Anisotropic Scaling of Random Fields with Long-Range Dependence

Scaling Limits of Random Fields with Applications

Anisotropic Scaling of Random Fields with Long-Range Dependence

Scaling Limits of Random Fields with Applications

Donatas Surgailis
Vilnius University, Lithuania

World Scientific

NEW JERSEY • LONDON • SINGAPORE • BEIJING • SHANGHAI • HONG KONG • TAIPEI • CHENNAI • TOKYO

Published by

World Scientific Publishing Co Pte Ltd
5 Toh Tuck Link, Singapore 596224
USA office: 27 Warren Street, Suite 401-402, Hackensack, NJ 07601
UK office: 57 Shelton Street, Covent Garden, London WC2H 9HE

Library of Congress Cataloging-in-Publication Data
Names: Surgailis, Donatas author
Title: Anisotropic scaling of random fields with long-range dependence : scaling limits of random fields with applications / author, Donatas Surgailis, (Vilnius University, Lithuania).
Description: Hackensack, NJ : World Scientific Publishing Co Pte Ltd, [2027] | Includes bibliographical references and index.
Identifiers: LCCN 2026013444 | ISBN 9789811249419 hardcover | ISBN 9789811249426 ebook | ISBN 9789811249433 ebook other
Subjects: LCSH: Random fields | Scaling laws (Statistical physics) | Stochastic processes
Classification: LCC QA274.45 .S874 2027
LC record available at https://lccn.loc.gov/2026013444

British Library Cataloguing-in-Publication Data
A catalogue record for this book is available from the British Library.

For any available supplementary material, please visit
https://www.worldscientific.com/worldscibooks/10.1142/12646#t=suppl

Desk Editors: Murali Appadurai/Cian Sacker Ooi

Typeset by Stallion Press
Email: enquiries@stallionpress.com

To my parents Elena and Henrikas

Preface

My interest in random fields started nearly 50 years ago when I attended a semester devoted to probability theory at the Banach International Mathematical Center in Warsaw. That year (1976), the semester was guided by Zbigniew Ciesielski, who was very friendly and supportive, and proposed to me an interesting problem on Gaussian Markov fields. Spatial dependence is a fascinating field of stochastic processes, accommodating a huge variety of diverse models and applications. Markov dependence (a special form of short-range dependence) is antipodal to long-range dependence (LRD). LRD is the main subject of this book, introduced to me by Roland Dobrushin around 1980. At present, LRD is a large area of probabilistic and statistical research, documented in several monographs, albeit most work in this area concern random processes with one-dimensional time parameter. This book deals with spatial LRD, which is more diverse and complex than temporal LRD, and therefore requires different approaches.

The idea of writing a book occurred to me around 10 years ago while working with PhD students Donata Puplinskaitė and Vytautė Pilipauskaitė on anisotropic scaling limits for random fields with long-range dependence. Rather unexpectedly, we found that summing up a random field over anisotropically increasing rectangles may lead to different partial sum limits depending on the mutual increase rates of the sides; moreover, these limits exhibit a trichotomy and have a degenerate dependence structure except for a critical rate ratio. It turned out that similar, albeit different, limits and trichotomies appear in telecommunications and econometrics and may reflect a universal trait of anisotropic scaling under long-range dependence.

This book reflects the long professional endeavour of the author in stochastic processes, which relied on the collaboration and exchange of ideas with colleagues, students, and other researchers. I want to thank Roland Dobrushin, Liudas Giraitis, Hira Koul, Murad Taqqu, Stanislav Molchanov, Wojbor Woyczynski, Remigijus Leipus, Vygantas Paulauskas, Peter Robinson, Anne Philippe, Paul Doukhan, Marie-Claude Viano, Jean-Marc Bardet, and, especially, Vytautė Pilipauskaitė (the co-author of most joint references in this book). Remigijus Lapinskas helped me with literature and figures. I also acknowledge the support from the Institute of Applied Mathematics, Vilnius University.

Finally, I am grateful to my wife Rūta for her love and patience.

Vilnius, October 2025
Donatas Surgailis

About the Author

Donatas Surgailis is Professor Emeritus at Vilnius University, Member Emeritus at the Lithuanian Academy of Sciences, and former Head of the Department of Stochastic Processes, Institute of Mathematics and Informatics at Vilnius University (2006–2016). He is the author of 150 research papers and the co-author (with Liudas Giraitis and Hira Koul) of the monograph "Large Sample Inference for Long Memory Processes", Imperial College Press, 2012. His main research interests are long-range dependence (including statistical inference), empirical processes, regression models, limit theorems, self-similar and fractal random processes, fractional integration, ARCH models, random fields and their applications in natural sciences, aggregation of spatio-temporal models, Bayesian networks, and stochastic geometry.

Contents

Notation and Conventions

$\mathbb{R} := (-\infty, \infty),\ \mathbb{Z} := \{0, \pm 1, \pm 2, \dots\}$

$\mathbb{R}_+ := (0, \infty),\ \mathbb{Z}_+ = \mathbb{N} := \{0, 1, \dots\}$

$\mathbb{R}^\nu := \{\boldsymbol{x} = (x_1, \dots, x_\nu) : x_i \in \mathbb{R}, i = 1, \dots, \nu\}$

$\mathbb{Z}^\nu := \{\boldsymbol{x} = (x_1, \dots, x_\nu) : x_i \in \mathbb{Z}, i = 1, \dots, \nu\}$

$\mathbb{R}^\nu_+ := \{\boldsymbol{x} = (x_1, \dots, x_\nu) : x_i \in \mathbb{R}_+, i = 1, \dots, \nu\}$

$\mathbb{Z}^\nu_+ := \{\boldsymbol{x} = (x_1, \dots, x_\nu) : x_i \in \mathbb{Z}_+, i = 1, \dots, \nu\}$

$\mathbf{1} := (1, \dots, 1) \in \mathbb{R}^\nu$, $\mathbf{0} := (0, \dots, 0) \in \mathbb{R}^\nu$, $\boldsymbol{e}_i := (\underbrace{0, \dots, 0}_{i-1}, 1, 0, \dots, 0) \in \mathbb{R}^\nu$

$\mathbb{R}^\nu_0 := \mathbb{R}^\nu \setminus \{\mathbf{0}\}$, $\mathbb{R}_0 := \mathbb{R} \setminus \{0\}$

$x \wedge y := \min(x, y),\ x \vee y := \max(x, y)\ (x, y \in \mathbb{R})$

$[x]$ denotes the integer part of $x \in \mathbb{R}$

$\mathrm{Leb}_\nu(A) :=$ the ν-dimensional Lebesgue measure of a set $A \subset \mathbb{R}^\nu$

$|\boldsymbol{x}| := (|x_1|^2 + \cdots + |x_\nu|^2)^{1/2},\ \boldsymbol{x} = (x_1, \dots, x_\nu) \in \mathbb{R}^\nu$

$\langle \boldsymbol{x}, \boldsymbol{y} \rangle := \sum_{i=1}^\nu x_i y_i,\quad \boldsymbol{x} = (x_1, \dots, x_\nu) \in \mathbb{R}^\nu,\ \boldsymbol{y} = (y_1, \dots, y_\nu) \in \mathbb{R}^\nu$

$]\boldsymbol{x}, \boldsymbol{y}] := \{\boldsymbol{z} = (z_1, \dots, z_\nu) \in \mathbb{R}^\nu : x_i < z_i \le y_i\},\ \boldsymbol{x} = (x_1, \dots, x_\nu) \in \mathbb{R}^\nu,\ \boldsymbol{y} = (y_1, \dots, y_\nu) \in \mathbb{R}^\nu$

$\|f\|_p := \big(\int_{\mathcal{X}} |f(x)| \mathrm{d}\mu(x)\big)^{1/p}$ $(p > 0)$, where $f : \mathcal{X} \to \mathbb{R}$ is a measurable function on a measure space $(\mathcal{X}, \mu)$

$\Psi(z) := \mathrm{e}^{\mathrm{i}z} - 1 - \mathrm{i}z,\ z \in \mathbb{R}$

$L^p(\mathbb{R}^\nu) :=$ the class of Borel functions $f : \mathbb{R}^\nu \to \mathbb{R}$ with $\|f\|_p = (\int_{\mathbb{R}^\nu} |f(\boldsymbol{x})|^p \mathrm{d}\boldsymbol{x})^{1/p} < \infty$

$L^p(\mathbb{Z}^\nu)$:= the class of functions $g : \mathbb{Z}^\nu \to \mathbb{R}$ with $\|g\|_p = (\sum_{\boldsymbol{x}\in\mathbb{Z}^\nu} |g(\boldsymbol{x})|^p)^{1/p} < \infty$

$\mathrm{diag}(\gamma_1, \ldots, \gamma_\nu)$ denotes the diagonal $\nu \times \nu$ matrix with entries $\gamma_1, \ldots, \gamma_\nu$ on the diagonal

$\mathbb{I}(A)$ stands for the indicator function of a set A

C denote generic positive constants which may be different at different locations

We write $\xrightarrow{\mathrm{d}}$, $\overset{\mathrm{d}}{=}$, $\overset{\mathrm{d}}{\neq}$ ($\xrightarrow{\mathrm{fdd}}$, $\overset{\mathrm{fdd}}{=}$, $\overset{\mathrm{fdd}}{\neq}$) for the weak convergence, equality, and inequality of (finite-dimensional) distributions

r.v. := random variable

i.i.d. := independent and identically distributed

l.h.s. := left-hand side, r.h.s. := right-hand side

w.r.t. := with respect to

d.f. := distribution function

ch.f. := characteristic function

w.l.g. := without loss of generality

$N(\mu, \sigma^2)$:= a normal distribution with mean μ and variance σ^2

CLT := the central limit theorem

DCT := the dominated convergence theorem

RF := random field, RG := random grain

BM := Brownian motion, FBM := fractional Brownian motion, FBS := fractional Brownian sheet

SS := self-similar, SI := stationary increment

MSS := multi-self-similar

LRD := long-range dependent, SRD:= short-range dependent, ND := negatively dependent

AR := autoregressive, MA := moving average, RC:= random coefficient

VO := variable order

All equalities and inequalities between random variables are assumed to hold almost surely (a.s.).

All statements (theorems, lemmas, propositions, corollaries, remarks, and examples) use the same consecutive enumeration.

Introduction

While the word 'scaling' has many meanings, in mathematics and physics it usually stands for 'zooming out', or getting a distant view of the object. As in the real world, looking at an object from far away, 'small-scale' details or distinctions disappear, but 'large-scale' ones may prevail, resulting in a robust or schematic image capturing only the essential features of the object and ignoring less important local and possibly random details. In probability theory and mathematical statistics, scaling is closely related to limit theorems, particularly to the fundamental central limit theorem (CLT) and Donsker's invariance principle. Indeed, with any sample $\{X(1), \ldots, X(n)\}$, we can associate a random process,

$$S_n(x) := \sum_{t=1}^{[nx]} X(t), \quad x \in [0,1], \tag{1}$$

on the interval $x \in [0,1]$, called the *partial sums process*. Since numerous statistics are written as functionals of the partial sums process, the scaling limit of S_n plays a vital role in the asymptotic statistical analysis. If the observation process $\{X(t)\}$ is a sequence of i.i.d. r.v.s with zero mean and variance $\sigma^2 < \infty$, then by the classical CLT,

$$n^{-1/2} S_n(x) \xrightarrow{\text{fdd}} \sqrt{\sigma^2} B(x), \tag{2}$$

where $B = \{B(x); x \in [0,1]\}$ is a standard Brownian motion (BM). Moreover, (2) extends to the functional convergence in the Skorohod space $D[0,1]$, as well as to a wide class of weakly dependent stationary processes; see, e.g., the monograph by Billingsley [42] (in the latter case, σ^2 in (2) is defined as $\sigma^2 = \sum_{t\in\mathbb{Z}} \mathrm{Cov}(X(0), X(t))$, often termed the *long-run variance of* X). We may say that the scaling limit σB in (2) provides a 'large-scale summary' for a vast class of weakly dependent processes X, which include Markov chains with an arbitrary number of values, autoregressive processes, various queueing models, and many other random processes; most of the concrete details and differences determining these evolutions disappear in the scaling limit, which depends only on the single parameter σ^2. The characteristic property $\sum_{t\in\mathbb{Z}} |\mathrm{Cov}(X(0), X(t))| < \infty, \sum_{t\in\mathbb{Z}} \mathrm{Cov}(X(0), X(t)) > 0$ of a stationary process X with finite variance is called *short-range dependence* (SRD), to be distinguished from *negative dependence* (ND) defined as $\sum_{t\in\mathbb{Z}} |\mathrm{Cov}(X(0), X(t))| < \infty, \sum_{t\in\mathbb{Z}} \mathrm{Cov}(X(0), X(t)) = 0$ and *long-range dependence* (LRD) $\sum_{t\in\mathbb{Z}} |\mathrm{Cov}(X(0), X(t))| = \infty$. Under LRD, the convergence in (2) with normalisation $n^{1/2}$ does not hold, and the partial sums process may grow faster than $n^{1/2}$, with a suitable normalised limit of $S_n(x)$ being different from BM. In the general case, the scaling limit of a stationary process X is defined as

$$d_n^{-1} S_n(x) \xrightarrow{\mathrm{fdd}} V(x), \quad n \to \infty, \tag{3}$$

where $d_n \to \infty$ is a normalisation and $V = \{V(x); x > 0\}$ is a nontrivial random process. We note that the definition in (3) does not assume the existence of covariance or any other finite moments of X. It turns out that under a mild stochastic continuity assumption on V, (3) implies that the normalisation d_n grows as n^H (modulo a slowly varying factor) for some $H > 0$, and V is H*-self-similar* (H-SS):

$$V(\lambda x) \overset{\mathrm{fdd}}{=} \lambda^H V(x) \quad \text{for any } \lambda > 0. \tag{4}$$

These facts were discovered by Lamperti [146] and constitute what is known as the Lamperti theorem [72]. An obvious implication of (3) is the limit distribution of the sample mean $\bar{X}_n := n^{-1} \sum_{t=1}^n X(t)$:

$$n d_n^{-1} \bar{X}_n \xrightarrow{\mathrm{d}} V(1),$$

which is of primary interest in statistics. However, the 'large-scale summary' V provides much more information about X, which is useful in statistical inference, including nonlinear statistics. The limit in (3) also gives rise to a useful strong dependence concept, different from the SRD/LRD covariance classification above. There are two possibilities: either V has independent increments or the increments of V are dependent, meaning that for some $0 \le x_1 < x_2 \le x_3$, r.v.s $V(x_2) - V(x_1)$ and $V(x_4) - V(x_3)$ are not independent. We say that the process X exhibits the *distributional SRD* property if V has independent increments and the *distributional LRD* property if V has dependent increments.

The concept of distributional SRD/LRD, introduced by Cox [62] and Dehling and Philipp [74], has several advantages. As noted above, it does not depend on any moment assumptions, so finite and infinite variance processes are treated from the same angle. A faster growth of normalisation d_n means stronger dependence; therefore, H is a quantitative indicator of the degree of dependence of $\{X(t)\}$. The characterisation of SRD through (3) is highly robust and essentially reduces to Lévy stable behaviour since all SS processes with stationary independent increments are stable Lévy processes. On the other hand, the class of all SS processes V in (3) with dependent increments is very large and does not allow for a judicious description. Fortunately, stochastically continuous Gaussian H-SS processes with stationary increments are well known, and the covariance function is written as

$$\mathrm{Cov}(V(x), V(y)) = (\sigma^2/2)(x^{2H} + y^{2H} - |x-y|^{2H}), \quad x, y \ge 0, \quad (5)$$

determined by two parameters: $H \in (0, 1], \sigma > 0$. The standardised Gaussian process $V = B_H$, corresponding to $\sigma = 1$ in (5), is called the *fractional Brownian motion* (FBM). Thus, the existence of the Gaussian scaling limit $V = \sigma B_H$ in (3) essentially clarifies the large-scale dependence structure of X, which resembles the dependence structure of the differenced process $\{B_H(t) - B_H(t-1); t \in \mathbb{Z}\}$, called the *fractional Gaussian noise* (FGN), and is completely determined by the (Hurst) parameter H.

The scaling operation in multidimensional space $\mathbb{R}^\nu$ ($\nu \ge 2$) can be defined in many ways. A general form of scaling is associated with a linear operator (matrix), $\Gamma : \mathbb{R}^\nu \to \mathbb{R}^\nu$, as

$$\boldsymbol{x} \mapsto \lambda^\Gamma \boldsymbol{x}, \quad \lambda > 0, \ \boldsymbol{x} \in \mathbb{R}^\nu. \quad (6)$$

In what follows, we use the notation λ for the scaling parameter, which takes values in $\mathbb{R}_+$ instead of $n \in \mathbb{N}_+$ in (1). The matrix exponential, defined as $\lambda^\Gamma \boldsymbol{x} = \mathrm{e}^{\Gamma \log \lambda} \boldsymbol{x} = \sum_{k=0}^\infty \frac{(\log \lambda)^k}{k!} \Gamma^k \boldsymbol{x}$, satisfies $\lambda_1^\Gamma \lambda_2^\Gamma = (\lambda_1 \lambda_2)^\Gamma$, and hence (6) forms a transformation group. The above group carries over to the group $T_\lambda, \lambda > 0$, of transformations

$$T_\lambda f(\boldsymbol{x}) = f(\lambda^\Gamma \boldsymbol{x}), \quad \boldsymbol{x} \in \mathbb{R}^\nu,$$

acting on functions $f = f(\boldsymbol{x}); \boldsymbol{x} \in \mathbb{R}^\nu$ and hence on the random field (RF) $Y = \{Y(\boldsymbol{x}); \boldsymbol{x} \in \mathbb{R}^\nu\}$ as well, indexed by points $\boldsymbol{x} \in \mathbb{R}^\nu$. Following Biermé *et al.* [37], we call an RF Y an *operator scaling with index* $H \in \mathbb{R}$ (H-OSRF) if

$$T_\lambda Y(\boldsymbol{x}) \stackrel{\mathrm{fdd}}{=} \lambda^H Y(\boldsymbol{x})$$

holds for any $\lambda > 0$. Obviously, the transformation in (6) and the class of H-OSRFs strongly depend on the matrix Γ. The simplest and most natural scaling transformation corresponds to the unit matrix $\Gamma = I := \mathrm{diag}(1, \ldots, 1)$. In the sequel, we refer to (6) with $\Gamma = I$ as the *isotropic scaling* and to (6) with $\Gamma \neq \gamma I, \gamma > 0$ as *non-isotropic scaling*. A particular case of non-isotropic scaling is given by a diagonal matrix,

$$\Gamma = \mathrm{diag}(\gamma_1, \ldots, \gamma_\nu), \tag{7}$$

where $\gamma_i > 0, i = 1, \ldots, \nu$ are positive and not all equal to each other. We refer to (6)–(7) as *anisotropic scaling*, this book being limited to scaling with a diagonal matrix.

Scaling limits for RFs with continuous or discrete ν-dimensional argument are usually defined as the limits in the distribution of integrals (sums) of the RF over rescaled increasing sets. More generally, they refer to the limit distribution of integrals

$$X_{\lambda,\Gamma}(\phi) := \int_{\mathbb{R}^\nu} X(\boldsymbol{t}) \phi(\lambda^{-\Gamma} \boldsymbol{t}) \mathrm{d}\boldsymbol{t}, \quad \text{as } \lambda \to \infty, \tag{8}$$

(or the respective sums in the discrete argument case), where $X = \{X(\boldsymbol{t}); \boldsymbol{t} \in \mathbb{R}^\nu\}$ is a given stationary RF, for each ϕ from a class of (test) functions, $\Phi = \{\phi : \mathbb{R}^\nu \to \mathbb{R}\}$. A suitably normalised limit of (8) is an RF $V_\Gamma(\phi)$ indexed by $\phi \in \Phi$, called the scaling limit of X under scaling (6). In the case of $\Gamma = I$, the integral in (8) is written as

$$X_\lambda(\phi) := \int_{\mathbb{R}^\nu} X(\boldsymbol{t})\phi(\boldsymbol{t}/\lambda)\mathrm{d}\boldsymbol{t}, \tag{9}$$

and the isotropic scaling limit $V = \{V(\phi); \phi \in \Phi\}$ is defined as the limit in distribution

$$d_\lambda^{-1} X_\lambda(\phi) \xrightarrow{\mathrm{d}} V(\phi), \quad \lambda \to \infty, \quad \forall\, \phi \in \Phi \tag{10}$$

under a normalisation, $d_\lambda \to \infty\,(\lambda \to \infty)$. The above approach is common in the theory of generalised RFs and is discussed in Dobrushin [79], together with various classes of Φ. For $\phi(\boldsymbol{t}) = \mathbb{I}(\boldsymbol{t} \in A)$, where A is a bounded Borel subset of $\mathbb{R}^\nu$, (9) is the integral (sum) of $X(\boldsymbol{t})$'s over $\boldsymbol{t} \in \lambda A$, whose limit distribution for A with irregular boundary may be of interest but is not simple; see e.g., Lahiri and Robinson [145]. For the class $\Phi = \Phi_{\mathrm{rec},\nu} := \{\mathbb{I}(\boldsymbol{t} \in]\boldsymbol{0}, \boldsymbol{x}]); \boldsymbol{x} \in \mathbb{R}^\nu_+\}$ of ν-dimensional 'rectangles' $]\boldsymbol{0}, \boldsymbol{x}] := \prod_{i=1}^\nu]0, x_i]$, (8) presents a ν-dimensional analogue,

$$S_{\lambda,\Gamma}(\boldsymbol{x}) := \int_{]\boldsymbol{0},\lambda^\Gamma \boldsymbol{x}]} X(\boldsymbol{t})\mathrm{d}\boldsymbol{t}, \quad \boldsymbol{x} \in \mathbb{R}^\nu_+, \tag{11}$$

of the partial sum process in (1), leading to a limit RF indexed by $\boldsymbol{x} \in \mathbb{R}^\nu_+$.

As in the one-dimensional case $\nu = 1$, there is no general scaling theory for dependent RFs since there is no possibility of describing such objects. For stationary RF with finite variance, dependence is usually associated with the covariance function or spectral density. Particularly, the classification into (covariance) LRD/SRD/ND classes for RFs on $\mathbb{Z}^\nu$ is completely analogous to the case $\nu = 1$ above. This book focuses on *linear* RFs,

$$X(\boldsymbol{t}) = \sum_{\boldsymbol{u} \in \mathbb{Z}^\nu} a(\boldsymbol{t} - \boldsymbol{u})\varepsilon(\boldsymbol{u}) \quad \text{and} \quad X(\boldsymbol{t}) = \int_{\mathbb{R}^\nu} a(\boldsymbol{t} - \boldsymbol{u})M(\mathrm{d}\boldsymbol{u}), \tag{12}$$

written as moving averages (MAs) in i.i.d. r.v.s $\{\varepsilon(\boldsymbol{s})\}$ or stochastic integrals with respect to an independently scattered Lévy random measure $M(\mathrm{d}\boldsymbol{u})$ and their subordinated nonlinear RFs $Y(\boldsymbol{t}) = G(X(\boldsymbol{t}))$, where $G : \mathbb{R} \to \mathbb{R}$ is a given nonlinear function.

Dependence in such RFs is determined by the MA kernel $a(\boldsymbol{t})$ and the nonlinear function G, resulting in a variety of Gaussian and non-Gaussian scaling limits under the LRD/SRD/ND conditions. The linearity of (12) and (9) warrants the use of the powerful method of characteristic functions in limit theorems of probability theory, widely applied in this book. Subordinated RFs, $Y(\boldsymbol{t}) = G(X(\boldsymbol{t}))$, are treated via the polynomial expansion of G and multiple polynomial forms generalising the Hermite expansion and the scaling limits of Gaussian subordinated RF in the seminal works of Dobrushin and Major [80] and Breuer and Major [47].

A related class of linear and nonlinear RFs is given by *random grain* (RG),

$$X(\boldsymbol{t}) = \sum_{j=1}^{\infty} \mathbb{I}(\boldsymbol{t} \in (\boldsymbol{u}_j + \Xi_j)), \quad \boldsymbol{t} \in \mathbb{R}^{\nu}, \tag{13}$$

and its subordinated models. Here, $\{\boldsymbol{u}_j\}$ is a stationary Poisson point process in $\mathbb{R}^{\nu}$ and $\{\Xi_j\}$ are random sets (i.i.d. copies of a generic *grain set*, $\Xi \subset \mathbb{R}^{\nu}$), independent of $\{\boldsymbol{u}_j\}$. By definition, the sum in (13) counts the number of RGs, $\boldsymbol{u}_j + \Xi_j$, covering $\boldsymbol{t}$. Isotropic scaling limits of a subordinated RF, $Y(\boldsymbol{t}) = G(X(\boldsymbol{t}))$, of the RG model in (13) are derived using the Charlier expansion of G and Mehler's formula for bivariate Poisson distribution.

Let us consider anisotropic scaling limits for linear and nonlinear RFs under the LRD/SRD/ND dependence conditions. This question is more difficult than isotropic scaling since the answer should depend on Γ in (7), which is supposed to be arbitrary. In this book, this question is essentially treated for $\nu = 2, 3$ and 'rectangular' test functions only, as in (11). For $\nu = 2, \Gamma = \mathrm{diag}(1, \gamma)$, and an RF X indexed by $\mathbb{Z}^2$, anisotropic scaling limits are defined as

$$d_{\lambda,\gamma}^{-1} \sum_{\boldsymbol{t} \in K_{[\lambda x_1, \lambda^{\gamma} x_2]}} X(\boldsymbol{t}) \xrightarrow{\mathrm{fdd}} V_{\gamma}(\boldsymbol{x}), \quad \boldsymbol{x} \in \mathbb{R}_+^2, \quad \lambda \to \infty, \tag{14}$$

where $K_{[\lambda x_1, \lambda^{\gamma} x_2]} := \{\boldsymbol{t} \in \mathbb{Z}^2 : 1 \le t_1 \le \lambda x_1, 1 \le t_2 \le \lambda^{\gamma} x_2\} \subset \mathbb{Z}^2$ is a family of rectangles, whose sides increase at possibly different rates λ and λ^{γ}, and $d_{\lambda,\gamma} \to \infty$ is a normalisation. The surprising fact first noted in Ref. [221] is that for a large class of RFs, nontrivial scaling limits in (14) *exist for any* $\gamma > 0$. In such a case, with a given RF X, one can associate a one-parameter family $\{V_\gamma; \gamma > 0\}$ of the scaling limits in (14), termed the *scaling diagram of* X, which characterises the dependence structure and large sample properties of X. What is more surprising, a large class of linear and nonlinear LRD/ND RFs on $\mathbb{Z}^2$ exhibit a *scaling transition*, i.e., a dramatic change in their scaling behaviour at a critical point, $\gamma_{\rm cr} > 0$, determined by internal parameters of the RF, in the sense that the limits $V_\gamma = V_\pm$ do not depend on γ for $\gamma > \gamma_{\rm cr}$ or $\gamma < \gamma_{\rm cr}$, and V_+, V_- are different RFs.

A basic model with the above behaviour, discussed in Section 4 of Chapter 6, is an MA RF, with MA coefficients decaying at different rates in the horizontal and vertical directions:

$$X(\boldsymbol{t}) = \sum_{\boldsymbol{u} \in \mathbb{Z}^2} a(\boldsymbol{t} - \boldsymbol{u}) \varepsilon(\boldsymbol{u}), \tag{15}$$

where $\varepsilon(\boldsymbol{u}), \boldsymbol{u} \in \mathbb{Z}^2$ are standardised i.i.d. r.v.s, and

$$a(\boldsymbol{t}) \sim \text{const}\, \rho(\boldsymbol{t})^{-1} \quad \text{as} \quad |\boldsymbol{t}| \to \infty, \tag{16}$$

where $\rho(\boldsymbol{t}) := |t_1|^{q_1} + |t_2|^{q_2}$, and the parameters $q_1, q_2 > 0$ satisfy

$$Q := \frac{1}{q_1} + \frac{1}{q_2} \in (0, 2). \tag{17}$$

Note that $a(t_1, 0) = O(|t_1|^{-q_1})$, $a(0, t_2) = O(|t_2|^{-q_2})$ decay at a different rates in the horizontal and vertical directions when $q_1 \neq q_2$, in which case the RF in (15) exhibits strong anisotropy. For $1 < Q < 2$, X in (15) is LRD, with covariance function $r(\boldsymbol{t}) = \mathrm{E} X(\mathbf{0}) X(\boldsymbol{t})$ decaying as

$$r(\boldsymbol{t}) \sim L(\boldsymbol{t})\, \tilde\rho(\boldsymbol{t})^{-1}, \quad |\boldsymbol{t}| \to \infty,$$

where

$$\tilde\rho(\boldsymbol{t}) := |t_1|^{\tilde q_1} + |t_2|^{\tilde q_2} \quad \text{and} \quad \tilde q_i := q_i(2 - Q), \quad i = 1, 2, \tag{18}$$

and $L(\boldsymbol{t})$ is a strictly positive and bounded 'angular function'. Particularly,

$$\gamma_{\rm cr} := \tilde q_1 / \tilde q_2 = q_1 / q_2 \tag{19}$$

can be considered the *internal scale ratio* of X in (15), or the ratio of the (reciprocal) asymptotic power decay exponents of the covariance function in the vertical and horizontal directions:

$$\gamma_{\rm cr} = \frac{\lim_{|t_2|\to\infty}(\log|t_2|)/(\log r(0,t_2))}{\lim_{|t_1|\to\infty}(\log|t_1|)/(\log r(t_1,0))}. \tag{20}$$

On the other hand, $\gamma = \log\lambda^\gamma/\log\lambda$ can be viewed as the *external scale ratio*, defined as the ratio of the power exponents of λ of the scaling factors in the vertical and horizontal directions in (14). The disagreement between the two ratios for X in (15), $1 < Q < 2$, has critical consequences for the dependence structure of the 'unbalanced' scaling limits in (14):

$$V_\pm = B_{H_1^\pm, H_2^\pm}, \tag{21}$$

given by a *fractional Brownian sheet (FBS) with one of the parameters* $H_1^\pm$ *and* $H_2^\pm$ *equal to* 1 *or* 1/2. The above facts suggest that 'unbalanced' scaling 'destroys' the dependence structure of X, the 'destruction', consisting in the fact that rectangular increments of $V_\pm$ in one of the two directions on the plane are either completely independent or completely dependent. A similar but different scaling transition occurs for linear RF X in (15), with coefficients as in (16), $0 < Q < 1$, and satisfying the ND condition $\sum_{t\in\mathbb{Z}^2} a(\boldsymbol{t}) = 0$, in which case $V_\pm$ are FBS, as in (19), except that *one of the parameters* $H_1^\pm$ *and* $H_2^\pm$ *is equal to* 0 *or* 1/2. FBS is one of the most important limit RFs in this book; for $\nu = 2$, the two-parameter FBS family $B_{H_1,H_2}, (H_1, H_2) \in [0,1]^2$ is defined as a zero-mean Gaussian RF on $\mathbb{R}_+^2$ with covariance function

$$\mathrm{E}B_{H_1,H_2}(\boldsymbol{x})B_{H_1,H_2}(\boldsymbol{y}) = \prod_{i=1}^{2} \mathrm{E}B_{H_i}(x_i)B_{H_i}(y_i), \quad \boldsymbol{x},\boldsymbol{y} \in \mathbb{R}_+^2,$$

where $B_H = \{B_H(x); x > 0\}, H \in (0,1]$ is FBM with covariance in (5) ($\sigma = 1$) and $B_0 = \{B_0(x); x > 0\}$ is a non-separable Gaussian process with

$$\mathrm{E}B_0(x)B_0(y) := \lim_{H\downarrow 0} \mathrm{E}B_H(x)B_H(y) = \mathbb{I}(x = y) + \frac{1}{2}\mathbb{I}(x \neq y), \quad x, y > 0. \tag{22}$$

The FBS B_{H_1,H_2} with $H_1 \wedge H_2 = 0$ are extremely singular RFs featuring complete negative dependence in one direction, caused by 'edge effects' at the boundaries of the rectangle $K_{[\lambda x_1, \lambda^\gamma x_2]}$ under ND. All FBS B_{H_1,H_2}, $(H_1, H_2) \in [0,1]^2$ satisfy the important *multi-self-similarity* (MSS) property, which seems to be characteristic of unbalanced scaling limits in dimension $\nu = 2$ and occurs in other models discussed in this book.

Section 6 of Chapter 6 shows that scaling transition is not limited to linear RFs and may occur in subordinated LRD RFs, $Y(\boldsymbol{t}) = G(X(\boldsymbol{t}))$, where X is a moving-average RF in (15)–(16) and G is a nonlinear function as well. Particularly, when $G = A_k$ is the Appell polynomial of degree

$$2 \le k < \tilde{Q} := \frac{1}{\tilde{q}_1} + \frac{1}{\tilde{q}_2},$$

$\tilde{q}_i, i = 1, 2$, as in (18), the subordinated RF $Y = A_k(X)$ exhibits scaling transition at the same point $\gamma_{\rm cr}$ in (19), which can be defined as in (20), with $r(\boldsymbol{t})$ replaced by $\mathrm{Cov}(Y(\boldsymbol{0}), Y(\boldsymbol{t}))$, with the difference that the unbalanced limits with completely dependent rectangular increments are (non-Gaussian) 'generalised Hermite slides'.

Needless to say, the results in Chapter 6 strongly rely on the form of the MA coefficients in (16). Indeed, if these coefficients asymptotically factorise as

$$a(\boldsymbol{t}) \sim \mathrm{const} \prod_{i=1}^{2} |t_i|^{-q_i}, \quad |\boldsymbol{t}| \to \infty, \tag{23}$$

with $q_i \in (\frac{1}{2}, 1), i = 1, 2$, the linear RF X in (15) and (23) is well defined and LRD but the scaling transition does not occur since all anisotropic limits in (14) agree with FBS $B_{H_1,H_2}, H_i := \frac{3}{2} - q_i, i = 1, 2$. The asymptotic forms in (16) and (23) are very different and might be dubbed 'elliptic' and 'hyperbolic', respectively, following the form of the level sets of functions $\rho(\boldsymbol{t})$ and $\prod_{i=1}^{2} |t_1|^{q_1} |t_2|^{q_2}$. A common feature of (16) and (23) with $q_1 \neq q_2$ is that the *dependence axis of* X (defined as the direction in the plane along which the MA coefficients decay at the smallest rate) agrees with the horizontal axis if $q_1 < q_2$ and with the vertical axis if $q_1 > q_2$. Since the scaling in (14) is parallel to the coordinate axes, we may say in both cases that *the scaling is congruous with the dependence axis of* X. This raises the question of what happens if the dependence axis of X does not agree with any of the two coordinate axes (the case of *incongruous scaling*). A linear RF with an 'oblique' dependence axis can be obtained by 'rotation', $a(B\boldsymbol{t})$, of $a(\boldsymbol{t})$ (16) or (23) with a non-degenerate 2×2 matrix B. The results under incongruous scaling in Section 5 of Chapter 6 are less complete than in the case of congruous scaling; however, they suggest that the presence of an 'oblique' dependence axis dramatically changes the scaling transition, which occurs at $\gamma_{\mathrm{cr}} = 1$ for $a(B\boldsymbol{t})$ in both cases (16) and (23) independently of the parameters q_1 and q_2.

Whereas the most attention in this book is given to planar RFs, Section 7 of Chapter 6 discusses anisotropic scaling of RFs in higher dimensions, particularly anisotropic scaling limits on 'rectangles' $]\mathbf{0}, \lambda^\Gamma \boldsymbol{x}]$, $\Gamma = \mathrm{diag}(\gamma_1, \gamma_2, \gamma_3)$ for a class of LRD MA RFs X on $\mathbb{Z}^3$ and coefficients $a(\boldsymbol{t})$ decaying as in (23), with $\rho(\boldsymbol{t}) = \sum_{i=1}^3 |t_i|^{q_i}$ depending on three parameters $q_i > 0, i = 1, 2, 3$, for $Q := \sum_{i=1}^3 \frac{1}{q_i} \in (1, 2)$. For the above X, the corresponding scaling limits

$$d_{\lambda,\gamma}^{-1} \sum_{\boldsymbol{t} \in]\mathbf{0}, \lambda^\Gamma \boldsymbol{x}]} X(\boldsymbol{t}) \xrightarrow{\mathrm{fdd}} V_\gamma(\boldsymbol{x}), \quad \boldsymbol{x} \in \mathbb{R}_+^3, \quad \lambda \to \infty, \tag{24}$$

exist for any $\boldsymbol{\gamma} = (\gamma_1, \gamma_2, \gamma_3) \in \mathbb{R}_+^3$ and are Gaussian RFs as in the case $\nu = 2$; however, the scaling diagram is more complicated and contains more than three elements, which crucially depend on the fulfilment or violation of the *balance conditions*:

$$\gamma_i / \gamma_j = q_j / q_i, \quad 1 \le i < j \le 3. \tag{25}$$

The limits in (24) are FBS $B_{H_1,H_2,H_3}, H_i \in [1/2, 1], i = 1, 2, 3$, if and only if *all balance conditions in* (25) *are violated.* Extending (24) to ND and/or higher dimensions $\nu \ge 4$ remains an open problem.

For functions and RFs indexed by continuous argument $\boldsymbol{x} \in \mathbb{R}^\nu$, scaling in (10) and (14) as $\lambda \to \infty$ can be 'reversed' and replaced by *small-scale* limits as $\lambda \to 0$. The latter limits usually involve increments of an RF: $Y = \{Y(\boldsymbol{t}); \boldsymbol{t} \in \mathbb{R}^\nu\}$. The two basic types of increments of Y at a given point, $\boldsymbol{t}_0 \in \mathbb{R}^\nu$, are the *(ordinary) increment* $Y(\boldsymbol{t}_0 + \boldsymbol{t}) - Y(\boldsymbol{t}_0)$ and the *rectangular increment* $Y(]\boldsymbol{t}_0, \boldsymbol{t}_0 + \boldsymbol{t}])$ on 'rectangle' $]\boldsymbol{t}_0, \boldsymbol{t}_0 + \boldsymbol{t}] \subset \mathbb{R}^\nu$, for $\nu = 2$ the latter increment is given by the double difference

$$\begin{aligned} Y(]\boldsymbol{t}_0, \boldsymbol{t}_0 + \boldsymbol{t}]) = Y(t_{01} + t_1, t_{02} + t_2) - Y(t_{01}, t_{02} + t_2) \\ - Y(t_{01} + t_1, t_{02}) + Y(t_{01}, t_{02}). \end{aligned} \tag{26}$$

Chapter 7 discusses small-scale anisotropic scaling limits for ordinary and rectangular increments for a class of Lévy-driven fractional RFs Y on $\mathbb{R}^2$, written as stochastic integrals w.r.t. an independently scattered and homogeneous Lévy random measure, $M(\mathrm{d}\boldsymbol{u})$, featuring an α-stable $(0 < \alpha \le 2)$ small-scale the behaviour specified in Assumption $(\mathrm{M})_\alpha$ of Section 6 of Chapter 2. Particularly, the distribution of rectangular increments of Y does not depend on $\boldsymbol{t}_0$ and coincides with that of stochastic integral

$$Y(]\mathbf{0},\boldsymbol{t}]) = \int_{\mathbb{R}^2} g(]-\boldsymbol{u},\boldsymbol{t}-\boldsymbol{u}])M(\mathrm{d}\boldsymbol{u}), \tag{27}$$

involving rectangular increments of deterministic kernel g, which behaves at the origin as a (generalised) homogeneous function

$$g(\boldsymbol{t}) \sim \rho(\boldsymbol{t})^{\chi}, \quad \boldsymbol{t} \to \mathbf{0}, \tag{28}$$

where $\rho(\boldsymbol{t})$ in (16) and $q_1 > 0, q_2 > 0, \kappa \ne 0$, are parameters, with MA coefficients $a(\boldsymbol{t})$ as $|\boldsymbol{t}| \to \infty$. The class of the above Y is quite large and contains many fractional RFs, *Fractional Lévy RFs* (Example 3 of Chapter 7), and *Matérn RFs* (Example 4 of Chapter 7). We show that under certain conditions on q_1, q_2, κ, rescaled rectangular increments in (27) display a scaling transition at the same $\gamma_{\mathrm{cr}} = q_1/q_2$ as in (19), and the scaling behaviour

$$d_{\lambda,\gamma}^{-1} Y(]\mathbf{0}, \lambda^{\Gamma}\boldsymbol{t}]) \xrightarrow{\mathrm{fdd}} V_\gamma(\boldsymbol{t}), \quad \lambda \to 0, \quad \Gamma := \mathrm{diag}(1,\gamma)$$

is somewhat similar to the large-scale behaviour in (14) and (21) of X in (15), except that $V_\pm$ are α-stable RFs $(0 < \alpha \le 2)$, which satisfy the $(H_1^\pm, H_2^\pm)$-MSS property with *one of the two indices* $H_i^\pm, i = 1, 2$ *being equal to 0 or 1.*

Let us turn to the application part of this book. The interest in limit theorems and scaling limits in dimension $\nu = 1$ is largely due to statistical analysis of time series – observations indexed by the time parameter from $\mathbb{Z}$ or $\mathbb{R}$ arising in various natural sciences (hydrology, econometrics, network traffic, etc.) – which may display LRD/ND properties and specific dependence structures. Much of the work in time series inference is devoted to regression estimates and empirical processes. The limit results under strong dependence use Hermite expansions or causal martingale methods. See the monographs by

Beran [30], Beran *et al.* [31], and Giraitis *et al.* [103], dedicated to statistics of long-memory time series.

Spatial inference faces two major distinctions from time series. The first one is the lack of causality, which is often unnatural for spatial observations. The second distinction is that the observation set is often highly irregular (e.g., not a ν-dimensional 'cube' or any other 'regular' set). The application in Section 4 of Chapter 5 to the *Kolmogorov–Smirnov statistic*

$$F_\lambda(y) = \mathrm{Leb}_\nu(\lambda A)^{-1} \sum_{\boldsymbol{t}\in\lambda A} \mathbb{I}(X(\boldsymbol{t}) \le y), \quad \text{as } \lambda \to \infty$$

(a consistent estimator of the marginal distribution function $F(y) = \mathrm{P}(X(\boldsymbol{t}) \le y)$ of the stationary linear RF X in (12)), avoids both causality and regularity limitations since the result (a functional CLT for $F_\lambda(y) - F(y)$) applies to an arbitrary Borel set A and a noncausal RF X under SRD and LRD assumptions. For the same X and A, we obtain the limit distribution for a class of M estimators of regression coefficients $\boldsymbol{\beta} \in \mathbb{R}^p$ in a linear regression model:

$$Y_\lambda(\boldsymbol{t}) = \langle \boldsymbol{\beta}, \boldsymbol{v}_\lambda(\boldsymbol{t})\rangle + X(\boldsymbol{t}), \quad \boldsymbol{t} \in \lambda A,$$

with the deterministic vector-valued regression function $\boldsymbol{v}_\lambda(\boldsymbol{t}) \in \mathbb{R}^p$.

Section 5 of Chapter 5 discusses the application of estimating the volume fraction of the Boolean set

$$\mathcal{X} = \bigcup_{j=1}^{\infty} (\boldsymbol{u}_j + \Xi_j) \subset \mathbb{R}^\nu,$$

defined as the union of all grains in the Poisson RG model (13). The Boolean model is fundamental in stochastic geometry and stereology. The *volume fraction* $p \in (0,1)$ is the mean fraction of the volume (the Lebesgue measure) occupied by a region of unit volume and one of the most important parameters of a Boolean set. A natural estimator of p from the observation of $\mathcal{X}$ on a large 'inflated' set, $\lambda A \subset \mathbb{R}^\nu$, is the ratio

$$\widehat{p}_{\lambda,A} := \frac{\widehat{X}_\lambda(A)}{\mathrm{Leb}_\nu(\lambda A)}, \tag{29}$$

called the *sample volume fraction*, where $\widehat{X}_\lambda(A) := \mathrm{Leb}_\nu(\mathcal{X} \cap \lambda A)$ is the volume of the intersection of the Boolean set with λA.

Since $\hat{X}_\lambda(A) = \int_{\mathbb{R}^\nu} \mathbb{I}(\boldsymbol{t}/\lambda \in A)(X(\boldsymbol{t}) \wedge 1)\mathrm{d}\boldsymbol{t}$ is written as the integral of the nonlinear function $G(x) = x \wedge 1$ of the RG model in (12) and the results of Section 5 of Chapter 5 on isotropic scaling of subordinated RFs apply to (29), it yields α-stable $(1 < \alpha < 2)$ or the Gaussian limit of the sample volume fraction under LRD/SRD conditions. A similar approach extends to estimates of the volume fraction on ν_0-dimensional sets $\lambda A \subset H_{\nu_0}$, where $H_{\nu_0} \subset \mathbb{R}^\nu$ is a hyperplane of dimension $1 \le \nu_0 < \nu$. The third application in Chapter 5 concerns the limit behaviour under parabolic scaling of solutions of Burgers' equation:

$$\partial \vec{v}(t, \boldsymbol{x})/\partial t + \langle \vec{v}(t, \boldsymbol{x}), \nabla \rangle \vec{v}(t, \boldsymbol{x}) = \frac{1}{2}\kappa \Delta \vec{v}(t, \boldsymbol{x}),$$

with random initial data $\vec{v}(0, \boldsymbol{x}) = -\nabla \xi(\boldsymbol{x}) = -(\partial \xi(\boldsymbol{x})/\partial x_1, \ldots, \partial \xi(\boldsymbol{x})/\partial x_\nu)$, where $\xi = \{\xi(\boldsymbol{x});\ \boldsymbol{x} \in \mathbb{R}^\nu\}$ is a scalar-valued shot-noise or RG RF, which may also depend on the scaling parameter $\lambda > 0$. The study of random solutions of Burgers' equation (known as *Burgers's turbulence problem*) is a broad field of mathematical and physical research; see Burgers [52] and Woyczynski [268].

Surprisingly, scaling transition occurs in spatio-temporal RF models or panel data,

$$X_i(t), \quad t = 1, \ldots, T, \quad i = 1, \ldots, M,$$

with 'spatial independence' and 'temporal LRD', consisting of M *independent* copies taken from an LRD random process X with discrete or continuous one-dimensional time parameters. The quantity of interest is the limit distribution of the *aggregated sum*:

$$A(T, M) := \sum_{i=1}^{M} \sum_{t=1}^{T} X_i(t), \tag{30}$$

as T and M increase jointly to infinity, possibly at different rates. Such problems appear under joint temporal and contemporaneous aggregation of network traffic and panel data in telecommunication and econometrics. By assuming that $T = [x\lambda], M = [y\lambda^\gamma]$ increase jointly with $\lambda \to \infty$ for some $\gamma > 0$, the aggregated sum $A_{\lambda,\gamma}(x, y) := A([x\lambda], [y\lambda^\gamma])$ becomes an RF indexed by $(x, y) \in \mathbb{R}_+^2$, and the limit problem is written as (large-scale) anisotropic scaling on rectangles of RF $X(t,s) \equiv X_s(t)$ with a two-dimensional argument (t,s). In Chapters 8 and 9, anisotropic scaling limits are derived for the centred 'aggregate' $A_{\lambda,\gamma}(x,y)$ formed by two general classes of LRD 'inputs' X. Class (I) comprises Poisson shot-noise processes with random 'pulse' of a general form. Class (II) is formed by regenerative X with a random pulse starting anew at points of a heavy-tailed renewal process. The above classes contain the M/G/∞ queue and the ON/OFF process studied in the important works of Mikosch *et al.* [185], Gaigalas and Kaj [93], and Kaj and Taqqu [130].

Chapter 10 discusses the scaling limits of aggregated sums (30) for *random-coefficient AR(1)* (RCAR(1)) 'inputs'

$$X_i(t) = a_i X_i(t-1) + \varepsilon_i(t), \qquad (31)$$

with i.i.d. random coefficients $a_i \in (0,1)$, independent of 'innovations' $\{\varepsilon_i(t)\}$. In econometric applications, $X_i(t)$ in (31) and the aggregated series $\sum_{i=1}^{M} X_i(t)$ may represent time evolution of individual 'micro-economic agents' and the 'macro-economic' variable. Following Robinson [223] and Granger [107], the aggregation of RCAR(1) processes with random coefficient having a sufficient mass near 'unit root' $a = 1$ has been widely accepted as an explanation of the long-memory phenomenon in observed macroeconomic time series. Under suitable assumptions on the latter distribution, the aggregated sums $A_{\lambda,\gamma}(x,y)$ of the RCAR(1) model in (31) feature a scaling transition with Gaussian and stable 'unbalanced' limits $V_\pm$ and an *intermediate Ornstein–Uhlenbeck'* limit V_{cr}, different from the corresponding limit (the *Telecom RF*) in Chapters 8 and 9, unsurprisingly since the 'memory mechanism' of a nonergodic process in (31) is very different from that of the telecommunication models.

Random fields is a very large area of research, due to the variety of mathematical models arising in natural and applied sciences (statistical physics, statistical hydrodynamics, astronomy, geophysics, image analysis, etc.). There are excellent texts (e.g., Adler and Taylor [2], Cressie [63], Guyon [108], and Preston [217]) devoted to path properties, geometry of RFs, spatial statistics, autoregressive models, Gibbsian RFs, and other areas of RFs that are not covered here.

This book focuses on limit theorems for RFs with strong dependence. Related topics treated here more or less systematically include spatial fractional integration, multiple stochastic integrals, local scaling limits of Lévy-driven RFs with continuous argument, renewal theory, self-similarity concepts for RFs, 'intermediate' RFs, and stochastic integral representations. The RF models discussed here are rather simple, with the dependence in the RF essentially determined by MA coefficients or their 'randomised' counterparts. The author views the mathematical level as moderate and accessible to doctoral students with a probability theory and stochastic processes background. The proofs are often reduced to classical central limit theorems for i.i.d. r.v.s, including stable and general infinitely divisible limit laws. Familiarity with stochastic integrals with deterministic integrands is required. Another prerequisite for the accessible reading of this text is a good command of the Fourier transform and characteristic functions.

A less pleasant feature of anisotropic scaling is the necessity of dealing with multifold cases and sub-cases corresponding to different parameter regions. The author attempted a reasonable compromise between readability and tedious technical details of some proofs, which the interested reader can find in the cited references.

Chapter 1

Random Fields and Long-Range Dependence

1 Concepts of Strong/Weak Dependence

We start with some terminology. A random field (RF) on T is a family $X = \{X(\boldsymbol{t}); \boldsymbol{t} \in T\}$ of r.v.s indexed by elements of a set T (where, usually, T is a subset of $\mathbb{R}^\nu, \nu \geq 2$). The distribution or the law of X is completely determined by finite-dimensional distributions, $\mathrm{P}(X(\boldsymbol{t}_1) \in A_1, \ldots, X(\boldsymbol{t}_n) \in A_n)$, for arbitrary $n \geq 1, \boldsymbol{t}_i \in T$ and Borel sets $A_i \subset \mathbb{R}, 1 \leq i \leq n$. In turn, a finite-dimensional distribution is completely specified by its characteristic function $(\theta_1, \ldots, \theta_n) \mapsto \mathrm{E}\exp\{\mathrm{i}\sum_{i=1}^n \theta_i X(\boldsymbol{t}_i)\} : \mathbb{R}^n \to \mathbb{C}$. An RF $X = \{X(\boldsymbol{t}); \boldsymbol{t} \in T\}, \boldsymbol{t} = (t_1, \ldots, t_\nu)$ on $T = \mathbb{Z}^\nu$ or $T = \mathbb{R}^\nu$ is called *stationary* if X and $U_{\boldsymbol{a}}X := \{X(\boldsymbol{a}+\boldsymbol{t}); \boldsymbol{t} \in T\}$ have the same distribution for any $\boldsymbol{a} \in T$. In the case when X has a finite second moment, $\mathrm{E}X(\boldsymbol{t})^2 < \infty\,(\forall\,\boldsymbol{t} \in T)$, stationarity of X implies that the mean $\mathrm{E}X(\boldsymbol{t})$ is constant (i.e., not dependent on $\boldsymbol{t} \in T$) and the covariance function $r(\boldsymbol{t}, \boldsymbol{s}) = \mathrm{Cov}(X(\boldsymbol{t}), X(\boldsymbol{s}))$ depends only on the difference $\boldsymbol{t} - \boldsymbol{s}$

$$r(\boldsymbol{t}, \boldsymbol{s}) = r(\boldsymbol{t} - \boldsymbol{s}), \quad \forall\,\boldsymbol{t}, \boldsymbol{s} \in T. \tag{1}$$

An RF is called *weakly stationary* if $\mathrm{E}X(\boldsymbol{t})^2 < \infty, \mathrm{E}X(\boldsymbol{t}) = \mathrm{E}X(\boldsymbol{s})\,(\forall \boldsymbol{t}, \boldsymbol{s} \in T)$ and (1) hold. A stationary RF with finite variance is weakly stationary, but the converse statement is not true. By Bochner's theorem, the covariance function of a weakly stationary

RF X on $\mathbb{Z}^\nu$ has a spectral representation:

$$r(\boldsymbol{t}) = \int_{\Pi^\nu} \mathrm{e}^{\mathrm{i}\langle \boldsymbol{t}, \boldsymbol{x}\rangle} F(\mathrm{d}\boldsymbol{x}), \qquad \boldsymbol{t} \in \mathbb{Z}^\nu, \tag{2}$$

where $F(\mathrm{d}x)$ is a nonnegative finite measure on Π^ν, called the *spectral measure of* X. In most cases of interest, the spectral measure is absolutely continuous w.r.t. the Lebesgue measure, which is determined by its Radon–Nikodym derivative $f(x), x \in \Pi^\nu$, called the *spectral density of* X. A *weak white noise* is an orthogonal sequence $\{\varepsilon(\boldsymbol{t}); \boldsymbol{t} \in \mathbb{Z}^\nu\}$ with zero mean and $\mathrm{E}\varepsilon(\boldsymbol{t})\varepsilon(\boldsymbol{s}) = \mathbb{I}(\boldsymbol{t} = \boldsymbol{s})$. A weak white noise is weakly stationary and has a constant spectral density $f_\varepsilon(x) = 1/(2\pi)^\nu$:

$$(2\pi)^{-\nu} \int_{\Pi^\nu} \mathrm{e}^{\mathrm{i}\langle \boldsymbol{t}, \boldsymbol{x}\rangle} \mathrm{d}\boldsymbol{x} \;=\; \mathbb{I}(\boldsymbol{t} = \boldsymbol{0}), \quad \boldsymbol{t} \in \mathbb{Z}^\nu. \tag{3}$$

A standardised i.i.d. sequence $\{\varepsilon(\boldsymbol{t}); \boldsymbol{t} \in \mathbb{Z}^\nu\}$ with $\mathrm{E}\varepsilon(\boldsymbol{t}) = 0, \mathrm{E}\varepsilon(\boldsymbol{t})^2 = 1$ is a particular case of weak white noise.

Definition 1. Let $X = \{X(\boldsymbol{t}); \boldsymbol{t} \in \mathbb{Z}^\nu\}$ be a weakly stationary RF on $\mathbb{Z}^\nu$. We say that X is:

(i) *covariance LRD* if $\sum_{\boldsymbol{t}\in\mathbb{Z}^\nu} |r(\boldsymbol{t})| = \infty$;
(ii) *covariance SRD* if $\sum_{\boldsymbol{t}\in\mathbb{Z}^\nu} |r(\boldsymbol{t})| < \infty$ and $\sum_{\boldsymbol{t}\in\mathbb{Z}^\nu} r(\boldsymbol{t}) \neq 0$;
(iii) *covariance ND* if $\sum_{\boldsymbol{t}\in\mathbb{Z}^\nu} |r(\boldsymbol{t})| < \infty$ and $\sum_{\boldsymbol{t}\in\mathbb{Z}^\nu} r(\boldsymbol{t}) = 0$.

Synonymous terminology for LRD is *long memory*, although it is more common in time series analysis and econometrics for processes indexed by $\mathbb{Z}$; see Definition 3.1.2 in Ref. [103]. Related definitions in terms of spectrum are as follows.

Definition 2. Let $X = \{X(\boldsymbol{t}); \boldsymbol{t} \in \mathbb{Z}^\nu\}$ be a weakly stationary RF on $\mathbb{Z}^\nu$ with finite second moment and spectral density f which is bounded on $\{\boldsymbol{x} \in \Pi^\nu : |\boldsymbol{x}| > \epsilon\}$ for any $\epsilon > 0$. We say that X has:

(i) *spectrum LRD* if $\lim_{\boldsymbol{x}\to\boldsymbol{0}} |f(\boldsymbol{x})| = \infty$;
(ii) *spectrum SRD* if $\lim_{\boldsymbol{x}\to\boldsymbol{0}} f(\boldsymbol{x}) > 0$;
(iii) *spectrum ND* if $\lim_{\boldsymbol{x}\to\boldsymbol{0}} f(\boldsymbol{x}) = 0$.

Definitions 1 and 2 are not equivalent, even when assuming the existence of a spectral density. Note that the covariance SRD property implies the existence of a spectral density f, given by the Fourier series

$$f(\boldsymbol{x}) = (2\pi)^{-\nu} \sum_{\boldsymbol{t}\in\mathbb{Z}^\nu} \mathrm{e}^{-\mathrm{i}\langle \boldsymbol{t},\boldsymbol{x}\rangle} r(\boldsymbol{t}), \tag{4}$$

and $f(\boldsymbol{x}), \boldsymbol{x} \in \Pi^\nu$, is a bounded and continuous function with $f(\mathbf{0}) = (2\pi)^{-\nu} \sum_{\boldsymbol{t}\in\mathbb{Z}^\nu} r(\boldsymbol{t}) \neq 0$. On the other hand, the fact that the spectral density f is unbounded implies that X has covariance LRD since the convergence of the covariance series results in a bounded spectral density.

The above classification is helpful in limit theorems. Consider the sum $S_\lambda = \sum_{\boldsymbol{t}\in K_\lambda} X(\boldsymbol{t})$ of the values of X over a large 'sampling region' $K_\lambda \subset \mathbb{Z}^\nu$, monotonically increasing to the whole lattice $\mathbb{Z}^\nu$ and such that $|K_\lambda| := \sum_{\boldsymbol{t}\in K_\lambda} 1 \to \infty \, (\lambda \to \infty)$. The sample mean $\hat{\mu}_\lambda := |K_\lambda|^{-1} S_\lambda$ consistently estimates the (unknown) mean $\mu = \mathrm{E}X(\boldsymbol{t})$ under general conditions on X and K_λ. The convergence rate of this estimator, or the difference $\hat{\mu}_\lambda - \mu$, can be assessed by knowing the growth rate of the variance $\mathrm{Var}(S_\lambda)$ as $\lambda \to \infty$. By stationarity,

$$\mathrm{Var}(S_\lambda) = \sum_{\boldsymbol{t}\in K_\lambda} \left(\sum_{\boldsymbol{s}\in K_\lambda} r(\boldsymbol{s}-\boldsymbol{t}) \right),$$

where the second sum tends to $\sigma^2 := \sum_{\boldsymbol{s}\in\mathbb{Z}^\nu} r(\boldsymbol{s}) < \infty$ under covariance SRD. In the latter case, $\mathrm{Var}(\hat{\mu}_\lambda) = O(1/|K_\lambda|)$, as in the case of i.i.d. observations. Under covariance LRD and $r(\boldsymbol{t}) \geq 0$, we have that $\mathrm{Var}(S_\lambda)/|K_\lambda| \to \infty$, and the convergence rate of $\hat{\mu}_\lambda$ is worse than $O_p(|K_\lambda|^{-1/2})$ whereas under ND, $\mathrm{Var}(S_\lambda)$ may grow much slower than $|K_\lambda|$ (– as slow as the 'volume' $|\partial K_\lambda|$ of the boundary $\partial K_\lambda := \{\boldsymbol{t} \notin \mathbb{Z}^\nu : \mathrm{dist}(\boldsymbol{t}, K_\lambda) \leq 1\}$, or even slower than $|\partial K_\lambda|$). Definitions 1–3 can be extended to stationary RFs with a continuous parameter $\boldsymbol{t} \in \mathbb{R}^\nu$.

Despite several attempts, there are no satisfactory analogues of covariance and Definitions 1/2 for infinite variance processes. Numerical characteristics of dependence for such processes include *codifference* [231] and *spectral covariance* [68]; however, these characteristics are not easily determined in concrete cases and/or apply to special infinite variance processes. In the literature, the most attention has been given to linear or moving-average (MA) processes with infinite variance innovations, for which the concepts of LRD/SRD/ND can be defined through MA coefficients. Consider an MA RF,

$$X(\boldsymbol{t}) = \sum_{\boldsymbol{u}\in\mathbb{Z}^\nu} a(\boldsymbol{t}-\boldsymbol{u})\varepsilon(\boldsymbol{u}), \quad \boldsymbol{t}\in\mathbb{Z}^\nu, \tag{5}$$

where $\varepsilon(\boldsymbol{u}), \boldsymbol{u} \in \mathbb{Z}^\nu$ are i.i.d. r.v.s, possibly with infinite variance, and $a(\boldsymbol{t}), \boldsymbol{t} \in \mathbb{Z}^\nu$ are deterministic coefficients such that

$$\sum_{\boldsymbol{t}\in\mathbb{Z}^\nu} |a(\boldsymbol{t})|^p < \infty, \qquad \mathrm{E}|\varepsilon(0)|^p < \infty, \quad \text{and} \quad \mathrm{E}\varepsilon(0) = 0$$
$$\text{for some } p \in (1,2]. \tag{6}$$

It is well known that (6) imply the L^p-convergence of the series in (5) for any $\boldsymbol{t} \in \mathbb{Z}^\nu$ and define a stationary RF with zero mean and finite pth moment.

Definition 3. Let X be an MA RF in (5) satisfying (6). We say that X is:

(i) coefficient LRD if $\sum_{\boldsymbol{t}\in\mathbb{Z}^\nu} |a(\boldsymbol{t})| = \infty$;
(ii) coefficient SRD if $\sum_{\boldsymbol{t}\in\mathbb{Z}^\nu} |a(\boldsymbol{t})| < \infty$ and $\sum_{\boldsymbol{t}\in\mathbb{Z}^\nu} a(\boldsymbol{t}) \neq 0$;
(iii) coefficient ND if $\sum_{\boldsymbol{t}\in\mathbb{Z}^\nu} |a(\boldsymbol{t})| < \infty$ and $\sum_{\boldsymbol{t}\in\mathbb{Z}^\nu} a(\boldsymbol{t}) = 0$.

In the finite variance case of $p = 2$, Definition 3 roughly agrees with Definition 1; particularly, $\sum_{\boldsymbol{t}\in\mathbb{Z}^\nu} |a(\boldsymbol{t})| < \infty$ implies $\sum_{\boldsymbol{t}\in\mathbb{Z}^\nu} |r(\boldsymbol{t})| \le \mathrm{E}\varepsilon^2(0)\big(\sum_{\boldsymbol{t}\in\mathbb{Z}^\nu} |a(\boldsymbol{t})|\big)^2 < \infty$ and $\sum_{\boldsymbol{t}\in\mathbb{Z}^\nu} r(\boldsymbol{t}) = \mathrm{E}\varepsilon^2(0)\big(\sum_{\boldsymbol{t}\in\mathbb{Z}^\nu} a(\boldsymbol{t})\big)^2 = 0$ whenever $\sum_{\boldsymbol{t}\in\mathbb{Z}^\nu} a(\boldsymbol{t}) = 0$. On the other hand, the above classification is very rough since the limit results require much more detailed information about the dependence structure of RF as seen in the following sections. Under LRD and ND, these results rely on a particular asymptotic form of the covariance function or MA coefficients, which may decay at different rates in different directions in $\mathbb{Z}^\nu$. In the 'isotropic' case, this rate is the same for all directions, as in the following assumption. Let $C(S_{\nu-1})$ denote the class of all nontrivial bounded continuous function on $S_{\nu-1} = \{\boldsymbol{t} \in \mathbb{R}^\nu : |\boldsymbol{t}| = 1\}$.

Assumption (A)$(d;\mathbb{Z}^\nu)$**.** Let $a(\boldsymbol{t}), \boldsymbol{t} \in \mathbb{Z}^\nu$ be a sequence of real numbers satisfying the following properties:

(i) Let $0 < d < \nu/4$. Then,

$$a(\boldsymbol{t}) = \frac{1}{|\boldsymbol{t}|^{\nu-2d}} \left(\ell\left(\frac{\boldsymbol{t}}{|\boldsymbol{t}|}\right) + o(1) \right), \quad |\boldsymbol{t}| \to \infty, \tag{7}$$

where $\ell(\cdot) \in C(S_{\nu-1})$.

(ii) Let $-\nu/4 < d < 0$. Then, $a(\boldsymbol{t})$ satisfy (7) with the same $\ell(\boldsymbol{t})$ and, moreover, $\sum_{\boldsymbol{t}\in\mathbb{Z}^\nu} a(\boldsymbol{t}) = 0$.

(iii) Let $d = 0$. Then, $\sum_{\boldsymbol{t}\in\mathbb{Z}^\nu} |a(\boldsymbol{t})| < \infty$ and $\sum_{\boldsymbol{t}\in\mathbb{Z}^\nu} a(\boldsymbol{t}) \neq 0$, i.e., $a(\boldsymbol{t})$ satisfies the coefficient SRD property.

By easy observation, Assumption $(\mathrm{A})(d;\mathbb{Z}^\nu)$ implies the coefficient LRD property for $d > 0$ and the coefficient ND property for $d < 0$. Following the terminology in time series [103], the parameter d in (7) may be called the *memory parameter* of the linear RF X in (5), except that for $\nu = 1$, the memory parameter is usually defined as $2d \in (-1/2, 1/2)$.

Linear RFs are the basic models of RFs studied in this book, except that they are not limited to MAs on $\mathbb{Z}^\nu$, as in (5), and include MA RFs on $\mathbb{R}^\nu$ written as stochastic integrals and shot-noise RFs containing several functional parameters. The dependence structure in such RFs is often determined by a kernel, $a(\boldsymbol{t}), \boldsymbol{t} \in \mathbb{R}^\nu$. The following assumption is a counterpart of Assumption $(\mathrm{A})(d;\mathbb{Z}^\nu)$ for continuous-argument kernels.

Assumption (A)$(d;\mathbb{R}^\nu)$**.** $a(\cdot) \in L^2(\mathbb{R}^\nu)$ is a real-valued function satisfying the following properties:

(i) Let $0 < d < \nu/4$. Then, $a(\boldsymbol{t})$ satisfies (7).
(ii) Let $-\nu/4 < d < 0$. Then, $a(\boldsymbol{t})$ satisfies (7) and $\int_{\mathbb{R}^\nu} a(\boldsymbol{t})\mathrm{d}\boldsymbol{t} = 0$.
(iii) Let $d = 0$. Then, $\int_{\mathbb{R}^\nu} |a(\boldsymbol{t})|\mathrm{d}\boldsymbol{t} < \infty$ and $\int_{\mathbb{R}^\nu} a(\boldsymbol{t})\mathrm{d}\boldsymbol{t} \neq 0$.

The covariance function of the linear RF X in (5) with standardised i.i.d. innovations is written as

$$r(\boldsymbol{t}) = \sum_{\boldsymbol{u}\in\mathbb{Z}^\nu} a(\boldsymbol{u})a(\boldsymbol{t}+\boldsymbol{u}), \quad \boldsymbol{t} \in \mathbb{Z}^\nu,$$

or the lattice convolution of $a(\boldsymbol{t})$ with itself. We use the notation $[a_1 \star a_2]$ for lattice convolution and $(a_1 \star a_2)$ for the continuous convolution:

$$[a_1 \star a_2](\boldsymbol{t}) := \sum_{\boldsymbol{u} \in \mathbb{Z}^\nu} a_1(\boldsymbol{u}) a_2(\boldsymbol{t}+\boldsymbol{u}), \quad \boldsymbol{t} \in \mathbb{Z}^\nu, \tag{8}$$

$$(a_1 \star a_2)(\boldsymbol{t}) := \int_{\mathbb{R}^\nu} a_1(\boldsymbol{u}) a_2(\boldsymbol{t}+\boldsymbol{u}) \mathrm{d}\boldsymbol{u}, \quad \boldsymbol{t} \in \mathbb{R}^\nu,$$

which are well defined for any $a_i \in L^2(\mathbb{Z}^\nu), i = 1,2$ (respectively, for any $a_i \in L^2(\mathbb{R}^\nu)$, $i = 1,2$).

Proposition 4. *(i) Let $a_i \in L^2(\mathbb{Z}^\nu)$ satisfy Assumption (A)($d; \mathbb{Z}^\nu$), with $0 < d < \nu/4$ and some $\ell_i \in C(S_{\nu-1}), i = 1,2$. Then,*

$$[a_1 \star a_2](\boldsymbol{t}) = |\boldsymbol{t}|^{4d-\nu} \left(L_{12}\left(\frac{\boldsymbol{t}}{|\boldsymbol{t}|}\right) + o(1) \right), \qquad |\boldsymbol{t}| \to \infty, \tag{9}$$

where the (angular) function $L_{12}(\cdot) \in C(S_{\nu-1})$ is given by

$$L_{12}(\boldsymbol{t}) := \int_{\mathbb{R}^\nu} \frac{\ell_1(\boldsymbol{s}/|\boldsymbol{s}|)\ell_2((\boldsymbol{t}-\boldsymbol{s})/|\boldsymbol{t}-\boldsymbol{s}|)}{|\boldsymbol{s}|^{\nu-2d}|\boldsymbol{t}-\boldsymbol{s}|^{\nu-2d}} \mathrm{d}\boldsymbol{s}, \qquad \boldsymbol{t} \in S_{\nu-1}. \tag{10}$$

(ii) Let $a_i \in L^2(\mathbb{R}^\nu)$ satisfy Assumption (A)($d; \mathbb{R}^\nu$), with $0 < d < \nu/4$ and some $\ell_i \in C(S_{\nu-1}), i = 1,2$. Then,

$$(a_1 \star a_2)(\boldsymbol{t}) = |\boldsymbol{t}|^{4d-\nu} \left(L_{12}\left(\frac{\boldsymbol{t}}{|\boldsymbol{t}|}\right) + o(1) \right), \qquad |\boldsymbol{t}| \to \infty, \tag{11}$$

with $L_{12}(\cdot) \in C(S_{\nu-1})$ as given in (10). Moreover,

$$(|a_1| \star |a_2|)(\boldsymbol{t}) \le C(|\boldsymbol{t}| \vee 1)^{4d-\nu}, \quad \boldsymbol{t} \in \mathbb{R}^\nu. \tag{12}$$

Proof. (i) The existence and continuity of L_{12} follow from the finiteness of integrals $\int_{|\boldsymbol{s}|<1} |\boldsymbol{s}|^{2d-\nu} \mathrm{d}\boldsymbol{s} < \infty$, $\int_{|\boldsymbol{s}|>1} |\boldsymbol{s}|^{2(2d-\nu)} \mathrm{d}\boldsymbol{s} < \infty$. For (9), it suffices to show that

$$|\boldsymbol{t}|^{\nu-4d}[a_1 \star a_2](\boldsymbol{t}) - L_{12}(\boldsymbol{t}/|\boldsymbol{t}|) \to 0, \qquad |\boldsymbol{t}| \to \infty. \tag{13}$$

Let $|\boldsymbol{t}|_+ := |\boldsymbol{t}| \vee 1$ and $a_i^0(\boldsymbol{t}) := |\boldsymbol{t}|_+^{2d-\nu} \ell_i(\boldsymbol{t}/|\boldsymbol{t}|_+), a_i^1(\boldsymbol{t}) := a_i(\boldsymbol{t}) - a_i^0(\boldsymbol{t}) = o(|\boldsymbol{t}|^{2d-\nu}), i = 1,2$; see (7). Then, $[a_1 \star a_2](\boldsymbol{t}) = \sum_{i,j=0}^1 [a_1^i \star$

$a_2^j](\boldsymbol{t})$. Clearly, (13) follows from

$$|\boldsymbol{t}|^{\nu-4d}[a_1^0 \star a_2^0](\boldsymbol{t}) - L_{12}(\boldsymbol{t}/|\boldsymbol{t}|) \to 0, \qquad |\boldsymbol{t}| \to \infty \tag{14}$$

and

$$[a_1^i \star a_2^j](\boldsymbol{t}) = o(|\boldsymbol{t}|^{4d-\nu}), \quad |\boldsymbol{t}| \to \infty, \quad (i,j) \neq (0,0), \ i,j = 0,1. \tag{15}$$

To prove (14), rewrite $[a_1^0 \star a_2^0](\boldsymbol{t}) = \int_{\mathbb{R}^\nu} a_1^0([\boldsymbol{u}])a_2^0(\boldsymbol{t} + [\boldsymbol{u}])\mathrm{d}\boldsymbol{u}$ as an integral and change the variable $\boldsymbol{u} \to |\boldsymbol{t}|\boldsymbol{u}$ in it. This leads to $|\boldsymbol{t}|^{\nu-4d}[a_1^0 \star a_2^0](\boldsymbol{t}) = \widetilde{L}_{\boldsymbol{t}}(\boldsymbol{t}/|\boldsymbol{t}|)$, where

$$\widetilde{L}_{\boldsymbol{t}}(\boldsymbol{z}) := \int_{\mathbb{R}^\nu} a_{1,\boldsymbol{t}}(\tilde{\boldsymbol{u}})a_{2,\boldsymbol{t}}(\boldsymbol{z} + \tilde{\boldsymbol{u}})\mathrm{d}\boldsymbol{u}, \qquad \boldsymbol{z} \in S_{\nu-1}, \tag{16}$$

where

$$a_{i,\boldsymbol{t}}(\tilde{\boldsymbol{u}}) := \frac{1}{(|\boldsymbol{t}|^{-1} \vee |\tilde{\boldsymbol{u}}|)^{\nu-2d}}\ell_i\left(\frac{\tilde{\boldsymbol{u}}}{|\boldsymbol{t}|^{-1} \vee |\tilde{\boldsymbol{u}}|}\right), \qquad \tilde{\boldsymbol{u}} := \frac{[|\boldsymbol{t}|\boldsymbol{u}]}{|\boldsymbol{t}|}.$$

Relation (14) follows once we prove the uniform convergence $\sup_{\boldsymbol{z}\in S_\nu} |\widetilde{L}_{\boldsymbol{t}}(\boldsymbol{z}) - L_{12}(\boldsymbol{z})| \to 0\,(|\boldsymbol{t}| \to \infty)$. Since $S_{\nu-1}$ is a compact set and L_2 is continuous, the last relation is implied by the sequential convergence

$$|\widetilde{L}_{\boldsymbol{t}}(\boldsymbol{z}_{\boldsymbol{t}}) - L_{12}(\boldsymbol{z})| \to 0 \qquad (|\boldsymbol{t}| \to \infty) \tag{17}$$

for any $\boldsymbol{z} \in S_{\nu-1}$ and any $\{\boldsymbol{z}_{\boldsymbol{t}}\}$ convergent to $\boldsymbol{z}$: $|\boldsymbol{z}_{\boldsymbol{t}} - \boldsymbol{z}| \to 0\,(|\boldsymbol{t}| \to \infty)$. The proof of (17) uses the bound

$$|a_{i,\boldsymbol{t}}(\tilde{\boldsymbol{u}})| \le C|\boldsymbol{u}|^{2d-\nu}, \qquad \boldsymbol{u} \in \mathbb{R}^\nu, \ i = 1,2, \tag{18}$$

which follows from the boundedness of ℓ_i and $|\boldsymbol{u}| \le |\tilde{\boldsymbol{u}}| + |\boldsymbol{u} - \tilde{\boldsymbol{u}}|$, with $|\boldsymbol{u} - \tilde{\boldsymbol{u}}| \le \nu^{1/2}/|\boldsymbol{t}|$, and hence $|\boldsymbol{u}| \le \nu^{1/2}(|\tilde{\boldsymbol{u}}| + |\boldsymbol{t}|^{-1}) \le 2\nu^{1/2}(|\tilde{\boldsymbol{u}}| \vee |\boldsymbol{t}|^{-1})$. Note that $a_{1,\boldsymbol{t}}(\tilde{\boldsymbol{u}})a_{2,\boldsymbol{t}}(\boldsymbol{z} + \tilde{\boldsymbol{u}}) \to a_1^0(\boldsymbol{u})a_2^0(\boldsymbol{z} + \boldsymbol{u})\,(|\boldsymbol{t}| \to \infty)$ for any $\boldsymbol{u} \neq \boldsymbol{0}, \boldsymbol{z}$ and $|a_{1,\boldsymbol{t}}(\tilde{\boldsymbol{u}})a_{2,\boldsymbol{t}}(\boldsymbol{z}+\tilde{\boldsymbol{u}})| \le C|\boldsymbol{u}|^{2d-\nu}|\boldsymbol{z}+\boldsymbol{u}|^{2d-\nu}$ according to (18). Since $h(\boldsymbol{u}) := C|\boldsymbol{u}|^{2d-\nu}|\boldsymbol{z} + \boldsymbol{u}|^{2d-\nu}$ does not depend on $\boldsymbol{t}$ and $\int_{\mathbb{R}^\nu} h(\boldsymbol{u})\mathrm{d}\boldsymbol{u} < \infty$, Pratt's lemma applies to the integral in (16), resulting in (17) and (14). As the proof of (15) is similar and simpler, it is omitted.

(ii) Using the same decomposition $a_i(\boldsymbol{t}) = a_i^0(\boldsymbol{t}) + a_i^1(\boldsymbol{t}), i = 1, 2$ as in part (i), the proof of $|\boldsymbol{t}|^{\nu-4d}(a_1^0 \star a_2^0)(\boldsymbol{t}) - L_{12}(\boldsymbol{t}/|\boldsymbol{t}|) \to 0$ is analogous to that of (14). Let us prove $(a_1^1 \star a_2)(\boldsymbol{t}) = o(|\boldsymbol{t}|^{4d-\nu}), |\boldsymbol{t}| \to \infty$. For a fixed but large $K > 0$, split $(a_1^1 \star a_2)(\boldsymbol{t}) = \sum_{i=1}^3 J_{\boldsymbol{t},i}$, where $J_{\boldsymbol{t},1} := \int_{|\boldsymbol{u}|\le K}(a_1(\boldsymbol{u}) - a_1^0(\boldsymbol{u}))a_2(\boldsymbol{t}+\boldsymbol{u})\mathrm{d}\boldsymbol{u}$, $J_{\boldsymbol{t},2} := \int_{|\boldsymbol{t}+\boldsymbol{u}|\le K}(a_1(\boldsymbol{u}) - a_1^0(\boldsymbol{u}))a_2(\boldsymbol{t}+\boldsymbol{u})\mathrm{d}\boldsymbol{u}$, and $J_{\boldsymbol{t},3} := \int_{|\boldsymbol{u}|>K, |\boldsymbol{t}+\boldsymbol{u}|>K}(a_1(\boldsymbol{u}) - a_1^0(\boldsymbol{u}))a_2(\boldsymbol{t}+\boldsymbol{u})\mathrm{d}\boldsymbol{u}$. Then, $|J_{\boldsymbol{t},1}| \le C|\boldsymbol{t}|^{-(\nu-2d)}\int_{|\boldsymbol{u}|<K}(|a_1(\boldsymbol{u})|+1)\mathrm{d}\boldsymbol{u} = o(|\boldsymbol{t}|^{4d-\nu})$. Similarly, $|J_{\boldsymbol{t},2}| = o(|\boldsymbol{t}|^{4d-\nu})$. On the other hand, since $|a_2(\boldsymbol{t}+\boldsymbol{u})| \le C|\boldsymbol{t}+\boldsymbol{u}|^{2d-\nu}$ and $|a_1(\boldsymbol{u}) - a_1^0(\boldsymbol{u})| \le \delta_K|\boldsymbol{u}|^{2d-\nu}$ hold for all $|\boldsymbol{t}+\boldsymbol{u}| > K, |\boldsymbol{u}| > K$ with $\delta_K > 0$ arbitrarily small, it follows that

$$|J_{\boldsymbol{t},3}| \le C\,\delta_K \int_{|\boldsymbol{u}|\ge K} |\boldsymbol{u}|^{2d-\nu}|\boldsymbol{t}+\boldsymbol{u}|^{2d-\nu}\mathrm{d}\boldsymbol{u} \le C\delta_K|\boldsymbol{t}|^{4d-\nu}; \tag{19}$$

therefore, $|\boldsymbol{t}|^{\nu-4d}|(a_1^1 \star a_2)(\boldsymbol{t})|$ can be made small uniformly in $\boldsymbol{t}$ by choosing K large enough.

Consider (12). Since $|(|a_1| \star |a_2|)(\boldsymbol{t})| \le \prod_{i=1}^2 \|a_i\|_{L^2(\mathbb{R}^\nu)} < C$, it suffices to prove (12) for $|\boldsymbol{t}|$ large enough. Similarly as above, split $|(|a_1| \star |a_2|)(\boldsymbol{t})| = \sum_{i=1}^3 \bar{J}_{\boldsymbol{t},i}$, where $\bar{J}_{\boldsymbol{t},1} := \int_{|\boldsymbol{u}|\le K}|a_1(\boldsymbol{u})a_2(\boldsymbol{t}+\boldsymbol{u})|\mathrm{d}\boldsymbol{u}$, $\bar{J}_{\boldsymbol{t},2} := \int_{|\boldsymbol{t}+\boldsymbol{u}|\le K}|a_1(\boldsymbol{u})a_2(\boldsymbol{t}+\boldsymbol{u})|\mathrm{d}\boldsymbol{u}$, $\bar{J}_{\boldsymbol{t},3} := \int_{|\boldsymbol{u}|>K,|\boldsymbol{t}+\boldsymbol{u}|>K}|a_1(\boldsymbol{u})a_2(\boldsymbol{t}+\boldsymbol{u})|\mathrm{d}\boldsymbol{u}$. Then, $\bar{J}_{\boldsymbol{t},1} \le C|\boldsymbol{t}|^{-(\nu-2d)}\int_{|\boldsymbol{u}|\le K}|a_1(\boldsymbol{u})|\mathrm{d}\boldsymbol{u} \le C|\boldsymbol{t}|^{-(\nu-2d)}\|a_1\|_{L^2(\mathbb{R}^\nu)} = o(|\boldsymbol{t}|^{-(\nu-4d)})$ and $\bar{J}_{\boldsymbol{t},3} = o(|\boldsymbol{t}|^{-(\nu-4d)})$. Finally,

$$|\bar{J}_{\boldsymbol{t},3}| \le C\int_{|\boldsymbol{u}|\ge K,|\boldsymbol{t}+\boldsymbol{u}|\ge K}|\boldsymbol{u}|^{2d-\nu}|\boldsymbol{t}+\boldsymbol{u}|^{2d-\nu}\mathrm{d}\boldsymbol{u} \le C|\boldsymbol{t}|^{4d-\nu},$$

as in (19). Proposition 4 is proved. □

2 Scaling and Self-Similarity Concepts for RFs

Scaling operation is usually defined as a group of scaling transformations, $\{X \mapsto T_\lambda X; \lambda > 0\}$, acting in a given class of RFs, X, with the group property meaning that $T_{\lambda_1}T_{\lambda_2}X \overset{\mathrm{fdd}}{=} T_{\lambda_1\lambda_2}X\ \forall \lambda_1, \lambda_2 > 0$ and $T_1X \overset{\mathrm{fdd}}{=} X$. Clearly, for any $H \ge 0$, the family $\{X \mapsto \lambda^{-H}T_\lambda X; \lambda > 0\}$ preserves the group property. An RF Y which is invariant (a fixed point) w.r.t. the above transformation group, viz., $\lambda^{-H}T_\lambda Y \overset{\mathrm{fdd}}{=} Y \ \forall\, \lambda > 0$, is called *self-similar with (self-similarity) index* H (H-SS for short). (For reasons explained in Introduction, we include $H = 0$ in the above definition, albeit most authors *a priori* restrict to $H > 0$.) The group property is sometimes relaxed; for instance, in statistical physics, the famous renormalisation theory (see Ref. [79]) defines the scaling transformation through sums of RF with a discrete argument, $\boldsymbol{t} \in \mathbb{Z}^2$, on blocks of size n. This operation does not have an inverse, resulting in a semigroup of transformations, $n^{-H}T_n$, indexed by $n = 1, 2, \ldots$; however, its fixed point – H-SS RF – is defined as above. Scaling operations can be also defined for generalised RFs, which do not exist in the usual sense; see, e.g., Refs. [78, 79], in which case they act in a given class of test functions.

A very general form of scaling acting on the RF $X = \{X(\boldsymbol{t}); \boldsymbol{t} \in \mathbb{R}^\nu\}$ is associated with a linear operator (matrix), $\Gamma : \mathbb{R}^\nu \to \mathbb{R}^\nu$, as

$$T_\lambda X(\boldsymbol{t}) := X(\lambda^\Gamma \boldsymbol{t}), \quad \boldsymbol{t} \in \mathbb{R}^\nu. \tag{20}$$

The matrix exponential (defined as the linear operator $\boldsymbol{t} \mapsto \lambda^\Gamma \boldsymbol{t} = \mathrm{e}^{\Gamma \log \lambda}\boldsymbol{t} = \sum_{k=0}^\infty \frac{(\log \lambda)^k}{k!}\Gamma^k \boldsymbol{t} : \mathbb{R}^\nu \to \mathbb{R}^\nu$) satisfies $\lambda_1^\Gamma \lambda_2^\Gamma = (\lambda_1\lambda_2)^\Gamma$, and hence (20) forms a transformation group. A real-valued RF X satisfying $\lambda^{-H}T_\lambda X(\boldsymbol{t}) \overset{\mathrm{fdd}}{=} X(\boldsymbol{t})$ ($\forall\, \lambda > 0$) is called *operator scaling* RF (OSRF) with index H (H-OSRF). In this book, we almost exclusively consider *scaling with a diagonal matrix:*

$$\Gamma = \mathrm{diag}(\gamma_1, \ldots, \gamma_\nu) \quad \text{is a diagonal matrix with entries} \quad \gamma_i > 0,$$
$$i = 1, \ldots, \nu. \tag{21}$$

An OSRF X with Γ in (21) is said to be $(H, \boldsymbol{\gamma})$*-SS (self-similar)* ($H \in \mathbb{R}, \boldsymbol{\gamma} = (\gamma_1, \ldots, \gamma_\nu) \in R_+^\nu$):

$$\{X(\lambda^{\gamma_1}t_1, \ldots, \lambda^{\gamma_\nu}t_\nu); \boldsymbol{t} \in \mathbb{R}^\nu\} \overset{\mathrm{fdd}}{=} \{\lambda^H X(\boldsymbol{t}); \boldsymbol{t} \in \mathbb{R}^\nu\} \quad (\forall\, \lambda > 0). \tag{22}$$

Since $\lambda^\Gamma \boldsymbol{t} \in \mathbb{R}_+^\nu$ ($\boldsymbol{t} \in \mathbb{R}_+^\nu$), the $(H, \boldsymbol{\gamma})$-SS property in (22) also applies to the RF $\{X(\boldsymbol{t}); \boldsymbol{t} \in \mathbb{R}_+^\nu\}$ defined on $\mathbb{R}_+^\nu$. For $\boldsymbol{\gamma} = \mathbf{1} = (1, \ldots, 1)$,

(22) agrees with the usual definition of H-SS RF:

$$\{X(\lambda \boldsymbol{t}); \boldsymbol{t} \in \mathbb{R}^\nu\} \overset{\text{fdd}}{=} \{\lambda^H X(\boldsymbol{t}); \boldsymbol{t} \in \mathbb{R}^\nu\} \quad (\forall\, \lambda > 0). \tag{23}$$

An RF X is said to be *multi-self-similar with index* $\boldsymbol{H} = (H_1, \ldots, H_\nu) \in \mathbb{R}^\nu$ ($\boldsymbol{H}$-MSS) if

$$\{X(\lambda_1 t_1, \ldots, \lambda_\nu t_\nu); \boldsymbol{t} \in \mathbb{R}^\nu\} \overset{\text{fdd}}{=} \left\{ \left(\prod_{i=1}^{\nu} \lambda_i^{H_i} \right) X(\boldsymbol{t}); \boldsymbol{t} \in \mathbb{R}^\nu \right\}$$
$$(\forall\, \lambda_1, \ldots, \lambda_\nu > 0). \tag{24}$$

Contrary to some other works, our definitions in (22)– (24) of various self-similarity properties *do not assume stochastic continuity of* X and are formulated in terms of finite-dimensional distributions alone. Clearly, *an* $\boldsymbol{H}$*-MSS RF is* $(H, \boldsymbol{\gamma})$*-SS for any* $\boldsymbol{\gamma} \in \mathbb{R}^\nu_+$, with

$$H = \sum_{i=1}^{\nu} H_i \gamma_i. \tag{25}$$

An RF $X = \{X(\boldsymbol{t}); \boldsymbol{t} \in \mathbb{R}^\nu_+\}$ is said to be *non-trivial* if $\mathrm{P}(X(\boldsymbol{t}) \neq 0) > 0$ for some $\boldsymbol{t} \in \mathbb{R}^\nu_+$. For $\boldsymbol{\lambda} = (\lambda_1, \ldots, \lambda_\nu) \in \mathbb{R}^\nu_+$, let $\boldsymbol{\Lambda} = \mathrm{diag}(\lambda_1, \ldots, \lambda_\nu)$, and we write $\boldsymbol{\lambda} \to \infty$ (respectively, $\boldsymbol{\lambda} \to \boldsymbol{0}$) whenever $\min_{1 \le i \le \nu} \lambda_i \to \infty$ (respectively, $\max_{1 \le i \le \nu} \lambda_i \to 0$).

Definition 5. Let $Y = \{Y(\boldsymbol{t}); \boldsymbol{t} \in \mathbb{R}^\nu_+\}$ be an RF, $\Gamma, \boldsymbol{\gamma} = (\gamma_1, \ldots, \gamma_\nu) \in \mathbb{R}^\nu_+$, as in (21). We say that:

(i) Y has a large-scale $\boldsymbol{\gamma}$-scaling limit (respectively, a small-scale $\boldsymbol{\gamma}$-scaling limit), $V_{\boldsymbol{\gamma}} = \{V_{\boldsymbol{\gamma}}(\boldsymbol{t}); \boldsymbol{t} \in \mathbb{R}^\nu_+\}$, if

$$d_{\lambda,\boldsymbol{\gamma}}^{-1} Y(\lambda^\Gamma \boldsymbol{t}) \xrightarrow{\text{fdd}} V_{\boldsymbol{\gamma}}(\boldsymbol{t}) \quad \text{as} \quad \lambda \to \infty \quad (\text{respectively,} \quad \lambda \to 0), \tag{26}$$

where $d_{\lambda,\boldsymbol{\gamma}} > 0$ is a normalisation tending to ∞ as $\lambda \to \infty$ (respectively, tending to 0 as $\lambda \to 0$). The family $\{V_{\boldsymbol{\gamma}}; \boldsymbol{\gamma} \in \mathbb{R}^\nu_+\}$ of all non-trivial $\boldsymbol{\gamma}$-scaling limits in (26) is called the large-scale (respectively,small-scale) scaling diagram of the RF Y.

(ii) Y has a large-scale joint scaling limit (respectively, small-scale joint scaling limit) $V = \{V(\boldsymbol{t}); \boldsymbol{t} \in \mathbb{R}_+^\nu\}$ if

$$d_{\boldsymbol{\lambda}}^{-1} Y(\boldsymbol{\Lambda t}) \xrightarrow{\text{fdd}} V(\boldsymbol{t}) \quad \text{as} \quad \boldsymbol{\lambda} \to \infty \quad (\text{respectively,} \quad \boldsymbol{\lambda} \to \boldsymbol{0}), \tag{27}$$

where $d_{\boldsymbol{\lambda}} > 0$ is a normalisation tending to ∞ as $\boldsymbol{\lambda} \to \infty$ (respectively, tending to 0 as $\boldsymbol{\lambda} \to \boldsymbol{0}$).

The scaling limit in (26) is said to be *isotropic* if $\gamma_1 = \cdots = \gamma_\nu =: \gamma > 0$ or $\boldsymbol{\gamma} = \gamma \boldsymbol{1}$ and *anisotropic* otherwise. Let us remark that the above self-similarity concepts do not assume (and sometimes exclude) the stationarity of RF. For *nonstationary* RFs, dependence can be characterised by their increments, provided the latter is stationary. Given an RF $X = \{X(\boldsymbol{t}); \boldsymbol{t} \in \mathbb{R}^\nu\}$, the two basic types of its increment at a point $\boldsymbol{t}_0 \in \mathbb{R}^\nu$ are the *(ordinary) increment* $X(\boldsymbol{t}_0 + \boldsymbol{t}) - X(\boldsymbol{t}_0)$ and the *rectangular increment*:

$$X(]\boldsymbol{t}_0, \boldsymbol{t}_0 + \boldsymbol{t}]) := \sum_{k_1,\ldots,k_\nu=0,1} (-1)^{k_1+\cdots+k_\nu} X\left(\boldsymbol{t}_0 + \sum_{i=1}^{\nu} k_i t_i \boldsymbol{e}_i\right),$$
$$\boldsymbol{t} \in \mathbb{R}_+^\nu. \tag{28}$$

Particularly, for $\nu = 2$,

$$\begin{aligned} X(]\boldsymbol{t}_0, \boldsymbol{t}_0 + \boldsymbol{t}]) &= X(t_{01} + t_1, t_{02} + t_2) - X(t_{01}, t_{02} + t_2) \\ &\quad - X(t_{01} + t_1, t_{02}) + X(t_{01}, t_{02}). \end{aligned} \tag{29}$$

Definition 6. Let $X = \{X(\boldsymbol{t}); \boldsymbol{t} \in \mathbb{R}^\nu\}$ be an RF. We say that

(i) X *has stationary increments* if $\{X(\boldsymbol{t}_0 + \boldsymbol{t}) - X(\boldsymbol{t}_0); \boldsymbol{t} \in \mathbb{R}^\nu\} \overset{\text{fdd}}{=} \{X(\boldsymbol{t}) - X(\boldsymbol{0}); \boldsymbol{t} \in \mathbb{R}^\nu\}$ for any $\boldsymbol{t}_0 \in \mathbb{R}^\nu$;
(ii) X *has stationary rectangular increments* if $(X(]\boldsymbol{t}_0, \boldsymbol{t}_0 + \boldsymbol{t}]); \boldsymbol{t} \in \mathbb{R}_+^\nu\} \overset{\text{fdd}}{=} \{X(]\boldsymbol{0}, \boldsymbol{t}]); \boldsymbol{t} \in \mathbb{R}_+^\nu\}$ for any $\boldsymbol{t}_0 \in \mathbb{R}^\nu$;
(iii) X *has independent rectangular increments* if, for any disjoint rectangles $K_i :=]\boldsymbol{t}_{0i}, \boldsymbol{t}_{0i} + \boldsymbol{t}_i], i = 1, \ldots, m, K_i \cap K_j = \emptyset\, (i \neq j), m = 1, 2, \ldots$, the rectangular increments $X(K_i), i = 1, \ldots, m$ are independent.

If X has stationary increments, then it also has stationary rectangular increments since (28) can be written as a sum of ordinary increments; see (29). However, the converse implication is not true. For instance, a Brownian sheet (BS) B (see Example 11, (45)) has stationary rectangular increments, but the variance $\mathrm{Var}(B(\boldsymbol{t}_0+\boldsymbol{t})-B(\boldsymbol{t}_0)) = \mathrm{Leb}_\nu(]\mathbf{0},\boldsymbol{t}_0+\boldsymbol{t}]) - \mathrm{Leb}_\nu(]\mathbf{0},\boldsymbol{t}_0]), \boldsymbol{t}\in\mathbb{R}^\nu_+$ of the ordinary increment depends on $\boldsymbol{t}_0$ when $\nu \geq 2$.

In the literature, the isotropic local scaling limit of the increment RF $Y(\boldsymbol{t}) = X(\boldsymbol{t}_0+\boldsymbol{t}) - X(\boldsymbol{t}_0)$ is often called a *tangent* RF; see Refs. [58, 88]. We introduce the following terminology, with 'rectangent' as an abridged form of 'rectangular tangent'.

Definition 7. Let $X = \{X(\boldsymbol{t}); \boldsymbol{t}\in\mathbb{R}^\nu\}$ be an RF, $\boldsymbol{t}_0\in\mathbb{R}^\nu, \boldsymbol{\gamma}\in\mathbb{R}^\nu_+$. We say that X has a (i) $\boldsymbol{\gamma}$-tangent RF $T_{\boldsymbol{\gamma}} = \{T_{\boldsymbol{\gamma}}(\boldsymbol{t}); \boldsymbol{t}\in\mathbb{R}^\nu\}$ at $\boldsymbol{t}_0$ if

$$d^{-1}_{\lambda,\boldsymbol{\gamma}}(X(\boldsymbol{t}_0+\lambda^\Gamma\boldsymbol{t}) - X(\boldsymbol{t}_0)) \overset{\text{fdd}}{\longrightarrow} T_{\boldsymbol{\gamma}}(\boldsymbol{t}), \qquad \lambda\to 0, \tag{30}$$

where $d_{\lambda,\boldsymbol{\gamma}} \to 0$ $(\lambda\to 0)$ is a normalisation. (ii) $\boldsymbol{\gamma}$-rectangent RF $V_{\boldsymbol{\gamma}} = \{V_{\boldsymbol{\gamma}}(\boldsymbol{t}); \boldsymbol{t}\in\mathbb{R}^\nu_+\}$ at $\boldsymbol{t}_0$ if

$$d^{-1}_{\lambda,\boldsymbol{\gamma}}X(]\boldsymbol{t}_0,\boldsymbol{t}_0+\lambda^\Gamma\boldsymbol{t}]) \overset{\text{fdd}}{\longrightarrow} V_{\boldsymbol{\gamma}}(\boldsymbol{t}), \qquad \lambda\to 0, \tag{31}$$

where $d_{\lambda,\boldsymbol{\gamma}} \to 0$ $(\lambda\to 0)$ is a normalisation.

Needless to say, the limits in (30) and (31) generally depend on $\boldsymbol{t}_0$, except for the case when X has stationary increments and stationary rectangular increments as in Chapter 7, where these limits are studied for a particular class of RFs X on $\mathbb{R}^2$.

Let us turn to self-similarity properties of scaling limits in (26) and (27). The following theorem deals with large-scale limits, but it also applies to small-scale limits with obvious changes.

Theorem 8. *Let an RF $Y = \{Y(\boldsymbol{t}); \boldsymbol{t}\in\mathbb{R}^\nu_+\}$ have a large-scale $\boldsymbol{\gamma}$-scaling limit $V_{\boldsymbol{\gamma}}$ for some $\boldsymbol{\gamma}\in\mathbb{R}^\nu_+$.*

(i) *Let*

$$d_{\lambda,\boldsymbol{\gamma}} = \mathcal{L}(\lambda)\lambda^H, \tag{32}$$

where $H > 0$ and $\mathcal{L}:\mathbb{R}_+\to\mathbb{R}_+$ is slowly varying at infinity. Then, $V_{\boldsymbol{\gamma}}$ is $(H,\boldsymbol{\gamma})$-SS. Moreover, if Y has stationary rectangular increments, then $V_{\boldsymbol{\gamma}}$ also has stationary rectangular increments.

(ii) *Assume that $\mathrm{P}(V_{\boldsymbol{\gamma}}(\boldsymbol{t})\neq 0) > 0\,\forall\boldsymbol{t}\in\mathbb{R}^\nu_+$ and $\nu\leq 2$. Then, (32) is true.*

Proof. (i) Fix $\lambda > 0$, and let $\tilde{\mu} := \mu\lambda, \mu > 0$. Then, $\mathcal{L}(\mu)/\mathcal{L}(\tilde{\mu}) \to 1$, $\mu \to \infty$ and

$$\begin{aligned} V_\gamma(\lambda^\Gamma \boldsymbol{t}) &= \operatorname*{fdd-lim}_{\mu\to\infty} \frac{1}{d_{\mu,\gamma}} Y((\lambda\mu)^\Gamma \boldsymbol{t}) \\ &= \operatorname*{fdd-lim}_{\tilde{\mu}\to\infty} \frac{\mathcal{L}(\tilde{\mu})}{\mathcal{L}(\mu)} \frac{\lambda^H}{d_{\tilde{\mu},\gamma}} Y(\tilde{\mu}^\Gamma \boldsymbol{t}) \overset{\mathrm{fdd}}{=} \lambda^H V_\gamma(\boldsymbol{t}). \end{aligned}$$

(ii) Let $\nu = 2$ since the case $\nu = 1$ uses the same argument. W.l.g., we can set $\gamma_1 = 1, \gamma_2 = \gamma$ and write $d_{\lambda,\gamma} = d_{\lambda,\gamma}$. For a fixed $t_1 > 0$ and $\Gamma = \operatorname{diag}(1,\gamma)$, we have $Y(\lambda, \lambda^\gamma) = Y(\lambda^\Gamma \mathbf{1}) = Y(\tilde{\lambda}^\Gamma \tilde{\boldsymbol{t}})$, where $\tilde{\lambda} := t_1\lambda$, $\tilde{\boldsymbol{t}} := (1/t_1, 1/t_1^\gamma)$. Then, $d_{\lambda,\gamma}^{-1} Y(\lambda^\Gamma \mathbf{1}) \xrightarrow{\mathrm{d}} V_\gamma(\mathbf{1})$, $d_{\tilde{\lambda},\gamma}^{-1} Y(\tilde{\lambda}^\Gamma \tilde{\boldsymbol{t}}) = d_{\tilde{\lambda},\gamma}^{-1} Y(\lambda^\Gamma \mathbf{1}) \xrightarrow{\mathrm{d}} V_\gamma(\tilde{\boldsymbol{t}})$ and the non-degeneracy of the limit distribution imply that

$$\lim_{\lambda\to\infty} \frac{d_{\lambda t_1,\gamma}}{d_{\lambda,\gamma}} = t_1^H \tag{33}$$

for some $H > 0$ [146]. Thus, $\lambda \mapsto d_{\lambda,\gamma}$ is regularly varying with index H, and hence $d_{\lambda,\gamma} = \mathcal{L}(\lambda)\lambda^H$, with $\mathcal{L}$ as in (32). □

In Theorem 8 (ii), the case of $\nu = 1$ is a simplified version of Lamperti's theorem [146, Thm. 2]. We note that Refs. [72, 146] and most other studies on Lamperti's theorem and its extensions assume the limit RF V_γ to be stochastically continuous instead of $\mathrm{P}(V_\gamma(\boldsymbol{t}) \neq 0) > 0, \forall \boldsymbol{t} \in \mathbb{R}_+^2$. However, some scaling limits in this book are not stochastically continuous, and the formulation as in Theorem 8 is preferable.

Remark 9. In Chapters 7–11, scaling operations are applied to *partial sums (integrals)* RFs of the form

$$Y(\boldsymbol{t}) = \sum_{\boldsymbol{s}\in]\mathbf{0},\boldsymbol{t}]\cap\mathbb{Z}^\nu} X(\boldsymbol{s}) \quad \text{or} \quad Y(\boldsymbol{t}) = \int_{]\mathbf{0},\boldsymbol{t}]\cap\mathbb{R}^\nu} X(\boldsymbol{s})\mathrm{d}\boldsymbol{s}, \qquad \boldsymbol{t} \in \mathbb{R}_+^\nu, \tag{34}$$

for stationary RF X on $\mathbb{Z}^\nu$ or $\mathbb{R}^\nu$. In such cases, the terminology in Definitions 5 and 18 is used respectively for RF X instead of Y.

Scaling of RFs indexed by test functions. Equation (34) naturally extends to integral functionals:

$$X(\phi) := \int_{\mathbb{R}^\nu} X([\boldsymbol{s}])\phi(\boldsymbol{s})\mathrm{d}\boldsymbol{s} \quad \text{or} \quad X(\phi) := \int_{\mathbb{R}^\nu} X(\boldsymbol{s})\phi(\boldsymbol{s})\mathrm{d}\boldsymbol{s}, \quad \phi \in \Phi, \tag{35}$$

where $X(\boldsymbol{s})$ is as in (34) and ϕ belongs to a linear space Φ of (test) functions $\phi : \mathbb{R}^\nu \to \mathbb{R}$. (34) corresponds to (35) and $\phi(\boldsymbol{s}) = \mathbb{I}(\boldsymbol{s} \in]\mathbf{0}, \boldsymbol{t}])$, in which case Φ can be identified with the linear space of indicator functions from $\Phi_{\mathrm{rec},\nu} := \{\phi(\boldsymbol{s}) = \mathbb{I}(\boldsymbol{s} \in]\mathbf{0}, \boldsymbol{t}]); \boldsymbol{t} \in \mathbb{R}^\nu_+\}$. Shift and scaling operations for $X(\phi)$ in (35) can be defined through test functions, provided Φ is closed under such operations. In such cases, following Definition 5, we can define a rescaled RF in (35) as

$$X_{\lambda,\Gamma}(\phi) := \int_{\mathbb{R}^\nu} \phi(\lambda^{-\Gamma}\boldsymbol{t})X(\boldsymbol{t})\mathrm{d}\boldsymbol{t}, \quad \phi \in \Phi \tag{36}$$

for Γ as in Definition 5, and the corresponding scaling limits

$$d^{-1}_{\lambda,\Gamma} X_{\lambda,\Gamma}(\phi) \xrightarrow{\mathrm{d}} V_\Gamma(\phi), \quad \forall\, \phi \in \Phi, \tag{37}$$

as λ tends to ∞ or 0. The scaling in (36) is said to be *isotropic* if $\Gamma = I$, the unit matrix, or $\gamma_1 = \cdots = \gamma_\nu$, and *anisotropic* otherwise. The approach through test functions is common in the theory of *generalised RFs* [78, 79, 94]. The choice of the set Φ of test functions is very important for the existence or non-existence of the limits in (37). In most works on generalised RFs, Φ is considered as the Schwartz space $\mathcal{S}(\mathbb{R}^\nu)$ or $\mathcal{D}(\mathbb{R}^\nu)$ of infinitely differentiable functions (rapidly decreasing or with compact support). The above classes do not include indicator functions from $\Phi_{\mathrm{rec},\nu}$, and we do not use them as a rule. The main class of test functions in this book is

$$\Phi = L^1(\mathbb{R}^\nu) \cap L^\infty(\mathbb{R}^\nu), \tag{38}$$

or the class of all (measurable) functions $\phi : \mathbb{R}^\nu \to \mathbb{R}$, with $\int_{\mathbb{R}^\nu} |\phi(\boldsymbol{s})|\mathrm{d}\boldsymbol{s} + \sup_{\boldsymbol{s}\in\mathbb{R}^\nu} |\phi(\boldsymbol{s})| < \infty$. Φ in (38) is closed under shift and scaling operations, and it contains the indicator functions $\phi(\boldsymbol{s}) = \mathbb{I}(\boldsymbol{s} \in A)$ of the arbitrary Borel set $A \subset \mathbb{R}^\nu$, with $\mathrm{Leb}_\nu(A) < \infty$.

Scaling with aggregation. Let $X_i, i = 1, \dots, M$ be independent copies of RF X and

$$X_{\lambda,M,\Gamma}(\phi) := \sum_{i=1}^{M} \int_{\mathbb{R}^\nu} \phi(\lambda^{-\Gamma}\boldsymbol{t}) X_i(\boldsymbol{t}) \mathrm{d}\boldsymbol{t}, \quad \phi \in \Phi, \tag{39}$$

be the 'aggregated sum' of (36). The suitably normalised limit of 'aggregates' $X_{\lambda,M,\Gamma}(\phi)$, in distribution, as $\lambda \to \infty$ and M increases with λ at a certain rate, is called the *scaling limit with aggregation.* By concretising M as $M = [\lambda^\gamma y]$ for some $\gamma, y > 0$, (39) can be regarded as a discretised version of the integral in (36) for stationary RF X' on $\mathbb{R}^\nu \times \mathbb{Z} = \{\boldsymbol{t}' = (\boldsymbol{t}, s), \boldsymbol{t} \in \mathbb{R}^\nu, s \in \mathbb{Z}\}$, given by $X'(\boldsymbol{t}') = X_s(\boldsymbol{t})$, corresponding to the scaling $\boldsymbol{t}' := (\boldsymbol{t}, s) \to (\lambda^{-\Gamma}\boldsymbol{t}, \lambda^{-\gamma}s) = \lambda^{-\Gamma'}\boldsymbol{t}'$ in $\mathbb{R}^{\nu+1}$, with diagonal matrix $\Gamma' := \mathrm{diag}(\gamma_1, \dots, \gamma_\nu, \gamma)$ and test function $\phi'(s, \boldsymbol{t}) := \phi(\boldsymbol{t})\mathbb{I}(s \in]0, y])$. In the case when $\Gamma = \mathrm{diag}(1, \dots, 1)$ is the unit matrix, we denote (39) by $X_{\lambda,M}(\phi)$ and call the corresponding limit the *isotropic scaling limit of X with aggregation.*

Example 10 (Lévy Sheet). A Lévy Sheet $W_\alpha(\boldsymbol{x}) = \{W_\alpha(\boldsymbol{x}); \boldsymbol{x} \in \mathbb{R}^\nu_+\}, \alpha \in (0, 2]$ is an RF such that $W_\alpha(\boldsymbol{x}) = 0$ for $\boldsymbol{x} = (x_1, \dots, x_\nu), \prod_{j=1}^{\nu} x_j = 0$ and for any *disjoint* rectangles $K_j =]\boldsymbol{y}_j, \boldsymbol{x}_j], j = 1, \dots, m, K_i \cap K_j = \emptyset \, (i \neq j)$, the rectangular increments $W_\alpha(K_j), j = 1, \dots, m$ are independent with characteristic function

$$\mathrm{E}\mathrm{e}^{\mathrm{i}\theta W_\alpha(K_j)} = \mathrm{e}^{-|\theta|^\alpha \omega_\alpha(\theta) \mathrm{Leb}_\nu(K_j)}, \qquad \theta \in \mathbb{R}, \tag{40}$$

where

$$\omega_\alpha(\theta) := \begin{cases} \sigma^2/2, & \alpha = 2, \\ \frac{\Gamma(2-\alpha)}{1-\alpha}\big((c_+ + c_-)\cos\left(\frac{\pi\alpha}{2}\right) & 0 < \alpha < 2, \alpha \neq 1, \\ -\mathrm{i}(c_+ - c_-)\mathrm{sgn}(\theta)\sin\left(\frac{\pi\alpha}{2}\right)\big), & \\ (c_+ + c_-)\frac{\pi}{2}, & \alpha = 1, c_+ = c_-, \end{cases} \tag{41}$$

and $\sigma > 0, c_+ \geq 0, c_- \geq 0, c_+ + c_- > 0$ are parameters. Equations (40)– (41) mean that $W_\alpha(K_j)$ has α-stable distribution (where, for $\alpha = 1$, this distribution is symmetric). The fact that (40)–(41) determine the distribution of an arbitrary linear combination $\sum_{j=1}^{m} \theta_j W_\alpha(\boldsymbol{x}_j)$ follows, as the latter can be represented as $\sum_{j=1}^{m'} \theta'_j W_\alpha(K_j)$ with disjoint K_j and some $\theta'_j, j = 1, \dots, m' \geq m$

written in terms of $\theta_j, j = 1, \ldots, m$. The two properties of the Lévy sheet often used in this book are as follows:

- W_α has stationary and independent rectangular increments;
- W_α is $\boldsymbol{H}$-MSS with $\boldsymbol{H} = (1/\alpha, \ldots, 1/\alpha) \in \mathbb{R}^\nu_+$.

Indeed, the stationarity of increments holds since shifts in $\mathbb{R}^\nu$ do not change $\mathrm{Leb}_\nu(K_j)$. Similarly, $\mathrm{Leb}_\nu(\Lambda K_j) = (\prod_{i=1}^\nu \lambda_i)\mathrm{Leb}_\nu(K_j)$ and $|\theta|^\alpha \omega_\alpha(\theta)\mathrm{Leb}_\nu(\Lambda K_j) = |\theta'|^\alpha \omega_\alpha(\theta')\mathrm{Leb}_\nu(K_j)$, $\theta' := \theta \prod_{i=1}^\nu \lambda_i^{1/\alpha}$ imply the second property.

Example 11 (Fractional Brownian sheet (FBS)). An FBS $B_{\boldsymbol{H}}$ with (Hurst) parameter vector $\boldsymbol{H} = (H_1, \ldots, H_\nu) \in [0,1]^\nu$ is defined as a Gaussian process on $\mathbb{R}^\nu_+$ with zero mean and covariance

$$\mathrm{E}B_{\boldsymbol{H}}(\boldsymbol{x})B_{\boldsymbol{H}}(\boldsymbol{y}) = \prod_{i=1}^{\nu} r_{H_i}(x_i, y_i), \quad \boldsymbol{x} = (x_1, \ldots, x_\nu),$$
$$\boldsymbol{y} = (y_1, \ldots, y_\nu), \tag{42}$$

where $r_H(x,y), x, y \in \mathbb{R}_+$ is the covariance of FBM $B_H(x), x \in \mathbb{R}$ with Hurst parameter $H \in [0,1]$, or

$$r_H(x,y) = (1/2)(|x|^{2H} + |y|^{2H} - |x-y|^{2H}), \quad 0 < H \le 1,$$
$$r_0(x,y) = \lim_{H \downarrow 0} r_H(x,y) = \begin{cases} 1, & x = y, \\ 1/2, & x \ne y. \end{cases} \tag{43}$$

Also, note that $r_1(x,y) = xy$ and $B_1(x) \overset{\mathrm{fdd}}{=} Zx, Z \sim N(0,1)$ is a Gaussian line process. A similar fact is true for FBS when (some) of its Hurst parameters equal 1, namely

$$B_{\boldsymbol{H}}(\boldsymbol{x}) \overset{\mathrm{fdd}}{=} \begin{cases} \left(\prod_{i \in N_1} x_i\right) B_{\hat{\boldsymbol{H}}}(\hat{\boldsymbol{x}}), & \nu_1 < \nu, \\ Z \prod_{i=1}^\nu x_i, & \nu_1 = \nu, \end{cases} \tag{44}$$

where $N_1 := \{i : 1 \le i \le \nu, H_i = 1\}, \nu_1 := \mathrm{Card}(N_1) \le \nu, \hat{\boldsymbol{H}} := (H_i, i \notin N_1) \in [0,1)^{\nu-\nu_1}, \hat{\boldsymbol{x}} := (x_i, i \notin N_1) \subset \mathbb{R}^{\nu-\nu_1}$. Particularly, $B_{\mathbf{1}}(\boldsymbol{x}) \overset{\mathrm{fdd}}{=} Z \prod_{i=1}^\nu x_i$. An important special case of FBS is the *BS*

$$B(\boldsymbol{x}) := B_{1/2,\ldots,1/2}(\boldsymbol{x}), \tag{45}$$

whose covariance function equals the product $\mathrm{E}B(\boldsymbol{x})B(\boldsymbol{y}) = \prod_{i=1}^{\nu}(x_i \wedge y_i)$ of covariance functions of standard Brownian motion. Note that $B \overset{\mathrm{fdd}}{=} \sigma L_2$ for L_2, defined in Example 10. For parameter values $\boldsymbol{H} \in (0,1]^\nu$, the RF $B_{\boldsymbol{H}}(\boldsymbol{x})$ has a.s. continuous paths in $\mathbb{R}_+^\nu$. On the other side, FBS with $\min_{1\le i\le \nu} H_i = 0$ are highly unusual and extremely singular RFs. They can be defined by Kolmogorov's theorem as the Gaussian probability measures on the space $\mathbb{R}^{\mathbb{R}_+^\nu}$ of *all* functions $\mathbb{R}_+^\nu \to \mathbb{R}$, equipped with the σ-field generated by cylinder sets. These RFs have highly irregular (nonseparable) paths since their covariance functions are discontinuous at $\boldsymbol{x} = \boldsymbol{y}$.

Proposition 12. *FBS $B_{\boldsymbol{H}}, \boldsymbol{H} = (H_1, \dots, H_\nu) \in [0,1]^\nu$ satisfies the following properties:*

(i) it has stationary rectangular increments;
(ii) it is $\boldsymbol{H}$-MSS;
(iii) the BS B in (45) *has independent rectangular increments.*

Proof. (i) By Gaussianity, this follows from the fact that $\mathrm{E}B_{\boldsymbol{H}}(]\boldsymbol{y}, \boldsymbol{y}+\boldsymbol{x}])B_{\boldsymbol{H}}(]\boldsymbol{y}, \boldsymbol{y}+\boldsymbol{x}']), \boldsymbol{x}, \boldsymbol{x}' \in \mathbb{R}_+^\nu$ does not depend on the shift $\boldsymbol{y} = (y_1, \dots, y_\nu) \in \mathbb{R}_+^\nu$. Any such shift can be written as a superposition of ν 'one-dimensional' shifts in direction $\boldsymbol{e}_i, i = 1, \dots, \nu$. Hence, it suffices to consider the case $\boldsymbol{y} = y\boldsymbol{e}_i, y > 0$, reducing the proof to that of the stationarity of increments of FBM. For $H \in [0,1]$, we see that $\mathrm{E}B_H(]y, y+x])B_H(]y, y+x']) = r_H(y+x, y+x') - r_H(y, y+x') - r_H(y+x, y) + r_H(y, y) = (1/2)(x^{2H} + x'^{2H} - |x-x'|^{2H})$ does not depend on $y > 0$. Then,

$$\begin{aligned}
&\mathrm{E}B_{\boldsymbol{H}}(]y\boldsymbol{e}_i, y\boldsymbol{e}_i + \boldsymbol{x}])B_{\boldsymbol{H}}(]y\boldsymbol{e}_i, y\boldsymbol{e}_i + \boldsymbol{x}']) \\
&\quad = (r_{H_i}(y+x_i, y+x_i') - r_{H_i}(y, y+x_i') - r_{H_i}(y+x_i, y) \\
&\qquad + r_{H_i}(y, y)) \prod_{j\ne i} r_{H_j}(x_j, x_j')
\end{aligned}$$

also does not depend on y, proving part (i).
(ii) This follows from (42) and $r_H(\lambda x, \lambda y) = \lambda^{2H} r_H(x, y)\, \forall \lambda > 0$.
(iii) It suffices to show that $\mathrm{E}B(K)B(K') = 0$ for any disjoint rectangles $K =]\boldsymbol{y}, \boldsymbol{x}], K' =]\boldsymbol{y}', \boldsymbol{x}'], K \cap K' = \emptyset$. This implies $y_j < x_j \le y_j' < x_j'$ for some $1 \le j \le \nu$. By (i), we can assume $y_j = 0$, and let $j = 1$ w.l.g. Then, $\mathrm{E}B(K)B(K') = 0$ holds since

$\mathrm{E}B(x_1, \tilde{\boldsymbol{x}})(B(x'_1, \tilde{\boldsymbol{x}}') - B(y'_1, \tilde{\boldsymbol{x}}')) = x_1 \prod_{i=2}^{\nu}(\tilde{x}_i \wedge \tilde{x}'_i) - x_1 \prod_{i=2}^{\nu}(\tilde{x}_i \wedge \tilde{x}'_i) = 0$ for any $\tilde{\boldsymbol{x}}, \tilde{\boldsymbol{x}}' \in \mathbb{R}_+^{\nu-1}$. □

Example 13 ((Discrete) Fractional Gaussian noise (FGN)). We define the FGN $B^{\square}_{\boldsymbol{H}} = \{B^{\square}_{\boldsymbol{H}}(\boldsymbol{t}); \boldsymbol{t} \in \mathbb{Z}^{\nu}\}$, as the rectangular increment RF of FBS:

$$B^{\square}_{\boldsymbol{H}}(\boldsymbol{t}) := B_{\boldsymbol{H}}(]\boldsymbol{t}, \boldsymbol{t}+\boldsymbol{1}]), \quad \boldsymbol{t} \in \mathbb{Z}^{\nu}. \tag{46}$$

The RF $B^{\square}_{\boldsymbol{H}}$ is Gaussian, stationary, has zero mean, and has covariance function

$$\begin{aligned} \mathrm{E}B^{\square}_{\boldsymbol{H}}(\boldsymbol{t})B^{\square}_{\boldsymbol{H}}(\boldsymbol{t}') &= \prod_{j=1}^{\nu}(2r_{H_j}(t_j, t'_j) - r_{H_j}(t_j+1, t'_j) - r_{H_j}(t_j, t'_j+1)) \\ &= 2^{-\nu}\prod_{j=1}^{\nu}(|t_j - t'_j + 1|^{2H_j} + |t_j - t'_j - 1|^{2H_j} \\ &\quad - 2|t - t'_j|^{2H_j}), \qquad \boldsymbol{t} \in \mathbb{Z}^{\nu}, \end{aligned} \tag{47}$$

with the convention $|t|^0 := 1\ (t \neq 0), := 0\ (t = 0)$, as in (43). Particularly, for $\nu = 1$,

$$\mathrm{E}B^{\square}_0(t)B^{\square}_0(t') = \begin{cases} 1, & t = t', \\ -1/2, & |t - t'| = 1, \quad t, t' \in \mathbb{Z}. \\ 0, & |t - t'| \geq 2, \end{cases} \tag{48}$$

We see from (47) that the covariance function of RF $B^{\square}_{\boldsymbol{H}}$ factorises into the product of the covariance functions of the FGN $B^{\square}_H(t), t \in \mathbb{Z}$ with $H = H_j$. The spectral density of the last process is

$$f_H(x) = \begin{cases} (H\Gamma(2H)\sin(H\pi)/2)\{\frac{|\mathrm{e}^{\mathrm{i}x}-1|^2}{|x|^{2H+1}} + \tilde{f}_H(x)\}, & H \in (0,1), \\ |\mathrm{e}^{\mathrm{i}x} - 1|^2, & H = 0, \end{cases} \tag{49}$$

where $\tilde{f}_H(x) := |\mathrm{e}^{\mathrm{i}x} - 1|^2 \sum_{k \in \mathbb{Z}, k \neq 0} |x + 2\pi k|^{-1-2H}$ is bounded and continuous in $x \in \Pi$ when $H > 0$. The fact $\lim_{H \downarrow 0} f_H(x) = f_0(x)$ follows from (49) and relation $\lim_{H\downarrow 0} 2H \sum_{k\in\mathbb{Z}} |x + 2\pi k|^{-1-2H} = 1\ (x \in \Pi, x \neq 0)$. (For $H = 1$, $B^{\square}_1(t) \overset{\mathrm{fdd}}{=} Z \in N(0,1), t \in \mathbb{Z}$ is a random constant and does not have a spectral density.) We see that for any $H \in [0, 1)$, the spectral density $f_H(x)$ is bounded and continuous on $|x| > \delta(\forall \delta > 0), x \in \Pi$ and

$$f_H(x) \sim \begin{cases} (H\Gamma(2H)\sin(H\pi)/\pi) & H \in (0,1), \\ |x|^{1-2H}, & H \neq 1/2, \\ 1/2\pi, & H = 1/2, \\ |x|^2, & H = 0, \end{cases} \quad x \to 0. \tag{50}$$

From (47), it follows that the spectral density $f_{\boldsymbol{H}}(\boldsymbol{x})$ of RF $B^{\square}_{\boldsymbol{H}}$ exists for $\boldsymbol{H} \in [0, 1)^{\nu}$ and factorises as

$$f_{\boldsymbol{H}}(\boldsymbol{x}) = \prod_{j=1}^{\nu} f_{H_j}(x_j). \tag{51}$$

We conclude from (47), (50), and (51) that the FGN $B^{\square}_{H}$ has:

- the covariance LRD property for $\boldsymbol{H} \in [0, 1]^{\nu}$, $\max_{1 \le j \le \nu} H_j > 1/2$;
- the spectrum LRD property for $\boldsymbol{H} \in [0, 1)^{\nu}$, $\max_{1 \le j \le \nu} H_j > 1/2$;
- the covariance and spectrum SRD properties for $H_1 = \cdots = H_{\nu} = 1/2$;
- the covariance and spectrum ND properties for $\boldsymbol{H} \in [0, 1]^{\nu}$, $0 \le \min_{1 \le i \le \nu} H_j < \max_{1 \le i \le \nu} H_j \le 1/2$.

Remark 14. One may ask whether the FBS in Example 11 exhausts *all* Gaussian $\boldsymbol{H}$-MSS RFs with zero mean and stationary rectangular increments. Surprisingly, this is not true. References [174,175] proved that the general form of the covariance function of such RFs with $0 < H_i < 1, H_i \neq 1/2, i = 1, \dots, \nu$ is given by

$$\mathrm{E}V(\boldsymbol{x})V(\boldsymbol{y}) = \frac{1}{4}\prod_{i=1}^{\nu}(x_i^{2H_i} + y_i^{2H_i} - |x_i - y_i|^{2H_i}) + \frac{c}{4}\prod_{i=1}^{\nu}\tan(\pi H_i) \\ \times (-x_i^{2H_i} + y_i^{2H_i} + \mathrm{sign}(x_i - y_i)|x_i - y_i|^{2H_i}), \tag{52}$$

$\boldsymbol{x}, \boldsymbol{y} \in \mathbb{R}^{\nu}_{+}$, where $c \in [-1, 1]$ is arbitrary (for $H_i = 1/2\,(\exists 1 \le i \le \nu)$, Ref. [175] provides a somewhat different representation of the covariance function).

Remark 15 (Scaling diagram of i.i.d. RF). The scaling diagram of i.i.d. RF $\{X(\boldsymbol{t}); \boldsymbol{t} \in \mathbb{Z}^\nu\}$ (or RF Y given by the sum in (34)) with zero mean and finite variance consists of a single element – a multiple σB of the Brownian Sheet in (45). In the case of i.i.d. RF with infinite variance, the scaling diagram can be empty. If the distribution of the r.v. $X(\mathbf{0}) \equiv X$ belongs to the domain of attraction of a stable law with index $0 < \alpha \le 2$ and the centring constant is zero, the scaling diagram of i.i.d. RF consists of a single element – the Lévy sheet W_α in Example 10 with the characteristic function given by (40)– (41). Necessary and sufficient conditions on the d.f. of X for this are well known [89, 122]. In this book, we use a simpler version of the latter conditions guaranteeing the convergence

$$\lim_{n\to\infty} (\mathrm{E}\mathrm{e}^{\mathrm{i}\theta/n^{1/\alpha}X})^n = \mathrm{E}\mathrm{e}^{\mathrm{i}\theta W_\alpha} = \mathrm{e}^{-|\theta|^\alpha \omega_\alpha(\theta)}, \quad \theta \in \mathbb{R} \tag{53}$$

towards α-stable law $W_\alpha = W_\alpha(\mathbf{1})$, formulated in what follows as Assumption $(\mathrm{D})_\alpha$.

Assumption $(\mathbf{D})_\alpha$. An r.v. ξ satisfies one of the following two conditions (i)/(ii):

(i) $\alpha = 2$ and $\sigma^2 := \mathrm{E}\xi^2 < \infty$; or
(ii) $0 < \alpha < 2$, and there exist $c_+, c_- \ge 0, c_+ + c_- > 0$ such that

$$\lim_{x\to\infty} x^\alpha \mathrm{P}(\xi > x) = c_+, \qquad \lim_{x\to-\infty} |x|^\alpha \mathrm{P}(\xi \le x) = c_-. \tag{54}$$

Moreover, $\mathrm{E}\xi = 0$ for $1 < \alpha \le 2$, and for $\alpha = 1$, the d.f. of ξ is symmetric.

Assumption $(\mathrm{D})_\alpha$ (abbreviated as $\xi \in (D)_\alpha$) implies that the ch.f. of ξ resembles the stable ch.f. in (53) at the origin [122, Thm. 2.6.5]: there exists $\epsilon > 0$ such that

$$\mathrm{E}\mathrm{e}^{\mathrm{i}\theta\xi} = \mathrm{e}^{-|\theta|^\alpha \omega_\alpha(\theta) h(\theta)} \quad \text{for any } \theta \in \mathbb{R},\ |\theta| < \epsilon, \tag{55}$$

where $h(\theta)$ is a positive function tending to 1 as $\theta \to 0$.

3 Further Concepts of Dependence in Dimension $\nu = 2$: Scaling Transition

Clearly, the covariance LRD property does not exclude the possibility that the covariance decays rapidly and sums up to zero in some directions or hyperplanes in $\mathbb{Z}^\nu$, leading to more detailed classification of the LRD/SRD/ND properties. The latter classifications and differences appear in the context of anisotropic RFs and anisotropic scaling limits in Chapter 3. In this section, we introduce these concepts for RFs on $\mathbb{Z}^2$ since this case is simpler, and this book mostly deals with planar RFs.

Definition 16. Let $X = \{X(\boldsymbol{t}); \boldsymbol{t} = (t_1, t_2) \in \mathbb{Z}^2\}$ be a stationary RF on $\mathbb{Z}^2$ with finite second moment. We say that X has:

(i) *covariance vertical LRD* property if $\sum_{t\in\mathbb{Z}} |r(0,t)| = \infty$;
(ii) *covariance vertical SRD* property if $\sum_{t\in\mathbb{Z}} |r(0,t)| < \infty$ and $\sum_{t\in\mathbb{Z}} r(0,t) \neq 0$;
(iii) *covariance vertical ND* property if $\sum_{t\in\mathbb{Z}} |r(0,t)| < \infty$ and $\sum_{t\in\mathbb{Z}} r(0,t) = 0$.

Analogously, we can define covariance horizontal LRD, SRD, and ND properties. What can be said about the spectral counterparts of these properties? Definition 16 refers to the dependence properties of the stationary sectional process $Y = \{Y(t); t \in \mathbb{Z}\} = \{X(0,t); t \in \mathbb{Z}\}$ arising on the vertical axis. Obviously, $r_Y(t) = \mathrm{Cov}(Y(0), Y(t)) = r(0,t)$ is the restriction of the covariance of X on the vertical axis. We also see that

$$r(0,t) = \int_{\Pi^2} \mathrm{e}^{\mathrm{i}tx_2} f(x_1,x_2)\mathrm{d}x_1\mathrm{d}x_2 = \int_{\Pi} \mathrm{e}^{\mathrm{i}tx} f_Y(x)\mathrm{d}x, \tag{56}$$

with

$$f_Y(x) := \int_{\Pi} f(x_1, x)\mathrm{d}x, \quad x \in \Pi. \tag{57}$$

In other words, if the RF X on $\mathbb{Z}^2$ has a spectral density f, then the sectional process Y has a spectral density f_Y in (57). Then, following Definitions 16, we can introduce the *spectrum vertical LRD/SRD/ND* properties by requiring that the sectional process Y has the respective spectrum LRD/SRD/ND properties as defined in Definition 2. Particularly, the spectral vertical ND property holds if f_Y is continuous at 0 and

$$f_0(0) = \int_{\Pi} f(0,x)\mathrm{d}x = 0, \tag{58}$$

which is true when f is continuous on Π^2 and vanishes on the vertical axis: $f(0, x) = 0, x \in \Pi$.

Sectional LRD/SRD/ND properties, as in Definition 16, are loosely related to the LRD/SRD/ND properties of RF X, allowing for various combinations. For instance, a covariance LRD RF X may feature not only the covariance vertical LRD but also the covariance vertical ND since the latter property allows for $\sum_{\boldsymbol{t}\in\mathbb{Z}^2} |r(\boldsymbol{t})| = \infty$. On the other hand, if $\sum_{\boldsymbol{t}\in\mathbb{Z}^2} |r(\boldsymbol{t})| < \infty$, then the covariance vertical ND property implies the covariance ND property, as shown in the following proposition.

Proposition 17. *Let RF $X = \{X(\boldsymbol{t}); \boldsymbol{t} \in \mathbb{Z}^2\}$ be a covariance stationary RF, with $\sum_{\boldsymbol{t}\in\mathbb{Z}^2} |r(\boldsymbol{t})| < \infty$ having the covariance vertical ND property. Then, $\sum_{t_2\in\mathbb{Z}} r(t_1, t_2) = 0$ for any $t_1 \in \mathbb{Z}$. Particularly, X is a covariance ND RF.*

Proof. By the stationarity of X,

$$
\begin{aligned}
0 \le N^{-1}\mathrm{E}\left(\sum_{t_2=1}^{N}(X(0, t_2) - X(t_1, t_2))\right)^2 \\
= 2N^{-1}\sum_{t_2, t_2'=1}^{N}(r(0, t_2 - t_2') - r(t_1, t_2 - t_2')) \\
= 2(S_N(0) - S_N(t_1)),
\end{aligned}
$$

where $S_N(t_1) := \sum_{|t_2|\le N-1} r(t_1, t_2)(1 - \frac{|t_2|}{N}) \to S(t_1) := \sum_{t_2\in\mathbb{Z}} r(t_1, t_2)$ as $N \to \infty$ follows by the DCT. Therefore, $S(t_1) \le S(0) = 0$. Similarly, $0 \le N^{-1}\mathrm{E}\big(\sum_{t_2=1}^{N}(X(0, t_2) + X(t_1, t_2))\big)^2 = 2(S_N(0) + S_N(t_1))$ implies $-S(t_1) \le S(0) = 0$, or $S(t_1) \ge 0$. Hence, $S(t_1) = 0$ for any $t_1 \in \mathbb{Z}$. □

There are other concepts of strong dependence apart from covariance or coefficient LRD properties which are limited to finite variance or MA processes. A useful strong dependence property in dimension $\nu = 1$ can be defined in terms of the partial sums limit.

The concept of scaling transition, defined as follows, is a more concrete interpretation (although limited to planar RFs) of this 'otherwise'. W.l.g., in dimension $\nu = 2$, we take $\boldsymbol{\gamma} = (1, \gamma), \gamma_1 = 1$,

$\gamma_2 = \gamma, \Gamma = \mathrm{diag}(1, \gamma)$ in Definition 5 and everywhere in the subsequent text, which allows for dealing with the scalar scaling exponent $\gamma \in \mathbb{R}_+$ instead of $\boldsymbol{\gamma} \in \mathbb{R}^2_+$.

Definition 18. Let $Y = \{Y(\boldsymbol{t}); \boldsymbol{t} \in \mathbb{R}^2_+\}$ be an RF with non-trivial large-scale γ-scaling limits V_γ (respectively, non-trivial small-scale γ-scaling limits V_γ) for any $\gamma > 0$. We say that Y exhibits large-scale scaling transition (respectively, small-scale scaling transition) if there exists $\gamma_{\mathrm{cr}} > 0$ such that V_γ do not depend on γ for $\gamma > \gamma_{\mathrm{cr}}$ and $\gamma < \gamma_{\mathrm{cr}}$:

$$V_\gamma \overset{\mathrm{fdd}}{=} \begin{cases} V_+, & \gamma > \gamma_{\mathrm{cr}}, \\ V_-, & \gamma < \gamma_{\mathrm{cr}}, \\ V_{\mathrm{cr}}, & \gamma = \gamma_{\mathrm{cr}} \end{cases} \tag{59}$$

and $V_+ \overset{\mathrm{fdd}}{\neq} aV_-$ $(\forall\, a > 0)$.

The condition $V_+ \overset{\mathrm{fdd}}{\neq} aV_-$ $(\forall\, a > 0)$ excludes the trivial scaling transition arising from a linear change of normalisation at $\gamma = \gamma_0$. We use the terminology *well-balanced limit* for V_0 and *unbalanced limits* for $V_\pm$ in (59). In the case of (59), the respective large-scale and small-scale scaling diagrams of Y consist of at most three elements: $\{V_\gamma, \gamma > 0\} = \{V_+, V_-, V_{\mathrm{cr}}\}$. The word 'unbalanced' suggests a lack of 'balance' between vertical and horizontal scaling, in which case we may expect that the limit RFs feature certain 'degenerate' properties in one direction, which are formalised in Definition 19.

Definition 19. Let $V = \{V(\boldsymbol{x}); \boldsymbol{x} \in \mathbb{R}^2_+\}$ be an RF with stationary rectangular increments. We say that V has:

(i) independent rectangular increments in the horizontal (respectively, vertical) direction if $V(K)$ and $V(K')$ are independent for any disjoint rectangles $K, K' \subset \mathbb{R}^2_+$ such that $K' = K + (x_1, 0)$ for some $x_1 \in \mathbb{R}$ (respectively, $K' = K + (0, x_2)$ for some $x_2 \in \mathbb{R}$);
(ii) shift-invariant rectangular increments in the horizontal (respectively, vertical) direction if $V(K) = V(K')$ for any two rectangles $K, K' \subset \mathbb{R}^2_+$ such that $K' = K + (x_1, 0)$ for some $x_1 \in \mathbb{R}$ (respectively, $K' = K + (0, x_2)$ for some $x_2 \in \mathbb{R}$);

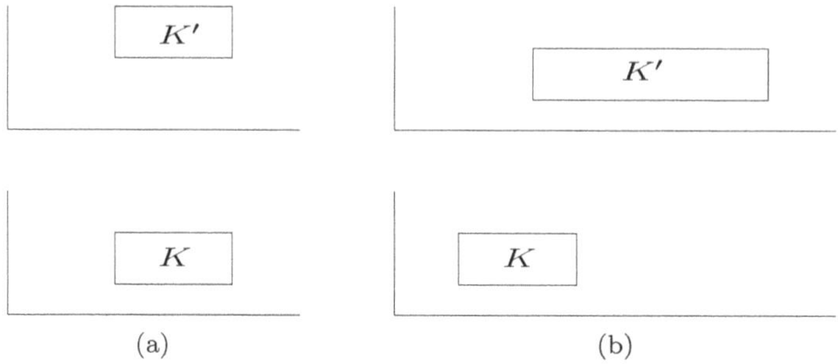

Figure 1. (a) Rectangle K shifted in the vertical direction. (b) Rectangle K scaled in the horizontal direction by a factor of 2.

(iii) scale-invariant rectangular increments in the horizontal (respectively, vertical) direction if $V(K) \stackrel{\mathrm{d}}{=} V(K')$ for any two rectangles $K = (x_1, y_1] \times (x_2, y_2]$, $K' = (x'_1, y'_1] \times (x'_2, y'_2]$, $K, K' \subset \mathbb{R}^2_+$ such that $x'_1 = \lambda x_1, y'_1 = \lambda y_1, x'_2 = x_2, y'_2 = y_2$ for some $\lambda > 0$; (respectively, $x'_2 = \lambda x_2, y'_2 = \lambda y_2, x'_1 = x_1, y'_1 = y_1$ for some $\lambda > 0$);

Figure 1 illustrates the transformation of the rectangle $K \to K'$ in Definition 19 (shift in (i) and (ii) and scaling in (iii)). Properties (i)–(iii) in Definition 19 can be verified in the case of FBS, where they correspond to special values of the parameters $\boldsymbol{H} = (H_1, H_2) \in [0, 1]^2$. Namely, an FBS B_{H_1,H_2} has:

(i) independent rectangular increments in the horizontal (vertical) direction if $H_1 = 1/2$ ($H_2 = 1/2$);
(ii) shift-invariant rectangular increments in the horizontal (vertical) direction if $H_1 = 1$ ($H_2 = 1$);
(iii) scale-invariant rectangular increments in the horizontal (vertical) direction if $H_1 = 0$ ($H_2 = 0$).

Properties (i) and (iii) follow from the respective properties of the covariance function of FBS as in Proposition 12, while (ii) is obvious from (44). These properties are reflected in the FGN $B^{\square}_{H_1,H_2}$ with covariance in (47). Let $B^{\square,(j)}_{H}, j \in \mathbb{Z}$ denote independent copies of FGN $B^{\square}_{H}(t), t \in \mathbb{Z}$. By symmetry, we formulate these facts only for $H_1 = 1/2, 1, 0$:

(j) $B^{\square}_{1/2,H_2}(\boldsymbol{t}) \overset{\text{fdd}}{=} B^{\square,(t_1)}_{H_2}(t_2), \boldsymbol{t} \in \mathbb{Z}^2$, meaning that the 'vertical sections' of $B^{\square}_{1/2,H_2}$ are i.i.d. FGN with $H = H_2$.
(jj) $B^{\square}_{1,H_2}(\boldsymbol{t}) \overset{\text{fdd}}{=} B^{\square}_{H_2}(t_2), \boldsymbol{t} \in \mathbb{Z}^2$, meaning that the 'vertical sections' of $B^{\square}_{1,H_2}$ are identical FGN with $H = H_2$.
(jjj) $B^{\square}_{0,H_2}(\boldsymbol{t}) \overset{\text{fdd}}{=} (1/\sqrt{2})(B^{\square,(t_1+1)}_{H_2}(t_2) - B^{\square,(t_1)}_{H_2}(t_2)), \boldsymbol{t} \in \mathbb{Z}^2$, meaning that the 'vertical sections' of $B^{\square}_{0,H_2}$ are the differences of i.i.d. FGN with $H = H_2$.
(jv) $B^{\square}_{0,0}(\boldsymbol{t}) \overset{\text{fdd}}{=} (1/2)(Z(t_1+1, t_2+1) - Z(t_1, t_2+1) - Z(t_1+1, t_2) + Z(t_1, t_2)) = (1/2)Z(]\boldsymbol{t}, \boldsymbol{t}+\mathbf{1}]), \boldsymbol{t} \in \mathbb{Z}^2$, where $\{Z(\boldsymbol{t}); \boldsymbol{t} \in \mathbb{Z}^2\}$ is an i.i.d. RF with generic distribution $Z \sim N(0,1)$.

Properties (j)–(jv) follow from (47) by checking that the covariance functions on both sides of these relations coincide. For instance, the covariance $(1/2)\mathrm{E}(B^{\square,(t_1+1)}_{H_2}(t_2) - B^{\square,(t_1)}_{H_2}(t_2))(B^{\square,(t'_1+1)}_{H_2}(t'_2) - B^{\square,(t'_1)}_{H_2}(t'_2))$ of the RF on the r.h.s. of (jjj) for $t_1 = t'_1$ equals $\mathrm{E}B^{\square}_{H_2}(t_2)B^{\square}_{H_2}(t'_2) = \mathrm{E}B^{\square}_{0,H_2}(t_1,t_2)B^{\square}_{0,H_2}(t_1,t'_2)$; see (48), whereas for $t'_1 = t_1+1$, it equals $(-1/2)\mathrm{E}B^{\square}_{H_2}(t_2)B^{\square}_{H_2}(t'_2) = \mathrm{E}B^{\square}_{0,H_2}(t_1,t_2)B^{\square}_{0,H_2}(t_1+1,t'_2)$, and for $t' \geq t_1+2$, it equals 0 by independence.

For the RF $B^{\square}_{H_1,H_2}$ on $\mathbb{Z}^2$, the dependence concepts in Definition 16 can be verified. Indeed, $\mathrm{E}B^{\square}_{\boldsymbol{H}}(0,0)B^{\square}_{\boldsymbol{H}}(0,t) = (1/2)(|t+1|^{2H_2} + |t-1|^{2H_2} - 2|t|^{2H_2})$ for any $\boldsymbol{H} = (H_1,H_2) \in [0,1]^2, t \in \mathbb{Z}$. Using Taylor's expansion, we see that $B^{\square}_{\boldsymbol{H}}$ satisfies the covariance vertical LRD (respectively, SRD or ND) property if and only if $H_2 > 1/2$ (respectively, $H_2 = 1/2$ or $H_2 < 1/2$).

We end this section with a discussion about the definition of the finite-dimensional distributions of FBS with $H_1 \wedge H_2 = 0$ using uncountable families of one-dimensional FBM. For instance, the (joint) distribution of $B_{H_1,0}(\boldsymbol{x}_i), \boldsymbol{x}_i = (x_{i1}, x_{i2}) \in \mathbb{R}^2_+, i = 1,\ldots,m, 0 < x_{12} < \cdots < x_{m2}$ coincides with that of the random vector

$$\frac{1}{\sqrt{2}}\big(B^{(0)}_{H_1}(x_{11}) - B^{(1)}_{H_1}(x_{11}), B^{(0)}_{H_1}(x_{21}) - B^{(2)}_{H_1}(x_{21}), \ldots, B^{(0)}_{H_1}(x_{m1}) \\ - B^{(m)}_{H_1}(x_{m1})\big),$$

where $B^{(i)}_{H_1} = \{B^{(i)}_{H_1}(x); x > 0\}, i = 0, 1, \ldots, m$ are *independent* FBM. Note that for each $x_1 > 0$ fixed,

$$\{B_{H_1,0}(x_1, x_2); x_2 > 0\} \overset{\text{fdd}}{=} \frac{x_1^{H_1}}{\sqrt{2}} \{\zeta(0) - \zeta(x_2); x_2 > 0\}, \tag{60}$$

where $\zeta(x) \sim N(0,1)\ (\forall\, x \ge 0)$, $\zeta(x); x > 0$ are independent Gaussian r.v.s. Even more singular is $B_{0,0}$ with the representation

$$\{B_{0,0}(\boldsymbol{x}); \boldsymbol{x} \in \mathbb{R}^2_+\} \overset{\text{fdd}}{=} \frac{1}{2}\{\zeta(0,0) - \zeta(0, x_2) - \zeta(x_1, 0) + \zeta(x_1, x_2); \boldsymbol{x} = (x_1, x_2) \in \mathbb{R}^2_+\},$$

where $\zeta(\boldsymbol{x}); \boldsymbol{x} \in \bar{\mathbb{R}}^2_+$ is a family of independent standard normal variables $\zeta(\boldsymbol{x}) \in N(0,1)$ with $\mathrm{E}\zeta(\boldsymbol{x})\zeta(\boldsymbol{y}) = 0\, (\boldsymbol{x} \ne \boldsymbol{y})$. For $H = 0$, we define $B_0 = \{B_0(x); x > 0\}$ as a Gaussian process $B_0(x) = \frac{1}{\sqrt{2}}(\zeta(0) - \zeta(x))$ in (60). Note that the FBM B_0 has stationary and scale-invariant increments since for any $y, \lambda > 0$, we have $\{B_0(y + x) - B_0(y); x > 0\} \overset{\text{fdd}}{=} \frac{1}{\sqrt{2}}\{\zeta(y) - \zeta(x + y); x > 0\} \overset{\text{fdd}}{=} \frac{1}{\sqrt{2}}\{\zeta(y) - \zeta(\lambda x + y); x > 0\} \overset{\text{fdd}}{=} \{B_0(y + \lambda x) - B_0(y); x > 0\}$.

4 Notes

1. A comprehensive treatment of correlation theory and spectral representations for stationary and stationary increment RFs can be found in Yaglom [271]. RFs with (wide-sense) stationary rectangular increments and their spectral representations are discussed in Basse-O'Connor *et al.* [25].

2. Definitions 1–3 extend similar concepts in one dimension, $\nu = 1$, in Giraitis *et al.* [103, Chapter 3], which uses somewhat different terminology. Particularly, the term 'long memory' used by many authors is avoided in this book since it is more appropriate for time evolution. Lahiri and Robinson [145] use 'positive strong dependence' instead of 'coefficient LRD'. Pipiras and Taqqu [215] provide several non-equivalent formulations of the one-dimensional ($\nu = 1$) LRD property

and discuss the relationships between them. Samorodnitsky [230] discusses LRD properties from the point of view of ergodic theory and mixing. Heyde and Yang [116] define an LRD property (for the case $\nu = 1$) via the so-called *Allen variance*. The latter concept seems rather robust and is extended to nonstationary periodically correlated processes in Philippe *et al.* [203]. Daley and Vesilo [66,67] define an LRD property of point processes on $\mathbb{R}$. See Samorodnitsky and Taqqu [231], Paulauskas [198], Kulik and Spodarev [144], and Makogin and Spodarev [176] for dependence characteristics and LRD/SRD concepts for random processes and RFs with infinite variance.

3. The definitions of SRD given in this book should be distinguished from weak dependence conditions defined through mixing used in the literature. A detailed description of multiple concepts of weak dependence based on mixing can be found in Dedecker *et al.* [73]. For weakly dependent RFs, anisotropic scaling does not seem very interesting since for such RFs, summation domains may have a very general shape, and the scaling diagram usually consists of a single element (white noise) or is empty. Various central limit theorems for weakly dependent RFs can be found in Ref. [73], Bolthausen [45], Kuensch [143], Bulinski *et al.* [51], and other texts. Particularly, the class of $\Phi_{\mathrm{rec},\nu}$ of ν-dimensional rectangles satisfies van Hove's condition [143].

4. Real data with LRD from various scientific areas (from hydrology and astrophysics to economics and telecommunications) are discussed in many publications, including Beran [30], Doukhan *et al.* [82], Giraitis *et al.* [103], Park and Willinger [197], and Teyssière and Kirman [262].

5. The literature on self-similarity, scaling, and related topics is huge, see e.g., the reference list in Pipiras and Taqqu [215], albeit most works refer to the one-dimensional case $\nu = 1$. The origins of this large field of probabilistic and applied research can be traced back to the famous papers of Kolmogorov [137] and Mandelbrot and Van Ness [178]. The classical reference on scaling limits for LRD MA processes on $\mathbb{Z}$ is Davydov [71]. Spatial self-similarity seems to be primarily associated with Kolmogorov's turbulence theory [138], Mandelbrot's fractals [177] and the renormalisation group theory of phase transitions in statistical physics (the latter association is well explained for the probabilistic audience in Dobrushin [79]).

Dobrushin [79] also provides a very general framework of isotropic ($\Gamma = \mathrm{diag}(\mathbf{1})$) scaling transformations acting in the space of *states* or probability measures corresponding to (generalised) RFs indexed by (test) functions from a linear topological space. The classical reference on the theory of generalised RFs is Gel'fand and Vilenkin [94].

6. The basic definition of SS RFs appeared in the seminal paper by Lamperti [146], albeit Lamperti's definition and terminology are somewhat different. Gaussian generalised H-SS RFs with stationary increments of arbitrary order are fully described in Dobrushin [78]. Roughly, the H-SS property for such RFs X is equivalent to the spectral measure $F(\mathrm{d}\boldsymbol{x})$ of X being homogeneous w.r.t scaling $\boldsymbol{x} \to \lambda x, \lambda > 0$, see Dobrushin [78, Thm. 3.2]. The notion of OSRF with general (non-diagonal) matrix Γ is due to Biermé *et al.* [37]. The last paper constructs some classes of stable (Gaussian) OSRFs using MA or harmonisable representations. Clausel and Vedel [61] construct a class of Gaussian OSRFs. The related concept of *generalised scale invariance* was introduced by Schertzer and Lovejoy [233]. Pecknold *et al.* [200] discuss real data and simulations of such RFs. Quoting the abstract of the last paper, 'The analysis of various geophysical fields has shown that although they are typically scaling multifractals, they are indeed anisotropic involving differential rotation, stratification, and more complex scale changes'.

7. MSS RFs were introduced by Genton *et al.* [95]. The last paper also extended the work of Lamperti [146] by establishing a one-to-one correspondence between the class of all $\boldsymbol{H}$-MSS RFs X on $\mathbb{R}_+^\nu$ and the class of all stationary RFs Y on $\mathbb{R}^\nu$, given by *Lamperti's transformation*:

$$Y(\boldsymbol{t}) := \mathrm{e}^{-\sum_{j=1}^{\nu} t_j H_j} X(\mathrm{e}^{t_1}, \ldots, \mathrm{e}^{t_\nu}) = \mathrm{e}^{-\boldsymbol{t}\cdot\boldsymbol{H}} X(\mathrm{e}^{\mathrm{diag}(\boldsymbol{t})}\mathbf{1}), \qquad \boldsymbol{t} \in \mathbb{R}_+^\nu. \tag{61}$$

Makogin and Mishura [175] used (61) to characterise all Gaussian $\boldsymbol{H}$-MSS RFs with stationary rectangular increments and $\boldsymbol{H} \in (0,1)^\nu$. The transformation in (61) also extends to $\boldsymbol{H}$-MSS RF X, with $\boldsymbol{H} \in [0,1]^\nu$ yielding a stationary RF Y without spectral representation when $\min_{1 \le j \le \nu} H_j = 0$. For instance, for $\nu = 1$ and $X = B_0$, the covariance $\mathrm{E}Y(0)Y(t) = (1/2)(1 + \mathbb{I}(t = 0))$ is discontinuous at $t = 0$, and Bochner's theorem does not apply [89, Chapter 19]. We remark that the above B_0 is different from *regularided FBM with index* $H = 0$ introduced by Fyodorov, Khoruzhenko, and Simm [91], with the latter process being a.s. continuous but lacking the SS property. Davydov and Paulauskas [72] discussed various generalisations of Lamperti's theorem and the difficulties arising in the multidimensional case of $\nu > 1$m including scaling with a general (non-diagonal) matrix Γ.

8. An FBS $B_{\boldsymbol{H}}$ with $\boldsymbol{H} \in (0,1]^\nu$ was introduced by Kamont [131]. The extension to $\boldsymbol{H} \in [0,1]^\nu$ appeared in Surgailis [251]. Path properties (Hölder regularity and Hausdorff dimension) of FBS were studied by Ayache *et al.* [11] and Ayache and Xiao [14]. Ayache *et al.* [12,13] introduced and studied a class of α-stable ($0 < \alpha < 2$) $\boldsymbol{H}$-MSS RFs, called *linear fractional stable sheets* and containing the Lévy sheet in Example 10, systematically investigated in Ehm [86]. Properties of FGN are discussed in Beran [30], Taqqu [261], Ayache *et al.* [11], and other texts. The formula in (49) is due to Sinai [238].

9. Necessary and sufficient conditions for weak convergence of normalised sums of i.i.d. r.v.s with a generic distribution ξ towards a stable law can be found in the monographs by Gnedenko and Kolmogorov [104], Ibragimov and Linnik [122], and elsewhere. Assumption $(\mathrm{D})_\alpha$ for $\alpha \neq 1$ is necessary for the latter convergence with zero centring and normalisation $n^{1/\alpha}$.

10. Classifications in Definitions 16 and 19 appeared in Pilipauskaitė and Surgailis [209] and Puplinskaitė and Surgailis [221]. The notion of distributional LRD (long memory) is due to Dehling and Philipp [74] (an earlier version of which appeared in Cox [62]).

The concept of scaling transition for planar RFs appeared in Puplinskaitė and Surgailis [220, 221]. A weaker version of it, requiring that V_γ is independent of γ for $\gamma > \gamma_{\mathrm{cr}}$ and $\gamma < \gamma_{\mathrm{cr}}$ in a (small) neighbourhood of γ_{cr}, was proposed by Damarackas and Paulauskas [69]. The last paper showed that there exist linear RFs on $\mathbb{Z}^2$ which exhibit scaling transition in the above weaker sense at several points, $\gamma > 0$.

11. Samorodnitsky [230, Chapter 9] proposed a different point of view on LRD as a *phase transition* or an abrupt change (arising at the boundary between LRD and SRD) in the behaviour of partial sums or partial maxima of a 'parametric' stationary sequence.

A similar approach may be pursued in many concrete cases studied in this book by fixing $\gamma > 0$ and considering the change in the scaling limit as parameters of the RF change. Since the phase transition is a fundamental and well-developed concept of statistical physics (see e.g., Sinai [239]), we do not use it to avoid ambiguity.

Chapter 2

Tools: Cumulants, Random Measures, and Stochastic Integrals

The definition and basic properties of cumulants are provided, followed by cumulant-based criteria for the central limit theorem. The stochastic integration of deterministic functions is discussed, together with convergence in the distribution of stochastic integrals and linear and polynomial forms.

1 Cumulants and the Central Limit Theorem

A family $\{\xi_\lambda; \lambda > 0\}$ of r.v.s is said to *converge in distribution* or *weakly* to an r.v. ξ as $\lambda \to \infty$, and we write $\xi_\lambda \xrightarrow{\mathrm{d}} \eta$ if $\lim_{\lambda\to\infty} \mathrm{P}(\xi_\lambda \le x) = \mathrm{P}(\xi \le x)$ at each continuity point, $x \in \mathbb{R}$, of the distribution function $\mathrm{P}(\xi \le x)$. Convergence in distribution is equivalent to the point-wise convergence of the corresponding characteristic functions

$$\lim_{\lambda\to\infty} \mathrm{E}e^{i\theta\xi_\lambda} = \mathrm{E}e^{i\theta\xi}, \quad \forall\, \theta \in \mathbb{R}.$$

Convergence in distribution can be proved by the method of moments provided all the moments exist. Suppose that $\mathrm{E}|\xi_\lambda|^k < \infty, \mathrm{E}|\xi|^k < \infty$, $k \in \mathbb{N}, \lambda > 0$, and the distribution of ξ is uniquely determined by its moments $\mathrm{E}\xi^k, k \in \mathbb{N}$. Then,

$$\lim_{\lambda\to\infty} \mathrm{E}\xi_\lambda^k = \mathrm{E}\xi^k, \quad \forall\, k \in \mathbb{N}, \tag{1}$$

implies $\xi_\lambda \xrightarrow{\mathrm{d}} \eta$.

For the Gaussian limit r.v. $\xi \in N(0, \sigma^2)$, convergence of moments in (1) can be replaced by convergence of cumulants, which is often more convenient since cumulants of order $k \le 3$ of a Gaussian r.v. are zero. For any r.v. ξ with all moments finite, its kth-order cumulant, for $k \in \mathbb{N}$, is defined as

$$\mathrm{Cum}_k(\xi) := (-\mathrm{i})^k \mathrm{d}^k \log \mathrm{E}\mathrm{e}^{\mathrm{i}\theta\xi}/\mathrm{d}\theta^k\big|_{\theta=0}. \tag{2}$$

In particular,

$$\mathrm{Cum}_0(\xi) = 0, \ \mathrm{Cum}_1(\xi) = \mathrm{E}\xi, \quad \mathrm{Cum}_2(\xi) = \mathrm{Var}(\xi),$$
$$\mathrm{Cum}_3(\xi) = \mathrm{E}\xi^3 - 3\mathrm{E}(\xi - \mathrm{E}\xi)^2\mathrm{E}\xi - (\mathrm{E}\xi)^3.$$

Due to the one-to-one correspondence between moments and cumulants, the convergence of moments in (1) is equivalent to the convergence of cumulants

$$\lim_{\lambda\to\infty} \mathrm{Cum}_k(\xi_\lambda) = \mathrm{Cum}_k(\xi), \quad \forall\, k \in \mathbb{N}. \tag{3}$$

For $\xi \in N(0, \sigma^2)$, we have $\mathrm{Cum}_2(\xi) = \sigma^2, \mathrm{Cum}_k(\xi) = 0 (k \ne 2, k \in \mathbb{N})$. The above facts lead to the following simple cumulant-based criterion of asymptotic normality.

Corollary 1. *Suppose the r.v.s $\xi_\lambda, \lambda > 0$ have all moments finite. If, in addition,*

$$\lim_{\lambda\to\infty} \mathrm{E}\xi_\lambda = 0, \quad \lim_{\lambda\to\infty} \mathrm{Var}(\xi_\lambda) = \sigma^2,$$
$$\lim_{\lambda\to\infty} \mathrm{Cum}_k(\xi_\lambda) = 0, \quad k = 3, 4, \dots,$$

then $\xi_\lambda \xrightarrow{\mathrm{d}} N(0, \sigma^2)$.

Applying the above cumulant criterion to the sums of r.v.s, one needs to express the cumulant of the sum in terms of the joint cumulants of the summands. Let $\{\xi_1, \dots, \xi_m\}$ be a set of r.v.s with $\mathrm{E}|\xi_j|^m < \infty$, for all $j = 1, \dots, m$. The *joint cumulant* of $\{\xi_1, \dots, \xi_m\}$

is defined by

$$\begin{aligned}
&\mathrm{Cum}(\xi_1, \dots, \xi_m) \\
&\quad := (-\mathrm{i})^m \partial^m \log \mathrm{E} \exp \left\{ \mathrm{i} \sum_{j=1}^m \theta_j \xi_j \right\} / \partial\theta_1 \cdots \partial\theta_m \Big|_{\theta_1 = \cdots = \theta_m = 0}. \quad (4)
\end{aligned}$$

In particular,

$$\mathrm{Cum}(\xi_1) = \mathrm{E}\xi_1, \quad \mathrm{Cum}(\xi_1, \xi_2) = \mathrm{Cov}(\xi_1, \xi_2),$$
$$\mathrm{Cum}(\xi_1, \dots, \xi_m) = \mathrm{Cum}_m(\xi), \quad \text{if } \xi_1 = \cdots = \xi_m = \xi.$$

Some simple properties of the joint cumulant are listed in the following proposition.

Proposition 2. (i) *(Symmetry.)* $\mathrm{Cum}(\xi_1, \dots, \xi_m)$ *is symmetric under permutations of* $\xi_1, \dots, \xi_m$.
(ii) *(Multilinearity.) If* $\xi_j = \sum_{i=1}^n a_{ij}\eta_i$*, where* η_i *are r.v.s and* a_{ij}, $i = 1, \dots, m$, $j = 1, \dots, n$, *are real constants, then*

$$\mathrm{Cum}(\xi_1, \dots, \xi_m) = \sum_{j_1, \dots, j_m = 1}^n a_{1j_1} \cdots a_{mj_m} \mathrm{Cum}(\eta_{j_1}, \dots, \eta_{j_m}).$$

(iii) $\mathrm{Cum}(\xi_1, \dots, \xi_n, \xi_{n+1}, \dots, \xi_m) = 0$ *if the sub-collections* $\{\xi_1, \dots, \xi_n\}$ *and* $\{\xi_{n+1}, \dots, \xi_m\}$ *are independent, for some* $1 \le n < m$.
(iv) *If the r.v.s* $\{\xi_1, \dots, \xi_m\}$ *are jointly Gaussian and* $m \ge 3$*, then* $\mathrm{Cum}(\xi_1, \dots, \xi_m) = 0$.

Joint cumulants can be also expressed in terms of joint moments, and *vice versa.* These relations (usually referred to as *Leonov–Shiryaev formulas*) are well known [103, Prop. 4.2.2] but are not used here. For our purposes, a rough bound

$$|\mathrm{Cum}_k(\xi)| \le C\mathrm{E}|\xi|^k, \quad |\mathrm{Cum}(\xi_1, \dots, \xi_m)| \le C \max_{1 \le i \le k} \mathrm{E}|\xi_i|^k, \quad (5)$$

suffices, with $C = C_k$ depending on k but not on $\xi, \xi_1, \dots, \xi_m$.
We use cumulants to prove Lyapunov's criterion for the asymptotic normality of sums of independent r.v.s.

Proposition 3. *Suppose that $\xi_{in}, 1 \le i \le n$ are independent r.v.s with zero mean* $\mathrm{E}\xi_{in} = 0$ *and* $\mathrm{E}|\xi_{in}|^{2+\delta} < \infty\,(\exists 0 < \delta < 1)$*. Assume that*

$$\lim_{n\to\infty} \sum_{i=1}^{n} \mathrm{E}\left|\frac{\xi_{in}}{\sigma_n}\right|^{2+\delta} = 0, \tag{6}$$

where $\sigma_n^2 = \mathrm{E}(\sum_{i=1}^n \xi_{in})^2 = \sum_{i=1}^n \mathrm{E}\xi_{in}^2$*. Then,* $\sigma_n^{-1}\sum_{i=1}^n \xi_{in} \xrightarrow{\mathrm{d}} N(0,1)$.

Proof. Split $S_n := \sum_{i=1}^n \xi_{in} = S_n^- + S_n^+, S_n^\pm := \sum_{i=1}^n \xi_{in}^\pm$, $\xi_{in}^- := \xi_{in}\mathbb{I}(|\xi_{in}| \le \sigma_n) - \mathrm{E}\xi_{in}\mathbb{I}(|\xi_{in}| \le \sigma_n)$, $\xi_{in}^+ := \xi_{in}\mathbb{I}(|\xi_{in}| > \sigma_n) - \mathrm{E}\xi_{in}\mathbb{I}(|\xi_{in}| > \sigma_n)$. It suffices to show that

$$\sigma_n^{-1} S_n^- \xrightarrow{\mathrm{d}} N(0,1) \quad \text{and} \quad \sigma_n^{-1} S_n^+ \xrightarrow{\mathrm{d}} 0. \tag{7}$$

In view of (6),

$$\begin{aligned}\mathrm{E}(\sigma_n^{-1}S_n^+)^2 &= \sigma_n^{-2}\sum_{i=1}^n \mathrm{E}(\xi_{in}^+)^2 \le 2\sum_{i=1}^n \mathrm{E}\left|\frac{\xi_{in}}{\sigma_n}\right|^2 \mathbb{I}\left(\left|\frac{\xi_{in}}{\sigma_n}\right| > 1\right)\\ &\le 2\sum_{i=1}^n \mathrm{E}\left|\frac{\xi_{in}}{\sigma_n}\right|^{2+\delta} \to 0,\end{aligned}$$

and the second relation in (7) holds. Next,

$$\begin{aligned}\mathrm{Cum}_k(\sigma_n^{-1}S_n^-) &= \sum_{i=1}^n \mathrm{Cum}_k\left(\frac{\xi_{in}^-}{\sigma_n}\right) \le C\sum_{i=1}^n \mathrm{E}\left|\frac{\xi_{in}}{\sigma_n}\right|^k \mathbb{I}\left(\left|\frac{\xi_{in}}{\sigma_n}\right| \le 1\right)\\ &\le C\sum_{i=1}^n \mathrm{E}\left|\frac{\xi_{in}}{\sigma_n}\right|^{2+\delta} \to 0\end{aligned}$$

for any $k \ge 3$. Finally, $\mathrm{Cov}(S_n^-, S_n^+) = 0$, and hence $\mathrm{Cum}_2(\sigma_n^{-1}S_n^-) = 1 - \mathrm{E}(\sigma_n^{-1}S_n^+)^2 \to 1$, and the first relation in (7) holds by Corollary 1. □

An RF $X = \{X(\boldsymbol{t}); \boldsymbol{t} \in \mathbb{Z}^\nu\}$ is said to be *m-dependent* $(0 < m < \infty)$ if, for any subsets $A, B \subset \mathbb{Z}^\nu$, the families $\{X(\boldsymbol{t}); \boldsymbol{t} \in A\}$ and $\{X(\boldsymbol{s});\ \boldsymbol{s} \in B\}$ of r.v.s are independent, provided the distance between A and B is greater than m: $\inf\{|\boldsymbol{t} - \boldsymbol{s}| : \boldsymbol{t} \in A, \boldsymbol{s} \in B\} > m$.

The definition of the m-dependent RF $X = \{X(\boldsymbol{t}); \boldsymbol{t} \in \mathbb{R}^\nu\}$ is analogous except that in the latter case, A and B are arbitrary Borel sets from $\mathbb{R}^\nu$.

Proposition 4. *Let X be a stationary m-dependent RF on $\mathbb{Z}^\nu$ (respectively, on $\mathbb{R}^\nu$) with finite variance. Then, for any $\phi \in \Phi$,*

$$\lambda^{-\nu/2}(X_\lambda(\phi) - \mathrm{E}X_\lambda(\phi)) \xrightarrow{\mathrm{d}} \sigma \int_{\mathbb{R}^\nu} \phi(\boldsymbol{u}) W(\mathrm{d}\boldsymbol{u}), \tag{8}$$

where $\sigma^2 := \sum_{\boldsymbol{t}\in\mathbb{Z}^\nu} \mathrm{Cov}(X(\mathbf{0}), X(\boldsymbol{t})) \geq 0$ (respectively, $\sigma^2 := \int_{\mathbb{R}^\nu} \mathrm{Cov}(X(\mathbf{0}), X(\boldsymbol{t}))\mathrm{d}\boldsymbol{t} \geq 0$).

Proof. We prove (8) for the RF on $\mathbb{Z}^\nu$ only since the proof for the RF on $\mathbb{R}^\nu$ is completely analogous. Split $X(\boldsymbol{t}) = X(\boldsymbol{t})\mathbb{I}(|X(\boldsymbol{t})| \leq \log\lambda) + X(\boldsymbol{t})\mathbb{I}(|X(\boldsymbol{t})| > \log\lambda) =: X_1(\boldsymbol{t}) + X_2(\boldsymbol{t})$ and $X_\lambda(\phi) = X_{1,\lambda}(\phi) + X_{2,\lambda}(\phi)$ accordingly. By m-dependence, $r_i(\boldsymbol{t}) := \mathrm{Cov}(X_i(\mathbf{0}), X_i(\boldsymbol{t})) = 0\,(|\boldsymbol{t}| > m, i = 1, 2)$ and $r_1(\boldsymbol{t}) \to r(\boldsymbol{t}) := \mathrm{Cov}(X(\mathbf{0}), X(\boldsymbol{t}))\,(\forall \boldsymbol{t} \in \mathbb{Z}^\nu, \lambda \to \infty)$. Relation (8) follows from

$$\lambda^{-\nu/2}(X_{1,\lambda}(\phi) - \mathrm{E}X_{1,\lambda}(\phi)) \xrightarrow{\mathrm{d}} \sigma \int_{\mathbb{R}^\nu} \phi(\boldsymbol{u}) W(\mathrm{d}\boldsymbol{u}),$$
$$\lambda^{-\nu/2}(X_{2,\lambda}(\phi) - \mathrm{E}X_{2,\lambda}(\phi)) \xrightarrow{\mathrm{d}} 0. \tag{9}$$

Consider first the convergence of the corresponding variances:

$$\lambda^{-\nu}\mathrm{Var}(X_{1,\lambda}(\phi)) \to \sigma^2 \int_{\mathbb{R}^\nu} \phi(\boldsymbol{u})^2 \mathrm{d}\boldsymbol{u}, \quad \lambda^{-\nu}\mathrm{Var}(X_{2,\lambda}(\phi)) \to 0, \tag{10}$$

whose proof follows that of (8) later with minor changes. Clearly, (10) implies the second relation in (8). To show the first relation, we use the cumulant criterion of Corollary 1. Accordingly, it suffices to show that

$$\mathrm{Cum}_k(X_{1,\lambda}(\phi)) = o(\lambda^{k\nu/2}), \quad \forall\, k = 3, 4, \ldots. \tag{11}$$

Using $X_{1,\lambda}(\phi) = \sum_{\boldsymbol{s}\in\mathbb{Z}^\nu} X_1(\boldsymbol{s}) \int_{[\boldsymbol{s},\boldsymbol{s}+\mathbf{1}]} \phi(\boldsymbol{t}/\lambda)\mathrm{d}\boldsymbol{t}$ and the multilinearity property of cumulants, the l.h.s. of (11) can be written as $\sum_{(\boldsymbol{s})_k\in\mathbb{Z}^{k\nu}} \prod_{i=1}^k b_\lambda(\boldsymbol{s}_i)\mathrm{Cum}(X_1(\boldsymbol{s}_1), \ldots, X_1(\boldsymbol{s}_k))$, where $b_\lambda(\boldsymbol{s}) := \int_{[\boldsymbol{s},\boldsymbol{s}+\mathbf{1}[} \phi(\boldsymbol{t}/\lambda)\mathrm{d}\boldsymbol{t}$ satisfy $|b_\lambda(\boldsymbol{s})| < C$ and $\sum_{\boldsymbol{s}\in\mathbb{Z}^\nu} |b_\lambda(\boldsymbol{s})| \leq$

$\int_{\mathbb{R}^\nu} |\phi(\boldsymbol{t}/\lambda)|\mathrm{d}\boldsymbol{t} \le C\lambda^\nu$ since $\phi \in \Phi$ is bounded and integrable. From properties (i)–(iii) of cumulants, we also get that

$$|\mathrm{Cum}(X_1(\boldsymbol{s}_1), \ldots, X_1(\boldsymbol{s}_k))| \le \begin{cases} 0, & \text{if } |\boldsymbol{s}_i - \boldsymbol{s}_j| > m \, (\exists i, j = 1, \ldots, k), \\ C(\log \lambda)^k, & \text{otherwise.} \end{cases}$$

The above facts entail the bound $|\mathrm{Cum}_k(X_{1,\lambda}(\phi))| \le C(\log \lambda)^k \sum_{(\boldsymbol{s})_k \in \mathbb{Z}^{k\nu}: |\boldsymbol{s}_i - \boldsymbol{s}_j| \le m, i,j=1,\ldots,k} |b_\lambda(\boldsymbol{s}_1)| \le C(\log \lambda)^k \lambda^\nu$, proving (11) and the proposition. □

Remark 5. We remark that the 'long-run variance' σ^2 in (8) may vanish (as it happens for ND RFs), in which case (8) yields the trivial limit $\lambda^{-\nu/2}(X_\lambda(\phi) - \mathrm{E}X_\lambda(\phi)) \xrightarrow{\mathrm{d}} 0$. Of course, this fact does not exclude the possibility of a nontrivial limit under a normalisation growing slower than $O(\lambda^{\nu/2})$.

The cumulant criterion and other facts will be used to prove the limit results for weighted sums of i.i.d. r.v.s. (Propositions 7, 9 and 26). Let

$$S(g) := \sum_{\boldsymbol{u} \in \mathbb{Z}^\nu} g(\boldsymbol{u})\varepsilon(\boldsymbol{u}) \tag{12}$$

be a linear form (a weighted sum) in i.i.d. r.v.s $\{\varepsilon, \varepsilon(\boldsymbol{u}); \boldsymbol{u} \in \mathbb{Z}^\nu\}$ with deterministic coefficients $g(\boldsymbol{u})$. The convergence and existence of higher-order moments of sums in (12) are provided in the following application of von Bahr–Esséen and Rosenthal's inequalities (see Refs. [17, 110] and [103, Lemma 2.5.2]).

Proposition 6. *Let* $\mathrm{E}|\varepsilon|^p < \infty \, (p > 0)$ *and* $\mathrm{E}\varepsilon = 0 \, (p > 1)$. *Then,*

$$\mathrm{E}|S(g)|^p \le C_p \mathrm{E}|\varepsilon|^p \begin{cases} \sum_{\boldsymbol{u} \in \mathbb{Z}^\nu} |g(\boldsymbol{u})|^p, & \text{`}0 < p \le 2, \\ \left(\sum_{\boldsymbol{u} \in \mathbb{Z}^\nu} |g(\boldsymbol{u})|^2\right)^{p/2}, & p > 2, \end{cases} \tag{13}$$

where C_p *depends on* p *but not on* g *and the distribution of* ε. *The series in (12) converges in* L^p *and a.s. provided the r.h.s. of (13) is finite.*

Proposition 7. *Assume that* $\varepsilon \in (D)_2, d_\lambda^2 := \mathrm{E}S(g_\lambda)^2 = \sum_{\boldsymbol{u}\in\mathbb{Z}^\nu} g_\lambda^2(\boldsymbol{u}) \to \infty$ *and*

$$|g_\lambda|_\infty := \sup_{\boldsymbol{u}\in\mathbb{Z}^\nu} |g_\lambda(\boldsymbol{u})| = o(d_\lambda), \quad \lambda \to \infty. \tag{14}$$

Then, $d_\lambda^{-1} S(g_\lambda) \xrightarrow{\mathrm{d}} N(0,1), \ \lambda \to \infty.$

Proof. Assume first that $\mathrm{E}|\varepsilon|^k < \infty \, \forall\, k \in \mathbb{N}$. Since $\mathrm{E} d_\lambda^{-1} S(g_\lambda) = 0, \mathrm{E}(d_\lambda^{-1} S(g_\lambda))^2 = 1$, by Corollary 1 it suffices to show that

$$\mathrm{Cum}_k(d_\lambda^{-1} S(g_\lambda)) \to 0, \quad \forall\, k \geq 3. \tag{15}$$

Let $c_\lambda := \frac{|g_\lambda|_\infty}{d_\lambda}$. By properties (i)–(iii) of joint cumulants in Proposition 2 and condition (14), for $k \geq 3$,

$$\begin{aligned}
|\mathrm{Cum}_k(d_\lambda^{-1} S(g_\lambda))| &= \Big|\mathrm{Cum}_k(\varepsilon) \sum_{\boldsymbol{u}\in\mathbb{Z}^\nu} \Big(\frac{g_\lambda(\boldsymbol{u})}{d_\lambda}\Big)^k \Big| \\
&\leq \big|\mathrm{Cum}_k(\varepsilon|c_\lambda|^{k-2} \sum_{\boldsymbol{v}\in\mathbb{Z}^\nu} \frac{g_\lambda^2(\boldsymbol{v})}{d_\lambda^2} \\
&\leq C|c_\lambda|^{k-2} \to 0,
\end{aligned} \tag{16}$$

proving (15) and the statement of the proposition when all the moments of ε are finite.

Consider now the general case, when some higher moments of $\varepsilon(\boldsymbol{t})$ do not exist. Let

$$\varepsilon_1(\boldsymbol{t}) := \varepsilon(\boldsymbol{t})\mathbb{I}\big(|\varepsilon(\boldsymbol{t})| \leq |\log c_\lambda|\big), \quad \varepsilon_2(\boldsymbol{t}) := \varepsilon(\boldsymbol{t})\mathbb{I}\big(|\varepsilon(\boldsymbol{t})| > |\log c_\lambda|\big),$$

$\varepsilon_i \overset{\mathrm{d}}{=} \varepsilon_i(\boldsymbol{0}), i = 1, 2$, where $|\log c_\lambda| \to \infty$. Then,

$$\begin{aligned}
S(g_\lambda) &= S_1(g_\lambda) + S_2(g_\lambda), \\
S_i(g_\lambda) &:= \sum_{\boldsymbol{u}\in\mathbb{Z}^\nu} g_\lambda(\boldsymbol{u})(\varepsilon_i(\boldsymbol{u}) - \mathrm{E}\varepsilon_i(\boldsymbol{u})), \quad i = 1, 2.
\end{aligned}$$

It suffices to show that

$$d_\lambda^{-1} S_1(g_\lambda) \xrightarrow{\mathrm{d}} N(0,1), \quad d_\lambda^{-1} S_2(g_\lambda) \xrightarrow{\mathrm{d}} 0. \tag{17}$$

Let $d^2_{\lambda,i} := \mathrm{Var}(S_i(g_\lambda)),\ i = 1, 2$. Then, $d^2_{\lambda,1}/d^2_\lambda = \mathrm{Var}(\varepsilon_1)/\mathrm{Var}(\varepsilon) \to 1$, implying $\mathrm{Var}(d_\lambda^{-1} S_1(g_\lambda)) \to 1$. From the bound (5) and the definition of $\varepsilon_1(\boldsymbol{t})$, one can estimate

$$|\mathrm{Cum}_k(\varepsilon_1)| \le C\mathrm{E}|\varepsilon_1|^k \le C|\log c_\lambda|^k$$

with a constant C, depending on k but not on λ. Then, as in (16), for $k \ge 3$,

$$|\mathrm{Cum}_k(d_\lambda^{-1} S_1(g_\lambda))| \le C\big|\mathrm{Cum}_k(\varepsilon_1)|c_\lambda|^{k-2} \le C|c_\lambda|^{k-2}|\log c_\lambda|^k \to 0,$$

which proves the first convergence in (17). The second convergence holds because $d^2_{\lambda,2}/d^2_\lambda = \mathrm{Var}(\varepsilon_2)/\mathrm{Var}(\varepsilon) \to 0$. This completes the proof of Proposition 7. □

2 Random Measures and Stochastic Integrals

As a rule, the limit RFs in this book are written as stochastic integrals with respect to some random measure and a deterministic integrand. Therefore, the knowledge of basic facts about such integrals presented in this section is recommended for further reading. Particularly, we focus on the criteria of weak convergence,

$$\int_{\mathfrak{X}} f_\lambda(x) M_\lambda(\mathrm{d}x) \xrightarrow{\mathrm{d}} \int_{\mathfrak{X}} f(x) M(\mathrm{d}x), \tag{18}$$

of a family of stochastic integrals towards a limit stochastic integral when both the random measures M_λ and the integrands f_λ depend on $\lambda > 0$ tending to ∞ or 0. Since the limit in (18) often has a Gaussian or stable distribution, these criteria provide a convenient way for verifying the Lindeberg or central limit theorem in the case when the l.h.s. of (18) is a linear form in i.i.d. r.v.s written as a 'discrete' stochastic integral with a measure $\mathcal{M}_n$ concentrated on a countable subset of $\mathfrak{X}$.

The notion of a random measure is not universally agreed upon in the probabilistic literature. For instance, in the point process theory, random measure is usually assumed to be σ-additive a.s., which is not true in our case. Given a measurable space $\mathfrak{X}$ with a σ-finite measure μ defined on a σ-algebra $\mathcal{F}(\mathfrak{X})$ of measurable subsets of $\mathfrak{X}$, by *random measure* we understand a family $M = \{M(A); A \in \mathcal{F}_0(\mathfrak{X})\}$ of r.v.s

$M(A)$ defined for $A \in \mathcal{F}_0(\mathfrak{X}) := \{A \in \mathcal{F}_0(\mathfrak{X}) : \mu(A) < \infty\}$ with the following properties:

(i) for any disjoint $A_i \in \mathcal{F}_0(\mathfrak{X}), A_i \cap A_j = \emptyset \ (i = j)$ such that $\cup_{i=1}^{\infty} A_i \in \mathcal{F}_0(\mathfrak{X})$,

$$M\left(\bigcup_{i=1}^{\infty} A_i\right) = \sum_{i=1}^{\infty} M(A_i), \tag{19}$$

where the sum on the r.h.s. converges in probability;

(ii) for any disjoint $A_i \in \mathcal{F}_0(\mathfrak{X}), i = 1, 2, A_1 \cap A_2 = \emptyset$, the r.v.s $M(A_1)$ and $M(A_2)$ are independent;

(iii) for any $A_i \in \mathcal{F}_0(\mathfrak{X}), i = 1, 2$, with $\mu(A_1) = \mu(A_2)$, the r.v.s $M(A_1)$ and $M(A_2)$ are identically distributed.

Because of assumption (iii) (which is usually not imposed in other texts but satisfied in many concrete cases), the measure μ can be appropriately termed the *control measure* of the random measure M. The two basic examples of random measure are the Gaussian random measure (where $M(A)$ has the Gaussian distribution $N(0, \mu(A)) \ \forall A \in \mathcal{F}_0(\mathfrak{X})$, and often referred to as *Gaussian white noise with zero mean and variance* μ) and the Poisson random measure ($M(A)$ has the Poisson distribution with mean $\mu(A)$ for any $A \in \mathcal{F}_0(\mathfrak{X})$). Another example of a random measure is a Lévy random measure on $X = \mathbb{R}^\nu$ with $\mu = \mathrm{Leb}_\nu$ the Lebesgue measure on $\mathbb{R}^d$ and characteristic function

$$\mathrm{E}\mathrm{e}^{\mathrm{i}\theta M(A)} = \exp\Bigg\{\mathrm{Leb}_\nu(A)\Bigg(\mathrm{i}\theta a - \frac{\theta^2\sigma^2}{2} + \int_{\mathbb{R}} (\mathrm{e}^{\mathrm{i}\theta z} - 1 - \mathrm{i}\theta \mathbb{I}(|\theta| \le 1))\nu(\mathrm{d}z)\Bigg)\Bigg\}, \quad \theta \in \mathbb{R}, \tag{20}$$

where $a \in \mathbb{R}, \sigma \ge 0$, and ν is a measure on $\mathbb{R}$ with $\nu(\{0\}) = 0$ and $\int_{\mathbb{R}} (z^2 \wedge 1)\nu(\mathrm{d}z) < \infty$, called a *Lévy measure*. In the latter case, $M(A)$ has an infinitely divisible distribution, and (a, σ, ν) is called the *characteristic triplet of* M.

The definition of the stochastic integral $\int_{\mathfrak{X}} f(x) M(\mathrm{d}x)$ w.r.t. the random measure M proceeds similarly to the case of the Lebesgue integral. Namely, for a class of *simple* functions $f(x) = \sum_{j=1}^n f_j \mathbb{I}(x \in A_j)$ taking constant values on disjoint sets $A_j \in \mathcal{F}_0(\mathfrak{X}), j = 1, \ldots, n,$

the integral is defined as the sum $\int_{\mathfrak{X}} f(x)M(\mathrm{d}x) := \sum_{j=1}^{n} f_j M(A_j)$. The class $L_0(\mathfrak{X})$ of all simple functions is dense in the Banach space $L^p(\mathfrak{X}), p \geq 1$ of all measurable functions $f : \mathfrak{X} \to \mathbb{R}$ with finite norm $\|f\|_p := \big(\int_{\mathfrak{X}} |f(x)|^p \mu(\mathrm{d}x)\big)^{1/p} < \infty$. Then, the definition of the stochastic integral is extended to a suitable class of integrable functions on $\mathfrak{X}$, which depends on the mode of the convergence of r.v.s and distributional properties of M.

Consider the stochastic integral for the Gaussian white noise, usually denoted W in this book. For any $f \in L_0(\mathfrak{X})$, $\int_{\mathfrak{X}} f(x)W(\mathrm{d}x)$ has a Gaussian distribution with mean zero and variance $\|f\|_2^2$. Therefore, $\int_{\mathfrak{X}} f(x)W(\mathrm{d}x)$ is well defined for any $f \in L^2(\mathfrak{X})$, as a limit in mean square of integral sums, and has the same properties:

$$\int_{\mathfrak{X}} f(x)W(\mathrm{d}x) \sim N(0, \|f\|_2^2), \quad \mathrm{E}\int_{\mathfrak{X}} f(x)W(\mathrm{d}x) = 0,$$
$$\mathrm{E}\left(\int_{\mathfrak{X}} f(x)W(\mathrm{d}x)\right)^2 = \|f\|_2^2. \tag{21}$$

Let us discuss the definition of Poisson stochastic integrals. Let $\mathcal{N}$ be a Poisson random measure on $\mathfrak{X}$ with mean μ and $\widetilde{\mathcal{N}} := \mathcal{N} - \mu$ as the centred Poisson random measure. The stochastic integrals $\int_{\mathfrak{X}} f(x)\mathcal{N}(\mathrm{d}x)$ and $\int_{\mathfrak{X}} f(x)\widetilde{\mathcal{N}}(\mathrm{d}x)$ are defined for any measurable function $f : \mathfrak{X} \to \mathbb{R}$ satisfying

$$\int_{\mathfrak{X}} (1 \wedge |f(x)|)\mu(\mathrm{d}x) < \infty \quad \text{and} \quad \int_{\mathfrak{X}} (|f(x)| \wedge |f(x)|^2)\mu(\mathrm{d}x) < \infty, \tag{22}$$

respectively, as limits in probability of suitable integral sums, and their characteristic functions are given by

$$\mathrm{E}\mathrm{e}^{\mathrm{i}\theta \int_{\mathfrak{X}} f(x)\mathcal{N}(\mathrm{d}x)} = \exp\left\{\int_{\mathfrak{X}} (\mathrm{e}^{\mathrm{i}\theta f(x)} - 1)\mu(\mathrm{d}x)\right\},$$
$$\mathrm{E}\mathrm{e}^{\mathrm{i}\theta \int_{\mathfrak{X}} f(x)\widetilde{\mathcal{N}}(\mathrm{d}x)} = \exp\left\{\int_{\mathfrak{X}} (\mathrm{e}^{\mathrm{i}\theta f(x)} - 1 - \mathrm{i}\theta f(x))\mu(\mathrm{d}x)\right\}, \quad \theta \in \mathbb{R}. \tag{23}$$

We have $\mathrm{E}\int_{\mathfrak{X}} f(x)\widetilde{\mathcal{N}}(\mathrm{d}x) = 0$, $f \in L^1(\mathfrak{X})$ and $\mathrm{E}\big(\int_{\mathfrak{X}} f(x)\widetilde{\mathcal{N}}(\mathrm{d}x)\big)^2 = \|f\|_2^2$, $f \in L^2(\mathfrak{X})$. Moreover,

$$\mathrm{E}\left|\int_{\mathfrak{X}} f(x)\widetilde{\mathcal{N}}(\mathrm{d}x)\right|^p < \infty$$

$$\Leftrightarrow \begin{cases} \int_{\mathfrak{X}}(|f(x)|^2 \wedge |f(x)|^p)\mu(\mathrm{d}x) < \infty, & 1 \le p \le 2, \\ \int_{\mathfrak{X}}(|f(x)|^2 \vee |f(x)|^p)\mu(\mathrm{d}x) < \infty, & p \ge 2 \end{cases} \tag{24}$$

(see Ref. [222]), while

$$\mathrm{E}\left|\int_{\mathfrak{X}} f(x)\mathcal{N}(\mathrm{d}x)\right|^p \le C(p)\|f\|_p^p, \quad p \in (0,1],$$

$$\mathrm{E}\left|\int_{\mathfrak{X}} f(x)\widetilde{\mathcal{N}}(\mathrm{d}x)\right|^p \le C(p)\|f\|_p^p, \quad p \in [1,2], \tag{25}$$

where the constant $C(p) < \infty$ depends only on p. For step functions $f = \sum_{j=1}^{\infty} f_j \mathbb{I}(x \in A_j) \in L_0(\mathfrak{X})$, the first inequality of (25) with $C(p) = 1$ follows from $\big|\sum_{j=1}^{\infty} f_j \mathcal{N}(A_j)\big|^p \le \sum_{j=1}^{\infty} |f_j|^p |\mathcal{N}(A_j)|^p$ and $\mathrm{E}|\mathcal{N}(A_j)|^p \le \mathrm{E}[\mathcal{N}(A_j) = \mu(A_j)$. The second inequality in (25), with $C(p) = 2^{2-p} \le 2$, is obtained in Ref. [244, Thm. 3.1] by interpolation between $L^1(\mathfrak{X})$ and $L^2(\mathfrak{X})$. It also follows from the inequality due to von Bahr and Esséen [17], according to which $\mathrm{E}\big|\sum_{j=1}^{\infty} f_j \widetilde{\mathcal{N}}(A_j)\big|^p \le 2\sum_{j=1}^{\infty} |f_j|^p \mathrm{E}|\widetilde{\mathcal{N}}(A_j)|^p$ and $\mathrm{E}|\widetilde{\mathcal{N}}(A)|^p \le 3\mu(A)$ for $\mu(A) \le 1$. We also note that $\mathrm{E}\mathrm{e}^{\int_{\mathfrak{X}} f(x))\widetilde{\mathcal{N}}(\mathrm{d}x)} < \infty$, provided

$$\int_{\mathfrak{X}} |\mathrm{e}^{f(x)} - 1 - f(x)|\mu(\mathrm{d}x) < \infty, \tag{26}$$

and the useful formulas

$$\mathrm{E}\mathrm{e}^{\theta \int_{\mathfrak{X}} f(x)\widetilde{\mathcal{N}}(\mathrm{d}x)} = \exp\left\{\int_{\mathfrak{X}} (\mathrm{e}^{\theta f(x)} - 1 - \theta f(x))\mu(\mathrm{d}x)\right\},$$

$$\mathrm{E}\left[\mathrm{e}^{\int_{\mathfrak{X}} f(x)\widetilde{\mathcal{N}}(\mathrm{d}x)} \int_{\mathfrak{X}} h(y)\widetilde{\mathcal{N}}(\mathrm{d}y)\right]$$

$$= \mathrm{E}\mathrm{e}^{\int_{\mathfrak{X}} f(x)\widetilde{\mathcal{N}}(\mathrm{d}x)} \int_{\mathfrak{X}} (\mathrm{e}^{f(y)} - 1)h(y)\mu(\mathrm{d}y), \tag{27}$$

which hold for any $f, h \in f \in L^2(\mathfrak{X})$, f satisfying (26) and any $0 \le \theta \le 1$; see Ref. [246, Prop. 2.1]. Using (26) and (27), we can derive an elegant expression for the covariance of the exponential functions of Poisson stochastic integrals, which will be used in Section 5 of Chapter 5. We write $\int_{\mathfrak{X}} f(x)\widetilde{\mathcal{N}}(\mathrm{d}x) \equiv \int_{\mathfrak{X}} f \mathrm{d}\widetilde{\mathcal{N}}$ for short.

Proposition 8. *Let $f_i \in L^2(\mathfrak{X}), i = 1, 2$, satisfy*

$$\int_{\mathfrak{X}} \mathrm{e}^{2f_i(x)} \mathbb{I}(f_i(x) > 1)\mu(\mathrm{d}x) < \infty, \quad i = 1, 2. \tag{28}$$

Then,

$$\mathrm{Cov}\big(\mathrm{e}^{\int_{\mathfrak{X}} f_1 \mathrm{d}\widetilde{\mathcal{N}}}, \mathrm{e}^{\int_{\mathfrak{X}} f_2 \mathrm{d}\widetilde{\mathcal{N}}}\big) = a_1 a_2 \big(\mathrm{e}^{\int_{\mathfrak{X}} (\mathrm{e}^{f_1(x)} - 1)(\mathrm{e}^{f_2(x)} - 1)\mu(\mathrm{d}x)} - 1\big), \tag{29}$$

$$\begin{aligned}
&\mathrm{Cov}\left(\mathrm{e}^{\int_{\mathfrak{X}} f_1 \mathrm{d}\widetilde{\mathcal{N}}} - a_1 \int_{\mathfrak{X}} f_1 \mathrm{d}\widetilde{\mathcal{N}},\, \mathrm{e}^{\int_{\mathfrak{X}} f_2 \mathrm{d}\widetilde{\mathcal{N}}} - a_2 \int_{\mathfrak{X}} f_2 \mathrm{d}\widetilde{\mathcal{N}}\right) \\
&= a_1 a_2 \left\{ \widetilde{\Psi}\left(\int_{\mathfrak{X}} (\mathrm{e}^{f_1(x)} - 1)(\mathrm{e}^{f_2(x)} - 1)\mu(\mathrm{d}x) \right) \right. \\
&\quad \left. + \int_{\mathfrak{X}} \widetilde{\Psi}(f_1(x)) \widetilde{\Psi}(f_2(x)) \mu(\mathrm{d}x) \right\},
\end{aligned} \tag{30}$$

where $\widetilde{\Psi}(z) := \mathrm{e}^z - 1 - z, z \in \mathbb{R}, a_i := \mathrm{E}\mathrm{e}^{\int_{\mathfrak{X}} f_i \mathrm{d}\widetilde{\mathcal{N}}}, i = 1, 2.$

Proof. Note that all expectations and integrals in (29) are well defined by (26) due to (28) and other assumptions on $f_i, i = 1, 2$. Relation (29) follows from (27), where the difference on the r.h.s. can be written as $\mathrm{e}^{\int_{\mathfrak{X}} (\mathrm{e}^{f_1(x)} - 1)(\mathrm{e}^{f_2(x)} - 1)\mu(\mathrm{d}x)} - 1 = \widetilde{\Psi}\big(\int_{\mathfrak{X}} (\mathrm{e}^{f_1(x)} - 1)(\mathrm{e}^{f_2(x)} - 1)\mu(\mathrm{d}x) + \int_{\mathfrak{X}} (\mathrm{e}^{f_1(x)} - 1)(\mathrm{e}^{f_2(x)} - 1)\mu(\mathrm{d}x)$. A similar computation of the covariances on the l.h.s. of (30) results in this identity. □

3 Linear Forms in i.i.d. r.v.s Tending towards Gaussian Stochastic Integral

Proposition 7 is simple and general; however, it does not provide a stochastic integral representation of the limit Gaussian distribution or the covariance structure when applied to partial sums of a linear RF on $\mathbb{Z}^\nu$. In order to obtain it, assume that for some $\boldsymbol{\gamma} = (\gamma_1, \dots, \gamma_\nu) \in \mathbb{R}^\nu_+$, the rescaled kernel

$$g^\dagger_\lambda(\boldsymbol{u}) := \lambda^{-\sum_{j=1}^\nu \gamma_j/2} g_\lambda([\lambda^\Gamma \boldsymbol{u}]), \quad \boldsymbol{u} \in \mathbb{R}^\nu, \quad \lambda^\Gamma = \mathrm{diag}(\lambda^{\gamma_1}, \dots, \lambda^{\gamma_\nu}), \tag{31}$$

tends to a limit function, $h = h(\boldsymbol{u}), \boldsymbol{u} \in \mathbb{R}^\nu$, in the sense of the convergence in $L^2(\mathbb{R}^\nu)$. (The last fact holds, for instance, when $S(g_\lambda) = \lambda^{-\sum_{j=1}^\nu \gamma_j/2} \sum_{\boldsymbol{u} \in]\mathbf{0}, \lambda^\Gamma \boldsymbol{x}]} \varepsilon(\boldsymbol{u})$ is the normalised sum of $\varepsilon(\boldsymbol{u})$'s on the rectangle $]\mathbf{0}, \lambda^\Gamma \boldsymbol{x}]$.) Then, with $\Lambda := \mathrm{diag}(\lambda^{\gamma_1}, \dots, \lambda^{\gamma_\nu})$,

$$S(g_\lambda) = \int_{\mathbb{R}^\nu} g^\dagger_\lambda(\boldsymbol{u}) M_\lambda(\mathrm{d}\boldsymbol{u})$$

can be written as a stochastic integral of $g^\dagger_\lambda$ in (31) w.r.t. the random measure

$$M_\lambda(A) := \det(\Lambda)^{-1/2} \sum_{\Lambda^{-1}\boldsymbol{u} \in A} \varepsilon(\boldsymbol{u}) \tag{32}$$

concentrated on the points of the rescaled lattice $\Lambda^{-1}\mathbb{Z}^\nu = \{(u_1/\lambda^{\gamma_1}, \dots, u_\nu/\lambda^{\gamma_\nu}) : \boldsymbol{u} \in \mathbb{Z}^\nu\}$. The random measure in (32) obviously satisfies conditions (i)–(iii) with the control measure $\mu_\lambda(A) = \det(\Lambda)^{-1/2} \sum_{\Lambda^{-1}\boldsymbol{u} \in A} 1$ concentrated on $\Lambda^{-1}\mathbb{Z}^\nu$ and tending to the Lebesgue measure $\mathrm{Leb}_\nu(A) = \int_A \mathrm{d}\boldsymbol{u}$ on each bounded open subset $A \subset \mathbb{R}^\nu$. Since $M_\lambda(A) \xrightarrow{\mathrm{d}} W(A)$, where $W(\mathrm{d}u)$ is the Gaussian white noise with zero mean and variance $\mathrm{d}\boldsymbol{u}$, we can expect that

$$S(g_\lambda) \xrightarrow{\mathrm{d}} \int_{\mathbb{R}^\nu} h(\boldsymbol{u}) W(\mathrm{d}\boldsymbol{u}). \tag{33}$$

Relation (33) is formally proved later, being a particular case of more general Propositions 9 and 21.

Proposition 9. *For $\lambda > 0$, let $S(g_\lambda), g_\lambda \in L^2(\mathbb{Z}^\nu)$ be a linear form in (12) in i.i.d.r.v.s with zero mean and unit variance. Assume that for some $\nu \times \nu$ non-degenerate matrix A and $\Lambda = \mathrm{diag}(l_1, \dots, l_\nu)$ with $l_i = l_i(\lambda) \to \infty$, $i = 1, \dots, \nu$ as $\lambda \to \infty$, the functions*

$$g^\dagger_\lambda(\boldsymbol{u}) := |\det(A\Lambda)|^{1/2} g_\lambda(\lceil A\Lambda\boldsymbol{u} \rceil), \quad \boldsymbol{u} \in \mathbb{R}^\nu, \quad \lambda > 0, \tag{34}$$

tend to a limit h in $L^2(\mathbb{R}^\nu)$, i.e.,

$$\|g^\dagger_\lambda - h\|_2^2 = \int_{\mathbb{R}^\nu} |g^\dagger_\lambda(\boldsymbol{u}) - h(\boldsymbol{u})|^2 \mathrm{d}\boldsymbol{u} \to 0, \quad \lambda \to \infty. \tag{35}$$

Then,

$$S(g_\lambda) \xrightarrow{\mathrm{d}} I(h) = \int_{\mathbb{R}^\nu} h(\boldsymbol{u}) W(\mathrm{d}\boldsymbol{u}), \quad \lambda \to \infty. \tag{36}$$

Proof. The convergence in (33) follows from (36) by taking $l_i(\lambda) = \lambda^{\gamma_i}$ and A as the unit matrix. Denote by $\mathcal{S}_0(\mathbb{R}^\nu)$ the set of step functions $f(\boldsymbol{u}) = \sum_{\boldsymbol{m} \in \mathbb{Z}^\nu} \tilde{f}(\square^K_{\boldsymbol{m}}) \mathbb{I}(\boldsymbol{u} \in \square^K_{\boldsymbol{m}})$, $\boldsymbol{u} \in \mathbb{R}^\nu$, which are finite linear combinations of the indicator functions of disjoint squares

$$\square^K_{\boldsymbol{m}} := \prod_{i=1}^{\nu} (m_i/K, (m_i+1)/K] \subset \mathbb{R}^\nu, \tag{37}$$

with real coefficients $\tilde{f}(\square^K_{\boldsymbol{m}})$, $\boldsymbol{m} = (m_1, \ldots, m_\nu) \in \mathbb{Z}^\nu$, $K \in \mathbb{N}$. The set $\mathcal{S}_0(\mathbb{R}^\nu)$ is dense in $L^2(\mathbb{R}^\nu)$: given $f \in L^2(\mathbb{R}^\nu)$, for every $\epsilon > 0$, there exists $f_\epsilon \in \mathcal{S}_0(\mathbb{R}^\nu)$ such that $\|f - f_\epsilon\|_2 < \epsilon$. Equation (36) follows once we show that for every $\epsilon > 0$, there exists $h_\epsilon \in \mathcal{S}_0(\mathbb{R}^\nu)$ such that as $\lambda \to \infty$, the following relations (j)–(jjj) hold: (j) $\mathrm{E}|S(g_\lambda) - S(h_{\epsilon,\lambda})|^2 < \epsilon$, (jj) $S(h_{\epsilon,\lambda}) \xrightarrow{\mathrm{d}} I(h_\epsilon)$, (jjj) $\mathrm{E}|I(h_\epsilon) - I(h)|^2 < \epsilon$, where

$$h_{\epsilon,\lambda}(\boldsymbol{s}) := |\det(A\Lambda)|^{-1/2} h_\epsilon((A\Lambda)^{-1}\boldsymbol{s}), \quad \boldsymbol{s} \in \mathbb{Z}^\nu, \quad \lambda > 0. \tag{38}$$

As for (j), note that

$$\begin{aligned}
&\mathrm{E}|S(g_\lambda) - S(h_{\epsilon,\lambda})|^2 \\
&= \int_{\mathbb{R}^\nu} |g_\lambda(\lceil \boldsymbol{s} \rceil) - h_{\epsilon,\lambda}(\lceil \boldsymbol{s} \rceil)|^2 \mathrm{d}\boldsymbol{s} \\
&= |\det(A\Lambda)| \int_{\mathbb{R}^\nu} |g_\lambda(\lceil A\Lambda\boldsymbol{u} \rceil) - h_{\epsilon,\lambda}(\lceil A\Lambda\boldsymbol{u} \rceil)|^2 \mathrm{d}\boldsymbol{u} = \|g^\dagger_\lambda - h^\dagger_{\epsilon,\lambda}\|_2^2,
\end{aligned}$$

where $h^\dagger_{\epsilon,\lambda}$ is derived from $h_{\epsilon,\lambda}$ in the same way as $g^\dagger_\lambda$ is derived from g_λ in (34). To prove (j), we need to find suitable $h_\epsilon \in \mathcal{S}_0(\mathbb{R}^\nu)$ and thus $h_{\epsilon,\lambda}$ in (38). By (35), there exists $\lambda_0 > 0$ such that $\|g^\dagger_\lambda - h\|_2 < \epsilon/4$,

$\forall \lambda \ge \lambda_0$. Given $g^\dagger_{\lambda_0} \in L^2(\mathbb{R}^\nu)$, there exists $h_\epsilon \in \mathcal{S}_0(\mathbb{R}^\nu)$ such that $\|g^\dagger_{\lambda_0} - h_\epsilon\|_2 < \epsilon/4$. Note that

$$\|h_\epsilon - h^\dagger_{\epsilon,\lambda}\|_2^2 = \int_{\mathbb{R}^\nu} |h_\epsilon(\boldsymbol{u}) - h_\epsilon((A\Lambda)^{-1}\lceil A\Lambda \boldsymbol{u}\rceil)|^2 \mathrm{d}\boldsymbol{u} \to 0, \quad \lambda \to \infty,$$

follows from $|(A\Lambda)^{-1}\lceil A\Lambda\boldsymbol{u}\rceil - \boldsymbol{u}| = |(A\Lambda)^{-1}(\lceil A\Lambda\boldsymbol{u}\rceil - A\Lambda\boldsymbol{u})| \le C\min(l_j, j = 1,\dots,\nu)^{-1} = o(1)$ uniformly in $\boldsymbol{u} \in \mathbb{R}^\nu$ and the fact that h_ϵ is bounded and has compact support. Thus, there exists $\lambda_1 > 0$ such that $\|h_\epsilon - h^\dagger_{\epsilon,\lambda}\|_2 < \epsilon/4$, $\forall \lambda \ge \lambda_1$. Hence,

$$\begin{aligned}\|g^\dagger_\lambda - h^\dagger_{\epsilon,\lambda}\|_2 &\le \|g^\dagger_\lambda - h\|_2 + \|h - g^\dagger_{\lambda_0}\|_2 + \|g^\dagger_{\lambda_0} - h_\epsilon\|_2 \\ &\quad + \|h_\epsilon - h^\dagger_{\epsilon,\lambda}\|_2 < \epsilon, \quad \forall \lambda \ge \lambda_0 \vee \lambda_1,\end{aligned}$$

completing the proof of (j). The above reasoning also implies (jjj) since $(\mathrm{E}|I(h_\epsilon) - I(h)|^2)^{1/2} = \|h_\epsilon - h\|_2 \le \|h_\epsilon - g^\dagger_{\lambda_0}\|_2 + \|g^\dagger_{\lambda_0} - h\|_2 < \epsilon/2$.

It remains to prove (jj). The step function h_ϵ in the above proof of (i) can be written as $h_\epsilon(\boldsymbol{u}) = \sum_{\boldsymbol{m}\in\mathbb{Z}^\nu} \tilde{h}_\epsilon(\square^K_{\boldsymbol{m}})\mathbb{I}(\boldsymbol{u} \in \square^K_{\boldsymbol{m}})$, $\boldsymbol{u} \in \mathbb{R}^2$, $(\exists K \in \mathbb{N})$, where $\tilde{h}_\epsilon(\square^K_{\boldsymbol{m}}) = 0$, except for a finite number of $\boldsymbol{m} \in \mathbb{Z}^\nu$. Then, by (38),

$$h_{\epsilon,\lambda}(\boldsymbol{s}) = |\det(A\Lambda)|^{-1/2} \sum_{\boldsymbol{m}\in\mathbb{Z}^\nu} \tilde{h}_\epsilon(\square^K_{\boldsymbol{m}})\mathbb{I}(\boldsymbol{s} \in A\Lambda\square^K_{\boldsymbol{m}}), \quad \boldsymbol{s} \in \mathbb{Z}^\nu,$$

and $S(h_{\epsilon,\lambda}) = \sum_{\boldsymbol{m}\in\mathbb{Z}^\nu} \tilde{h}_\epsilon(\square^K_{\boldsymbol{m}}) W_\lambda(\square^K_{\boldsymbol{m}})$, where

$$M_\lambda(\square^K_{\boldsymbol{m}}) := |\det(A\Lambda)|^{-1/2} \sum_{\boldsymbol{s}\in A\Lambda\square^K_{\boldsymbol{m}}} \varepsilon(\boldsymbol{s}), \quad \boldsymbol{m} \in \mathbb{Z}^\nu. \tag{39}$$

Since the r.v.s $\varepsilon(\boldsymbol{s})$, $\boldsymbol{s} \in \mathbb{Z}^\nu$, are i.i.d. with $\mathrm{E}\varepsilon(\mathbf{0}) = 0$, $\mathrm{E}\varepsilon(\mathbf{0})^2 = 1$ and the parallelograms $A\Lambda\square^K_{\boldsymbol{m}}$, $\boldsymbol{k} \in \mathbb{Z}^\nu$, are disjoint, the r.v.s $M_\lambda(\square^K_{\boldsymbol{m}})$, $\boldsymbol{m} \in \mathbb{Z}^\nu$, are independent and satisfy $\mathrm{E}M_\lambda(\square^K_{\boldsymbol{m}}) = 0$, $\mathrm{E}|M_\lambda(\square^K_{\boldsymbol{m}})|^2 = |\det(A\Lambda)|^{-1} \sum_{\boldsymbol{s}\in\mathbb{Z}^\nu} \mathbb{I}(\boldsymbol{s} \in A\Lambda\square^K_{\boldsymbol{m}}) \to \int_{\square^K_{\boldsymbol{m}}} \mathrm{d}\boldsymbol{u}$, $\lambda \to \infty$. Hence, by the classical CLT, for every $J \in \mathbb{N}$,

$$\begin{aligned}&\{M_\lambda(\square^K_{\boldsymbol{m}}),\ \boldsymbol{m} \in \{-J,\dots,J\}^\nu\} \\ &\quad \xrightarrow{\mathrm{d}} \{W(\square^K_{\boldsymbol{m}}),\ \boldsymbol{m} \in \{-J,\dots,J\}^\nu\}, \quad \lambda \to \infty,\end{aligned} \tag{40}$$

implying the convergence $S(h_{\epsilon,\lambda}) \xrightarrow{\mathrm{d}} \sum_{\boldsymbol{m}\in\mathbb{Z}^\nu} \tilde{h}_\epsilon(\Box^K_{\boldsymbol{m}}) W(\Box^K_{\boldsymbol{m}}) = I(h_\epsilon)$, $\lambda \to \infty$, or part (jj), and completing the proof of the proposition. □

4 Hermite, Charlier, and Appell Polynomials

This section summarises some properties of the Hermite, Charlier, and Appell polynomials, which play an important role in the study of scaling limits of nonlinear functions $G(X)$ of a linear LRD RF X. These facts will be used in Chapters 5 and 6. The interest in the above polynomials can be roughly explained by the fact that $G(X) - \mathrm{E}G(X) = \sum_{k=1}^\infty c_k P_k(X)$ can be expanded into polynomials $P_k(X)$ with the LRD properties of $G(X)$ determined by the lowest non-zero expansion term $c_{k^*} P_{k^*}(X)$, with $c_{k^*} \neq 0, c_k = 0, 1 \le k < k^*$. In many cases of interest, the above approximation with $k^* = 1$ allows us to reduce the scaling limit of $G(X)$ to that of the linear RF X.

Hermite polynomials. Hermite polynomials, $H_k(x), x \in \mathbb{R}, k \in \mathbb{N}$, are related to the standard Gaussian distribution $Z \sim N(0,1)$ with density function $\phi(x) = (2\pi)^{-1/2}\mathrm{e}^{-x^2/2}$ and are usually defined as

$$\begin{aligned} H_k(x) &:= (-1)^k \phi(x)^{-1} \mathrm{d}^k \phi(x)/\mathrm{d}x^k \\ &= (-1)^k \mathrm{e}^{x^2/2} \mathrm{d}^k \mathrm{e}^{-x^2/2}/\mathrm{d}x^k, \quad x \in \mathbb{R},\ k \in \mathbb{N}. \end{aligned} \tag{41}$$

They can also be defined by the power series (generating function)

$$\psi(x,u) := \mathrm{e}^{ux - u^2/2} = \sum_{k=0}^\infty \frac{u^k}{k!} H_k(x), \quad x, u \in \mathbb{R}, \tag{42}$$

and arise as the coefficients of the powers of u. This sum converges for all values of x and u. To see the equivalence of definitions in (41) and (42), note that

$$H_k(x) = \mathrm{d}^k \psi(x,u)/\mathrm{d}u^k\big|_{u=0} = \mathrm{e}^{x^2/2} \mathrm{d}^k \mathrm{e}^{-(u-x)^2/2}/\mathrm{d}u^k\big|_{u=0} \tag{43}$$

$$= (-1)^k \mathrm{e}^{x^2/2} \mathrm{d}^k \mathrm{e}^{-(u-x)^2/2}/\mathrm{d}x^k\big|_{u=0} \tag{44}$$

$$= (-1)^k \mathrm{e}^{x^2/2} \mathrm{d}^k \mathrm{e}^{-x^2/2}/\mathrm{d}x^k. \tag{45}$$

Since $u \mapsto \mathrm{e}^{ux-u^2/2} : \mathbb{C} \to \mathbb{C}$ is analytic on the whole complex plane $\mathbb{C}$, the power series in (42) can be extended to imaginary $\mathrm{i}u \in \mathbb{C}$ or

$$\mathrm{e}^{\mathrm{i}ux}/\mathrm{E}\mathrm{e}^{\mathrm{i}uZ} = \mathrm{e}^{\mathrm{i}ux+u^2/2} = \sum_{k=0}^{\infty} \frac{(\mathrm{i}u)^k}{k!} H_k(x), \quad x, u \in \mathbb{R}. \tag{46}$$

The r.v.s $H_k(Z), k \geq N$ are orthogonal, i.e.,

$$\begin{aligned} &\mathrm{E}H_k(Z) = 0, \quad \mathrm{E}H_k(Z)^2 = k!, \quad k = 1, 2, \ldots, \\ &\mathrm{E}H_k(Z)H_\ell(Z) = 0, \quad k \neq \ell = 0, 1, \ldots. \end{aligned} \tag{47}$$

Facts (47) follow from multiplying the series in (46) at the points u and v, taking the expectation of the product

$$\begin{aligned} \sum_{k,\ell=0}^{\infty} \frac{(\mathrm{i}u)^k (\mathrm{i}v)^\ell}{k!\ell!} \mathrm{E}H_k(Z)H_\ell(Z) &= \mathrm{E}\mathrm{e}^{\mathrm{i}(u+v)Z} \mathrm{e}^{(u^2+v^2)/2} \\ &= \mathrm{e}^{-uv} = \sum_{k=0}^{\infty} \frac{(-1)^k u^k v^k}{k!}, \end{aligned} \tag{48}$$

and equating the coefficients of $u^k v^\ell, k, \ell \in \mathbb{N}$ of the power series on both sides.

Relation (47) means that $H_k(x)/\sqrt{k!}, k = 0, 1, \ldots$ form an orthonormal system in the space $L^2(\Phi)$ of all measurable functions $G : \mathbb{R} \to \mathbb{R}$, with $\mathrm{E}G(Z)^2 = \int_{\mathbb{R}} G(x)^2 \mathrm{d}\Phi(x) < \infty$, where $\Phi(x) := \mathrm{P}(Z \leq x)$ is the d.f. of $Z \sim N(0,1)$. The above system is complete and forms an orthogonal basis in $L^2(\Phi)$, meaning that any function $G \in L^2(\Phi)$ can be expanded into Hermite polynomials:

$$G(x) = \sum_{j=0}^{\infty} \frac{h_G(j)}{j!} H_j(x), \quad \text{with} \tag{49}$$

$$h_G(j) := \mathrm{E}G(Z)H_j(Z) = \int_{\mathbb{R}} G(x)H_j(x)\mathrm{d}\Phi(x), \quad j = 0, 1, \ldots.$$

Note that $h_G(0) = \mathrm{E}G(Z)$ and

$$\mathrm{E}G(Z)^2 = \sum_{j=0}^{\infty} \frac{h_G^2(j)}{j!}. \tag{50}$$

Example 10. (i) Let $G(x) = \mathrm{e}^{ax}, x, a \in \mathbb{R}$. Then, $h_G(j) = \mathrm{E}G^{(j)}(Z) = a^j \mathrm{E}\mathrm{e}^{aZ} = a^j \mathrm{e}^{a^2/2}$, yielding the convergent Hermite expansion of the exponential function:

$$\mathrm{e}^{ax} = \mathrm{e}^{a^2/2} \sum_{j=0}^{\infty} \frac{a^j}{j!} H_j(x). \tag{51}$$

(ii) Let $G(x) = \mathbb{I}(x \le y), x, y \in \mathbb{R}$. Then, $h_G(j) = (-1)^j \mathrm{E}G(Z)$ $\Phi^{(j+1)}(Z) = (-1)^k \Phi^{(j)}(y)$, yielding the convergent Hermite expansion of the exponential function:

$$\mathbb{I}(x \le y) = \sum_{j=0}^{\infty} \frac{(-1)^j \Phi^{(j)}(y)}{j!} H_j(x). \tag{52}$$

Proposition 11. *Let (Z_1, Z_2) have a bivariate normal distribution with mean zero, unit variances, and correlation coefficient $\rho \in (-1, 1)$, with the joint density*

$$\phi(x, y) = (2\pi\sqrt{1-\rho^2})^{-1} \times \exp\left\{-\frac{1}{2(1-\rho^2)}(x^2 + y^2 - 2\rho xy)\right\}, \quad x, y \in \mathbb{R}. \tag{53}$$

(i) (Orthogonality property): *For any $k, \ell \in \mathbb{N}$,*

$$\mathrm{E}H_k(Z_1)H_\ell(Z_2) = \begin{cases} 0, & k \ne \ell, \\ \rho^k k!, & k = \ell, \end{cases} \tag{54}$$

with the convention $0^0 := 1$.
(ii) *Let $G_i = G_i(x), x \in \mathbb{R}, i = 1, 2$ be given functions such that*

$$\mathrm{E}G_i^2(Z_i) < \infty, \quad i = 1, 2. \tag{55}$$

Then,

$$\mathrm{E}G_1(Z_1)G_2(Z_2) = \sum_{k=0}^{\infty} \frac{h_{G_1}(k) h_{G_2}(k)}{k!} \rho^k. \tag{56}$$

(iii) (Mehler's formula):

$$\phi(x,y) = \sum_{k=0}^{\infty} \frac{\rho^k}{k!} \phi^{(k)}(x)\phi^{(k)}(y) \tag{57}$$

$$= \phi(x)\phi(y) \sum_{k=0}^{\infty} \frac{\rho^k}{k!} H_k(x)H_k(y).$$

Proof. (i) Similarly to (48), multiply the generating functions $\mathrm{e}^{\mathrm{i}uZ_1+u^2/2}$ and $\mathrm{e}^{\mathrm{i}vZ_2+v^2/2}$, use (46), and take the expectation to obtain

$$\sum_{k,\ell=0}^{\infty} \frac{(\mathrm{i}u)^k(\mathrm{i}v)^\ell}{k!\ell!} \mathrm{E}H_k(Z_1)H_\ell(Z_2) = \mathrm{E}\mathrm{e}^{\mathrm{i}(uZ_1+vZ_2)}\mathrm{e}^{(u^2+v^2)/2}$$

$$= \mathrm{e}^{-\rho uv} = \sum_{k=0}^{\infty} \frac{(-1)^k}{k!} \rho^k u^k v^k.$$

Hence, (54) follows from this equality by comparing the coefficients of powers $u^k v^\ell$ on both sides.

(ii) Immediate from (49) and (54).

(iii) By definition,

$$\int_{\mathbb{R}^2} \mathrm{e}^{\mathrm{i}(xu+yv)}\phi(x,y)\mathrm{d}x\mathrm{d}y = \mathrm{e}^{-(u^2-2\rho uv+v^2)/2}.$$

It suffices to show that the last equality remains valid, with $\phi(x,y)$ replaced by the r.h.s. of (57), denoted by $\tilde{\phi}(x,y)$. We have

$$I := \int_{\mathbb{R}^2} \mathrm{e}^{\mathrm{i}(xu+yv)}\tilde{\phi}(x,y)\mathrm{d}x\mathrm{d}y$$

$$= \sum_{k=0}^{\infty} \frac{\rho^k}{k!} \int_{\mathbb{R}} \mathrm{e}^{\mathrm{i}xu}\phi^{(k)}(x)\mathrm{d}x \int_{\mathbb{R}} \mathrm{e}^{\mathrm{i}yv}\phi^{(k)}(y)\mathrm{d}y.$$

Integrating by parts, $\int_{\mathbb{R}} \mathrm{e}^{\mathrm{i}xu}\phi^{(k)}(x)\mathrm{d}x = (\mathrm{i}u)^k \int_{\mathbb{R}} \mathrm{e}^{\mathrm{i}xu}\phi(x)\mathrm{d}x = (\mathrm{i}u)^k\mathrm{e}^{-u^2/2}$, and hence

$$I = \mathrm{e}^{-(u^2+v^2)/2} \sum_{k=0}^{\infty} \frac{\rho^k}{k!} (\mathrm{i}u)^k(\mathrm{i}v)^k = \mathrm{e}^{-(u^2-2\rho uv+v^2)/2},$$

which completes the proof of the first equality in (57). The second equality in (57) follows from (41). □

The *Hermite rank* $k_H(G)$ of $G \in L^2(\Phi)$ is the index of the first non-zero coefficient in the Hermite expansion (49): $G(x) - \mathrm{E}G(Z) = \sum_{j=k_H(G)}^{\infty} h_G(j)H_j(x)/j!$, where $h_G(k_H(G)) \neq 0$. For instance, the Hermite rank of $G(x) = x^2 - c$ is 2 if $c = 1$ or $G(x) = H_2(x)$ and 1 if $c \neq 1$.

Corollary 12. *Let $G_i, i = 1, 2$, be as in (55) and $k_H^* := k_H(G_1) \vee k_H(G_2)$. Then,*

$$\begin{aligned} |\mathrm{Cov}(G_1(Z_1), G_2(Z_2))| &\le |\rho|^{k_H^*} \sum_{k=k_H^*}^{\infty} \frac{|h_{G_1}(k)h_{G_2}(k)|}{k!} \qquad (58) \\ &\le |\rho|^{k_H^*} \left(\sum_{k=k_H^*}^{\infty} \frac{h_{G_1}^2}{k!} \right)^{1/2} \left(\sum_{k=k_H^*}^{\infty} \frac{h_{G_2}^2}{k!} \right)^{1/2} \\ &\le |\rho|^{k_H^*} \sqrt{\mathrm{Var}(G_1(Z_1))\mathrm{Var}(G_2(Z_2))}. \end{aligned}$$

Particularly, $\sup\{|\mathrm{Cov}(G_1(Z_1), G_2(Z_2))| : \mathrm{Var}(G_i(Z_i)) = 1, i = 1, 2\} = |\rho|$, *and the last supremum is attained by* $G_i(z) = z, i = 1, 2$.

Remark 13. A natural extension of standard Hermite polynomials depending on the variance parameter $\sigma^2 > 0$ is defined through the generating function

$$\sum_{k=0}^{\infty} \frac{(\mathrm{i}u)^k}{k!} H_k(x; \sigma^2) = \mathrm{e}^{\mathrm{i}ux + (1/2)\sigma^2 u^2} = \frac{\mathrm{e}^{\mathrm{i}ux}}{\mathrm{E}\mathrm{e}^{\mathrm{i}u\sigma Z}}, \qquad u, x \in \mathbb{R}, \qquad (59)$$

or directly as $H_k(x; \sigma^2) := \sigma^k H_k(x/\sigma), k \in \mathbb{N}$.

Charlier polynomials. Because of many similarities, Charlier polynomials can be regarded as 'Poisson counterparts' of Hermite polynomials. The Charlier polynomials $P_k(x; \mu), k \in \mathbb{N}$ of discrete variable $x \in \mathbb{N}$ depending on parameter $\mu > 0$ can be defined through the generating function

$$\mathcal{P}(u; x, \mu) := \sum_{k=0}^{\infty} \frac{u^k}{k!} P_k(x; \mu) = (1 + u)^x \mathrm{e}^{-u\mu}, \qquad (60)$$

where the series is convergent for any $x \in \mathbb{N}, \mu > 0$ and any (complex) $u \in \mathbb{C}$. We have $P_0(x; \mu) = 1, P_1(x; \mu) = x - \mu, P_2(x; \mu) =$

$x^2 - (2\mu + 1)x + \mu^2$ and

$$P_k(x;\mu) = (-1)^k \mu^k p(x;\mu)^{-1} D_-^k p(x;\mu), \quad k \in \mathbb{N}, \tag{61}$$

where $D_-^k := D_- D_-^{k-1}$ is the backward difference operator, $D_- G(x) := G(x) - G(x-1)\mathbb{I}(x \ge 1), D_-^0 G(x) = G(x)$, and

$$p(x;\mu) := \mathrm{e}^{-\mu} \frac{\mu^x}{x!}, \quad x \in \mathbb{N}, \tag{62}$$

is the distribution of Poisson r.v. N with mean μ. Relation (61) for $\mu = 1$ resembles (41) for Hermite polynomials, with D_-^k being the discrete analogue of $\mathrm{d}^k/\mathrm{d}x^k$, and can be verified as follows. Denote $\tilde{P}_k(x;\mu)$ the r.h.s. of (61) and check $\tilde{P}_k(x;\mu) = P_k(x;\mu)$ for $k = 0, 1, 2$. By induction, verify the recurrence relation $\tilde{P}_{k+1}(x;\mu) = (x + \mu - k)\tilde{P}_k(x;\mu) - \mu k \tilde{P}_{k-1}(x;\mu), k = 1, 2, \ldots$. The same recurrence relation is satisfied by the polynomials $P_k(x;\mu)$ with the generating function in (60) using the identity $(1+u)\partial \mathcal{P}(u;x,\mu)/\partial u = (x - (1+u)\mu)\mathcal{P}(u;x,\mu)$ [246]. We have

$$\begin{aligned} &\mathrm{E}P_k(N;\mu) = 0, \quad \mathrm{E}P_k(N;\mu)^2 = k!\mu^k, \quad k = 1, 2, \ldots, \\ &\mathrm{E}P_k(N;\mu)P_\ell(N;\mu) = 0, \quad k \ne \ell = 0, 1, \ldots. \end{aligned} \tag{63}$$

Facts (63) follow from multiplying the series in (60) at the points u and v, taking the expectation of the product

$$\begin{aligned} \sum_{k,\ell=0}^{\infty} \frac{u^k v^\ell}{k!\ell!} \mathrm{E}P_k(N;\mu)P_\ell(N;\mu) &= \mathrm{e}^{-(u+v)\mu} \mathrm{E}((1+u)(1+v))^N \\ &= \mathrm{e}^{\mu uv} = \sum_{k=0}^{\infty} \frac{(\mu uv)^k}{k!}, \end{aligned}$$

and equating the coefficients of $u^k v^\ell, k, \ell \in \mathbb{N}$ of the power series on both sides.

Any $G = G(x), x \in \mathbb{N}$ with $\mathrm{E}G(N)^2 < \infty$ can be uniquely expanded into Charlier polynomials:

$$G(x) = \sum_{k=0}^{\infty} \frac{c_G(k;\mu)}{k!} P_k(x;\mu), \quad x \in \mathbb{N}, \tag{64}$$

with the *Charlier coefficients* $c_G(k;\mu)$ of G

$$c_G(k;\mu) := \mu^{-k}\mathrm{E}G(N)P_k(N;\mu), \quad k \in \mathbb{N} \tag{65}$$

satisfying

$$\sum_{k=0}^{\infty} \frac{c_G^2(k;\mu)\mu^k}{k!} = \mathrm{E}G(N)^2 < \infty; \tag{66}$$

see (63). Relation (61) and summation by parts yields another expression for these coefficients:

$$c_G(k;\mu) = \mathrm{E}D_+^k G(N), \quad k \in \mathbb{N}, \tag{67}$$

where $D_+^k := D_+D_+^{k-1}$ is the forward difference operator, $D_+^0G(x) := G(x), D_+G(x) := G(x+1) - G(x), D_+^2G(x) = D_+G(x+1) - D_+G(x) = G(x+2) - 2G(x+1) + G(x)$, etc. Equations (65) and (63) yield the bound

$$|c_G(k;\mu)| \le \mu^{-k}\sqrt{\mathrm{E}[G^2(N)]\mathrm{E}[P_k^2(N;\mu)]} = C(k!/\mu^k)^{1/2},$$
$$C = \sqrt{\mathrm{E}G(N)^2}. \tag{68}$$

Let us present two concrete examples of Charlier expansions which are used in Section 4 and 5.

Example 14. Let $G(x) = x \wedge 1 = 1 - \mathbb{I}(x = 0)$ $(x \in \mathbb{N})$. Then, $D_+^kG(x) = (-1)^{k+1}\mathbb{I}(x = 0), k \ge 1$. According to (67), $c_G(0;\mu) = \mathrm{P}(N \ge 1) = 1 - \mathrm{e}^{-\mu}, c_G(k;\mu) = (-1)^{k+1}\mathrm{P}(N = 0) = (-1)^{k+1}\mathrm{e}^{-\mu}$ $(k \ge 1)$. Thus,

$$x \wedge 1 = 1 - \mathrm{e}^{-\mu} + \mathrm{e}^{-\mu}\sum_{k=1}^{\infty}\frac{(-1)^{k+1}}{k!}P_k(x;\mu), \quad x \in \mathbb{N}. \tag{69}$$

Example 15. Let $G(x) = \mathrm{e}^{ax}$ $(x \in \mathbb{N}, a \in \mathbb{R})$. Then $D_+^kG(x) = (\mathrm{e}^a - 1)^k\mathrm{e}^{ax}$ and $c_G(k;\mu) = (\mathrm{e}^a - 1)^k\mathrm{e}^{(\mathrm{e}^a-1)\mu}$ $(k \ge 0)$, yielding

$$\mathrm{e}^{ax} = \mathrm{e}^{(\mathrm{e}^a-1)\mu}\sum_{k=0}^{\infty}\frac{(\mathrm{e}^a-1)^k}{k!}P_k(x;\mu), \quad x \in \mathbb{N}. \tag{70}$$

Lemma 16. *Let $M_i, i = 1, 2, 3$, be independent Poisson distributed r.v.s with means* $\mathrm{E}M_1 = \mu_1 - \mu_3, \mathrm{E}M_2 = \mu_2 - \mu_3.\mathrm{E}M_3 = \mu_3$, $\mu_3 \le \mu_1 \wedge \mu_2$, *and*

$$N_i := M_i + M_3, \quad i = 1, 2. \tag{71}$$

Let

$$p(x, y; \mu_1, \mu_2, \mu_3) := \mathrm{P}(N_1 = x, N_2 = y), \quad (x, y) \in \mathbb{N}^2, \tag{72}$$

denote the joint distribution of (N_1, N_2).
(i) (Orthogonality property): *For any* $k, \ell \in \mathbb{N}$,

$$\mathrm{E}P_k(N_1; \mu_1)P_\ell(N_2; \mu_2) = \begin{cases} 0, & k \ne \ell, \\ \mu_3^k k!, & k = \ell, \end{cases} \tag{73}$$

with the convention $0^0 := 1$.
(ii) Let $G_i = G_i(x), x \in \mathbb{N}, i = 1, 2$ *be given functions such that*

$$\mathrm{E}G_i^2(N_i) < \infty, \quad i = 1, 2. \tag{74}$$

Then,

$$\mathrm{E}G_1(N_1)G_2(N_2) = \sum_{k=0}^{\infty} \frac{c_{G_1}(k; \mu_1)c_{G_2}(k; \mu_2)}{k!} \mu_3^k. \tag{75}$$

(iii) (Mehler's formula):

$$p(x, y; \mu_1, \mu_2, \mu_3) = \sum_{k=0}^{\infty} \frac{\mu_3^k}{k!} D_-^k p(x; \mu_1) D_-^k p(y; \mu_2) \tag{76}$$

$$= p(x; \mu_1)p(y; \mu_2) \sum_{k=0}^{\infty} \frac{\rho_{12}^k}{k!} P_k(x; \mu_1)P_k(y; \mu_2),$$

where $\rho_{12} := \mu_3/\sqrt{\mu_1\mu_2} = \mathrm{Corr}(N_1, N_2)$ *is the correlation coefficient.*

Proof. (i) The proof of (73) using the generating function in (61) is similar as in the univariate case of (63). Consider the expectation

$$\begin{aligned} &\mathrm{E}\mathcal{P}(u; N_1, \mu_1)\mathcal{P}(v; N_2, \mu_2) \\ &= \mathrm{e}^{-u\mu_1 - v\mu_2} \mathrm{E}[(1+u)^{N_1}(1+v)^{N_2}] \end{aligned} \tag{77}$$

$$= \mathrm{e}^{-u\mu_1 - v\mu_2}\mathrm{E}[(1+u)^{M_1}]\mathrm{E}[(1+v)^{M_2}]\mathrm{E}[((1+u)(1+v))^{M_3}]$$
$$= \mathrm{e}^{-u\mu_1 - v\mu_2}\mathrm{e}^{(\mu_1-\mu_3)u}\mathrm{e}^{(\mu_2-\mu_3)v}\mathrm{e}^{((1+u)(1+v)-1)\mu_3}$$
$$= \mathrm{e}^{uv\mu_3} = \sum_{k=0}^{\infty}\frac{(uv\mu_3)^k}{k!}.$$

On the other hand,

$$\mathrm{E}\mathcal{P}(u; N_1, \mu_1)\mathcal{P}(v; N_2, \mu_2) = \sum_{k,\ell=0}^{\infty}\frac{u^k v^\ell}{k!\ell!}\mathrm{E}[P_k(N_1;\mu_1)P_\ell(N_2;\mu_2)],$$

and (73) follows by equating the coefficients of $u^k v^\ell, k, \ell \in \mathbb{N}$ of the power series on both sides.
(ii) Immediate from (64) and (73).
(iii) Apply (75) to $G_1(x) := \mathbb{I}(x = n), G_2(x) := \mathbb{I}(x = m)$, for given $n, m \in \mathbb{N}$. By (66) and (61), $c_{G_1}(k;\mu_1) = \mathrm{E}[D_+^k \mathbb{I}(N_1 = n)] = D_-^k p(n;\mu_1) = (-1)^k \mu_1^{-k} P_k(n;\mu_1)p(n;\mu_1)$, $c_{G_2}(k;\mu_2) = \mathrm{E}[D_+^k \mathbb{I}(N_2 = m)] = D_-^k p(m;\mu_2) = (-1)^k \mu_2^{-k} P_k(m;\mu_2)p(m;\mu_2)$. This proves (76) and the lemma as well. □

The *Charlier rank* $k_C(G;\mu) \geq 1$ of $G(x), x \in \mathbb{N}, \mathrm{E}G(N)^2 < \infty$ is the index of the first non-zero coefficient in the Charlier expansion (64): $G(x) - \mathrm{E}G(N) = \sum_{j=k_C(G;\mu)}^{\infty}\frac{c_G(j;\mu)}{j!}P_j(x;\mu)$, where $c_G(k_C(G;\mu);\mu) \neq 0$. For instance, the Charlier rank of the second-order polynomial $G(x) = x^2 + b_1 x + b_2$ is

$$k_C(G;\mu) = \begin{cases} 1, & b_1 + 2\mu + 1 \neq 0, \\ 2, & b_1 + 2\mu + 1 = 0. \end{cases}$$

Lemma 16 (75) and the bound in (68) imply the following.

Corollary 17. *Let $G_i, N_i, i = 1, 2$ be as in Lemma 16. Then,*

$$\mathrm{Cov}(G_1(N_1), G_2(N_2)) = \sum_{k=k_{12}^*}^{\infty}\frac{c_{G_1}(k;\mu_1)c_{G_2}(k;\mu_2)}{k!}\mu_3^k$$
$$= \frac{c_{G_1}(k_{12}^*;\mu_1)c_{G_2}(k_{12}^*;\mu_2)}{k_{12}^*!}\mu_3^{k^*} + R(k^*),$$

where $k_{12}^* := k_C(G_1;\mu_1) \vee k_C(G_2;\mu_2)$ *and*

$$|R(k_{12}^*)| \le \frac{(\mu_3/\sqrt{\mu_1\mu_2})^{k_{12}^*+1}}{1-(\mu_3/\sqrt{\mu_1\mu_2})} \prod_{i=1}^{2} \mathrm{E}^{1/2} G(N_i)^2.$$

Corollary 18. *Let* $G_i, N_i, i = 1, 2$ *be as in (74),* $\rho_{12} = \mathrm{Corr}(N_1, N_2)$, $k_{12}^* = k_C(G_1;\mu_1) \vee k_C(G_2;\mu_2)$. *Then,*

$$\begin{aligned}
&|\mathrm{Cov}(G_1(N_1), G_2(N_2))| \\
&\le \sum_{k=k_{12}^*}^{\infty} |\rho_{12}|^k \frac{|c_{G_1}(k;\mu_1)c_{G_2}(k;\mu_2)|(\mu_1\mu_2)^{k/2}}{k!} \\
&\le |\rho_{12}|^{k_{12}^*} \left(\sum_{k=k_C^*}^{\infty} \frac{c_{G_1}^2(k;\mu_1)\mu_1^k}{k!}\right)^{1/2} \left(\sum_{k=k_{12}^*}^{\infty} \frac{c_{G_2}^2(k;\mu_2)\mu_2^k}{k!}\right)^{1/2} \\
&\le |\rho_{12}|^{k_{12}^*} \sqrt{\mathrm{Var}(G_1(N_1))\mathrm{Var}(G_2(N_2))}. \qquad (78)
\end{aligned}$$

Particularly, $\sup\{|\mathrm{Cov}(G_1(N_1), G_2(N_2))| : \mathrm{Var}(G_i(N_i)) = 1, i = 1, 2\} = |\rho_{12}|$, *and the last supremum is attained by the linear functions* $G_i(x) = x/\sqrt{\mu_i}, i = 1, 2$.

Remark 19. Let $X = \{X(t); t \in \mathbb{Z}\}$ be a stationary Markov process on $\mathbb{N}$ with marginal Poisson distribution $\mathrm{P}(X(t) = x) = p(x;\mu)$ and transition probabilities

$$p(y|x;\mu) := \frac{p(x,y;\mu)}{p(x;\mu)}, \quad x, y \in \mathbb{N}, \qquad (79)$$

where $p(x,y;\mu) := p(x,y;\mu,\mu,\mu_3)$ is the joint distribution in (72) with $\mu_1 = \mu_2 =: \mu$. This process is well known in the literature as the *Poisson AR(1)* or *INAR(1)* and is related to the M/M/∞ queueing system; see e.g., Ref. [182]. Substitution of (76) into (79) yields an expansion of (79) into a series of Charlier polynomials. Since the transition probability of the Poisson INAR(1) process is usually written using a different expansion, the coincidence with (79) can be easily verified through the bivariate generating function of $(X(1), X(2))$ in (77); see Ref. [182, (9)]. We remark that the Poisson INAR(1) process is a particular case of stationary Markov evolutions of non-interacting particle systems with Poisson marginal distribution discussed in Ref. [246], closely related to chaos expansions in multiple Poisson stochastic integrals.

The Poisson INAR(1) process has an elegant representation as a *Poisson trawl process with an exponential trawl function*, as defined in what follows. Let $\mathcal{N}(s, \mathrm{d}u)$ be a Poisson random measure on $\mathbb{Z} \times \mathbb{R}_+$ with $\mathrm{E}\mathcal{N}(s, \mathrm{d}u) = c\,\mathrm{d}u$ and

$$X(t) := \mathcal{N}(A_t), \quad \text{where } A_t := \{(s,u) \in \mathbb{Z} \times \mathbb{R}_+ : s \le t, u \le a^{t-s)}\}, \tag{80}$$

with $c > 0, 0 \le a < 1$ being the parameters. Thus, $X(t)$ counts the number of Poisson points ('seeds') in the 'trawl set' A_t. The process $X = \{X(t); t \in \mathbb{Z}\}$ in (80) is stationary and has a marginal Poisson distribution with mean

$$\mu = \mathrm{E}X(t) = c \sum_{s \le t} a^{-(t-s)} = \frac{c}{1-a}$$

and the covariance function

$$\mathrm{Cov}(X(0), X(t)) = c \sum_{s \le 0} a^s \wedge a^{-(t-s)} = \mu\, a^t, \quad t = 0, 1, \ldots. \tag{81}$$

The bivariate generating function

$$\begin{aligned} \mathrm{E}\,(1+u_1)^{X(1)}(1+u_2)^{X(2)} &= \mathrm{E}\,(1+u_1)^{\mathcal{N}(A_1)}(1+u_2)^{\mathcal{N}(A_2)} \\ &= \mathrm{e}^{\mu(u_1+u_2+u_1u_2a)} \end{aligned}$$

is computed as in (77), with $N_i = \mathcal{N}(A_i), i = 1,2$ and $M_3 = \mathcal{N}(A_1 \cap A_2)$. The above computation can be extended to the *multivariate generating function* of X in (80):

$$\begin{aligned} \mathrm{E}\prod_{t=1}^{n}(1+u_t)^{X(t)} &= \mathrm{E}\prod_{t=1}^{n}(1+u_t)^{\mathcal{N}(A_t)} \\ &= \exp\left\{\mu \sum_{1 \le s \le t \le n} a^{t-s} u_s(1+u_{s+1}) \cdots \right. \\ &\qquad \left. \times\ (1+u_{t-1})u_t \right\}, \end{aligned} \tag{82}$$

with the convention that $u_s(1+u_{s+1})\cdots(1+u_{t-1})u_t := u_t$ when $s=t$. The multivariate generating function uniquely determines the law of X and coincides with the corresponding generating function of the Poisson INAR(1) process [19, Prop. 2.1].

Appell polynomials. Given an r.v. ξ with d.f. $F = F(x) = \mathrm{P}(\xi \le x), x \in \mathbb{R}$ and finite kth moment $\mathrm{E}|\xi|^k < \infty$ $(k = 1, 2, \dots)$, the kth Appell polynomial $A_k(x; F), x \in \mathbb{R}$ is defined by

$$A_k(x;F) := (-\mathrm{i})^k \mathrm{d}^k(\mathrm{e}^{\mathrm{i}xu}/\mathrm{E}\mathrm{e}^{\mathrm{i}u\xi})/\mathrm{d}u^k\big|_{u=0}. \tag{83}$$

If ξ has all finite moments, the above definition agrees with the definition of Appell polynomials through a formal power series:

$$\sum_{k=0}^{\infty} \frac{(\mathrm{i}u)^k}{k!} A_k(x;F) = \frac{\mathrm{e}^{\mathrm{i}ux}}{\mathrm{E}\mathrm{e}^{\mathrm{i}u\xi}}. \tag{84}$$

The convergence of the series in (84) can be justified under strong additional assumptions on F or its characteristic function, so that we use only (83) here. It follows that for $\xi = Z \sim N(0,1)$, $A_k(x;F) = H_k(x)$ agree with Hermite polynomials. For any $k \ge 0$, $A_k(x;F)$ is a polynomial of degree k in x whose coefficients are expressed in terms of moments $m_j := \mathrm{E}\xi^j, j \le k$. In particular,

$$\begin{aligned}
&A_0(x;F) = 1, \quad A_1(x;F) = x - m_1,\\
&A_2(x;F) = x^2 - 2m_1 x + 2m_1^2 - m_2,\\
&A_3(x;F) = x^3 - 3m_1 x^2 + 3x(2m_1^2 - m_2) + 6m_1 m_2 - m_2 - 6m_1^2.
\end{aligned}$$

We note the property

$$\mathrm{E}A_k(\xi;F) = \int_{\mathbb{R}} A_k(x;F)\mathrm{d}F(x) = 0, \quad k = 1, 2, \dots, \tag{85}$$

which follows from (83) by taking the expectation of $A_k(\xi;F)$ and exchanging it with differentiation on the r.h.s. Another useful property is the differentiation rule

$$\mathrm{d}A_k(x;F)/\mathrm{d}x = kA_{k-1}(x;F), \quad k = 1, 2, \dots, \tag{86}$$

which follows similarly from (83) by exchanging the order of differentiation w.r.t x and u. The class of Appell polynomials contains many classical polynomials (Euler and Bernoulli) but lacks the orthogonality property except for Hermite polynomials [237].

Any polynomial $G(x), x \in \mathbb{R}$ of degree $n \geq 0$ can be expanded using Appell polynomials:

$$G(x) = \sum_{k=0}^{n} a_{k,F} A_k(x; F), \tag{87}$$

with coefficients

$$\begin{aligned} a_{k,F} &= \mathrm{E}G^{(k)}(\xi) = G_\infty^{(k)}(0), \\ G_\infty(x) := \mathrm{E}G(\xi + x) &= \int_{\mathbb{R}} G(x+y)\mathrm{d}F(y), \end{aligned} \tag{88}$$

where $h^{(k)}(x) := \mathrm{d}^k h(x)/\mathrm{d}x^k$ denotes the derivative. Equations (87)–(88) follow by taking the expectation of the derivatives of both sides in (87) and using the properties in (85)–(86). For more general (non-polynomial) G, one can expect an Appell expansion as in (87)–(88) of infinite order $n = \infty$, viz.,

$$G(x) = \sum_{k=0}^{\infty} a_{k,F} A_k(x; F), \tag{89}$$

but the convergence of the series in (89) is difficult to justify due to the lack of orthogonality mentioned above. Therefore, infinite-order expansion (89) is not very useful. Moreover, Appell polynomials exist under suitable moment assumptions only. In the sequel, we shall refer to (89) as a *formal Appell expansion of* G. Nevertheless, in limit theorems involving nonlinear functions G of stationary linear LRD process X, the *finite Appell expansion* $G(X) - \mathrm{E}G(X) = \sum_{k=1}^{n} a_{k,F} A_{k,F}(X) + R_n(X)$ is useful, with the property that the lowest non-zero coefficient $a_{k,F} \neq 0$ determines the limit distribution, whereas the 'remainder term' $R_n(X)$ is negligible.

The Hermite expansion in (49) is a special case of Gaussian chaos expansions discussed in many texts (e.g., Refs. [78, 80, 103, 117, 172, 199, 266]) and in the following section. Similarly, Charlier expansion is a special case of chaos expansion on Poisson spaces [117, 125, 199, 246]. Appell polynomials are also related to Wiener–Itô integrals, although this relationship is less straightforward. Let $X(\boldsymbol{t}) = \sum_{\boldsymbol{u} \in \mathbb{Z}^\nu} a(\boldsymbol{t} - \boldsymbol{u})\varepsilon(\boldsymbol{u})$ be a stationary linear RF in (2) of Chapter 3 with innovations having finite $2k$-moment ($k \ge 1$) and square summable coefficients $\sum_{\boldsymbol{t}\in\mathbb{Z}^\nu} a(\boldsymbol{t})^2 < \infty$. Let $A_k(x;F)$ be an Appell polynomial relative to the marginal distribution $F(x) = \mathrm{P}(X(\boldsymbol{t}) \le x), x \in \mathbb{R}$, which does not depend on $\boldsymbol{t}$ by stationarity. Then, $A_k(X(\boldsymbol{t});F)$ can be represented using the *Wick products* of noise variables:

$$A_k(X(\boldsymbol{t});F) = \sum_{(\boldsymbol{u})_k\in\mathbb{Z}^{\nu k}} a(\boldsymbol{t}-\boldsymbol{u}_1)\cdots a(\boldsymbol{t}-\boldsymbol{u}_k) :\varepsilon(\boldsymbol{u}_1)\times\cdots\times\varepsilon(\boldsymbol{u}_k):, \tag{90}$$

the sum being taken over all collections $(\boldsymbol{u})_k = (\boldsymbol{u}_1,\ldots,\boldsymbol{u}_k) \in \mathbb{Z}^{2\nu}$ of points $\boldsymbol{u}_j \in \mathbb{Z}^\nu, 1 \le j \le \nu$. By definition, the Wick product $:\xi_1 \times\cdots\times \xi_k:$ of the r.v.s $\xi_j, \mathrm{E}|\xi_j|^k < \infty, j = 1,\ldots,k$ is given by

$$:\xi_1\times\cdots\times\xi_k: := \frac{(-\mathrm{i})^k\partial^k}{\partial u_1\cdots\partial u_k}\left(\frac{\exp\{\mathrm{i}\sum_{j=1}^k u_j\xi_j\}}{\mathrm{E}\exp\{\mathrm{i}\sum_{j=1}^k u_j\xi_j\}}\right)\Bigg|_{u_1=\cdots=u_k=0}. \tag{91}$$

The Wick product in (91) is a polynomial form in $\xi_j, j = 1,\ldots,k$ with coefficients expressed in terms of the moments or cumulants of these r.v.s. For $\xi_1 = \cdots = \xi_k =: \xi$, (91) coincides with $A_k(\xi;F)$ in (83). For Gaussian r.v.s, the Wick products are called *multivariate Hermite polynomials.* The notation in (91) is common in mathematical physics and has the advantage that it does not need to indicate the distribution of $\xi_j, j = 1,\ldots,k$. We also use the shorthand

$$:\xi^k: \equiv :\xi\times\cdots\times\xi := \begin{cases} A_k(\xi), & \\ H_k(\xi), & \text{if } \xi \sim N(0,1). \end{cases} \tag{92}$$

The Wick product is different from the usual multiplication of r.v.s but shares some properties with the latter. These and other simple properties of the Wick product are summarised in the following proposition.

Proposition 20. (i) *(Symmetry.) The polynomial form* $:\xi_1 \times \cdots\times\xi_k:$ *is invariant under permutations of* $\xi_1,\ldots,\xi_k$.

(ii) *(Multilinearity.) If* $\xi_i = \sum_{j=1}^n a_{ij}\eta_j$, *where* $\eta_j, j = 1,\dots,n$ *are r.v.s and* $a_{ij} \in \mathbb{R}$, *then*

$$:\xi_1 \times \cdots \times \xi_k := \sum_{j_1,\dots,j_k=1}^{n} a_{1j_1} \cdots a_{kj_k} :\eta_{j_1} \times \cdots \times \eta_{j_k}: .$$

(iii) *If* $(\xi_1,\dots,\xi_m)$ *and* $(\xi_{m+1},\dots,\xi_k)$ *are independent,* $1 \le m < k$, *then*

$$:\xi_1 \times \cdots \times \xi_k: = :\xi_1 \times \cdots \times \xi_m: :\xi_{m+1} \times \cdots \times \xi_k:$$

equals the (usual) product of the r.v.s $:\xi_1 \times \cdots \times \xi_m:$ *and* $:\xi_{m+1} \times \cdots \times \xi_k:$.

(iv) *If* $\xi_j, j = 1,\dots,k$ *are independent r.v.s with zero mean* $\mathrm{E}\xi_j = 0$, *then their Wick product is the usual product:*

$$:\xi_1 \times \cdots \times \xi_k := \prod_{j=1}^{k} \xi_j.$$

Returning to the representation in (90), we see that for $a(\boldsymbol{t})$ having finite support (i.e., $a(\boldsymbol{t}) = 0$ for $|\boldsymbol{t}| > K$ and some $K > 0$), (90) follows from the multilinearity property of Proposition 20. In the general case of square-summable $a(\boldsymbol{t})$, (90) follows by approximating both sides in the mean square by corresponding sums with truncated coefficients. See Ref. [10]. By properties (iii)–(iv), for *distinct* points $\boldsymbol{u}_j \ne \boldsymbol{u}_{j'}$ $(j \ne j', 1 \le j, j' \le i)$, the Wick product

$$\begin{aligned}
&:\varepsilon(\boldsymbol{u}_1)^{k_1} \times \cdots \times \varepsilon(\boldsymbol{u}_i)^{k_i}: \\
&=: \underbrace{\varepsilon(\boldsymbol{u}_1) \times \cdots \times \varepsilon(\boldsymbol{u}_1)}_{k_1} \times \cdots \times \underbrace{\varepsilon(\boldsymbol{u}_i) \times \cdots \times \varepsilon(\boldsymbol{u}_i)}_{k_i}: \\
&= \prod_{j=1}^{i} \widetilde{A}_{k_j}(\varepsilon(\boldsymbol{u}_j))
\end{aligned} \tag{93}$$

equals the product of *independent* r.v.s $\widetilde{A}_{k_j}(\varepsilon(u_j, v_j)), 1 \le j \le i$, where

$$\widetilde{A}_k(x) := A_k(x; F_\epsilon), \quad k \ge 0, \quad F_\epsilon(x) := \mathrm{P}(\varepsilon(\boldsymbol{u}) \le x) \tag{94}$$

is the Appell polynomials corresponding to the innovation distribution F_ε. This leads to the decomposition of (90) into the 'off-diagonal' and 'diagonal' parts:

$$A_k(X(\boldsymbol{t})) = X^{\bullet k}(\boldsymbol{t}) + \mathcal{Z}(\boldsymbol{t}), \tag{95}$$

where

$$X^{\bullet k}(\boldsymbol{t}) := \sum_{(\boldsymbol{u})_k}^{\bullet} a(\boldsymbol{t} - \boldsymbol{u}_1) \cdots a(\boldsymbol{t} - \boldsymbol{u}_k) \varepsilon(\boldsymbol{u}_1) \cdots \varepsilon(\boldsymbol{u}_k), \tag{96}$$

and the sum $\sum_{(\boldsymbol{u})_k}^{\bullet}$ is taken over all $(\boldsymbol{u})_k = (\boldsymbol{u}_1, \ldots, \boldsymbol{u}_k) \in \mathbb{Z}^{\nu k}$ such that $\boldsymbol{u}_i \neq \boldsymbol{u}_j$ $(i \neq j)$ (the set of such $(\boldsymbol{u})_k \in \mathbb{Z}^{\nu k}$ will be denoted by $\mathbb{Z}^{\bullet \nu k}$). By definition, the 'diagonal' part $\mathcal{Z}(\boldsymbol{t})$ in (95) is given by the r.h.s. of (90), with $(\boldsymbol{u})_k \in \mathbb{Z}^{\nu k}$ replaced by $(\boldsymbol{u})_k \in \mathbb{Z}^{\nu k} \setminus \mathbb{Z}^{\bullet \nu k}$. In most of the limit results, $\mathcal{Z}(\boldsymbol{t})$ is negligible, and $X^{\bullet k}(\boldsymbol{t})$ is the main term, which is easier to handle compared to $A_k(X(\boldsymbol{t}))$ in (95).

5 Multilinear Forms and Multiple Wiener–Itô Stochastic Integral

In this section, we extend Proposition 9 to multilinear forms in i.i.d. r.v.s tending to a multiple Wiener–Itô stochastic integral w.r.t. Gaussian white noise on $\mathbb{R}^\nu$. We start with the definition of the above integral. Let $L^2(\mathbb{R}^{\nu k})$ denote the Hilbert space of real-valued functions $h = h((\boldsymbol{u})_k), (\boldsymbol{u})_k := (\boldsymbol{u}_1, \ldots, \boldsymbol{u}_k) \in \mathbb{R}^{\nu k}$, with finite norm $\|h\|_{2,k} := \{\int_{\mathbb{R}^{\nu k}} h((\boldsymbol{u})_k)^2 \mathrm{d}(\boldsymbol{u})_k\}^{1/2}$, $\mathrm{d}(\boldsymbol{u})_k = \mathrm{d}\boldsymbol{u}_1 \cdots \mathrm{d}\boldsymbol{u}_k$. Let $W = \{W(\mathrm{d}\boldsymbol{u});\ \boldsymbol{u} \in \mathbb{R}^\nu\}$ denote a real-valued Gaussian white noise on $\mathbb{R}^\nu$ with zero mean and variance $\mathrm{E}W(\mathrm{d}\boldsymbol{u})^2 = \mathrm{d}\boldsymbol{u}$, as in Section 1 of Chapter 2. Define the symmetrisation of $h \in L^2(\mathbb{R}^{\nu k})$ by

$$(\operatorname{sym} h)((\boldsymbol{u})_k) := \frac{1}{k!} \sum_{(\pi)_k} h(\boldsymbol{u}_{\pi_1}, \ldots, \boldsymbol{u}_{\pi_k}), \tag{97}$$

where the sum is taken over all permutations $(\pi)_k = (\pi_1, \ldots, \pi_k)$ of the integers $1, 2, \ldots, k$. For given integers $K \geq 1, k \geq 1$ and a collection $(\boldsymbol{m})_k = (\boldsymbol{m}_1, \ldots, \boldsymbol{m}_k), \boldsymbol{m}_i \in \mathbb{Z}^\nu, i = 1, \ldots, k$, let

$$\square^K_{(\boldsymbol{m})_k} := \square^K_{\boldsymbol{m}_1} \times \cdots \times \square^K_{\boldsymbol{m}_k} \subset \mathbb{R}^{\nu k} \tag{98}$$

denote the Cartesian product of squares in (37) (a 'cube' in $\mathbb{R}^{\nu k}$ with side length $1/K$). Denote by $\mathcal{S}_{0,K}(\mathbb{R}^{\nu k})$ the set of step functions $h : \mathbb{R}^{\nu k} \to \mathbb{R}$, taking a finite number of non-zero constant values $\tilde{h}(\Box^K_{(\boldsymbol{m})_k})$ on 'cubes' (98) and vanishing on 'diagonals': $h((\boldsymbol{u})_k) = \tilde{h}(\Box^K_{(\boldsymbol{m})_k}) = 0, (\boldsymbol{u})_k \in \Box^K_{(\boldsymbol{m})_k}$, provided $\boldsymbol{u}_i = \boldsymbol{u}_j$ for some $i \neq j$, $i,j = 1,\dots,k$. For $h \in \mathcal{S}_{0,K}(\mathbb{R}^{\nu k})$, the k-tuple Wiener–Itô integral $I_k(h)$ is defined as the sum

$$I_k(h) := \sum_{(\boldsymbol{m})_k \in \mathbb{Z}^{\nu k}} \tilde{h}(\Box^K_{(\boldsymbol{m})_k}) W(\Box^K_{\boldsymbol{m}_1}) \cdots W(\Box^K_{\boldsymbol{m}_k}) \tag{99}$$

over all 'cubes' in (98), which is finite as h vanishes except for a finite number of 'cubes'. The following properties of such integral sums follow easily: for any $k, l = 1, 2, \dots$ and any $h \in \mathcal{S}_{0,K}(\mathbb{R}^{\nu k}), g \in \mathcal{S}_{0,K}(\mathbb{R}^{\nu l})$,

$$\begin{aligned} I_k(h) &= I_k(\mathrm{sym}\, h), \quad \mathrm{E} I_k(h) = 0, \\ \mathrm{E} I_k(h) I_l(g) &= \begin{cases} k!\langle h, \mathrm{sym}\, g\rangle_{2,k}, & k = l, \\ 0, & k \neq l, \end{cases} \end{aligned} \tag{100}$$

where $\langle h, g\rangle_{2,k} := \int_{\mathbb{R}^{\nu k}} h((\boldsymbol{u})_k) g((\boldsymbol{u})_k) \mathrm{d}(\boldsymbol{u})_k$ is the scalar product in $L^2(\mathbb{R}^{\nu k})$. The union $\mathcal{S}_0(\mathbb{R}^{\nu k}) := \cup_{K \geq 1} \mathcal{S}_{0,K}(\mathbb{R}^{\nu k})$ of step functions, being a dense set in $L^2(\mathbb{R}^{\nu k})$, the above definition of $I_k(h)$ extends to any $h \in L^2(\mathbb{R}^{\nu k})$ in a standard way, with preservation of the properties in (100). Namely, for any $h \in L^2(\mathbb{R}^{\nu k})$, there exists a sequence $h_n \in \mathcal{S}_0(\mathbb{R}^{\nu k}), n = 1, 2, \dots$, such that $\|h - h_n\|_{2,k} \to 0$ $(n \to \infty)$, and then the Wiener–Itô integral $I_k(h)$ is defined as the limit in mean square:

$$I_k(h) := \lim_{n\to\infty} I_k(h_n). \tag{101}$$

We use the notation

$$I_k(h) = \int_{\mathbb{R}^{\nu k}} h(\boldsymbol{u}_1, \dots, \boldsymbol{u}_k) W(\mathrm{d}\boldsymbol{u}_1) \cdots W(\mathrm{d}\boldsymbol{u}_k) \equiv \int_{\mathbb{R}^{\nu k}} h \,\mathrm{d}^k W. \tag{102}$$

Consequently, and from (100), it follows that

$$\mathrm{E}\left(\int_{\mathbb{R}^{\nu k}} h\,\mathrm{d}^k W\right)^2 \le k!\|h\|_{2,k}^2, \tag{103}$$

$$= k!\|h\|_{2,k}^2, \quad \text{if } h \text{ is symmetric: } h = \operatorname{sym} h.$$

We note the following formulas for a change of variables in integral (102). Write $(U_{\boldsymbol{y}}h)(\boldsymbol{u}_1,\ldots,\boldsymbol{u}_k) := h(\boldsymbol{u}_1+\boldsymbol{y},\ldots,\boldsymbol{u}_k+\boldsymbol{y})$ $(\boldsymbol{y}\in\mathbb{R}^\nu)$, $(T_{\boldsymbol{\lambda}}h)(\boldsymbol{u}_1,\ldots,\boldsymbol{u}_k) := h(\Lambda\boldsymbol{u}_1,\ldots,\Lambda\boldsymbol{u}_k)$ $(\Lambda = \operatorname{diag}(\lambda_1,\ldots,\lambda_\nu), \boldsymbol{\lambda} = (\lambda_1,\ldots,\lambda_\nu)\in\mathbb{R}_+^\nu)$. Then, for any $h\in L^2(\mathbb{R}^{\nu k}), k\ge 1$, and any $\boldsymbol{y}\in\mathbb{R}^\nu, \boldsymbol{\lambda}\in\mathbb{R}_+^\nu$,

$$I_k(U_{\boldsymbol{y}}h) \stackrel{\mathrm{d}}{=} I_k(h), \quad I_k(T_{\boldsymbol{\lambda}}h) \stackrel{\mathrm{d}}{=} \det(\Lambda)^{-k/2} I_k(h). \tag{104}$$

The above formulas are easily obtained as consequences of defining the Wiener–Itô integral through step functions. Indeed, as linear combinations of indicator functions

$$h((\boldsymbol{u})_k) = \prod_{i=1}^k \mathbb{I}(\boldsymbol{u}_i \in]\boldsymbol{v}_i', \boldsymbol{v}_i'']) \tag{105}$$

of disjoint 'rectangles' $]\boldsymbol{v}_i',\boldsymbol{v}_i''] \subset \mathbb{R}^\nu,]\boldsymbol{v}_i',\boldsymbol{v}_i'']\cap]\boldsymbol{v}_j',\boldsymbol{v}_j''] = \emptyset$ $(i,j = 1,\ldots,k, i\ne j)$ are dense in $L^2(\mathbb{R}^{\nu k})$, it suffices to verify (104) for h in (105). But for such h, $\int_{\mathbb{R}^{\nu k}} h\,\mathrm{d}^k W = \prod_{i=1}^k W(]\boldsymbol{v}_i',\boldsymbol{v}_i''])$ and $\int_{\mathbb{R}^{\nu k}} T_{\boldsymbol{\lambda}}h\,\mathrm{d}^k W = \prod_{i=1}^k W(]\Lambda^{-1}\boldsymbol{v}_i',\Lambda^{-1}\boldsymbol{v}_i''])$ $\stackrel{\mathrm{d}}{=} \prod_{i=1}^k W(]\boldsymbol{v}_i',\boldsymbol{v}_i''])/\det(\Lambda)^{1/2}$, yielding the second relation in (104). Similarly, one can verify the first relation in (104).

Any r.v. ζ with $\mathrm{E}\zeta^2<\infty$ measurable w.r.t. to the σ-field generated by $W(A), A\in\mathcal{B}_0(\mathbb{R}^\nu)$ has a 'chaotic' expansion in the Wiener–Itô integrals

$$\zeta = \sum_{k=0}^\infty \frac{1}{k!}\int_{\mathbb{R}^{\nu k}} h_k\,\mathrm{d}^k W, \tag{106}$$

convergent in mean square. In particular, if $\zeta = G(\xi), G\in L^2(\Phi), \xi = \int_{\mathbb{R}^\nu} h(\boldsymbol{u})\mathrm{d}W, h\in L^2(\mathbb{R}^\nu)$ and $\|h\|_2^2 = \mathrm{E}\xi^2 = 1$, then $h_k((\boldsymbol{u})_k) =$

$h(\boldsymbol{u}_1)\cdots h(\boldsymbol{u}_k)$ the multiple integrals $\int_{\mathbb{R}^{\nu k}} h_k \mathrm{d}^k W$ in (106) coincide with $H_k(\xi)$:

$$H_k\left(\int_{\mathbb{R}^\nu} h(\boldsymbol{u})\mathrm{d}W\right) = \int_{\mathbb{R}^{\nu k}} h(\boldsymbol{u}_1)\cdots h(\boldsymbol{u}_k)W(\mathrm{d}\boldsymbol{u}_1)\cdots W(\mathrm{d}\boldsymbol{u}_k), \tag{107}$$

and (106) becomes the Hermite expansion in (49).

Now, we turn to the convergence in the distribution of off-diagonal multilinear forms towards a multiple Wiener–Itô integral. Let $L^2(\mathbb{Z}^{\bullet\nu k})$ be the class of all real functions $g = g((\boldsymbol{u})_k)$, $(\boldsymbol{u})_k \in \mathbb{Z}^{\nu k}$, with $\sum_{(\boldsymbol{u})_k\in\mathbb{Z}^{\nu k}} g((\boldsymbol{u})_k)^2 < \infty$, which vanish on 'diagonals' $\mathbb{Z}^{\nu k}\setminus\mathbb{Z}^{\bullet\nu k}$, and a polynomial form

$$\begin{aligned} S_k(g) &:= \sum_{(\boldsymbol{u})_k}^{\bullet} g((\boldsymbol{u})_k)\varepsilon(\boldsymbol{u}_1)\cdots\varepsilon(\boldsymbol{u}_k) \\ &= \sum_{(\boldsymbol{u})_k\in\mathbb{Z}^{\nu k}} g((\boldsymbol{u})_k)\varepsilon(\boldsymbol{u}_1)\cdots\varepsilon(\boldsymbol{u}_k), \quad g \in L^2(\mathbb{Z}^{\bullet\nu k}); \end{aligned} \tag{108}$$

see (96) for notation. The r.v.s $\{\varepsilon(\boldsymbol{u})\}$ in (108) are i.i.d. and have zero mean, unit variance. Then, $\mathrm{E}S_k(g) = 0$ and

$$\mathrm{E}S_k^2(g) \le k! \sum_{(\boldsymbol{u})_k\in\mathbb{Z}^{\nu k}} g((\boldsymbol{u})_k)^2, \tag{109}$$

with equality in (109) when $g(\boldsymbol{u}_1,\ldots,\boldsymbol{u}_k)$ is symmetric under permutations of its arguments. Given $\Lambda = \mathrm{diag}(\lambda^{\gamma_1},\ldots,\lambda^{\gamma_\nu})$, $\lambda > 0$, $\boldsymbol{\gamma} = (\gamma_1,\ldots,\gamma_\nu) \in \mathbb{R}_+^\nu$ and $g_\lambda \in L^2(\mathbb{Z}^{\bullet\nu k})$, define a step function $g_\lambda^\dagger \in L^2(\mathbb{R}^{\nu k})$ by

$$g_\lambda^\dagger((\boldsymbol{u})_k) := \det(\Lambda)^{k/2} g_\lambda([\Lambda\boldsymbol{u}_1],\ldots,[\Lambda\boldsymbol{u}_k]), \quad (\boldsymbol{u})_k \in \mathbb{R}^{\nu k}. \tag{110}$$

For $k = 1$ and $A = I$ (equal to the identity matrix), Propositions 9 and 21 agree.

Proposition 21. *Assume that there exists* $h \in L^2(\mathbb{R}^{\nu k})$ *such that* $\lim_{\lambda\to\infty} \|g_\lambda^\dagger - h\|_{2,k} = 0$. *Then,*

$$S_k(g_\lambda) \xrightarrow{\mathrm{d}} \int_{\mathbb{R}^{\nu k}} h\,\mathrm{d}^k W \quad (\lambda\to\infty). \tag{111}$$

Proof. We use a three-step procedure similar to the proof of Proposition 9. Accordingly, we show that $S_k(g_\lambda)$ and $I_k(h) = \int_{\mathbb{R}^{\nu k}} h\, \mathrm{d}^k W$ can be approximated in mean square by 'step sums' $S_k(h_{\epsilon,\lambda})$ and $I_k(h_\epsilon), \epsilon > 0$ uniformly in λ and that for each $\epsilon > 0$ $S_k(h_{\epsilon,\lambda}) \xrightarrow{\mathrm{d}} I_k(h_\epsilon)$ as $\lambda \to \infty$ using the classical multidimensional CLT.

Recall the definition $\mathcal{S}_0(\mathbb{R}^{\nu k}) = \cup_{K\ge 1}\mathcal{S}_{0,K}(\mathbb{R}^{\nu k})$ as a dense subset of step functions in $L^2(\mathbb{R}^{\nu k})$. Since $\lim_{\lambda\to\infty} \|g^\dagger_\lambda - h\|_{2,k} = 0$, for any $\delta > 0$, there exist $\lambda_0 > 0, \epsilon > 0, K \ge 1$, and $h_\epsilon \in \mathcal{S}_{0,K}(\mathbb{R}^{\nu k})$ such that

$$\|h - h_\epsilon\|_{2,k} + \|h_\epsilon - g^\dagger_\lambda\|_{2,k} < \delta, \quad \forall\, \lambda \ge \lambda_0. \tag{112}$$

Define the 'discretisation' of h_ϵ on $\mathbb{Z}^{\nu k}$ by

$$\begin{aligned} h_{\epsilon,\lambda}((\boldsymbol{u})_k) &:= \det(\Lambda)^{-k/2} h_\epsilon(\Lambda^{-1}\boldsymbol{u}_1, \ldots, \Lambda^{-1}\boldsymbol{u}_k), \\ (\boldsymbol{u})_k &= (\boldsymbol{u}_1, \ldots, \boldsymbol{u}_k) \in \mathbb{Z}^{\nu k}. \end{aligned} \tag{113}$$

Clearly, for such $h_{\epsilon,\lambda}$, the polynomial form $S_k(h_{\epsilon,\lambda})$ is well defined and

$$\begin{aligned} &\mathrm{E}(S_k(h_{\epsilon,\lambda}) - S_k(g_\lambda))^2 \\ &\le k! \sum_{(\boldsymbol{u})_k}^{\bullet} \left(\det(\Lambda)^{-k/2} h_\epsilon(\Lambda^{-1}\boldsymbol{u}_1, \ldots, \Lambda^{-1}\boldsymbol{u}_k) - g_\lambda(\boldsymbol{u}_1, \ldots, \boldsymbol{u}_k)\right)^2 \\ &\le k! \|h_\epsilon - g^\dagger_\lambda\|^2_{2,k}; \end{aligned}$$

see (113), (110) and (109). By (112), this implies

$$\mathrm{E}(S_k(h_{\epsilon,\lambda}) - S_k(g_\lambda))^2 < k!\delta, \quad \forall\, \lambda \ge \lambda_0. \tag{114}$$

On the other hand, since $h_\epsilon((\boldsymbol{u})_k) = \tilde{h}(\square^K_{(\boldsymbol{m})_k}), (\boldsymbol{u})_k \in \square^K_{(\boldsymbol{m})_k})$,

$$\begin{aligned} S_k(h_{\epsilon,\lambda}) = &\sum_{(\boldsymbol{m})_k \in \mathbb{Z}^{\nu k}} \tilde{h}(\square^K_{(\boldsymbol{m})_k}) \\ &\times \det(\Lambda)^{k/2} \sum_{(\boldsymbol{u})_k \in \mathbb{Z}^{\nu k}} \varepsilon(\boldsymbol{u}_1) \cdots \varepsilon(\boldsymbol{u}_k) \\ &\times \mathbb{I}(\Lambda^{-1}\boldsymbol{u}_1 \in \square^K_{\boldsymbol{m}_1}, \ldots, \Lambda^{-1}\boldsymbol{u}_k \in \square^K_{\boldsymbol{m}_k}) \\ = &\sum_{(\boldsymbol{m})_k \in \mathbb{Z}^{\nu k}} \tilde{h}(\square^K_{(\boldsymbol{m})_k}) M_\lambda(\square^K_{\boldsymbol{m}_1}) \cdots M_\lambda(\square^K_{\boldsymbol{m}_k}), \end{aligned}$$

where M_λ is the 'discrete' random measure in (32) corresponding to $A = I$, the identity matrix. From the joint convergence in (40), we obtain

$$S_k(h_{\epsilon,\lambda}) \xrightarrow{\mathrm{d}} \sum_{(\boldsymbol{m})_k \in \mathbb{Z}^{\nu k}} \tilde{h}(\Box^K_{(\boldsymbol{m})_k})\, W(\Box^K_{\boldsymbol{m}_1}) \cdots W(\Box^K_{\boldsymbol{m}_k}) = I_k(h_\epsilon), \tag{115}$$

referring to (99). We also have from (112) that $\mathrm{E}(I_k(h) - I_k(h_\epsilon))^2 \le k! \|h - h_\epsilon\|^2_{2,k} < k!\delta$. This, together with (114) and (115), completes the proof of Proposition 21. □

A very simple version of Proposition 21 given as follows concerns the weak convergence of rescaled multiple Wiener–Itô integrals. Given $g_\lambda \in L^2(\mathbb{R}^{\nu k})$ and a diagonal $\nu \times \nu$ matrix Λ as above, define $g^\dagger_\lambda \in L^2(\mathbb{R}^{\nu k})$ by

$$g^\dagger_\lambda((\boldsymbol{u})_k) := \det(\Lambda)^{k/2} g_\lambda(\Lambda \boldsymbol{u}_1, \ldots, \Lambda \boldsymbol{u}_k), \quad (\boldsymbol{u})_k \in \mathbb{R}^{\nu k}, \tag{116}$$

analogously to (110).

Proposition 22. *Assume that there exists $h \in L^2(\mathbb{R}^{\nu k})$ such that $\lim_{\lambda \to \infty} \|g^\dagger_\lambda - h\|_{2,k} = 0$. Then,*

$$\int_{\mathbb{R}^{\nu k}} g_\lambda \, \mathrm{d}^k W \xrightarrow{\mathrm{d}} \int_{\mathbb{R}^{\nu k}} h \, \mathrm{d}^k W \quad (\lambda \to \infty). \tag{117}$$

Proof. By the change of variable formula in (104), $\int_{\mathbb{R}^{\nu k}} g_\lambda \, \mathrm{d}^k W \stackrel{\mathrm{d}}{=} \int_{\mathbb{R}^{\nu k}} g^\dagger_\lambda \, \mathrm{d}^k W$. Hence, (117) follows by $\mathrm{E}\big(\int_{\mathbb{R}^{\nu k}} g^\dagger_\lambda \, \mathrm{d}^k W - \int_{\mathbb{R}^{\nu k}} h \, \mathrm{d}^k W\big)^2 \le k! \|g^\dagger_\lambda - h\|^2_{2,k} \to 0$. □

6 Lévy Random Measures and Convergence of Stochastic Integrals

In this section, we consider a class of Lévy random measures with α-stable small-scale behaviour and the convergence of stochastic integrals. Let $M = \{M(A); A \in \mathcal{B}_0(\mathbb{R}^\nu)\}$ be a Lévy random measure on

$\mathbb{R}^\nu$ having the characteristic function

$$\mathrm{E}\mathrm{e}^{\mathrm{i}\theta M(A)} = \exp\left\{\mathrm{Leb}_\nu(A)\left(-\frac{\theta^2\sigma^2}{2}\right.\right.$$
$$\left.\left. + \int_{\mathbb{R}} (\mathrm{e}^{\mathrm{i}\theta z} - 1 - \mathrm{i}\theta_\alpha \tau(z))\pi(\mathrm{d}z)\right)\right\}, \quad \theta \in \mathbb{R}, \tag{118}$$

where $\sigma \geq 0, \pi$ is a Lévy measure on $\mathbb{R}$, satisfying Assumption $(\mathrm{M})_\alpha$, and $\tau_\alpha(z)$ is the centring function

$$\tau_\alpha(z) := \begin{cases} z, & 1 < \alpha \leq 2, \\ z\mathbb{I}(|z| \leq 1), & \alpha = 1, \\ 0, & 0 < \alpha < 1. \end{cases} \tag{119}$$

Assumption $(\mathbf{M})_\alpha$. The Lévy characteristics σ, π in (118) satisfy one of the two following conditions (i)/(ii):

(i) $\alpha = 2, \sigma > 0$ and $\int_{\mathbb{R}} z^2\pi(\mathrm{d}z) < \infty$;

(ii) $0 < \alpha < 2, \sigma = 0$, and there exist $c_\pm \geq 0, c_+ + c_- > 0$ such that

$$\lim_{r\downarrow 0} r^\alpha \pi([r,\infty)) = c_+, \quad \lim_{r\downarrow 0} r^\alpha \pi((-\infty,-r]) = c_- \tag{120}$$

and

$$\sup r^\alpha \pi(|z| > r) < \infty. \tag{121}$$

Moreover, if $\alpha = 1$, then $\pi(\mathrm{d}z) = \pi(-\mathrm{d}z), z > 0$ is symmetric.

Essentially, $(\mathrm{M})_\alpha$ (ii) pertains to the behaviour of the Lévy measure at the origin and does not exclude $\int_{\mathbb{R}} z^2\pi(\mathrm{d}z) < \infty$ or $\mathrm{E}M(A)^2 < \infty$. An important class of Lévy random measures satisfying Assumption $(\mathrm{M})_\alpha$ constitute *α-stable random measures* W_α $(0 < \alpha \leq 2)$, corresponding to $\sigma > 0, \pi = 0$ $(\alpha = 2)$ and $\sigma = 0, \pi = \pi_\alpha$ $(0 < \alpha < 2)$, where

$$\pi_\alpha(\mathrm{d}z) := (c_+\mathbb{I}(z > 0) + c_-\mathbb{I}(z < 0))\frac{\alpha \mathrm{d}z}{|z|^{\alpha+1}}, \quad z \in \mathbb{R}, \tag{122}$$

where $c_\pm \ge 0, c_+ + c_- > 0$ are parameters, $c_+ = c_-$ if $\alpha = 1$. The characteristic function of W_α is written as

$$\mathrm{E}\mathrm{e}^{\mathrm{i}\theta W_\alpha(A)} = \mathrm{e}^{-\mathrm{Leb}_\nu(A)|\theta|^\alpha \omega_\alpha(\theta)}, \quad \theta \in \mathbb{R}, \quad A \in \mathcal{B}_0(\mathbb{R}^\nu), \quad (123)$$

where

$$\omega_\alpha(\theta) := \begin{cases} \frac{1}{2}\sigma^2, & \alpha = 2, \\ \frac{\Gamma(2-\alpha)}{1-\alpha}\left((c_+ + c_-)\cos\left(\frac{\pi\alpha}{2}\right)\right. & \\ \qquad \left.- \mathrm{i}(c_+ - c_-)\mathrm{sgn}(\theta)\sin\left(\frac{\pi\alpha}{2}\right)\right), & 0 < \alpha < 2, \alpha \ne 1, \\ (c_+ + c_-)\frac{\pi}{2}, & \alpha = 1, c_+ = c_-; \end{cases} \quad (124)$$

see Ref. [122, Chapter 2, &2]. Relation (124) is a consequence of the useful formula

$$\int_{\mathbb{R}} (\mathrm{e}^{\mathrm{i}\theta z} - 1 - \mathrm{i}\theta\tau_\alpha(z))\pi_\alpha(\mathrm{d}z) := -|\theta|^\alpha \omega_\alpha(\theta), \quad \forall\, \theta \in \mathbb{R}. \quad (125)$$

Assumption $(\mathrm{M})_\alpha$ means that M belongs to the domain of local attraction of α-stable random measure W_α (see (132)), but M need not be α-stable, as it may have finite variance or even all moments when $\alpha < 2$, as in the case of the 'truncated' stable case

$$\pi(\mathrm{d}z) = \mathbb{I}(|z| \le 1)|z|^{-1-\alpha}\mathrm{d}z. \quad (126)$$

Stochastic integrals w.r.t. M can be defined by reducing them to Poisson stochastic integrals on the product space $\mathbb{R}^\nu \times \mathbb{R}$ discussed above. Let $\mathcal{N}(\mathrm{d}\boldsymbol{u}, \mathrm{d}z)$ be a Poisson random measure on $\mathbb{R}^\nu \times \mathbb{R}$ with mean $\mu(\mathrm{d}\boldsymbol{u}, \mathrm{d}z) = \mathrm{d}\boldsymbol{u}\pi(\mathrm{d}z)$, $\widetilde{\mathcal{N}}(\mathrm{d}\boldsymbol{u}, \mathrm{d}z) = \mathcal{N}(\mathrm{d}\boldsymbol{u}, \mathrm{d}z) - \mu(\mathrm{d}\boldsymbol{u}, \mathrm{d}z)$. We use the fact that for M satisfying Assumption $(\mathrm{M})_\alpha$ and any $A \in \mathcal{B}_0(\mathbb{R}^2)$,

$$M(A) = \begin{cases} \int_{A\times\mathbb{R}} z\widetilde{\mathcal{N}}(\mathrm{d}\boldsymbol{u}, \mathrm{d}z), & 1 < \alpha < 2, \\ \int_{A\times\mathbb{R}} z\mathcal{N}(\mathrm{d}\boldsymbol{u}, \mathrm{d}z), & 0 < \alpha < 1, \\ \int_{A\times\mathbb{R}} z(\mathbb{I}(|z| \le 1)\widetilde{\mathcal{N}}(\mathrm{d}\boldsymbol{u}, \mathrm{d}z) & \\ \qquad + \mathbb{I}(|z| > 1)\mathcal{N}(\mathrm{d}\boldsymbol{u}, \mathrm{d}z)), & \alpha = 1, \end{cases} \quad (127)$$

where the stochastic integrals are well defined using the criterion in (22). Consequently, we can define the stochastic integral w.r.t. to such M as

$$\int_{\mathbb{R}^2} f(\boldsymbol{u})M(\mathrm{d}\boldsymbol{u}) \tag{128}$$
$$:= \begin{cases} \int_{\mathbb{R}^\nu\times\mathbb{R}} f(\boldsymbol{u})z\widetilde{\mathcal{N}}(\mathrm{d}\boldsymbol{u},\mathrm{d}z), & 1<\alpha<2,\\ \int_{\mathbb{R}^\nu\times\mathbb{R}} f(\boldsymbol{u})z\mathcal{N}(\mathrm{d}\boldsymbol{u},\mathrm{d}z), & 0<\alpha<1,\\ \int_{\mathbb{R}^\nu\times\mathbb{R}} f(\boldsymbol{u})z(\mathbb{I}(|z|\le 1)\widetilde{\mathcal{N}}(\mathrm{d}\boldsymbol{u},\mathrm{d}z) & \\ \quad +\mathbb{I}(|z|>1)\mathcal{N}(\mathrm{d}\boldsymbol{u},\mathrm{d}z)), & \alpha=1, \end{cases}$$

whereas for $\alpha = 2$, the integral $\int_{\mathbb{R}^\nu} f(\boldsymbol{u})M(\mathrm{d}\boldsymbol{u})$ is well defined for any $f \in L^2(\mathbb{R}^\nu)$ and has zero mean and variance $C\|f\|_2^2$, $C = \sigma^2 + \int_{\mathbb{R}}|z|^2\pi(\mathrm{d}z)$. Using (121) and integrating by parts, we also see that for the Poisson integral in (128), conditions (22) hold when $f \in L^\alpha(\mathbb{R}^\nu)$. For instance, when $1 < \alpha < 2$, we have that $\int_{\mathbb{R}^\nu\times\mathbb{R}}(|zf(\boldsymbol{u})| \wedge |zf(\boldsymbol{u})|^2)\mathrm{d}\boldsymbol{u}\pi(\boldsymbol{u}) = \int_{\mathbb{R}^\nu}|f(\boldsymbol{u})|\int_{|z|>1/|f(\boldsymbol{u})|}|z|\pi(\mathrm{d}z) + \int_{\mathbb{R}^\nu}|f(\boldsymbol{u})|^2\int_{|z|\le 1/|f(\boldsymbol{u})|}|z|^2\pi(\mathrm{d}z) \le C\|f\|_\alpha^\alpha$ since $\int_{|z|>1/|f(\boldsymbol{u})|}|z|\pi(\mathrm{d}z) \le C\int_{1/|f(\boldsymbol{u})}z^{-\alpha}\mathrm{d}z = C|f(\boldsymbol{u})|^{\alpha-1}$ and $\int_{|z|\le 1/|f(\boldsymbol{u})|}|z|^2\pi(\mathrm{d}z) \le C\int_0^{1/|f(\boldsymbol{u})}z^{1-\alpha}\mathrm{d}z = C|f(\boldsymbol{u})|^{\alpha-2}$.

For the reader's convenience, we summarise the facts about such integrals in the following proposition.

Proposition 23. (i) *Let M be a Lévy random measure on $\mathbb{R}^\nu$ with characteristic function in* (118) *and satisfying Assumption $(M)_\alpha$, $0 < \alpha \le 2$. Then, the stochastic integral $\int_{\mathbb{R}^\nu} f(\boldsymbol{u})M(\mathrm{d}\boldsymbol{u})$ is well defined for any $f \in L^\alpha(\mathbb{R}^\nu)$ and has an infinitely divisible distribution with characteristic function*

$$\mathrm{E}\mathrm{e}^{\mathrm{i}\theta\int_{\mathbb{R}^\nu} f(\boldsymbol{u})M(\mathrm{d}\boldsymbol{u})} \tag{129}$$
$$= \exp\left\{\int_{\mathbb{R}^\nu}\left(-\frac{\theta^2\sigma^2}{2}|f(\boldsymbol{u})|^2 \right.\right.$$
$$\left.\left. + \int_{\mathbb{R}}(\mathrm{e}^{\mathrm{i}\theta z f(\boldsymbol{u})} - 1 - \mathrm{i}\theta\tau_\alpha(z)f(\boldsymbol{u}))\pi(\mathrm{d}z)\right)\mathrm{d}\boldsymbol{u}\right\},\ \theta\in\mathbb{R}.$$

Particularly, $\mathrm{E}\int_{\mathbb{R}^\nu} f(\boldsymbol{u})M(\mathrm{d}\boldsymbol{u}) = 0$ *when* $1 < \alpha \le 2$ *and* $\mathrm{E}(\int_{\mathbb{R}^\nu} f(\boldsymbol{u})M(\mathrm{d}\boldsymbol{u}))^2 = C\|f\|_2^2 < \infty$, $C = \sigma^2 + \int_{\mathbb{R}} |z|^2\pi(\mathrm{d}z)$ *when* $\alpha = 2$.

(ii) *Let* W_α *be an* α*-stable random measure on* $\mathbb{R}^\nu$, $0 < \alpha \le 2$ *with characteristic function in* (124). *Then, the stochastic integral* $\int_{\mathbb{R}^\nu} f(\boldsymbol{u})W_\alpha(\mathrm{d}\boldsymbol{u})$ *is well defined for any* $f \in L^\alpha(\mathbb{R}^\nu)$ *and has* α*-stable distribution with*

$$\mathrm{E}\mathrm{e}^{\mathrm{i}\theta \int_{\mathbb{R}^\nu} f(\boldsymbol{u})W_\alpha(\mathrm{d}\boldsymbol{u})} \exp\left\{-|\theta|^\alpha \int_{\mathbb{R}^\nu} |f(\boldsymbol{u})|^\alpha \omega_\alpha(\theta f(\boldsymbol{u}))\mathrm{d}\boldsymbol{u}\right\}, \qquad \theta \in \mathbb{R}. \tag{130}$$

Given a Lévy random measure M and an α-stable random measure W_α on $\mathbb{R}^\nu$, with respective characteristic functions in (118) and (124), we can define two RFs on $\mathbb{R}^\nu_+$ termed *Lévy Sheet* and α*-stable Sheet*

$$M(]\mathbf{0}, \boldsymbol{x}]), \quad W_\alpha(]\mathbf{0}, \boldsymbol{x}]), \quad \boldsymbol{x} \in \mathbb{R}^\nu_+. \tag{131}$$

For $\alpha = 2$, $W_2(]\mathbf{0}, \boldsymbol{x}]) \overset{\mathrm{fdd}}{=} \sigma B_{1/2,1/2}(\boldsymbol{x})$ agrees with the Brownian sheet defined in Example 11 of Chapter 1. Assumption $(\mathrm{M})_\alpha$ means that M belongs to the domain of local attraction of the α-stable random measure:

$$(\lambda_1\lambda_2)^{-1/\alpha} M(]\mathbf{0}, \Lambda\boldsymbol{x}]) \overset{\mathrm{fdd}}{\longrightarrow} W_\alpha(]\mathbf{0}, \boldsymbol{x}]), \quad \lambda_i \downarrow 0, \ i = 1, 2, \tag{132}$$

where $\Lambda = \mathrm{diag}(\lambda_1, \lambda_2)$ and W_α is an α-stable sheet with characteristic function as in (124). Relation (132) follows from Proposition 24, where the convergence towards stable limit is established for stochastic integrals w.r.t. M. The RFs in (131) have independent rectangular increments in both (horizontal and vertical) directions (Definition 19 of Chapter 1) and constitute direct two-dimensional analogues of the classical homogeneous Lévy process and α-stable process with one-dimensional time and independent increments [231, 232]. The α-stable sheet is (H_1, H_2)-MSS with $H_1 = H_2 = 1/\alpha$, viz., for any $\lambda_1, \lambda_2 > 0, \Lambda = \mathrm{diag}(\lambda_1, \lambda_2)$,

$$W_\alpha(]\mathbf{0}, \Lambda\boldsymbol{x}]) \overset{\mathrm{fdd}}{=} (\lambda_1\lambda_2)^{1/\alpha} W_\alpha(]\mathbf{0}, \boldsymbol{x}]). \tag{133}$$

The scaling property in (133) extends to arbitrary sets in $\mathcal{B}_0(\mathbb{R}^2)$, which is often written as $W_\alpha(\mathrm{d}\Lambda\boldsymbol{x}) \overset{\mathrm{fdd}}{=} (\lambda_1\lambda_2)^{1/\alpha} W_\alpha(\mathrm{d}\boldsymbol{x})$.

Consider the weak convergence of stochastic integrals in (128) w.r.t. to M towards stochastic integrals w.r.t. W_α.

For any $0 < \alpha \le 2, \lambda > 0$, and $\boldsymbol{\gamma} = (\gamma_1, \dots, \gamma_\nu) \in \mathbb{R}^\nu_+$ and any measurable function $f = f_\lambda : \mathbb{R}^\nu \to \mathbb{R}$ depending on λ, define the rescaled function $f^\dagger_\lambda : \mathbb{R}^\nu \to \mathbb{R}$ by

$$f^\dagger_\lambda(\boldsymbol{u}) := \det(\Lambda)^{1/\alpha} f_\lambda(\Lambda \boldsymbol{u}), \quad \boldsymbol{u} \in \mathbb{R}^\nu, \quad \Lambda = \operatorname{diag}(\lambda^{\gamma_1}, \dots, \lambda^{\gamma_\nu}). \tag{134}$$

Proposition 24 is similar in spirit to Proposition 9 except that now $\lambda \to 0$ and the convergence (135) extends to the $L^\alpha(\mathbb{R}^\nu)$-norm instead of the $L^2(\mathbb{R}^\nu)$-norm; moreover, its proof does not use approximation of f_λ by step functions due to the explicit form of the characteristic function in (129).

Proposition 24. *Assume that the Lévy basis M satisfies Assumption $(M)_\alpha$ with some $0 < \alpha \le 2$. Let $f_\lambda \in L^\alpha(\mathbb{R}^\nu), \lambda > 0$ be given and there exists $h \in L^\alpha(\mathbb{R}^\nu)$ such that*

$$\lim_{\lambda \to 0} \|f^\dagger_\lambda - h\|_\alpha = 0, \tag{135}$$

for some $\boldsymbol{\gamma} \in \mathbb{R}^\nu_+$. Then,

$$\int_{\mathbb{R}^\nu} f_\lambda(\boldsymbol{u}) M(\mathrm{d}\boldsymbol{u}) \xrightarrow{\mathrm{d}} \int_{\mathbb{R}^\nu} h(\boldsymbol{u}) W_\alpha(\mathrm{d}\boldsymbol{u}), \quad \lambda \to 0. \tag{136}$$

Proof. Write $\int_{\mathbb{R}^\nu} = \int$ for short. It suffices to prove the convergence of characteristic functions:

$$\begin{aligned} C_\lambda(\theta) &:= \mathrm{E}\exp\left\{\mathrm{i}\theta \int f_\lambda(\boldsymbol{u}) M(\mathrm{d}\boldsymbol{u})\right\} \to C_0(\theta) \\ &:= \mathrm{E}\exp\left\{\mathrm{i}\theta \int h(\boldsymbol{u}) W_\alpha(\mathrm{d}\boldsymbol{u})\right\}, \quad \forall\, \theta \in \mathbb{R}. \end{aligned} \tag{137}$$

W.l.g., assume additionally that

$$f^\dagger_\lambda(\boldsymbol{u}) \to h(\boldsymbol{u}) \quad \text{for a.e. } \boldsymbol{u} \in \mathbb{R}^\nu. \tag{138}$$

(Indeed, suppose *ad absurdum* that (137) does not hold. Then, there exist $\theta \in \mathbb{R}, \epsilon > 0$ and a sequence $\lambda_n \to 0 (n \in \mathbb{N}, n \to \infty)$ such that $|C_{\lambda_n}(\theta) - C_0(\theta)| > \epsilon$ ($\forall\ n \in \mathbb{N}$). Due to (135), we can find a subsequence $\{n_k\} \subset \mathbb{N}$ such that (138) and (137) hold for $\lambda_{n_k} \to 0$ $(k \to \infty)$, which is a contradiction.) Split $M(\mathrm{d}\boldsymbol{u}) = M_1(\mathrm{d}\boldsymbol{u}) + M_2(\mathrm{d}\boldsymbol{u})$, where M_1 (respectively, M_2) is a Lévy random measure with characteristics $(0, \pi)$ (respectively, with $(\sigma, 0)$) and M_1, M_2 being independent. Accordingly, $C_\lambda(\theta) = C_{\lambda,1}(\theta) + C_{\lambda,2}(\theta)$, where $C_{\lambda,i}(\theta) := \mathrm{E}\exp\{\mathrm{i}\theta \int f_\lambda(\boldsymbol{u}) M_i(\mathrm{d}\boldsymbol{u})\}, i = 1, 2$. The subsequent proof is split into four parts depending on the value of α. Let

$$\tilde{\lambda} := \det(\Lambda) = \prod_{j=1}^{\nu} \lambda^{\gamma_j},$$

$$\Psi_\alpha(z) := \mathrm{e}^{\mathrm{i}z} - 1 - \mathrm{i}\tau_\alpha(z) \quad (0 < \alpha \le 2, z \in \mathbb{R}). \tag{139}$$

Case $\alpha = 2$. Then, $\sigma > 0$ and $C_{\lambda,2}(\theta) \to C_0(\theta)$ is immediate from (135), and hence (136) follows from $C_{\lambda,1}(\theta) \to 1$. We have $C_{\lambda,1}(\theta) = \mathrm{e}^{I_\lambda}$, where

$$\begin{aligned} I_\lambda &:= \int_{\mathbb{R}^\nu \times \mathbb{R}} \Psi_2(\theta z f_\lambda(\boldsymbol{u}))\mathrm{d}\boldsymbol{u}\, \pi(\mathrm{d}z) \\ &= \tilde{\lambda} \int_{\mathbb{R}^\nu \times \mathbb{R}} \Psi_2\big(\theta z \tilde{\lambda}^{-\frac{1}{2}} f_\lambda^\dagger(\boldsymbol{u})\big)\mathrm{d}\boldsymbol{u}\, \pi(\mathrm{d}z) \to 1 \end{aligned}$$

follows by (Pratt's) Lemma 30. Indeed, using $|\Psi_2(z)| \le 2(|z| \wedge |z|^2)$, for θ fixed, we have $\big|\tilde{\lambda}\Psi_2\big(\theta z \tilde{\lambda}^{-\frac{1}{2}} f_\lambda^\dagger(\boldsymbol{u})\big)\big| \le C\tilde{\lambda}^{1/2}|z f_\lambda^\dagger(\boldsymbol{u})| \to 0$ for any $z \in \mathbb{R}$ and a.e. $\boldsymbol{u} \in \mathbb{R}^\nu$ satisfying (138); moreover, $\big|\tilde{\lambda}\Psi_2\big(\theta z \tilde{\lambda}^{-\frac{1}{2}} f_\lambda^\dagger(\boldsymbol{u})\big)\big| \le C|z f_\lambda^\dagger(\boldsymbol{u})|^2$, with $\int_{\mathbb{R}^\nu \times \mathbb{R}} |z f_\lambda^\dagger(\boldsymbol{u})|^2 \mathrm{d}\boldsymbol{u}\, \pi(\mathrm{d}z) \to \int_{\mathbb{R}^\nu \times \mathbb{R}} |z h(\boldsymbol{u})|^2 \mathrm{d}\boldsymbol{u}\, \pi(\mathrm{d}z)$, proving (136) for $\alpha = 2$.

Case $1 < \alpha < 2$. In this case, $\sigma = 0$. Let $\bar{\pi}_+(z) := \pi([z, \infty)), \bar{\pi}_-(z) := \pi((-\infty, -z]), z > 0$. Then, $C_\lambda(\theta) = \exp\Big\{\tilde{\lambda} \int_{\mathbb{R}^\nu \times \mathbb{R}} \Psi_\alpha\big(\theta z \tilde{\lambda}^{-\frac{1}{\alpha}} f_\lambda^\dagger(\boldsymbol{u})\big)\mathrm{d}\boldsymbol{u}\, \pi(\mathrm{d}z)\Big\} = \mathrm{e}^{I_\lambda}$, where $I_\lambda = \int_{\mathbb{R}^\nu \times \mathbb{R}} \Phi_{\tilde{\lambda}}(\theta f_\lambda^\dagger(\boldsymbol{u}), z)\mathrm{d}\boldsymbol{u}\, \mathrm{d}z$ follows upon integrating by parts with

$$\begin{aligned} \Phi_{\tilde{\lambda}}(u, z) &:= \tilde{\lambda}\, \bar{\pi}_{\mathrm{sgn}(z)}\left(\frac{|z|\tilde{\lambda}^{1/\alpha}}{|u|}\right)(\mathrm{e}^{\mathrm{i}\, z\, \mathrm{sgn}(u)} - 1)\mathrm{i}\, \mathrm{sgn}(uz) \\ &\to |u|^\alpha \frac{c_{\mathrm{sgn}(z)}}{|z|^\alpha}(\mathrm{e}^{\mathrm{i}\, z\, \mathrm{sgn}(u)} - 1)\mathrm{i}\, \mathrm{sgn}(uz) =: \Phi_0(u, z), \end{aligned}$$

and hence $\Phi_{\tilde{\lambda}}(\theta f_\lambda^\dagger(\boldsymbol{u}), z) \to \Phi_0(\theta h(\boldsymbol{u}), z)$ for any $(\boldsymbol{u}, z) \in \mathbb{R}^\nu \times \mathbb{R}$, such that the convergence in (138) holds. Therefore, we can expect that

$$I_\lambda \to \int_{\mathbb{R}^\nu \times \mathbb{R}} \Phi_0(\theta h(\boldsymbol{u}), z)\mathrm{d}\boldsymbol{u}\mathrm{d}z \ = -|\theta|^\alpha \int_{\mathbb{R}^\nu} |h(\boldsymbol{u})|^\alpha \omega_\alpha(\theta h(\boldsymbol{u}))\mathrm{d}\boldsymbol{u}; \tag{140}$$

see (125). To justify the convergence of the integrals in (140), by $\bar{\nu}_\pm(z) \le Cz^{-\alpha}, z > 0$, we have $|\Phi_{\tilde{\lambda}}(u,z)| \le C|u|^\alpha(|z| \wedge 1)|z|^{-\alpha} := \bar{\Phi}(u,z)$, implying $|\Phi_{\tilde{\lambda}}(\theta f^\dagger_\lambda(\boldsymbol{u}), z)| \le \bar{\Phi}(\theta f^\dagger_\lambda(\boldsymbol{u}), z)$, where $\bar{\Phi}(\theta f^\dagger_{\boldsymbol{\lambda}}(\boldsymbol{u}), z) \to \bar{\Phi}(\theta h(\boldsymbol{u}), z)$ for a.e. $(\boldsymbol{u}, z) \in \mathbb{R}^\nu \times \mathbb{R}$ and $\int_{\mathbb{R}^\nu \times \mathbb{R}} \bar{\Phi}(\theta f^\dagger_\lambda(\boldsymbol{u}), z)\mathrm{d}\boldsymbol{u}\, \mathrm{d}z \to \int_{\mathbb{R}^\nu \times \mathbb{R}} \bar{\Phi}(\theta h(\boldsymbol{u}), z)\mathrm{d}\boldsymbol{u}\mathrm{d}z < \infty$ so that (140) follows by Lemma 30, as in the case of $\alpha = 2$ above. Hence, $C_\lambda(\theta) \to C_0(\theta)$ $(\forall \theta \in \mathbb{R})$.

Case $0 < \alpha < 1$. Then, $C_\lambda(\theta) = \exp\{\int J_{\tilde{\lambda}}(\theta f^\dagger_\lambda(\boldsymbol{u}))\mathrm{d}\boldsymbol{u}\} = \mathrm{e}^{I_\lambda}$, where for a large $K > 0$, we split $J_{\tilde{\lambda}}(u) = J_{\tilde{\lambda},K,0}(u) + J_{\tilde{\lambda},K,1}(u)$, with

$$J_{\tilde{\lambda},K,0}(u) := \tilde{\lambda} \int_{|z| \le \frac{\tilde{\lambda}^{\frac{1}{\alpha}} K}{|u|}} (\mathrm{e}^{\mathrm{i}\tilde{\lambda}^{-\frac{1}{\alpha}} uz} - 1)\pi(\mathrm{d}z),$$

$$J_{\tilde{\lambda},K,1}(u) := \tilde{\lambda} \int_{|z| > \frac{\tilde{\lambda}^{\frac{1}{\alpha}} K}{|u|}} (\mathrm{e}^{\mathrm{i}\tilde{\lambda}^{-\frac{1}{\alpha}} uz} - 1)\pi(\mathrm{d}z). \tag{141}$$

Integrating by parts, we rewrite $J_{\tilde{\lambda},K,0}(u) = \Phi_{\tilde{\lambda}}(u,K) - \Phi_{\tilde{\lambda}}(u,-K) - \int_{|z|\le K} \phi_{\tilde{\lambda}}(u,z)\mathrm{d}z$, with

$$\Phi_{\tilde{\lambda}}(u,z) := -\tilde{\lambda}\bar{\nu}_{\mathrm{sgn}(z)}\left(\frac{\tilde{\lambda}^{\frac{1}{\alpha}}|z|}{|u|}\right)\mathrm{sgn}(z)(\mathrm{e}^{\mathrm{i}z\,\mathrm{sgn}(u)} - 1),$$

$$\phi_{\tilde{\lambda}}(u,z) := -\tilde{\lambda}\bar{\nu}_{\mathrm{sgn}(z)}\left(\frac{\tilde{\lambda}^{\frac{1}{\alpha}}|z|}{|u|}\right)\mathrm{i}\,\mathrm{sgn}(uz)\mathrm{e}^{\mathrm{i}z\,\mathrm{sgn}(u)}.$$

In the corresponding decomposition $I_\lambda = I_{\lambda,K,0} + I_{\lambda,K,1}$, with $I_{\lambda,K,0} := \int J_{\tilde{\lambda},K,0}(\theta f^\dagger_{\boldsymbol{\lambda}}(\boldsymbol{u}))\mathrm{d}\boldsymbol{u}$, we use $\Phi_{\tilde{\lambda}}(\theta f^\dagger_\lambda(\boldsymbol{u}), z) \to \Phi_0(\theta h(\boldsymbol{u}), z)$ and $\phi_{\tilde{\lambda}}(\theta f^\dagger_\lambda(\boldsymbol{u}), z) \to \phi_0(\theta h(\boldsymbol{u}), z)$ for any $(\boldsymbol{u}, z) \in \mathbb{R}^\nu \times \mathbb{R}$ such that (138) holds, where

$$\Phi_0(u,z) := -|u|^\alpha \frac{c_{\mathrm{sgn}(z)}}{|z|^\alpha}\mathrm{sgn}(z)(\mathrm{e}^{\mathrm{i}z\,\mathrm{sgn}(u)} - 1),$$

$$\phi_0(u,z) := -|u|^\alpha \frac{c_{\mathrm{sgn}(z)}}{|z|^\alpha}\mathrm{i}\,\mathrm{sgn}(uz)\mathrm{e}^{\mathrm{i}z\,\mathrm{sgn}(u)}.$$

Arguing as in the proof of (140), we obtain $I_{\lambda,K,0} \to I_{K,0} := \int J_{K,0}(\theta h(\boldsymbol{u}))\mathrm{d}\boldsymbol{u}$, with

$$
\begin{aligned}
J_{K,0}(u) &:= \Phi_0(u,K) - \Phi_0(u,-K) - \int_{|z|\le K} \phi_0(u,z)\mathrm{d}z \\
&= |u|^\alpha \int_{|z|\le K} \frac{\alpha c_{\mathrm{sgn}(z)}}{|z|^{\alpha+1}} (\mathrm{e}^{\mathrm{i}\,\mathrm{sgn}(u)z} - 1)\mathrm{d}z,
\end{aligned}
$$

where $I_{K,0} \to -|\theta|^\alpha \int |h(\boldsymbol{u})|^\alpha \omega_\alpha(\theta h(\boldsymbol{u}))\mathrm{d}\boldsymbol{u}$ as $K \to \infty$. The fact that $\int J_{\tilde\lambda,K,1}(\theta f^\dagger_\lambda(\boldsymbol{u}))\mathrm{d}\boldsymbol{u} =: I_{\lambda,K,1}$ is negligible follows from $|I_{\lambda,K,1}| \le 2\tilde\lambda \int \nu(|z| > \frac{\tilde\lambda^{\frac{1}{\alpha}} K}{|\theta f^\dagger_\lambda(\boldsymbol{u})|})\mathrm{d}\boldsymbol{u} \le \frac{C}{K^\alpha}\int |\theta f^\dagger_\lambda(\boldsymbol{u})|^\alpha \mathrm{d}\boldsymbol{u} \le \frac{C}{K^\alpha}$ by choosing K large enough. This proves $C_\lambda(\theta) \to C_0(\theta)$ $(\forall \theta \in \mathbb{R})$ in the case of $0 < \alpha < 1$.

Case $\alpha = 1$. By symmetry of π and W_1, $C_\lambda(\theta) = \mathrm{e}^{2I_\lambda}, C_0(\theta) = \mathrm{e}^{2I_0}$, where

$$
\begin{aligned}
I_\lambda &:= \tilde\lambda \int_{\mathbb{R}^\nu \times \mathbb{R}_+} (\cos(\theta f^\dagger_\lambda(\boldsymbol{u})\tilde\lambda^{-1} z) - 1)\mathrm{d}\boldsymbol{u}\nu(\mathrm{d}z), \\
I_0 &:= c_+ \int_{\mathbb{R}^\nu \times \mathbb{R}_+} (\cos(\theta h(\boldsymbol{u})z) - 1)z^{-2}\mathrm{d}\boldsymbol{u}\mathrm{d}z = -c_+|\theta|\|h\|_1 J,
\end{aligned}
$$

where $J := \int_0^\infty (1-\cos z)z^{-2}\mathrm{d}z = \int_0^\infty \frac{\sin z}{z}\mathrm{d}z = \frac{\pi}{2}$. Similarly to (141), for a large $K > 0$, split $I_\lambda = I_{\lambda,K,0} + I_{\lambda,K,1}$, where

$$
I_{\lambda,K,0} := \tilde\lambda \int_{\mathbb{R}^\nu \times \left\{0<z\le \frac{\tilde\lambda K}{|\theta f^\dagger_\lambda(\boldsymbol{u})|}\right\}} \cdots, \qquad I_{\lambda,K,1} := \tilde\lambda \int_{\mathbb{R}^\nu \times \left\{z> \frac{\tilde\lambda K}{|\theta f^\dagger_\lambda(\boldsymbol{u})|}\right\}} \cdots.
$$

Then, integrating by parts and using $(\mathrm{M})_1$, Lemma 30 similarly as in the case $0 < \alpha < 1$, we obtain

$$
I_{\lambda,K,0} \to c_+ \int |\theta h(\boldsymbol{u})|\mathrm{d}\boldsymbol{u} \int_0^K (\cos z - 1)z^{-2}\mathrm{d}z =: I_{K,0},
$$

where $I_{K,0} \to I_0$ $(K \to \infty)$, while $I_{\lambda,K,1}$ can be made arbitrarily small uniformly in $\lambda > 0$ by choosing $K > 0$ large enough. Proposition 24 is proved. □

Remark 25. The conditions of Proposition 24 are not necessary for the convergence of the stochastic integrals in (136). Necessary and sufficient conditions for the convergence in the distribution of infinitely divisible r.v.s in terms of their Lévy triplets are given in Sato [232] and other texts. However, the verification of these conditions is facilitated by the application of Proposition 24.

Proposition 24 admits a 'large-scale' counterpart implying the convergence

$$\int_{\mathbb{R}^\nu} f_\lambda(\boldsymbol{u}) M(\mathrm{d}\boldsymbol{u}) \xrightarrow{\mathrm{d}} \int_{\mathbb{R}^\nu} h(\boldsymbol{u}) W_\alpha(\mathrm{d}\boldsymbol{u}) \quad \text{as} \quad \lambda \to \infty \tag{142}$$

under analogous conditions $\lim_{\lambda\to\infty} \|f_\lambda^\dagger - h\|_\alpha = 0$ on $f_\lambda^\dagger$ in (138) and the behaviour of the Lévy measure π at infinity. We omit the formulation of this counterpart since it is almost identical to that of Proposition 24 and is not used in this book. Instead, we provide an extension of the convergence criterion in Proposition 9 for linear forms in i.i.d. r.v.s belonging to the domain of attraction of the α-stable law.

Proposition 26. *Let $0 < \alpha < 2$ and $S(g_\lambda) = \sum_{\boldsymbol{u}\in\mathbb{Z}^\nu} g_\lambda(\boldsymbol{u})\varepsilon(\boldsymbol{u})$ be a linear form in i.i.d. r.v.s $\varepsilon(\boldsymbol{u}), \varepsilon \in D(\alpha)$ with real coefficients $g_\lambda(\boldsymbol{u}), \boldsymbol{u} \in \mathbb{Z}^\nu$. Assume that there exists $0 \le p < \alpha < p'$ and a function $h \in L^p(\mathbb{R}^\nu) \cap L^{p'}(\mathbb{R}^\nu)$ such that the functions*

$$g_\lambda^\dagger(\boldsymbol{u}) := \det(\Lambda)^{1/\alpha} g_\lambda([\Lambda\boldsymbol{u}]), \quad \boldsymbol{u} \in \mathbb{R}^\nu, \quad \Lambda = \mathrm{diag}(\lambda^{\gamma_1}, \ldots, \lambda^{\gamma_\nu}). \tag{143}$$

satisfy

$$\|g_\lambda^\dagger - h\|_p + \|g_\lambda^\dagger - h\|_{p'} \to 0, \quad \lambda \to \infty. \tag{144}$$

Then,

$$S(g_\lambda) \xrightarrow{\mathrm{d}} \int_{\mathbb{R}^\nu} h(\boldsymbol{u}) W_\alpha(\mathrm{d}\boldsymbol{u}), \tag{145}$$

where $W_\alpha(\mathrm{d}\boldsymbol{u})$ is an α-stable random measure with ch.f. in (123).

Proof. Note first that $h \in L^p(\mathbb{R}^\nu) \cap L^{p'}(\mathbb{R}^\nu)$ implies $h \in L^\alpha(\mathbb{R}^\nu)$, and hence the stochastic integral $\int_{\mathbb{R}^\nu} h(\boldsymbol{x}) W_\alpha(\mathrm{d}\boldsymbol{x}) =: I_\alpha(h)$ in (145) is well defined. To prove (145), we follow the proof of Proposition 9. The set $\mathcal{S}_0(\mathbb{R}^\nu)$ of step functions $f : \mathbb{R}^\nu \to \mathbb{R}$ defined in the latter proof is dense in $L^p(\mathbb{R}^\nu) \cap L^{p'}(\mathbb{R}^\nu)$: given $h \in L^p(\mathbb{R}^\nu) \cap L^{p'}(\mathbb{R}^\nu)$, for every $\epsilon > 0$, there exists $h_\epsilon \in \mathcal{S}_0(\mathbb{R}^\nu)$ such that

$$\|h - h_\epsilon\|_p + \|h - h_\epsilon\|_{p'} < \epsilon. \tag{146}$$

As in the proof of Proposition 9, (145) follows once we show that for every $\epsilon > 0$, there exists $h_\epsilon \in \mathcal{S}_0(\mathbb{R}^\nu)$ such that as $\lambda \to \infty$, the following relations (j)–(jjj) hold: (j) $\mathrm{E}|S(g_\lambda) - S(h_{\epsilon,\lambda})|^p < \epsilon$, (jj) $S(h_{\epsilon,\lambda}) \xrightarrow{\mathrm{d}} I_\alpha(h_\epsilon)$, (jjj) $\mathrm{E}|I_\alpha(h_\epsilon) - I_\alpha(h)|^p < \epsilon$, where

$$h_{\epsilon,\lambda}(\boldsymbol{u}) := |\det(\Lambda)|^{-1/\alpha} h_\epsilon(\Lambda^{-1}\boldsymbol{u}), \quad \boldsymbol{u} \in \mathbb{Z}^\nu, \quad \lambda > 0.$$

We only prove (j) since facts (jj) and (jjj) follow from the properties of the stochastic integral I_α discussed above.

Towards this end, split $S(g_\lambda) - S(h_{\epsilon,\lambda}) = S(g_\lambda - h_{\epsilon,\lambda}) = S_-(g_\lambda - h_{\epsilon,\lambda}) + S_+(g_\lambda - h_{\epsilon,\lambda})$, where

$$S_\pm(g) := \sum_{\boldsymbol{u} \in \mathbb{Z}^\nu} g(\boldsymbol{u})\varepsilon_\pm(\boldsymbol{u}),$$

$$\varepsilon_+(\boldsymbol{u}) := \varepsilon(\boldsymbol{u})\mathbb{I}(|\varepsilon(\boldsymbol{u})| > \tilde{\lambda}^{1/\alpha}) - \mathrm{E}[\varepsilon(\boldsymbol{u})\mathbb{I}(|\varepsilon(\boldsymbol{u})| > \tilde{\lambda}^{1/\alpha})]\mathbb{I}(\alpha > 1),$$
$$\varepsilon_-(\boldsymbol{u}) := \varepsilon(\boldsymbol{u})\mathbb{I}(|\varepsilon(\boldsymbol{u})| \le \tilde{\lambda}^{1/\alpha}) - \mathrm{E}[\varepsilon(\boldsymbol{u})\mathbb{I}(|\varepsilon(\boldsymbol{u})| \le \tilde{\lambda}^{1/\alpha})]\mathbb{I}(\alpha > 1).$$

and $\tilde{\lambda} := \det(\Lambda)$. By the moment inequality of Proposition 6 with $0 < p < \alpha < p' \le 2$,

$$\mathrm{E}|S_+(g_\lambda - h_{\epsilon,\lambda})|^p \le C\mathrm{E}|\varepsilon_+|^p \|g_\lambda - h_{\epsilon,\lambda}\|_p^p, \tag{147}$$
$$\mathrm{E}|S_-(g_\lambda - h_{\epsilon,\lambda})|^{p'} \le C\mathrm{E}|\varepsilon_-|^{p'} \|g_\lambda - h_{\epsilon,\lambda}\|_{p'}^{p'},$$

By the change of variable $\boldsymbol{u} \to \Lambda\boldsymbol{u}$, for any $p > 0$, we see that

$$\|g_\lambda - h_{\epsilon,\lambda}\|_p^p = \int_{\mathbb{R}^\nu} |g_\lambda(\boldsymbol{u}) - h_{\epsilon,\lambda}(\boldsymbol{u})|^p \mathrm{d}\boldsymbol{u}$$
$$= \tilde{\lambda}^{1-p/\alpha} \|g_\lambda^\dagger(\boldsymbol{u}) - h_\epsilon(\boldsymbol{u})\|_p^p \le C\epsilon\tilde{\lambda}^{1-p/\alpha} \tag{148}$$

holds thanks to (146) for all λ large enough. On the other hand, using the condition $\varepsilon \in D(\alpha)$ and integration by parts, we obtain that $\{\varepsilon_\pm, \varepsilon_\pm(\boldsymbol{u})\}$ are i.i.d. sequences satisfying

$$\begin{aligned}
\mathrm{E}|\varepsilon_+|^p &\le C\mathrm{E}|\varepsilon|^p \mathbb{I}(|\varepsilon| > \tilde\lambda^{1/\alpha}) \\
&= C\int_{\tilde\lambda^{1/\alpha}}^{\infty} z^p \mathrm{dP}(|\varepsilon| \le z) \le C\tilde\lambda^{p/\alpha - 1}, \quad p < \alpha, \qquad (149) \\
\mathrm{E}|\varepsilon_-|^{p'} &\le C\mathrm{E}|\varepsilon|^{p'} \mathbb{I}(|\varepsilon| \le \tilde\lambda^{1/\alpha}) \\
&= C\int_0^{\tilde\lambda^{1/\alpha}} z^{p'} \mathrm{dP}(|\varepsilon| \le z) \le C\tilde\lambda^{p'/\alpha - 1}, \quad \alpha < p' \le 2.
\end{aligned}$$

The inequalities in (147), (148), and (149) prove relation (j) and the proposition as well. □

Remark 27. Condition (144) of Proposition 26 cannot be weakened to $\|g^\dagger_\lambda - h\|_p \to 0$, $p < \alpha$ in general. Indeed, let $\varepsilon \overset{\mathrm{d}}{=} W_\alpha(\mathbf{1})$ have an α-stable distribution of Example 10 of Chapter 1 and $g_\lambda(\boldsymbol{u}) = g(\boldsymbol{u}) = \mathbb{I}(\boldsymbol{u} \in]\mathbf{0}, \boldsymbol{x}])$ do not depend on λ. Then, $S(g_\lambda) = W_\alpha(]\mathbf{0}, \boldsymbol{x}])$ and (145) holds with $h(\boldsymbol{u}) = \mathbb{I}(\boldsymbol{u} \in]\mathbf{0}, \boldsymbol{x}])$. On the other hand, $\|g^\dagger_\lambda\|_p^p = \det(\Lambda)^{p/\alpha - 1} \to 0$ leads to $h(\boldsymbol{u}) = 0$, which is a contradiction. Note that $\|g^\dagger_\lambda\|_{p'}^{p'} = \det(\Lambda)^{p'/\alpha - 1} \to \infty$ in this example, meaning that (144) is not satisfied for any $p < \alpha < p'$.

Example 28. *Fractional Linear Stable Motion* $L_{\alpha,d} = \{L_{\alpha,d}(\tau); \tau \ge 0\}$, depending on the parameters $\alpha \in (0, 2], -1/\alpha < d < 1 - \frac{1}{\alpha}, d \ne 0$, is defined as the stochastic integral

$$L_{\alpha,d}(\tau) = \int_{\mathbb{R}} ((\tau - s)_+^d - (-s)_+^d) W_\alpha(\mathrm{d}s) \qquad (150)$$

w.r.t. an α-stable random measure W_α on $\mathbb{R}$. The above process is H-SS with $H = d + \frac{1}{\alpha} \in (0, 1)$ and has stationary increments. For $\alpha = 2$, (150) is a multiple of FBM. Various properties and generalisations of (23) are discussed in the monographs [215, 230, 231].

Let $X = \{X(t); t \in \mathbb{Z}\}$ be a stationary MA process,

$$X(t) = \sum_{s \le t} a(t - s)\varepsilon(s), \qquad (151)$$

in i.i.d. innovations with generic $\varepsilon \in D(\alpha)$, and deterministic coefficients satisfying $\sum_{s=0}^\infty |a(s)|^p < \infty$ for some $0 < p < \alpha$. The above class is well defined by Proposition 6 and includes the case of regularly decaying coefficients,

$$a(s) \sim c_0 \, s^{d-1}, \quad s \to \infty. \tag{152}$$

The process X in (151)–(152) is well defined for $d < 1 - \frac{1}{\alpha}$ by applying Proposition 6 since $\sum_{s=0}^{\infty} |a(s)|^p \le C \sum_{s=0}^{\infty} s_+^{p(d-1)} < \infty$ holds if $p < \alpha$ is sufficiently close to α. For $\alpha \in (1, 2]$, the above X satisfies the coefficient LRD property when $0 < d < 1 - \frac{1}{\alpha}$ and the coefficient ND property when $d < 0$ and $\sum_{s=0}^{\infty} a(s) = 0$ hold. An application of Proposition 26, we derive the partial sums limit of X in the LRD/SRD/ND cases.

Theorem 29 ([8]). *Let X in be as in (151), $\varepsilon \in D(\alpha), 0 < \alpha < 2$. In addition, consider the following:*

(i) *Let* (152) *hold with $-\frac{1}{\alpha} < d < 1 - \frac{1}{\alpha}, d \neq 0$. In addition, suppose $\sum_{s=0}^{\infty} a(s) = 0$ when $d < 0$. Then,*

$$\lambda^{-d-\frac{1}{\alpha}} \sum_{t=1}^{[\lambda\tau]} X(t) \xrightarrow{\text{fdd}} (c_0/d) L_{\alpha,d}(\tau), \tag{153}$$

where $\{L_{\alpha,d}(\tau); \tau > 0\}$ is the fractional linear stable motion in (150).

(ii) *Let $\sum_{j=0}^{\infty} |a(s)| < \infty$. Then,*

$$\lambda^{-\frac{1}{\alpha}} \sum_{t=1}^{[\lambda\tau]} X(t) \xrightarrow{\text{fdd}} A W_\alpha(\tau), \tag{154}$$

where $A := \sum_{s=0}^{\infty} a(s)$ and $\{W_\alpha(\tau) = W_\alpha(]0, \tau]); \tau > 0\}$ is the same Lévy process as in (150).

Proof. (i) Write the l.h.s. of (153) as the linear form $S(g_\lambda)$, with $g_\lambda(s) = \lambda^{-d-\frac{1}{\alpha}} \sum_{t=1}^{[\lambda\tau]} a(t-s), s \in \mathbb{Z}$. For $0 < d < 1 - \frac{1}{\alpha}$, the rescaled weights can be written as the integral

$$g_\lambda^\dagger(s) = \lambda^{1/\alpha} g_\lambda([\lambda s]) = \int_0^\tau a_\lambda(s,t) \mathrm{d}t, \quad s \in \mathbb{R},$$

where

$$a_\lambda(t,s) := \lambda^{1-d}a([\lambda t]-[\lambda s]) \to c_0(t-s)_+^{d-1} =: a_\infty(t-s), \qquad \lambda\to\infty \tag{155}$$

point-wise for any $t \neq s$ according to (152). We also see that $|a_\lambda(t-s)| \le C\lambda^{1-d}|1 \vee ([\lambda t]-[\lambda s])|^{d-1} = C|(1/\lambda) \vee ([\lambda t]/\lambda - [\lambda s]/\lambda)|^{d-1}$, where $(1/\lambda) \vee ([\lambda t]/\lambda - [\lambda s]/\lambda) \ge (t-s)/2$ for any $t \ge s, \lambda > 0$. Therefore, $|g_\lambda^\dagger(s)| \le \bar g(s) := C\int_0^\tau (t-s)_+^{d-1}\mathrm{d}s$ is dominated by $\bar g$ independent of λ, with $\|\bar g\|_p^p \le C\int_{\mathbb{R}}\big(\int_0^\tau (t-s)_+^{d-1}\mathrm{d}t\big)^p\mathrm{d}s = C\int_{\mathbb{R}}\big((\tau-s)_+^d - (-s)_+^d\big)^p\mathrm{d}s < \infty$ for $0<d<1-\frac{1}{p}$, the last inequality satisfied in both cases $p<\alpha$ and $p>\alpha$ when p is sufficiently close to α. Therefore, condition (144) holds with $h(s) = \int_0^\tau a_\infty(t-s)\mathrm{d}t$, implying one-dimensional convergence in (153) in view of (145). The general case of (153) follows analogously by the Cramér–Wold device and taking linear combinations of the above $g_\lambda(s)$'s.

Next, consider the ND case of $-\frac{1}{\alpha}<d<0$. Here, we represent the l.h.s. of (153) as $S(g_\lambda)$ with $g_\lambda(s) = -\lambda^{-d-\frac{1}{\alpha}}\sum_{t\in\mathbb{Z},t\notin[1,[\lambda\tau]]} a(t-s)$, $s\in\mathbb{Z}$ and $g_\lambda^\dagger(s) = \lambda^{1/\alpha}g_\lambda([\lambda s]) = -\int_{\mathbb{R}\setminus]0,\tau]} a_\lambda(s,t)\mathrm{d}t$, $t,s\in\mathbb{R}$, with the same $a_\lambda(s,t)$ as in (155). Then, analogously as in the case $d>0$, we get that

$$\begin{aligned} g_\lambda^\dagger(s) \to h(s) &= -c_0\int_{\mathbb{R}\setminus]0,\tau]}((t-s)_+^{d-1} - (-s)_+^{d-1})\mathrm{d}t \\ &= (c_0/d)((\tau-s)_+^d - (-s)_+^d) \end{aligned}$$

converge point-wise for any $s\in\mathbb{R}, s\neq 0,\tau$ and satisfy the conditions of Proposition 26, proving (153) for $d<0$.

(ii) We have $\lambda^{-\frac{1}{\alpha}}\sum_{t=1}^{[\lambda\tau]} X(t) = S(g_\lambda)$, with $g_\lambda(s) = \lambda^{-\frac{1}{\alpha}}\sum_{t=1}^{[\lambda\tau]} a(t-s)$ and

$$g_\lambda^\dagger(s) = \sum_{t=1}^{[\lambda\tau]} a(t-[\lambda s]) \to A\,\mathbb{I}(s\in]0,\tau]) =: h(s), \tag{156}$$

tending point-wise for any $s\in\mathbb{R}, s\neq 0,\tau$.

Let us show (144). Split $\int_{\mathbb{R}}|g_\lambda^\dagger(s)-h(s)|^p\mathrm{d}s = \int_{-\infty}^0|g_\lambda^\dagger(s)|^p\mathrm{d}s + \int_0^\tau|g_\lambda^\dagger(s)-h(s)|^p\mathrm{d}s =: I_\lambda^- + I_\lambda^+$. Then, $\lim_{\lambda\to\infty} I_\lambda^+ = 0$ for any $p>0$

follows from (156) by $\sum_{s=0}^{\infty} |a(s)| < \infty$ and the DCT. It remains to show that

$$\lim_{\lambda\to\infty} I_\lambda^- = 0 \tag{157}$$

for some $p \neq \alpha$ sufficiently close to α. Since $g_\lambda^\dagger$ in (156) is bounded, it suffices to prove (157) for $p \le 1$. Towards this end, let $\tau = 1$ and $\lambda, \mu \in \mathbb{N}_+$, $\mu = o(\lambda)$. Then, $I_\lambda^- = \lambda^{-1} \sum_{s=0}^{\infty} \big| \sum_{t=1}^{\mu} a(t+s) + \sum_{t=\mu+1}^{\lambda} a(t+s)\big|^p$, where

$$\lambda^{-1} \sum_{s=0}^{\infty} \Big| \sum_{t=1}^{\mu} a(t+s)\Big|^p \le C\lambda^{-1} \sum_{s=0}^{\infty} \sum_{t=1}^{\mu} |a(t+s)|^p = O(\mu/\lambda) = o(1)$$

since $\sum_{s=0}^{\infty} |a(s)|^p < \infty$. Next,

$$\lambda^{-1} \sum_{s=0}^{\infty} \Big| \sum_{t=\mu+1}^{\lambda} a(t+s)\Big|^p \le C\lambda^{-1} \sum_{t=1}^{\lambda} \sum_{s=0}^{\infty} |a(\mu+t+s)|^p = o(1)$$

follows by the DCT since $\sum_{s=0}^{\infty} |a(\mu+t+s)|^p$ is bounded and vanishes with $\mu \to \infty$ for any $t \in \mathbb{N}_+$. Theorem 29 is proved. □

We remark that similar convergence and DCT arguments as in the above theorem are used in the following sections to prove the scaling limits of linear RFs.

Lemma 30 ([216]). *Let $(\mathcal{X}, \mu)$ be a measure space and $L^p(\mathcal{X})$ be the space of all measurable functions $f : \mathcal{X} \to \mathbb{R}$, with $\int |f|^p \equiv \int_{\mathcal{X}} |f(x)|^p \mu(\mathrm{d}x) < \infty$, $p > 0$.*
(i) *Let $f_\lambda, g_\lambda \in L^p(\mathcal{X}), \lambda > 0$ be such that*
(j) $\lim_{\lambda\to\infty} f_\lambda(x) = f(x)$, $\lim_{\lambda\to\infty} g_\lambda(x) = g(x)$ *exists a.e. in $\mathcal{X}$,*
(jj) $|f_\lambda(x)| \le |g_\lambda(x)|$ *a.e. in $\mathcal{X}$, $\lambda > 0$, and*
(jjj) $\lim_{\lambda\to\infty} \int |g_\lambda|^p = \int |g|^p < \infty$.
Then, $f \in L^p(\mathcal{X})$ and

$$\lim_{\lambda\to\infty} \int |f_\lambda - f|^p = 0. \tag{158}$$

(ii) *$f_\lambda, f \in L^p(\mathcal{X}), \lambda > 0$ and (158) implies* $\lim_{\lambda\to\infty} \int |f_\lambda|^p = \int |f|^p$.

Proof. By Fatou's lemma, $\int \liminf_{\lambda\to\infty} h_\lambda \le \liminf_{\lambda\to\infty} \int h_\lambda$ for any $0 \le h_\lambda \in L^1(\mathcal{X})$. Hence, $\int |f|^p \le \liminf_{\lambda\to\infty} \int |f_\lambda|^p \le \liminf_{\lambda\to\infty} \int |g_\lambda|^p < \infty$. Similarly, by applying Fatou's lemma for $h_\lambda(x) := |g_\lambda(x)|^p - |f_\lambda(x)|^p \ge 0$ and using (j)–(jjj), we obtain $\int |g|^p - \int |f|^p \le \liminf_{\lambda\to\infty}(\int |g_\lambda|^p - \int |f|^p) = \int |g|^p - \limsup_{\lambda\to\infty} \int |f_\lambda|^p$, proving $\limsup_{\lambda\to\infty} \int |f_\lambda|^p \le \int |f|^p \le \liminf_{\lambda\to\infty} \int |f_\lambda|^p$ or $\lim_{\lambda\to\infty} \int |f_\lambda|^p = \int |f|^p < \infty$.

It remains to show (158). Using the inequality $|a-b|^p \le 2^p(|a|^p + |b|^p)$, $a, b \in \mathbb{R}$, apply Fatou's lemma to

$$h_\lambda(x) := 2^p(|f(x)|^p + |f_\lambda(x)|^p) - |f(x) - f_\lambda(x)|^p \ge 0,$$

with $\lim_{\lambda\to\infty} h_\lambda(x) = 2^{p+1}|f(x)|^p$ a.e. in $\mathcal{X}$. Since $\lim_{\lambda\to\infty} \int |f_\lambda|^p = \int |f|^p$, this yields $2^{p+1} \int |f|^p \le 2^{p+1} \int |f|^p - \limsup_{\lambda\to\infty} \int |f - f_\lambda|^p$, or $\limsup_{\lambda\to\infty} \int |f - f_\lambda|^p = 0$, proving (158) and part (i).

Part (ii) follows by the triangle inequality and Fatou's lemma, implying $\limsup_{\lambda\to\infty} \int |f_\lambda|^p \le \int |f|^p$ and $\int |f|^p \le \liminf_{\lambda\to\infty} \int |f_\lambda|^p$, respectively. □

7 Notes

1. Various properties of cumulants and diagram formulas can be found in Malyshev and Minlos [173], Giraitis and Surgailis [99], Breuer and Major [47], Shiryaev [236], Giraitis *et al.* [103], and Peccati and Taqqu [199], among others.

2. Most facts in Section 2 of Chapter 3 stem from Rajput and Rosinski [222]. Proposition 9 appeared in Pilipauskaitė and Surgailis [210].

3. The discussion of Hermite and Charlier polynomials follows Giraitis *et al.* [103] and Surgailis [253]. Probabilistic aspects of Appell polynomials are described in Giraitis and Surgailis [99], Giraitis *et al.* [103], and Avram and Taqqu [10].

4. The multiple Wiener–Itô integral was introduced by Wiener [266] and Itô [125]. A detailed discussion of such integrals and their properties can be found in Dobrushin [78], Dobrushin and Major [80], Peccati and Taqqu [199], and other texts. Multiple stochastic integrals can also be defined w.r.t. to Poisson and other random measures; see Itô [126], as well as Surgailis [244, 246], Peccati and

Taqqu [199], and Samorodnitsky [230]. Poisson multiple stochastic integrals are related to Charlier polynomials and useful in the study of polynomial forms of random grain and other integer-valued processes.

5. Proposition 24 appeared in Pilipauskaitė and Surgailis [211]. The convergence towards fractional linear stable motion in Theorem 29 (i) was first discussed by Astrauskas [8]. For a more detailed treatment and extensions of issues pertaining to this theorem, see Kasahara and Maejima [132], Astrauskas *et al.* [9], and Samorodnisky and Taqqu [231]. Lemma 30 is due to Pratt [216].

Chapter 3

Moving-Average RFs

This chapter discusses normalised limits of integrals in (1) for moving-average (MA) RFs satisfying Assumptions (A)($d;\mathbb{Z}^\nu$)/(A)($d;\mathbb{R}^\nu$) of Section 1 of Chapter 1. As shown in Theorems 4 and 5, these limits can be written as stochastic integrals w.r.t. the Gaussian white noise with integrands taking a different form in the LRD ($d > 0$), SRD ($d = 0$), and ND ($d < 0$) cases. We present concrete examples and parametric classes of MA RFs and kernels $a(\boldsymbol{t})$ with the above properties involving fractional integration/differentiation operators $(I - T)^d$ on $\mathbb{Z}^\nu$, defined via the binomial expansion of the transition operator T of a random walk on $\mathbb{Z}^\nu$. The main result of Section 2 is Theorem 8, which provides a sufficient condition for the invertibility of $(I - T)^d$ in terms of the characteristic function of the random walk. The asymptotic behaviour of fractional MA coefficients is identified by following a local limit theorem for random walks. Examples involving fractional Laplace and heat operators are studied in detail. Simultaneous and conditional representations of fractionally integrated RFs are discussed.

1 Isotropic Scaling Limits

Let X be a stationary RF on $\mathbb{Z}^\nu$ or $\mathbb{R}^\nu$. By the *isotropic scaling limit of* X, we mean the limit distribution of integral functionals

$$X_\lambda(\phi) := \int_{\mathbb{R}^\nu} X([\boldsymbol{t}])\phi(\boldsymbol{t}/\lambda)\mathrm{d}\boldsymbol{t} \quad \text{or } X_\lambda(\phi) := \int_{\mathbb{R}^\nu} X(\boldsymbol{t})\phi(\boldsymbol{t}/\lambda)\mathrm{d}\boldsymbol{t} \quad (1)$$

as $\lambda \to \infty$, for each ϕ from a class Φ of test functions $\phi : \mathbb{R}^\nu \to \mathbb{R}$; see (35), (38) of Chapter 1. This section discusses the above limits for the MA RF

$$X(\boldsymbol{t}) = \begin{cases} \sum_{\boldsymbol{u}\in\mathbb{Z}^\nu} a(\boldsymbol{t}-\boldsymbol{u})\varepsilon(\boldsymbol{u}), & \boldsymbol{t}\in\mathbb{Z}^\nu, \\ \int_{\mathbb{R}^\nu} a(\boldsymbol{t}-\boldsymbol{u})M(\mathrm{d}\boldsymbol{u}), & \boldsymbol{t}\in\mathbb{R}^\nu, \end{cases} \tag{2}$$

which can be written as sums or integrals over the standardised i.i.d. r.v.s $\varepsilon(\boldsymbol{s}), \boldsymbol{s}\in\mathbb{Z}^\nu$ and a Lévy random measure $M(\mathrm{d}\boldsymbol{u}), \boldsymbol{u}\in\mathbb{R}^\nu$ with zero mean and variance $\mathrm{E}M(\mathrm{d}\boldsymbol{u})^2 = \mathrm{d}\boldsymbol{u}$. The MA kernel $a(\boldsymbol{t})$ in (2) is a deterministic function satisfying the respective Assumptions (A)$(d;\mathbb{Z}^\nu)$ /(A)$(d;\mathbb{R}^\nu)$ of Section 1 of Chapter 1.

The first question concerning the limit of the integral in (1) concerns the asymptotics of its variance, which can be written as the double integral

$$\mathrm{Var}(X_\lambda(\phi)) = \int_{\mathbb{R}^{2\nu}} \begin{Bmatrix} \mathrm{Cov}(X([\boldsymbol{t}_1]), X([\boldsymbol{t}_2])) \\ \mathrm{Cov}(X(\boldsymbol{t}_1), X(\boldsymbol{t}_2)) \end{Bmatrix} \phi(\boldsymbol{t}_1/\lambda)\phi(\boldsymbol{t}_2/\lambda \mathrm{d}\boldsymbol{t}_1\mathrm{d}\boldsymbol{t}_2 \tag{3}$$

in the respective discrete and continuous argument cases, in terms of the covariance function of X. For the MA RF X satisfying Assumptions (A)$(d;\mathbb{Z}^\nu)$ /(A)$(d;\mathbb{R}^\nu)$ with $0 < d < \nu/4$, the covariance can be written as the convolution $[a\star a](\boldsymbol{t})/(a\star a)(\boldsymbol{t})$ (see (8)), whose asymptotic behaviour was obtained in Proposition 4 of Chapter 1. The following Proposition 1 obtains the asymptotics of the double integral in (3) for RFs indexed by $\mathbb{Z}^\nu$, which is used in the limit Theorem 4. Although analogous relations as in (6) and (8) hold for the covariances indexed by $\mathbb{R}^\nu$, they are not included in Proposition 1 since the scaling limits of Lévy MAs in Theorem 5 are handled directly via the characteristic function.

Proposition 1. *Let* $\phi_i \in \Phi, i = 1, 2$.

(i) For any $\beta > 0$ *as* $\lambda \to \infty$,

$$\int_{\mathbb{R}^{2\nu}} |\phi_1(\boldsymbol{t}_1/\lambda)\phi_2(\boldsymbol{t}_2/\lambda)|(1\wedge|\boldsymbol{t}_1-\boldsymbol{t}_2|^{-\beta})\mathrm{d}\boldsymbol{t}_1\mathrm{d}\boldsymbol{t}_2$$

$$= \begin{cases} O(\lambda^\nu), & \beta > \nu, \\ O(\lambda^{2\nu-\beta}), & \beta < \nu, \\ O(\lambda^\nu \log\lambda), & \beta = \nu. \end{cases} \tag{4}$$

(ii) Let $r(\cdot) \in L^{\infty}(\mathbb{Z}^{\nu})$ *satisfy*

$$r(\boldsymbol{t}) = |\boldsymbol{t}|^{4d-\nu} \left(L\left(\frac{\boldsymbol{t}}{|\boldsymbol{t}|}\right) + o(1) \right), \quad |\boldsymbol{t}| \to \infty, \tag{5}$$

with $0 < d < \nu/4, L(\cdot) \in C(S_{\nu-1})$. *Then,*

$$\lim_{\lambda\to\infty} \lambda^{-\nu-4d} \int_{\mathbb{R}^{2\nu}} \phi_1(\boldsymbol{t}_1/\lambda)\phi_2(\boldsymbol{t}_2/\lambda) r([\boldsymbol{t}_1] - [\boldsymbol{t}_2]) \mathrm{d}\boldsymbol{t}_1 \mathrm{d}\boldsymbol{t}_2 = c(\phi_1, \phi_2), \tag{6}$$

where

$$c(\phi_1, \phi_2) := \int_{\mathbb{R}^{2\nu}} \phi_1(\boldsymbol{t}_1)\phi_2(\boldsymbol{t}_2) L\left(\frac{\boldsymbol{t}_1 - \boldsymbol{t}_2}{|\boldsymbol{t}_1 - \boldsymbol{t}_2|}\right) \frac{\mathrm{d}\boldsymbol{t}_1 \mathrm{d}\boldsymbol{t}_2}{|\boldsymbol{t}_1 - \boldsymbol{t}_2|^{\nu-4d}}. \tag{7}$$

(iii) Let $r(\cdot) \in L^1(\mathbb{Z}^{\nu})$. *Then,*

$$\lim_{\lambda\to\infty} \lambda^{-\nu} \int_{\mathbb{R}^{2\nu}} \phi_1(\boldsymbol{t}_1/\lambda)\phi_2(\boldsymbol{t}_2/\lambda) r([\boldsymbol{t}_1] - [\boldsymbol{t}_2]) \mathrm{d}\boldsymbol{t}_1 \mathrm{d}\boldsymbol{t}_2$$

$$= \int_{\mathbb{R}^{\nu}} \phi_1(\boldsymbol{t})\phi_2(\boldsymbol{t}) \mathrm{d}\boldsymbol{t} \times \sum_{\boldsymbol{s}\in\mathbb{Z}^{\nu}} r(\boldsymbol{s}). \tag{8}$$

Proof. (i) Write $I_{\lambda,\beta}$ for the l.h.s. of (4). First, let $\beta > \nu$. Then, $I_{\lambda,\beta} \le C\int_{\mathbb{R}^{\nu}} |\phi_1(\boldsymbol{t}_1/\lambda)| \mathrm{d}\boldsymbol{t}_1 \times \int_{\mathbb{R}^{\nu}} 1 \wedge |\boldsymbol{t}_2 - \boldsymbol{t}_1|^{-\beta} \, \mathrm{d}\boldsymbol{t}_2 \le C\int_{\mathbb{R}^{\nu}} |\phi_1(\boldsymbol{t}_1/\lambda)| \mathrm{d}\boldsymbol{t}_1 = C\lambda^{\nu} \int_{\mathbb{R}^{\nu}} |\phi_1(\boldsymbol{t})| \mathrm{d}\boldsymbol{t} = O(\lambda^{\nu})$ as $\int_{\mathbb{R}^{\nu}} 1 \wedge |\boldsymbol{t}|^{-\beta} \mathrm{d}\boldsymbol{t} < \infty$. Next, let $\beta < \nu$. Then, $I_{\lambda,\beta} \le \lambda^{2\nu-\beta} J_{\beta}$, where $J_{\beta} := \int_{\mathbb{R}^{2\nu}} |\phi_1(\boldsymbol{t}_1)\phi_2(\boldsymbol{t}_2)||\boldsymbol{t}_1 - \boldsymbol{t}_2|^{-\beta} \mathrm{d}\boldsymbol{t}_1 \mathrm{d}\boldsymbol{t}_2 < \infty$ follows by $J_{\beta} \le C\int_{\mathbb{R}^{\nu}} \phi_1(\boldsymbol{t}_1)|\mathrm{d}\boldsymbol{t}_1 \int_{|\boldsymbol{t}_2-\boldsymbol{t}_1|\le 1} |\boldsymbol{t}_2 - \boldsymbol{t}_1|^{-\beta} \mathrm{d}\boldsymbol{t}_2 + \int_{\mathbb{R}^{2\nu}} |\phi_1(\boldsymbol{t}_1)\phi_2(\boldsymbol{t}_2)| \mathrm{d}\boldsymbol{t} \mathrm{d}\boldsymbol{t}_2 < \infty$. Finally, for $\beta = \nu$, we have $I_{\lambda,\nu} = \lambda^{\nu} J_{\lambda,\nu}$, where $J_{\lambda,\nu} := \int_{\mathbb{R}^{2\nu}} |\phi_1(\boldsymbol{t}_1)\phi_2(\boldsymbol{t}_2)|(\lambda^{-1} \wedge |\boldsymbol{t}_1 - \boldsymbol{t}_2|)^{-\nu} \mathrm{d}\boldsymbol{t}_1 \mathrm{d}\boldsymbol{t}_2 = O(\log \lambda)$ follows similarly.

(ii) The convergence of the integral in (7) follows from that of J_{β} in part (i), with $\beta = \nu - 4d$. Let $c_{\lambda}(\phi_1, \phi_2)$ denote the integral on the l.h.s. of (6). By change of variables,

$$\frac{c_{\lambda}(\phi_1, \phi_2)}{\lambda^{\nu+4d}} = \int_{\mathbb{R}^{2\nu}} \frac{\phi_1(\boldsymbol{t}_1)\phi_2(\boldsymbol{t}_2)}{|\boldsymbol{t}_1 - \boldsymbol{t}_2|^{\nu-4d}} \tilde{L}_{\lambda}(\boldsymbol{t}_1, \boldsymbol{t}_2) \mathrm{d}\boldsymbol{t}_1 \mathrm{d}\boldsymbol{t}_2,$$

where $\tilde{L}_\lambda(\boldsymbol{t}_1, \boldsymbol{t}_2) \to L_2(\frac{\boldsymbol{t}_1-\boldsymbol{t}_2}{|\boldsymbol{t}_1-\boldsymbol{t}_2|})$ $(\lambda \to \infty)$ for any $\boldsymbol{t}_1 \neq \boldsymbol{t}_2$. Using Pratt's lemma [216], it suffices to prove (6) for $L_2 \equiv 1$. In the latter case, with $\tilde{\boldsymbol{t}}_i := [\lambda \boldsymbol{t}_i]/\lambda, i = 1, 2$, we see that $|\tilde{L}_\lambda(\boldsymbol{t}_1, \boldsymbol{t}_2)| \le C\big(|\boldsymbol{t}_1 - \boldsymbol{t}_2|/(|\tilde{\boldsymbol{t}}_1 - \tilde{\boldsymbol{t}}_2| \vee (1/\lambda))\big)^{\nu-2d} \le C$, as in the proof of Proposition 4 of Chapter 1. Thus, (6) follows from the dominated convergence theorem (DCT).

(iii) Let $c_\lambda(\phi_1, \phi_2)$ be the same as in the proof of (ii). For a large $K > 0$, write $c_\lambda(\phi_1, \phi_2) = \sum_{i=1}^3 c_{i,\lambda}$, where $c_{1,\lambda} := \int_{|\boldsymbol{t}_1-\boldsymbol{t}_2|>K} \phi_1(\boldsymbol{t}_1/\lambda)\phi_2(\boldsymbol{t}_2/\lambda) r([\boldsymbol{t}_1] - [\boldsymbol{t}_2]) \mathrm{d}\boldsymbol{t}_1 \mathrm{d}\boldsymbol{t}_2$, $c_{2,\lambda} := \int_{|\boldsymbol{t}_1-\boldsymbol{t}_2|\le K} \phi_1(\boldsymbol{t}_1/\lambda) \times \phi_2(\boldsymbol{t}_1/\lambda) r([\boldsymbol{t}_1] - [\boldsymbol{t}_2]) \mathrm{d}\boldsymbol{t}_1 \mathrm{d}\boldsymbol{t}_2$, and $c_{3,\lambda} := \int_{|\boldsymbol{t}_1-\boldsymbol{t}_2|\le K} \phi_1(\boldsymbol{t}_1/\lambda)(\phi_2(\boldsymbol{t}_2/\lambda) - \phi_2(\boldsymbol{t}_1/\lambda)) r([\boldsymbol{t}_1] - [\boldsymbol{t}_2]) \mathrm{d}\boldsymbol{t}_1 \mathrm{d}\boldsymbol{t}_2$. Here, $\lambda^{-\nu}|c_{1,K}| \le C\lambda^{-\nu} \int_{\mathbb{R}^\nu} |\phi_1(\boldsymbol{t}/\lambda)| \mathrm{d}\boldsymbol{t} \sum_{|\boldsymbol{s}|>K} |r(\boldsymbol{s})| \le C \sum_{|\boldsymbol{s}|>K} |r(\boldsymbol{s})|$ can be made arbitrarily small uniformly in $\lambda \ge 1$ by choosing K large enough. Next,

$$\lambda^{-\nu}|c_{3,\lambda}| \le C \int_{\mathbb{R}^\nu} |\phi_1(\boldsymbol{t})| \mathrm{d}\boldsymbol{t} \int_{|\boldsymbol{s}|\le K} \left|\phi_2\left(\boldsymbol{t} + \frac{\boldsymbol{s}}{\lambda}\right) - \phi_2(\boldsymbol{t})\right| \mathrm{d}\boldsymbol{s}.$$

By the boundedness of ϕ_2, we see that the integral $\int_{|\boldsymbol{s}|\le K} |\phi_2(\boldsymbol{t} + \frac{\boldsymbol{s}}{\lambda}) - \phi_2(\boldsymbol{t})| \mathrm{d}\boldsymbol{s} \to 0$ $(\lambda \to \infty)$ a.e. in $\mathbb{R}^\nu$, and is bounded in $\boldsymbol{t} \in \mathbb{R}^\nu$. Then, since $\phi_1 \in L^1(\mathbb{R}^\nu)$, we conclude $\lim_{\lambda\to\infty} \lambda^{-\nu}|c_{3,\lambda}| = 0$ by the DCT. Finally, $\lambda^{-\nu} c_{2,\lambda} = \int_{\mathbb{R}^\nu} \phi_1(\boldsymbol{t})\phi_2(\boldsymbol{t}) \mathrm{d}\boldsymbol{t} \int_{|\boldsymbol{s}+[\lambda\boldsymbol{t}]-\lambda\boldsymbol{t}|\le K} r([\boldsymbol{s}]) \mathrm{d}\boldsymbol{s}$, and we can replace the last integral by the r.h.s. of (8) uniformly in λ provided K is large enough. □

Proposition 1 does not apply to ND covariances satisfying (5) with negative $d < 0$. This case is more delicate since it requires additional regularity conditions on the test functions and the occurrence of 'edge effects'. A detailed analysis of this issue in dimension $\nu = 2$ and for the indicator (test) functions of rectangles in $\mathbb{R}^2_+$ is provided in Chapter 6. In the following, we present a result in this direction and sufficient conditions on $d, \phi_i, i = 1, 2$, when the limits take a form similar to (6). Introduce a subclass of test functions:

$$\Phi_- := \Bigg\{\phi \in \Phi,\ \phi(\cdot) \text{ a.e. continuous on } \mathbb{R}^\nu,$$
$$\int_{\mathbb{R}^\nu} \left(\int_{\mathbb{R}^\nu} |\phi(\boldsymbol{t}+\boldsymbol{s}) - \phi(\boldsymbol{s})|^2 \mathrm{d}\boldsymbol{s}\right)^{1/2} |\boldsymbol{t}|^{2d-\nu} \mathrm{d}\boldsymbol{t} < \infty\Bigg\}. \quad (9)$$

Proposition 2. *Let $a \in L^2(\mathbb{Z}^\nu)$ satisfy Assumption (A)$(d;\mathbb{Z}^\nu)$ with $-\nu/4 < d < 0$. Then, for any $\phi_i \in \Phi_-, i = 1,2$, we have that*

$$\lim_{\lambda\to\infty} \lambda^{-\nu-4d} \int_{\mathbb{R}^{2\nu}} \phi_1(\boldsymbol{t}_1/\lambda)\phi_2(\boldsymbol{t}_2/\lambda)[a\star a]([\boldsymbol{t}_1]-[\boldsymbol{t}_2])\mathrm{d}\boldsymbol{t}_1\mathrm{d}\boldsymbol{t}_2$$
$$= c_-(\phi_1,\phi_2), \tag{10}$$

where

$$c_-(\phi_1,\phi_2)$$
$$:= \int_{\mathbb{R}^\nu}\prod_{i=1}^{2}\int_{\mathbb{R}^\nu}\left(\int_{\mathbb{R}^\nu}(\phi_i(\boldsymbol{t}+\boldsymbol{s})-\phi_i(\boldsymbol{s}))|\boldsymbol{t}|^{2d-\nu}\ell\left(\frac{\boldsymbol{t}}{|\boldsymbol{t}|}\right)\mathrm{d}\boldsymbol{t}\right)\mathrm{d}\boldsymbol{s}. \tag{11}$$

Proof. The convergence of the integral on the r.h.s. of (11) follows from (9) and Minkowski's integral inequality: $\big\{\int_{\mathbb{R}^\nu}\big(\int_{\mathbb{R}^\nu}|\phi(\boldsymbol{t}+\boldsymbol{s})-\phi(\boldsymbol{s})||\boldsymbol{t}|^{2d-\nu}\mathrm{d}\boldsymbol{t}\big)^2\mathrm{d}\boldsymbol{s}\big\}^{1/2} \le \int_{\mathbb{R}^\nu}\|\phi(\boldsymbol{t}+\cdot)-\phi(\cdot)\|_{L^2(\mathbb{R}^\nu)}\times|\boldsymbol{t}|^{2d-\nu}\mathrm{d}\boldsymbol{t}$.

The proof of the convergence in (10) resembles that of (6). Write $c_\lambda(\phi_1,\phi_2)$ for the integral on the l.h.s. of (10). Using $\sum_{\boldsymbol{s}\in\mathbb{Z}^\nu} a(\boldsymbol{s}) = 0$, we rewrite $\int_{\mathbb{R}^\nu}\phi_i(\boldsymbol{t}_i)a([\lambda\boldsymbol{t}_i] - `[\lambda\boldsymbol{s})\mathrm{d}\boldsymbol{t}_i = \int_{\mathbb{R}^\nu}\big(\phi_i(\boldsymbol{t}_i+\frac{\boldsymbol{s}}{\lambda})-\phi_i(\frac{\boldsymbol{s}}{\lambda})\big)\,a([\lambda\boldsymbol{t}_i])\mathrm{d}\boldsymbol{t}_i, i=1,2, \boldsymbol{s}\in\mathbb{Z}^\nu$ and

$$\frac{c_\lambda(\phi_1,\phi_2)}{\lambda^{\nu+4d}} = \int_{\mathbb{R}^\nu}\mathrm{d}\boldsymbol{s}\prod_{i=1}^{2}\int_{\mathbb{R}^\nu}\left(\phi_i\left(\boldsymbol{t}_i+\frac{[\lambda\boldsymbol{s}]}{\lambda}\right)\right.$$
$$\left. - \phi_i\left(\frac{[\lambda\boldsymbol{s}]}{\lambda}\right)\right)\lambda^{\nu-2d}a([\lambda\boldsymbol{t}_i])\mathrm{d}\boldsymbol{t}_i,$$

where the inner integrals tend to those on the r.h.s. of (11) at each continuity point $\boldsymbol{s}$ of ϕ_i; see (12). The remaining details are similar as in (6) and hence omitted. □

Remark 3. The restriction $d > -\nu/4$ in Proposition 2 is not necessary for (11). Indeed, if $\phi \in \Phi$ satisfies the uniform Lipschitz condition $|\phi(\boldsymbol{t}+\boldsymbol{s})-\phi(\boldsymbol{s})| < C|\boldsymbol{t}|$ ($|\boldsymbol{t}| < 1, \boldsymbol{s}\in\mathbb{R}^\nu$), then the integral in (9) converges for $0 > d > -\nu/2$, implying $\phi \in \Phi_-$. On the other hand, for the indicator functions $\phi(\boldsymbol{t}) = \mathbb{I}(\boldsymbol{t}\in A)$ of a bounded Borel set $A \subset \mathbb{R}^\nu$ with a 'regular' boundary, we typically have $\|\phi(\boldsymbol{t}+\cdot) - \phi(\cdot)\|_{L^2(\mathbb{R}^\nu)} = O(|\boldsymbol{t}|^{1/2})$, leading to $d > -\nu/4$.

Relation (7) of Chapter 1 entails the existence of the scaling limit

$$\lim_{\lambda\to\infty} \lambda^{\nu-2d} a([\lambda \boldsymbol{t}]) = a_\infty(\boldsymbol{t}) := |\boldsymbol{t}|^{2d-\nu} \ell\left(\frac{\boldsymbol{t}}{|\boldsymbol{t}|}\right), \quad \lambda \to \infty, \quad \forall \boldsymbol{t} \in \mathbb{R}^\nu_0, \tag{12}$$

which is a continuous homogeneous function on $\mathbb{R}^\nu$: for any $\lambda > 0$, we have that

$$a_\infty(\lambda \boldsymbol{t}) = \lambda^{2d-\nu} a_\infty(\boldsymbol{t}), \quad \boldsymbol{t} \in \mathbb{R}^\nu_0. \tag{13}$$

With the limit function in (12), we associate a Gaussian RF:

$$W_d(\phi) := \begin{cases} \int_{\mathbb{R}^\nu} (a_\infty \star \phi)(\boldsymbol{u}) W(\mathrm{d}\boldsymbol{u}), & 0 < d < \nu/4, \phi \in \Phi, \\ \int_{\mathbb{R}^\nu} (a_\infty \star \phi)_{\mathrm{reg}}(\boldsymbol{u}) W(\mathrm{d}\boldsymbol{u}), & -\nu/4 < d < 0, \phi \in \Phi_-, \\ \int_{\mathbb{R}^\nu} \phi(\boldsymbol{u}) W(\mathrm{d}\boldsymbol{u}), & d = 0, \phi \in \Phi, \end{cases} \tag{14}$$

where $W(\mathrm{d}\boldsymbol{u})$ is a real-valued Gaussian white noise with zero mean and variance $\mathrm{d}\boldsymbol{u}$, $(a_\infty \star \phi)(\boldsymbol{u}) = \int_{\mathbb{R}^\nu} a_\infty(\boldsymbol{t})\phi(\boldsymbol{t}+\boldsymbol{u})\mathrm{d}\boldsymbol{t}$ is the usual, and

$$(a_\infty \star \phi)_{\mathrm{reg}}(\boldsymbol{u}) := \int_{\mathbb{R}^\nu} a_\infty(\boldsymbol{t})(\phi(\boldsymbol{t}+\boldsymbol{u}) - \phi(\boldsymbol{u}))\mathrm{d}\boldsymbol{t}, \quad \boldsymbol{u} \in \mathbb{R}^\nu, \tag{15}$$

is the 'regularised' convolution. For the indicator test function $\phi(\boldsymbol{t}) = \mathbb{I}(\boldsymbol{t} \in B)$ of a Borel set $B \subset \mathbb{R}^\nu$ (belonging to Φ_-), we see that the latter convolution equals

$$(a_\infty \star \phi)_{\mathrm{reg}}(\boldsymbol{u}) = \begin{cases} \int_B a_\infty(\boldsymbol{t}-\boldsymbol{u})\mathrm{d}\boldsymbol{t}, & \boldsymbol{u} \notin B, \\ -\int_{\mathbb{R}^\nu \setminus B} a_\infty(\boldsymbol{t}-\boldsymbol{u})\mathrm{d}\boldsymbol{t}, & \boldsymbol{u} \in B. \end{cases} \tag{16}$$

The existence of the stochastic integrals in (14) follows from Propositions 1 and 2. Particularly, the variances $\mathrm{E}W_d^2(\phi) = c(\phi,\phi)$ $(0 < d < \nu/4)$ and $\mathrm{E}W_d^2(\phi) = c_-(\phi,\phi)$ $(-\nu/4 < d < 0)$ agree with (7) and (11). The RF in (14) appears as scaling limits in the following theorem.

Theorem 4. *Let X be an MA RF satisfying Assumption (A)$(d;\mathbb{Z}^\nu)$. Then,*

$$\lambda^{-(\nu+4d)/2}X_\lambda(\phi) \xrightarrow{\mathrm{d}} \begin{cases} W_d(\phi), & 0<d<\nu/4,\ \forall\,\phi\in\Phi, \\ W_d(\phi), & -\nu/4<d<0,\ \forall\,\phi\in\Phi_-, \\ \sigma_0 W_0(\phi), & d=0,\ \forall\,\phi\in\Phi, \end{cases} \tag{17}$$

where $\sigma_0^2 := \big(\sum_{\boldsymbol{t}\in\mathbb{Z}^\nu} a(\boldsymbol{t})\big)^2$.

Proof. Since (1) can be written as a linear form $X_\lambda(\phi) = \sum_{\boldsymbol{u}\in\mathbb{Z}^\nu} \varepsilon(\boldsymbol{u}) \int_{\mathbb{R}^\nu} \phi(\boldsymbol{t}/\lambda)a([\boldsymbol{t}]-\boldsymbol{u})\mathrm{d}\boldsymbol{t}$ in i.i.d. r.v.s, we can apply the Lindeberg criterion in Proposition 7, Chapter 2. Accordingly, it suffices to show that

$$\sup_{\boldsymbol{u}\in\mathbb{Z}^\nu} \Big| \int_{\mathbb{R}^\nu} \phi(\boldsymbol{t}/\lambda)a([\boldsymbol{t}]-\boldsymbol{u})\mathrm{d}\boldsymbol{t}\Big| = o(\sqrt{\mathrm{Var}(X_\lambda(\phi))}), \quad \lambda\to\infty, \tag{18}$$

holds in each case of $d>0, d<0$, and $d=0$ of the theorem. The behaviour of the last variance is detailed in Propositions 1 and 2 tending to infinity in each case of d. On the other hand, the l.h.s. of (18) does not exceed $\|\phi\|_{L^\infty(\mathbb{R}^\nu)}\|a\|_{L^1(\mathbb{Z}^\nu)}$, which is bounded in cases $d<0$ and $d=0$. Finally, in the case of $d>0$, we see that the l.h.s. of (18) does not exceed $\|\phi(\cdot/\lambda)\|_{L^2(\mathbb{R}^\nu)}\|a\|_{L^2(\mathbb{Z}^\nu)} = O(\lambda^{\nu/2})$, and (18) holds since $d>0$. □

Let us turn to scaling limits of the Lévy MA RF in (2). From the characteristic function of the stochastic integral w.r.t. $M\in(M)_2$ in (129), we have that

$$\mathrm{E}\mathrm{e}^{\mathrm{i}\theta X(\phi)} = \exp\Big\{-(\sigma^2\theta^2/2)\|a\star\phi\|^2_{L^2(\mathbb{R}^\nu)} + \int_{\mathbb{R}\times\mathbb{R}^\nu} \Psi(\theta v(a\star\phi)(\boldsymbol{u}))\pi(\mathrm{d}v)\mathrm{d}\boldsymbol{u}\Big\}, \tag{19}$$

where σ,π are the Lévy characteristics of M and $\Psi(x)=\mathrm{e}^{\mathrm{i}x}-1-\mathrm{i}x, x\in\mathbb{R}$. Particularly, when $\pi=0, \sigma=1$, (2) is a Gaussian RF which can be written as the stochastic integral $X(\boldsymbol{t}) = \int_{\mathbb{R}^\nu} a(\boldsymbol{t}-\boldsymbol{s})W(\mathrm{d}\boldsymbol{s})$ w.r.t. the Gaussian white noise as in (14). We see that the limit results for the Lévy MAs in (2) are almost identical to those in Theorem 4.

Theorem 5. *Let X be a Lévy MA in (2) with the kernel $a(\cdot)$ satisfying Assumption (A)$(d;\mathbb{R}^\nu)$. Then, the convergences in (17) hold, with σ_0^2 in (17) replaced by $\sigma_0^2 := \big(\int_{\mathbb{R}^\nu} a(\boldsymbol{t})\mathrm{d}\boldsymbol{t}\big)^2$.*

Proof. The convergence of the variances in (17) follows as in Propositions 1 and 2. Instead of the Lindeberg criterion (which does not directly apply to stochastic integrals), the limits in (17) can be obtained with the help of the characteristic function in (19). Particularly, for $0 < d < \nu/4$, $\lambda^{-(\nu+4d)}\|a \star \phi(\cdot/\lambda\|^2_{L^2(\mathbb{R}^\nu)} \to \|a_\infty \star \phi\|^2_{L^2(\mathbb{R}^\nu)}$ uses only the variance argument and $\int_{\mathbb{R}^\nu} \Psi\big(\theta v\lambda^{-(\nu+4d)/2}(a\star \phi(\cdot/\lambda))(\boldsymbol{u})\big)\mathrm{d}\boldsymbol{u} = \int_{\mathbb{R}^\nu} \lambda^\nu\Psi\big(\theta v\lambda^{-\nu/2}\lambda^{\nu-2d}(a \star \phi))(\boldsymbol{u})\big)\mathrm{d}\boldsymbol{u}$, where $\lambda^{\nu-2d}(a\star\phi))(\boldsymbol{u}) \to (a_\infty\star\phi)(\boldsymbol{u})$ and $\lambda^\nu\Psi\big(\theta v\lambda^{-\nu/2}\lambda^{\nu-2d}(a\star\phi))(\boldsymbol{u})\big) \to -(\theta^2v^2/2)(a_\infty \star \phi)(\boldsymbol{u})$, for any $\boldsymbol{u} \in \mathbb{R}^\nu, v \in \mathbb{R}, \phi \in \Phi$, according to (7) of Chapter 1 and $\Psi(x) \sim -x^2/2 (x \to 0)$. Moreover, $\big|\lambda^\nu\Psi\big(\theta v\lambda^{-\nu/2}\lambda^{\nu-2d}(a \star \phi))(\boldsymbol{u})\big)\big| \le Cv^2\big|\lambda^{\nu-2d}(a \star \phi)(\boldsymbol{u})\big|^2$, where

$$\int_{\mathbb{R}\times\mathbb{R}^\nu} \big|\lambda^{\nu-2d}(a\star\phi)(\boldsymbol{u})\big|^2\pi(\mathrm{d}v)\mathrm{d}\boldsymbol{u} \to \int_{\mathbb{R}} v^2\pi(\mathrm{d}v)\int_{\mathbb{R}^\nu} |(a_\infty\star\phi)(\boldsymbol{u})|^2\mathrm{d}\boldsymbol{u}$$

in view of the above variance argument. Then, $\log \mathrm{E}\mathrm{e}^{\mathrm{i}\theta\lambda^{-(\nu+4d)/2}X_\lambda(\phi)} \to -(\theta 2/2)(\sigma^2 + \int_{\mathbb{R}} v^2\pi(\mathrm{d}v)) \times \|a_\infty \star \phi\|^2_{L^2(\mathbb{R}^\nu)} = \log \mathrm{E}\mathrm{e}^{\mathrm{i}\theta W_d(\phi)}$ by (Pratt's) Lemma 30 of Chapter 2. In the case of $-\nu/4 < d < 0$, we replace $(a \star \phi)(\boldsymbol{u})$ on the r.h.s. of (19) by $(a_\infty \star \phi)_{\mathrm{reg}}(\boldsymbol{u})$ of (15) and proceed with the proof of (17) as above. Finally, in the case of $d = 0$, use $|\Psi(x) + (1/2)x^2| \le C(|x|^3 \wedge x^2)\,(x \in \mathbb{R})$ to conclude that $\mathrm{E}\mathrm{e}^{\mathrm{i}\theta\lambda^{-\nu/2}X_\lambda(\phi)} = \exp\big\{-(\theta^2/2)\lambda^{-\nu}\mathrm{E}X^2_\lambda(\phi) + R_\lambda\big\}$, where $\lambda^{-\nu}\mathrm{E}X^2_\lambda(\phi) \to \sigma_0^2\mathrm{E}W_0^2(\phi)$ and $|R_\lambda| \le C\lambda^{-3\nu/2}\int_{\mathbb{R}\times\mathbb{R}^\nu}\Big(\big|v(a\star \phi(\cdot/\lambda))(\boldsymbol{u})\big|^3 \wedge \big|v(a\star\phi(\cdot/\lambda))(\boldsymbol{u})\big|^2\Big)\pi(\mathrm{d}v)\mathrm{d}\boldsymbol{u} = O(\lambda^{-\nu/2}) = o(1)$. □

2 Fractional Integration/Differentiation on $\mathbb{Z}^\nu$

Fractional operators on $\mathbb{Z}$. Consider the fractional differentiation/integration operators acting on functions $g = g(t), t \in \mathbb{Z}$, where t can be interpreted as 'time'. Fractional differentiation operators $(I - T)^d, 0 < d < 1$, form a class of difference operators which interpolate between the identity $Ig(t) = g(t)$ and the 'discrete derivative' $g(t) - g(t-1) = (I - T)g(t)$, where $Tg(t) = g(t-1)$ is the backward shift. Analogously, the fractional integration operators $(I - T)^{-d}, 0 < d < 1$ interpolate between the identity and the 'discrete integral' $(I - T)^{-1}g(t) := \sum_{s=-\infty}^{t} g(s)$. These operators can be defined by substituting $z = T$ in the binomial expansion

$$(1 - z)^d = \sum_{j=0}^{\infty} \psi_j(d) z^j, \quad \psi_j(d) := \frac{\Gamma(j - d)}{\Gamma(j + 1)\Gamma(-d)}, \quad z \in \mathbb{C}, \ |z| < 1, \tag{20}$$

which is important for the subsequent discussion. The representation in (20) can be more explicitly written as

$$\psi_0(d) = 1, \qquad \psi_j(d) = \prod_{k=1}^{j} \frac{k - 1 - d}{k} \qquad (j = 1, 2, \ldots). \tag{21}$$

The application of Stirling's formula to the Gamma functions in the numerator and the denominator in (20) yields the asymptotics of the binomial coefficients:

$$\psi_j(d) \sim \Gamma(-d)^{-1} j^{-d-1}, \qquad j \to \infty, \quad 0 < |d| < 1 \tag{22}$$

It easily follows from (21) and (22) that

$$\begin{aligned} &\psi_j(d) < 0 \quad (j \geq 1), \quad \sum_{j=0}^{\infty} \psi_j(d) = 0 \qquad \text{if} \quad 0 < d < 1, \\ &\psi_j(d) > 0 \quad (j \geq 1), \quad \sum_{j=0}^{\infty} \psi_j(d) = \infty \qquad \text{if} \quad -1 < d < 0. \end{aligned} \tag{23}$$

The fractional power of order d of $I - T$ is defined as the difference operator

$$(I - T)^d g(t) := \sum_{j=0}^{\infty} \psi_j(d) T^j g(t) = \sum_{j=0}^{\infty} \psi_j(d) g(t - j), \quad t \in \mathbb{Z}, \tag{24}$$

provided the last series converges. The identity $(1 - z)^d (1 - z)^{-d} = 1$ leads to

$$1 = \sum_{j,k=0}^{\infty} \psi_j(d) \psi_l(-d) z^{j+k} = \sum_{n=0}^{\infty} z^n \sum_{j=0}^{n} \psi_j(d) \psi_{n-j}(-d)$$

and the orthogonality relations between the binomial coefficients:

$$\sum_{j=0}^{n} \psi_j(d)\psi_{n-j}(-d) = \mathbb{I}(n=0), \quad n \in \mathbb{N}. \tag{25}$$

Let $\{\varepsilon(t); t \in \mathbb{Z}\}$ be a discrete weak white noise (an uncorrelated sequence of r.v.s with zero mean and unit variance). An autoregressive fractionally integrated MA ARFIMA$(0, d, 0)$ process $X = \{X(t); t \in \mathbb{Z}\}$ is defined as a stationary solution of the stochastic difference equation

$$(I - T)^d X(t) = \sum_{j=0}^{\infty} \psi_j(d) X(t-j) = \varepsilon(t), \qquad t \in \mathbb{Z}. \tag{26}$$

For $d \in (-1/2, 1/2)$, the solution of (26) is obtained by applying the inverse operator:

$$X(t) = (I - T)^{-d}\varepsilon(t) = \sum_{j=0}^{\infty} \psi_j(d)\varepsilon(t-j), \quad t \in \mathbb{Z}. \tag{27}$$

Since (22) implies $\sum_{j=0}^{\infty} \psi_j(d)^2 < \infty$ $(|d| < 1/2)$, we see that (27) is well defined as a stationary MA process with zero mean and finite variance. It is worth noting that (27) has an explicit covariance function,

$$\mathrm{Cov}(X(0), X(t)) = \frac{\Gamma(t+d)}{\Gamma(t-d+1)} \frac{\Gamma(1-2d)}{\Gamma(1-d)\Gamma(d)}, \quad t \in \mathbb{N}, \tag{28}$$

and a spectral density,

$$f(x) = (2\pi)^{-1}|1 - \mathrm{e}^{-\mathrm{i}x}|^{-2d}, \quad x \in \Pi, \tag{29}$$

which explodes or vanishes at the origin $x = 0$ as $(2\pi)^{-1}|x|^{-2d}$ depending on the sign of d.

Invertibility and properties of fractional operators on $\mathbb{Z}^\nu$. The definition in (26) can be extended to the case of $\nu \geq 1$ by replacing the shift $Tg(t) = g(t-1)$ by the transition operator

$$Tg(\boldsymbol{t}) = \sum_{\boldsymbol{u} \in \mathbb{Z}^\nu} g(\boldsymbol{t} + \boldsymbol{u})p(\boldsymbol{u}) = \mathrm{E}g(S_1 + \boldsymbol{t}), \quad \boldsymbol{t} \in \mathbb{Z}^\nu, \tag{30}$$

of a general random walk, $\{S_j; j \ge 0\}$, on $\mathbb{Z}^\nu$, starting at $S_0 = \mathbf{0}$, with the (1-step) probabilities $p = \{p(\boldsymbol{u}) := \mathrm{P}(S_1 = \boldsymbol{u}); \boldsymbol{u} \in \mathbb{Z}^\nu\}$. The operator $T - I$ is called the generator of the above random walk and has the property that $(T - I)g = 0$ for a constant $g = \text{const}$. Clearly, $T^j g(\boldsymbol{t}) = \sum_{\boldsymbol{u} \in \mathbb{Z}^\nu} g(\boldsymbol{t} + \boldsymbol{u}) p_j(\boldsymbol{u}) = \mathrm{E} g(S_j + \boldsymbol{t}), j = 1, 2, \ldots$, where $p_j(\boldsymbol{u}) := \mathrm{P}(S_j = \boldsymbol{u})$ are the j-step probabilities. Similarly to (26), we define the fractional powers $(I - T)^d, -1 < d < 1, d \neq 0$ of $I - T$ in (30) through the binomial expansion in (20), yielding

$$(I - T)^d g(\boldsymbol{t}) = \sum_{j=0}^{\infty} \psi_j(d) T^j g(\boldsymbol{t}) = \sum_{\boldsymbol{u} \in \mathbb{Z}^\nu} \tau(\boldsymbol{u}; d) g(\boldsymbol{t} - \boldsymbol{u}), \quad (31)$$

with coefficients

$$\tau(\boldsymbol{u}; d) := \sum_{j=0}^{\infty} \psi_j(d) p_j(\boldsymbol{u}), \quad (32)$$

expressed through the binomial coefficients $\psi_j(d)$ and random walk probabilities $p_j(\boldsymbol{u}) := \mathrm{P}(S_j = \boldsymbol{u})$; $p_0(\boldsymbol{u}) := \mathbb{I}(\boldsymbol{u} = \mathbf{0})$.

The first question concerning the fractional operators in (31) is the convergence of the series in (32). For $d > 0$, the latter series converges absolutely due to (22) and $p_j(\boldsymbol{s}) \le 1$. On the other hand, for $d < 0$, the series in (32) may diverge, and its convergence imposes certain conditions on d and the random walk.

Lemma 6. *(i) Let $0 < d < 1$. Then, the series in (32) converges for any $\boldsymbol{u} \in \mathbb{Z}^\nu$ and*

$$\tau(\mathbf{0}; d) > 0, \quad \tau(\boldsymbol{u}; d) < 0 \ (\boldsymbol{u} \neq \mathbf{0}) \quad \text{and} \quad \sum_{\boldsymbol{u} \in \mathbb{Z}^\nu} \tau(\boldsymbol{u}; d) = 0. \quad (33)$$

(ii) Let $-1 < d < 0$. Then, all summands on the r.h.s. of (32) are nonnegative and $\tau(\boldsymbol{u}; d) \in (0, \infty]$. Moreover, $\tau(\mathbf{0}; d) < \infty$ implies $\tau(\boldsymbol{u}; d) < \infty$ and

$$\sum_{\boldsymbol{u} \neq \mathbf{0}} \tau(\boldsymbol{u}; d) |\tau(-\boldsymbol{u}; -d)| \le \tau(\mathbf{0}; d) < \infty. \quad (34)$$

(iii) Let $0 < d < 1$ and $\tau(\mathbf{0}; -d) < \infty$. Then,

$$\sum_{\boldsymbol{s} \in \mathbb{Z}^\nu} \tau(\boldsymbol{s}; d) \tau(\boldsymbol{t} - \boldsymbol{s}; -d) = \mathbb{I}(\boldsymbol{t} = \mathbf{0}), \quad \boldsymbol{t} \in \mathbb{Z}^\nu. \quad (35)$$

Proof. (i) From (32) and (23), we get

$$\tau(\mathbf{0};d) = 1 + \sum_{j=1}^{\infty} \psi_j(d)p_j(\mathbf{0}) > 1 + \sum_{j=1}^{\infty} \psi_j(d) = 0 \tag{36}$$

since $p_j(\mathbf{0}) = 1 (\forall j \geq 1)$ is not possible. On the other hand, for $\boldsymbol{u} \neq \mathbf{0}$, we have $p_0(\boldsymbol{u}) = 0$ and

$$\tau(\boldsymbol{u};d) = \sum_{j=1}^{\infty} \psi_j(d)p_j(\boldsymbol{u}) < 0 \tag{37}$$

in view of (23).

(ii) Since $\psi_j(d)p_j(\boldsymbol{u}) \geq 0$ is obvious from (23), it suffices to show (34) since it implies $\tau(\boldsymbol{u};d) < \infty$ by (33). We have

$$\begin{aligned}
\Sigma_0 &:= \sum_{\boldsymbol{u}\neq\mathbf{0}} \tau(\boldsymbol{u};d)|\tau(-\boldsymbol{u};-d)) \\
&= \sum_{\boldsymbol{u}\neq\mathbf{0}} \sum_{j,k=1}^{\infty} \psi_j(d)(-\psi_k(-d))p_j(\boldsymbol{u})p_k(-\boldsymbol{u}) \\
&= \sum_{n=2}^{\infty}\sum_{j=1}^{n-1} \psi_j(d)(-\psi_{n-j}(-d)) \sum_{\boldsymbol{u}\neq\mathbf{0}} p_j(\boldsymbol{u})p_{n-j}(-\boldsymbol{u}),
\end{aligned}$$

where exchanging the order of summation is legitimate, as all the summands are nonnegative. Hence, using $\sum_{\boldsymbol{u}\neq\mathbf{0}} p_j(\boldsymbol{u})p_{n-j}(-\boldsymbol{u}) \leq p_n(\mathbf{0})$ and (25), we get

$$\begin{aligned}
\Sigma_0 &\leq \sum_{n=2}^{\infty} p_n(\mathbf{0}) \sum_{j=1}^{n-1} \psi_j(d)(-\psi_{n-j}(-d)) \\
&= \sum_{n=2}^{\infty} p_n(\mathbf{0})(\psi_n(d) + \psi_n(-d)) \\
&\leq \sum_{n=2}^{\infty} p_n(\mathbf{0})\psi_n(d) < \tau(\mathbf{0};d),
\end{aligned}$$

proving part (ii).

(iii) The convergence of the series in (35) and the equality follow as in (34):

$$\begin{aligned}\sum_{s\in\mathbb{Z}^\nu}\tau(\boldsymbol{s};d)\tau(\boldsymbol{t}-\boldsymbol{s};-d) &= \sum_{j,k=0}^{\infty}\psi_j(d)\psi_k(-d)\sum_{\boldsymbol{s}\in\mathbb{Z}^\nu}p_j(\boldsymbol{s})p_k(\boldsymbol{t}-\boldsymbol{s})\\ &= \sum_{n=0}^{\infty}p_n(\boldsymbol{t})\sum_{j=0}^{n}\psi_j(d)\psi_{n-j}(-d)\\ &= p_0(\boldsymbol{t}) = \mathbb{I}(\boldsymbol{t}=\boldsymbol{0}).\end{aligned}$$

Lemma 6 is proved. □

Remark 7. Let $0 < d < 1$. Then, the inequalities are strict: $\tau(\boldsymbol{u};d) < 0$ and $\tau(\boldsymbol{u};-d) > 0$ if $p_j(\boldsymbol{u}) > 0$ for some j, i.e., $\boldsymbol{u}$ is accessible from state $\mathbf{0}$. Moreover, if state $\mathbf{0}$ is transient, i.e., the probability of eventual return to $\mathbf{0}$ is strictly less than 1, which is equivalent to $\sum_{j=0}^{\infty}p_j(\mathbf{0}) < \infty$, then $\tau(\mathbf{0};-d) < \infty$.

As noted above, the main result of Section 3 of Chapter 3 is Theorem 8, which provides the conditions for square summability of the fractional coefficients in (32), in terms of the characteristic function of the random walk

$$\widehat{p}(\boldsymbol{x}) := \mathrm{E}\exp\{\mathrm{i}\langle\boldsymbol{x},S_1\rangle\},\quad \boldsymbol{x}\in[-\pi,\pi]^\nu\equiv\Pi^\nu.$$

Write $\widehat{f}$ for the Fourier transform of a function, $f:\mathbb{Z}^\nu\to\mathbb{R}$. For $r\in(0,1), d\in(-1,1)$, introduce the *tempered fractional operators*

$$(I-rT)^d g(\boldsymbol{t}) = \sum_{j=0}^{\infty}r^j\psi_j(d)T^j g(\boldsymbol{t}) = \sum_{\boldsymbol{u}\in\mathbb{Z}^\nu}\tau_r(\boldsymbol{u};d)g(\boldsymbol{t}+\boldsymbol{u}),\quad \boldsymbol{t}\in\mathbb{Z}^\nu,$$

with coefficients

$$\tau_r(\boldsymbol{u};d) := \sum_{j=0}^{\infty}r^j\psi_j(d)p_j(\boldsymbol{u}), \tag{38}$$

and the Fourier transform $\widehat{\tau}_r(\boldsymbol{x};d) = (1-r\widehat{p}(\boldsymbol{x}))^d$.

Theorem 8. *For* $-1 < d < 1$, *the following conditions are equivalent:*

$$\int_{\Pi^\nu} |1 - \widehat{p}(\boldsymbol{x})|^{-2|d|} \mathrm{d}\boldsymbol{x} < \infty, \tag{39}$$

$$\sum_{\boldsymbol{u} \in \mathbb{Z}^\nu} \tau(\boldsymbol{u}; -|d|)^2 < \infty. \tag{40}$$

Either of these conditions implies

$$\widehat{\tau}(\cdot; -|d|) = (1 - \widehat{p}(\cdot))^{-|d|} \quad \text{in } L^2(\Pi^\nu). \tag{41}$$

Moreover, for $0 < d < 1$, *the above conditions (39), (40), and (41) hold with* d *in place of* $-|d|$.

Proof. Let $0 < d < 1$. Firstly, we consider $\tau(\boldsymbol{u}; d)$ in (32). They satisfy $\sum_{\boldsymbol{u} \in \mathbb{Z}^\nu} |\tau(\boldsymbol{u}; d)| \le \sum_{j=0}^{\infty} |\psi_j(d)| < \infty$ because of (22) and $\sum_{\boldsymbol{u} \in \mathbb{Z}^\nu} p_j(\boldsymbol{u}) = 1$, with $0 \le p_j(\boldsymbol{u}) \le 1$. Then, $\sum_{\boldsymbol{u} \in \mathbb{Z}^\nu} \tau(\boldsymbol{u}; d)^2 < \infty$ is immediate. Moreover, we have the Fourier transform $\widehat{\tau}(\boldsymbol{x}; d) = \sum_{j=0}^{\infty} \psi_j(d) \widehat{p}_j(\boldsymbol{x})$, where $\widehat{p}_j(\boldsymbol{x}) = \widehat{p}(\boldsymbol{x})^j$ satisfies $|\widehat{p}(\boldsymbol{x})| \le 1$. We see that

$$\widehat{\tau}(\boldsymbol{x}; d) = (1 - \widehat{p}(\boldsymbol{x}))^d, \quad \boldsymbol{x} \in \Pi^\nu, \tag{42}$$

belongs to $L^2(\Pi^\nu)$.

Now, let us prove the implication (39) $\Rightarrow$ (40). We use the approximation by tempered fractional coefficients $\tau_r(\boldsymbol{u}; -d)$ in (38) as $r \nearrow 1$. We have that $\widehat{\tau}_r(\boldsymbol{x}; -d) = (1 - r\widehat{p}(\boldsymbol{x}))^{-d} \to (1 - \widehat{p}(\boldsymbol{x}))^{-d}$ a.e. as $r \nearrow 1$. Next, for $z \in \mathbb{C}$, $|z| \le 1$, $0 < r < 1$, the inequality $|1 - z| \le |1 - rz| + |rz - z| \le |1 - rz| + 1 - r$, where $1 - r \le 1 - |rz| \le |1 - rz|$, becomes $|1 - z| \le 2|1 - rz|$. Using it, we get the domination for all $0 < r < 1$, $\boldsymbol{x} \in \Pi^\nu$,

$$|\widehat{\tau}_r(\boldsymbol{x}; -d)| \le \frac{1}{|1 - r\widehat{p}(\boldsymbol{x})|^d} \le \frac{2^d}{|1 - \widehat{p}(\boldsymbol{x})|^d}$$

by a function in $L^2(\Pi^\nu)$ according to (39). Hence, by the DCT, $\widehat{\tau}_r(\cdot; -d) \to (1 - \widehat{p}(\cdot))^{-d}$, as $r \nearrow 1$ holds in $L^2(\Pi^\nu)$. As a consequence, $\widehat{\tau}_r(\cdot; -d)$, $0 < r < 1$, is a Cauchy sequence in $L^2(\Pi^\nu)$. By Parseval's theorem, the inverse Fourier transforms

$$\tau_r(\boldsymbol{u};-d) = \frac{1}{(2\pi)^\nu}\int_{\Pi^\nu} \mathrm{e}^{-\mathrm{i}\langle \boldsymbol{u},\boldsymbol{x}\rangle}\widehat{\tau}_r(\boldsymbol{x};-d)\mathrm{d}\boldsymbol{x}, \quad \boldsymbol{u}\in\mathbb{Z}^\nu,\ 0<r<1,$$

is a Cauchy sequence in $L^2(\mathbb{Z}^\nu)$, and so $\tau_r(\cdot;-d)$ converges in $L^2(\mathbb{Z}^\nu)$ to some $f\in L^2(\mathbb{Z}^\nu)$ as $r\nearrow 1$. This f must be $\tau(\cdot;-d)$ because $\tau_r(\boldsymbol{u};-d)\nearrow\tau(\boldsymbol{u};-d)$ as $r\nearrow 1$ for all $\boldsymbol{u}$. We conclude that $\tau(\cdot;-d)\in L^2(\mathbb{Z}^\nu)$, or (40).

Let us turn to the implication (40) $\Rightarrow$ (39). From (40) and $\tau_r(\boldsymbol{u};-d)\nearrow\tau(\boldsymbol{u};-d)$ for all $\boldsymbol{u}$, it follows that $\tau_r(\cdot;-d)\to\tau(\cdot;-d)$, as $r\nearrow 1$ holds in $L^2(\mathbb{Z}^\nu)$. By Parseval's theorem, $\widehat{\tau}_r(\cdot;-d) = (1-r\widehat{p}(\cdot))^{-d}$, $0<r<1$, is a Cauchy sequence in $L^2(\Pi^\nu)$. It follows that $\lim_{r\nearrow 1}\int_{\Pi^\nu}|\widehat{\tau}_r(\boldsymbol{x};-d)-g(\boldsymbol{x})|^2\mathrm{d}\boldsymbol{x}=0$ for some $g\in L^2(\Pi^\nu)$. We also have that $\lim_{r\nearrow 1}(1-r\widehat{p}(\boldsymbol{x}))^{-d}=(1-\widehat{p}(\boldsymbol{x}))^{-d}$ for each $\boldsymbol{x}\in\Pi^\nu$ such that $\widehat{p}(\boldsymbol{x})\neq 1$. Since $\mathrm{Leb}_\nu(\boldsymbol{x}\in\Pi^\nu:\widehat{p}(\boldsymbol{x})=1)=0$, referring to Lemma 2.3.2(a) in Ref. [149], we conclude that $g(\cdot)=(1-\widehat{p}(\cdot))^{-d}$ a.e., proving (39).

The above argument also proves (41). On the one hand, $\widehat{\tau}(\cdot;-d)$ is the limit of $\widehat{\tau}_r(\cdot;-d)$ in $L^2(\Pi^\nu)$ as $r\nearrow 1$ because $\tau_r(\cdot;-d)$ converges in $L^2(\mathbb{Z}^\nu)$ to $\tau(\cdot;-d)$ as $r\nearrow 1$. On the other hand, $\widehat{\tau}_r(\cdot;-d) = (1-r\widehat{p}(\cdot))^{-d}\to(1-\widehat{p}(\cdot))^{-d}$ in $L^2(\Pi^\nu)$ as $r\nearrow 1$. We conclude that $\widehat{\tau}(\cdot;-d)=(1-\widehat{p}(\cdot))$ a.e. Theorem 8 is proved. □

Asymptotics of fractional coefficients. Next, we turn to the asymptotics of 'fractional coefficients' $\tau(\boldsymbol{u};d)$ in (32) as $|\boldsymbol{u}|\to\infty$. We use the local limit theorem in Ref. [149] for the random walk probabilities $p_j(\boldsymbol{u})=\mathrm{P}(S_j=\boldsymbol{u})$. Following the latter work, we assume that

$$\mathrm{E}\mathrm{e}^{c|S_1|}<\infty \quad (\exists\, c>0) \quad \text{and } \{S_j\} \text{ is zero mean, aperiodic, and irreducible.} \tag{43}$$

For example, if S_1 is symmetric, i.e., $S_1\overset{\mathrm{d}}{=}-S_1$, and moreover, has finite support which contains $\mathbf{0}$, $\boldsymbol{e}_i$, $i=1,\dots,\nu$, the random walk satisfies our assumption (43). The conditions in (43) imply that the random walk has zero mean, $\mathrm{E}S_1=\sum_{\boldsymbol{u}\in\mathbb{Z}^\nu}\boldsymbol{u}p(\boldsymbol{u})=\mathbf{0}$, and an invertible covariance matrix,

$$\Gamma := \mathrm{E}S_1S_1'.$$

According to the classical (integral) CLT, the normalised sum $S_j/\sqrt{j}, j\to\infty$ approaches a Gaussian distribution on $\mathbb{R}^\nu$ with the density

$$\phi(\boldsymbol{z}) := \frac{1}{(2\pi)^{\nu/2}\sqrt{\det\Gamma}} \mathrm{e}^{-\langle \boldsymbol{z}, \Gamma^{-1}\boldsymbol{z}\rangle/2}, \quad \boldsymbol{z} \in \mathbb{R}^\nu.$$

Denote

$$\bar{p}_j(\boldsymbol{u}) := \frac{1}{(2\pi j)^{\nu/2}\sqrt{\det\Gamma}} \mathrm{e}^{-\langle \boldsymbol{u}, \Gamma^{-1}\boldsymbol{u}\rangle/2j}, \quad \boldsymbol{u} \in \mathbb{R}^\nu.$$

Lemma 9 ([149, Thm. 2.3.11]). *Under conditions (43), there exists $C > 0$ such that*

$$|p_j(\boldsymbol{u}) - \bar{p}_j(\boldsymbol{u})| \le C\bar{p}_j(\boldsymbol{u}) \left(\frac{1}{j^{1/2}} + \frac{|\boldsymbol{u}|^3}{j^2}\right), \quad \forall\ |\boldsymbol{u}| < j^2,\ \boldsymbol{u} \in \mathbb{Z}^\nu. \tag{44}$$

For 'very atypical' values $|S_j| > j$, we use the following bound [149, Prop. 2.1.2]: for any $k \ge 1$, there exists $C > 0$ such that

$$\mathrm{P}(|S_j| > z\sqrt{j}) \le Cz^{-k}, \quad \forall\ z > 0. \tag{45}$$

Proposition 10. *Let $p = \{p(\boldsymbol{u}); \boldsymbol{u} \in \mathbb{Z}^\nu\}$ satisfy (43). The coefficients in (32) are well defined for any $-(1 \wedge \frac{\nu}{2}) < d < 1, d \ne 0$ and satisfy*

$$\tau(\boldsymbol{u}; d) = (B_1(d) + o(1))\langle \boldsymbol{u}, \Gamma^{-1}\boldsymbol{u}\rangle^{-(\nu/2)-d}, \quad |\boldsymbol{u}| \to \infty, \tag{46}$$

where

$$B_1(d) := \frac{2^d\Gamma(d + (\nu/2))}{\pi^{\nu/2}\Gamma(-d)\sqrt{\det\Gamma}}.$$

Proof. Let us prove (46). Since Γ is positive-definite, $|\boldsymbol{u}|_\Gamma := \sqrt{\langle \boldsymbol{u}, \Gamma^{-1}\boldsymbol{u}\rangle}$, $\boldsymbol{u} \in \mathbb{R}^\nu$, is a norm. Note that it is equivalent to the Euclidean norm: $c_-|\boldsymbol{u}| \le |\boldsymbol{u}|_\Gamma \le c_+|\boldsymbol{u}|$ for some constants $0 < c_- < c_+ < \infty$ since any two norms are equivalent on a finite-dimensional real vector space. Using (32) for a large $K > 0$, decompose $|\boldsymbol{u}|_\Gamma^{\nu+2d}\tau(\boldsymbol{u};d) = \sum_{i=1}^3 J_i(\boldsymbol{u})$, where

$$J_1(\boldsymbol{u}) := |\boldsymbol{u}|_\Gamma^{\nu+2d}\Gamma(-d)^{-1} \sum_{j>|\boldsymbol{u}|_\Gamma^2/K} j^{-d-1}p_j(\boldsymbol{u}),$$

$$J_2(\boldsymbol{u}) := |\boldsymbol{u}|_\Gamma^{\nu+2d} \sum_{j>|\boldsymbol{u}|_\Gamma^2/K} (\psi_j(d) - \Gamma(-d)^{-1}j^{-d-1})p_j(\boldsymbol{u}),$$

$$J_3(\boldsymbol{u}) := |\boldsymbol{u}|_\Gamma^{\nu+2d} \sum_{0\le j\le|\boldsymbol{u}|_\Gamma^2/K} \psi_j(d)p_j(\boldsymbol{u}).$$

It suffices to show that

$$\lim_{K\to\infty}\lim_{|\boldsymbol{u}|\to\infty} J_1(\boldsymbol{u}) = B_1(d), \qquad \lim_{K\to\infty}\limsup_{|\boldsymbol{u}|\to\infty} J_i(\boldsymbol{u}) = 0, \quad i = 2,3. \tag{47}$$

To show the first relation in (47), use (44). We have $J_1(\boldsymbol{u}) = J_1'(\boldsymbol{u}) + J_1''(\boldsymbol{u})$, where, for each $K > 0$ fixed, the main term $J_1'(\boldsymbol{u})$ and the remainder term $J_1''(\boldsymbol{u})$ asymptotically behave when $|\boldsymbol{u}| \to \infty$ as

$$\begin{aligned}
J_1'(\boldsymbol{u}) &:= |\boldsymbol{u}|_\Gamma^{\nu+2d}\Gamma(-d)^{-1} \sum_{j>|\boldsymbol{u}|_\Gamma^2/K} j^{-d-1}\bar{p}_j(\boldsymbol{u}) \\
&= \frac{|\boldsymbol{u}|_\Gamma^{\nu+2d}}{(2\pi)^{\nu/2}\Gamma(-d)\sqrt{\det\Gamma}} \\
&\quad \times \int_0^\infty \mathbb{I}(|\boldsymbol{u}|_\Gamma^2/K < [y])[y]^{-d-1-(\nu/2)}\mathrm{e}^{-|\boldsymbol{u}|_\Gamma^2/2[y]}\mathrm{d}y \\
&\sim \frac{1}{(2\pi)^{\nu/2}\Gamma(-d)\sqrt{\det\Gamma}} \int_{1/K}^\infty x^{-d-1-(\nu/2)}\mathrm{e}^{-1/2x}\mathrm{d}x
\end{aligned}$$

and, for some constants $C, c > 0$,

$$\begin{aligned}
|J_1''(\boldsymbol{u})| &\le C|\boldsymbol{u}|_\Gamma^{\nu+2d}K^{3/2} \sum_{j>|\boldsymbol{u}|_\Gamma^2/K} j^{-d-3/2}\bar{p}_j(\boldsymbol{u}) \\
&\le C|\boldsymbol{u}|_\Gamma^{-1}K^{3/2} \int_0^\infty x^{-d-(3/2)-(\nu/2)}\mathrm{e}^{-c/x}\mathrm{d}x = o(1).
\end{aligned}$$

Hence, the first relation in (47) follows using $\int_0^\infty x^{-1-\tau}\mathrm{e}^{-1/x}dx = \Gamma(\tau), \tau > 0$. In view of (22), the same argument also proves the second relation in (47) for $i = 2$.

Consider (47) for $i = 3$. Split $J_3(\boldsymbol{u}) = J_3''(\boldsymbol{u}) + J_3'(\boldsymbol{u})$ into two sums over $j > 0$, where $j^2 \le |\boldsymbol{u}|$ and $j^2 > |\boldsymbol{u}|$, respectively. In the sum $J_3'(\boldsymbol{u})$, we also have that $j \le |\boldsymbol{u}|_\Gamma^2/K \le |\boldsymbol{u}|^2$, and Lemma 9 entails the bound

$$p_j(\boldsymbol{u}) \le C\bar{p}_j(\boldsymbol{u})\left(\frac{|\boldsymbol{u}|^3}{j^2}\right) \le C|\boldsymbol{u}|^3 j^{-(\nu/2)-2}\mathrm{e}^{-c|\boldsymbol{u}|^2/j}(j)^{3/2}$$

for some constants $C, c > 0$. Hence,

$$\begin{aligned}|J_3'(\boldsymbol{u})| &\le C|\boldsymbol{u}|^{\nu+2d+3}\int_0^{|\boldsymbol{u}|^2}[y]^{-d-3-(\nu/2)}\mathrm{e}^{-c|\boldsymbol{u}|^2/[y]}dy\\ &\le C|\boldsymbol{u}|^{-1}\int_0^1 x^{-d-3-(\nu/2)}\mathrm{e}^{-c/x}\mathrm{d}x = o(1)\end{aligned}$$

since the last integral converges for any d. Finally, by (45), given large enough $k > 0$, there exists $C > 0$ such that $p_j(\boldsymbol{u}) \le Cj^{k/2}/|\boldsymbol{u}|^k$, which implies $J_3''(\boldsymbol{u}) = o(1)$. This proves (47) and completes the proof of Proposition 10. □

Lemma 9 does not apply to the simple random walk (which is not aperiodic), in which case the local CLT takes a somewhat different form; see Ref. [149, Thm. 2.1.3]. The application of the latter result and the argument in the proof of Proposition 10 yields the following result.

Proposition 11. *Let $p(\boldsymbol{e}_j) = p(-\boldsymbol{e}_j) = \frac{1}{2\nu}$, $j = 1, \ldots, \nu$. The coefficients in (32) are well defined for any $-(1 \wedge \frac{\nu}{2}) < d < 1, d \ne 0$ and satisfy*

$$\tau(\boldsymbol{u}; d) = (B(d) + o(1))|\boldsymbol{u}|^{-\nu-2d}, \quad |\boldsymbol{u}| \to \infty,$$

where

$$B(d) := \frac{2^d\Gamma(d + (\nu/2))}{\nu^d\Gamma(-d)}.$$

Proposition 10 and Lemma 9 do not apply to random walks with non-zero mean as in Example 19 (fractional heat operator), in which case the fractional coefficients exhibit an anisotropic behaviour different from (46). Such behaviour is described in the following proposition. We assume that the underlying random walk factorises into a deterministic drift by 1 in the direction of $-\boldsymbol{e}_1$ and a random walk on $\mathbb{Z}^{\nu-1}$ as in Lemma 9:

$$p(\boldsymbol{u}) = \begin{cases} 1-\theta, & \boldsymbol{u} = -\boldsymbol{e}_1, \\ \theta\tilde{q}(\tilde{\boldsymbol{u}}), & \boldsymbol{u} = -\boldsymbol{e}_1 + (0, \tilde{\boldsymbol{u}}), \end{cases} \tag{48}$$

where $\theta \in (0,1)$ and $\tilde{q}(\tilde{\boldsymbol{u}})$ is a probability distribution concentrated on $\boldsymbol{u} = (u_2, \dots, u_\nu) \in \mathbb{Z}^{\nu-1}$ such that $\tilde{\boldsymbol{u}} \neq \mathbf{0}$. Write $\{S_j; j \geq 0\}$ for the random walk starting at $\mathbf{0}$ with the j-step probabilities $\mathrm{P}(\tilde{S}_j = \tilde{\boldsymbol{u}} | \tilde{S}_0 = \mathbf{0}) =: \tilde{q}_j(\tilde{\boldsymbol{u}})$, $j = 0, 1, \dots$, such that $\tilde{q}_1(\tilde{\boldsymbol{u}}) := \tilde{q}(\tilde{\boldsymbol{u}})$, $\tilde{\boldsymbol{u}} \in \mathbb{Z}^{\nu-1}$. In order to apply Lemma 9, we make a similar assumption for (43):

$$\mathrm{E}e^{c|\tilde{S}_1|} < \infty \quad (\exists\, c > 0) \quad \text{and } \{\tilde{S}_j\} \text{ is zero mean and irreducible} \tag{49}$$

and we denote $\tilde{\Gamma} := \mathrm{E}\tilde{S}_1\tilde{S}_1'$ as the respective covariance matrix. Let

$$\rho(\boldsymbol{x}) := \left(x_1^2 + \langle \tilde{\boldsymbol{x}}, \tilde{\Gamma}^{-1}\tilde{\boldsymbol{x}}\rangle^2\right)^{1/2}, \quad \boldsymbol{x} = (x_1, \tilde{\boldsymbol{x}}) \in \mathbb{R}^\nu,$$

be a positive function on $\mathbb{R}^\nu$ satisfying the homogeneity property: $\rho(\lambda x_1, \lambda^{1/2}\tilde{\boldsymbol{x}}) = \lambda\rho(\boldsymbol{x})$, $\forall \lambda > 0$. As in Example 19, fractional coefficients for $p(\boldsymbol{u})$ in (48) can be written as

$$\tau(-\boldsymbol{u}; d) = \psi_{u_1}(d) p_{u_1}(-\boldsymbol{u}) \mathbb{I}(u_1 \geq 0), \quad \boldsymbol{u} = (u_1, \tilde{\boldsymbol{u}}) \in \mathbb{Z}^\nu. \tag{50}$$

Proposition 12. *Let $p = \{p(\boldsymbol{u}); \boldsymbol{u} \in \mathbb{Z}^\nu\}$ satisfy (49) and $\theta \in (0,1)$. Then,*

$$\tau(-\boldsymbol{u}; d) = \frac{u_1^{-d-(\nu+1)/2}}{\Gamma(-d)(2\pi\theta)^{(\nu-1)/2}\sqrt{\det\tilde{\Gamma}}} \exp\left\{ -\frac{\langle \tilde{\boldsymbol{u}}, \tilde{\Gamma}^{-1}\tilde{\boldsymbol{u}}\rangle}{2\theta u_1} \right\} (1 + o(1)) \tag{51}$$

as $u_1 \to \infty$ and $|\tilde{\boldsymbol{u}}| \to \infty$, $|\tilde{\boldsymbol{u}}| = o(u_1^{2/3})$. We also have that

$$\tau(-\boldsymbol{u}; d) = \rho(\boldsymbol{u})^{-d-(\nu+1)/2} \left(L_0\left(\frac{u_1}{\rho(\boldsymbol{u})}\right) + o(1) \right), \quad |\boldsymbol{u}| \to \infty, \tag{52}$$

where $\ell_0(z), z \in [-1,1]$ *is a continuous function on* $[-1,1]$ *given by*

$$\ell_0(z) := \frac{z^{-d-(\nu+1)/2}}{\Gamma(-d)(2\pi\theta)^{(\nu-1)/2}\sqrt{\det\tilde{\Gamma}}} \exp\left\{-(1/2\theta)\sqrt{(1/z)^2-1}\right\}$$

for $z \in (0,1]$ *and equal 0 for* $z \in [-1,0]$.

Proof. Consider the following j-step probabilities of a random walk on $\mathbb{Z}^{\nu-1}$, starting at $\mathbf{0}$: $q_j(\tilde{\boldsymbol{u}}) := p_j(\boldsymbol{u})$, where $\boldsymbol{u} = (-j, \tilde{\boldsymbol{u}})$ for $\tilde{\boldsymbol{u}} \in \mathbb{Z}^{\nu-1}$, $j = 0,1,\ldots$. Let us estimate them by $\bar{q}_j(\tilde{\boldsymbol{u}}) := (2\pi j)^{-(\nu-1)/2}(\det\Gamma)^{-1/2} \times \exp\{-\langle\tilde{\boldsymbol{u}}, \Gamma^{-1}\tilde{\boldsymbol{u}}\rangle/2j\}$, where Γ is the covariance matrix of the 1-step distribution $q_1(\tilde{\boldsymbol{u}})$, $\boldsymbol{u} \in \mathbb{Z}^{\nu-1}$. Note that $\Gamma = \theta\tilde{\Gamma}$. By Lemma 9,

$$|q_j(\tilde{\boldsymbol{u}}) - \bar{q}_j(\tilde{\boldsymbol{u}})| \le C\bar{q}_j(\tilde{\boldsymbol{u}})\left(\frac{1}{j^{1/2}} + \frac{|\tilde{\boldsymbol{u}}|^3}{j^2}\right), \quad \forall\ |\tilde{\boldsymbol{u}}| < j^2,\ \tilde{\boldsymbol{u}} \in \mathbb{Z}^{\nu-1}. \tag{53}$$

Relation (51) follows directly from (50), (53), and (22). Relation (52) can be written as

$$\rho(\boldsymbol{u})^{d+(\nu+1)/2}\tau(-\boldsymbol{u}; d) - \ell_0\left(\frac{u_1}{\rho(\boldsymbol{u})}\right) \to 0, \quad |\boldsymbol{u}| \to \infty. \tag{54}$$

The asymptotics in (54) is immediate from (51) for $|\boldsymbol{u}|$ tending to ∞ as in (51). The general case of (54) also follows from (51) using the continuity of L_0. For $\nu = 2$, the details can be found in Ref. [209, proof of Prop. 4.1]. □

Remark 13. The approximation in (51) is compared with the kernel

$$h_{c,-d}(\boldsymbol{t}) = c_1\, t_1^{-d-\frac{1+\nu}{2}} \exp\left\{-ct_1 - \frac{|\tilde{\boldsymbol{t}}|^2}{4t_1}\right\}\mathbb{I}(t_1 > 0), \quad \boldsymbol{t} = (t_1, \tilde{\boldsymbol{t}}) \in \mathbb{R}^\nu \tag{55}$$

of the fractional heat operator $(c + \partial_1 - \widetilde{\Delta})^d, \partial_1 - \widetilde{\Delta} := \partial/\partial t_1 - \sum_{i=2}^{\nu} \partial^2/\partial t_i^2$ for all $c > 0, d < 0$ and some $c_1 \in \mathbb{R}$. For $\nu = 2$, Ref. [211, (3.7)] has recently derived the analytic form in (55) of the kernel from the absolute square of its Fourier transform:

$$|\widehat{h}_{c,-d}(\boldsymbol{z})|^2 = \left|\int_{\mathbb{R}^\nu} \mathrm{e}^{\mathrm{i}\langle \boldsymbol{z},\boldsymbol{t}\rangle} h_{c,-d}(\boldsymbol{t})\mathrm{d}\boldsymbol{t}\right|^2$$
$$= c_1^2(4\pi)^{\nu-1}\Gamma(-d)^2(z_1^2 + (c+|\tilde{\boldsymbol{z}}|^2)^2)^d,$$
$$\boldsymbol{z} = (z_1, \tilde{\boldsymbol{z}}) \in \mathbb{R}^\nu, \tag{56}$$

which is an implicit definition of this kernel in Ref. [134]. Similarly to the derivations in Ref. [211], for $\nu \geq 2$, the table of integrals [106, 3.944.5-6] gives

$$\widehat{h}_{c,-d}(\boldsymbol{z}) = c_1 \int_0^\infty \mathrm{e}^{\mathrm{i}z_1t_1 - ct_1} t_1^{d-\frac{1+\nu}{2}} \mathrm{d}t_1 \int_{\mathbb{R}^{\nu-1}} \exp\left\{\mathrm{i}\langle \tilde{\boldsymbol{z}}, \tilde{\boldsymbol{t}}\rangle - \frac{|\tilde{\boldsymbol{t}}|^2}{4t_1}\right\} \mathrm{d}\tilde{\boldsymbol{t}}$$
$$= c_1(4\pi)^{\frac{\nu-1}{2}} \int_0^\infty \mathrm{e}^{\mathrm{i}z_1t_1 - t_1(c+|\tilde{\boldsymbol{z}}|^2)} t_1^{-d-1}\mathrm{d}t_1$$
$$= c_1(4\pi)^{\frac{\nu-1}{2}}\Gamma(-d)\left(z_1^2 + (c+|\tilde{\boldsymbol{z}}|^2)^2\right)^{\frac{d}{2}}$$
$$\times \exp\left\{-\mathrm{i}d \arctan\left(\frac{z_1}{c+|\tilde{\boldsymbol{z}}|^2}\right)\right\},$$

yielding (56).

Finally, the tempered fractional coefficients in (38) are summable: $\sum_{\boldsymbol{u}\in\mathbb{Z}^\nu} |\tau_r(\boldsymbol{u};d)| \leq \sum_{j=0}^\infty r^j|\psi_j(d)| \leq 2(1-r)^{-|d|} < \infty$ for any $d \in (-1,1), r \in (0,1)$ and *any* random walk $\{S_j\}$. Assuming the existence of an exponential moment, $\mathrm{E}\mathrm{e}^{\kappa|S_1|} < \infty$ for some $\kappa > 0$, (38) decay exponentially:

$$|\tau_r(\boldsymbol{u};d)| \leq C\mathrm{e}^{-c|\boldsymbol{u}|}, \quad \boldsymbol{u} \in \mathbb{Z}^\nu, \tag{57}$$

for some $C, c > 0$. Indeed, Markov's inequality gives $r^j|\psi_j(d)|p_j(\boldsymbol{u}) \leq \mathrm{P}(|S_j| \geq |\boldsymbol{u}|) \leq \mathrm{e}^{-\kappa|\boldsymbol{u}|}\mathrm{E}\mathrm{e}^{\kappa|S_j|} \leq \mathrm{e}^{-\kappa|\boldsymbol{u}|}(\mathrm{E}\mathrm{e}^{\kappa|S_1|})^j \leq \mathrm{e}^{-(\kappa/2)|\boldsymbol{u}|}$ for any $0 \leq j < c|\boldsymbol{u}|$ and large enough $|\boldsymbol{u}|$. Moreover, $\sum_{j\geq c|\boldsymbol{u}|} r^j|\psi_j(d)|p_j(\boldsymbol{u}) \leq \sum_{j\geq c|\boldsymbol{u}|} r^j = r^{c|\boldsymbol{u}|}/(1-r)$, proving (57).

3 Fractionally Integrated Random Fields on $\mathbb{Z}^\nu$: Examples

A fractionally integrated MA RF X on $\mathbb{Z}^\nu$ is defined as a stationary solution of the difference equation

$$(I - T)^d X(\boldsymbol{t}) = \varepsilon(\boldsymbol{t}), \quad \boldsymbol{t} \in \mathbb{Z}^\nu. \tag{58}$$

Corollary 15 obtains the conditions for the existence of a stationary solution of (58) given by the inverse operator $X(\boldsymbol{t}) = (I - T)^{-d}\varepsilon(\boldsymbol{t})$, which are detailed in Examples 18 and 19 for the fractional Laplacian and fractional heat operators.

Let $\{\varepsilon(\boldsymbol{t}); \boldsymbol{t} \in \mathbb{Z}^\nu\}$ be a weak white noise; see Section 1 of Chapter 1, (3). Consider fractional equations on $\mathbb{Z}^\nu$:

$$(I - T)^d X(\boldsymbol{t}) = \sum_{\boldsymbol{s} \in \mathbb{Z}^\nu} \tau(\boldsymbol{s}; d) X(\boldsymbol{t} + \boldsymbol{s}) = \varepsilon(\boldsymbol{t}), \tag{59}$$

$$(I - rT)^d X(\boldsymbol{t}) = \sum_{\boldsymbol{s} \in \mathbb{Z}^\nu} \tau_r(\boldsymbol{s}; d) X(\boldsymbol{t} + \boldsymbol{s}) = \varepsilon(\boldsymbol{t}), \quad 0 < r < 1, \quad \boldsymbol{t} \in \mathbb{Z}^\nu, \tag{60}$$

whose solutions are obtained by inverting these operators, as described later. In the case when $\{\varepsilon(\boldsymbol{t}); \boldsymbol{t} \in \mathbb{Z}^\nu\}$ is a standardised i.i.d. sequence, these solutions are written as MA RFs with discrete arguments; however, the discussion in this section uses only second-moment properties of RFs and allows for weak white noise in equations (59)–(60).

Definition 14. Let $d \in (-1, 1)$ and $\tau(\boldsymbol{u}; \pm d)$ in (32) be well defined. By a stationary solution of equation (59) (respectively, (60)), we mean a stationary RF X such that for each $\boldsymbol{t} \in \mathbb{Z}^\nu$, the series in (59) converges in mean square and (59) holds (respectively, the series in (60) converges in mean square and (60) holds).

Corollary 15. *(i) Let $-1 < d < 1$ and (39) hold. Then,*

$$X(\boldsymbol{t}) = (I - T)^{-d}\varepsilon(\boldsymbol{t}) = \sum_{\boldsymbol{u} \in \mathbb{Z}^\nu} \tau(\boldsymbol{u}; -d)\varepsilon(\boldsymbol{t} + \boldsymbol{u}), \quad \boldsymbol{t} \in \mathbb{Z}^\nu, \tag{61}$$

is a stationary solution of equation (59). The RF in (61) has zero mean and spectral density

$$f(\boldsymbol{x}) = (2\pi)^{-\nu}|1 - \widehat{p}(\boldsymbol{x})|^{-2d}, \quad \boldsymbol{x} \in \Pi^{\nu}. \tag{62}$$

For $0 < d < 1$, *(39) is also necessary for the existence of the above* X.
(ii) Let $0 < d < 1$ *and (39) hold. Then,* X *in (61) has the coefficient LRD property. Moreover, it has nonnegative covariance function* $\mathrm{Cov}(X(\mathbf{0}), X(\boldsymbol{t})) \geq 0$ *and* $\sum_{\boldsymbol{t}\in\mathbb{Z}^{\nu}} \mathrm{Cov}(X(\mathbf{0}), X(\boldsymbol{t})) = \infty$.

(iii) Let $-1 < d < 0$ *and (39) hold. Then,* X *in (61) has the coefficient ND property and* $\sum_{\boldsymbol{t}\in\mathbb{Z}^{\nu}} \mathrm{Cov}(X(\mathbf{0}), X(\boldsymbol{t})) = 0$.

(iv) Let $-1 < d < 1, 0 < r < 1$. *Then,*

$$X(\boldsymbol{t}) = (I - rT)^{-d}\varepsilon(\boldsymbol{t}) = \sum_{\boldsymbol{u}\in\mathbb{Z}^{\nu}} \tau_r(\boldsymbol{u}; -d)\varepsilon(\boldsymbol{t} + \boldsymbol{u}), \quad \boldsymbol{t} \in \mathbb{Z}^{\nu}, \tag{63}$$

is a stationary solution of equation (60). Moreover, X *in (63) has the coefficient SRD property and* $\sum_{\boldsymbol{t}\in\mathbb{Z}^{\nu}} |\mathrm{Cov}(X(\mathbf{0}), X(\boldsymbol{t}))| < \infty$, $\sum_{\boldsymbol{t}\in\mathbb{Z}^{\nu}} \mathrm{Cov}(X(\mathbf{0}), X(\boldsymbol{t})) = (1 - r)^{-2d} > 0$.

Proof. (i) Let $0 < d < 1$. X in (61) is well defined if and only if (40) holds, which is therefore a necessary condition. Let us show that X in (61) is a stationary solution of (59). We use the spectral representation of white noise

$$\varepsilon(\boldsymbol{t}) = \int_{\Pi^{\nu}} \mathrm{e}^{\mathrm{i}\langle \boldsymbol{t}, \boldsymbol{x}\rangle} Z(\mathrm{d}\boldsymbol{x}), \quad \boldsymbol{t} \in \mathbb{Z}^{\nu}, \tag{64}$$

where $Z(\mathrm{d}\boldsymbol{x})$ is a random complex-valued spectral measure on Π^{ν} with zero mean and variance $\mathrm{E}|Z(\mathrm{d}\boldsymbol{x})|^2 = \mathrm{d}\boldsymbol{x}/(2\pi)^{\nu}$. Then, $X(\boldsymbol{t})$ is written as

$$X(\boldsymbol{t}) = \int_{\Pi^{\nu}} \mathrm{e}^{\mathrm{i}\langle \boldsymbol{t}, \boldsymbol{x}\rangle} \widehat{\tau}(\boldsymbol{x} \ - d) Z(\mathrm{d}\boldsymbol{x}) = \int_{\Pi^{\nu}} \mathrm{e}^{\mathrm{i}\langle \boldsymbol{t}, \boldsymbol{x}\rangle} \frac{Z(\mathrm{d}\boldsymbol{x})}{(1 - \widehat{p}(\boldsymbol{x}))^{d}}; \tag{65}$$

see (41). Then, $(I - T)^d X(\boldsymbol{t}) = \int_{\Pi^{\nu}} \mathrm{e}^{\mathrm{i}\langle \boldsymbol{t}, \boldsymbol{x}\rangle} \sum_{\boldsymbol{s}\in\mathbb{Z}^{\nu}} \tau(\boldsymbol{s}; d) \mathrm{e}^{\mathrm{i}\langle \boldsymbol{s}, \boldsymbol{x}\rangle}(1 - \widehat{p}(\boldsymbol{x}))^{-d} Z(\mathrm{d}\boldsymbol{x}) = \varepsilon(\boldsymbol{t})$ follows by (42) and the absolute summability $\sum_{\boldsymbol{s}\in\mathbb{Z}^{\nu}} |\tau(\boldsymbol{s}; d)| < \infty$; see (33) and (37).

Next, let $-1 < d < 0$. Then, X in (61) is well defined and written as (65) due to $\sum_{\boldsymbol{s}\in\mathbb{Z}^\nu} |\tau(\boldsymbol{s}; -d)| < \infty$. We need to show that the series in (59) converges in mean square towards $\varepsilon(\boldsymbol{t})$ if and only if (39) or (40) hold. The latter convergence is written as

$$\lim_{M\to\infty} \mathrm{E}|s_M - \varepsilon(\boldsymbol{t})|^2 = 0, \quad \text{where} \quad s_M := \sum_{|\boldsymbol{s}|\le M} \tau(\boldsymbol{s}; d) X(\boldsymbol{t}+\boldsymbol{s}).$$

From (65),

$$\begin{aligned}
&\mathrm{E}|s_M - \varepsilon(\boldsymbol{t})|^2 \\
&= (2\pi)^{-\nu} \int_{\Pi^\nu} \Big| \sum_{|\boldsymbol{s}|\le M} \mathrm{e}^{\mathrm{i}\langle \boldsymbol{s},\boldsymbol{x}\rangle} \tau(\boldsymbol{s}; d) - (1-\widehat{p}(\boldsymbol{x}))^d \Big|^2 |1-\widehat{p}(\boldsymbol{x})|^{2|d|} \mathrm{d}\boldsymbol{x} \\
&\le C \int_{\Pi^\nu} \Big| \sum_{|\boldsymbol{s}|\le M} \mathrm{e}^{\mathrm{i}\langle \boldsymbol{s},\boldsymbol{x}\rangle} \tau(\boldsymbol{s}; d) - (1-\widehat{p}(\boldsymbol{x}))^d \Big|^2 \mathrm{d}\boldsymbol{x} \\
&= C \int_{\Pi^\nu} \Big| \sum_{|\boldsymbol{s}|> M} \mathrm{e}^{\mathrm{i}\langle \boldsymbol{s},\boldsymbol{x}\rangle} \tau(\boldsymbol{s}; d) \Big|^2 \mathrm{d}\boldsymbol{x} \\
&= C \sum_{|\boldsymbol{s}|>M} \tau(\boldsymbol{s}; -|d|)^2 \to 0 \quad (M\to\infty)
\end{aligned}$$

in view of (40). The spectral density in (62) is a consequence of (42) or (65). This proves part (i).

(ii) From (23) and (32), we see that $\tau(\boldsymbol{s}; -d) \ge 0$ are nonnegative and $\sum_{\boldsymbol{s}\in\mathbb{Z}^\nu} \tau(\boldsymbol{s}; -d) = \sum_{j=0}^\infty \psi_j(-d) = \infty$. Thus, $\mathrm{Cov}(X(\mathbf{0}), X(\boldsymbol{t})) = \sum_{\boldsymbol{s}\in\mathbb{Z}^\nu} \tau(\boldsymbol{s}; -d)\tau(\boldsymbol{t}+\boldsymbol{s}; -d) \ge 0$ and $\sum_{\boldsymbol{t}\in\mathbb{Z}^\nu} \mathrm{Cov}(X(\mathbf{0}), X(\boldsymbol{t})) = \infty$.

(iii) As in the proof of (i), we get that $\sum_{\boldsymbol{s}\in\mathbb{Z}^\nu} |\tau(\boldsymbol{s}; -d)| \le 1 + \sum_{j=1}^\infty \sum_{\boldsymbol{s}\in\mathbb{Z}^\nu} |\psi_j(-d)| p_j(\boldsymbol{s}) = 1 + \sum_{j=1}^\infty |\psi_j(-d)| = 2$, referring to (23), and $\sum_{\boldsymbol{s}\in\mathbb{Z}^\nu} \tau(\boldsymbol{s}; -d) = 0$, implying $\sum_{\boldsymbol{t}\in\mathbb{Z}^\nu} \mathrm{Cov}(X(\mathbf{0}), X(\boldsymbol{t})) = \sum_{\boldsymbol{t},\boldsymbol{s}\in\mathbb{Z}^\nu} \tau(\boldsymbol{s}; -d)\tau(\boldsymbol{t}+\boldsymbol{s}; -d) = 0$.

(iv) Using $\sum_{\boldsymbol{u}\in\mathbb{Z}^\nu} |\tau_r(\boldsymbol{u}; d)| < \infty$, $\sum_{\boldsymbol{u}\in\mathbb{Z}^\nu} \tau_r(\boldsymbol{u}; d) = \sum_{j=0}^\infty r^j \psi_j(d) = (1-r)^d$, the proof is similar as above. Corollary 15 is proved. □

The ARFIMA(0, d, 0) equation (26) is autoregressive since the best linear predictor (or conditional expectation in the Gaussian case) of $X(t)$ given the 'past' $X(s), s < t$ is a linear combination, $\sum_{j=1}^{\infty} \psi_j(d)X(t-j)$, of the 'past' observations due to the fact that $\mathrm{Cov}(X(s), \varepsilon(t)) = 0\,(s < t)$. For spatial equations as in (59) or (60), an analogous property given the 'past' $X(\boldsymbol{s}), \boldsymbol{s} \neq \boldsymbol{t}$ does not hold since $\mathrm{Cov}(X(\boldsymbol{s}), \varepsilon(\boldsymbol{t})) \neq 0\,(\boldsymbol{s} \neq \boldsymbol{t})$ as a rule. This issue is important in spatial statistics and has been discussed in the literature (see Refs. [34,35] and the references therein), distinguishing between 'simultaneous' and 'conditional autoregressive schemes'. The recent work Ref. [90] discusses some conditional autoregressive models with the LRD property. Recall $\mathbb{Z}_0^\nu := \mathbb{Z}^\nu \setminus \{\mathbf{0}\}$.

Definition 16. Let X be an RF with $\mathrm{E}X(\boldsymbol{t})^2 < \infty$ for each $\boldsymbol{t} \in \mathbb{Z}^\nu$. We say that X has:

(i) a simultaneous autoregressive representation with coefficients $b(\boldsymbol{s}), \boldsymbol{s} \in \mathbb{Z}_0^\nu$ if for each $\boldsymbol{t} \in \mathbb{Z}^\nu$,

$$X(\boldsymbol{t}) = \sum_{\boldsymbol{s} \in \mathbb{Z}_0^\nu} b(\boldsymbol{s})X(\boldsymbol{t} - \boldsymbol{s}) + \xi(\boldsymbol{t}),$$

where the series converges in mean square and the r.v.s $\xi(\boldsymbol{t}), \boldsymbol{t} \in \mathbb{Z}^\nu$ satisfy $\mathrm{Cov}(\xi(\boldsymbol{t}), \xi(\boldsymbol{s})) = 0\,(\forall\, \boldsymbol{s} \neq \boldsymbol{t})$.

(ii) a conditional autoregressive representation with coefficients $c(\boldsymbol{s}), \boldsymbol{s} \in \mathbb{Z}_0^\nu$ if for each $\boldsymbol{t} \in \mathbb{Z}^\nu$,

$$X(\boldsymbol{t}) = \sum_{\boldsymbol{s} \in \mathbb{Z}_0^\nu} c(\boldsymbol{s})X(\boldsymbol{t} - \boldsymbol{s}) + \eta(\boldsymbol{t}), \tag{66}$$

where the series converges in mean square and the r.v.s $\eta(\boldsymbol{t}), \boldsymbol{t} \in \mathbb{Z}^\nu$ satisfy $\mathrm{Cov}(\eta(\boldsymbol{t}), X(\boldsymbol{s})) = 0\,(\forall\, \boldsymbol{s} \neq \boldsymbol{t})$.

Corollary 17. *(i) Let $d \in (-1, 1)$ and X be a fractionally integrated RF in (61) and (39) holds. Then, X has a simultaneous autoregressive representation with coefficients $b(\boldsymbol{s}) = -\tau(-\boldsymbol{s}; d)/\tau(\mathbf{0}; d), \boldsymbol{s} \in \mathbb{Z}_0^\nu$ and $\xi(\boldsymbol{s}) = \varepsilon(\boldsymbol{s})/\tau(\mathbf{0}; d), \boldsymbol{s} \in \mathbb{Z}^\nu$.*

(ii) Let $d \in (0, 1)$, X be a fractionally integrated RF in (61) and (39) holds. Then, X has a conditional autoregressive representation with coefficients $c(\boldsymbol{s}) = -\gamma^*(\boldsymbol{s})/\gamma^*(\mathbf{0}), \boldsymbol{s} \in \mathbb{Z}_0^\nu$ and $\eta(\boldsymbol{s}) = \int_{\Pi^\nu} \mathrm{e}^{\mathrm{i}\langle \boldsymbol{s}, \boldsymbol{x}\rangle}(1 - \widehat{p}(-\boldsymbol{x}))^d Z(\mathrm{d}\boldsymbol{x})/\gamma^*(\mathbf{0})$, where $Z(\mathrm{d}\boldsymbol{x})$ is a

complex-valued random measure given in (64) with zero mean and variance $\mathrm{E}|Z(\mathrm{d}\boldsymbol{x})|^2 = \mathrm{d}\boldsymbol{x}/(2\pi)^\nu$ *and*

$$\gamma^*(\boldsymbol{s}) := \frac{1}{(2\pi)^\nu}\int_{\Pi^\nu} \mathrm{e}^{-\mathrm{i}\langle \boldsymbol{s},\boldsymbol{x}\rangle}|1-\widehat{p}(\boldsymbol{x})|^{2d}\mathrm{d}\boldsymbol{x}, \quad \boldsymbol{s}\in\mathbb{Z}^\nu.$$

(iii) Let $d\in(-1,1), 0<r<1$ *and* X *be a (tempered) fractionally integrated RF in (63). Then,* X *has a simultaneous autoregressive representation with* $b(\boldsymbol{s}) = -\tau_r(-\boldsymbol{s};d)/\tau_r(\boldsymbol{0};d), \xi(\boldsymbol{t}) = \varepsilon(\boldsymbol{t})/\tau_r(\boldsymbol{0};d)$ *and a conditional autoregressive representation with* $c(\boldsymbol{s}) = -\gamma_r^*(\boldsymbol{s})/\gamma_r^*(\boldsymbol{0})$, $\eta(\boldsymbol{t}) = \int_{\Pi^\nu}\mathrm{e}^{\mathrm{i}\langle \boldsymbol{t},\boldsymbol{x}\rangle}(1-r\widehat{p}(-\boldsymbol{x}))^d Z(\mathrm{d}\boldsymbol{x})/\gamma_r^*(\boldsymbol{0})$, *with the same* $Z(\mathrm{d}\boldsymbol{x})$ *as in part (ii) and*

$$\gamma_r^*(\boldsymbol{s}) := \frac{1}{(2\pi)^\nu}\int_{\Pi^\nu} \mathrm{e}^{-\mathrm{i}\langle \boldsymbol{s},\boldsymbol{x}\rangle}|1-r\widehat{p}(\boldsymbol{x})|^{2d}\mathrm{d}\boldsymbol{x}, \quad \boldsymbol{s}\in\mathbb{Z}^\nu.$$

Proof. (i) is obvious from Corollary 15 and (59), $\tau(\boldsymbol{0};d)\neq 0$.
(ii) By (39), $c(\boldsymbol{s})$ and $\eta(\boldsymbol{t})$ are well defined, $\eta(\boldsymbol{t})\in\mathbb{R}$ and $\mathrm{E}\eta(\boldsymbol{t})^2<\infty$. The orthogonality relation $\mathrm{E}X(\boldsymbol{t})\eta(\boldsymbol{s}) = 0\,(\boldsymbol{t}\neq\boldsymbol{s})$ follows from the spectral representations in (65), (64):

$$\begin{aligned}\mathrm{E}X(\boldsymbol{t})\eta(\boldsymbol{s}) &= \frac{1}{(2\pi)^\nu\gamma^*(\boldsymbol{0})}\int_{\Pi^\nu}\mathrm{e}^{\mathrm{i}\langle \boldsymbol{t}-\boldsymbol{s},\boldsymbol{x}\rangle}\frac{\overline{(1-\widehat{p}(-\boldsymbol{x}))^d}}{(1-\widehat{p}(\boldsymbol{x}))^d}\mathrm{d}\boldsymbol{x}\\ &= \frac{1}{(2\pi)^\nu\gamma^*(\boldsymbol{0})}\int_{\Pi^\nu}\mathrm{e}^{\mathrm{i}\langle \boldsymbol{t}-\boldsymbol{s},\boldsymbol{x}\rangle}\mathrm{d}\boldsymbol{x} = 0 \quad (\boldsymbol{t}\neq\boldsymbol{s}).\end{aligned}$$

It remains to show (66), including the convergence of the series. In view of the definition of $c(\boldsymbol{s})$, this amounts to showing

$$\sum_{\boldsymbol{s}\in\mathbb{Z}^\nu} X(\boldsymbol{t}-\boldsymbol{s})\gamma^*(\boldsymbol{s}) = \gamma^*(\boldsymbol{0})\eta(\boldsymbol{t}),$$

or, in spectral terms, the convergence of the Fourier series

$$\frac{1}{(1-\widehat{p}(\boldsymbol{x}))^d}\sum_{\boldsymbol{s}\in\mathbb{Z}^\nu}\mathrm{e}^{-\mathrm{i}\langle \boldsymbol{x},\boldsymbol{s}\rangle}\gamma^*(\boldsymbol{s}) = (1-\widehat{p}(-\boldsymbol{x}))^d = \frac{|1-\widehat{p}(-\boldsymbol{x})|^{2d}}{(1-\widehat{p}(\boldsymbol{x}))^d} \tag{67}$$

in $L^2(\Pi^\nu)$. Note that $\gamma^*(\boldsymbol{s}) = \mathrm{Cov}(X^*(\boldsymbol{0}), X^*(\boldsymbol{s}))$, where the RF $X^*(\boldsymbol{t}) := (1-T)^d\varepsilon(\boldsymbol{t})$, $\boldsymbol{t}\in\mathbb{Z}^\nu$, results from the application of the inverse operator. Since X^* has negative dependence, referring to (65) and the proof of Corollary 15 (iii), the covariances $\gamma^*(\boldsymbol{s}), \boldsymbol{s} \in \mathbb{Z}^\nu$ are absolutely summable. Therefore, the Fourier series on the l.h.s. of (67) converges uniformly in $\boldsymbol{x} \in \Pi^\nu$ to $|1 - \widehat{p}(-\boldsymbol{x})|^{2d}$, proving (67).

(iii) The proof is analogous to those of (and simpler than) (i)–(ii), using $\sum_{\boldsymbol{u} \in \mathbb{Z}^\nu} |\tau_r(\boldsymbol{u}; d)| < \infty$. □

Example 18 (Fractional Laplace RF). The (lattice) Laplace operator on $\mathbb{Z}^\nu$ is defined as

$$[\Delta]g(\boldsymbol{t}) := \frac{1}{2\nu} \sum_{j=1}^{\nu} (g(\boldsymbol{t} + \boldsymbol{e}_j) + g(\boldsymbol{t} - \boldsymbol{e}_j) - 2g(\boldsymbol{t})), \quad \boldsymbol{t} \in \mathbb{Z}^\nu \tag{68}$$

so that $[\Delta] = T - I$, where $Tg(\boldsymbol{t}) = \frac{1}{2\nu} \sum_{j=1}^{\nu} (g(\boldsymbol{t} + \boldsymbol{e}_j) + g(\boldsymbol{t} - \boldsymbol{e}_j)$ is the transition operator of the simple random walk $\{S_j; j = 0, 1, \ldots\}$ on $\mathbb{Z}^\nu$ with equal 1-step transition probabilities $1/2\nu$ to the nearest-neighbours $\boldsymbol{t} \to \boldsymbol{t} \pm \mathrm{e}_j, j = 1, \ldots, \nu$. Fractional Laplace RF can be defined as a stationary solution of the difference equation

$$(-[\Delta])^d X(\boldsymbol{t}) = \varepsilon(\boldsymbol{t}), \quad \boldsymbol{t} \in \mathbb{Z}^\nu, \tag{69}$$

with weak white noise on the r.h.s., written as an MA RF:

$$X(\boldsymbol{t}) = (-[\Delta])^{-d} \varepsilon(\boldsymbol{t}) = \sum_{\boldsymbol{u} \in \mathbb{Z}^\nu} \tau(\boldsymbol{u}; -d) \varepsilon(\boldsymbol{t} - \boldsymbol{u}). \tag{70}$$

We find that $\widehat{p}(\boldsymbol{x}) = (1/\nu) \sum_{j=1}^{\nu} \cos(x_j), \boldsymbol{x} = (x_1, \ldots, x_\nu) \in \Pi^\nu$ and

$$1 - \widehat{p}(\boldsymbol{x}) = \tfrac{1}{\nu} \textstyle\sum_{j=1}^{\nu} (1 - \cos(x_j)) \geq C|\boldsymbol{x}|^2$$

for some $C > 0$ and $1 - \widehat{p}(\boldsymbol{x}) \sim (1/2\nu)|\boldsymbol{x}|^2$ $(|\boldsymbol{x}| \to 0)$. Hence, condition (39) for (68) translates to

$$\int_{\Pi^\nu} \frac{\mathrm{d}\boldsymbol{x}}{|1 - \widehat{p}(\boldsymbol{x})|^{2|d|}} < \infty \quad \Longleftrightarrow \quad |d| < \frac{\nu}{4}. \tag{71}$$

The asymptotic behaviour of $\tau(\boldsymbol{u}; -d)$ is given in Proposition 11 and written as

$$\tau(\boldsymbol{u}; -d) \sim B(-d)|\boldsymbol{u}|^{2d-\nu-2d}, \qquad |\boldsymbol{u}| \to \infty,$$

with the r.h.s. representing the kernel of the fractional Laplacian $(-\Delta)^{-d}$ in $\mathbb{R}^\nu$ (see Example 22).

Example 19 (Fractional heat operator RF). For $\nu \geq 2$ and a (diffusive) parameter $0 < \theta < 1$, the (lattice) heat operator on $\mathbb{Z}^\nu$ is defined as [209]

$$\begin{aligned}\Delta_{1,2} g(\boldsymbol{t}) := g(\boldsymbol{t}) - g(\boldsymbol{t} - \boldsymbol{e}_1) - \frac{\theta}{2(\nu-1)} \sum_{j=2}^{\nu} (g(\boldsymbol{t} - \boldsymbol{e}_1 + \boldsymbol{e}_j) \\ + g(\boldsymbol{t} - \boldsymbol{e}_1 - \boldsymbol{e}_j) - 2g(\boldsymbol{t} - \boldsymbol{e}_1)). \qquad (72)\end{aligned}$$

Thus, $T - I = \Delta_{1,2}$ corresponds to the random walk on $\mathbb{Z}^\nu$ with distribution $p(-\boldsymbol{e}_1) = 1 - \theta, p(-\boldsymbol{e}_1 \pm \boldsymbol{e}_j) = \frac{\theta}{2(\nu-1)}, j = 2, \ldots, \nu$ (a particular case of the random walk in (48)). The fractional heat operator RF can be defined as a stationary solution of the difference equation

$$\Delta_{1,2}^d X(\boldsymbol{t}) = \varepsilon(\boldsymbol{t}), \quad \boldsymbol{t} \in \mathbb{R}^\nu, \qquad (73)$$

with weak white noise on the r.h.s., written as an MA RF:

$$X(\boldsymbol{t}) = \Delta_{1,2}^{-d} \varepsilon(\boldsymbol{t}) = \sum_{\boldsymbol{u} \in \mathbb{Z}^\nu} \tau(\boldsymbol{u}; -d) \varepsilon(\boldsymbol{t} - \boldsymbol{u}). \qquad (74)$$

We find that

$$\begin{aligned}|1 - \widehat{p}(\boldsymbol{x})|^2 = \theta^2 \left(\frac{1}{\nu - 1} \sum_{j=2}^{\nu} (1 - \cos(x_j)) \right)^2 \\ + 2(1 - \cos(x_1)) \left(1 - \frac{\theta}{\nu - 1} \sum_{j=2}^{\nu} (1 - \cos(x_j)) \right), \\ \boldsymbol{x} = (x_1, \ldots, x_\nu) \in \Pi^\nu.\end{aligned}$$

Using the Taylor expansion,

$$|1 - \widehat{p}(\boldsymbol{x})|^2 \sim \left(\frac{\theta}{2(\nu - 1)} \right)^2 |\tilde{\boldsymbol{x}}|^4 + x_1^2, \quad \boldsymbol{x} \to \boldsymbol{0}, \quad \tilde{\boldsymbol{x}} := (0, x_2, \ldots, x_\nu).$$

We also find that $|1-\widehat{p}(\boldsymbol{x})|$ is separated from 0 except for the origin: $|1-\widehat{p}(\boldsymbol{x})| \ge |\mathrm{Re}(1-\widehat{p}(\boldsymbol{x}))| = 1-\cos(x_1)\big(1-\frac{\theta}{\nu-1}\sum_{j=2}^{\nu}\cos(x_j)\big) \ge c|\boldsymbol{x}|^2, \boldsymbol{x}\in\Pi^\nu$ for some constant $c>0$. Therefore, with $y=|\tilde{\boldsymbol{x}}|$, we obtain that

$$\int_{\Pi^\nu}\frac{\mathrm{d}\boldsymbol{x}}{|1-\widehat{p}(\boldsymbol{x})|^{2d}} \le C\int_0^1\int_0^1\frac{y^{\nu-2}\mathrm{d}x\mathrm{d}y}{(x^2+y^4)^d}$$
$$\le C\int_0^1\frac{\mathrm{d}x}{x^{2d-(\nu-1)/2}} < \infty \quad \text{if} \quad d < \frac{\nu+1}{4}$$

and $\int_{\Pi^\nu}|1-\widehat{p}(\boldsymbol{x})|^{-2d}\mathrm{d}\boldsymbol{x} = \infty$ if $d \ge \frac{\nu+1}{4}$, providing the sufficient and necessary condition $|d| < \frac{\nu+1}{4}$ for the existence of a stationary solution of (73) given in (74). The asymptotics in (51) translates to

$$\tau(\boldsymbol{u};-d) \sim \frac{u_1^{d-\frac{1+\nu}{2}}}{\Gamma(-d)(2\pi\theta)^{(\nu-1)/2}}\exp\left\{-\frac{|\tilde{\boldsymbol{u}}|^2}{2\theta u_1}\right\}, \qquad u_1\to\infty, |\tilde{\boldsymbol{u}}|\to\infty,$$

where the r.h.s. represents the kernel of the fractional heat operator $(\partial/\partial_1-\theta\widetilde{\Delta})^d$ in $\mathbb{R}^\nu$; see Section 4.

Example 20 (Separately fractionally integrated RF). Let $T_i g(\boldsymbol{t}) := g(\boldsymbol{t}-\boldsymbol{e}_i), d_i \in (-1/2,1/2)\setminus\{0\}, i=1,\ldots,\nu$. We define a separately fractionally integrated RF $X=\{X(\boldsymbol{t});\boldsymbol{t}\in\mathbb{Z}^\nu\}$ as the stationary solution of the difference equation

$$(I-T_1)^{d_1}\cdots(I-T_2)^{d_\nu}X(\boldsymbol{t})$$
$$= \sum_{\boldsymbol{u}\in\mathbb{Z}_+^\nu}\prod_{j=1}^{\nu}\psi_{u_j}(d_j)X(\boldsymbol{t}-\boldsymbol{u}) = \varepsilon(\boldsymbol{t}), \quad \boldsymbol{t}\in\mathbb{Z}^\nu, \tag{75}$$

given by the unilateral MA representation

$$X(\boldsymbol{t}) = (I-T_1)^{-d_\nu}\cdots(I-T_2)^{-d_1}\varepsilon(\boldsymbol{t}) \tag{76}$$
$$= \sum_{\boldsymbol{u}\in\mathbb{Z}_+^\nu}\prod_{j=1}^{\nu}\psi_{u_j}(-d_j)\varepsilon(\boldsymbol{t}-\boldsymbol{u}), \quad \boldsymbol{t}\in\mathbb{Z}^\nu.$$

The RF in (76) has zero mean and explicit covariance function

$$\begin{aligned}\mathrm{Cov}(X(\boldsymbol{t}), X(\boldsymbol{0})) &= \prod_{j=1}^{\nu} \sum_{u_j \le t_j \wedge 0} \psi_{t_j - u_j}(-d_j)\psi_{-u_j}(-d_j) \\ &= \prod_{j=1}^{\nu} \frac{\Gamma(|t_j| + d_j)}{\Gamma(|t_j| - d_j + 1)} \frac{\Gamma(1 - 2d_j)}{\Gamma(1 - d_j)\Gamma(d_j)}, \qquad (77)\end{aligned}$$

which factorises into the product of ν ARFIMA(0, d, 0) covariance functions in (28). It follows that the separately fractionally integrated RF X in (76) has coefficient LRD if $\bigvee_{i=1}^{\nu} d_i > 0$ and ND if $\bigvee_{i=1}^{\nu} d_i < 0$.

Example 21 (Fractionally integrated time series models (case $\nu = 1$)). As noted above, the ARFIMA$(0, d, 0)$ process is a particular case of (61) corresponding to the backward shift $Tg(t) = g(t-1)$ or the deterministic random walk $t \to t-1$. Another fractionally integrated time series model is given in Example 18 and corresponds to the symmetric nearest-neighbour random walk on $\mathbb{Z}$ with probabilities 1/2. It is of interest to compare these two processes and their properties. Let $T_1 g(t) := g(t-1), T_2 g(t) := (1/2)(g(t+1) + g(t-1)), t \in \mathbb{Z}$ be the corresponding operators,

$$\begin{aligned} X_1(t) &:= (I - T_1)^{-d_1}\varepsilon(t) = \sum_{u=0}^{\infty} \psi_u(-d_1)\varepsilon(t-u), \\ X_2(t) &:= (I - T_2)^{-d_2}\varepsilon(t) = \sum_{u \in \mathbb{Z}} \tau(\boldsymbol{u}, -d_2)\varepsilon(t-u), \quad t \in \mathbb{Z}. \end{aligned}$$

Stationary processes X_1 and X_2 are well defined for $d_1 < 1/2$ and $d_2 < 1/4$; moreover, for $0 < d_1 < 1/2$ and $0 < d_2 < 1/4$, they are stationary solutions of respective equations $(I - T_1)^{d_1} X(t) = \varepsilon(t)$ and $(I - T_2)^{d_2} X(t) = \varepsilon(t)$. The spectral densities of X_1 and X_2 are given by

$$\begin{aligned} f_1(x) &= \frac{1}{|1 - \mathrm{e}^{-\mathrm{i}x}|^{2d_1}} = \frac{1}{2^{d_1}|1 - \cos(x)|^{d_1}}, \\ f_2(x) &= \frac{1}{|1 - (1/2)(\mathrm{e}^{-\mathrm{i}x} + \mathrm{e}^{\mathrm{i}x})|^{2d_2}} = \frac{1}{|1 - \cos(x)|^{2d_2}}. \end{aligned}$$

We see that when $d_1 = 2d_2$ processes X_1 and X_2 have the same second-order properties up to a multiplicative constant so that X_2 is a noncausal representation of ARFIMA$(0, 2d_2, 0)$.

4 Fractionally Integrated Random Fields on $\mathbb{R}^\nu$. Examples

The discussion in this section of fractionally integrated RF with continuous argument $\boldsymbol{t} \in \mathbb{R}^\nu$ is less systematic and focuses on classical partial differential operators – the Laplace operator $\Delta = \sum_{j=1}^{\nu} \partial^2/\partial^2 t_j$ and the Helmholtz operator $c^2 - \Delta, c > 0$. The well-known *fractional Lévy RF* $\mathcal{L}_H$ and *Matérn Lévy RF* $\mathcal{M}_{c,H}$ depending on parameter $H \in (0,1)$ are defined in Examples 22 and 23 and can be identified with fractional powers of these operators applied to the Lévy random measure M. We note that the RF $\mathcal{L}_H$ is nonstationary, whereas $\mathcal{M}_{c,H}$ is SRD. MA LRD RFs $X = (I - T_B)^\kappa \mathcal{L}_H$ and $X = (I - T_B)^{-\kappa} \mathcal{M}_{c,H}$ satisfying Assumption (A)$(d; \mathbb{R}^\nu)$, $(0 < d < \nu/4)$ are defined in Examples 25 and 27 by applying 'discrete fractional operators' analogous to (31) with T replaced by the transition operator T_B of the discrete-time standard Brownian random walk $\{B_j; j \in \mathbb{N}\}$, $\{B_t; t \geq 0\}$ a Brownian motion in $\mathbb{R}^\nu$. Conditions on parameters κ, H, c for the existence of the above 'discretely fractionally integrated' RFs and their properties are given in Propositions 26 and 28.

We start with three important models of fractional RFs with continuous argument $\boldsymbol{t} \in \mathbb{R}^\nu$.

Example 22. *Fractional Lévy RF* on $\mathbb{R}^\nu$ is defined as stochastic integral

$$\mathcal{L}_H(\boldsymbol{t}) := \int_{\mathbb{R}^\nu} \big(|\boldsymbol{t} + \boldsymbol{u}|^{H - \frac{\nu}{2}} - |\boldsymbol{u}|^{H - \frac{\nu}{2}}\big) M(\mathrm{d}\boldsymbol{u}), \quad \boldsymbol{t} \in \mathbb{R}^\nu, \tag{78}$$

w.r.t. the Lévy random measure M as in (2), where $0 < H < 1, H \neq \frac{\nu}{2}$ is a parameter. The fact that the integral in (78) is well defined and has finite variance, viz.,

$$\mathrm{E}|\mathcal{L}_H(\boldsymbol{t})|^2 = \int_{\mathbb{R}^\nu} \big||\boldsymbol{t} + \boldsymbol{u}|^{H - \frac{\nu}{2}} - |\boldsymbol{u}|^{H - \frac{\nu}{2}}\big|^2 \mathrm{d}\boldsymbol{u} < \infty,$$

follows from $\int_{|\boldsymbol{u}|<1} |\boldsymbol{u}|^{2(H-\frac{\nu}{2})} \mathrm{d}\boldsymbol{u} < \infty$ when $H > 0$, whereas for (large) $|\boldsymbol{u}| \ge 2|\boldsymbol{t}|$, we have $\big||\boldsymbol{t}+\boldsymbol{u}|^{H-\frac{\nu}{2}} - |\boldsymbol{u}|^{H-\frac{\nu}{2}}\big| \le C|\boldsymbol{t}||\boldsymbol{u}|^{H-1-\frac{\nu}{2}}$ by the Taylor expansion, where the exponent $2(\frac{\nu}{2}+1-H) = \nu + 2(1-H) > \nu$. In the Gaussian case of $M = W$, (78) is called a *fractional Brownian RF*, usually denoted by $\mathcal{B}_H(\boldsymbol{t}) := \int_{\mathbb{R}^\nu} \big(|\boldsymbol{t} + \boldsymbol{u}|^{H-\frac{\nu}{2}} - |\boldsymbol{u}|^{H-\frac{\nu}{2}}\big) W(\mathrm{d}\boldsymbol{u})$. Since the kernel and the noise in (78) are isotropic, it follows that the fractional Lévy RF is *isotropic*:

$$\{\mathcal{L}_H(\mathcal{O}\boldsymbol{t}); \boldsymbol{t} \in \mathbb{R}^\nu\} \overset{\mathrm{fdd}}{=} \{\mathcal{L}_H(\boldsymbol{t}); \boldsymbol{t} \in \mathbb{R}^\nu\} \tag{79}$$

for any orthogonal linear transformation $\mathcal{O} : \mathbb{R}^\nu \to \mathbb{R}^\nu$. The covariance of the fractional Lévy RF has a particularly simple and elegant form

$$\begin{aligned}
&\mathrm{E}\mathcal{L}_H(\boldsymbol{t})\mathcal{L}_H(\boldsymbol{s}) \\
&\quad = \int_{\mathbb{R}^\nu} \big(|\boldsymbol{t}+\boldsymbol{u}|^{H-\frac{\nu}{2}} - |\boldsymbol{u}|^{H-\frac{\nu}{2}}\big)\big(|\boldsymbol{s}+\boldsymbol{u}|^{H-\frac{\nu}{2}} - |\boldsymbol{u}|^{H-\frac{\nu}{2}}\big) \mathrm{d}\boldsymbol{u} \\
&\quad = c_{1,H}^2 \big(|\boldsymbol{t}|^{2H} + |\boldsymbol{s}|^{2H} - |\boldsymbol{t}-\boldsymbol{s}|^{2H}\big), \quad \boldsymbol{t}, \boldsymbol{s} \in \mathbb{R}^\nu,
\end{aligned} \tag{80}$$

with $c_{1,H}^2 := \mathrm{E}\mathcal{L}_H^2(\boldsymbol{e}_1)/2 < \infty$. Equation (80) agrees with the covariance of the fractional Brownian motion in the case of $\nu = 1$. To show (80), note that for $\boldsymbol{t} = \boldsymbol{s}$, it follows by the change of variable $\boldsymbol{u} \to |\boldsymbol{t}|\boldsymbol{u}$ in the integral and isometry in (79). In the general case, we have $\mathrm{E}\mathcal{L}_H(\boldsymbol{t})\mathcal{L}_H(\boldsymbol{s}) = \frac{1}{2}\big(\mathrm{E}\mathcal{L}_H^2(\boldsymbol{t}) + \mathrm{E}\mathcal{L}_H^2(\boldsymbol{s}) - \mathrm{E}(\mathcal{L}_H(\boldsymbol{t}) - \mathcal{L}_H(\boldsymbol{s}))^2\big)$, where $\mathrm{E}(\mathcal{L}_H(\boldsymbol{t}) - \mathcal{L}_H(\boldsymbol{s}))^2 = \mathrm{E}\mathcal{L}_H^2(\boldsymbol{t}-\boldsymbol{s}) = \mathrm{E}\mathcal{L}_H^2(\boldsymbol{e}_1)|\boldsymbol{t}-\boldsymbol{s}|^{2H}$ by the stationarity of increments property. The covariance in (80) admits a spectral representation:

$$\begin{aligned}
\mathrm{Cov}(\mathcal{L}_H(\boldsymbol{t}), \mathcal{L}_H(\boldsymbol{s})) = c_{2,H}^2 \int_{\mathbb{R}^\nu} (\mathrm{e}^{\mathrm{i}\langle \boldsymbol{t}, \boldsymbol{z}\rangle} - 1)(\mathrm{e}^{-\mathrm{i}\langle \boldsymbol{s}, \boldsymbol{z}\rangle} - 1)|\boldsymbol{z}|^{-2H-\nu} \mathrm{d}\boldsymbol{z}, \\
\boldsymbol{t}, \boldsymbol{s} \in \mathbb{R}^\nu,
\end{aligned} \tag{81}$$

with $c_{i,H}^2, i = 1, 2$ in (80) and (81) related by $c_{1,H}^2 = c_{2,H}^2 \int_{\mathbb{R}^\nu} |\mathrm{e}^{\mathrm{i}\langle \boldsymbol{e}_1, \boldsymbol{z}\rangle} - 1|^2 |\boldsymbol{z}|^{-2H-\nu} \mathrm{d}\boldsymbol{z}$. Since $\mathrm{E}\mathcal{L}_H(\lambda\boldsymbol{t})\mathcal{L}_H(\lambda\boldsymbol{s}) = \lambda^{2H}\mathrm{E}\mathcal{L}_H(\boldsymbol{t})\mathcal{L}_H(\boldsymbol{s}) \, \forall \lambda > 0$, in the Gaussian case of $\mathcal{L}_H = \mathcal{B}_H$, this implies that the fractional Brownian RF is H-self-similar. A fractional Lévy RF satisfies the fractional equation

$$(-\Delta)^{\frac{H}{2}+\frac{\nu}{4}}\mathcal{L}_H(\boldsymbol{t}) = c_{3,H}\dot{M}(\boldsymbol{t}) \tag{82}$$

with Laplacian $\Delta = \sum_{j=1}^{\nu} \partial^2/\partial^2 t_j$, $\dot{M}(\boldsymbol{t}) = M(\mathrm{d}\boldsymbol{t})/\mathrm{d}\boldsymbol{t}$ defined as a generalised process with $\dot{M}(\phi) = \int_{\mathbb{R}^\nu} \phi(\boldsymbol{t})M(\mathrm{d}\boldsymbol{t}), \phi \in L^2(\mathbb{R}^\nu)$, and some constant $c_{3,H} > 0$. We refer to Ref. [169] for a rigorous definition of the fractional Laplace operator and the various mathematical and probabilistic aspects of a fractional Brownian field. The constants $c_{i,H}, i = 1, 2, 3$ in (80)–(82) are explicitly written in terms of H and ν, particularly $c_{1,H}^2 = c_{3,H}2^{-2H-\nu}\pi^{-\nu/2}\Gamma(-H)/\Gamma(H+\frac{\nu}{2})$ [169, Thm. 3.3].

Example 23. The stationary *Matérn RF* on $\mathbb{R}^\nu$ is defined by

$$\mathcal{M}_{c,H}(\boldsymbol{t}) := \int_{\mathbb{R}^\nu} m_{c,H}(\boldsymbol{t}-\boldsymbol{u})M(\mathrm{d}\boldsymbol{u}), \quad \boldsymbol{t} \in \mathbb{R}^\nu, \tag{83}$$

where

$$m_{c,H}(\boldsymbol{t}) := \mu_{1,c,H}|c\boldsymbol{t}|^{\frac{H}{2}-\frac{\nu}{4}}K_{\frac{H}{2}-\frac{\nu}{4}}(c|\boldsymbol{t}|), \quad \boldsymbol{t} \in \mathbb{R}^\nu, \tag{84}$$

and $c, H > 0$ are parameters, $\mu_{1,c,H} := \frac{2^{1+\frac{H}{2}-\frac{\nu}{4}}c^{\frac{\nu}{2}-H}}{\Gamma(\frac{\nu}{4}-\frac{H}{2})}$, $K_\tau, \tau \in \mathbb{R}$ denotes the modified Bessel function of the second kind. Using the asymptotics $K_\tau(x) = K_{-\tau}(x) \sim \Gamma(\tau)2^{\tau-1}x^{-\tau}$ as $x \downarrow 0$ $(\tau > 0)$, we see that

$$m_{c,H}(\boldsymbol{t}) \sim |\boldsymbol{t}|^{H-\frac{\nu}{2}}, \quad \text{as} \quad c \downarrow 0,$$

and hence for $0 < H < 1$, the fractional Lévy RF is the limiting case of the increment $\mathcal{M}_{c,H}(\boldsymbol{t}) - \mathcal{M}_{c,H}(\boldsymbol{0})$ of the Matérn RF as $c \downarrow 0$. The RF in (83) is well defined for any $c, H > 0$ and has an explicit (Matérn) covariance function,

$$\begin{aligned} \mathrm{E}\mathcal{M}_{c,H}(\boldsymbol{0})\mathcal{M}_{c,H}(\boldsymbol{t}) &= \mu_{2,c,H}(c|\boldsymbol{t}|)^H K_H(c|\boldsymbol{t}|), \\ \mu_{2,c,H} &:= \mathrm{E}\mathcal{M}_{c,H}^2(\boldsymbol{0})\left(\frac{2^{1-H}}{\Gamma(H)}\right), \end{aligned} \tag{85}$$

which is widely used in spatial applications; see Refs. [44, 109, 114, 127, 265] and the references therein. Similarly to (82), the RF in (83) can be regarded as a solution to the fractional equation

$$(c^2 - \Delta)^{\frac{H}{2}+\frac{\nu}{4}} \mathcal{M}_{c,H}(\boldsymbol{t}) = \mu_{3,c,H} \dot{M}(\boldsymbol{t})$$

and has a bounded spectral density,

$$f(\boldsymbol{z}) = (2\pi)^{-\nu} |\widehat{m}_{c,H}(\boldsymbol{z})|^2 = \frac{\mu_{4,c,H}}{(c^2 + |\boldsymbol{z}|^2)^{H+\frac{\nu}{2}}}, \qquad \boldsymbol{z} \in \mathbb{R}^\nu, \tag{86}$$

with some constants $\mu_{i,c,H} > 0, i = 2, 3, 4$, which can be explicitly written in terms of c and H. Since $K_\tau(x)$ is positive and decays exponentially, $K_\tau(x) \sim \sqrt{\pi/2x}\, \mathrm{e}^{-x}$ $(x \to \infty)$, the Matérn RF is SRD, and the coefficients $m_{c,H}(\cdot)$ satisfy (A)($\mathbb{R}^\nu; 0$) property for any $c, H > 0$.

Example 24. *Fractional heat operator RF* on $\mathbb{R}^\nu$ can be defined as an MA RF in

$$\mathcal{H}_{c,d}(\boldsymbol{t}) := \int_{\mathbb{R}^\nu} h_{c,d}(\boldsymbol{t} - \boldsymbol{u}) M(\mathrm{d}\boldsymbol{u}), \qquad \boldsymbol{t} \in \mathbb{R}^\nu, \tag{87}$$

with spectral density

$$f(\boldsymbol{z}) = (2\pi)^{-\nu} |\widehat{h}_{c,d}(\boldsymbol{z})|^2 = \frac{1}{(2\pi)^\nu (z_1^2 + (c + |\tilde{\boldsymbol{z}}|^2)^2)^d}, \quad \boldsymbol{z} = (z_1, \tilde{\boldsymbol{z}}) \in \mathbb{R}^\nu, \tag{88}$$

where $c > 0$ and

$$d > \frac{\nu + 1}{4} \tag{89}$$

are parameters. (We use the notation $\tilde{\boldsymbol{z}} = (z_2, \dots, z_\nu) \in \mathbb{R}^{\nu-1}$ and identify $\tilde{\boldsymbol{z}}$ with $(0, z_2, \dots, z_\nu) \in \mathbb{R}^{\nu-1}$ by writing $\boldsymbol{z} = (z_1, \tilde{\boldsymbol{z}})$ as in Proposition 12.) Condition (89) is necessary and sufficient for the integrability of $f(\cdot)$ in (88). The RF in (87) can be regarded as a solution to the stochastic equation

$$(c + \partial_1 - \widetilde{\Delta})^d \mathcal{H}_{c,d}(\boldsymbol{t}) = \dot{M}(\boldsymbol{t}), \tag{90}$$

involving a fractional power of the heat operator $\partial_1 - \widetilde{\Delta} := \partial/\partial t_1 - \sum_{i=2}^{\nu} \partial^2/\partial t_i^2$; see Ref. [134]. The MA kernel

$$h_{c,d}(\boldsymbol{t}) = c_1\, t_1^{d-\frac{1+\nu}{2}} \exp\left\{-ct_1 - \frac{|\tilde{\boldsymbol{t}}|^2}{4t_1}\right\} \mathbb{I}(t_1 > 0), \quad \boldsymbol{t} = (t_1, \tilde{\boldsymbol{t}}) \in \mathbb{R}^\nu, \tag{91}$$

in (87)–(88) is identified from Remark 13, (55). We note that the kernel in (55) is bounded, nonnegative, and integrable:

$$\begin{aligned} \int_{\mathbb{R}^\nu} h_{c,d}(\boldsymbol{t})\mathrm{d}\boldsymbol{t} &= c_1(4\pi)^{\frac{\nu-1}{2}} \int_0^\infty t^{d-1}\mathrm{e}^{-ct}\mathrm{d}t \\ &= (c_1/c^d)(4\pi)^{\frac{\nu-1}{2}}\Gamma(d) \ \in \ (0,\infty), \end{aligned}$$

and hence the RF in (87) is SRD and satisfies (A)($\mathbb{R}^\nu$; 0) property. Contrary to Examples 22 and 23, the RF in (87) is anisotropic and has very different dependence properties in t_1 and $\tilde{\boldsymbol{t}}$.

As noted above, the RFs in Examples 22–24 are either SRD or nonstationary. A natural question is the existence of parametric fractionally integrated linear RFs with LRD/ND and continuous argument. We show that such stationary and isotropic Lévy MAs

$$X(\boldsymbol{t}) = \int_{\mathbb{R}^\nu} a(\boldsymbol{t} - \boldsymbol{u}) M(\mathrm{d}\boldsymbol{u}), \quad \boldsymbol{t} \in \mathbb{R}^\nu, \tag{92}$$

as in (2) can be constructed by applying to RFs in Examples 22 and 23 the isotropic 'discrete' fractional operator $(I - T_B)^\kappa$, where

$$T_B g(\boldsymbol{t}) := \int_{\mathbb{R}^\nu} p_1(\boldsymbol{s} - \boldsymbol{t}) g(\boldsymbol{s})\mathrm{d}\boldsymbol{s}, \quad \boldsymbol{t} \in, \mathbb{R}^\nu \tag{93}$$

is the transition operator of a (discrete-time) standard Brownian random walk $\{B_j; j \in \mathbb{N}\}$ on $\mathbb{R}^\nu$ with j-step transition probabilities

$$p_j(\boldsymbol{s} - \boldsymbol{t}) := (2\pi j)^{-\nu/2}\mathrm{e}^{-|\boldsymbol{s}-\boldsymbol{t}|^2/2j}, \quad \boldsymbol{t}, \boldsymbol{s} \in \mathbb{R}^\nu.$$

The operator T_B in (93) is well defined for each $g \in L^p(\mathbb{R}^\nu), p \ge 1$, and satisfies $T_B^j g(\boldsymbol{x}) := \int_{\mathbb{R}^\nu} p_j(\boldsymbol{y}-\boldsymbol{x}) g(\boldsymbol{y}) \mathrm{d}\boldsymbol{y}$. Thus,

$$(I - T_B)^\kappa g(\boldsymbol{t}) = \int_{\mathbb{R}^\nu} \tau_B(\boldsymbol{s};\kappa) g(\boldsymbol{s}+\boldsymbol{t}) \mathrm{d}\boldsymbol{s}, \quad \boldsymbol{t} \in \mathbb{R}^\nu, \tag{94}$$

where the kernel in (94) is given by the series

$$\tau_B(\boldsymbol{s};\kappa) := \textstyle\sum_{j=0}^\infty \psi_j(\kappa) p_j(\boldsymbol{s}) \tag{95}$$

involving the binomial coefficients $\psi_j(\kappa) = \Gamma(j-\kappa)/\Gamma(j+1)\Gamma(-\kappa), j \ge 0$ in (20). The 'continuous' kernel in (95) satisfies similar properties as the 'discrete' one in Propositions 10 and 11, viz., for any $-(1 \wedge \frac{\nu}{2}) < \kappa < 1, \kappa \neq 0$

$$\tau_B(\boldsymbol{s};\kappa) = (B(\kappa) + o(1))|\boldsymbol{s}|^{-\nu-2\kappa}, \quad |\boldsymbol{s}| \to \infty, \tag{96}$$

and $\int_{\mathbb{R}^\nu} \tau_B(\boldsymbol{s};\kappa) \mathrm{d}\boldsymbol{s} = 0, \kappa > 0$. Note that $\tau_B(\boldsymbol{s};\kappa)$ is an isotropic function on $\mathbb{R}^\nu$, depending only on the norm $|\boldsymbol{s}|$. Its Fourier transform is given by

$$\widehat{\tau}_B(\boldsymbol{z};\kappa) = \sum_{j=0}^\infty \psi_j(\kappa) \mathrm{e}^{-j|\boldsymbol{z}|^2/2} = (1 - \mathrm{e}^{-|\boldsymbol{z}|^2/2})^\kappa, \quad \boldsymbol{z} \in \mathbb{R}^\nu. \tag{97}$$

Example 25. *Discretely fractionally differenced fractional Lévy RF* is defined as

$$\begin{aligned} X(\boldsymbol{t}) &:= (I - T_B)^\kappa \mathcal{L}_H(\boldsymbol{t}) \\ &= \int_{\mathbb{R}^\nu} \tau_B(\boldsymbol{s};\kappa) \mathcal{L}_H(\boldsymbol{s}+\boldsymbol{t}) \mathrm{d}\boldsymbol{s}, \quad \boldsymbol{t} \in \mathbb{R}^\nu, \end{aligned} \tag{98}$$

where $\kappa, H > 0$ are parameters satisfying (100). Substituting the stochastic integral in (78) into (98), we obtain the stochastic integral representation (92) of $X(\boldsymbol{t})$ with the isotropic kernel

$$a(\boldsymbol{t}) := \int_{\mathbb{R}^\nu} \tau_B(\boldsymbol{s};\kappa)(|\boldsymbol{s}+\boldsymbol{t}|^{H-\frac{\nu}{2}} - |\boldsymbol{t}|^{H-\frac{\nu}{2}}) \mathrm{d}\boldsymbol{s}, \quad \boldsymbol{t} \in \mathbb{R}^\nu. \tag{99}$$

The existence and LRD properties of (98) are given in the following proposition. It is limited to $\nu \ge 2$ since for $\nu = 1$, condition (100) takes a somewhat different form.

Proposition 26. *Let* $\nu \geq 2$ *and*

$$0 < H < 2\kappa < 1. \tag{100}$$

Then, we have the following:
(i) The function $a(\cdot)$ *in (99) is well defined, belongs to* $L^2(\mathbb{R}^\nu)$*, and satisfies*

$$a(\boldsymbol{t}) = \frac{c_{1,a}}{|\boldsymbol{t}|^{\frac{\nu}{2}+2\kappa-H}}\big(1+o(1)\big), \quad |\boldsymbol{t}| \to \infty, \tag{101}$$

with $c_{1,a} := B(\kappa)\int_{\mathbb{R}^\nu}|\boldsymbol{s}|^{-\nu-2\kappa}(|\boldsymbol{e}_1+\boldsymbol{s}|^{H-\frac{\nu}{2}}-1)\mathrm{d}\boldsymbol{s}$*. Particularly, the kernel in (99) satisfies Assumption (A)*$(d;\mathbb{R}^\nu)$ *with* $d = (\nu+2H-4\kappa)/4 \in (0,\nu/4)$ *and a constant angular function* $\ell(\boldsymbol{t}) = c_a, \boldsymbol{t} \in S_{\nu-1}$*.*
(ii) The MA RF X *in (92) with kernel* $a(\cdot)$ *in (99) and Lévy measure* $M \in (M)_2$ *is well defined, has zero mean, and covariance function* $r(\boldsymbol{t}) = \mathrm{E}X(\boldsymbol{s})X(\boldsymbol{t}+\boldsymbol{s})$ *given by*

$$\begin{aligned} r(\boldsymbol{t}) &= -c_{1,H}^2\int_{\mathbb{R}^{2\nu}}\tau_B(\boldsymbol{s}_1;\kappa)\tau_B(\boldsymbol{s}_2;\kappa)|\boldsymbol{t}+\boldsymbol{s}_1-\boldsymbol{s}_2|^{2H}\mathrm{d}\boldsymbol{s}_1\mathrm{d}\boldsymbol{s}_2 \\ &= \frac{c_{1,r}}{|\boldsymbol{t}|^{4\kappa-2H}}\big(1+o(1)\big), \quad |\boldsymbol{t}| \to \infty, \end{aligned} \tag{102}$$

where $c_{1,r} := c_{1,a}^2\int_{\mathbb{R}^\nu}(|\boldsymbol{s}||\boldsymbol{e}_1+\boldsymbol{s}|)^{H-2\kappa-\frac{\nu}{2}}\mathrm{d}\boldsymbol{s}$*. Moreover, the above RF has a spectral density,*

$$f(\boldsymbol{z}) = \frac{(1-\mathrm{e}^{-|\boldsymbol{z}|^2/2})^{2\kappa}}{|\boldsymbol{z}|^{\nu+2H}}, \quad \boldsymbol{z} \in \mathbb{R}^\nu, \tag{103}$$

which explodes as $f(\boldsymbol{z}) = \frac{1}{4|\boldsymbol{z}|^{\nu+2H-4\kappa}}\big(1+o(1)\big)$ *when* $|\boldsymbol{z}| \to 0$*.*

Proof. (i) Note that the integral in the definition (99) of $a(\boldsymbol{t})$ is well defined for $H < 2\kappa + \frac{\nu}{2}$ and any $\boldsymbol{t} \in \mathbb{R}^\nu$. By change of variables $\boldsymbol{s} \to |\boldsymbol{t}|\boldsymbol{s}$ in this integral, we get that $|\boldsymbol{t}|^{\frac{\nu}{2}+2\kappa-H}a(\boldsymbol{t}) = c_{1,a,\boldsymbol{t}}$, where

$$c_{1,a,\boldsymbol{t}} := \int_{\mathbb{R}^\nu}(|\boldsymbol{t}|^{\nu+2\kappa}\tau_B(|\boldsymbol{t}|\boldsymbol{s}))|\boldsymbol{s}|^{-\nu-2\kappa}\big(|\boldsymbol{e}_1+\boldsymbol{s}|^{H-\frac{\nu}{2}}-1\big)\mathrm{d}\boldsymbol{s} \to c_{1,a}$$

as $|\boldsymbol{t}| \to \infty$ since $|\boldsymbol{t}|^{\nu+2\kappa}\tau_B(|\boldsymbol{t}|\boldsymbol{s}) \to B(\kappa)$ $(|\boldsymbol{t}| \to \infty)$ for each $\boldsymbol{s} \neq \boldsymbol{0}$ in view of (96). Hence, relation (101) follows by the DCT, proving part (i).
(ii) Facts (102) and (102) are consequences of (101), (81), and (97). □

Discrete fractional differentiation in (98) with $\kappa > 0$ transforms the nonstationary FBF $\mathcal{B}_H$ into a stationary LRD RF X in Proposition 26. We may expect that an analogous operation with $\kappa < 0$ (discrete fractional integration) can transform a stationary SRD RF into a stationary LRD RF, as in the following example.

Example 27. *Discretely fractionally integrated Matérn RF* is defined as

$$X(\boldsymbol{t}) := (I - T_B)^{-\kappa}\mathcal{M}_{c,H}(\boldsymbol{t}) = \int_{\mathbb{R}^\nu} \tau_B(\boldsymbol{s}; -\kappa)\mathcal{M}_{c,H}(\boldsymbol{s} + \boldsymbol{t})\mathrm{d}\boldsymbol{s}, \quad \boldsymbol{t} \in \mathbb{R}^\nu, \tag{104}$$

where $c, \kappa, H > 0$ are parameters satisfying (106). From (83)–(84), we obtain the stochastic integral representation (92) of $X(\boldsymbol{t})$ with the isotropic kernel

$$a(\boldsymbol{t}) := \mu_{1,c,H} \int_{\mathbb{R}^\nu} \tau_B(\boldsymbol{t} + \boldsymbol{s}; -\kappa)(c|\boldsymbol{s}|)^{\frac{H}{2} - \frac{\nu}{4}} K_{\frac{H}{2} - \frac{\nu}{4}}(c|\boldsymbol{s}|)\mathrm{d}\boldsymbol{s}, \quad \boldsymbol{t} \in \mathbb{R}^\nu. \tag{105}$$

Proposition 28. *Let*

$$H, c > 0, \quad 0 < \kappa < \frac{\nu}{4}. \tag{106}$$

Then, we have the following:

(i) The function $a(\cdot)$ in (105) is well defined, belongs to $L^2(\mathbb{R}^\nu)$. and satisfies

$$a(\boldsymbol{t}) = \frac{c_{2,a}}{|\boldsymbol{t}|^{\nu - 2\kappa}}(1 + o(1)), \quad |\boldsymbol{t}| \to \infty, \tag{107}$$

with $c_{2,a} := B(-\kappa)\mu_{1,c,H} \int_{\mathbb{R}^\nu} (c|\boldsymbol{s}|)^{\frac{H}{2} - \frac{\nu}{4}} K_{\frac{H}{2} - \frac{\nu}{4}}(c|\boldsymbol{s}|)\mathrm{d}\boldsymbol{s}$. Particularly, the kernel in (99) satisfies Assumption (A)$(d; \mathbb{R}^\nu)$. with $d = \kappa \in (0, \nu/4)$ and a constant angular function $\ell(\boldsymbol{t}) = c_{2,a}, \boldsymbol{t} \in S_{\nu-1}$.

(ii) The MA RF X in (92) with kernel $a(\cdot)$ in (107) and the Lévy measure $M \in (M)_2$ is well defined, has zero mean, and covariance

function $r(\boldsymbol{t}) = \mathrm{E}X(\boldsymbol{s})X(\boldsymbol{t}+\boldsymbol{s})$ *given by*

$$\begin{aligned} r(\boldsymbol{t}) &= \mu_{2,c,H} \int_{\mathbb{R}^{2\nu}} \tau_B(\boldsymbol{s}_1; -\kappa)\tau_B(\boldsymbol{s}_2; -\kappa)(c|\boldsymbol{t}+\boldsymbol{s}_1-\boldsymbol{s}_2|)^H \\ &\quad \times K_H(c|\boldsymbol{t}+\boldsymbol{s}_1-\boldsymbol{s}_2|)\mathrm{d}\boldsymbol{s}_1\mathrm{d}\boldsymbol{s}_2 \\ &= \frac{c_{2,r}}{|\boldsymbol{t}|^{\nu-4\kappa}}(1+o(1)), \quad |\boldsymbol{t}| \to \infty, \end{aligned}$$

where $c_{2,r} := c_{2,a}^2 \int_{\mathbb{R}^\nu} (|\boldsymbol{s}||\boldsymbol{e}_1+\boldsymbol{s}|)^{2\kappa-\nu}\mathrm{d}\boldsymbol{s}$. *Moreover, the above RF has a spectral density,*

$$f(\boldsymbol{z}) = \frac{\mu_{4,c,H}}{(1-\mathrm{e}^{-|\boldsymbol{z}|^2/2})^{2\kappa}(c^2+|\boldsymbol{z}|^2)^{H+\frac{\nu}{2}}}, \quad \boldsymbol{z} \in \mathbb{R}^\nu,$$

which explodes as $f(\boldsymbol{z}) = O(|\boldsymbol{z}|^{-4\kappa})$ *when* $|\boldsymbol{z}| \to 0$.

Proof. (i) We have

$$|\boldsymbol{t}|^{\nu-2\kappa}a(\boldsymbol{t}) = \mu_{1,c,H} \int_{\mathbb{R}^\nu} \phi_{\boldsymbol{t}}(\boldsymbol{s})(c|\boldsymbol{s}|)^{\frac{H}{2}-\frac{\nu}{4}} K_{\frac{H}{2}-\frac{\nu}{4}}(c|\boldsymbol{s}|)\mathrm{d}\boldsymbol{s},$$

where $\phi_{\boldsymbol{t}}(\boldsymbol{s}) := |\boldsymbol{t}|^{\nu-2\kappa}\tau_B(\boldsymbol{t}+\boldsymbol{s}; -\kappa) \to B(-\kappa)$ $(|\boldsymbol{t}| \to \infty)$ according to (96). The passage to the limit as $|\boldsymbol{t}| \to \infty$ in the last integral can by justified using (Pratt's) Lemma 30 of Chapter 2 and the bound $0 \le \tau_B(\boldsymbol{s}; -\kappa) \le C\int_0^\infty r^{\kappa-1}\mathrm{e}^{-|\boldsymbol{s}|^2/2j} = C|\boldsymbol{s}|^{2\kappa-\nu}$ or $\phi_{\boldsymbol{t}}(\boldsymbol{s})$ replaced by $\bar{\phi}_{\boldsymbol{t}}(\boldsymbol{s}) := C|\boldsymbol{t}|^{\nu-2\kappa}/|\boldsymbol{t}+\boldsymbol{s}|^{\nu-2\kappa}$, in which case this passage follows by standard argument. Part (ii) follows from Proposition 12 and relations (107), (83), (84), and (97). □

5 Notes

1. Theorem 4 is proved in Pilipauskaitė and Surgailis [212]; for $0 < d < \nu/4$ and indicator functions $\phi(\boldsymbol{s}) = \mathbb{I}(\boldsymbol{s} \in]0, \boldsymbol{t}])$; this result goes back to Surgailis [245]. Lahiri and Robinson [145] studied the case of indicator functions $\phi(\boldsymbol{s}) = \mathbb{I}(\boldsymbol{s} \in A)$ of a star-like region, $A \subset \mathbb{R}^\nu$, having an irregular boundary, under LRD, SRD, and ND conditions on MA coefficients $a(\boldsymbol{t})$. Lavancier [147] discussed partial sum limits for a class of nonisotropic MA RFs on $\mathbb{Z}^\nu$.

2. Fractional calculus is a large field of mathematical and applied research as documented in several texts and monographs Oldham

and Spanier [191], Samko *et al.* [227], Martinez and Sanz [179]. A considerable part of this research deal with fractional differentiation and integration on the real line or its interval. Multidimensional fractional calculus is closely related to partial differential and singular operators on Euclidean spaces. A general fractional theory of bounded operators was developed in Hille [118].

3. Fractional operators and properties of the ARFIMA(p,d,q) times series are discussed in detail in many texts, e.g., Brockwell and Davis [48], Giraitis *et al.* [103], and Samorodnitsky and Taqqu [231]. The discussion in Sections 2 and 3 of Chapter 4 follows Pilipauskaitė and Surgailis [212]. For other approaches towards multi-dimensional fractional integration, see the review paper by Kostić [139]. Berger [33] discussed local limit theorems for random walk on $\mathbb{Z}^\nu$ with non-zero mean and/or infinite variance.

4. Regularity and other properties of fractional Brownian/Lévy RFs were studied by Ciesielski and Kamont [55], Cohen [58], Cohen and Istas [59], and other works.

5. Leonenko *et al.* [163] discussed several classes of fractional RFs on $\mathbb{R}^\nu$, solving a linear partial or fractional partial differential equation with *fractional Gaussian noise* $\dot{B}_{\boldsymbol{H}}$ (a 'continuous' or 'infinitesimal' version of the discrete-time FGN of Example 13) of Chapter 1. The RFs solving the above equations are defined in Leonenko *et al.* [163] via Fourier transform and the spectral representation of $\dot{B}_{\boldsymbol{H}}$. Leonenko *et al.* [163] also reviewed numerous earlier works and the literature on stochastic fractional PDEs.

6. Fractional powers of Markov semigroup. Let $M = \{M_j; j = 0,1,\ldots\}$ be a homogeneous Markov process on a countable set, $S = \{s\}$, with transition operator

$$\begin{aligned} Tg(t) &:= \mathrm{E}[g(M_1)|M_0 = t] = \sum_{u\in S} g(u)p_1(t,u), \\ p_1(t,u) &:= \mathrm{P}(M_1 = u|M_0 = t). \end{aligned} \tag{108}$$

The operators $T_j, j = 0,1,\ldots,$ recursively defined by $T_j := TT_{j-1}, T_0 := I$ form a Markov semigroup:

$$T_j g(s) = T^j g(s) = \sum_{u\in S} g(u)p_j(s,u), \quad p_j(s,u) := \mathrm{P}(M_j = u|M_0 = s),$$

which follows from the Kolmogorov–Chapman identity:

$$\sum_{s\in S} p_j(t,s)p_\ell(s,u) = p_{j+\ell}(t,s), \quad t,s \in S, \quad j,\ell = 0,1,\dots.$$

The fractional powers $(I-T)^d$ of $I-T$ in (108) can be formally defined as in (31):

$$(I-T)^d g(t) := \sum_{j=0}^{\infty} \psi_j(d) T^j g(t) = \sum_{s\in S} \tau(t,s;d) g(s), \tag{109}$$

with coefficients

$$\tau(t,s;d) := \sum_{j=0}^{\infty} \psi_j(d) p_j(t,s), \quad t,u \in S.$$

The (formal) identity $(I-T)^d(I-T)^{-d} = I$ follows similarly to the case of (35):

$$\begin{aligned}\sum_{s\in S} \tau(t,s;d)\tau(s,u;-d) &= \sum_{j,\ell=0}^{\infty} \psi_j(d)\psi_\ell(-d) \sum_{s\in S} p_j(t,s)p_\ell(s,u)\\ &= \sum_{n=0}^{\infty} p_n(t,u) \sum_{j=0}^{n} \psi_j(d)\psi_{n-j}(-d)\\ &= \mathbb{I}(t=u), \quad t,u \in S.\end{aligned}$$

It is notable that (109) apply to a very general index set S (possibly, a fractal set) and to a general Markov operator T which need not be a random walk when $S = \mathbb{Z}^\nu$ as in (31).

7. Fractional operators of variable order (VO). Philippe *et al.* [202–204] introduced and studied two classes of VO fractional operators on $\mathbb{Z}$, defined as

$$A_{d(\cdot)} g(t) := \sum_{j=0}^{\infty} a_j(t) g(t-j), \quad B_{d(\cdot)} g(t) := \sum_{j=0}^{\infty} b_j(t) g(t-j),$$
$$t \in \mathbb{Z}, \tag{110}$$

with coefficients

$$
\begin{aligned}
a_j(t) &:= \frac{d_{t-1}}{1} \cdot \frac{d_{t-2}+1}{2} \cdot \frac{d_{t-3}+1}{3} \cdots \frac{d_{t-j}+j-1}{j}, \\
b_j(t) &:= \frac{d_{t-1}}{1} \cdot \frac{d_{t-2}+j-1}{2} \cdot \frac{d_{t-2}+j-2}{3} \cdots \frac{d_{t-j}+1}{j},
\end{aligned} \tag{111}
$$

$a_0(t) = b_0(t) := 1, a_1(t) = b_1(t) := d_{t-1}$. Here, $d(\cdot) = \{d_t, t \in \mathbb{Z}\}$ is a given sequence of real numbers (a 'VO fractional parameter'). For constant $d_t \equiv d$, (110) agree with fractional operators in (24): $A_{d(\cdot)}g(t) = B_{d(\cdot)}g(t) = (I-T)^d g(t) = \sum_{j=0}^{\infty} \psi_j(d) g(t-j)$. A remarkable property of (110) is invertibility:

$$
B_{-d(\cdot)} A_{d(\cdot)} = A_{-d(\cdot)} B_{d(\cdot)} = I, \tag{112}
$$

where $-d(\cdot) := \{-d_t; t \in \mathbb{Z}\}$. Property (112) leads to explicit solutions of the VO fractional equations

$$
A_{d(\cdot)} X(t) = \varepsilon(t), \quad B_{d(\cdot)} Y(t) = \varepsilon(t) \tag{113}
$$

with white noise $\{\varepsilon(t)\}$ on the r.h.s., generalising the ARFIMA(0, d, 0) class in (26) to nonstationary processes with LRD. Various LRD properties of (113), including partial sums limits, nonlinear functions, and infinite-variance innovations were studied by Philippe *et al.* [202–204], Doukhan *et al.* [84], and Bružaitė *et al.* [49]. We note that these works essentially assume only asymptotic conditions on the functional parameter $d(\cdot)$ and no restrictions on d_t for finite t.

It is of interest to extend VO operators in (110) to higher dimensions $\boldsymbol{t} \in \mathbb{Z}^\nu$. Let $\boldsymbol{d}(\cdot) = \{d(\boldsymbol{t}); \boldsymbol{t} \in \mathbb{Z}^\nu\}$ be a sequence of real numbers $d(\boldsymbol{t}) \in \mathbb{R}$ and $p = \{p(\boldsymbol{u}) := \mathrm{P}(S_1 = \boldsymbol{u}); \boldsymbol{u} \in \mathbb{Z}^\nu\}$ be a one-step random walk probability distribution as in (30). Generalising (110) and $(I-T)^d$ in (30), we may define

$$
A_{\boldsymbol{d}(\cdot)} g(\boldsymbol{t}) := \sum_{\boldsymbol{s} \in \mathbb{Z}^\nu} a(\boldsymbol{t}, \boldsymbol{s}) g(\boldsymbol{s}), \quad B_{\boldsymbol{d}(\cdot)} g(\boldsymbol{t}) := \sum_{\boldsymbol{s} \in \mathbb{Z}^\nu} b(\boldsymbol{t}, \boldsymbol{s}) g(\boldsymbol{s}), \tag{114}
$$

with coefficients

$$a(\boldsymbol{t},\boldsymbol{s}) := \sum_{j=0}^{\infty} a_j(\boldsymbol{t},\boldsymbol{s}), \quad b(\boldsymbol{t},\boldsymbol{s}) := \sum_{j=0}^{\infty} b_j(\boldsymbol{t},\boldsymbol{s}), \quad \boldsymbol{t},\boldsymbol{s} \in \mathbb{Z}^\nu,$$

$$a_j(\boldsymbol{t},\boldsymbol{s}) := \sum_{\boldsymbol{u}_1,\dots,\boldsymbol{u}_{j-1}\in\mathbb{Z}^\nu} \frac{d(\boldsymbol{u}_1)p(\boldsymbol{u}_1-\boldsymbol{t})}{1}\frac{(d(\boldsymbol{u}_2)+1)p(\boldsymbol{u}_2-\boldsymbol{u}_1)}{2}$$
$$\cdots\frac{(d(\boldsymbol{s})+j-1)p(\boldsymbol{s}-\boldsymbol{u}_{j-1})}{j},$$

$$b_j(\boldsymbol{t},\boldsymbol{s}) := \sum_{\boldsymbol{u}_1,\dots,\boldsymbol{u}_{j-1}\in\mathbb{Z}^\nu} \frac{d(\boldsymbol{u}_1)p(\boldsymbol{u}_1-\boldsymbol{t})}{1}\frac{(d(\boldsymbol{u}_2)+j-1)p(\boldsymbol{u}_2-\boldsymbol{u}_1)}{2}$$
$$\cdots\frac{(d(\boldsymbol{s})+1)p(\boldsymbol{s}-\boldsymbol{u}_{j-1})}{j}, \tag{115}$$

$a(\boldsymbol{t},\boldsymbol{t}) = b(\boldsymbol{t},\boldsymbol{t}) := 1, a_1(\boldsymbol{t},\boldsymbol{s}) = b_1(\boldsymbol{t},\boldsymbol{s}) := d(\boldsymbol{s})p(\boldsymbol{s}-\boldsymbol{t})$. We see that $A_{\boldsymbol{d}(\cdot)} = A_{d(\cdot)}, B_{\boldsymbol{d}(\cdot)} = B_{d(\cdot)}$ in (110) for $\nu = 1$ and $p(1) = 1$, and $A_{\boldsymbol{d}(\cdot)} = B_{d(\cdot)} = (I-T)^d$ in (30) for constant $d(\boldsymbol{t}) = d$ and any $\nu \ge 2$ and random walk $p = \mathrm{P}(S_1 = \boldsymbol{u}); \boldsymbol{u} \in \mathbb{Z}^\nu\}$.

The following notes ignore the convergence issues of the infinite series in (114), (115), and other properties of these operators, aiming at formal (algebraic) justification of their invertibility:

$$B_{-\boldsymbol{d}(\cdot)}A_{\boldsymbol{d}(\cdot)} = A_{-\boldsymbol{d}(\cdot)}B_{\boldsymbol{d}(\cdot)} = I, \tag{116}$$

where $-\boldsymbol{d}(\boldsymbol{t}) = \{-d(\boldsymbol{t}); \boldsymbol{t} \in \mathbb{Z}^\nu\}$. In terms of the coefficients, (116) amounts to obtaining the identities:

$$\Delta'(\boldsymbol{t},\boldsymbol{w}) := \sum_{\boldsymbol{s}\in\mathbb{Z}^\nu} b^-(\boldsymbol{t},\boldsymbol{s})a(\boldsymbol{s},\boldsymbol{w}) = \mathbb{I}(\boldsymbol{t}=\boldsymbol{w}) \tag{117}$$

and

$$\Delta''(\boldsymbol{t},\boldsymbol{w}) := \sum_{\boldsymbol{s}\in\mathbb{Z}^\nu} a^-(\boldsymbol{t},\boldsymbol{s})b(\boldsymbol{s},\boldsymbol{w}) = \mathbb{I}(\boldsymbol{t}=\boldsymbol{w}), \tag{118}$$

where $a^-(\boldsymbol{t},\boldsymbol{s})b^-(\boldsymbol{t},\boldsymbol{s})$ are obtained by replacing $\boldsymbol{d}(\cdot)$ by $-\boldsymbol{d}(\cdot)$ in (115). From the definitions in (114)–(115), we have that $\Delta'(\boldsymbol{t},\boldsymbol{w}) =$

$\sum_{n=0}^{\infty} \Delta'_n(\boldsymbol{t}, \boldsymbol{w})$, where $\Delta'_n(\boldsymbol{t}, \boldsymbol{w}) := \sum_{j=0}^{n} \Delta'_{j,n-j}(\boldsymbol{t}, \boldsymbol{w})$ and

$$\begin{aligned}
\Delta'_{j,n-j}(\boldsymbol{t}, \boldsymbol{w}) &:= \sum_{\boldsymbol{s}} b_j^-(\boldsymbol{t}, \boldsymbol{s}) a_{n-j}(\boldsymbol{s}, \boldsymbol{w}) \\
&= \frac{1}{j!(n-j)!} \sum_{\boldsymbol{s}} \sum_{\boldsymbol{v}_1, \dots, \boldsymbol{v}_{j-1}} \sum_{\boldsymbol{u}_1, \dots, \boldsymbol{u}_{n-j-1}} \\
&\quad \times (-d(\boldsymbol{v}_1)) p(\boldsymbol{v}_1 - \boldsymbol{t})(-d(\boldsymbol{v}_2) + j - 1) p(\boldsymbol{v}_2 - \boldsymbol{v}_1) \\
&\quad \cdots (-d(\boldsymbol{s}) + 1) p(\boldsymbol{s} - \boldsymbol{v}_{j-1}) \\
&\quad \times d(\boldsymbol{u}_1) p(\boldsymbol{u}_1 - \boldsymbol{s})(d(\boldsymbol{u}_2) + 1) p(\boldsymbol{u}_2 - \boldsymbol{u}_1) \\
&\quad \cdots (d(\boldsymbol{w}) + n - j - 1) p(\boldsymbol{w} - \boldsymbol{u}_{n-k-1}).
\end{aligned}$$

By changing the variables in the above sum as $\boldsymbol{v}_1 =: \boldsymbol{z}_1, \boldsymbol{v}_2 =: \boldsymbol{z}_2, \dots, \boldsymbol{v}_{j-1} =: \boldsymbol{z}_{j-1}, \boldsymbol{s} =: \boldsymbol{z}_j, \boldsymbol{u}_1 =: \boldsymbol{z}_{j+1}, \dots, \boldsymbol{u}_{n-j-1} =: \boldsymbol{z}_{n-1}$, it can be written as

$$\begin{aligned}
\Delta'_{j,n-j}(\boldsymbol{t}, \boldsymbol{w}) = \frac{1}{j!(n-j)!} &\sum_{\boldsymbol{z}_1, \dots, \boldsymbol{z}_{n-1}} p(\boldsymbol{z}_1 - \boldsymbol{t}) p(\boldsymbol{z}_2 - \boldsymbol{z}_1) \\
&\cdots p(\boldsymbol{z}_{n-1} - \boldsymbol{z}_{n-2}) p(\boldsymbol{w} - \boldsymbol{z}_{n-1})(-d(\boldsymbol{z}_1)) \\
&\times (-d(\boldsymbol{z}_2) + j - 1) \cdots (-d(\boldsymbol{z}_j) + 1) \\
&\times d(\boldsymbol{z}_{j+1}(d(\boldsymbol{z}_{j+2} + 1) \cdots (d(\boldsymbol{w}) + n - j - 1),
\end{aligned}$$

implying

$$\begin{aligned}
\Delta'_n(\boldsymbol{t}, \boldsymbol{w}) = \sum_{j=0}^{n} \Delta'_{j,n-j}(\boldsymbol{t}, \boldsymbol{w}) &\sum_{\boldsymbol{z}_1, \dots, \boldsymbol{z}_{n-1}} j'_n(\boldsymbol{t}, \boldsymbol{w}, \boldsymbol{z}_1, \dots, \boldsymbol{z}_{n-1}) \\
&\times p(\boldsymbol{z}_1 - \boldsymbol{t}) p(\boldsymbol{z}_2 - \boldsymbol{z}_1) \cdots p(\boldsymbol{z}_{n-1} - \boldsymbol{z}_{n-2}) p(\boldsymbol{w} - \boldsymbol{z}_{n-1}),
\end{aligned}$$

with

$$\begin{aligned}
&j'_n(\boldsymbol{t}, \boldsymbol{w}, \boldsymbol{z}_1, \dots, \boldsymbol{z}_{n-1}) \\
&\quad := d(\boldsymbol{z}_1) \sum_{j=0}^{n} \frac{(-1)^j}{j!(n-j)!} (d(\boldsymbol{z}_2) - j + 1) \cdots (d(\boldsymbol{z}_j) - 1) d(\boldsymbol{z}_{j+1}) \\
&\qquad \times (d(\boldsymbol{z}_{j+2}) + 1) \cdots (d(\boldsymbol{w}) + n - j - 1). \qquad (119)
\end{aligned}$$

The formal proof of (117) reduces to the identity

$$j_n'(\boldsymbol{t}, \boldsymbol{w}, \boldsymbol{z}_1, \ldots, \boldsymbol{z}_{n-1}) \equiv 0 \quad (n \geq 1) \tag{120}$$

and the observation that the r.h.s. of (119) *does not involve random walk probabilities and applies to an arbitrary sequence,* $\boldsymbol{d}(\cdot)$. Hence, (117) follows from the analogous identity for the coefficients in (111); see Philippe *et al.* [202, (3.3)]. The formal proof of (118) is similar and reduces to the identity in Philippe *et al.* [202, (3.4)].

The summability and other properties of VO operators in (114) strongly depend on random walk probabilities p and are generally open. These properties are important for the existence and scaling properties of the solutions of fractional VO equations

$$A_{\boldsymbol{d}(\cdot)}X(\boldsymbol{t}) = \varepsilon(\boldsymbol{t}), \quad B_{\boldsymbol{d}(\cdot)}Y(\boldsymbol{t}) = \varepsilon(\boldsymbol{t}), \quad \boldsymbol{t} \in \mathbb{Z}^\nu, \tag{121}$$

analogous to (113). The case of the simple random walk p on $\mathbb{Z}^\nu$ may be of particular interest. Following Doukhan *et al.* [84], the functional parameter $\boldsymbol{d}(\cdot)$ may be *random* (e.g., an i.i.d. sequence), leading to *stationary* solutions X and Y of equations in (121).

There is considerable literature on VO fractional operators. Samko and Ross [229] introduced VO Liouville fractional operators on the real line. Samko [228] and Ortigueira *et al.* [194] presented overviews of the existing works and various approaches to VO fractional integration and differentiation operators. Ortigueira *et al.* [194] noted that they lack the mutual invertibility property: the derivative is not the left inverse of the integral. The latter property is satisfied by VO operators introduced by Surgailis [249] and defined as the limits of discrete VO operators in (110) indexed by $t/n, t \in \mathbb{Z}$ as $n \to \infty$. An interesting question is whether the approach in Surgailis [249] may work in higher dimensions and for the VO operators $A_{d(\cdot)}, B_{d(\cdot)}$ in (113). These and some other open problems in spatial fractional integration are discussed in Surgailis [254].

Chapter 4

Random Grain Model

A random grain (RG) model is an RF $X = \{X(\boldsymbol{t}); \boldsymbol{t} \in \mathbb{R}^\nu\}$ counting the number of uniformly scattered i.i.d. random sets (random grains) $\Xi_j, j \geq 1$, which cover a given point $\boldsymbol{t}$. The closely related (Poisson) Boolean model is defined as the union of all random grains. The RF RG X exhibits the LRD or SRD property depending on whether the Lebesgue measure of the generic grain set Ξ has an infinite or finite second moment. The main result of this section is Theorem 6, dealing with Gaussian/stable isotropic scaling limits of RG RF X under respective SRD/LRD conditions on Ξ. Theorem 10 provides scaling limits of the RG model with aggregation and a a trichotomy of the limit distribution, including Gaussian, stable, and intermediate RFs.

1 Definition and General Properties of RG Model: The Boolean Model

A moving-average RF $X = \{X(\boldsymbol{t}) = \sum_{\boldsymbol{u} \in \mathbb{Z}^\nu} a(\boldsymbol{t} - \boldsymbol{u})\varepsilon(\boldsymbol{u}); \boldsymbol{t} \in \mathbb{Z}^\nu\}$ in (2) of Chapter 3 can be viewed as a sum or superposition of 'pulses' $a(\boldsymbol{t} - \boldsymbol{u})$ 'located' at lattice sites $\boldsymbol{u} \in \mathbb{Z}^\nu$ and having random 'amplitudes' $\varepsilon(\boldsymbol{u})$. Intuitively, dependence in such an RF occurs because different points $\boldsymbol{t}_1 \neq \boldsymbol{t}_2$ can be 'reached' by the same 'pulse' originating at $\boldsymbol{u}$, with $a(\boldsymbol{t}_i - \boldsymbol{u}) \neq 0, i = 1, 2$. This remark also extends to Lévy moving averages X in (2) of Chapter 3 with continuous argument $\boldsymbol{t} \in \mathbb{R}^\nu$, where the role of the 'amplitude' of the 'pulse' $a(\boldsymbol{t} - \boldsymbol{u})$ is played by the Lévy measure $M(\mathrm{d}\boldsymbol{u})$. It is particularly apparent in a *shot-noise* RF,

$$X(\boldsymbol{t}) = \sum_{j=1}^{\infty} \zeta_j a(\boldsymbol{t} - \boldsymbol{u}_j), \quad \boldsymbol{t} \in \mathbb{R}^\nu, \tag{1}$$

written as a sum of 'pulses' $a(\boldsymbol{t} - \boldsymbol{u}_j)$ 'located' at points $\boldsymbol{u}_j$ of a (stationary) Poisson point process $\{\boldsymbol{u}_j; j \geq 1\} \subset \mathbb{R}^\nu$ with i.i.d. 'amplitudes' ζ_j, independent of the Poisson process. When the ζ_js have zero mean and finite variance, (1) becomes a special case of a moving-average RF in (92) of Chapter 3, with $M(\mathrm{d}\boldsymbol{u}) = \int_{\mathbb{R}} v\mathcal{N}(\mathrm{d}\boldsymbol{u}, \mathrm{d}v) = \int_{\mathbb{R}} v\widetilde{\mathcal{N}}(\mathrm{d}\boldsymbol{u}, \mathrm{d}v)$ being a compound Poisson measure and $\mathcal{N}(\mathrm{d}\boldsymbol{u}, \mathrm{d}v)$ being a Poisson random measure on $\mathbb{R}^\nu \times \mathbb{R}$ with mean $\mathrm{E}\mathcal{N}(\mathrm{d}\boldsymbol{u}, \mathrm{d}v) = \mathrm{d}\boldsymbol{u}\mathrm{P}(\zeta \in \mathrm{d}v)$, $\widetilde{\mathcal{N}}(\mathrm{d}\boldsymbol{u}, \mathrm{d}v) = \mathcal{N}(\mathrm{d}\boldsymbol{u}, \mathrm{d}v) - \mathrm{E}\mathcal{N}(\mathrm{d}\boldsymbol{u}, \mathrm{d}v)$. The covariance of the shot-noise RF in (1) can be written as a convolution, $\mathrm{Cov}(X(\boldsymbol{0}), X(\boldsymbol{t})) = \mathrm{E}\zeta^2(a \star a)(\boldsymbol{t})$, and its decay rate and LRD property are determined by the properties of the 'pulse' (moving-average kernel) $a(\cdot)$. As seen in Proposition 5 of Chapter 3, the scaling behaviour of the above shot-noise RF is Gaussian and quite similar to that of Gaussian RFs with the same covariance structure, except for the fact that the marginal distribution of (1) may have a much heavier tail than Gaussian.

A far-reaching generalisation of the shot-noise model of (1) is written as a sum,

$$X(\boldsymbol{t}) := \sum_{j=1}^{\infty} W_j(\boldsymbol{t} - \boldsymbol{u}_j), \quad \boldsymbol{t} \in \mathbb{R}^\nu, \tag{2}$$

of i.i.d. *pulse RFs* $W_j = \{W_j(\boldsymbol{u}); \boldsymbol{u} \in \mathbb{R}^\nu\}, j = 1, 2, \ldots,$ 'located' at points of a stationary Poisson point process. We assume that the generic pulse $W = \{W(\boldsymbol{u}); \boldsymbol{u} \in \mathbb{R}^\nu\}$ is a measurable RF independent of $\{\boldsymbol{u}_j; j \geq 1\}$ and defined on a probability space (Ω, P). Then, (2) can be rewritten as a Poisson stochastic integral,

$$X(\boldsymbol{t}) = \int_{\mathbb{R}^\nu \times \Omega} w(\boldsymbol{t} - \boldsymbol{u})\mathcal{N}(\mathrm{d}\boldsymbol{u}, \mathrm{d}w), \tag{3}$$

w.r.t. to a Poisson random measure $\mathcal{N}(\mathrm{d}\boldsymbol{u}, \mathrm{d}w)$ on the product space $\mathbb{R}^\nu \times \Omega$ equipped with the measure $\mathrm{E}\mathcal{N}(\mathrm{d}\boldsymbol{u}, \mathrm{d}w) = \mathrm{d}\boldsymbol{u}\mathrm{P}(W(\cdot) \in \mathrm{d}w)$. From the definition and properties of Poisson stochastic integral listed in (22)–(27) of Chapter 2, we get that (3) and (2) are well-defined if

$$\int_{\mathbb{R}^\nu} \mathrm{E}[|W(\boldsymbol{u})| \wedge 1]\mathrm{d}\boldsymbol{u} < \infty. \tag{4}$$

Finally, for the pulse RF with

$$\int_{\mathbb{R}^\nu} \mathrm{E}W(\boldsymbol{u})^2\mathrm{d}\boldsymbol{u} < \infty, \tag{5}$$

the RF in (3) has finite variance, a covariance function

$$\mathrm{Cov}(X(\mathbf{0}), X(\boldsymbol{t})) = \int_{\mathbb{R}^\nu} \mathrm{E}[W(\boldsymbol{u})W(\boldsymbol{t}+\boldsymbol{u})]\mathrm{d}\boldsymbol{u}, \quad \boldsymbol{t} \in \mathbb{R}^\nu, \tag{6}$$

and a representation

$$X(\boldsymbol{t}) = \mathrm{E}X(\boldsymbol{t}) + \int_{\mathbb{R}^\nu \times \Omega} w(\boldsymbol{t}-\boldsymbol{u})\widetilde{\mathcal{N}}(\mathrm{d}\boldsymbol{u}, \mathrm{d}w), \quad \boldsymbol{t} \in \mathbb{R}^\nu \tag{7}$$

w.r.t. the centered Poisson random measure $\widetilde{\mathcal{N}}(\mathrm{d}\boldsymbol{u}, \mathrm{d}w) = \mathcal{N}(\mathrm{d}\boldsymbol{u}, \mathrm{d}w) - \mathrm{E}\mathcal{N}(\mathrm{d}\boldsymbol{u}, \mathrm{d}w)$. The RFs in (2) and (7) are stationary and have infinitely divisible finite-dimensional distributions. In particular, the (joint) characteristic function of the RF in (7) is written as

$$\mathrm{E}\mathrm{e}^{\mathrm{i}\sum_{j=1}^n \theta_j X(\boldsymbol{t}_j)} = \exp\left\{\int_{\mathbb{R}^\nu} \mathrm{E}\Psi\left(\sum_{j=1}^n \theta_j W(\boldsymbol{t}_j - \boldsymbol{u})\right)\mathrm{d}\boldsymbol{u} + \mathrm{i}m\sum_{j=1}^n \theta_j\right\} \tag{8}$$

for any $\boldsymbol{t}_j \in \mathbb{R}^\nu, \theta_j \in \mathbb{R}^\nu, j = 1, \ldots, n, n \in \mathbb{N}_+$.

A shot-noise RF in (2) takes integer values if the pulse RF does. In particular, the pulse RF may take two values, 0 and 1, viz.,

$$W(\boldsymbol{u}) = \mathbb{I}(\boldsymbol{u} \in \Xi), \quad \boldsymbol{u} \in \mathbb{R}^\nu, \tag{9}$$

where $\Xi \subset \mathbb{R}^\nu$ is a *random set* ('generic grain'). Although the latter notion is not unanimous, in the literature, it often refers to a measurable map (random element) $\Xi : \Omega \to \mathbb{F}(\mathbb{R}^\nu)$, taking values from the set $\mathbb{F}(\mathbb{R}^\nu)$ of all closed compact subsets $\xi \subset \mathbb{R}^\nu$, endowed with the σ-field $\mathcal{F}(\mathbb{R}^\nu)$ generated by all $\xi \cap K \neq \emptyset, \xi, K \in \mathbb{F}(\mathbb{R}^\nu)$. In this book, we use the same terminology for a random set, identifying it with a probability distribution on $(\mathbb{F}(\mathbb{R}^\nu), \mathcal{F}(\mathbb{R}^\nu))$. In the case of (9), the corresponding shot-noise RF is called a *random grain (RG) model*,

$$X(\boldsymbol{t}) = \sum_{j=1}^{\infty} \mathbb{I}(\boldsymbol{t} \in (\boldsymbol{u}_j + \Xi_j)), \quad \boldsymbol{t} \in \mathbb{R}^\nu, \tag{10}$$

counting, for each given point $\boldsymbol{t} \in \mathbb{R}^\nu$, the number of RGs $\boldsymbol{u}_j + \Xi_j$ covering $\boldsymbol{t}$. A closely related object, the *Boolean set* $\mathcal{X}$, is defined as the union of all grains:

$$\mathcal{X} := \bigcup_{j=1}^{\infty} (\boldsymbol{u}_j + \Xi_j) \subset \mathbb{R}^\nu. \tag{11}$$

The random set in (11) can be identified with its indicator function

$$\widehat{X}(\boldsymbol{t}) := \mathbb{I}(\boldsymbol{t} \in \mathcal{X}), \quad \boldsymbol{t} \in \mathbb{R}^\nu, \tag{12}$$

which is an RF on $\mathbb{R}^\nu$, taking two values: 0 and 1. In the sequel, we shall refer to the binary RF in (12) as the *Boolean model.* The relation between this model and the RG model in (10) is very simple:

$$\widehat{X}(\boldsymbol{t}) = X(\boldsymbol{t}) \wedge 1, \tag{13}$$

indicating that the Boolean model can be viewed as a nonlinear transformation of the linear RF in (10). The LRD property of the Boolean model will be studied in Section 5 of Chapter 5.

Let us turn to the existence and LRD properties of the RG model. From (9), $\int_{\mathbb{R}^\nu} (|W(\boldsymbol{u})| \wedge 1) \mathrm{d}\boldsymbol{u} = \mathrm{Leb}_\nu(\Xi)$, and condition (4) translates to

$$\mu := \mathrm{E}\,\mathrm{Leb}_\nu(\Xi) < \infty. \tag{14}$$

We see that under (14), the RG model in (10) is well-defined and admits a Poisson integral representation in (7), or

$$\begin{aligned} X(\boldsymbol{t}) &= \int_{\mathbb{R}^\nu \times \mathbb{F}(\mathbb{R}^\nu)} \mathbb{I}(\boldsymbol{t} \in (\boldsymbol{u} + \xi)) \mathcal{N}(\mathrm{d}\boldsymbol{u}, \mathrm{d}\xi) \qquad (15) \\ &= \mathrm{E}\,\mathrm{Leb}_\nu(\Xi) + \int_{\mathbb{R}^\nu \times \mathbb{F}(\mathbb{R}^\nu)} \mathbb{I}(\boldsymbol{t} \in (\boldsymbol{u} + \xi)) \widetilde{\mathcal{N}}(\mathrm{d}\boldsymbol{u}, \mathrm{d}\xi), \quad \boldsymbol{t} \in \mathbb{R}^\nu, \end{aligned}$$

where $\mathcal{N}(\mathrm{d}\boldsymbol{u}, \mathrm{d}\xi)$ is a Poisson random measure with intensity $\mathrm{d}\boldsymbol{u}\mathrm{P}(\Xi \in \mathrm{d}\xi)$, $\widetilde{\mathcal{N}}(\mathrm{d}\boldsymbol{u}, \mathrm{d}\xi) = \widetilde{\mathcal{N}}(\mathrm{d}\boldsymbol{u}, \mathrm{d}\xi) - \mathrm{E}\widetilde{\mathcal{N}}(\mathrm{d}\boldsymbol{u}, \mathrm{d}\xi)$. The RF in (15) has a marginal Poisson distribution, $P(X(\boldsymbol{t}) = k) = \mathrm{e}^{-\mu} k^{\mu}/k!, k = 0, 1, \dots,$ with mean $\mu := \mathrm{E}\,\mathrm{Leb}_{\nu}(\Xi)$ and a nonnegative covariance function

$$\begin{aligned}\mathrm{Cov}(X(\mathbf{0}), X(\boldsymbol{t})) &= \mathrm{E}\int_{\mathbb{R}^{\nu}} \mathbb{I}(\boldsymbol{s} \in \Xi, \boldsymbol{s} + \boldsymbol{t} \in \Xi)\mathrm{d}\boldsymbol{s} \\ &= \mathrm{E}\mathrm{Leb}_{\nu}(\Xi \cap (\Xi - \boldsymbol{t})),\end{aligned} \tag{16}$$

in agreement with (6). In a similar way,

$$\begin{aligned}\int_{\mathbb{R}^{\nu}} \mathrm{Cov}(X(\mathbf{0}), X(\boldsymbol{t}))\mathrm{d}\boldsymbol{t} &= \mathrm{E}\int_{\mathbb{R}^{\nu}} \mathbb{I}(\boldsymbol{s} \in \Xi)\mathrm{d}\boldsymbol{s} \int_{\mathbb{R}^{\nu}} \mathbb{I}(\boldsymbol{s} + \boldsymbol{t} \in \Xi)\mathrm{d}\boldsymbol{t} \\ &= \mathrm{E}\mathrm{Leb}_{\nu}(\Xi)^2.\end{aligned} \tag{17}$$

The latter fact is emphasised in the following corollary.

Corollary 1. *An RG model in* (15) *has covariance LRD (respectively, covariance SRD) if and only if*

$$\mathrm{E}\mathrm{Leb}_{\nu}(\Xi)^2 = \infty \tag{18}$$

(respectively, $\mathrm{E}\mathrm{Leb}_{\nu}(\Xi)^2 < \infty$*).*

2 Randomly Homothetic RG Model: Examples

Probably, the simplest RG model is the *random balls* model, with grain

$$\Xi = \{|\boldsymbol{t}| \le R^{1/\nu}\} = R^{1/\nu} B_1 \tag{19}$$

being a ball of random radius $R^{1/nu}$. Since $\mathrm{Leb}_{\nu}(\Xi) = R\mathrm{Leb}_{\nu}(B_1)$, (14) is equivalent to $\mathrm{E}R < \infty$, whereas (18) translates to $\mathrm{E}R^2 = \infty$. A similar fact holds if the unit ball B_1 in (19) is replaced by an arbitrary compact *deterministic* set $\Xi^0 \subset \mathbb{R}^{\nu}$, viz., $\Xi = R^{1/\nu}\,\Xi^0$, in which case $\mathrm{Leb}_{\nu}(\Xi) = R\mathrm{Leb}_{\nu}(\Xi^0)$ is again proportional to R. In the above case, all grains $\Xi_j = R_j^{1/\nu}\,\Xi^0$ in the sum (10) are homothetic dilations of Ξ^0 by random factors $R_j^{1/\nu}$. This form of grain may be unrealistic in applications. We extend the above model by allowing Ξ^0 to be *random*. The extension is formalised as follows.

Definition 2. A random set $\Xi \subset \mathbb{R}^\nu$ is said to be a *random homothetic grain* if it can be represented as

$$\Xi = R^{1/\nu}\,\Xi^0, \tag{20}$$

where Ξ^0 is a bounded random set with $\mathrm{Leb}_\nu(\Xi^0) > 0$, $\Xi^0 \subset \{|\boldsymbol{t}| < K\}$ for some nonrandom $K > 0$ and $R > 0$ is an r.v., independent of Ξ^0.

In the following, we shall refer to the corresponding RG RF, viz.,

$$X(\boldsymbol{t}) = \sum_{j=1}^{\infty} \mathbb{I}(\boldsymbol{t} \in (\boldsymbol{u}_j + R_j^{1/\nu}\Xi_j^0)), \quad \boldsymbol{t} \in \mathbb{R}^\nu, \tag{21}$$

as a *randomly homothetic RG model.* For (20), we have $\mathrm{Leb}_\nu(\Xi) = R\mathrm{Leb}_\nu(\Xi^0) \le CR$ due to the boundedness of Ξ^0 so that the finiteness of the expectation $\mathrm{E}R < \infty$ is necessary and sufficient for the existence of (21). For a randomly homothetic RG model, condition (14) (respectively, (18)) simplifies to $\mathrm{E}R < \infty$ (respectively, $\mathrm{E}R^2 = \infty$).

Example 3 (Hard balls grain). Let $\{\boldsymbol{u}_j^0\}$ be a Poisson point process on $B_1 := \{|\boldsymbol{u}| \le 1\} \subset \mathbb{R}^\nu$ with Lebesgue intensity. From each point $\boldsymbol{u}_j^0$, a hard ball starts growing at unit rate, and the growth stops after it hits another ball or the boundary $\partial B_1 = \{|\boldsymbol{u}| = 1\}$. The set Ξ^0 is defined as the union of all such balls.

Example 4 (Cluster Boolean grain). Let $\{(\boldsymbol{u}_j^0, y_j^0)\}$ be a Poisson point process on $B_1\times]0,1] \subset \mathbb{R}^\nu \times \mathbb{R}_+$ with intensity $\mu^0(\mathrm{d}\boldsymbol{u}, \mathrm{d}y)$, $\int_{B_1\times]0,1]} y\mu^0(\mathrm{d}\boldsymbol{u}, \mathrm{d}y) < \infty$, and

$$\Xi^0 = \bigcup_{j=1}^{\infty} (\boldsymbol{u}_j^0 + y_j^{1/\nu} B_1).$$

Then, $\Xi^0 \subset B_2$ is a.s. bounded and is a union of infinite number of balls unless the intensity measure $\mu^0(B_1\times]0,1]) < \infty$ is bounded.

Fig. 1 provides simulated trajectories of RG models in Examples 3 and 4. (the thick dots refer to points of the Poisson process $\{\boldsymbol{u}_j\}$ of 'grain centers'.)

Proposition 5. *Let* $\{X(\boldsymbol{t}); \boldsymbol{t} \in \mathbb{R}^\nu\}$ *be a randomly homothetic RG model in* (21), *with* (R, Ξ^0) *satisfying the following two conditions:*

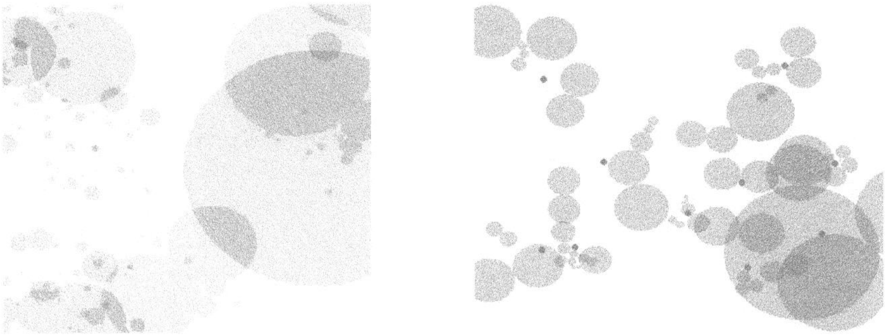

Figure 1. Trajectories of RG models with grain set Ξ^0 in Examples 3 (left) and 4 (right).

(i) *the function* $\boldsymbol{t} \to \mathrm{ELeb}_\nu(\Xi^0 \cap (\Xi^0 - \boldsymbol{t}))$ *is continuous on* $\mathbb{R}^\nu_0$, *and*

(ii) $R > 0$ *is an r.v. with a probability density of* $f(r), r > 0$, *such that*

$$f(r) \sim c_f r^{-1-\alpha}, \qquad r \to \infty \qquad (\exists\, c_f > 0, \quad \alpha \in (1,2)). \quad (22)$$

Then,

$$\mathrm{P}(\mathrm{Leb}_\nu(\Xi) > x) \sim c_\Xi x^{-\alpha} \ (x \to \infty), \quad c_\Xi := (c_f/\alpha)\mathrm{ELeb}_\nu(\Xi^0)^\alpha, \quad (23)$$

whereas $r(\boldsymbol{t}) = \mathrm{Cov}(X(\boldsymbol{0}), X(\boldsymbol{t})) = \mathrm{ELeb}_\nu((R^{1/\nu}\Xi^0) \cap (R^{1/\nu}\Xi^0 - \boldsymbol{t}))$ *has the asymptotics*

$$r(\boldsymbol{t}) = |\boldsymbol{t}|^{-\nu(\alpha-1)}\left(L\left(\frac{\boldsymbol{t}}{|\boldsymbol{t}|}\right) + o(1)\right), \quad |\boldsymbol{t}| \to \infty, \quad (24)$$

with $L(\cdot) \in S_{\nu-1}$ *given by*

$$L(\boldsymbol{z}) := c_f \int_{\mathbb{R}_+} \mathrm{ELeb}_\nu(\Xi^0 \cap (\Xi^0 - r^{-1/\nu}\boldsymbol{z}))r^{-\alpha}\mathrm{d}r. \quad (25)$$

Proof. The statement in (23) is a direct consequence of $\mathrm{Leb}_\nu(\Xi) = R\mathrm{Leb}_\nu(\Xi^0)$ (22) and Breiman's lemma. To prove (24), write

$$\begin{aligned}|\boldsymbol{t}|^{\nu(\alpha-1)}r(\boldsymbol{t}) &= |\boldsymbol{t}|^{\nu(\alpha-1)}\int_{\mathbb{R}_+} \mathrm{ELeb}_\nu(r^{1/\nu}\Xi^0 \cap (r^{1/\nu}\Xi^0 - \boldsymbol{t}))f(r)\mathrm{d}r \\ &= \int_{\mathbb{R}_+} \mathrm{ELeb}_\nu\left(\Xi^0 \cap \left(\Xi^0 - r^{-1/\nu}\frac{\boldsymbol{t}}{|\boldsymbol{t}|}\right)\right)\tilde{f}(r;|\boldsymbol{t}|^\nu)r^{-\alpha}\mathrm{d}r,\end{aligned}$$

where

$$\tilde{f}(r;\lambda) := (\lambda r)^{1+\alpha} f(\lambda r) \to c_f \quad (\lambda \to \infty) \tag{26}$$

for any $r > 0$. By the boundedness of Ξ^0, $\mathrm{Leb}_\nu(\Xi^0 \cap (\Xi^0 - r^{-1/\nu}\boldsymbol{z}))$ is bounded and vanishes for small enough $r > 0$ uniformly in $|\boldsymbol{z}| = 1$. Thus, (24) follows from (26) and the DCT. The properties of $L(\boldsymbol{z})$ follow from its definition and assumption (i). □

3 Scaling Limits of Random Grain Model

The tail behaviour of the distribution

$$\mathrm{P}(\mathrm{Leb}_\nu(\Xi) > x), \quad \text{as } x \to \infty, \tag{27}$$

of the Lebesgue measure $\mathrm{Leb}_\nu(\Xi)$ of generic grain Ξ is very important in the scaling limit of the RG and Boolean models, as well as their statistical inference, as explained later in Section 3 of Chapter 5. Indeed, the number of grains (points of the Poisson process) in a large 'inflated' set λA is approximately $\mathrm{Leb}_\nu(\lambda A) = \lambda^\nu \mathrm{Leb}_\nu(A)$, so that, in rough terms, the integral $\int_{\lambda A} X(\boldsymbol{t})\mathrm{d}\boldsymbol{t}$ can be viewed as a sum, $\sum_{j=1}^{[\lambda^\nu \mathrm{Leb}_\nu(A)]} \mathrm{Leb}_\nu(\Theta_j)$, of i.i.d. r.v.s with distribution (27). If the probability in (27) decays regularly with exponent $\alpha \in (1,2)$ (as in Proposition 5), we can expect that $\int_{\lambda A}(X(\boldsymbol{t}) - \mathrm{E}X(\boldsymbol{t}))\mathrm{d}\boldsymbol{t}$ tends to an α-stable distribution under normalisation, which grows as $\lambda^{\nu/\alpha}$ when $\lambda \to \infty$.

The above α-stable scaling behaviour applies to randomly homothetic RG models of Proposition 5 but is not true under condition (23) alone. Indeed, a more exact approximation $\int_{\lambda A} X(\boldsymbol{t})\mathrm{d}\boldsymbol{t}$ by the sum of $[\lambda^\nu \mathrm{Leb}_\nu(A)]$ independent r.v.s should use a 'truncated' Lebesgue measure, $\mathrm{Leb}_\nu(\Theta_j \cap (\lambda A))$, which may have a crucial effect on the limit $\lambda \to \infty$ if the grains have an 'elongated' or 'dispersed' shape. This 'truncation' effect does not occur for randomly homothetic RG model because of the special form of Ξ. On the other hand, isotropic dilation by a heavy-tailed R in (20) is a very particular class of random sets. It is of interest to find general conditions for the stable scaling behaviour of RG models, which is done in Theorem 6 later.

Theorem 6. *Let X be a random grain model in* (10) *with generic grain set Ξ satisfying the following conditions: either*

(i) $\alpha = 2$ *and* $\mathrm{ELeb}_\nu(\Xi)^2 < \infty$ *or*
(ii) $1 < \alpha < 2$,

$$\mathrm{P}(\mathrm{Leb}_\nu(\Xi) > x) \sim c_\Xi x^{-\alpha}, \quad x \to \infty \qquad (\exists\, c_\Xi > 0) \tag{28}$$

and

$$\mathrm{ELeb}_\nu\big(\Xi \cap \{|\boldsymbol{t}| > \lambda\}\big) = o(\lambda^{(1-\alpha)\nu/\alpha}), \qquad \lambda \to \infty. \tag{29}$$

Then, for any $\phi \in \Phi$,

$$\lambda^{-\nu/\alpha}(X_\lambda(\phi) - \mathrm{E}X_\lambda(\phi)) \xrightarrow{\mathrm{d}} \begin{cases} \sigma_\alpha L_\alpha(\phi), & 1 < \alpha < 2, \\ \sigma_2 W(\phi), & \alpha = 2, \end{cases} \tag{30}$$

where

$$\sigma_\alpha^\alpha := \begin{cases} c_\Xi \Gamma(2-\alpha)\cos(\pi\alpha)/(1-\alpha), & 1 < \alpha < 2, \\ \mathrm{ELeb}_\nu(\Xi)^2, & \alpha = 2. \end{cases} \tag{31}$$

Before turning to the proof of the above theorem, we present two examples of nonhomothetic RG models that explain the need for condition (29) and show that the scaling behaviour without it can be 'nontypical' and quite complex.

Example 7. Let $\nu = 2$ and

$$\Xi =]0,1] \times]0,R] \subset \mathbb{R}^2, \tag{32}$$

where $R > 0$ is an r.v. with distribution as in Proposition 9(i). The RG model X with 'rectangular' Ξ (32) appears in network traffic models studied in Refs. [130, 159, 185] and elsewhere. According to these works, for indicator functions $\phi(\boldsymbol{t}) = \mathbb{I}(\boldsymbol{t} \in]\mathbf{0}, \boldsymbol{x}]), \boldsymbol{x} \in \mathbb{R}^2_+$, $\lambda^{-(4-\alpha)/2}(X_\lambda(\phi) - \mathrm{E}X_\lambda(\phi))$ is asymptotically normal; hence, they do not satisfy Theorem 6(ii). It is easy to see that (32) violates condition (29): for large λ. we have $\mathrm{ELeb}_2\big(\Xi \cap \{|\boldsymbol{t}| > \lambda\}\big) \sim \mathrm{E}(R-\lambda)_+ = O(\lambda^{-(\alpha-1)}) \gg O(\lambda^{-2(\alpha-1)/\alpha})$ since $\alpha < 2$.

Example 8 ('Deterministically related transmission rate and duration' model; see Refs. [159, 208]). Let

$$\Xi =]0, R^{1-p}] \times]0, R^p] \subset \mathbb{R}^2, \tag{33}$$

where $p \in (0,1)$ is a (shape) parameter and $R > 0$ is the same as in (33). By symmetry, it suffices to consider the case of $p \geq 1/2$. According to these works, the limit distribution of $X_\lambda(\phi)$ for indicator functions ϕ is Gaussian if $\alpha > 2p$ and α-stable if $\alpha < 2p$; for $\alpha = 2p$, this limit is an 'intermediate' one, written as a Poisson stochastic integral. We have $\mathrm{Leb}_2(\Xi) = R$ and $\mathrm{Leb}_2(\Xi \cap \{|\boldsymbol{t}| > \lambda\}) < 2(R^p - \lambda)R^{1-p}$, so that $\mathrm{E}\mathrm{Leb}_2(\Xi \cap \{|\boldsymbol{t}| > \lambda\}) = O(\lambda^{-(\alpha-1)/p}) = o(\lambda^{-(\alpha-1)(2/\alpha)})$ when $2p < \alpha$, and (29) holds. Similarly, one can check that for $2p \ge \alpha$, condition (29) is not satisfied and Theorem 6(ii) for (33) does not apply.

As noted above, Theorem 6 applies to randomly homothetic RG models.

Corollary 9. *Let X be a randomly homothetic RG model in* (21) *satisfying the conditions of Corollary* 5. *Then,* $\Xi = R^{1/\nu}\Xi^0$ *satisfies conditions* (28) *and* (29) *of Theorem* 6, *and the convergence in* (30) *holds, with* σ_α^α *in* (31) *and* $c_\Xi = c_f \mathrm{Leb}_\nu^\alpha(\Xi^0)/\alpha$.

Proof. It suffices to check only (29). Since $\Xi^0 \subset B_1$ is bounded, $\mathrm{Leb}_\nu(\Xi \cap \{|\boldsymbol{t}| > \lambda\}) \le \mathrm{Leb}_\nu(\{\lambda < |\boldsymbol{t}| < R^{1/\nu}\}) \le C(R - \lambda^\nu) \vee 0$ and, therefore,

$$
\begin{aligned}
\mathrm{E}\mathrm{Leb}_\nu(\Xi \cap \{|\boldsymbol{t}| > \lambda\}) &\le C \int_{\lambda^\nu}^{\infty} \mathrm{P}(R > r)\mathrm{d}r \\
&= O(\lambda^{-\nu(\alpha-1)}) = o(\lambda^{-\nu(\alpha-1)/\alpha})
\end{aligned}
$$

since $\alpha > 1$. □

Proof of Theorem 6. *(ii) ($1 < \alpha < 2$).* Let $j_\lambda(\theta) := \log \mathrm{E} \exp\{\mathrm{i}\theta\lambda^{-\nu/\alpha}(X_\lambda(\phi) - \mathrm{E}X_\lambda(\phi))\}, \theta \in \mathbb{R}$ denote the log-characteristic functional of the l.h.s. of (30). We need to show that it converges to the corresponding functional $j(\theta) := \log \mathrm{E}\mathrm{e}^{\mathrm{i}\theta L_\alpha(\phi)}$ of the r.h.s. as $\lambda \to \infty$. It is convenient to write the limit functional as the double integral

$$
j(\theta) := \mathrm{i}c_\Xi \int_{\mathbb{R}^\nu} \theta\phi(\boldsymbol{s}) \left\{ \int_{\mathbb{R}_+} (\mathrm{e}^{\mathrm{i}\theta\phi(\boldsymbol{s})x} - 1)x^{-\alpha}\mathrm{d}x \right\} \mathrm{d}\boldsymbol{s}. \tag{34}
$$

We have

$$
\begin{aligned}
j_\lambda(\theta) &= \int_{\mathbb{R}^\nu} \mathrm{E}\Psi\left(\frac{\theta}{\lambda^{\nu/\alpha}} \int_{\mathbb{R}^\nu} \phi(\boldsymbol{t}/\lambda)\mathbb{I}(\boldsymbol{t} - \boldsymbol{s} \in \Xi)\mathrm{d}\boldsymbol{t}\right) \mathrm{d}\boldsymbol{s} \\
&= \lambda^\nu \int_{\mathbb{R}^\nu} \mathrm{E}\Psi\left(\frac{\theta}{\lambda^{\nu/\alpha}} \int_{\mathbb{R}^\nu} \phi\left(\frac{\boldsymbol{t}}{\lambda} + \boldsymbol{s}\right) \mathbb{I}(\boldsymbol{t} \in \Xi)\mathrm{d}\boldsymbol{t}\right) \mathrm{d}\boldsymbol{s}.
\end{aligned} \tag{35}
$$

The intuitive argument leading to $j_\lambda(\theta) \to j(\theta)$ uses the observation that

$$\int_{\mathbb{R}^\nu} \phi\left(\frac{\boldsymbol{t}}{\lambda} + \boldsymbol{s}\right) \mathbb{I}(\boldsymbol{t} \in \Xi)\mathrm{d}\boldsymbol{t} \ \to \ \phi(\boldsymbol{s})\mathrm{Leb}_\nu(\Xi) \tag{36}$$

a.s. at each continuity point $\boldsymbol{s}$ of $\phi(\cdot)$. Using integration by parts and the tail condition in (28), we see that

$$\begin{aligned}\lambda^\nu \mathrm{E}\Psi\left(\frac{\theta\phi(\boldsymbol{s})}{\lambda^{\nu/\alpha}}\mathrm{Leb}_\nu(\Xi)\right) &= \mathrm{i}\theta\phi(\boldsymbol{s})\lambda^\nu \int_{\mathbb{R}_+} (\mathrm{e}^{\mathrm{i}\theta\phi(\boldsymbol{s})x} - 1)\lambda^\nu \mathrm{P}(\mathrm{Leb}_\nu(\Xi) \\ &\qquad > x\lambda^{\nu/\alpha})\mathrm{d}x \\ &\sim \mathrm{i}c_\Xi\theta\phi(\boldsymbol{s}) \int_{\mathbb{R}_+} (\mathrm{e}^{\mathrm{i}\theta\phi(\boldsymbol{s})x} - 1)x^{-\alpha}\mathrm{d}x. \end{aligned} \tag{37}$$

Hence, if the inner integral in (35) can be replaced by the r.h.s. of (36), i.e., $j_\lambda(\theta)$ can be replaced by

$$\tilde{j}_\lambda(\theta) := \lambda^\nu \int_{\mathbb{R}^\nu} \mathrm{E}\Psi\left(\frac{\theta\phi(\boldsymbol{s})}{\lambda^{\nu/\alpha}}\mathrm{Leb}_\nu(\Xi)\right)\mathrm{d}\boldsymbol{s}, \tag{38}$$

the statement of the theorem will follow rather easily.

A rigorous justification of the above argument using condition (29) is somewhat involved. We face two difficulties. Firstly, ϕ need not be continuous; secondly, even if it is, the convergence in (36) need not hold for large $|\boldsymbol{t}| = O(\lambda)$.

The classical Lusin's theorem states that each (measurable) function ϕ is nearly continuous, or, in other words, for any $K > 0$ and $\epsilon > 0$, there is a measurable set $U_{\epsilon,K} \subset B_K := \{\boldsymbol{u} \in \mathbb{R}^\nu : |\boldsymbol{u}| < K\} \subset \mathbb{R}^\nu$ such that ϕ restricted to $U_{\epsilon,K}$ is continuous and $\mathrm{Leb}_\nu(B_K \setminus U_{\epsilon,K}) < \epsilon$. Accordingly, denote

$$\xi_\lambda(\boldsymbol{u}) := \lambda^{-\nu/\alpha} \int_{\mathbb{R}^\nu} \phi(\boldsymbol{t}/\lambda + \boldsymbol{u})\mathbb{I}(\boldsymbol{t} \in \Xi)\mathrm{d}\boldsymbol{t},$$

$$j_\lambda(\theta; U_{\epsilon,K}) := \lambda^\nu \int_{U_{\epsilon,K}} \mathrm{E}\Psi(\theta\xi_\lambda(\boldsymbol{u}))\mathrm{d}\boldsymbol{u}.$$

Note that $j_\lambda(\theta) - j_\lambda(\theta; U_{\epsilon,K}) = j_\lambda(\theta; U^c_{\epsilon,K})$ for $U^c_{\epsilon,K} := \mathbb{R}^\nu \setminus U_{\epsilon,K}$. Therefore, $j_\lambda(\theta) \to j(\theta)$ as $\lambda \to \infty$ follows, provided the following two relations hold:

$$\lim_{\epsilon\to 0,\, K\to\infty} \limsup_{\lambda\to\infty} |j_\lambda(\theta; U^c_{\epsilon,K})| = 0, \tag{39}$$

$$\forall \epsilon, K > 0, \qquad \lim_{\lambda\to\infty} j_\lambda(\theta; U_{\epsilon,K}) = j(\theta; U_{\epsilon,K}), \tag{40}$$

where

$$j(\theta; U_{\epsilon,K}) := \int_{U_{\epsilon,K}} \mathrm{i} c_\Xi \theta \phi(\boldsymbol{u}) \left\{ \int_{\mathbb{R}_+} (\mathrm{e}^{\mathrm{i}\theta\phi(\boldsymbol{u})x} - 1) x^{-\alpha} \mathrm{d}x \right\} \mathrm{d}\boldsymbol{u},$$

c.f. (34). To show (39), we recall that $\phi \in L^1(\mathbb{R}^\nu) \cap L^\infty(\mathbb{R}^\nu) = \Phi$ and introduce the integral

$$\Phi_p(\boldsymbol{t}) := \left(\int_{U^c_{\epsilon,K}} |\phi(\boldsymbol{t} + \boldsymbol{u})|^p \mathrm{d}\boldsymbol{u} \right)^{1/p}, \qquad p = 1, 2,$$

satisfying $(\Phi_2(\boldsymbol{t}))^2 \le \Phi_1(\boldsymbol{t}) \|\phi\|_\infty \le \|\phi\|_1 \|\phi\|_\infty$ for each $\boldsymbol{t} \in \mathbb{R}^\nu$, and in particular, for each $|\boldsymbol{t}| \le 1$,

$$\begin{aligned}\Phi_1(\boldsymbol{t}) = \int_{B_K \setminus U_{\epsilon,K}} |\phi(\boldsymbol{t} + \boldsymbol{u})| \mathrm{d}\boldsymbol{u} + \int_{B^c_K} |\phi(\boldsymbol{t} + \boldsymbol{u})| \mathrm{d}\boldsymbol{u} \le \epsilon \|\phi\|_\infty \\ + \int_{B^c_{K-1}} |\phi(\boldsymbol{u})| \mathrm{d}\boldsymbol{u},\end{aligned} \tag{41}$$

where we have used $\mathrm{Leb}_\nu(B_K \setminus U_{\epsilon,K}) < \epsilon$ and which yields

$$\sup_{|\boldsymbol{t}| \le 1,\, p=1,2} \Phi_p(\boldsymbol{t}) =: \delta \to 0 \qquad \text{as } \epsilon \to 0,\ K \to \infty. \tag{42}$$

Next, using $|\Psi(z)| \le (2|z|) \wedge (|z|^2/2)$, $z \in \mathbb{R}$, and the Minkowski inequality, we get

$$\begin{aligned}\int_{U^c_{\epsilon,K}} |\Psi(\theta\xi_\lambda(\boldsymbol{u}))| \mathrm{d}\boldsymbol{u} &\le C \int_{U^c_{\epsilon,K}} |\xi_\lambda(\boldsymbol{u})| \wedge |\xi_\lambda(\boldsymbol{u})|^2 \mathrm{d}\boldsymbol{u} \\ &\le C \left(\int_{U^c_{\epsilon,K}} |\xi_\lambda(\boldsymbol{u})| \mathrm{d}\boldsymbol{u} \right) \wedge \left(\int_{U^c_{\epsilon,K}} |\xi_\lambda(\boldsymbol{u})|^2 \mathrm{d}\boldsymbol{u} \right) \\ &\le C(\xi^c_{\lambda,1} \wedge (\xi^c_{\lambda,2})^2),\end{aligned}$$

with

$$\begin{aligned}\xi^c_{\lambda,p} &:= \lambda^{-\nu/\alpha}\int_{\mathbb{R}^\nu}\Phi_p(\boldsymbol{t}/\lambda)\mathbb{I}(\boldsymbol{t}\in\Xi)\mathrm{d}\boldsymbol{t}\\ &= \lambda^{-\nu/\alpha}\Big(\int_{\bar{B}_\lambda}\Phi_p(\boldsymbol{t}/\lambda)\mathbb{I}(\boldsymbol{t}\in\Xi)\mathrm{d}\boldsymbol{t}+\int_{\bar{B}^c_\lambda}\Phi_p(\boldsymbol{t}/\lambda)\mathbb{I}(\boldsymbol{t}\in\Xi)\mathrm{d}\boldsymbol{t}\Big)\\ &\le \lambda^{-\nu/\alpha}(\delta\mathrm{Leb}_\nu(\Xi)+C\mathrm{Leb}_\nu(\Xi\cap\bar{B}^c_\lambda)) =: \xi^c_\lambda, \qquad (43)\end{aligned}$$

where $\bar{B}_\lambda := \{\boldsymbol{t}\in\mathbb{R}^\nu : |\boldsymbol{t}|\le\lambda\}$ denotes a closed ball of radius λ and $\bar{B}^c_\lambda := \mathbb{R}^\nu\setminus\bar{B}_\lambda$ denotes its complement. Hence, $\int_{U^c_{\epsilon,K}}|\Psi(\theta\xi_\lambda(\boldsymbol{u}))|\mathrm{d}\boldsymbol{u}\le C(\xi^c_\lambda\wedge(\xi^c_\lambda)^2)$ and

$$\begin{aligned}|j_\lambda(\theta;U^c_{\epsilon,K})| &\le C\lambda^\nu(\mathrm{E}[(\xi^c_\lambda)^2\mathbb{I}(\xi^c_\lambda\le 1)]+\mathrm{E}[\xi^c_\lambda\mathbb{I}(\xi^c_\lambda>1)])\\ &\le C\lambda^\nu\left(\int_0^1 x\mathrm{P}(\xi^c_\lambda>x)\mathrm{d}x+\int_1^\infty\mathrm{P}(\xi^c_\lambda>x)\mathrm{d}x\right),\end{aligned}$$

where the last inequality follows using integration by parts. Here, $\mathrm{P}(\xi^c_\lambda > x) \le \mathrm{P}(\xi^{c,1}_\lambda > x/2)+\mathrm{P}(\xi^{c,2}_\lambda > x/2)$, where $\xi^{c,1}_\lambda := \lambda^{-\nu/\alpha}\delta\mathrm{Leb}_\nu(\Xi), \xi^{c,2}_\lambda := \lambda^{-\nu/\alpha}\mathrm{Leb}_\nu(\Xi\cap\bar{B}^c_\lambda)$. By condition (28), $\mathrm{P}(\xi^{c,1}_\lambda > x)\le C\lambda^{-\nu}\delta^\alpha x^{-\alpha}$ and, therefore, $\int_0^1 x\mathrm{P}(\xi^{c,1}_\lambda > x)\mathrm{d}x + \int_1^\infty\mathrm{P}(\xi^{c,1}_\lambda > x)\mathrm{d}x \le C\lambda^{-\nu}\delta^\alpha\big(\int_0^1 x^{1-\alpha}\mathrm{d}x+\int_1^\infty x^{-\alpha}\mathrm{d}x\big)\le C\lambda^{-\nu}\delta^\alpha$. Similarly, using condition (29), $\int_0^1 x\mathrm{P}(\xi^{c,2}_\lambda > x)\mathrm{d}x+\int_1^\infty\mathrm{P}(\xi^{c,2}_\lambda > x)\mathrm{d}x\le\int_0^\infty\mathrm{P}(\xi^{c,2}_\lambda > x)\mathrm{d}x = \mathrm{E}\xi^{2,c}_\lambda = \lambda^{-\nu/\alpha}o(\lambda^{(\nu/\alpha)(1-\alpha)} = o(\lambda^{-\nu})$. Therefore, $|j_\lambda(\theta;U^c_{\epsilon,r})|\le C\delta^\alpha+o(1)$, proving (39).

Consider (40). Recall by Lusin's theorem, there exists a continuous $\phi_{\epsilon,K}:\mathbb{R}^\nu\to\mathbb{R}$ with compact support in B_K such that $\phi_{\epsilon,K}=\phi$ on $U_{\epsilon,K}$; moreover, $\|\phi_{\epsilon,K}\|_\infty\le\|\phi\|_\infty$. Since $\mathrm{Leb}_\nu(B_K\setminus U_{\epsilon,K})<\epsilon$, it suffices to prove (40) for $\phi_{\epsilon,K}$ on B_K in place of ϕ on $U_{\epsilon,K}$. More specifically, it suffices to prove the relation (40) for integrals over the sets $B^+_K := \{\boldsymbol{u}\in B_K : \phi_{\epsilon,K}(\boldsymbol{u})\ge 0\}$ and $B^-_K := \{\boldsymbol{u}\in B_K : \phi_{\epsilon,K}(\boldsymbol{u})\le 0\}$. Assume w.l.g. that $\phi=\phi_{\epsilon,K}\ge 0$, so that $B^+_K = B_K$ and $j_\lambda(\theta;B_K)=j_\lambda(\theta)$ and $j(\theta;B_K)=j(\theta)$. Then, integrating by parts as in (37),

$$j_\lambda(\theta) = \mathrm{i}\theta\int_{B_K}\left\{\int_{\mathbb{R}_+}(\mathrm{e}^{\mathrm{i}\theta x}-1)\lambda^\nu\mathrm{P}(\xi_\lambda(\boldsymbol{u})>x)\mathrm{d}x\right\}\mathrm{d}\boldsymbol{u}.$$

Hence, $\lim_{\lambda\to\infty} j_\lambda(\theta) = j(\theta)$ follows, provided the following two relations hold: for all $x > 0$, $\boldsymbol{u} \in B_K$,

$$\lim_{\lambda\to\infty} \lambda^\nu x^\alpha \mathrm{P}(\xi_\lambda(\boldsymbol{u}) > x) = c_\Xi \phi(\boldsymbol{u})^\alpha \qquad (44)$$
$$\text{and } \lambda^\nu x^\alpha \mathrm{P}(\xi_\lambda(\boldsymbol{u}) > x) < C,$$

with C independent of $\lambda > 0$ and $x, \boldsymbol{u}$. The second relation in (44) follows from (28) since $\lambda^{-\nu/\alpha}\|\phi\|_\infty \mathrm{Leb}_\nu(\Xi) \ge \xi_\lambda(\boldsymbol{u})$ for all $\boldsymbol{u} \in B_K$. Consider the first one. Note that condition (28) implies that $\tilde{\xi}_\lambda(\boldsymbol{u}) := \lambda^{-\nu/\alpha}\phi(\boldsymbol{u})\mathrm{Leb}_\nu(\Xi)$ satisfies $\lim_{\lambda\to\infty} \lambda^\nu x^\alpha \mathrm{P}(\tilde{\xi}_\lambda(\boldsymbol{u}) > x) = c_\Xi \phi(\boldsymbol{u})^\alpha$ for all $x > 0$, $\boldsymbol{u} \in B_K$. Denote $\eta_\lambda(\boldsymbol{u}) := \xi_\lambda(\boldsymbol{u}) - \tilde{\xi}_\lambda(\boldsymbol{u})$. Then, for any $\gamma > 0$,

$$\begin{aligned} \mathrm{P}(\xi_\lambda(\boldsymbol{u}) > x) &\le \mathrm{P}(\tilde{\xi}_\lambda(\boldsymbol{u}) > (1-\gamma)x) + \mathrm{P}(|\eta_\lambda(\boldsymbol{u})| > \gamma x), \\ \mathrm{P}(\xi_\lambda(\boldsymbol{u}) > x) &\ge \mathrm{P}(\tilde{\xi}_\lambda(\boldsymbol{u}) > (1+\gamma)x) - \mathrm{P}(|\eta_\lambda(\boldsymbol{u})| > \gamma x). \end{aligned} \qquad (45)$$

It remains to prove that

$$\lim_{\lambda\to\infty} \lambda^\nu x^\alpha \mathrm{P}(|\eta_\lambda(\boldsymbol{u})| > \gamma x) = 0. \qquad (46)$$

Since ϕ is uniformly continuous, for any $\delta > 0$, there is a $\tau > 0$ such that $\sup_{|\boldsymbol{t}|\le\tau} |\phi(\boldsymbol{t}+\boldsymbol{u}) - \phi(\boldsymbol{u})| < \gamma\delta$ uniformly in $\boldsymbol{u} \in B_K$. Therefore,

$$\begin{aligned} |\eta_\lambda(\boldsymbol{u})| &\le \lambda^{-\nu/\alpha} \int_{\mathbb{R}^\nu} |\phi(\boldsymbol{t}/\lambda + \boldsymbol{u}) - \phi(\boldsymbol{u})| \mathbb{I}(\boldsymbol{t} \in \Xi) \mathrm{d}\boldsymbol{t} \\ &\le \lambda^{-\nu/\alpha}(\gamma\delta \mathrm{Leb}_\nu(\Xi) + C\mathrm{Leb}_\nu(\Xi \cap \bar{B}^c_{\tau\lambda})). \end{aligned}$$

Thus, the proof of (46) is completely analogous to that of the estimation of ξ^c_λ in (43), and we omit the details. This also completes the proof of (40) and hence part (ii) as well.

(i) ($\alpha = 2$). We use approximation by an m-dependent RG RF and the CLT of Proposition 4 of Chapter 2. The approximating RF X_m is defined as

$$X_m(\boldsymbol{t}) = \sum_{j=1}^{\infty} \mathbb{I}(\boldsymbol{t} \in (\boldsymbol{u}_j + \Xi_j \cap B_{m/2})), \quad \boldsymbol{t} \in \mathbb{R}^\nu,$$

with generic grain $\Xi \cap B_{m/2} \subset \{|\boldsymbol{t}| \le m/2\}$ belonging to the ball $B_{m/2}$ of radius $m/2$. Thus, $X_m(\boldsymbol{t}_1)$ and $X_m(\boldsymbol{t}_2)$ are independent when

$|\boldsymbol{t}_1 - \boldsymbol{t}_2| > m$, a fact which follows from the independence property of Poisson stochastic integrals with disjoint supports, as $\mathbb{I}(\boldsymbol{t}_i - \boldsymbol{u} \in \Xi \cap B_{m/2}, i = 1,2) \le \mathbb{I}(|\boldsymbol{t}_i - \boldsymbol{u}| \le m/2, i = 1,2) \le \mathbb{I}(|\boldsymbol{t}_1 - \boldsymbol{t}_2| \le m) = 0$ for any $\boldsymbol{u} \in \mathbb{R}^\nu$ by triangle inequality. Note that $r_{X_m}(\boldsymbol{t}) := \mathrm{Cov}(X_m(\mathbf{0}), X_m(\boldsymbol{t})) = \mathrm{E}\mathrm{Leb}_\nu((\Xi \cap B_{m/2}) \cap ((\Xi \cap B_{m/2}) - \boldsymbol{t}))$. By Proposition 4 of Chapter 2, for any $m > 0$,

$$\lambda^{-\nu/2}(X_m(\phi) - \mathrm{E}X_m(\phi)) \xrightarrow{\mathrm{d}} \sigma_{2,m} W(\phi), \tag{47}$$

where $\sigma_{2,m}^2 = \mathrm{E}\mathrm{Leb}_\nu(\Xi \cap B_{m/2})^2 \to \sigma_2^2 = E\mathrm{Leb}_\nu(\Xi)^2$ as $m \to \infty$. We also see that $\lambda^{-\nu/2}\mathrm{Var}(X(\phi) - X_m(\phi)) = \mathrm{E}\int_{\mathbb{R}^{2\nu}}(\phi \star \phi)(\frac{\boldsymbol{t}_1 - \boldsymbol{t}_2}{\lambda})\mathbb{I}(\boldsymbol{t}_i \in \Xi \cap B^c_{m/2}, i = 1,2)\mathrm{d}\boldsymbol{t}_1\mathrm{d}\boldsymbol{t}_2 \le C\mathrm{E}\int_{\mathbb{R}^{2\nu}} \mathbb{I}(\boldsymbol{t}_i \in \Xi \cap B^c_{m/2}, i = 1,2)\mathrm{d}\boldsymbol{t}_1\mathrm{d}\boldsymbol{t}_2 = C\mathrm{E}\mathrm{Leb}_\nu(\Xi \cap B^c_{m/2})^2 \to 0 (m \to \infty)$, where C is independent of λ, m and $B^c_{m/2} := \mathbb{R}^\nu \setminus B_{m/2}$. Hence,

$$\lim_{m\to\infty} \limsup_{\lambda\to\infty} \lambda^{-\nu/2}\mathrm{Var}(X(\phi) - X_m(\phi)) = 0. \tag{48}$$

Relations (47) and (48) prove (30) for $\alpha = 2$, thereby completing the proof of Theorem 6. □

Scaling limits with aggregation. Recall from Section 2 of Chapter 1 that isotropic scaling with aggregation refers to the limit distribution of sums,

$$X_{\lambda,M}(\phi) := \sum_{i=1}^{M} \int_{\mathbb{R}^\nu} \phi(\boldsymbol{t}/\lambda) X_i(\boldsymbol{t})\mathrm{d}\boldsymbol{t}, \quad \phi \in \Phi, \tag{49}$$

of i.i.d. copies of $X_\lambda(\phi)$ in (1) of Chapter 3 when M increases with λ at a given rate, say, $M = [\lambda^\gamma]$ for some $\gamma > 0$. In the case of Gaussian scaling limits in Propositions 4 and 5 of Chapter 3, aggregation leads to the same results and is not very interesting. On the other hand, the discrepancy between finite variance of the summands in (49) and infinite variance limits as in Theorem 6(ii) hints at the possibility that the growth rate of M may have a deciding effect on the limit behaviour of (49). Indeed, this happens for randomly homothetic X, leading to a trichotomy of the limit distribution depending on γ, as described in Theorem 10.

Recall the simple fact that for RG models based on Poisson processes, as in (15), summing over M independent copies reduces to multiplying by M the intensity of the underlying Poisson process. Accordingly, let the aggregate in (49) of the randomly homothetic RG model in Proposition 5 be represented as the stochastic integral

$$X_{\lambda,M}(\phi) = \int_{\mathbb{R}^\nu \times \mathbb{R}_+ \times \mathbb{F}(\mathbb{R}^\nu)} \left\{ \int_{\mathbb{R}^\nu} \mathbb{I}(\boldsymbol{t} - \boldsymbol{u} \in r^{1/\nu}\xi)\phi(\boldsymbol{t}/\lambda)\mathrm{d}\boldsymbol{t} \right\} \times \mathcal{N}_M(\mathrm{d}\boldsymbol{u}, \mathrm{d}r, \mathrm{d}\xi) \tag{50}$$

with respect to the Poisson random measure $\mathcal{N}_M(\mathrm{d}\boldsymbol{u}, \mathrm{d}r, \mathrm{d}\xi)$ on $\mathbb{R}^\nu \times \mathbb{R}_+ \times \mathbb{F}(\mathbb{R}^\nu)$ with intensity $M\mathrm{d}\boldsymbol{u} f(r)\mathrm{d}r\mathrm{P}(\Xi^0 \in \mathrm{d}\xi)$. We also introduce the stochastic integrals

$$\begin{aligned} B_\alpha(\phi) &:= \int_{\mathbb{R}^\nu \times \mathbb{R}_+ \times \mathbb{F}(\mathbb{R}^\nu)} W_\alpha(\mathrm{d}\boldsymbol{u}, \mathrm{d}r, \mathrm{d}\xi) \int_{\mathbb{R}^d} \phi(\boldsymbol{t})\mathbb{I}(\boldsymbol{t} - \boldsymbol{u} \in r^{1/\nu}\xi)\mathrm{d}\boldsymbol{t}, \\ J_\alpha(\phi) &:= \int_{\mathbb{R}^\nu \times \mathbb{R}_+ \times \mathbb{F}(\mathbb{R}^\nu)} \widetilde{\mathcal{N}}_\alpha(\mathrm{d}\boldsymbol{u}, \mathrm{d}r, \mathrm{d}\xi) \int_{\mathbb{R}^\nu} \phi(\boldsymbol{t})\mathbb{I}(\boldsymbol{t} - \boldsymbol{u} \in r^{1/\nu}\xi)\mathrm{d}\boldsymbol{t} \end{aligned} \tag{51}$$

w.r.t. the corresponding Gaussian white noise measure $W_\alpha(\mathrm{d}\boldsymbol{u}, \mathrm{d}r, \mathrm{d}\xi)$ and Poisson random measure $\widetilde{\mathcal{N}}_\alpha(\mathrm{d}\boldsymbol{u}, \mathrm{d}r, \mathrm{d}\xi) = \mathcal{N}_\alpha(\mathrm{d}\boldsymbol{u}, \mathrm{d}r, \mathrm{d}\xi) - \mathrm{E}\mathcal{N}_\alpha(\mathrm{d}\boldsymbol{u}, \mathrm{d}r, \mathrm{d}\xi)$ with zero mean and the same variance $\mathrm{E}W_\alpha(\mathrm{d}\boldsymbol{u}, \mathrm{d}r, \mathrm{d}\xi)^2 = \mathrm{E}\widetilde{\mathcal{N}}_\alpha(\mathrm{d}\boldsymbol{u}, \mathrm{d}r, \mathrm{d}\xi)^2 = \mathrm{E}\mathcal{N}_\alpha(\mathrm{d}\boldsymbol{u}, \mathrm{d}r, \mathrm{d}\xi) = c_f r^{-1-\alpha}\mathrm{d}r\mathrm{d}\boldsymbol{u}\mathrm{P}(\Xi^0 \in \mathrm{d}\xi)$, with $c_f > 0, \alpha \in (1, 2)$ as in Proposition 5. Therefore, $\mathrm{E}B_\alpha(\phi)^2 = \mathrm{E}J_\alpha(\phi)^2$ and, with $L(\boldsymbol{z})$ in (25),

$$\begin{aligned} \mathrm{E}B_\alpha(\phi)^2 &= c_f \int_{\mathbb{R}_+ \times \mathbb{R}^\nu} r^{-1-\alpha}\mathrm{d}r\mathrm{d}\boldsymbol{u}\mathrm{E}\Big(\int_{\mathbb{R}^\nu} \phi(\boldsymbol{t})\mathbb{I}(\boldsymbol{t} - \boldsymbol{u} \in r^{1/\nu}\Xi^0)\mathrm{d}\boldsymbol{t} \Big)^2 \\ &= c_f \int_{\mathbb{R}^{2\nu}} \phi(\boldsymbol{t}_1)\phi(\boldsymbol{t}_2)|\boldsymbol{t}_1 - \boldsymbol{t}_2|^{-\nu(\alpha-1)} L\left(\frac{\boldsymbol{t}_1 - \boldsymbol{t}_2}{|\boldsymbol{t}_1 - \boldsymbol{t}_2|} \right) \mathrm{d}\boldsymbol{t}_1\mathrm{d}\boldsymbol{t}_2 \end{aligned} \tag{52}$$

coincides with $c(\phi, \phi)$ in (7) of Chapter 3 for $c_f = 1$, provided the exponents $\nu(\alpha - 1) = \nu - 4d$ and angular functions $L(\boldsymbol{z})$ in these expressions coincide. Therefore, the Gaussian RFs $B_\alpha(\phi)$ in (51) and $W_d(\phi)$ in (14) of Chapter 3 have the same distribution: $\alpha = 2 - (4d/\nu)$.

Theorem 10. *Let X be a randomly homothetic RG model satisfying the conditions of Proposition* 5 *and $X_{\lambda,M}(\phi)$ as in* (50)*, $M = \lambda^\gamma$ ($\gamma > 0$). Then, for any $\phi \in \Phi$, as $\lambda \to \infty$,*

$$\lambda^{-H(\gamma)}(X_{\lambda,M}(\phi) - \mathrm{E}X_{\lambda,M}(\phi))$$
$$\xrightarrow{\mathrm{d}} \begin{cases} B_\alpha(\phi), & \gamma > \nu(\alpha-1),\ H(\gamma) = \frac{\gamma+(3-\alpha)\nu}{2}, \\ c_\alpha L_\alpha(\phi), & \gamma < \nu(\alpha-1),\ H(\gamma) = \frac{\gamma+\nu}{\alpha}, \\ J_\alpha(\phi), & \gamma = \nu(\alpha-1),\ H(\gamma) = \nu, \end{cases} \tag{53}$$

with the limit RFs B_α and J_α defined in (51) *and c_α and L_α as in* (30) *and Corollary* 9.

Proof. Using similar notation as in the proof of Theorem 6, consider the log-characteristic function $j_\lambda(\theta) := \log \mathrm{E}\exp\{\mathrm{i}\theta\lambda^{-H(\gamma)} (X_{\lambda,M}(\phi) - \mathrm{E}X_{\lambda,M}(\phi))\}, \theta \in \mathbb{R}$ of the l.h.s. in (53). We need to show that it converges to the corresponding log-characteristic function $j(\theta)$ of the r.h.s. as $\lambda \to \infty$, for each $\theta \in \mathbb{R}$ fixed.

Case: $\gamma > \nu(\alpha-1)$. Set $\tilde{j}_\lambda(\theta) := -(\theta^2/2)\mathrm{Var}(\lambda^{-H(\gamma)} X_{\lambda,M}(\phi))$, $j(\theta) := \log \mathrm{E}\exp\{\mathrm{i}\theta B_\alpha(\phi)\} = -(\theta^2/2)\mathrm{E}B_\alpha(\phi)^2$. Since $\mathrm{Var}(X_{\lambda,M}(\phi) = M\mathrm{Var}(X_\lambda(\phi))$, $\tilde{j}_\lambda(\theta) \to j(\theta)$ $(\lambda \to \infty)$ for any $\gamma > 0$ follows from the convergence of the variance $\mathrm{Var}(\lambda^{-H(0)} X_\lambda(\phi) \to \mathrm{E}B_\alpha(\phi)^2$; see (6), (7) of Chapter 3, and (52). Next,

$$j_\lambda(\theta) - \tilde{j}_\lambda(\theta) = \lambda^\gamma \int_{\mathbb{R}^\nu \times \mathbb{R}_+} \mathrm{d}\boldsymbol{u} f(r)\mathrm{d}r \mathrm{E}\Bigg\{\Psi\Bigg(\frac{\theta}{\lambda^{H(\gamma)}} \int_{\mathbb{R}^\nu} \phi(\boldsymbol{t}/\lambda)$$
$$\times\ \mathbb{I}(\boldsymbol{t}-\boldsymbol{u} \in r^{1/\nu}\Xi^0)\mathrm{d}\boldsymbol{t}\Bigg)$$
$$+(1/2)\left(\frac{\theta}{\lambda^{H(\gamma)}} \int_{\mathbb{R}^\nu} \phi(\boldsymbol{t}/\lambda)\mathbb{I}(\boldsymbol{t}-\boldsymbol{u} \in r^{1/\nu}\Xi^0)\mathrm{d}\boldsymbol{t}\right)^2\Bigg\}.$$

Using $|\Psi(z) + (1/2)z^2| \le |z|^3$ and (6) of Chapter 3, we get

$$|j_\lambda(\theta) - \tilde{j}_\lambda(\theta)| \le C\lambda^\gamma \int_{\mathbb{R}^\nu \times \mathbb{R}_+} \mathrm{E}\Bigg|\lambda^{-H(\gamma)} \int_{\mathbb{R}^\nu} \phi(\boldsymbol{t}/\lambda)\mathbb{I}$$
$$\times\ (\boldsymbol{t}-\boldsymbol{u} \in r^{1/\nu}\Xi^0)\mathrm{d}\boldsymbol{t}\Bigg|^3 \mathrm{d}\boldsymbol{u} f(r)\mathrm{d}r$$

$$\leq C\lambda^{-(\gamma/2)-(3/2)(3-\alpha)\nu}\left(\int_{\mathbb{R}^\nu}|\phi(\boldsymbol{t}/\lambda)|\mathrm{d}\boldsymbol{t}\right)\mathrm{Var}(X_M(\phi))$$
$$\leq C\lambda^{-(\gamma-\nu(\alpha-1))/2} = o(1)$$

since $|\int_{\mathbb{R}^\nu}\phi(\boldsymbol{t}/\lambda)\mathbb{I}(\boldsymbol{t}-\boldsymbol{u}\in r^{1/\nu}\Xi^0)\mathrm{d}\boldsymbol{t}| \leq \int_{\mathbb{R}^\nu}|\phi(\boldsymbol{t}/\lambda)|\mathrm{d}\boldsymbol{t} = \lambda^\nu\int_{\mathbb{R}^\nu}|\phi(\boldsymbol{t})|\mathrm{d}\boldsymbol{t}$ holds a.s. and uniformly in $\boldsymbol{u}\in\mathbb{R}^\nu$. Thus, $j_\lambda(\theta)\to j(\theta)$ $(\forall\theta\in\mathbb{R})$.

Case: $\gamma<\nu(\alpha-1)$. The proof is similar to that of Theorem 6(ii) but simpler due to the particular form $\Xi=R^{1/\nu}\Xi^0$ of the generic grain. By change of variables, we rewrite $j_\lambda(\theta)$ as

$$j_\lambda(\theta)=\int_{\mathbb{R}^\nu\times\mathbb{R}_+}\mathrm{E}\Psi\left(\theta\lambda'\int_{\mathbb{R}^\nu}\phi(\boldsymbol{t}+\boldsymbol{u})\mathbb{I}\left(\boldsymbol{t}\in\left(\frac{r}{\lambda'}\right)^{\frac{1}{\nu}}\Xi^0\right)\mathrm{d}\boldsymbol{t}\right)$$
$$\times\tilde{f}(r;\lambda'')\frac{\mathrm{d}\boldsymbol{u}\mathrm{d}r}{r^{\alpha+1}}, \tag{54}$$

with $\tilde{f}(r,\lambda)$ being as in (26) and

$$\lambda':=\lambda^{\nu-\frac{\gamma+\nu}{\alpha}},\quad \lambda'':=\lambda^{\frac{\gamma+\nu}{\alpha}} \tag{55}$$

depending on γ and satisfying $\lambda'\to\infty$, $\lambda''\to\infty$ and $(\lambda\to\infty)$ for $\gamma<\nu(\alpha-1)$. Note that

$$h_\lambda(\boldsymbol{u},r):=\lambda'\int_{\mathbb{R}^\nu}\phi(\boldsymbol{t}+\boldsymbol{u})\mathbb{I}\left(\boldsymbol{t}\in\left(\frac{r}{\lambda'}\right)^{\frac{1}{\nu}}\Xi^0\right)\mathrm{d}\boldsymbol{t}\to r\phi(\boldsymbol{u})\mathrm{Leb}_\nu(\Xi^0)$$
$$=: h(\boldsymbol{u},r) \tag{56}$$

a.s. at each continuity point $\boldsymbol{u}$ of ϕ. Let $U_{\epsilon,K}\subset\mathbb{R}^\nu$ be as in the proof of Theorem 6(ii), $U^c_{\epsilon,K}=\mathbb{R}^\nu\setminus U_{\epsilon,K}$, and

$$j_\lambda(\theta;\epsilon,K):=\int_{U_{\epsilon,K}\times\mathbb{R}_+}\mathrm{E}\Psi\big(\theta h_\lambda(\boldsymbol{u},r)\big)\tilde{f}(r;\lambda'')\mathrm{d}\boldsymbol{u}\mathrm{d}r/r^{\alpha+1},$$

$$j(\theta;\epsilon,K):=c_f\int_{U_{\epsilon,K}\times\mathbb{R}_+}\Psi\big(\theta h(\boldsymbol{u},r)\big)\mathrm{d}\boldsymbol{u}\mathrm{d}r/r^{\alpha+1},$$

and $j^c_\lambda(\theta;\epsilon,K):=j_\lambda(\theta)-j_\lambda(\theta;\epsilon,K), j^c(\theta;\epsilon,K):=j(\theta)-j(\theta;\epsilon,K)$. As in (39)–(40), it suffices to show

$$\lim_{\epsilon\to0,K\to\infty}\big(\limsup_{\lambda\to\infty}|j^c_\lambda(\theta;\epsilon,K)|+|j^c(\theta;\epsilon,K)|\big) = 0, \tag{57}$$

$$\lim_{\lambda\to\infty}j_\lambda(\theta;\epsilon,K) = j(\theta;\epsilon,K)\qquad(\forall\,\epsilon,K>0). \tag{58}$$

Then, $\int_{U^c_{\epsilon,K}} \big|\Psi\big(\theta h(\boldsymbol{u},r)\big)\big|\mathrm{d}\boldsymbol{u} \leq C\big(\int_{U^c_{\epsilon,K}}|h(\boldsymbol{u},r)|\mathrm{d}\boldsymbol{u}\big) \wedge \big(\int_{U^c_{\epsilon,K}} |h(\boldsymbol{u},r)|\mathrm{d}\boldsymbol{u}\big)^2$, where

$$\begin{aligned}\int_{U^c_{\epsilon,K}} |h(\boldsymbol{u},r)|\mathrm{d}\boldsymbol{u} &\leq \lambda' \int_{\mathbb{R}^\nu} \mathbb{I}\left(\boldsymbol{t} \in \left(\frac{r}{\lambda'}\right)^{\frac{1}{\nu}} \Xi^0\right)\mathrm{d}\boldsymbol{t} \int_{U^c_{\epsilon,K}} |\phi(\boldsymbol{t}+\boldsymbol{u})|\mathrm{d}\boldsymbol{u} \\ &= r\int_{\mathbb{R}^\nu} \mathbb{I}(\boldsymbol{t}\in\Xi^0)\Phi_1\left(\left(\frac{r}{\lambda'}\right)^{1/\nu}\boldsymbol{t}\right)\mathrm{d}\boldsymbol{t} \\ &\leq r\int_{\mathbb{R}^\nu} \mathbb{I}(|\boldsymbol{t}|\leq C)\Phi_1\left(\left(\frac{r}{\lambda'}\right)^{1/\nu}\boldsymbol{t}\right)\mathrm{d}\boldsymbol{t} \\ &\leq Cr(\delta+\mathbb{I}(r>\lambda')),\end{aligned}$$

with $\Phi_1(\boldsymbol{t})$ and δ as in (41) and (42). Therefore, $|j^c_\lambda(\theta;\epsilon,K)| \leq C(\int_0^1 r^{1-\alpha}(\delta+\mathbb{I}(r>\lambda'))\mathrm{d}r + \int_1^\infty r^{-\alpha}(\delta+\mathbb{I}(r>\lambda'))^2\mathrm{d}r)$, implying $\lim_{\epsilon\to 0,K\to\infty}\limsup_{\lambda\to\infty}|j^c_\lambda(\theta;\epsilon,K) = 0$. The rest of the proof of (57)–(58), viz., proving the convergence $j_\lambda(\theta)\to j(\theta)$ ($\forall\theta\in\mathbb{R}$) in the case of $\gamma<\nu(\alpha-1)$, is rather easy and omitted.

Case: $\gamma=\nu(\alpha-1)$. The expression in (54) for $j_\lambda(\theta)$ is valid, with $\lambda'=1, \lambda''=\lambda^\nu$ (see (55)), yielding

$$j_\lambda(\theta) = \int_{\mathbb{R}^\nu\times\mathbb{R}_+} \mathrm{E}\Psi\left(\theta\int_{\mathbb{R}^\nu}\phi(\boldsymbol{t}+\boldsymbol{u})\mathbb{I}(\boldsymbol{t}\in r^{\frac{1}{\nu}}\,\Xi^0)\mathrm{d}\boldsymbol{t}\right)\tilde{f}(r;\lambda'')\frac{\mathrm{d}\boldsymbol{u}\mathrm{d}r}{r^{\alpha+1}}, \tag{59}$$

whereas $j(\theta)$ is written as an integral on the r.h.s. of (59), with $\tilde{f}(r;\lambda'')$ replaced by its limit c_f, viz.,

$$j(\theta) = c_f\int_{\mathbb{R}^\nu\times\mathbb{R}_+} \mathrm{E}\Psi\left(\theta\int_{\mathbb{R}^\nu}\phi(\boldsymbol{t}+\boldsymbol{u})\mathbb{I}(\boldsymbol{t}\in r^{\frac{1}{\nu}}\,\Xi^0)\mathrm{d}\boldsymbol{t}\right)\frac{\mathrm{d}\boldsymbol{u}\mathrm{d}r}{r^{\alpha+1}};$$

see (51) and (23) of Chapter 2 for the definition of $J_\alpha(\phi)$ in (51) and the characteristic function of the Poisson stochastic integral. The proof of $j_\lambda(\theta)\to j(\theta)$ reduces to justifying the passage to the limit in the integral on the r.h.s. of (59). A simple proof of the last fact, making use of (Pratt's) Lemma 30 of Chapter 2, goes as follows. Since $|\Psi(z)|\leq |z|^2/2$, the integrand in (59) is dominated by $(1/2)\big|\theta\int_{\mathbb{R}^\nu}\phi(\boldsymbol{t}+\boldsymbol{u})\mathbb{I}(\boldsymbol{t}\in r^{\frac{1}{\nu}}\,\Xi^0)\mathrm{d}\boldsymbol{t}\big|^2\tilde{f}(r;\lambda'')/r^{\alpha+1} =: \bar{g}_\lambda(\boldsymbol{u},r,\Xi^0)$, which tends to $(c_f/2)\big|\theta \int_{\mathbb{R}^\nu} \phi(\boldsymbol{t}+\boldsymbol{u})\mathbb{I}(\boldsymbol{t} \in r^{\frac{1}{\nu}}\Xi^0)\mathrm{d}\boldsymbol{t}\big|^2/r^{\alpha+1} =: \bar g(\boldsymbol{u}, r, \Xi^0)$ a.s. for any $(\boldsymbol{u}, r) \in \mathbb{R}^\nu \times \mathbb{R}_+$. We also have that

$$\int_{\mathbb{R}^\nu\times\mathbb{R}_+} \mathrm{E}\bar g_\lambda(\boldsymbol{u}, r, \Xi^0)\mathrm{d}\boldsymbol{u}\mathrm{d}r = (\theta^2/2)\mathrm{Var}(\lambda^{-\nu/\alpha}X_\lambda(\phi))$$

$$\to (\theta^2/2)\mathrm{E}J_\alpha(\phi)^2 = \int_{\mathbb{R}^\nu\times\mathbb{R}_+} \mathrm{E}\bar g(\boldsymbol{u}, r, \Xi^0)\mathrm{d}\boldsymbol{u}\mathrm{d}r,$$

as in the case of $\gamma > \nu(\alpha - 1)$ above. Theorem 10 is proved. □

4 Notes

1. Shot-noise and RG models belong to the large class of stationary infinitely divisible RFs. Many properties of infinitely divisible random vectors and processes are given in Sato [232]. Rajput and Rosinski [222] provide a general stochastic integral representation of an infinitely divisible RF. The Boolean model is discussed in detail in Matheron [181] and Serra [234]. The Boolean model is a basic model in stereology and stochastic geometry Stoyan *et al.* [242].

2. *Poisson infinite-source and trawl processes.* The infinite-source Poisson process or M/G/∞ queue is an RG process on $\mathbb{R}$ with the generic grain $\Theta = \mathbb{I}(t \in]0, R]), t \in \mathbb{R}$, where $R > 0$ is an r.v. with $\mathrm{E}R < \infty$. The above process is a basic model in the queueing theory and broadband network traffic research and is discussed in Section 2 of Chapter 9. Accordingly, this process has a representation,

$$X(t) = \int_{(-\infty,t]\times\mathbb{R}_+} \mathbb{I}(r > t - u)\mathcal{N}(\mathrm{d}u, \mathrm{d}r), \quad t \in \mathbb{R}, \tag{60}$$

w.r.t. the Poisson random measure $\mathcal{N}(\mathrm{d}u, \mathrm{d}r)$ on $\mathbb{R}\times\mathbb{R}_+$, with mean $\mathrm{E}\mathcal{N}(\mathrm{d}u, \mathrm{d}r) = \mathrm{d}u\mathrm{d}F(r)$, where $F(r) = \mathrm{P}(R \le r)$ is the d.f. of the r.v. R. A related class of Poisson trawl processes with the (deterministic) trawl function $0 \le a(t) < \infty, t > 0$, is defined as a stochastic integral,

$$\tilde X(t) = \int_{(-\infty,t]\times\mathbb{R}_+} \mathbb{I}(z < a(t-u))\tilde{\mathcal{N}}(\mathrm{d}u, \mathrm{d}z), \quad t \in \mathbb{R}, \tag{61}$$

w.r.t. the Poisson random measure $\tilde{\mathcal{N}}(\mathrm{d}u, \mathrm{d}z)$ on $\mathbb{R}\times\mathbb{R}_+$, with mean $\mathrm{E}\tilde{\mathcal{N}}(\mathrm{d}u, \mathrm{d}z) = \mathrm{d}u\mathrm{d}z$. Poisson trawl processes are particular cases of integer-valued trawl processes, discussed in Barndorff-Nielsen *et al.* [24]. Through simple observation, random processes X in (60) and $\tilde{X}$ in (61) exhibit the same finite-dimensional distributions, provided $a(t)$ and $F(r)$ are related as

$$a(t) = 1 - F(t), \quad t > 0. \tag{62}$$

The definitions in (60) and (61) extend to more general integer-valued random measures (see Barndorff-Nielsen *et al.* [24] and Roueff *et al.* [226]), as well as to RF setups, suggesting a close relation between RG and trawl RFs.

As noted by McKenzie [182], the modelling of discrete variate time series is a challenging task due to the need for generating sequences of dependent random variates with a particular marginal distribution and correlation structure. A discrete-time Poisson trawl process can be as in (61), with the trawl function $a(t)$ taking constant values on intervals $t \in [j, j+1[, j \in \mathbb{N}$. The above trawl process can display a variety of SRD and LRD covariance structures; see Doukhan *et al.* [85, Prop. 2]. In particular, the covariance function

$$r(t) := \frac{\Gamma(t+d)}{\Gamma(t-d+1)} \frac{\Gamma(1-d)}{\Gamma(d)}, \quad t \in \mathbb{N} \tag{63}$$

of an ARFIMA$(0, d, 0)$ process with unit variance (see (63)) corresponds to the trawl function

$$a(t) = r(t) - r(t+1) = r(t)\frac{1-2d}{t+1-d}, \quad t \in \mathbb{N}. \tag{64}$$

A discrete-time M/G/∞ queueing process with (multiple) Poisson arrivals and service time distribution, $F(x) = \mathrm{P}(R \le x), x \in \mathbb{N}$, can be defined as

$$X(t) = \sum_{s \in \mathbb{N}} N_s \mathbb{I}(s < t \le s + R_s), \quad t \in \mathbb{N}, \tag{65}$$

where $\{N_s; s \in \mathbb{Z}\}$ are i.i.d. Poisson r.v.s with $\mathrm{E}N_s = 1$ and $\{R_s; s \in \mathbb{Z}\}$ are i.i.d. r.v.s with distribution F. Note that for F in (62) and $a(t)$ in (64), $\mathrm{E}R = \sum_{t=0}^{\infty} a(t) = r(0) = \Gamma(1-2d)/\Gamma(1-d)^2$.

The above facts and the discussion around (60)–(64) lead to the following conclusion.

Corollary 11. *Let the service times* $R_s \overset{\mathrm{d}}{=} R$ *in* (65) *take values in* $\mathbb{N}$ *with d.f.*

$$\mathrm{P}(R \le x) = 1 - a(x) = 1 - r(x)\frac{1-2d}{x+1-d},$$
$$x = 0, 1, \ldots, \qquad 0 < d < 1/2.$$

Then, the $M/G/\infty$ *queueing process* X *in* (65) *is stationary, having a marginal Poisson distribution with mean* $\mathrm{E}X(t) = \mathrm{E}R = \Gamma(1-2d)/\Gamma(1-d)^2$ *and ARFIMA*$(0,d,0)$ *covariance function* $r(t) = \mathrm{Cov}(X(0), X(t))$ *in* (63).

3. The scaling limits of the RG model (including aggregation) are discussed in Kaj *et al.* [129], Biermé *et al.* [38], and Surgailis [253]. Biermé *et al.* [38] include a study of small-scale scaling limits. Further scaling results for RG models can be found in Pilipauskaitė and Surgailis [208], and Biermé *et al.* [40, 41].

Chapter 5

Nonlinear Subordinated Random Fields

This chapter deals with isotropic scaling limits of integrals in (3) of nonlinear functions G of linear RFs X on $\mathbb{R}^\nu$ and $\mathbb{Z}^\nu$. This question is effectively answered in the case of Gaussian and random grain RFs X using Hermite and Charlier expansions. Particular attention is given to non-Gaussian MA RFs X and the first-order reduction principle, reducing the limit distribution of the nonlinear RF $G(X)$ to that of the linear RF X multiplied by the first Appell coefficient of G. Applications of the above results include the limit distribution of the empirical process and the Kolmogorov–Smirnov statistic from observations of X on a large spatial domain, limit distribution of volume fraction estimation of a random Boolean set, and the limit distribution of solutions of Burgers' equation with RF initial data.

1 Nonlinear Functions of Gaussian MA RFs

This section obtains isotropic scaling limits of instantaneous functions,

$$Y(\boldsymbol{t}) := G(X(\boldsymbol{t})), \tag{1}$$

of an MA Gaussian RF,

$$X(\boldsymbol{t}) = \int_{\mathbb{R}^\nu} a(\boldsymbol{t} - \boldsymbol{u}) W(\mathrm{d}\boldsymbol{u}), \quad \boldsymbol{t} \in \mathbb{R}^\nu, \tag{2}$$

w.r.t. Gaussian white noise $W(\boldsymbol{u})$ with zero mean and variance $\mathrm{d}\boldsymbol{u}$ and MA coefficients $a(\boldsymbol{t})$ satisfying Assumption (A)$(d;\mathbb{R}^\nu)$, $\mathrm{E}X(\boldsymbol{t})^2 = \int_{\mathbb{R}^\nu} a(\boldsymbol{t})^2 \mathrm{d}\boldsymbol{t} = 1$. We write $r_X(\boldsymbol{t}) = \mathrm{Cov}(X(\mathbf{0}), X(\boldsymbol{t}))$, $r_Y(\boldsymbol{t}) = \mathrm{Cov}(Y(\mathbf{0})$, and $Y(\boldsymbol{t}))$ for the respective covariance functions and

$$Y_\lambda(\phi) = \int_{\mathbb{R}^\nu} \phi(\boldsymbol{t}/\lambda) Y(\boldsymbol{t}) \mathrm{d}\boldsymbol{t} \tag{3}$$

for the integral of RF Y with rescaled test function $\phi \in \Phi$. Recall from Proposition 4 (11) of Chapter 1 that

$$r_X(\boldsymbol{t}) = |\boldsymbol{t}|^{4d-\nu}\left(L_X\left(\frac{\boldsymbol{t}}{|\boldsymbol{t}|}\right) + o(1)\right), \quad |\boldsymbol{t}| \to \infty, \tag{4}$$

where $L_X(\cdot) \in C(S_{\nu-1})$ is given by

$$L_X(\boldsymbol{t}) := \int_{\mathbb{R}^\nu} \frac{\ell(\boldsymbol{s}/|\boldsymbol{s}|)\ell((\boldsymbol{t}+\boldsymbol{s})/|\boldsymbol{t}+\boldsymbol{s}|)}{|\boldsymbol{s}|^{\nu-2d}|\boldsymbol{t}+\boldsymbol{s}|^{\nu-2d}} \mathrm{d}\boldsymbol{s}, \quad \boldsymbol{t} \in S_{\nu-1}. \tag{5}$$

Also, recall from (12) of Chapter 3 the notation $a_\infty(\boldsymbol{t}) := |\boldsymbol{t}|^{2d-\nu}\ell(\boldsymbol{t}/|\boldsymbol{t}|)$ and

$$(a_\infty \star a_\infty)(\boldsymbol{t}) = |\boldsymbol{t}|^{4d-\nu} L_X(\boldsymbol{t}), \quad \boldsymbol{t} \in \mathbb{R}^\nu, \tag{6}$$

with $L_X(\cdot)$ in (5). Any function G with $\mathrm{E}G(Z)^2 < \infty$ can be expanded in Hermite polynomials:

$$G(x) = \mathrm{E}G(Z) + \sum_{j=k}^{\infty} \frac{h_G(j)}{j!} H_j(x), \tag{7}$$

where the integer $k = k_H(G) \geq 1$ is called the Hermite rank of G; see Section 4 of Chapter 2.

Introduce a *Hermite RF of order* k indexed by functions $\phi \in \Phi$ defined as the k-tuple Wiener–Itô integral

$$\begin{aligned} \mathcal{H}_k(\phi) &:= \int_{\mathbb{R}^{\nu k}} \left\{ \int_{\mathbb{R}^\nu} \phi(\boldsymbol{t}) \prod_{i=1}^{k} a_\infty(\boldsymbol{t} - \boldsymbol{u}_i) \mathrm{d}\boldsymbol{t} \right\} W(\mathrm{d}\boldsymbol{u}_1) \cdots W(\mathrm{d}\boldsymbol{u}_k) \\ &\equiv \int_{\mathbb{R}^{\nu k}} h_k \mathrm{d}^k W, \end{aligned} \tag{8}$$

with integrand

$$h_k(\boldsymbol{u}_1, \ldots, \boldsymbol{u}_k) := \int_{\mathbb{R}^\nu} \phi(\boldsymbol{t}) \prod_{i=1}^{k} a_\infty(\boldsymbol{t} - \boldsymbol{u}_i) \mathrm{d}\boldsymbol{t}; \tag{9}$$

see Section 3 of Chapter 2, (102), for the definition of the Wiener–Itô integral. The existence of the stochastic integral in (8) for $k < \nu/(\nu - 4d)$ and any $\phi \in \Phi$ or the fact that $\|h_k\|_{2,k} \equiv \|h_k\|_{L^2(\mathbb{R}^{\nu k})} < \infty$ follows from (6) and

$$\begin{aligned}
\|h_k\|_{2,k}^2 &\le \int_{\mathbb{R}^{2\nu}} |\phi(\boldsymbol{t}_1)\phi(\boldsymbol{t}_2)| \prod_{i=1}^{k} \int_{\mathbb{R}^\nu} |a_\infty(\boldsymbol{t}_1 + \boldsymbol{u}_i) a_\infty(\boldsymbol{t}_2 + \boldsymbol{u}_i)| \mathrm{d}\boldsymbol{u}_i \\
&\le C \int_{\mathbb{R}^{2\nu}} |\phi(\boldsymbol{t}_1)\phi(\boldsymbol{t}_2)| |\boldsymbol{t}_1 - \boldsymbol{t}_2|^{-k(\nu - 4d)} \mathrm{d}\boldsymbol{t}_1 \mathrm{d}\boldsymbol{t}_2 \ < \infty
\end{aligned}$$

(with the last integral converging if $k(\nu - 4d) < \nu$). We also note the stationarity and self-similarity properties of the Hermite RF:

$$\begin{aligned}
&\mathcal{H}_k(\phi(\cdot + \boldsymbol{y})) \stackrel{\mathrm{d}}{=} \mathcal{H}_k(\phi), \quad \mathcal{H}_k(\phi(\cdot/\lambda)) \stackrel{\mathrm{d}}{=} \lambda^{\nu(1 - \frac{k}{2}) + 2kd} \mathcal{H}_k(\phi), \\
&\forall\, \boldsymbol{y} \in \mathbb{R}^\nu, \ \lambda > 0, \ \phi \in \Phi,
\end{aligned}$$

which follow from the change of variables formulas in (104) of Chapter 2, the scaling property of a_∞ in (13) of Chapter 3, and the fact that $\mathcal{H}_1(\phi) = W_d(\phi)$ coincides with the Gaussian RF in (14) of Chapter 3.

Theorem 1. *Let X be an MA Gaussian RF in (2) satisfying Assumption (A)($d; \mathbb{R}^\nu$) and $Y = G(X)$ be a subordinated RF in (1), where $G = G(x), x \in \mathbb{R}$ is a measurable function with $\mathrm{E}G(X(\boldsymbol{t}))^2 < \infty$, the Hermite rank $k = k_H(G) \ge 1$, and the Hermite expansion in (7). Then,*

$$\begin{aligned}
r_Y(\boldsymbol{t}) &= (h_G^2(k)/k!) r_X^k(\boldsymbol{t}) \big(1 + o(1)\big) \\
&= |\boldsymbol{t}|^{k(4d - \nu)} \left(L_Y\left(\frac{\boldsymbol{t}}{|\boldsymbol{t}|}\right) + o(1) \right), \quad (|\boldsymbol{t}| \to \infty),
\end{aligned} \tag{10}$$

where $L_Y(\boldsymbol{z}) := (h_G^2(k)/k!) L_X^k(\boldsymbol{z})$. In addition, consider the following:

(i) *Let $0 < d < \nu/4$. Then, for any $\phi \in \Phi$*

$$\lambda^{-2kd-\nu(1-\frac{k}{2})}(Y_\lambda(\phi) - \mathrm{E}Y_\lambda(\phi)) \xrightarrow{\mathrm{d}} h_G(k)\mathcal{H}_k(\phi),$$
$$\text{if } k < \frac{\nu}{\nu - 4d} \tag{11}$$

and

$$\lambda^{-\frac{\nu}{2}}(Y_\lambda(\phi) - \mathrm{E}Y_\lambda(\phi)) \xrightarrow{\mathrm{d}} \sigma W(\phi), \quad \text{if } k > \frac{\nu}{\nu - 4d}, \tag{12}$$

where $W(\phi) \overset{\mathrm{d}}{=} \int_{\mathbb{R}^\nu} \phi(\boldsymbol{u})W(\mathrm{d}\boldsymbol{u})$ and $\sigma^2 := \int_{\mathbb{R}^\nu} r_Y(\boldsymbol{t})\mathrm{d}\boldsymbol{t}$.

(ii) *Let $-\nu/4 < d \leq 0$. Then, (12) holds, with the same $\sigma^2 = \int_{\mathbb{R}^\nu} r_Y(\boldsymbol{t})\mathrm{d}\boldsymbol{t}$.*

Remark 2. By property (54),

$$\sigma^2 = \sum_{j=k}^{\infty} \frac{h_G^2(j)}{j!}\sigma_j^2, \tag{13}$$

where $\sigma_j^2 := \int_{\mathbb{R}^\nu} \mathrm{Cov}(H_j(X(\mathbf{0})), H_j(X(\boldsymbol{t})))\mathrm{d}\boldsymbol{t} \geq 0, j \geq 1$, due to the positive-definite property of the covariance function. We also have that $\sigma_j^2 > 0$ for *even* $j \geq 2$ since $\mathrm{Cov}(H_j(X(\mathbf{0})), H_j(X(\boldsymbol{t}))) = j! r_X^j(\boldsymbol{t}) \geq 0, j = 2, 4, \ldots$. Therefore, the 'long-range variance' $\sigma^2 > 0$ in the CLT statement (12) in the case of $d = 0$, whereas when $|d| > 0$, this is true unless all even coefficients $h_G(j) = 0, j = 2, 4, \ldots$, in the expansion (7) vanish. The limit in (12) being trivial for $\sigma = 0$ leads to the question about a nontrivial limit in the above relation under a normalisation which grows slower than $\lambda^{\nu/2}$. This question is rather delicate and is left open. See Ref. [98] for some related results when $G = H_3$ is the Hermite polynomial of order three.

Proof of Theorem 1. Relation (10) follows by (4) and property (54) of Chapter 2 since

$$\begin{aligned} r_Y(\boldsymbol{t}) &= \sum_{j=k}^{\infty} \frac{h_G^2(j)}{j!} r_X^j(\boldsymbol{t}) \\ &= (h_G^2(k)/k!) r_X^k(\boldsymbol{t}) \left(1 + (k!/h_G^2(k)) \sum_{j=k+1}^{\infty} \frac{h_G^2(j)}{j!} r_X^{j-k}(\boldsymbol{t})\right), \end{aligned}$$

where the last sum does not exceed $|r_X(\boldsymbol{t})|\mathrm{E}Y^2(\mathbf{0})$ and vanishes as $|\boldsymbol{t}| \to \infty$.

Proof of the 'noncentral limit theorem' in (i), (11). Recall the Hermite expansion in (49) of Chapter 2. W.l.g., let $\mathrm{E}Y(\boldsymbol{t}) = 0$, $h_G(k) = k!$. Then,

$$Y(\boldsymbol{t}) = H_k(X(\boldsymbol{t})) + \sum_{j=k+1}^{\infty} \frac{h_G(j)}{j!} H_j(X(\boldsymbol{t})) =: Y'(\boldsymbol{t}) + Y''(\boldsymbol{t}), \quad \text{where}$$

$$Y'(\boldsymbol{t}) = \int_{\mathbb{R}^{\nu k}} \prod_{i=1}^{k} a(\boldsymbol{t} - \boldsymbol{u}_i) \mathrm{d}^k W; \tag{14}$$

see (107) of Chapter 2. Let $Y'_\lambda(\phi) := \int_{\mathbb{R}^\nu} \phi(\boldsymbol{t}/\lambda) Y'(\boldsymbol{t}) \mathrm{d}\boldsymbol{t}$, $Y''_\lambda(\phi) := \int_{\mathbb{R}^\nu} \phi(\boldsymbol{t}/\lambda) Y''(\boldsymbol{t}) \mathrm{d}\boldsymbol{t}$. We show that $Y'_\lambda(\phi)$ is the main term in (11), whereas $Y''_\lambda(\phi)$ is negligible in this limit. Using (14) and the change of variables formula $\boldsymbol{u}_i \to \lambda \boldsymbol{u}_i$ in (104) of Chapter 2, we get that

$$\begin{aligned}
Y'_\lambda(\phi) &= \int_{\mathbb{R}^{\nu k}} \left\{ \int_{\mathbb{R}^\nu} \phi(\boldsymbol{t}/\lambda) \prod_{i=1}^{k} a(\boldsymbol{t} - \boldsymbol{u}_i) \mathrm{d}\boldsymbol{t} \right\} W(\mathrm{d}\boldsymbol{u}_1) \cdots W(\mathrm{d}\boldsymbol{u}_k) \\
&= \lambda^\nu \int_{\mathbb{R}^{\nu k}} \left\{ \int_{\mathbb{R}^\nu} \phi(\boldsymbol{t}) \prod_{i=1}^{k} a(\lambda(\boldsymbol{t} - \boldsymbol{u}_i)) \mathrm{d}\boldsymbol{t} \right\} W(\mathrm{d}\lambda \boldsymbol{u}_1) \cdots W(\lambda \mathrm{d}\boldsymbol{u}_k) \\
&\overset{\mathrm{d}}{=} d_\lambda \int_{\mathbb{R}^{\nu k}} g^\dagger_\lambda(\boldsymbol{u}_1, \ldots, \boldsymbol{u}_k) W(\mathrm{d}\boldsymbol{u}_1) \cdots W(\mathrm{d}\boldsymbol{u}_k) =: \tilde{Y}_\lambda(\phi),
\end{aligned} \tag{15}$$

where $d_\lambda := \lambda^{2kd + \nu(1 - \frac{k}{2})}$ is the normalisation in (11) and

$$g^\dagger_\lambda((\boldsymbol{u})_k) := \int_{\mathbb{R}^\nu} \phi(\boldsymbol{t}) \prod_{i=1}^{k} \lambda^{\nu - 2d} a(\lambda(\boldsymbol{t} - \boldsymbol{u}_i)) \mathrm{d}\boldsymbol{t}, \quad (\boldsymbol{u})_k \in \mathbb{R}^{\nu k}. \tag{16}$$

From the isometry property of the Wiener–Itô integral in (100) of Chapter 2, the relation

$$\mathrm{E}(d_\lambda^{-1} \tilde{Y}_\lambda(\phi) - \mathcal{H}_k(\phi))^2 \to 0 \tag{17}$$

is equivalent to

$$\lim_{\lambda \to \infty} \|g^\dagger_\lambda - h_k\|_{2,k} \; = \; 0 \tag{18}$$

for h_k in (9). Relation (12) of Chapter 3 implies point-wise convergence:

$$\lim_{\lambda \to \infty} \prod_{i=1}^{k} \lambda^{\nu - 2d} a(\lambda(\boldsymbol{t} - \boldsymbol{u}_i)) = \prod_{i=1}^{k} a_\infty(\boldsymbol{t} - \boldsymbol{u}_i), \quad \lambda \to \infty \tag{19}$$

for any $\boldsymbol{t}, \boldsymbol{u}_1, \dots, \boldsymbol{u}_k \in \mathbb{R}^\nu$ such that $\boldsymbol{t} \neq \boldsymbol{u}_i, i = 1, \dots, k$. The proof of (18) from (19) uses the DCT and the bound $|a_\infty(\boldsymbol{t}-\boldsymbol{u})| \le C|\boldsymbol{t}-\boldsymbol{u}|^{2d-\nu}$. Thus, $|g^\dagger_\lambda((\boldsymbol{u})_k - h((\boldsymbol{u})_k)| \le C \int_{\mathbb{R}^\nu} |\phi(\boldsymbol{t})| \prod_{i=1}^k |\boldsymbol{t}-\boldsymbol{u}_i|^{2d-\nu} \mathrm{d}\boldsymbol{t} =: \bar{h}((\boldsymbol{u})_k)$, where the dominated function is integrable: $\|\bar{h}\|_{2,k} < \infty$. This proves (18) and (17). Since $Y'_\lambda(\phi) \overset{\mathrm{d}}{=} \tilde{Y}_\lambda(\phi)$ have the same distribution, (17) implies $d_\lambda^{-1} Y'_\lambda(\phi) \xrightarrow{\mathrm{d}} \mathcal{H}_k(\phi)$. Thus, (11) reduces to

$$\begin{aligned} \mathrm{Var}(Y''_\lambda(\phi)) = \int_{\mathbb{R}^{2\nu}} &\phi(\boldsymbol{t}_1/\lambda)\phi(\boldsymbol{t}_2/\lambda))\mathrm{Cov}(Y''(\boldsymbol{t}_1), \\ Y''(\boldsymbol{t}_2))\mathrm{d}\boldsymbol{t}_1\mathrm{d}\boldsymbol{t}_2 = o(d_\lambda^2), \end{aligned} \tag{20}$$

which follows from $|\mathrm{Cov}(Y''(\boldsymbol{t}_1), Y''(\boldsymbol{t}_2))| \le C|r_X(\boldsymbol{t}_1-\boldsymbol{t}_2)|^k \le C(1 \vee |\boldsymbol{t}_1-\boldsymbol{t}_2|)^{-(k+1)(\nu-4d)}$ and (4) with $\beta = (k+1)(\nu-4d)$, proving (11).
Proof of the 'central limit theorem' in (12), *parts (i) and (ii).* By inequality (58) of Chapter 2,

$$|r_Y(\boldsymbol{t})| \le C \begin{cases} |r_X(\boldsymbol{t})|^k, & d > 0, \\ |r_X(\boldsymbol{t})|, & d \le 0, \end{cases} \tag{21}$$

and hence $\int_{\mathbb{R}^\nu} |r_Y(\boldsymbol{t})|\mathrm{d}\boldsymbol{t} < \infty$ holds in both cases $d > 0, k > \frac{\nu}{\nu-4d}$ and $d \le 0$. The proof of (12) uses the approximation of Y in (1) by an m-dependent RF Y_m and the CLT for such an RF in Proposition 4 of Chapter 2. The approximating RF is defined as

$$Y_m(\boldsymbol{t}) := G(X_m(\boldsymbol{t})), \quad \boldsymbol{t} \in \mathbb{R}^\nu, \tag{22}$$

where G is the same as in (1) and X_m is a 'truncated' Gaussian RF:

$$X_m(\boldsymbol{t}) = \int_{\mathbb{R}^\nu} a_m(\boldsymbol{t}-\boldsymbol{u})W(\mathrm{d}\boldsymbol{u}), \quad \text{with } a_m(\boldsymbol{t}) := c_m^{-1} a(\boldsymbol{t})\mathbb{I}(|\boldsymbol{t}| < m), \tag{23}$$

$c_m := \int_{\mathbb{R}^\nu} |a(\boldsymbol{t})|^2 \mathbb{I}(|\boldsymbol{t}| < m)\mathrm{d}\boldsymbol{t} \to 1$ $(m \to \infty)$. Clearly, X_m and Y_m are $2m$-dependent RFs with finite variance; therefore,

$$\lambda^{-\nu/2}(Y_{\lambda,m}(\phi) - \mathrm{E}Y_{\lambda,m}(\phi)) \xrightarrow{\mathrm{d}} \sigma_m W(\phi), \quad \lambda \to \infty$$

holds for $Y_{\lambda,m}(\phi) := \int_{\mathbb{R}^\nu} \phi(\boldsymbol{t}/\lambda) Y_m(\boldsymbol{t})\mathrm{d}\boldsymbol{t}$ and any $\phi \in \Phi$ in view of the above-mentioned CLT, with $\sigma_m^2 := \int_{\mathbb{R}^m} r_{Y_m}(\boldsymbol{t})\mathrm{d}\boldsymbol{t}$. The above facts reduce the proof of (12) to

$$\lim_{m\to\infty}\limsup_{\lambda\to\infty}\lambda^{-\nu}\mathrm{Var}(Y_\lambda(\phi)-Y_{\lambda,m}(\phi))=0 \quad \text{and} \quad \lim_{m\to\infty}\sigma_m^2=\sigma^2. \tag{24}$$

We have $\mathrm{Var}(Y_\lambda(\phi) - Y_{\lambda,m}(\phi)) = \int_{\mathbb{R}^{2\nu}} \phi(\boldsymbol{t}_1/\lambda)\phi(\boldsymbol{t}_2/\lambda)\rho_m(\boldsymbol{t}_1 - \boldsymbol{t}_2)\mathrm{d}\boldsymbol{t}_1\mathrm{d}\boldsymbol{t}_2$,, where $\rho_m(\boldsymbol{t}) := \mathrm{Cov}(Y(\boldsymbol{0})-Y_m(\boldsymbol{0}), Y(\boldsymbol{t})-Y_m(\boldsymbol{t}))$. Relation (24) follows from $\lim_{m\to\infty}\rho_m(\boldsymbol{t}) = 0\,(\forall \boldsymbol{t}\in\mathbb{R}^\nu)$ and the dominating bound

$$\begin{aligned}|\rho_m(\boldsymbol{t})| &\le |r_Y(\boldsymbol{t})| + |r_{Y_m}(\boldsymbol{t})| + |\mathrm{Cov}(Y(\boldsymbol{0}), Y_m(\boldsymbol{t})| \\ &\quad + |\mathrm{Cov}(Y(\boldsymbol{0}), Y_m(-\boldsymbol{t})| \ \le\ \bar\rho(\boldsymbol{t}), \quad \boldsymbol{t}\in\mathbb{R}^\nu,\end{aligned} \tag{25}$$

with the integrable $\bar\rho(\cdot)\in L^1(\mathbb{R}^\nu)\cap L^\infty(\mathbb{R}^\nu)$ independent of m. The proof of (25) follows from (58) of Chapter 2, as all these covariances can be written as the l.h.s. of (58) of Chapter 2, with $G_1 = G_2 = G$ having Hermite rank k (for $d > 0$) or 1 (for $d \le 0$) and $Z_i \sim N(0,1), i = 1,2$, with correlation coefficient

$$\begin{aligned}|\mathrm{Cov}(Z_1, Z_2)| &\le \max(|r_X(\boldsymbol{t})|, |r_{X_m}(\boldsymbol{t})|, |\mathrm{Cov}(X(\boldsymbol{0}), X_m(\boldsymbol{t}))|, \\ &\qquad \times |\mathrm{Cov}(X(\boldsymbol{0}), X_m(-\boldsymbol{t}))|) \\ &\le C\int_{\mathbb{R}^\nu} |a(\boldsymbol{u})|(|a(\boldsymbol{u}+\boldsymbol{t})| + |a(\boldsymbol{u}-\boldsymbol{t})|)\mathrm{d}\boldsymbol{u} \ =:\ \bar r(\boldsymbol{t}),\end{aligned}$$

satisfying $\bar r(\cdot) \le C(1\vee|\boldsymbol{t}|)^{4d-\nu}\,(d>0)$, $\bar\rho(\cdot)\in L^1(\mathbb{R}^\nu)\cap L^\infty(\mathbb{R}^\nu)\,(d<0)$. This ends the proof of (30) and Theorem 1 as well. □

We remark that a statement analogous to Theorem 1 is valid for nonlinear functions in (1) of Gaussian MA RF X on $\mathbb{Z}^\nu$ under Assumption (A)$(d;\mathbb{Z}^\nu)$, with the only difference that $\sigma \ge 0$ in (12) must be replaced by $\sigma^2 = \sum_{\boldsymbol{s}\in\mathbb{Z}^\nu} r_Y(\boldsymbol{s})$.

2 Polynomials of MA RF

Let $X(\boldsymbol{t}) = \sum_{\boldsymbol{s}\in\mathbb{Z}^\nu} a(\boldsymbol{t}-\boldsymbol{s})\varepsilon(\boldsymbol{s})$ be an MA RF as in (5) of Chapter 1 or (2) of Chapter 3, with (non-Gaussian) i.i.d innovations $\{\varepsilon(\boldsymbol{s})\}$, and $G = G(x), x\in\mathbb{R}$ be a polynomial of degree $k\ge 1$. We are interested in dependence properties of the subordinated RF $Y = G(X)$, particularly the behaviour of the covariance function

$r_Y(\boldsymbol{t}) := \mathrm{Cov}(Y(\mathbf{0}), Y(\boldsymbol{t}))$ and the isotropic scaling limits of the integrals $Y_\lambda(\phi) = \int_{\mathbb{R}^\nu} \phi(\boldsymbol{t}/\lambda) Y([\boldsymbol{t}])\mathrm{d}\boldsymbol{t}, \phi \in \Phi$. From Rosenthal's inequality, for any $p \geq 2$,

$$\mathrm{E}|X(\boldsymbol{t})|^p \leq C_p \left\{ \sum_{\boldsymbol{s}\in\mathbb{Z}^\nu} (\mathrm{E}|a(\boldsymbol{t}-\boldsymbol{s})\varepsilon(\boldsymbol{s})|^p)^{2/p} \right\}^{p/2}$$

$$= C_p \mathrm{E}|\varepsilon(\mathbf{0})|^p \left\{ \sum_{\boldsymbol{s}\in\mathbb{Z}^\nu} |a(\boldsymbol{s})|^2 \right\}^{p/2}.$$

It follows that for any polynomial G of degree $k \geq 1$, we have $\mathrm{E}|Y(\boldsymbol{t})|^2 < \infty$, provided

$$\mathrm{E}|\varepsilon(\mathbf{0})|^{2k} < \infty. \tag{26}$$

Recall from (87) of Chapter 2 that any such polynomial can be expanded in Appell polynomials relative to the marginal distribution F of X: $G(x) = \sum_{j=0}^k a_j A_j(x)$, with coefficients

$$a_j = \mathrm{E}G^{(j)}(X(\mathbf{0})), \qquad G^{(j)}(x) := \mathrm{d}^j G(x)/\mathrm{d}x^j. \tag{27}$$

As noted above, the LRD properties of $Y(\boldsymbol{t}) = \sum_{j=0}^n a_j A_j(X(\boldsymbol{t}))$ are determined by the Appell rank $k = k_A(G) \geq 1$ of G or the LRD properties of the Appell polynomial $A_k(X)$, with $a_k \neq 0, a_j = 0(1 \leq j < k$.

Theorem 3. *Let X be an RF on $\mathbb{Z}^\nu$ with the MA representation as in* (5) *of Chapter 1, standardised i.i.d. innovations, and coefficients satisfying Assumption (A)(d; $\mathbb{Z}^\nu$). Let* (26) *hold and $Y = A_k(X)$ be the Appell polynomial of degree $k \geq 1$ relative to the distribution of $X(\mathbf{0})$.*

(i) *Let $0 < d < \frac{\nu}{4}$ and $k < \frac{\nu}{\nu-4d}$. Then,*

$$r_Y(\boldsymbol{t}) = k! r_X^k(\boldsymbol{t})(1 + o(1))$$

$$= k!\, |\boldsymbol{t}|^{k(4d-\nu)} \left(L_X^k \left(\frac{\boldsymbol{t}}{|\boldsymbol{t}|} \right) + o(1) \right), \quad |\boldsymbol{t}| \to \infty \tag{28}$$

and

$$\lambda^{-2kd-\nu(1-\frac{k}{2})} Y_\lambda(\phi) \xrightarrow{\mathrm{d}} \mathcal{H}_k(\phi), \quad \forall\, \phi \in \Phi, \tag{29}$$

where $\mathcal{H}_k(\phi)$ is the Hermite RF in (8).

(ii) *Let either* $d > 0$ *and* $k > \frac{\nu}{\nu - 4d}$ *or* $d \le 0$ *hold. Then,* $\sum_{\boldsymbol{t}\in\mathbb{Z}^\nu} |r_Y(\boldsymbol{t})| < \infty$ *and*

$$\lambda^{-\frac{\nu}{2}} Y_\lambda(\phi) \xrightarrow{\mathrm{d}} \sigma W(\phi), \quad \sigma^2 := \sum_{\boldsymbol{t}\in\mathbb{Z}^\nu} r_Y(\boldsymbol{t}). \tag{30}$$

Proof. (i) We use the decomposition in (95) of Chapter 2,

$$A_k(X(\boldsymbol{t})) = X^{\bullet k}(\boldsymbol{t}) + \mathcal{Z}(\boldsymbol{t}), \tag{31}$$

of the Appell polynomial into the 'off-diagonal' polynomial form $X^{\bullet k}(\boldsymbol{t}) = \sum^*_{(\boldsymbol{u})_k} \prod_{i=1}^k a(\boldsymbol{t} - \boldsymbol{u}_i)\varepsilon(\boldsymbol{u}_1)\cdots\varepsilon(\boldsymbol{u}_k)$ (see (96) of Chapter 2) and a 'remainder' SRD RF, $\mathcal{Z}(\boldsymbol{t})$. According to (93) of Chapter 2, $\mathcal{Z}(\boldsymbol{t}) = \sum_{i=1}^{k-1} \mathcal{Z}_i(\boldsymbol{t})$ can be written as a sum of stationary 'off-diagonal' polynomial forms,

$$\mathcal{Z}_i(\boldsymbol{t}) = \sum_{(V)_i} \sum_{(\boldsymbol{u})_i}^{\bullet} a(\boldsymbol{t} - \boldsymbol{u}_1)^{|V_1|} \cdots a(\boldsymbol{t} - \boldsymbol{u}_i)^{|V_i|} \widetilde{A}_{|V_1|} \times (\varepsilon(\boldsymbol{u}_1)) \cdots \widetilde{A}_{|V_i|}(\varepsilon(\boldsymbol{u}_i)), \tag{32}$$

of order $i < k$ in zero-mean i.i.d. r.v.s $\widetilde{A}_{|V_\ell|}(\varepsilon(\boldsymbol{u}_\ell)), 1 \le \ell \le i$ with $\max(|V_1|, \ldots, |V_i|) \ge 2$, where $\widetilde{A}_j$ stands for the jth Appell polynomial relative to the noise distribution ε and the sum $\sum_{(V)_i}$ is taken over all partitions of $\{1, \ldots, k\}$ into i nonempty sets $V_1, \ldots, V_i$ having cardinality $|V_1| \ge 1, \ldots, |V_i| \ge 1, |V_1| + \cdots + |V_i| = k$. Note that $\mathrm{Cov}(\mathcal{Z}_i(\boldsymbol{t}), \mathcal{Z}_{i'}(\boldsymbol{t}')) = 0$ are uncorrelated for $i \ne i'$ and any $\boldsymbol{t}, \boldsymbol{t}'$, and hence $r_{\mathcal{Z}}(\boldsymbol{t}) := \mathrm{Cov}(\mathcal{Z}(\boldsymbol{0}), \mathcal{Z}(\boldsymbol{t}))$ is bounded by

$$|r_{\mathcal{Z}}(\boldsymbol{t})| \le \sum_{i=1}^{k-1} |\mathrm{Cov}(\mathcal{Z}_i(\boldsymbol{0}), \mathcal{Z}_i(\boldsymbol{t}))| \le C \sum_{i=1}^{k-1} \sum_{(v)_i, (v')_i} \prod_{\ell=1}^{i} [|a|^{v_\ell} \star |a|^{v'_\ell}](\boldsymbol{t}), \tag{33}$$

where the last sum is taken over all collections $(v)_i = (v_1, \ldots, v_i), (v')_i = (v'_1, \ldots, v'_i)$ of integers $v_\ell \ge 1, v'_\ell \ge 1$ with $\sum_{\ell=1}^i v_\ell = \sum_{\ell=1}^i v'_\ell = k$. (We also remind that $[|a|^v \star |a|^{v'}](\boldsymbol{t}) = \sum_{\boldsymbol{u}\in\mathbb{Z}^\nu} |a(\boldsymbol{u})|^v |a(\boldsymbol{t} + \boldsymbol{u})|^{v'}$ stands for 'discrete convolution'.) We use the following bound: for $0 < d < \nu/4$,

$$[|a|^{v} \star |a|^{v'}](\boldsymbol{t}) \le C \begin{cases} |\boldsymbol{t}|^{4d-\nu}, & \text{if } v = v' = 1, \\ |\boldsymbol{t}|^{2d-\nu}, & \text{if } v \ge 1, v' \ge 2 \text{ or } v \ge 2, v' \ge 1, \\ |\boldsymbol{t}|^{4d-2\nu}, & \text{if } v \ge 2, v' \ge 2. \end{cases} \tag{34}$$

Indeed, for $v = v' = 1$, (34) is a consequence of the asymptotics in (9) of Chapter 1, whereas for $v \vee v' \ge 2$, it follows by splitting the convolution as $[|a|^{v} \star |a|^{v'}](\boldsymbol{t}) = \sum_{|\boldsymbol{u}| \le |\boldsymbol{t}|/2} \cdots + \sum_{|\boldsymbol{u}| > |\boldsymbol{t}|/2} \cdots$ and using $|a(\boldsymbol{t}+\boldsymbol{u})|^{v'} \le C|\boldsymbol{t}|^{v'(2d-\nu)}$ ($|\boldsymbol{u}| \le |\boldsymbol{t}|/2$), $|a(\boldsymbol{u})|^{v} \le C|\boldsymbol{t}|^{v(2d-\nu)}$ ($|\boldsymbol{u}| > |\boldsymbol{t}|/2$). From (33), (34) and $\max_{1 \le \ell \le i} v_\ell \ge 2, \max_{1 \le \ell \le i} v'_\ell \ge 2$, we conclude that for $0 < d < \nu/4$,

$$r_{\mathcal{Z}}(\boldsymbol{t}) = O(|\boldsymbol{t}|^{4d-2\nu}), \quad |\boldsymbol{t}| \to \infty. \tag{35}$$

Since $2\nu - 4d > \nu$, this implies that $r_{\mathcal{Z}}(\boldsymbol{t})$ is summable:

$$\sum_{\boldsymbol{t} \in \mathbb{Z}^\nu} |r_{\mathcal{Z}}(\boldsymbol{t})| < \infty. \tag{36}$$

The same conclusion in (36) also holds for $d \le 0$ since in this case, $|a|^{v} \in L^1(\mathbb{Z}^\nu)\, \forall v \ge 1$, implying $[|a|^{v} \star |a|^{v'}] \in L^1(\mathbb{Z}^\nu)$ in (34).

Let us prove (28). By the orthogonality $\mathrm{Cov}(X^{\bullet k}(\boldsymbol{t}), \mathcal{Z}(\boldsymbol{t}')) = 0$ ($\boldsymbol{t}, \boldsymbol{t}' \in \mathbb{Z}^\nu$), we have that $r_Y(\boldsymbol{t}) = \mathrm{Cov}(X^{\bullet k}(\boldsymbol{0}), X^{\bullet k}(\boldsymbol{t})) + r_{\mathcal{Z}}(\boldsymbol{t})$, where $r_{\mathcal{Z}}(\boldsymbol{t})$ is negligible in view of (36), whereas the difference $k! r_X(\boldsymbol{t})^k - \mathrm{Cov}(X^{\bullet k}(\boldsymbol{0}), X^{\bullet k}(\boldsymbol{t})) = k!\{[a \star a](\boldsymbol{t})^k - \sum_{(\boldsymbol{u})_k}^{\bullet} a(\boldsymbol{u}_1) a(\boldsymbol{t}+\boldsymbol{u}_1) \cdots a(\boldsymbol{u}_k) a(\boldsymbol{t}+\boldsymbol{u}_k)\}$ satisfies the same bound as in (33); therefore, this difference is $O(|\boldsymbol{t}|^{4d-2\nu})$ as in (35) above, with $2\nu - 4d > \nu > k(\nu - 4d)$ for $k < \frac{\nu}{\nu - 4d}$, proving (28).

To prove the 'noncentral limit theorem' in (29), we use Proposition 21 of Chapter 2. Accordingly, we rewrite $Y_\lambda(\phi)$ as an 'off-diagonal' polynomial form in noise variables plus a remainder term which is asymptotically negligible. Using the decomposition in (31), we have that

$$d_\lambda^{-1} Y_\lambda(\phi) = S_k(g_\lambda) + d_\lambda^{-1} \mathcal{Z}_\lambda(\phi), \tag{37}$$

where $d_\lambda := \lambda^{2kd + \nu(1 - \frac{k}{2})}$ and

$$S_k(g_\lambda) := d_\lambda^{-1} \int_{\mathbb{R}^\nu} \phi(\boldsymbol{t}/\lambda) X^{\bullet k}([\boldsymbol{t}]) \mathrm{d}\boldsymbol{t}$$

$$= \sum_{(\boldsymbol{u})_k}^{\bullet} g_\lambda(\boldsymbol{u}_1, \ldots, \boldsymbol{u}_k)\varepsilon(\boldsymbol{u}_1)\cdots\varepsilon(\boldsymbol{u}_k),$$

$$\text{with} \quad g_\lambda(\boldsymbol{u}_1, \ldots, \boldsymbol{u}_k) := d_\lambda^{-1} \int_{\mathbb{R}^\nu} \phi(\boldsymbol{t}/\lambda) \prod_{i=1}^{k} a([\boldsymbol{t}] - \boldsymbol{u}_i)\mathrm{d}\boldsymbol{t} \tag{38}$$

and $\mathcal{Z}_\lambda(\phi) := \int_{\mathbb{R}^\nu} \phi(\boldsymbol{t}/\lambda)\mathcal{Z}([\boldsymbol{t}])\mathrm{d}\boldsymbol{t}$. Then, by (36) and Proposition 1 (8) of Chapter 3, $\mathrm{Var}(\mathcal{Z}_\lambda(\phi)) = O(\lambda^\nu)$, implying that $d_\lambda^{-1}\mathcal{Z}_\lambda(\phi) = o_p(1)$ is negligible, and relation (29) reduces to $S_k(g_\lambda) \xrightarrow{\mathrm{d}} \mathcal{H}_k(\phi)$. By Proposition 21 of Chapter 2, the last relation follows from

$$\lim_{\lambda\to\infty} \|g_\lambda^\dagger - h\|_{2,k} = 0 \tag{39}$$

for $g_\lambda^\dagger((\boldsymbol{u})_k), (\boldsymbol{u})_k = (\boldsymbol{u}_1, \ldots, \boldsymbol{u}_k) \in \mathbb{R}^{\nu k}$, defined in (110) from $g_\lambda((\boldsymbol{u})_k)$ in (38), with $\Lambda = \lambda I, \det(\Lambda) = \lambda^\nu$. More explicitly,

$$g_\lambda^\dagger((\boldsymbol{u})_k) = \int_{\mathbb{R}^\nu} \phi(\boldsymbol{t}) \prod_{i=1}^{k} \lambda^{\nu-2d} a([\lambda\boldsymbol{t}] - [\lambda\boldsymbol{u}_i])\mathrm{d}\boldsymbol{t}, \quad (\boldsymbol{u})_k \in \mathbb{R}^{\nu k}. \tag{40}$$

Relation (12) of Chapter 3 implies point-wise convergence

$$\lim_{\lambda\to\infty} \prod_{i=1}^{k} \lambda^{\nu-2d} a([\lambda\boldsymbol{t}] - [\lambda\boldsymbol{u}_i]) = \prod_{i=1}^{k} a_\infty(\boldsymbol{t} - \boldsymbol{u}_i), \quad \lambda \to \infty \tag{41}$$

for any $\boldsymbol{t}, \boldsymbol{u}_1, \ldots, \boldsymbol{u}_k \in \mathbb{R}^\nu$ such that $\boldsymbol{t} \neq \boldsymbol{u}_i, i = 1, \ldots, k$. The proof of (39) from (41) uses the DCT and the bounds $|a_\infty(\boldsymbol{t}-\boldsymbol{u})| \le C|\boldsymbol{t}-\boldsymbol{u}|^{2d-\nu}$, $\lambda^{\nu-2d}|a([\boldsymbol{t}] - [\boldsymbol{u}])| \le C|\boldsymbol{t}-\boldsymbol{u}|^{2d-\nu}$, the latter being similar to (18) of Chapter 1. Thus, $|g_\lambda^\dagger((\boldsymbol{u})_k - h((\boldsymbol{u})_k)| \le C\int_{\mathbb{R}^\nu} |\phi(\boldsymbol{t})| \prod_{i=1}^k |\boldsymbol{t} - \boldsymbol{u}_i|^{2d-\nu}\mathrm{d}\boldsymbol{t} =: \bar{h}((\boldsymbol{u})_k)$, where the dominated function is integrable: $\|\bar{h}\|_{2,k} < \infty$. This proves (39), (29) and part (i).
(ii) Relation $\sum_{\boldsymbol{t}\in\mathbb{Z}^\nu} |r_Y(\boldsymbol{t})| < \infty$ follows from (36) (valid also for $d \le 0$) and $|\mathrm{Cov}(X^{\bullet k}(\boldsymbol{0}), X^{\bullet k}(\boldsymbol{t}))| \le [|a| \star |a|]^k(\boldsymbol{t}) \in L^1(\mathbb{Z}^\nu)$. Consider the CLT in (30) whose proof uses the approximation of Y by m-dependent RF and is similar to that of (12) above. The approximating RF $Y_m(\boldsymbol{t})$ is defined by truncating MA coefficients in the

Wick representation (90) of Chapter 2 of $Y(\boldsymbol{t}) = A_k(X(\boldsymbol{t}))$:

$$Y_m(\boldsymbol{t}) := \sum_{(\boldsymbol{u})_k \in \mathbb{Z}^{\nu k}} a_m(\boldsymbol{t}-\boldsymbol{u}_1)\cdots a_m(\boldsymbol{t}-\boldsymbol{u}_k) :\varepsilon(\boldsymbol{u}_1)\times\cdots\times\varepsilon(\boldsymbol{u}_k):, \tag{42}$$

where

$$a_m(\boldsymbol{t}) := a(\boldsymbol{t})\mathbb{I}(|\boldsymbol{t}| \le m), \quad \boldsymbol{t} \in \mathbb{Z}^\nu. \tag{43}$$

Thus, $Y_m(\boldsymbol{t}) = A_{k,m}(X_m(\boldsymbol{t}))$, where $X_m(\boldsymbol{t}) := \sum_{\boldsymbol{u}\in\mathbb{Z}^\nu} a_m(\boldsymbol{t}-\boldsymbol{u})\varepsilon(\boldsymbol{u})$ and $A_{k,m}(x)$ is the kth Appell polynomial relative to the marginal distribution of the stationary RF X_m. Let $Y_{\lambda,m}(\phi) := \int_{\mathbb{R}^\nu}\phi(\boldsymbol{t}/\lambda)Y_m([\boldsymbol{t}])\mathrm{d}\boldsymbol{t}$. The aforementioned approximation leading to (30) amounts to showing the following relations:

$$\begin{aligned}&\lambda^{-\nu}\mathrm{Var}(Y_{\lambda,m}(\phi)) \to \sigma_m^2\int_{\mathbb{R}^\nu}\phi(\boldsymbol{u})^2\mathrm{d}\boldsymbol{u}, \quad \lambda^{-\nu/2}Y_{\lambda,m}(\phi) \to \sigma_m W(\phi),\\ &\lim_{m\to\infty}\limsup_{\lambda\to\infty}\lambda^{-\nu}\mathrm{Var}(Y_\lambda(\phi) - Y_{\lambda,m}(\phi)) = 0, \qquad \lim_{m\to\infty}\sigma_m^2 \to \sigma^2\end{aligned} \tag{44}$$

(c.f. (24)), where $\sigma_m^2 := \sum_{\boldsymbol{t}\in\mathbb{Z}^\nu} r_{Y_m}(\boldsymbol{t})$ is the sum of covariances of RF Y_m. The first two relations in (44) are rather easy in view of Propositions 1 (iv) of Chapter 3 and 4 of Chapter 2. In the more delicate case of $0 < d < \nu/4$, the proof of the main approximating step in (44) for $\mathrm{Var}(Y_\lambda(\phi) - Y_{\lambda,m}(\phi)) = \sum_{\boldsymbol{s}_1,\boldsymbol{s}_2\in\mathbb{Z}^\nu} b_\lambda(\boldsymbol{s}_1)b_\lambda(\boldsymbol{s}_2)\mathrm{Cov}(Y(\boldsymbol{s}_1) - Y_m(\boldsymbol{s}_1), Y(\boldsymbol{s}_2) - Y_m(\boldsymbol{s}_2)), b_\lambda(\boldsymbol{s}) := \int_{[\boldsymbol{s},\boldsymbol{s}+1[}\phi(\boldsymbol{t}/\lambda)\mathrm{d}\boldsymbol{t}$ uses a similar bound as in (34):

$$|r_Y(\boldsymbol{t})| \le C|\boldsymbol{t}|^{-(k(\nu-4d))\wedge(2\nu-4d)}, \tag{45}$$

implying the summability of these covariances over $\boldsymbol{t} \in \mathbb{Z}^\nu$, together with properties of $b_\lambda(\boldsymbol{s})$ noted in the proof of Proposition 4. Finally, the last relation in (44) is a consequence of (45), $\lim_{m\to\infty}\mathrm{Cov}(Y_m(\mathbf{0}), Y_m(\boldsymbol{t})) = \mathrm{Cov}(Y(\mathbf{0}), Y(\boldsymbol{t}))$, and the DCT. This ends the proof of (30) and Theorem 3 as well. □

Remark 4. We expect that Theorem 3 can be extended with minor changes to the Appell polynomial $Y = A_k(X)$ of the Lévy moving average RF $X(\boldsymbol{t}) = \int_{\mathbb{R}^\nu} a(\boldsymbol{t}-\boldsymbol{u})M(\mathrm{d}\boldsymbol{u}), \boldsymbol{t} \in \mathbb{R}^\nu$ in (1) of Chapter 3, with the coefficient satisfying Assumption (A)$(d; \mathbb{R}^\nu)$. In this case, the Wick decomposition in (31) can be replaced by the decomposition in Ref. [243, Prop. 2.5] of $A_k(X(\boldsymbol{t}))$ into a sum of Poisson multiple stochastic integrals. See Ref. [97] for some results in this direction.

3 First-Order Reduction Principle for Nonlinear Functions of Linear RFs

The natural question with regard to Theorem 3 is what can be said about the scaling limits of a non-polynomial subordinated RF $Y = G(X)$, unless X is Gaussian? Indeed, many statistical functionals of interest (such as the empirical distribution function) are not polynomials; therefore, this question deserves attention. For an LRD causal time series (case $\nu = 1$), it was studied in rather many works (see Refs. [103, 119, 120] and the references therein) using martingale methods. The latter methods are not well developed for multiparameter processes, making the study of non-polynomial LRD RFs much more difficult. The results obtained in this section are limited to G having Appell rank 1, or the first Appell coefficient as $a_G(1) \neq 0$. In this case, scaling of $Y = G(X)$ reduces to that of $a_G(1)X$, leading to the Gaussian limit $a_G(1)W_d(\phi)$. The above reduction is termed the *first-order reduction principle* and is rather technical, requiring additional conditions on X and G. As in Theorems 4 and 5 of Chapter 3, we discuss the MA RF X on $\mathbb{Z}^\nu$ and $\mathbb{R}^\nu$ satisfying the corresponding Assumptions (A)$(d; \mathbb{Z}^\nu)$ and (A)$(d; \mathbb{R}^\nu)$. Assuming that the marginal distribution of X has a differentiable probability density $f(x)$, the first Appell coefficient of G defined in (88) of Chapter 2 can be rewritten as

$$a_G(1) = -\int_{\mathbb{R}} G(x) f'(x) \mathrm{d}x. \tag{46}$$

Contrary to (27), the definition in (46) makes sense even if G is not differentiable, provided the last integral exists. The following Theorems 5 and 6 are rather similar but deal with RFs on $\mathbb{Z}^\nu$ and $\mathbb{R}^\nu$ and involve somewhat different assumptions on the underlying MA RF X.

Nonlinear function of linear RF on $\mathbb{Z}^\nu$. Let X be an MA RF on $\mathbb{Z}^\nu$ as in Theorem 3. The existence of smooth marginal density is a consequence of a mild regularity or decay condition of the characteristic function of the noise distribution:

$$|\phi(z)| \le C/(1+|z|)^{\tau}, \quad z \in \mathbb{R}, \quad (\exists\, C, \tau > 0),$$
$$\text{where} \quad \phi(z) := \mathrm{E}\mathrm{e}^{\mathrm{i}z\varepsilon}. \tag{47}$$

We shall assume that the innovations of X have 2ℓ finite moment:

$$\mathrm{E}|\varepsilon|^{2\ell} < \infty \quad (\exists \ell \ge 2, \ell \in \mathbb{N}), \tag{48}$$

whereas G satisfies the bound

$$|G(x)| \le C(1+|x|)^{\ell-2}, \quad x \in \mathbb{R} \tag{49}$$

with the same ℓ as in (48).

Theorem 5. *Let X be a linear RF on $\mathbb{Z}^\nu$ satisfying* (47), (48) *and Assumption (A)($d; \mathbb{Z}^\nu$), $0 < d < \nu/4$, and G be a measurable function on $\mathbb{R}$ such that* (49) *holds. Then,*

$$\lambda^{-(\nu+4d)/2} Y_\lambda(\phi) \xrightarrow{\mathrm{d}} a_G(1) W_d(\phi), \quad \forall \phi \in \Phi, \tag{50}$$

where $Y_\lambda(\phi) = \int_{\mathbb{R}^\nu} \phi(\mathbf{t}/\lambda) G(X([\mathbf{t}]))\mathrm{d}\mathbf{t}$ and Gaussian RF $W_d(\phi)$ is defined in (14) *of Chapter 3.*

Nonlinear function of Lévy MA RF on $\mathbb{R}^\nu$. Let $Y(\boldsymbol{t}) = G(X(\boldsymbol{t})), \boldsymbol{t} \in \mathbb{R}^\nu$, where

$$X(\boldsymbol{t}) = \int_{\mathbb{R}^\nu} a(\boldsymbol{t} - \boldsymbol{s}) M(\mathrm{d}\boldsymbol{s}), \quad \boldsymbol{t} \in \mathbb{R}^\nu, \tag{51}$$

is an MA in (2) of Chapter 3, written as a stochastic integral w.r.t. a Lévy random measure $M(\mathrm{d}\boldsymbol{u})$ on $\mathbb{R}^\nu$ with zero mean and variance $\mathrm{d}\boldsymbol{u}$. For brevity of exposition, we limit our discussion to the case when M has no Gaussian component, in which case its distribution is completely determined by the Lévy measure π; see (19) of Chapter 3. We assume that π satisfies the following two conditions:

$$\int_{\mathbb{R}} v^2 \mathbb{I}(|v| < r)\pi(\mathrm{d}v) \ \ge\ c_\pi r^{2-\alpha}, \quad \forall r \in (0,1) \tag{52}$$

and

$$\int_{\mathbb{R}} |v|^{2\ell} \pi(\mathrm{d}v) < \infty \tag{53}$$

for some $c_\pi > 0, \alpha \in (0,2), \ell \ge 2, \ell \in \mathbb{N}$. Condition (53) is a counterpart of the moment condition (48) and guarantees $\mathrm{E}|X(\boldsymbol{t})|^{2\ell} < \infty$, whereas (52) replaces the regularity condition in (47), implying the existence of a smooth (infinitely differentiable) marginal density $f(x)$.

Theorem 6. *Let X be an MA RF in* (51) *on $\mathbb{R}^\nu$ satisfying* (52), (53) *and Assumption* $(A)(d;\mathbb{R}^\nu)$, $0 < d < \frac{\nu}{4}$, *and G be a measurable function on $\mathbb{R}$ such that* (49) *holds. Then,*

$$\lambda^{-(\nu+4d)/2} Y_\lambda(\phi) \xrightarrow{\mathrm{d}} a_G(1) W_d(\phi), \quad \forall\, \phi \in \Phi, \tag{54}$$

where $Y_\lambda(\phi) = \int_{\mathbb{R}^\nu} \phi(\boldsymbol{t}/\lambda) Y(\boldsymbol{t}) \mathrm{d}\boldsymbol{t}$ and Gaussian RF $W_d(\phi)$ is defined in (14) *of Chapter 3.*

The main idea behind the proofs of the above theorems is the same and consists of showing that the difference

$$\mathcal{Z}(\boldsymbol{t}) := Y(\boldsymbol{t}) - a_G(1) X(\boldsymbol{t}) \tag{55}$$

has a covariance function $r_{\mathcal{Z}}(\boldsymbol{t})$ that decays faster than the covariance functions $r_X(\boldsymbol{t})$ and $r_Y(\boldsymbol{t})$ of the RFs X and Y. From (55),

$$\begin{aligned} r_Y(\boldsymbol{t}) = a_G^2(1) r_X(\boldsymbol{t}) + a_G(1)(\mathrm{Cov}(X(\boldsymbol{0}), \mathcal{Z}(\boldsymbol{t})) \\ + \mathrm{Cov}(X(\boldsymbol{0}), \mathcal{Z}(-\boldsymbol{t}))) + r_{\mathcal{Z}}(\boldsymbol{t}). \end{aligned} \tag{56}$$

Using $\mathrm{E}X(\boldsymbol{t}) = 0$ and (46), the covariances on the r.h.s. of (56) can be written through the bivariate probability density $f_{\boldsymbol{t}}(x_1, x_2)$ of $(X(\boldsymbol{s}), X(\boldsymbol{s}+\boldsymbol{t}))$ (which does not depend on $\boldsymbol{s}$ by stationarity) as integrals

$$\mathrm{Cov}(X(\boldsymbol{0}), \mathcal{Z}(\boldsymbol{t})) = \int_{\mathbb{R}^2} x_1 (G(x_2) - \mu_Y) g_{\boldsymbol{t}}(x_1, x_2) \mathrm{d}x_1 \mathrm{d}x_2, \tag{57}$$

$$r_{\mathcal{Z}}(\boldsymbol{t}) = \int_{\mathbb{R}^2} \prod_{j=1}^{2} (G(x_j) - \mu_Y - a_G(1) x_j) g_{\boldsymbol{t}}(x_1, x_2) \mathrm{d}x_1 \mathrm{d}x_2$$

involving the difference

$$\begin{aligned} g_{\boldsymbol{t}}(x_1, x_2) := f_{\boldsymbol{t}}(x_1, x_2) - f(x_1) f(x_2) - r_X(\boldsymbol{t}) f'(x_1) f'(x_2), \\ x_1, x_2 \in \mathbb{R} \end{aligned} \tag{58}$$

The existence of the above densities and a relevant bound on $g_{\boldsymbol{t}}(x_1, x_2)$ is established in Lemma 7.

Lemma 7. *Let X be as in Theorem 5 or Theorem 6. Then, for all $|\boldsymbol{t}|$ large enough, the densities and their derivatives in (58) exist and satisfy*

$$\sup_{x_1,x_2\in\mathbb{R}} \prod_{i=1}^{2}(1+|x_i|)^{\ell} f_{\boldsymbol{t}}(x_1,x_2) < C,$$

$$\sup_{x\in\mathbb{R}}(1+|x|)^{2\ell}(f(x)+|f'(x)|) \;<\; C, \tag{59}$$

$$\sup_{x_1,x_2\in\mathbb{R}} \prod_{i=1}^{2}(1+|x_i|)^{\ell}|g_{\boldsymbol{t}}(x_1,x_2)| \;\le\; C|\boldsymbol{t}|^{-(\nu-2d)\wedge(2\nu-8d)}, \tag{60}$$

with C independent of $\boldsymbol{t}$.

Using Lemma 7, the proofs of Theorems 5 and 6 are completely analogous and proceed as follows. Let us prove that $a_G(1)$ in (46) is well defined and, as $|\boldsymbol{t}|\to\infty$,

$$\begin{aligned} r_{\mathcal{Z}}(\boldsymbol{t}) &= O(|\boldsymbol{t}|^{-(\nu-2d)\wedge(2\nu-8d)+\epsilon}),\\ r_Y(\boldsymbol{t}) &= a_G^2(1)\, r_X(\boldsymbol{t}) + O(|\boldsymbol{t}|^{-(\nu-2d)\wedge(2\nu-8d)+\epsilon}), \end{aligned} \tag{61}$$

where $\epsilon>0$ is arbitrarily small. Indeed, $\int_{\mathbb{R}}|G(x)||f'(x)|\mathrm{d}x \le C\int_{\mathbb{R}}(1+|x|)^{\ell-2}(1+|x|)^{-2\ell}\mathrm{d}x<\infty$ according to (49) and (59). Consider (61). For $\ell\ge 3$, it easily follows with $\epsilon=0$ from (57), (49), and (60) since $|G(x)-\mu_Y-a_G(1)x|\le C(1+|x|)^{\ell-2}, x\in\mathbb{R}$. Let $\ell=2$. Then, $|G(x)|\le C$ and

$$\begin{aligned} |r_{\mathcal{Z}}(\boldsymbol{t})| &\le C\int_{\mathbb{R}^2}(1+|x_1)(1+|x_2|)|g_{\boldsymbol{t}}(x_1,x_2)|\mathrm{d}x_1\mathrm{d}x_2\\ &\le C(I_1(\boldsymbol{t})+I_2(\boldsymbol{t})), \end{aligned} \tag{62}$$

where $I_1(\boldsymbol{t}) := \int_{\mathbb{R}^2}|g_{\boldsymbol{t}}(x_1,x_2)|\mathrm{d}x_1\mathrm{d}x_2 \le C|\boldsymbol{t}|^{-(\nu-2d)\wedge(2\nu-8d)} =: C\rho_{\boldsymbol{t}}$ (see (60)), and

$$\begin{aligned} I_2(\boldsymbol{t}) := &\int_{[-K,K]^2}|x_1x_2||g_{\boldsymbol{t}}(x_1,x_2)|\mathrm{d}x_1\mathrm{d}x_2\\ &+\int_{\mathbb{R}^2\setminus[-K,K]^2}|x_1x_2||g_{\boldsymbol{t}}(x_1,x_2)|\mathrm{d}x_1\mathrm{d}x_2 =: I_2'(\boldsymbol{t})+I_2''(\boldsymbol{t}), \end{aligned}$$

where $K = |\boldsymbol{t}|^p$ grows with $|\boldsymbol{t}| \to \infty$ as a power function. In view of (60), we see that $I_2'(\boldsymbol{t}) \le C\rho_{\boldsymbol{t}}\left(\int_0^K \frac{x\mathrm{d}x}{1+x^2}\right)^2 \le C\rho_{\boldsymbol{t}} \log|\boldsymbol{t}|^{2p} = O(|\boldsymbol{t}|^{-(\nu-2d)\wedge(2\nu-8d)+\epsilon})$, for any $\epsilon > 0, p > 0$. On the other hand,

$$
\begin{aligned}
I_2''(\boldsymbol{t}) &\le C\int_{\mathbb{R}^2\setminus[-K,K]^2} |x_1x_2| \left(f_{\boldsymbol{t}}(x_1,x_2) + \prod_{i=1}^2 (1+|x_i|)^{-4}\right)\mathrm{d}x_1\mathrm{d}x_2 \\
&= O(K^{-2}) = O(|\boldsymbol{t}|^{-2p})
\end{aligned}
$$

follows as $\int_{\mathbb{R}^2\setminus[-K,K]^2} |x_1x_2| f_{\boldsymbol{t}}(x_1,x_2)\mathrm{d}x_1\mathrm{d}x_2 = \mathrm{E}|X(\mathbf{0})X(\boldsymbol{t})|\mathbb{I}(|X(\mathbf{0})| > K \mathrm{or} |X(\boldsymbol{t})| > K) \le 2\mathrm{E}X(\mathbf{0})^2\mathbb{I}(|X(\mathbf{0})| > K) \le K^{-2}\mathrm{E}X(\mathbf{0})^4$ and $\mathrm{E}X(\mathbf{0})^4 < \infty$ holds by $\mathrm{E}\zeta^4 < \infty$. Therefore, $I_2''(\boldsymbol{t}) = O(|\boldsymbol{t}|^{-(\nu-2d)\wedge(2\nu-8d)})$, provided $2p \ge (\nu-2d)\wedge(2\nu-8d)$. This proves the first relation in (61). A similar argument applies to the other two covariances on the r.h.s. of (56) represented as integrals in (57), yielding $|\mathrm{Cov}(X(\mathbf{0}), \mathcal{Z}(\boldsymbol{t}))| = O(|\boldsymbol{t}|^{-(\nu-2d)\wedge(2\nu-8d)+\epsilon})$ and the second relation in (61). Then, from (61) and Proposition 1 (i), we get $\mathrm{Var}(\mathcal{Z}_\lambda(\phi)) = o(\lambda^{\nu+4d})$, and the limit in (50) follows by writing $Y_\lambda(\phi) = a_G(1)X_\lambda(\phi) + \mathcal{Z}_\lambda(\phi)$, and using the respective Theorems 4 and 5 of Chapter 3.

The proof of Lemma 7 uses the Fourier transform or characteristic functions by representing the Fourier transform

$$
\begin{aligned}
\widehat{g}_{\boldsymbol{t}}(z_1,z_2) &:= \widehat{f}_{\boldsymbol{t}}(z_1,z_2) - \widehat{f}(z_1)\widehat{f}(z_2) + r_X(\boldsymbol{t})z_1z_2\widehat{f}(z_1)\widehat{f}(z_2), \\
&\quad z_1, z_2 \in \mathbb{R}
\end{aligned}
\tag{63}
$$

of $g_{\boldsymbol{t}}(x_1,x_2)$ in (58) through $\widehat{f}(z) = \mathrm{E}e^{\mathrm{i}zX(\mathbf{0})}$, $\widehat{f}_{\boldsymbol{t}}(z_1,z_2) = \mathrm{E}e^{\mathrm{i}(z_1X(\mathbf{0})+z_2X(\boldsymbol{t}))}$. In the case of the MA RF X on $\mathbb{Z}^\nu$, the latter characteristic functions are written as infinite products

$$
\widehat{f}(z) = \prod_{\boldsymbol{u}\in\mathbb{Z}^\nu} \phi(za(\boldsymbol{u})), \quad \widehat{f}_{\boldsymbol{t}}(z_1,z_2) = \prod_{\boldsymbol{u}\in\mathbb{Z}^\nu} \phi(z_1a(\boldsymbol{u}) + z_2a_2(\boldsymbol{t}+\boldsymbol{u}))
$$

of characteristic functions of the noise r.v. ε, whereas in the case of X in (51), we use the Lévy–Khinchine representation. Handling of infinite products in (64) is rather cumbersome, and hence the proof of Lemma 7 is limited to X in (51). It is convenient to extend this lemma to more general stochastic integrals:

$$X_i := \int_{\mathbb{R}\times\mathbb{R}^\nu} a_i(\boldsymbol{s}) M(\mathrm{d}\boldsymbol{s}), \quad i = 1, 2, \tag{64}$$

where $M(\mathrm{d}\boldsymbol{s})$ is the same as in (51) and $a_i(\cdot) \in L^2(\mathbb{R}^\nu) \cap L^\infty(\mathbb{R}^\nu), i = 1, 2$ are deterministic functions generally different from $a(\cdot)$ in (51). This is done by the following Lemma 8, from which Lemma 7 is derived afterwards.

From (19) of Chapter 3, the characteristic function of (X_1, X_2) is written as

$$\mathrm{E}\mathrm{e}^{\mathrm{i}(z_1X_1+z_2X_2)} = \exp\left\{\int_{\mathbb{R}^\nu\times\mathbb{R}} \Psi\big(v(z_1a_1(\boldsymbol{s}) + z_2a_2(\boldsymbol{s}))\big)\mathrm{d}\boldsymbol{s}\pi(\mathrm{d}v)\right\},$$
$$z_1, z_2 \in \mathbb{R}. \tag{65}$$

From (65) and the definition of cumulant in (4) of Chapter 2, we have that for $n + m \le 2\ell$, the joint cumulant $\kappa_{n,m} := \mathrm{Cum}(\underbrace{X_1, \dots, X_1}_{n}, \underbrace{X_2, \dots, X_2}_{m})$ exists and written as

$$\kappa_{n,m} = \int_{\mathbb{R}} v^{n+m}\pi(\mathrm{d}v) \int_{\mathbb{R}^\nu} a_1^n(\boldsymbol{s})a_2^m(\boldsymbol{s})\mathrm{d}\boldsymbol{s}. \tag{66}$$

We introduce a quantity,

$$\rho_{12} := \bar\kappa_{1,1}^2 + \bar\kappa_{2,2} + \bar\kappa_{2,1} + \bar\kappa_{1,2}, \tag{67}$$

characterising the dependence between X_1 and X_2, written in terms of integrals $\bar\kappa_{n,m} := \int_{\mathbb{R}} |v|^{n+m}\pi(\mathrm{d}v) \int_{\mathbb{R}^\nu} |a_1(\boldsymbol{s})|^n|a_2(\boldsymbol{s})|^m\mathrm{d}\boldsymbol{s}$ which are upper estimates of the cumulants $\kappa_{n,m}$ in (66). Since $\int_{\mathbb{R}} v^2\pi(\mathrm{d}v) = 1$, so

$$\kappa_{1,1} = r_{12} := \mathrm{Cov}(X_1, X_2) = \int_{\mathbb{R}^\nu} a_1(\boldsymbol{s})a_2(\boldsymbol{s})\mathrm{d}\boldsymbol{s}. \tag{68}$$

We denote by $f_{12}(x_1, x_2), f_i(x), i = 1, 2$ the bivariate and univariate probability densities of X_1 and X_2 in (64), and let

$$g_{12}(x_1, x_2) := f_{12}(x_1, x_2) - f_1(x_1)f_2(x_2) - r_{12}f_1'(x_1)f_2'(x_2)$$

be defined similarly to (58). Condition (52) guarantees the existence of infinitely differentiable marginal densities $f_i, i = 1, 2$ [232] but is obviously not enough for f_{12}. In order that the bivariate distribution of (X_1, X_2) be absolutely continuous, the integrands a_1 and a_2 must be 'different' or 'spread apart'. This feature of $a_i, i = 1, 2$, is formalised as condition (69) of Lemma 8, where α is the same as in (52). Equation (69) also implies that the quantities $\bar{\kappa}_{n,m}, n + m \geq 3$, and ρ_{12} in (67) are typically much smaller than the covariance r_{12} in (68), resulting in a crucial bound in (71).

Lemma 8. *Let $X_i, i = 1, 2$, be two stochastic integrals in* (64)*with a Lévy measure π fulfilling conditions* (52) *and* (53), *and integrands $a_i, i = 1, 2$, such that*

$$\int_{\mathbb{R}^\nu} |a_i(\boldsymbol{s})|^\alpha \mathbb{I}(|a_j(\boldsymbol{s})| < |a_i(\boldsymbol{s})|/2)\mathrm{d}\boldsymbol{s} > c_{12}, \quad i = 1, 2, j = i + 1 (\mathrm{mod}\, 2) \tag{69}$$

hold for some $c_{12} > 0$ (where the integral in (69) *may be infinite). Then, the probability densities $f_{12}(x_1, x_2), f_i(x), i = 1, 2$ exist, are infinitely differentiable, and satisfy the following relations*:

$$\sup_{x_1, x_2 \in \mathbb{R}} \prod_{i=1}^{2} (1 + |x_i|)^\ell f_{12}(x_1, x_2) < \infty,$$

$$\sup_{x \in \mathbb{R}} (1 + |x|)^{2\ell} (f_i(x) + |f_i'(x)|) < \infty, \tag{70}$$

$$\sup_{x_1, x_2 \in \mathbb{R}} \prod_{i=1}^{2} (1 + |x_i|)^\ell |g_{12}(x_1, x_2)| \ \leq \ C\rho_{12}, \tag{71}$$

with $C > 0$ in (71) *depending only on ℓ, $c_\pi > 0$ in* (52), *c_{12} in* (69), *and $\|a_i\|_{L^2(\mathbb{R}^\nu)}$, $\|a_i\|_{L^\infty(\mathbb{R}^\nu)}, i = 1, 2$.*

Proof. Let $\widehat{f}_i(z) := \mathrm{E}\mathrm{e}^{\mathrm{i}zX_i}, i = 1, 2, \widehat{f}_{12}(z_1, z_2) := \mathrm{E}\mathrm{e}^{\mathrm{i}(z_1 X_1 + z_2 X_2)}$, $z, z_1, z_2 \in \mathbb{R}$ denote the respective characteristic functions. We shall prove that these functions decay exponentially with $|z_i| \to \infty, i = 1, 2$, and hence the corresponding probability densities exist and coincide with the inverse Fourier transforms of the characteristic functions. Let

$$\widehat{g}_{12}(z_1, z_2) := \widehat{f}_{12}(z_1, z_2) - \widehat{f}_1(z_1)\widehat{f}_1(z_2) + r_{12} z_1 z_2 \widehat{f}_1(z_1)\widehat{f}_2(z_2). \tag{72}$$

Let $\partial_i^n := \partial^n/\partial z_i^n, n \in \mathbb{N}_+, \partial_i := \partial_i^1, i = 1, 2$ denote partial derivatives. The lemma follows from the bounds

$$|\partial_1^m \partial_2^n \widehat{f}_{12}(z_1, z_2)| + |\partial_1^{2m} \partial_2^{2n} \widehat{f}_1(z_1)\widehat{f}_2(z_2)| \\ \le C\mathrm{e}^{-c(z_1^2+z_2^2)^{\alpha/2}}, \quad z_1, z_2 \in \mathbb{R} \tag{73}$$

and

$$|\partial_1^m \partial_2^n \widehat{g}_{12}(z_1, z_2)| \le C\mathrm{e}^{-c(z_1^2+z_2^2)^{\alpha/2}} \rho_{12}, \quad z_1, z_2 \in \mathbb{R}, \tag{74}$$

for any $m, n \in \{0, 1, \dots, \ell\}$, with some constants $c > 0, C > 0$ detailed as follows and $\alpha \in (0, 2)$ in (52). Indeed, (73) implies $|x_1|^m |x_2|^n f_{12}(x_1, x_2) = C\big|\int_{\mathbb{R}^2} \mathrm{e}^{-\mathrm{i}(x_1 z_1 + x_2 z_2} \partial_1^m \partial_2^n \widehat{f}_{12}(z_1, z_2)\mathrm{d}z_1 \mathrm{d}z_2\big| \le C\int_{\mathbb{R}^2} |\partial_1^m \partial_2^n \widehat{f}_{12}(z_1, z_2)|\mathrm{d}z_1\mathrm{d}z_2 < C$. Similarly, $|x_1|^{2m}|x_2|^{2n}(f_1(x_1)$ $f_2(x_2) + |f_1'(x_1) f_2'(x_2)|) < C$, proving (70). Relation (71) follows from (74) completely analogously.

Consider (73) for $m = n = 0$. We have $\widehat{f}_{12}(z_1, z_2) = \mathrm{e}^{J_{12}(z_1,z_2)}$, $\widehat{f}_i(z) = \mathrm{e}^{J_i(z)}, i = 1, 2$, where

$$J_{12}(z_1, z_2) := \int_{\mathbb{R}\times\mathbb{R}^\nu} \Psi\Big(v\sum_{i=1}^2 z_i a_i(\boldsymbol{s})\big)\Big)\pi(\mathrm{d}v)\mathrm{d}\boldsymbol{s},$$

$$J_i(z) := \int_{\mathbb{R}\times\mathbb{R}^\nu} \Psi(vza_i(\boldsymbol{s}))\Big)\pi(\mathrm{d}v)\mathrm{d}\boldsymbol{s}.$$

Using $1 - \cos(x) \ge x^2/4, |x| \le 1$, $\mathrm{Re}(J_{12}(z_1, z_2)) = -\int_{\mathbb{R}\times\mathbb{R}^\nu}(1 - \cos\big(v\sum_{i=1}^2 z_i a_i(\boldsymbol{s})\big))\pi(\mathrm{d}v)\mathrm{d}\boldsymbol{s} \le -(1/4)I(z_1, z_2)$, where

$$I(z_1, z_2) := \int_{\mathbb{R}^\nu}\Big(\sum_{i=1}^2 z_i a_i(\boldsymbol{s})\Big)^2 \mathrm{d}\boldsymbol{s}\int_{\mathbb{R}} v^2 \mathbb{I}\Big(v\Big|\sum_{i=1}^2 z_i a_i(\boldsymbol{s})\Big| \le 1\Big)\pi(\mathrm{d}v),$$

and the last integral is larger or equal to $c_\pi|\sum_{i=1}^2 z_i a_i(\boldsymbol{s})|^{\alpha-2}\mathbb{I}(|\sum_{i=1}^2 z_i a_i(\boldsymbol{s})| > 1)$ according to assumption (52). Therefore, $I(z_1, z_2) \ge c_\pi \int_{\mathbb{R}^\nu}\big|\sum_{i=1}^2 z_i a_i(\boldsymbol{s})\big|^\alpha \mathbb{I}(|\sum_{i=1}^2 z_i a_i(\boldsymbol{s})| > 1)\mathrm{d}\boldsymbol{s} =: I'(z_1, z_2)$. In turn, for $|z_1| \le |z_2|$, $|a_2(\boldsymbol{s})| \le |a_1(\boldsymbol{s})|/2$, we have

$|\sum_{i=1}^2 z_i a_i(\boldsymbol{s})| \ge |z_1 a_1(\boldsymbol{s})|/2$, and hence

$$\begin{aligned} I'(z_1,z_2) &\ge c'_\pi \int_{\mathbb{R}^\nu} |z_1 a_1(\boldsymbol{s})|^\alpha \mathbb{I}(|z_1 a_1(\boldsymbol{s})| > 2, |a_2(\boldsymbol{s})| \le |a_1(\boldsymbol{s})|/2)\mathrm{d}\boldsymbol{s} \\ &= |z_1|^\alpha h(|z_1|), \end{aligned}$$

where $c'_\pi := c_\pi/2^\alpha$, the function

$$z \mapsto h(z) := c'_\pi \int_{\mathbb{R}^\nu} |a_1(\boldsymbol{s})|^\alpha \mathbb{I}(z|a_1(\boldsymbol{s})| > 2, |a_2(\boldsymbol{s})| \le |a_1(\boldsymbol{s})|/2)\mathrm{d}\boldsymbol{s}$$

is nondecreasing on $(0,\infty)$, and

$$\lim_{z\to\infty} h(z) = c'_\pi \int_{\mathbb{R}^\nu} |a_1(\boldsymbol{s})|^\alpha \mathbb{I}(|a_2(\boldsymbol{s})| < |a_1(\boldsymbol{s})|/2)\mathrm{d}\boldsymbol{s} \ge c'_\pi c_{12} \in (0,\infty] \tag{75}$$

due to (69). Hence, $I'(z_1,z_2) \ge c''|z_1|^\alpha, c'' := c'_\pi c_{12}/2$ for all $|z_1| > z_0, |z_2| \le |z_1|$ and some $z_0 > 0$. Similarly, $I'(z_1,z_2) \ge c''|z_2|^\alpha$ for all $|z_2| > z_0, |z_1| \le |z_2|$. As a consequence, $I'(z_1,z_2) \ge c(z_1^2+z_2^2)^{\alpha/2}$ and $|\widehat{f}_{12}(z_1,z_2)| \le C\mathrm{e}^{-c(z_1^2+z_2^2)^{\alpha/2}}$ for all $|z_1|+|z_2|$ large enough, with $c = c''/2^{\alpha/2} > 0$. Since $|\widehat{f}_{12}(z_1,z_2)| \le 1, (z_1,z_2) \in \mathbb{R}^2$, which proves $|\widehat{f}_{12}(z_1,z_2)| \le C\mathrm{e}^{-c(z_1^2+z_2^2)^{\alpha/2}}, (z_1,z_2) \in \mathbb{R}^2$. A similar bound $|\widehat{f}_1(z_1)\widehat{f}(z_2)| \le C\mathrm{e}^{-c(z_1^2+z_2^2)^{\alpha/2}}, (z_1,z_2) \in \mathbb{R}^2$ follows analogously, proving (73) for $m = n = 0$. These bounds also entail

$$|\widehat{g}_{12}(z_1,z_2)| \le C\mathrm{e}^{-(c/2)(z_1^2+z_2^2)^{\alpha/2}}, \quad (z_1,z_2) \in \mathbb{R}^2 \tag{76}$$

since $|z_1 z_2| \le (z_1^2+z_2^2) \le C\mathrm{e}^{\epsilon(z_1^2+z_2^2)^{\alpha/2}}$ with $\epsilon > 0$ arbitrarily small and some $C < \infty$.

The proof of (74) for $m = n = 0$ uses (76) and the Taylor expansion of the exponent function, as follows. We first note that by taking a smaller $c > 0$, ρ_{12} on the r.h.s. of (74) can be replaced by

$$\rho_{12}(z_1,z_2) := (1+z_1^2)(1+z_2^2)(\bar{\kappa}_{1,1}^2 + \bar{\kappa}_{2,2} + \bar{\kappa}_{2,1} + \bar{\kappa}_{1,2}), \tag{77}$$

as $\rho_{12}(z_1,z_2) \le C\mathrm{e}^{\epsilon(z_1^2+z_2^2)^{\alpha/2}}\rho_{12}$ with arbitrary $\epsilon > 0$. Next, due to (76), it suffices to show (74) with (77) for $\rho_{12}(z_1,z_2) \le 1$. From the definition in (63),

$$\widehat{g}_{12}(z_1,z_2) = \widehat{f}_1(z_1)\widehat{f}_2(z_2)(\mathrm{e}^{\widetilde{J}_{12}(z_1,z_2)} - 1 + z_1 z_2 r_{12}),$$

where

$$\begin{aligned}\widetilde{J}_{12}(z_1,z_2) &:= J_{12}(z_1,z_2) - J_1(z_1) - J_2(z_2)\\ &= \int_{\mathbb{R}\times\mathbb{R}^\nu} (\mathrm{e}^{\mathrm{i}vz_1a_1(\boldsymbol{s})} - 1)(\mathrm{e}^{\mathrm{i}vz_2a_2(\boldsymbol{s})} - 1)\pi(\mathrm{d}v)\mathrm{d}\boldsymbol{s}\end{aligned}$$

satisfies $|\widetilde{J}_{12}(z_1,z_2)| \le |z_1z_2|\bar{\kappa}_{1,1} \le |\rho_{12}(z_1,z_2)|^{1/2} \le 1$; see (77). Therefore, $|\widehat{g}_{12}(z_1,z_2)| \le |\widetilde{J}_{12}(z_1,z_2) + z_1z_2r_{12}| + (1/2)|\widetilde{J}_{12}(z_1,z_2)|^2$, where

$$\begin{aligned}\widetilde{J}_{12}(z_1,z_2) + z_1z_2r_{12} = \int_{\mathbb{R}\times\mathbb{R}^\nu} &\{(\mathrm{e}^{\mathrm{i}vz_1a_1(\boldsymbol{s})} - 1)(\mathrm{e}^{\mathrm{i}vz_2a_2(\boldsymbol{s})} - 1)\\ &+ z_1z_2v^2a_1(\boldsymbol{s})a_2(\boldsymbol{s})\}\pi(\mathrm{d}v)\mathrm{d}\boldsymbol{s}.\end{aligned}$$

Using the identity $(\mathrm{e}^{\mathrm{i}w_1} - 1)(\mathrm{e}^{\mathrm{i}w_2} - 1) + w_1w_2 = \Psi(w_1)\Psi(w_2) + \mathrm{i}w_1\Psi(w_2) + \mathrm{i}w_2\Psi(w_1)$, $w_1, w_2 \in \mathbb{R}$, together with $|\Psi(w)| \le w^2/2$ and the definitions in (66), (67) we get that

$$\begin{aligned}|\widetilde{J}_{12}(z_1,z_2) + z_1z_2r_{12}| &\le (1/2)(z_1^2|z_2|\bar{\kappa}_{2,1} + z_2^2|z_1|\bar{\kappa}_{1,2} + z_1^2z_2^2\bar{\kappa}_{2,2})\\ &\le \rho_{12}(z_1,z_2)\end{aligned} \tag{78}$$

and $|\widetilde{J}_{12}(z_1,z_2)|^2 \le \rho_{12}(z_1,z_2)$, proving (74) for $m = n = 0$.

Consider (73)–(74) in the general case $0 \le m,n \le \ell$. We have

$$\partial_1^n\partial_2^m\widehat{f}_{12}(z_1,z_2) = \widehat{f}_{12}(z_1,z_2)\sum_{i=0}^{n}\sum_{j=0}^{m}\binom{n}{i}\binom{m}{j}\partial_1^i\partial_2^jJ_{12}(z_1,z_2).$$

Then, $|\partial_1^i\partial_2^jJ_{12}(z_1,z_2)| \le (1 + |z_1| + |z_2|)\bar{\kappa}_{n,m}$, and hence $|\partial_1^n\partial_2^m\widehat{f}_{12}(z_1,z_2)| \le C|\widehat{f}_{12}(z_1,z_2)|(1 + |z_1| + |z_2|)$ and relation (73) follows as in the case of $m = n = 0$ with minor changes.

Consider (74) where we can replace ρ_{12} by $\rho_{12}(z_1,z_2)$ of (77) as before. Rewrite (3) as $\widehat{g}_{12}(z_1,z_2) = \widehat{f}_1(z_1)\widehat{f}_2(z_2)K_{12}(z_1,z_2)$, where

$K_{12}(z_1, z_2) := \mathrm{e}^{\widetilde{J}_{12}(z_1,z_2)} - 1 + z_1 z_2 r_{12}$. Then,

$$\partial_1^\ell \partial_2^m \widehat{g}_{12}(z_1, z_2) = \sum_{i=0}^{\ell} \sum_{j=0}^{m} \binom{\ell}{i} \binom{m}{j} \partial_1^{\ell-i} \widehat{f}_1(z_1) \partial_2^{m-j} \widehat{f}_2(z_2) \times \partial_1^i \partial_2^j K_{12}(z_1, z_2). \tag{79}$$

Applying the classical Faà di Bruno differentiation rule of composite function, the derivative $\partial_1^i \partial_2^j K_{12}(z_1, z_2)$ can be expressed as a polynomial in partial derivatives of $\widetilde{J}_{12}(z_1, z_2)$. Particularly, for $i \vee j \geq 2$ and omitting arguments, we obtain

$$\partial_1^i \partial_2^j K_{12} = \mathrm{e}^{\widetilde{J}_{12}} \sum_{\{V\}} \prod_{k=1}^{s} (\partial_1^{|V_k \cap T_1|} \partial_2^{|V_k \cap T_2|} \widetilde{J}_{12}),$$

where the sum is taken over all partitions $\{V\} = (V_1, \ldots, V_s), s = 1, \ldots, i+j$ of table $T = T_1 \cup T_2$ consisting of two rows $T_1 = \{1, \ldots, i\}, T_2 = \{1, \ldots, j\}$ by nonempty sets $V_k \subset T_{12}, V_k \cap V_{k'} = \emptyset \, (k \neq k'), \cup_{k=1}^s V_k = T_{12}$, and $|V|$ stands for the cardinality of $V \subset T_{12}$. The bound

$$|\partial_1^n \partial_2^m (\widetilde{J}_{12}(z_1, z_2) + z_1 z_2 r_{12})| \leq C \rho_{12}(z_1, z_2) \tag{80}$$

follows as in (78), for any $n, n \in \mathbb{N}$ fixed, leading to

$$|\partial_1^i \partial_2^j K_{12}(z_1, z_2)| \leq C \rho_{12}(z_1, z_2), \quad \text{for } \rho_{12}(z_1, z_2) \leq 1, \; i \vee j \geq 2. \tag{81}$$

In the remaining cases of $(i, j) = (1, 0), (0, 1), (1, 1)$, the bound in (81) can be directly verified, similarly as in the case of $(i, j) = (0, 0)$ above. Finally, (74) is a consequence of (79), (81), and (73), thereby completing the proof of Lemma 8. □

Proof of Lemma 6. Apply Lemma 8 to $X_1 = X(\mathbf{0}), X_2 = X(\boldsymbol{t})$, or $a_1(\boldsymbol{s}) = a(\boldsymbol{s}), a_2(\boldsymbol{s}) = a(\boldsymbol{t} + \boldsymbol{s})$. Let us show that for all $|\boldsymbol{t}|$ large enough, Assumption (A)$(d; \mathbb{R}^\nu)$ implies condition (69). Write $\boldsymbol{s} = r\boldsymbol{o} \in \mathbb{R}^\nu, r > 0, \boldsymbol{o} \in S_{\nu-1}$. Since $\ell(\boldsymbol{o}), \boldsymbol{o} \in S_{\nu-1}$ in (7) of Chapter 1 is nontrivial and continuous, there exist $\boldsymbol{o}_0 \in S_{\nu-1}$ and $\epsilon > 0$ such that $|\ell(\boldsymbol{o})| > 2\epsilon, |\boldsymbol{o} - \boldsymbol{o}_0| < \epsilon$. Therefore, $|a(\boldsymbol{s})| > \epsilon / r_0^{2d-\nu} =: \epsilon_1 > 0$

for $\boldsymbol{s} = r\boldsymbol{o}, r \in (r_0 - 1, r_0], |\boldsymbol{o} - \boldsymbol{o}_0| < \epsilon$ and some $r_0 > 1$ and $\int_{\mathbb{R}^\nu} |a(\boldsymbol{s})|^\alpha \mathbb{I}(r \in (r_0 - 1, r_0], |\boldsymbol{o} - \boldsymbol{o}_0| < \epsilon)\mathrm{d}\boldsymbol{s} =: c_{12} > 0$. Then, since $|a(\boldsymbol{t})| < C/|\boldsymbol{t}|^{\nu-2d}$ for all $|\boldsymbol{t}|$ large enough,

$$\begin{aligned}
&\int_{\mathbb{R}^\nu} |a_1(\boldsymbol{s})|^\alpha \mathbb{I}(r \in (r_0 - 1, r_0], |\boldsymbol{o} - \boldsymbol{o}_0| < \epsilon, |a_2(\boldsymbol{s})| < |a_1(\boldsymbol{s})|/2)\mathrm{d}\boldsymbol{s} \\
&\quad \geq \int_{\mathbb{R}^\nu} |a(\boldsymbol{s})|^\alpha \mathbb{I}(r \in (r_0 - 1, r_0], |\boldsymbol{o} - \boldsymbol{o}_0| < \epsilon, \\
&\quad\quad |a(\boldsymbol{t}+\boldsymbol{s})| < |a(\boldsymbol{s})|/2)\mathrm{d}\boldsymbol{s} \\
&\quad \geq \int_{\mathbb{R}^\nu} |a(\boldsymbol{s})|^\alpha \mathbb{I}\bigg(r \in (r_0 - 1, r_0], |\boldsymbol{o} - \boldsymbol{o}_0| < \epsilon, \\
&\quad\quad \frac{C}{|\boldsymbol{t}+r\boldsymbol{o}|^{\nu-2d}} < \epsilon_1/2\bigg)\mathrm{d}\boldsymbol{s} = c_{12}
\end{aligned}$$

for all $\boldsymbol{t} \in \mathbb{R}^\nu$ such that $\frac{C}{|\boldsymbol{t}+r\boldsymbol{o}|^{\nu-2d}} < \epsilon_1/2, r \in (r_0 - 1, r_0], |\boldsymbol{o} - \boldsymbol{o}_0| < \epsilon$ or $|\boldsymbol{t} + r\boldsymbol{o}| > (2C/\epsilon_1)^{1/(\nu-2d)}$ hold, which is obviously true when $|\boldsymbol{t}|$ is large enough. This proves (69) for $i = 1$ and $j = 2$, and the proof for $i = 2$ and $j = 1$ is the same. As a consequence, Lemma 8 and the bounds in (70) and (71) apply to the above $X_i, i = 1, 2$, yielding $\sup_{x_1,x_2\in\mathbb{R}} \prod_{i=1}^2 (1+|x_i|)^\ell |g_{\boldsymbol{t}}(x_1, x_2)|\mathrm{d}x_1\mathrm{d}x_2 \leq C\rho_{\boldsymbol{t}}$, with $g_{\boldsymbol{t}}(x_1, x_2)$ in (58) and $\rho_{12} \equiv \rho_{\boldsymbol{t}}$ defined as in (67), with

$$\bar{\kappa}_{n,m} := \int_{\mathbb{R}} |v|^{n+m}\pi(\mathrm{d}v) \int_{\mathbb{R}^\nu} |a(\boldsymbol{s})|^n |a(\boldsymbol{t}+\boldsymbol{s})|^m \mathrm{d}\boldsymbol{s}.$$

Thus, (60) follows from (71) and

$$\rho_{\boldsymbol{t}} \;=\; O(|\boldsymbol{t}|^{-(\nu-2d)\wedge(2\nu-8d)}). \tag{82}$$

Indeed, $\bar{\kappa}_{1,1}^2 = O(|\boldsymbol{t}|^{-2(\nu-4d)})$, whereas $\bar{\kappa}_{n,m} = O(|\boldsymbol{t}|^{-(\nu-2d)})$ for $m, n = 1, 2, m + n \geq 3$ can be verified as follows. Let $n = 2, m = 1$. For a fixed but large K, split $\int_{\mathbb{R}^\nu} |a(\boldsymbol{s})|^2 |a(\boldsymbol{t} + \boldsymbol{s})|\mathrm{d}\boldsymbol{s}$ over sets $\{|\boldsymbol{s}| \leq K\}, \{|\boldsymbol{t} + \boldsymbol{s}| \leq K\}$ and $\{|\boldsymbol{s}| > K, |\boldsymbol{t} + \boldsymbol{s}| > K\}$. Thus, $\bar{\kappa}_{2,1} \leq \sum_{j=1}^3 I_j(\boldsymbol{t})$, where $I_1(\boldsymbol{t}) \leq C \int_{|\boldsymbol{s}|\leq K} |a(\boldsymbol{s})|^2 |\boldsymbol{t}+\boldsymbol{s}|^{-(\nu-2d)}$ $\mathrm{d}\boldsymbol{s} \leq C|\boldsymbol{t}|^{-(\nu-2d)} \int_{\mathbb{R}^3} |a(\boldsymbol{s})|^2\mathrm{d}\boldsymbol{s} = O(|\boldsymbol{t}|^{-(\nu-2d)})$. Similarly, $I_2(\boldsymbol{t}) \leq C \int_{|\boldsymbol{t}+\boldsymbol{s}|\leq K} |\boldsymbol{s}|^{-2(\nu-2d)} |a(\boldsymbol{t} + \boldsymbol{s})|\mathrm{d}\boldsymbol{s} \leq C|\boldsymbol{t}|^{-2(\nu-2d)} \int_{|\boldsymbol{s}|\leq K} |a(\boldsymbol{s})|\mathrm{d}\boldsymbol{s} =$

$O(|\boldsymbol{t}|^{-2(\nu-2d)})$ since $\int_{|\boldsymbol{s}|\le K}|a(\boldsymbol{s})|\mathrm{d}\boldsymbol{s} \le C\big(\int_{|\boldsymbol{s}|\le K}|a(\boldsymbol{s})|^2\mathrm{d}\boldsymbol{s}\big)^{1/2} < C$. Finally, $I_3(\boldsymbol{t}) \le C\int_{|\boldsymbol{s}|>K}|\boldsymbol{s}|^{-2(\nu-2d)}|\boldsymbol{t}+\boldsymbol{s}|^{-(\nu-2d)}\mathrm{d}\boldsymbol{s} \le C\big(\int_{|\boldsymbol{t}|/2>|\boldsymbol{s}|>K}|\boldsymbol{s}|^{-2(\nu-2d)}|\boldsymbol{t}+\boldsymbol{s}|^{-(\nu-2d)}\mathrm{d}\boldsymbol{s} + \int_{|\boldsymbol{s}|>|\boldsymbol{t}|/2}|\boldsymbol{s}|^{-2(\nu-2d)}|\boldsymbol{t}+\boldsymbol{s}|^{-(\nu-2d)}\mathrm{d}\boldsymbol{s}\big) =: I_{3,1}(\boldsymbol{t}) + I_{3,2}(\boldsymbol{t})$, where $I_{3,1}(\boldsymbol{t}) \le C|\boldsymbol{t}|^{-(\nu-2d)}\int_{|\boldsymbol{s}|>K}|\boldsymbol{s}|^{-2(\nu-2d)}\mathrm{d}\boldsymbol{s} = O(|\boldsymbol{t}|^{-(\nu-2d)})$ and $I_{3,2}(\boldsymbol{t}) \le C|\boldsymbol{t}|^{\nu-3(\nu-2d)}\int_{|\boldsymbol{s}|>K/2}|\boldsymbol{s}|^{-2(\nu-2d)}|\frac{\boldsymbol{t}}{|\boldsymbol{t}|}+\boldsymbol{s}|^{-(\nu-2d)}\mathrm{d}\boldsymbol{s} = O(|\boldsymbol{t}|^{-(2\nu-6d)}) = o(|\boldsymbol{t}|^{-(2\nu-8d)})$, proving (82) and (60). □

Exponential function. Theorems 5 and 6 of Chapter 5 do not apply to nonlinear functions G, which grow faster than any polynomial. Particularly, the exponential function G is of interest since it appears in applications to Burgers' equation in Section 6 of Chapter 5. In what follows, we extend the first-order reduction principle to $G(x) = \mathrm{e}^x$ using a direct approach through the characteristic function of the Poisson stochastic integral. By definition (88) of Chapter 2, the first Appell coefficient of $G(x) = \mathrm{e}^x$ is

$$a_G(1) = \int_{\mathbb{R}}\mathrm{e}^x f(x)\mathrm{d}x = \mathrm{E}\mathrm{e}^{X(\mathbf{0})} \ne 0. \tag{83}$$

We can expect that scaling limits of RFs $Y(\boldsymbol{t}) = \mathrm{e}^{X(\boldsymbol{t})}$ and $a_G(1)X(\boldsymbol{t})$ are the same. A rigorous result is given in the following proposition. Unsurprisingly, dealing with the exponential function imposes strong boundedness conditions on the kernel and the Lévy measure in stochastic integral representation of X.

Proposition 9. *Let $Y(\boldsymbol{t}) = \mathrm{e}^{X(\boldsymbol{t})}$, where X is a Lévy MA RF in* (51) *on $\mathbb{R}^\nu$ with kernel $a(\cdot) \in L^2(\mathbb{R}^\nu)\cap L^\infty(\mathbb{R}^\nu)$ and the Lévy measure π having finite exponential moments: $\int_{\mathbb{R}}\mathrm{e}^{r|v|}\pi(\mathrm{d}v) < \infty\, (\forall r > 0)$. Then, $\mathrm{E}Y(\boldsymbol{t})^2 < \infty$. In addition, we have the following:*

(i) *Let $a(\cdot)$ satisfy Assumption $(A)(d;\mathbb{R}^\nu)$, $0 < d < \frac{\nu}{4}$. Then,*

$$r_{\mathcal{Z}}(\boldsymbol{t}) = O(|\boldsymbol{t}|^{-2(\nu-4d)}) = o(r_X(\boldsymbol{t})) \quad \text{and}$$
$$r_Y(\boldsymbol{t}) = a_G^2(1)(r_X(\boldsymbol{t}) + o(1)), \tag{84}$$

where $a_G(1) = \mathrm{E}Y(\mathbf{0}) = \mathrm{E}\mathrm{e}^{X(\mathbf{0})}$, $\mathcal{Z}(\boldsymbol{t}) = Y(\boldsymbol{t}) - a_G(1)X(\boldsymbol{t})$. Moreover,

$$\lambda^{-(\nu+4d)/2}Y_\lambda(\phi) \xrightarrow{\mathrm{d}} a_G(1)W_d(\phi), \quad \forall\,\phi\in\Phi, \tag{85}$$

where $Y_\lambda(\phi), W_d(\phi)$ are as in (54).

(ii) *Let* $a(\cdot) \in L^1(\mathbb{R}^\nu)$. *Then,* $\int_{\mathbb{R}^\nu} |r_Y(\boldsymbol{t})| \mathrm{d}\boldsymbol{t} < \infty$ *and*

$$\lambda^{-\nu/2} Y_\lambda(\phi) \xrightarrow{\mathrm{d}} \sigma_Y W(\phi), \quad \forall \phi \in \Phi, \quad \text{with}$$
$$\sigma_Y^2 := \int_{\mathbb{R}^\nu} r_Y(\boldsymbol{t}) \mathrm{d}\boldsymbol{t} > 0. \tag{86}$$

Proof. As in Theorem 6, we consider only the case of $\sigma = 0$. To show $\mathrm{E}Y(\boldsymbol{t})^2 < \infty$, we use Proposition 8 of Chapter 2 and the representation $X(\boldsymbol{t}) = \int_{\mathbb{R}^\nu \times \mathbb{R}} v a(\boldsymbol{t}-\boldsymbol{u}) \widetilde{\mathcal{N}}(\mathrm{d}\boldsymbol{u}, \mathrm{d}v)$ w.r.t. the centred Poisson measure $\widetilde{\mathcal{N}}(\mathrm{d}\boldsymbol{u}, \mathrm{d}v)$ with variance $\mathrm{d}\boldsymbol{u}\pi(\mathrm{d}v)$; see (128) of Chapter 2. Condition (28) of Chapter 2 for $f(\boldsymbol{u}, v) = va(\boldsymbol{t} - \boldsymbol{u})$ translates to

$$J := \int_{\mathbb{R}^\nu \times \mathbb{R}} \mathrm{e}^{2va(\boldsymbol{u})} \mathbb{I}(2va(\boldsymbol{u}) > 1) \pi(\mathrm{d}v) \mathrm{d}\boldsymbol{u} \ < \ \infty,$$

which can be verified as follows. Let $K := \sup_{\boldsymbol{u} \in \mathbb{R}^\nu} |a(\boldsymbol{u})|$. Then, $J \le \int_{\mathbb{R}^\nu} \mathrm{d}\boldsymbol{u} \int_{\mathbb{R}} \mathrm{e}^{r|v| - (r-2K)|v|} \, \mathbb{I}(2|v||a(\boldsymbol{u})| > 1) \pi(\mathrm{d}v)$, with $r - 2K \ge 1$. This yields $J \le \int_{\mathbb{R}^\nu} \mathrm{d}\boldsymbol{u} \int_{\mathbb{R}} \mathrm{e}^{r|v| - (r-2K)|v|} \mathbb{I}(2|a(\boldsymbol{u})||v| > 1) \pi(\mathrm{d}v) \le \int_{\mathbb{R}^\nu} \mathrm{e}^{-1/2|a(\boldsymbol{u})|} \mathrm{d}\boldsymbol{u} \int_{\mathbb{R}} \mathrm{e}^{r|v|} \pi(\mathrm{d}v) \le C \int_{\mathbb{R}^\nu} \mathrm{e}^{-1/2|a(\boldsymbol{u})|} \mathrm{d}\boldsymbol{u} < \infty$ since $a \in L^2(\mathbb{R}^\nu)$, proving $\mathrm{E}Y(\boldsymbol{t})^2 < \infty$.

(i) Since $\mathcal{Z}(\boldsymbol{t}) = \exp\{\int_{\mathbb{R}^\nu \times \mathbb{R}} f(\boldsymbol{u}, v) \mathrm{d}\widetilde{\mathcal{N}}\} - (\mathrm{E}\mathrm{e}^{X(\boldsymbol{t})}) \int_{\mathbb{R}^\nu \times \mathbb{R}} f(\boldsymbol{u}, v) \mathrm{d}\widetilde{\mathcal{N}}$, the covariance $r_{\mathcal{Z}}(\boldsymbol{t})$ is written as in (30):

$$r_{\mathcal{Z}}(\boldsymbol{t}) = a_{1,G}^2 \Bigg\{ \widetilde{\Psi}\bigg(\int_{\mathbb{R}^\nu \times \mathbb{R}} (\mathrm{e}^{va(\boldsymbol{u})} - 1)(\mathrm{e}^{va(\boldsymbol{t}+\boldsymbol{u})} - 1) \pi(\mathrm{d}v) \mathrm{d}\boldsymbol{u} \bigg)$$
$$+ \int_{\mathbb{R}^\nu \times \mathbb{R}} \widetilde{\Psi}(va(\boldsymbol{u})) \widetilde{\Psi}(va(\boldsymbol{t} + \boldsymbol{u})) \pi(\mathrm{d}v) \mathrm{d}\boldsymbol{u} \Bigg\}.$$

Denote $\bar{r}_i(\boldsymbol{t}) := \int_{\mathbb{R}^\nu} |a(\boldsymbol{u}) a(\boldsymbol{t} + \boldsymbol{u})|^i \mathrm{d}\boldsymbol{u}, i = 1, 2$. Using inequalities $|\mathrm{e}^z - 1| \le |z| \mathrm{e}^{|z|}, |\widetilde{\Psi}(z)| \le z^2 \mathrm{e}^{|z|}, z \in \mathbb{R}$ and the assumption $\|a(\boldsymbol{u})\|_{L^\infty(\mathbb{R}^\nu)} \le K$, we estimate the above integrals as $\left| \int_{\mathbb{R}^\nu \times \mathbb{R}} (\mathrm{e}^{va(\boldsymbol{u})} - 1)(\mathrm{e}^{va(\boldsymbol{t}+\boldsymbol{u})} - 1) \pi(\mathrm{d}v) \mathrm{d}\boldsymbol{u} \right| \le \int_{\mathbb{R}^\nu \times \mathbb{R}} v^2 |a(\boldsymbol{u}) a(\boldsymbol{t} +$

$\boldsymbol{u})|\mathrm{e}^{2Kv}\pi(\mathrm{d}v)\mathrm{d}\boldsymbol{u} \le C\bar{r}_1(\boldsymbol{t})$ and

$$|r_{\mathcal{Z}}(\boldsymbol{t})| \le C\left(\bar{r}_1^2(\boldsymbol{t}) + \int_{\mathbb{R}^\nu\times\mathbb{R}} v^4|a(\boldsymbol{u})a(\boldsymbol{t}+\boldsymbol{u})|^2\mathrm{e}^{2Kv}\pi(\mathrm{d}v)\mathrm{d}\boldsymbol{u}\right)$$
$$\le C(\bar{r}_1^2(\boldsymbol{t}) + \bar{r}_2(\boldsymbol{t})),$$

proving the first relation in (84) since $\bar{r}_2(\boldsymbol{t}) = O(|\boldsymbol{t}|^{-2(\nu-2d)}) = o(|\boldsymbol{t}|^{-2(\nu-4d)})$ $(|\boldsymbol{t}| \to \infty)$. The remaining facts in (84) and (85) are similar to those in Theorems 5 and 6.
(ii) From (29), $r_Y(\boldsymbol{t}) = (\mathrm{E}Y(\mathbf{0}))^2(\mathrm{e}^{J(\boldsymbol{t})} - 1)$, where $J(\boldsymbol{t}) := \int_{\mathbb{R}^\nu\times\mathbb{R}}(\mathrm{e}^{va(\boldsymbol{u})} - 1)(\mathrm{e}^{va(\boldsymbol{t}+\boldsymbol{u})} - 1)\pi(\mathrm{d}v)\mathrm{d}\boldsymbol{u}$ satisfies $|J(\boldsymbol{t})| \le C\bar{r}_1(\boldsymbol{t})$; see the proof of (i). Thus, $|r_Y(\boldsymbol{t})| \le C\bar{r}_1(\boldsymbol{t})$ and $\int_{\mathbb{R}^\nu}|r_Y(\boldsymbol{t})|\mathrm{d}\boldsymbol{t} \le C\big(\int_{\mathbb{R}^\nu}|a(\boldsymbol{t})|\mathrm{d}\boldsymbol{t}\big)^2 < \infty$. By inequality $\mathrm{e}^x - 1 \ge x, x \in \mathbb{R}$, we see that $\sigma_Y^2 \ge (\mathrm{E}Y(\mathbf{0}))^2\int_{\mathbb{R}^\nu}J(\boldsymbol{t})\mathrm{d}\boldsymbol{t} = (\mathrm{E}Y(\mathbf{0}))^2\int_{\mathbb{R}}\pi(\mathrm{d}v)\big(\int_{\mathbb{R}^\nu}(\mathrm{e}^{va(\boldsymbol{u})} - 1)\mathrm{d}\boldsymbol{u}\big)^2$. The function $g(v) := \int_{\mathbb{R}^\nu}(\mathrm{e}^{va(\boldsymbol{u})} - 1)\mathrm{d}\boldsymbol{u}, v \in \mathbb{R}$ is twice continuously differentiable and concave with $g(0) = 0, g''(v) = \int_{\mathbb{R}^\nu}a(\boldsymbol{u})^2\mathrm{e}^{va(\boldsymbol{u})}\mathrm{d}\boldsymbol{u} > 0\,(v \in \mathbb{R})$, implying $\int_{\mathbb{R}}g(v)^2\pi(\mathrm{d}v) > 0$ and, consequently, $\sigma_Y^2 > 0$.

Let us prove (86). We use the approximation of $Y(\boldsymbol{t})$ by the m-dependent RF $Y_m(\boldsymbol{t}) := \mathrm{e}^{X_m(\boldsymbol{t})}$ with $X_m(\boldsymbol{t}) := \int_{\mathbb{R}^\nu}a_m(\boldsymbol{t} - \boldsymbol{s})M(\mathrm{d}\boldsymbol{s}), a_m(\boldsymbol{t}) := a(\boldsymbol{t})\mathbb{I}(|\boldsymbol{t}| \le m)$ and the argument as in (44), which is often used in this book. The proof of the approximation, viz.,

$$\lim_{m\to\infty}\limsup_{\lambda\to\infty}\lambda^{-\nu}\mathrm{Var}(Y_\lambda(\phi) - Y_{\lambda,m}(\phi)) = 0,$$

is not difficult since the covariance of $Y(\boldsymbol{t}) - Y_m(\boldsymbol{t})$ is written explicitly using (29) of Chapter 2 and can be evaluated similarly to $r_Y(\boldsymbol{t})$ above. □

4 Application to Spatial Empirical Processes

Kolmogorov–Smirnov statistic. Let $A \subset \mathbb{R}^\nu$ be a bounded Borel set and $X = \{X(\boldsymbol{t}); \boldsymbol{t} \in \mathbb{Z}^\nu\}$ be a stationary RF on $\mathbb{Z}^\nu$. The ratio

$$F_\lambda(y) := \frac{\int_{\lambda A}\mathbb{I}(X([\boldsymbol{t}]) \le y)\mathrm{d}\boldsymbol{t}}{\mathrm{Leb}_\nu(\lambda A)}, \quad y \in \mathbb{R} \tag{87}$$

is the empirical distribution function (empirical process) and an estimator of the marginal distribution function $F(y) = \mathrm{P}(X(\boldsymbol{t}) \le y), y \in \mathbb{R}$ of RF X, with the property that

$$\mathrm{E}F_\lambda(y) = F(y), \quad \forall y \in \mathbb{R},$$

meaning $F_\lambda(y)$ is an unbiased estimator of $F(y)$. Analogously, given a stationary RF $X = \{X(\boldsymbol{t}); \boldsymbol{t} \in \mathbb{R}^\nu\}$ on $\mathbb{R}^\nu$ and $A \subset \mathbb{R}^\nu$ as above, we define the corresponding estimator as

$$F_\lambda(y) := \frac{\int_{\lambda A} \mathbb{I}(X(\boldsymbol{t}) \le y)\mathrm{d}\boldsymbol{t}}{\mathrm{Leb}_\nu(\lambda A)}, \quad y \in \mathbb{R}. \tag{88}$$

For fixed y, the numerator in (87)–(88) is a particular case of the integral $Y_\lambda(\phi)$ in Corollary 5, corresponding to indicator functions $\phi_A(\boldsymbol{t}) = \mathbb{I}(\boldsymbol{t} \in A)$ and $G_y(x) := \mathbb{I}(x \le y), x \in \mathbb{R}$. We can apply the results in Section 3 of Chapter 4 to study consistency and the limit distribution of the estimator in (87)–(88) as $\lambda \to \infty$. The object of interest is the classical *Kolmogorov–Smirnov statistic*

$$\mathcal{D}_\lambda := \sup_{y \in \mathbb{R}} |F_\lambda(y) - F(y)|. \tag{89}$$

For MA LRD RFs X in Theorems 5 and 6 and a fixed y, we can expect that the limit distribution of (87) and (88) is Gaussian and agrees with that of the linear statistic $-f(y)X_\lambda(A)/\mathrm{Leb}_\nu(A)$ since the first Appell coefficient

$$a_{G_y}(1) = -\int_{\mathbb{R}} G_y(x) f'(x)\mathrm{d}x = -\int_{-\infty}^{y} f'(x)\mathrm{d}x = -f(y) \tag{90}$$

coincides with the marginal density of X at point y. We also recall from Theorems 4 and 5 of Chapter 3 that for the above X,

$$\lambda^{-2d-\frac{\nu}{2}} X_\lambda(A) \xrightarrow{\mathrm{d}} W_d(A) := \int_{\mathbb{R}^\nu} \left\{ \int_A a_\infty(\boldsymbol{t} - \boldsymbol{u})\mathrm{d}\boldsymbol{t} \right\} W(\mathrm{d}\boldsymbol{u}) \tag{91}$$

tends to a Gaussian distribution with mean zero and variance

$$\mathrm{E}W_d(A)^2 = \sigma_A^2 := \int_A \int_A L\left(\frac{\boldsymbol{t}_1 - \boldsymbol{t}_2}{|\boldsymbol{t}_1 - \boldsymbol{t}_2|}\right) \frac{\mathrm{d}\boldsymbol{t}_1 \boldsymbol{t}_2}{|\boldsymbol{t}_1 - \boldsymbol{t}_2|^{\nu - 4d}}. \tag{92}$$

In view of the above facts, the question about the limit distribution of the Kolmogorov–Smirnov statistics in (89) basically reduces to the tightness of the 'remainder empirical process'

$$S_\lambda(y) := F_\lambda(y) - F(y) + f(y)X_\lambda(A)/\mathrm{Leb}_\nu(A), \quad y \in \mathbb{R}, \tag{93}$$

in the Skorohod space $D([-\infty, \infty])$.

Proposition 10. *Let X be as in Theorems* 5 *or* 6, $0 < d < \nu/4$, $\ell = 2$, *and* $S_\lambda(y)$ *be as in* (93), (87)–(88), *where* $A \subset \mathbb{R}^\nu$ *is an arbitrary bounded Borel set. Then,*

$$\sup_{y\in\mathbb{R}} |S_\lambda(y)| = o_p\left(\lambda^{2d-\frac{\nu}{2}}\right) \tag{94}$$

and

$$\lambda^{\frac{\nu}{2}-2d}\mathcal{D}_\lambda \xrightarrow{\mathrm{d}} \|f\|_\infty(\sigma_A/\mathrm{Leb}_\nu(A))|Z|, \quad Z \sim N(0,1), \tag{95}$$

where $\|f\|_\infty = \sup_y |f(y)|$ *and* $\sigma_A > 0$ *are defined in* (92).

Proof. The convergence in (95) is easy from (94) and the decomposition of $F_\lambda(y) - F(y)$ in (93), using (91) and the fact that $f(x)$ is bounded and continuous on $\mathbb{R}$. Therefore, we can focus on (94) in the remainder of the proof.

Towards this aim, we prove the bound on increment of (93): for any $-\infty \le y_1 < y_2 \le \infty$,

$$\mathrm{E}(S_\lambda(y_2) - S_\lambda(y_1))^2 \le \mu(y_1, y_2)\lambda^{4d-\nu-\delta}, \tag{96}$$

where μ is a *finite* measure on $\bar{\mathbb{R}} = [-\infty, \infty]$, $\delta > 0$. The proof of (94) from (96) uses a chaining argument which can be found in several texts [75, 141], [103, proof of Thm. 10.2.1]. We simply note that (96) agrees with (94) for a fixed y, concerning the rate of λ, whereas the presence of bounded μ on the r.h.s. of (96) allows us to build a chain connecting arbitrary y to a fixed finite partition of $\bar{\mathbb{R}}$ whose number of points grows logarithmically with λ.

Consider (96), whose proof uses the bounds (59) and (60). For $y_1 < y_2$, denote $S_\lambda(y_1, y_2) := S_\lambda(y_2) - S_\lambda(y_1)$, $G_{y_1,y_2}(x) := \mathbb{I}(y_1 < x \le y_2)$, $F(y_1, y_2) := F(y_2) - F(y_1)$, $f(y_1, y_2) := f(y_2) - f(y_1)$, $h_{y_1,y_2}(x) := G_{y_1,y_2}(x) - F(y_1, y_2) + f(y_1, y_2)x$ and

$$\begin{aligned} \mathcal{Z}_{y_1,y_2}(\boldsymbol{t}) := h_{y_1,y_2}(X(\boldsymbol{t})) &= \mathbb{I}(y_1 < X(\boldsymbol{t}) \le y_2) \\ &\quad - F(y_1, y_2) + f(y_1, y_2)X(\boldsymbol{t}), \end{aligned} \tag{97}$$

where $\boldsymbol{t}$ runs over points of $\mathbb{Z}^\nu$ or $\mathbb{R}^\nu$ for X in Theorems 4 and 5 of Chapter 3, respectively. Let X be as in the former theorem for concreteness. Then, $S_\lambda(y_1, y_2) = \int_{\mathbb{R}^\nu} \mathbb{I}(\boldsymbol{t} \in \lambda A)\mathcal{Z}_{y_1,y_2}([\boldsymbol{t}])\mathrm{d}\boldsymbol{t}/\mathrm{Leb}_\nu(\lambda A)$, with $\mathcal{Z}_{y_1,y_2}(\boldsymbol{t})$ being a particular case of $\mathcal{Z}(\boldsymbol{t})$ in (56) and corresponding to $G = G_{y_1,y_2}$. Accordingly, the covariance $\rho_{\boldsymbol{t}}(y_1, y_2) := \mathrm{Cov}(\mathcal{Z}_{y_1,y_2}(\mathbf{0}), \mathcal{Z}_{y_1,y_2}(\boldsymbol{t}))$ can be written as in (57):

$$\rho_{\boldsymbol{t}}(y_1, y_2) = \int_{\mathbb{R}^2} h_{y_1,y_2}(x_1) h_{y_1,y_2}(x_2) g_{\boldsymbol{t}}(x_1, x_2) \mathrm{d}x_1 \mathrm{d}x_2, \qquad (98)$$

with $g_{\boldsymbol{t}}(x_1, x_2)$ as in (58). Thus, (96) for $\mathrm{E}S^2_\lambda(y_1, y_2) = \int_{\mathbb{R}^{2\nu}} \mathbb{I}(\boldsymbol{t}_1, \boldsymbol{t}_2 \in \lambda A)\rho_{[\boldsymbol{t}_1]-[\boldsymbol{t}_2]}(y_1, y_2)\mathrm{d}\boldsymbol{t}_1\mathrm{d}\boldsymbol{t}_2 \times\lambda^{-2\nu}\mathrm{Leb}_\nu(A)^{-2}$ follows from Proposition 1 (4) of Chapter 3 and the bound

$$|\rho_{\boldsymbol{t}}(y_1, y_2)| \le \mu(y_1, y_2)(1 \vee |\boldsymbol{t}|)^{-\beta}, \quad \text{with}$$
$$\beta := (\nu - 2d) \wedge (2\nu - 8d) - \epsilon \qquad (99)$$

with arbitrary small $\epsilon > 0$ and $\delta > 0, \epsilon > 0$ related by $4d - \nu - \delta = -\beta = \epsilon - (\nu - 4d) \wedge (2\nu - 8d)$. For $|\boldsymbol{t}|$ large enough, relation (99) can be verified with the help of (60). Let $\rho_{\boldsymbol{t}} := |\boldsymbol{t}|^{-(\nu-2d)\wedge(2\nu-8d)}$. Following the proof of (61), case $\ell = 2$, split the integral in (98) over the sets $[-K, K]^2$ and $\mathbb{R}^2\backslash[-K, K]^2$, $K = |\boldsymbol{t}|^p$ and write $|\rho_{\boldsymbol{t}}(y_1, y_2)| \le C(\rho'_{\boldsymbol{t}}(y_1, y_2) + \rho''_{\boldsymbol{t}}(y_1, y_2))$ accordingly. Then, using (60), case $\ell = 2$, we see that

$$\begin{aligned}
&\rho'_{\boldsymbol{t}}(y_1, y_2) \\
&\quad \le C\rho_{\boldsymbol{t}}\left(\int_{[-K,K]} \frac{\mathbb{I}(y_1 < x \le y_2) + F(y_1, y_2) + |f(y_1, y_2)||x|}{1 + x^2}\mathrm{d}x\right)^2 \\
&\quad \le C\rho_{\boldsymbol{t}}(1 + \log K)(\mu(y_1, y_2))^2 \le \rho_{\boldsymbol{t}}(1 + \log|\boldsymbol{t}|)\mu(y_1, y_2),
\end{aligned}$$

where

$$\mu(y_1, y_2) := C\int_{y_1}^{y_2} \{(1 + x^2)^{-1} + f(x) + |f'(x)|\}\mathrm{d}x \qquad (100)$$

is a finite measure on $\bar{\mathbb{R}}$ since $|f'(x)| \le C(1 + |x|)^{-4}, x \in \mathbb{R}$; see (59). On the other hand, using (59), we get $\rho''_{\boldsymbol{t}}(y_1, y_2) \le CK^{-1}\mu(y_1, y_2)$ with the same μ in (100), and hence (99) when $p = \beta$. This proves (99) for large $|\boldsymbol{t}|$. It remains to show the last relation for bounded $|\boldsymbol{t}|$.

From (98) and the Cauchy inequality, we get

$$
\begin{aligned}
|\rho_{\boldsymbol{t}}(y_1, y_2)| &\le \int_{\mathbb{R}^2} h^2_{y_1,y_2}(x_1)|g_{\boldsymbol{t}}(x_1, x_2)|\mathrm{d}x_1\mathrm{d}x_2 \\
&\le 3\mathrm{E}\big(\mathbb{I}(y_1 < X(\mathbf{0}) \le y_2) + F(y_1, y_2)^2 \\
&\quad + |f(y_1, y_2)|^2|X(\mathbf{0})|^2\big) \le C\mu(y_1, y_2),
\end{aligned}
$$

with $\mu(y_1, y_2)$ as in (100). This proves (96), ending the proof of Proposition 10. □

M estimation in spatial linear regression. As another application of our limit results, we discuss the asymptotic distribution of regression estimates in the linear regression model:

$$
Y_\lambda(\boldsymbol{t}) = \langle \boldsymbol{\beta}, \boldsymbol{v}_\lambda(\boldsymbol{t})\rangle + X(\boldsymbol{t}), \quad \boldsymbol{t} \in \lambda A, \tag{101}
$$

where $\boldsymbol{\beta} = (\beta_1, \ldots, \beta_p) \in \mathbb{R}^p$ is an unknown vector parameter, $A \subset \mathbb{R}^\nu$ is a bounded Borel set as in (87)–(88), $\boldsymbol{v}_\lambda(\boldsymbol{t}) = (v_{1,\lambda}(\boldsymbol{t}), \ldots, v_{p,\lambda}(\boldsymbol{t})) \in \mathbb{R}^p$ is a known vector-valued regression function, possibly depending on λ, and defined, together with a zero mean 'error' RF $X(\boldsymbol{t})$, for all $\boldsymbol{t} \in \mathbb{Z}^\nu$ in the discrete argument case, or all $\boldsymbol{t} \in \mathbb{R}^\nu$ in the continuous argument case. The model in (101) corresponding to $p = 1, \boldsymbol{v}_\lambda(\boldsymbol{t}) \equiv 1$ includes the estimation of the unknown mean $\beta \in \mathbb{R}$ of RF $Y(\boldsymbol{t}) = \beta + X(\boldsymbol{t})$ observed in a large region λA, usually referred to as the *location model.* The error RF X in (101) has zero mean and a linear structure as in Proposition 10. We assume that the regressors in (101) have the form

$$
\boldsymbol{v}_\lambda([\boldsymbol{t}]) = \lambda^\nu \int_{]\frac{[\boldsymbol{t}]-\mathbf{1}}{\lambda}, \frac{[\boldsymbol{t}]}{\lambda}]} \boldsymbol{v}(\boldsymbol{s})\mathrm{d}\boldsymbol{s}, \quad \boldsymbol{t} \in \mathbb{R}^\nu \tag{102}
$$

so that in both cases, we deal with regression function $\boldsymbol{v}(\boldsymbol{t}) = (v_1(\boldsymbol{t}), \ldots, v_p(\boldsymbol{t}))$ of continuous argument $\boldsymbol{t} \in \mathbb{R}^\nu$ independent of λ, whose components are bounded and integrable. As usual, the positive definite $p \times p$ 'design matrix'

$$
\mathbf{V}_\lambda := \Big(\int_{\lambda A} v_{i,\lambda}([\boldsymbol{t}])v_{j,\lambda}([\boldsymbol{t}])\mathrm{d}\boldsymbol{t}\Big)_{i,j=1,\ldots,p},
$$

with $\langle \mathbf{V}_\lambda \boldsymbol{z}, \boldsymbol{z}\rangle = \int_{\lambda A}\langle \boldsymbol{v}_\lambda([\boldsymbol{t}]), \boldsymbol{z}\rangle^2\mathrm{d}\boldsymbol{t} \ge 0$, is assumed to be nondegenerate: $\langle \mathbf{V}_\lambda \boldsymbol{z}, \boldsymbol{z}\rangle > 0$ for any $\boldsymbol{z} \in \mathbb{R}^p \setminus \{\mathbf{0}\}$. In view of

(102), $\lambda^\nu \mathbf{V}_\lambda = \big(\int_A w_{i,\lambda}(\boldsymbol{t}) w_{j,\lambda}(\boldsymbol{t})\mathrm{d}\boldsymbol{t}\big)_{i,j=1,\dots,p}$, where $w_{i,\lambda}(\boldsymbol{t}) := \lambda^\nu \int_{]\frac{[\lambda \boldsymbol{t}]-\mathbf{1}}{\lambda}, \frac{[\lambda \boldsymbol{t}]}{\lambda}]} \boldsymbol{v}(\boldsymbol{s})\mathrm{d}\boldsymbol{s} \to v_i(\boldsymbol{t})\ (\lambda \to \infty)$ for a.e. $\boldsymbol{t} \in \mathbb{R}^\nu$, according to the Lebesgue–Vitali theorem [235, Chapters 4, 10, Thm. 1], implying

$$\lim_{\lambda\to\infty} \lambda^{-\nu}\mathbf{V}_\lambda = \mathbf{V}_\infty := \Big(\int_A v_i(\boldsymbol{t})v_j(\boldsymbol{t})\mathrm{d}\boldsymbol{t}\Big)_{i,j=1,\dots,p},$$

with $\langle \mathbf{V}_\infty \boldsymbol{z}, \boldsymbol{z}\rangle = \int_A \langle \boldsymbol{v}(\boldsymbol{t}), \boldsymbol{z}\rangle^2 \mathrm{d}\boldsymbol{t}$, so that the nondegeneracy condition for (102) is automatically satisfied unless $\boldsymbol{v}(\boldsymbol{t}) = 0, \boldsymbol{t} \in A$.

The least squares (LS) estimator $\widetilde{\boldsymbol{\beta}}_{\lambda,\mathrm{LS}}$ in the model (101) is defined as

$$\widetilde{\boldsymbol{\beta}}_{\lambda,\mathrm{LS}} := \operatorname{argmin}\Big\{\int_{\lambda A} \big(Y_\lambda([\boldsymbol{t}]) - \langle \boldsymbol{z}, \boldsymbol{v}_\lambda([\boldsymbol{t}])\rangle\big)^2 \mathrm{d}\boldsymbol{t} : \boldsymbol{z} \in \mathbb{R}^p\Big\}$$

in the discrete argument case, with obvious modification in the continuous argument case. It is well known that the corresponding minimiser solves a system of linear equation whose solution can be written as

$$\widetilde{\boldsymbol{\beta}}_{\lambda,\mathrm{LS}} = \mathbf{V}_\lambda^{-1} \int_{\lambda A} \boldsymbol{v}_\lambda([\boldsymbol{t}]) Y_\lambda([\boldsymbol{t}])\mathrm{d}\boldsymbol{t}.$$

Alternatively, the LS estimator can be defined by the relation

$$\widetilde{\boldsymbol{\beta}}_{\lambda,\mathrm{LS}} := \operatorname{argmin}\Big\{\Big|\mathbf{V}_\lambda^{-1/2} \int_{\lambda A} \boldsymbol{v}_\lambda([\boldsymbol{t}])\big(Y_\lambda([\boldsymbol{t}]) - \langle \boldsymbol{z}, \boldsymbol{v}_\lambda([\boldsymbol{t}])\rangle\big)\mathrm{d}\boldsymbol{t}\Big|^2 : \boldsymbol{z} \in \mathbb{R}^p\Big\}.$$

The idea behind M estimation is to replace the residuals $Y_\lambda([\boldsymbol{t}]) - \langle \boldsymbol{z}, \boldsymbol{v}_\lambda([\boldsymbol{t}])\rangle$ in the last integral by $\tau\big(Y_\lambda([\boldsymbol{t}]) - \langle \boldsymbol{z}, \boldsymbol{v}_\lambda([\boldsymbol{t}])\rangle\big)$, where $\tau(y), y \in \mathbb{R}$ is a *score function*, typically bounded or with the property that $\tau(y) = o(|y|), |y| \to \infty$. Accordingly, let

$$\mathcal{M}_\lambda(\boldsymbol{z}; \tau) := \mathbf{V}_\lambda^{-1/2} \int_{\lambda A} \boldsymbol{v}_\lambda([\boldsymbol{t}])\tau\big(Y_\lambda([\boldsymbol{t}]) - \langle \boldsymbol{z}, \boldsymbol{v}_\lambda([\boldsymbol{t}])\rangle\big)\mathrm{d}\boldsymbol{t}.$$

Then, a class of M estimators, one for each τ, of $\boldsymbol{\beta}$ is defined as

$$\widetilde{\boldsymbol{\beta}}_{\lambda,\mathrm{M}} := \operatorname{argmin}\big\{|\mathcal{M}_\lambda(\boldsymbol{z}; \tau)|^2 : \boldsymbol{z} \in \mathbb{R}^p\big\}.$$

Important examples of M estimators correspond to the Huber score $\tau(y) = y\mathbb{I}(|y| \le c) + c\mathbb{I}(|y| > c)\mathrm{sgn}(y)$, where $c > 0$ is a known positive number, and the least absolute deviation (LAD) score $\tau(y) = \mathrm{sgn}(y)$. The latter estimator is referred to as the LAD estimator, an analogue of the median in the location case. Introduce the following class of score functions:

$$\mathfrak{T} = \big\{\tau : \tau \text{ is nondecreasing right-continuous bounded function}$$
$$\text{from } \mathbb{R} \text{ to } \mathbb{R} \text{ and } \textstyle\int_{\mathbb{R}} \tau(y) f(y)\mathrm{d}y = 0, \int_{\mathbb{R}} \tau(y) f'(y)\mathrm{d}y \neq 0\big\},$$

where $f(y), y \in \mathbb{R}$ is the marginal probability density of error RF X. Note that $\int_{\mathbb{R}} \tau(y) f(y)\mathrm{d}y = a_\tau(0), -\int_{\mathbb{R}} \tau(y) f'(y)\mathrm{d}y = a_\tau(1)$ are the first two Appell coefficients of τ.

Theorem 11. *Consider the linear regression model in* (101) *with regression functions as in* (102), $v_i(\cdot) \in L^1(\mathbb{R}^\nu) \cap L^\infty(\mathbb{R}^\nu), i = 1, \ldots, p$, *where* $A \subset \mathbb{R}^\nu$ *is an arbitrary bounded Borel set. Let the error RF* X *in* (101) *be an MA as in Theorem* 5 *or* 6, $0 < d < \nu/4$, $\ell = 2$. *Then, the LS estimator in* (4) *is asymptotically normal:*

$$\lambda^{(\nu-4d)/2}(\widetilde{\boldsymbol{\beta}}_{\lambda,\mathrm{LS}} - \boldsymbol{\beta}) \xrightarrow{\mathrm{d}} \boldsymbol{W}_{d,\boldsymbol{v}}(A), \tag{103}$$

with the limit written as the stochastic integral

$$\boldsymbol{W}_{d,\boldsymbol{v}}(A) \overset{\mathrm{d}}{=} \int_{\mathbb{R}^\nu} \left\{ \int_A \mathbf{V}_\infty^{-1} \boldsymbol{v}(\boldsymbol{t}) a_\infty(\boldsymbol{t} - \boldsymbol{u}) \mathrm{d}\boldsymbol{t} \right\} W(\mathrm{d}\boldsymbol{u})$$

w.r.t. to the Gaussian white noise $W(\mathrm{d}\boldsymbol{u})$ *as in* (91). *Furthermore, every M estimator corresponding to a score function* $\tau \in \mathfrak{T}$ *is asymptotically first-order equivalent to the LS estimator:*

$$\lambda^{(\nu-4d)/2}(\widetilde{\boldsymbol{\beta}}_{\lambda,\mathrm{M}} - \boldsymbol{\beta}) = \lambda^{(\nu-4d)/2}(\widetilde{\boldsymbol{\beta}}_{\lambda,\mathrm{LS}} - \boldsymbol{\beta}) + o_p(1) \xrightarrow{\mathrm{d}} \boldsymbol{W}_{d,\boldsymbol{v}}(A).$$

The proof of the main result (104) of the above theorem uses the uniform reduction principle in (94) for the empirical process in (87) and its extension in for the *weighted empirical process*

$$\mathcal{F}_\lambda(y; \boldsymbol{z}) := \int_{\lambda A} \boldsymbol{v}_\lambda([\boldsymbol{t}]) \mathbb{I}(X([\boldsymbol{t}]) \le y + \langle \boldsymbol{v}_\lambda([\boldsymbol{t}]), \boldsymbol{z} - \boldsymbol{\beta} \rangle) \mathrm{d}\boldsymbol{t},$$
$$y \in \mathbb{R}, \ \ \boldsymbol{z} \in \mathbb{R}^p,$$

together with a general minimum dispersion argument developed in Ref. [140]. The details of this proof are rather involved and not included here, as they are mostly unrelated to RFs and similar to those in a time-series setting.

5 Nonlinear Functions of RG Model: Application to Volume Fraction Estimation

Recall from Section 2 of Chapter 4 (21) the randomly homothetic RG model

$$X(\boldsymbol{t}) := \sum_{j=1}^{\infty} \mathbb{I}(\boldsymbol{t} - \boldsymbol{u}_j \in R_j^{1/d}\Xi_j^0), \quad \boldsymbol{t} \in \mathbb{R}^\nu, \tag{104}$$

with the generic grain $R^{1/\nu}\Xi^0$ being an isotropic dilation by a random factor $R^{1/\nu} > 0$ of a random bounded set Ξ^0, independent of R. We assume that the distribution of (R, Ξ^0) satisfies the conditions of Proposition 5 of Chapter 4, according to which it has covariance LRD and the asymptotics

$$r_X(\boldsymbol{t}) := \mathrm{Cov}(X(\boldsymbol{0}), X(\boldsymbol{t})) = |\boldsymbol{t}|^{-\nu(\alpha-1)}\left(L\left(\frac{\boldsymbol{t}}{|\boldsymbol{t}|}\right) + o(1)\right), \quad |\boldsymbol{t}| \to \infty, \tag{105}$$

with $L(\cdot) \in S_{\nu-1}$ in (25) of Chapter 4.

This section discusses isotropic scaling limits of the nonlinear RF $Y(\boldsymbol{t}) = G(X(\boldsymbol{t}))$ subordinated to the RG model in (104), as well as scaling limits of nonlinear RFs

$$Y_M^{\mathrm{I}}(\boldsymbol{t}) := \sum_{j=1}^{M} G(X_j(\boldsymbol{t})) \quad \text{and}$$

$$Y_M^{\mathrm{II}}(\boldsymbol{t}) := G\big(M^{-1/2}(X_M(\boldsymbol{t}) - \mathrm{E}X_M(\boldsymbol{t}))\big), \tag{106}$$

involving independent copies X_j of X in (104), $X_M(\boldsymbol{t}) := \sum_{j=1}^{M} X_j(\boldsymbol{t})$, and representing the aggregated sum of a nonlinear function G of X and a nonlinear function G of the aggregated sum of X, respectively, when M increases with λ at a given rate of $M = [\lambda^\gamma]$, for some $\gamma > 0$. According to Theorem 12, the scaling limits of $Y_M^{\mathrm{I}}(\boldsymbol{t})$ in (106) and $X_M(\boldsymbol{t})$ are essentially the same (including the trichotomy of the limit distribution) *provided the Charlier rank of* G *is 1*, or

$$c_{G,\mu}(1) := \mathrm{E}[G(X(\boldsymbol{0}))(X(\boldsymbol{0}) - \mathrm{E}X(\boldsymbol{0}))] \neq 0. \tag{107}$$

A similar conclusion (part (ii) of Theorem 12) applies to the aggregated RF $Y_M^{\rm II}(\boldsymbol{t})$ *provided the Hermite rank of G is 1*, or

$$h_{G,\mu}(1) := \mu^{-1}\mathrm{E}[G(Z_\mu)Z_\mu] \neq 0, \quad Z_\mu \sim N(0,\mu), \tag{108}$$

commended by application $(X_M(\mathbf{0}) - \mathrm{E}X_M(\mathbf{0}))/M^{1/2} \xrightarrow{\rm d} Z_\mu\, (M \to \infty)$ of the central limit theorem. It follows that the difference between the scaling limits of $Y_M^{\rm I}$ and $Y_M^{\rm II}$ in (106) reduces to multiplicative factors in (107) and (108). The same condition (107) also guarantees the coincidence of the scaling limits of $Y(\boldsymbol{t}) = G(X(\boldsymbol{t}))$ and $X(\boldsymbol{t})$ up to the Charlier coefficient $c_{G,\mu}(1)$ (Corollary 14).

Theorem 12. *Let X be a randomly homothetic RG model in (104), satisfying the conditions of Proposition 5 of Chapter 4, and $Y_{\lambda,M}^{\rm I}(\phi) = \int_{\mathbb{R}^\nu}\phi(\boldsymbol{t}/\lambda)Y_M^{\rm I}(\boldsymbol{t})\mathrm{d}\boldsymbol{t}, Y_{\lambda,M}^{\rm II}(\phi) = \int_{\mathbb{R}^\nu}\phi(\boldsymbol{t}/\lambda)Y_M^{\rm II}(\boldsymbol{t})\mathrm{d}\boldsymbol{t}$ be as in (106), with $M = \lambda^\gamma$ for some $\gamma > 0$.*

(i) *Let $G : \mathbb{N} \to \mathbb{R}$ be any function with $\mathrm{E}G(X(\mathbf{0}))^2 < \infty$, with $c_{G,\mu}(1)$ in (107) being the first coefficient in the Charlier expansion of $G - \mathrm{E}G(X)$. Then, for any $\phi \in \Phi$ as $\lambda \to \infty$,*

$$\lambda^{-H(\gamma)}(Y_{\lambda,M}^{\rm I}(\phi) - \mathrm{E}Y_{\lambda,M}^{\rm I}(\phi))$$
$$\xrightarrow{\rm d} c_{G,\mu}(1)\begin{cases} B_\alpha(\phi), & \gamma > \nu(\alpha-1), \\ L_\alpha(\phi), & \gamma < \nu(\alpha-1), \\ J_\alpha(\phi), & \gamma = \nu(\alpha-1), \end{cases} \tag{109}$$

where $H(\gamma), B_\alpha(\phi), L_\alpha(\phi)$, and $J_\alpha(\phi)$ are the same as in Theorem 10, (53) of Chapter 4.

(ii) *Let $G : \mathbb{R} \to \mathbb{R}$ be a $\mathrm{d}x$-a.e. continuous function such that*

$$\lim_{M\to\infty} \mathrm{E}(Y_M^{\rm II}(\mathbf{0}))^2 = \mathrm{E}G(Z_\mu)^2 < \infty, \tag{110}$$

with $h_{G,\mu}(1)$ in (108) being the first coefficient in the Hermite expansion of $G - \mathrm{E}G(Z_\mu)$. Then, for any $\phi \in \Phi$ as $\lambda \to \infty$,

$$\lambda^{(\gamma/2)-H(\gamma)}(Y_{\lambda,M}^{\rm II}(\phi) - \mathrm{E}Y_{\lambda,M}^{\rm II}(\phi))$$
$$\xrightarrow{\rm d} h_{G,\mu}(1)\begin{cases} B_\alpha(\phi), & \gamma > \nu(\alpha-1), \\ L_\alpha(\phi), & \gamma < \nu(\alpha-1), \\ J_\alpha(\phi), & \gamma = \nu(\alpha-1), \end{cases} \tag{111}$$

where $H(\gamma), B_\alpha(\phi), L_\alpha(\phi)$, and $J_\alpha(\phi)$ are the same as in (109).

Proof. (i) Consider the decomposition $Y^{\mathrm{I}}_{\lambda,M}(\phi) = c_{G,\mu}(1)X_{\lambda,M}(\phi) + \mathcal{Z}^{\mathrm{I}}_{\lambda,M}(\phi)$, where $\mathcal{Z}^{\mathrm{I}}_{\lambda,M}(\phi) = \sum_{j=1}^{M} \mathcal{Z}_{j,\lambda}(\phi)$ is a sum of independent copies of integrals $\mathcal{Z}_\lambda(\phi) := \int_{\mathbb{R}^\nu} \mathcal{Z}(\boldsymbol{t})\phi(\boldsymbol{t}/\lambda)\mathrm{d}\boldsymbol{t}$ involving the stationary RF

$$\mathcal{Z}(\boldsymbol{t}) := G(X(\boldsymbol{t})) - \mathrm{E}G(X(\boldsymbol{t})) - c_{G,\mu}(1)X(\boldsymbol{t}), \quad \boldsymbol{t} \in \mathbb{R}^\nu.$$

The convergence in (109) is a consequence of the relation

$$\mathrm{Var}(\mathcal{Z}^{\mathrm{I}}_{\lambda,M}(\phi)) = o(\lambda^{2H(\gamma)}) \tag{112}$$

and (53) of Theorem 10 of Chapter 4. Since $\mathrm{Var}(\mathcal{Z}^{\mathrm{I}}_{\lambda,M}(\phi)) = M\mathrm{Var}(\mathcal{Z}_\lambda(\phi))$ and $|\mathrm{Cov}(\mathcal{Z}(\mathbf{0}), \mathcal{Z}(\boldsymbol{t}))| \le r^2_X(\boldsymbol{t})\mathrm{Var}(G(X(\mathbf{0})))$ (see Corollary 17 of Chapter 2), we get from Proposition 1 (4) of Chapter 3 that

$$\mathrm{Var}(\mathcal{Z}^{\mathrm{I}}_{\lambda,M}(\phi)) = \begin{cases} O(\lambda^{\nu+\gamma}), & \alpha > 3/2, \\ O(\lambda^{2\nu(2-\alpha)+\gamma}), & \alpha < 3/2, \\ O(\lambda^{\nu+\gamma}\log\lambda), & \alpha = 3/2, \end{cases}$$

and (112) holds due to

$$\max\{\nu, 2\nu(2-\alpha)\} < \min\left\{(3-\alpha)\nu, \frac{2(\gamma+\nu)}{\alpha} - \gamma\right\},$$
$$\forall \gamma > 0,\ \alpha \in (1,2). \tag{113}$$

(ii) We have $Y^{\mathrm{II}}_M(\boldsymbol{t}) = G_M(X_M(\boldsymbol{t}))$, where $G_M(x) := G((x - \mu M)/M^{1/2}), x \in \mathbb{N}$ and $X_M(\boldsymbol{t}) =: N_M$ has a Poisson distribution with mean μM. Following (64) of Chapter 2, consider the expansion

$$G_M(x) = \sum_{k=0}^{\infty} \frac{c_{G,\mu M}(k)}{k!} P_k(x;\mu M), \quad x \in \mathbb{N} \tag{114}$$

in the Charlier polynomials $P_k(x;\mu M), k \in \mathbb{N}$, where

$$c_{G,\mu M}(k) := (\mu M)^{-k}\mathrm{E}[G_M(N_M)P_k(N_M;\mu M)], \quad k \in \mathbb{N}.$$

Particularly,

$$M^{1/2}c_{G,\mu M}(1) = \frac{M^{1/2}}{\mu M}\mathrm{E}\left[G\left(\frac{N_M - M\mu}{M^{1/2}}\right)(N_M - M\mu)\right] \to h_{G,\mu}(1) \tag{115}$$

as $M \to \infty$, with the limit as in (108). The convergence in (115) under condition (110) can be verified with the help of (Pratt's) Lemma 30 of Chapter 2 as follows. Let $g(x) := G(x)x$, $x \in \mathbb{R}$, $\xi_M := \frac{N_M - M\mu}{M^{1/2}}$. Then, $M^{1/2} c_{G,\mu M}(1) = \mu^{-1}\mathrm{E}[g(\xi_M)\xi_M]$ and $\xi_M \xrightarrow{\mathrm{d}} Z_\mu \ (M \to \infty)$. Since G is a.e. continuous and Z_μ has a continuous distribution, this implies $(G(\xi_M), g(\xi_M)) \xrightarrow{\mathrm{d}} (G(Z_\mu), g(Z_\mu))$. By the Skorohod representation theorem, $(G(\xi_M), g(\xi_M)) \to_p (G(Z_\mu), g(Z_\mu))$ in probability. Moreover, $|g(\xi_M)| \le (1/2)((\xi_M)^2 + \xi_M^2) =: \bar{g}_M$, where $\bar{g}_M \to_p (1/2)(G(Z_\mu)^2 + Z_\mu^2) =: \bar{g}$ and $\mathrm{E}\bar{g}_M \to \mathrm{E}\bar{g}$ according to (110). This and the above-mentioned Lemma 30 of Chapter 2 prove (115).

In view of (114), we have the representation

$$
\begin{aligned}
Y^{\mathrm{II}}_{\lambda,M}(\phi) - \mathrm{E}Y^{\mathrm{II}}_{\lambda,M}(\phi) &= c_{G,\mu M}(1)(X_{\lambda,M}(\phi) \\
&\quad - \mathrm{E}X_{\lambda,M}(\phi)) + \mathcal{Z}^{\mathrm{II}}_{\lambda,M}(\phi), \quad \text{where} \\
\mathcal{Z}^{\mathrm{II}}_{\lambda,M}(\phi) &:= \int_{\mathbb{R}^\nu} \mathcal{Z}^{\mathrm{II}}_M(\boldsymbol{t})\phi(\boldsymbol{t}/\lambda)\mathrm{d}\boldsymbol{t}, \\
\mathcal{Z}^{\mathrm{II}}_M(\boldsymbol{t}) &:= \sum_{k=2}^{\infty} \frac{c_{G,\mu M}(k)}{k!} P_k\big(X_M(\boldsymbol{t}); \mu M\big). \qquad (116)
\end{aligned}
$$

The convergence in (111) follows from Theorem 10 of Chapter 4 and (115), once we show that $\mathcal{Z}^{\mathrm{II}}_{\lambda,M}(\phi)$ in (116) is negligible, or

$$
\mathrm{Var}(\mathcal{Z}^{\mathrm{II}}_{\lambda,M}(\phi)) = o(\lambda^{2H(\gamma)-\gamma}), \quad \lambda \to \infty \qquad (117)
$$

for any fixed $\gamma > 0$. Applying Corollary 17 of Chapter 2 with $k^* = 2, \mu_3 = \mathrm{Cov}(X_M(\boldsymbol{t}), X_M(\mathbf{0})) = Mr_X(\boldsymbol{t})$ and the bound $c^2_{G,\mu M}(k) \le (\mu M)^{-k} k! \mathrm{E}G_M(X_M(\mathbf{0}))^2$ (see (68)), we get

$$
\begin{aligned}
|\mathrm{E}\mathcal{Z}^{\mathrm{II}}_M(\boldsymbol{t})\mathcal{Z}^{\mathrm{II}}_M(\mathbf{0})| &\le \sum_{k=2}^{\infty} \frac{c^2_{G,M}(k)}{(k!)^2} |\mathrm{E}[P_k(X_M(\boldsymbol{t}); \mu M)P_k(X_M(\mathbf{0}); \mu M)]| \\
&\le \mathrm{E}G_M(X_M(\mathbf{0}))^2 \sum_{k=2}^{\infty} \frac{(r_X(\boldsymbol{t})M)^k}{(\mu M)^k} \\
&= \mathrm{E}G_M(X_M(\mathbf{0}))^2 \sum_{k=2}^{\infty} \left(\frac{r_X(\boldsymbol{t})}{r_X(\mathbf{0})}\right)^k \le C(1 \wedge r^2_X(\boldsymbol{t})).
\end{aligned} \qquad (118)
$$

Applying Proposition 1 (4) of Chapter 3, relation (117) follows by (113) as in part (i). □

Remark 13. Condition (110) on G involving only the convergence of the second moments is rather weak. Equation (110) can be replaced by a boundedness condition:

$$|G(x)| \le C_1 \mathrm{e}^{C_2|x|} \quad \text{or} \quad |G(x)| \le C_1(\mathrm{e}^{C_2 x} + \mathrm{e}^{-C_2 x}),$$
$$x \in \mathbb{R} \quad (\exists C_1, C_2 > 0). \tag{119}$$

Both conditions in (119) are equivalent. Verification of (110) for $G(x) = C_1 \mathrm{e}^{\pm C_2 x}$ is rather easy, as follows. Let $\widetilde{\Psi}(z) := \mathrm{e}^z - 1 - z$, $z \in \mathbb{R}$. We have

$$\begin{aligned}
\mathrm{E}\mathrm{e}^{2C_2(N_M - \mu M)/M^{1/2}} &= \mathrm{e}^{-M\mu} \sum_{k=0}^{\infty} \mathrm{e}^{2C_2(k-\mu M)/M^{1/2}} \frac{(M\mu)^k}{k!} \\
&= \exp\left\{ M\mu \widetilde{\Psi}\left(\frac{2C_2}{M^{1/2}}\right)\right\} \\
&\to \mathrm{e}^{\mu(2C_2)^2/\mu} = \mathrm{E}\mathrm{e}^{2C_2 Z_\mu} \quad (M \to \infty),
\end{aligned}$$

independently of the sign of C_2. The fact that (119) implies (110) for G in (119) follows from Lemma 30 of Chapter 2.

Part (i) of Theorem 12 also holds when $\gamma = 0$ or $M = 1$ is fixed. This case is separated in the following corollary.

Corollary 14. *Let $Y(\boldsymbol{t}) := G(X(\boldsymbol{t})), \boldsymbol{t} \in \mathbb{R}^\nu$, where X, G are as in Theorem 12 (i), $Y_\lambda(\phi) := \int_{\mathbb{R}^\nu} Y(\boldsymbol{t})\phi(\boldsymbol{t}/\lambda)\mathrm{d}\boldsymbol{t}, \phi \in \Phi$. Then,*

$$\lambda^{-d/\alpha}(Y_\lambda(\phi) - \mathrm{E}Y_\lambda(\phi)) \xrightarrow{\mathrm{d}} c_{G,\mu}(1) L_\alpha(\phi), \tag{120}$$

where $c_{G,\mu}(1), L_\alpha(\phi)$ are the same as in (109).

Proof. As in the proof of Theorem 12 (i), write $Y_\lambda(\phi) - \mathrm{E}Y_\lambda(\phi) = c_{G,\mu}(1)(X_\lambda(\phi) - \mathrm{E}X_\lambda(\phi)) + \mathcal{Z}_\lambda(\phi)$, where $\mathcal{Z}_\lambda(\phi) := \int_{\mathbb{R}^\nu} \mathcal{Z}(\boldsymbol{t})\phi(\boldsymbol{t}/\lambda)\mathrm{d}\boldsymbol{t}$ and $|\mathrm{E}\mathcal{Z}(\boldsymbol{t})\mathcal{Z}(\boldsymbol{0})| \le C(1 \wedge r_X^2(\boldsymbol{t}))$. Then, (120) follows as in the above-mentioned proof. □

Remark 15. Theorem 12 and Corollary 14 yield trivial limits if the respective coefficients $c_{G,\mu}(1)$, $h_{G,\mu}(1)$ vanish. The question of the limit distribution of $Y^{\mathrm{I}}_{\lambda,M}(\phi), Y^{\mathrm{II}}_{\lambda,M}(\phi)$ and $Y_\lambda(\phi)$ in such a case remains open.

Application to volume fraction estimation. The *volume fraction* of the Boolean set $\mathcal{X}$ in (11) of Chapter 4 is the mean of the 'volume' of $\mathcal{X}$ in the unit 'cube' $]\mathbf{0},\mathbf{1}]$:

$$p := \mathrm{E}\mathrm{Leb}_\nu(\mathcal{X}\cap]\mathbf{0},\mathbf{1}]) = \int_{]\mathbf{0},\mathbf{1}]} \mathrm{E}\widehat{X}(\boldsymbol{t})\mathrm{d}\boldsymbol{t} = \mathrm{E}\widehat{X}(\mathbf{0}) = 1 - \mathrm{P}(X(\mathbf{0}) = 0), \tag{121}$$

leading to

$$p = 1 - \mathrm{e}^{-\mu}, \tag{122}$$

as $X(\boldsymbol{t})$ has a Poisson distribution with mean $\mu = \mathrm{E}\mathrm{Leb}_\nu(\Xi)$ (14) of Chapter 4. By the stationarity of the RG model X, $p = \mathrm{E}\mathrm{Leb}_\nu(\mathcal{X}\cap A)/\mathrm{Leb}_\nu(A)$ for any Borel set $A, 0 < \mathrm{Leb}_\nu(A) < \infty$. The natural estimator of p from the observations of $\mathcal{X}$ on a large 'inflated' set $\lambda A \subset \mathbb{R}^\nu$ is the ratio

$$\widehat{p}_{\lambda,A} := \frac{\widehat{X}_\lambda(A)}{\mathrm{Leb}_\nu(\lambda A)}, \tag{123}$$

called the *sample volume fraction*, where $\widehat{X}_\lambda(A) := \mathrm{Leb}_\nu(\mathcal{X}\cap\lambda A)$ is the 'volume' of the intersection of the Boolean set with λA. Thus,

$$\lambda^\nu(\widehat{p}_{\lambda,A} - p) = \frac{\widehat{X}_\lambda(A) - \mathrm{E}\widehat{X}_\lambda(A)}{\mathrm{Leb}_\nu(A)}, \tag{124}$$

and finding the limit distribution of $\widehat{p}_{\lambda,A}$ reduces to that of the numerator of the last fraction. Since $\hat{X}(\boldsymbol{t}) = X(\boldsymbol{t}) \wedge 1$ corresponds to $Y(\boldsymbol{t}) = G(X(\boldsymbol{t}))$ with $G(x) = x \wedge 1$, $x \in \mathbb{N}$, we see that

$$\hat{X}_\lambda(A) = \int_{\mathbb{R}^\nu} \mathbb{I}(\boldsymbol{t}/\lambda \in A)\hat{X}(\boldsymbol{t})\mathrm{d}\boldsymbol{t} \tag{125}$$

is a particular case of $Y_\lambda(\phi)$ in (120), corresponding to the test function $\phi(\boldsymbol{x}) = \mathbb{I}(x \in A)$. As a consequence, (120) with $c_{G.\mu}(0) = 1 - \mathrm{e}^{-\mu}$ applies for X in Corollary 14, yielding an α-stable limit distribution for the estimator $\widehat{p}_{\lambda,A}$ in (124).

An interesting question related to stereological applications is the estimation of the volume fraction p of a Boolean set $\mathcal{X} \subset \mathbb{R}^\nu$ from observations

$$\mathcal{X} \cap (\lambda A) \cap H_{\nu_0} \tag{126}$$

on the intersection of $\mathcal{X} \cap (\lambda A)$ with a subspace $H_{\nu_0} \subset \mathbb{R}^\nu$ of dimension $1 \le \nu_0 \le \nu$. In the following, we take

$$H_{\nu_0} = \{\boldsymbol{t} \in \mathbb{R}^\nu : t_i = 0, \nu_0 < i \le \nu\}. \tag{127}$$

The corresponding estimator is defined as

$$\widehat{p}_{\lambda,A}(H_{\nu_0}) := \frac{\mathrm{Leb}_{\nu_0}(\mathcal{X} \cap H_{\nu_0} \cap (\lambda A))}{\mathrm{Leb}_{\nu_0}(H_{\nu_0} \cap (\lambda A))}. \tag{128}$$

Obviously, (128) and (123) coincide when $\nu_0 = \nu$: $\widehat{p}_{\lambda,A}(H_\nu) = \widehat{p}_{\lambda,A}$. The first question about (128) in the case of $\nu_0 < \nu$ is whether it estimates the same p as (123)? Indeed, $\mathrm{E}\widehat{p}_{\lambda,A}(H_{\nu_0}) = \mathrm{E}\mathrm{Leb}_{\nu_0}(\mathcal{X} \cap H_{\nu_0} \cap (\lambda A))/\mathrm{Leb}_{\nu_0}(H_{\nu_0} \cap (\lambda A)) = \mathrm{E}\hat{X}(\mathbf{0})\mathrm{Leb}_{\nu_0}(H_\nu) \cap (\lambda A))/\mathrm{Leb}_{\nu_0}(H_\nu) \cap (\lambda A)) = p$ by stationarity, referring to (121) and (122), so that (128) is an unbiased estimator of p for any $\nu_0 \in \{1, \dots, \nu\}$. The following theorem shows that the limit distribution of this estimator crucially depends on the subspace dimension ν_0, in the case of the Boolean set formed by a randomly homothetic RG model.

Theorem 16. *Let $\mathcal{X} = \{\boldsymbol{t} \in \mathbb{R}^\nu : X(\boldsymbol{t}) \ge 1\}$ be the Boolean set corresponding to a randomly homothetic RG model X satisfying the conditions of Corollary* 14, $1 < \alpha < 2$, *and $\widehat{p}_{\lambda,A}(H_{\nu_0})$ in* (128) *be the volume fraction estimator on the ν_0-dimensional subspace H_{ν_0} in* (127), *where $A \subset H_{\nu_0}$ is an arbitrary bounded Borel set,* $\mathrm{Leb}_{\nu_0}(A) > 0$.

(i) *Let $1 < \alpha < 1 + \frac{\nu_0}{\nu}$ and*

$$\alpha_0 := 1 + \frac{\nu}{\nu_0}(\alpha - 1). \tag{129}$$

Then,

$$\lambda^{\nu_0 - (\nu_0/\alpha_0)}(\widehat{p}_{\lambda,A}(H_{\nu_0}) - p) \xrightarrow{\mathrm{d}} \mathrm{e}^{-\mu} L_{\alpha_0}(A)/\mathrm{Leb}_{\nu_0}(A), \quad \lambda \to \infty, \tag{130}$$

where $L_{\alpha_0}(A) := \int_A L_{\alpha_0}(\mathrm{d}\tilde{\boldsymbol{t}})$ has a α_0-stable distribution.

(ii) *Let* $2 > \alpha > 1 + \frac{\nu_0}{\nu}$. *Then,*

$$\lambda^{\nu_0/2}(\widehat{p}_{\lambda,A}(H_{\nu_0}) - p) \xrightarrow{\mathrm{d}} \sigma W(A)/\mathrm{Leb}_{\nu_0}(A),$$
$$\lambda \to \infty \tag{131}$$

has a Gaussian distribution.

A sketch of the proof of the above theorem using the ideas from Theorem 6 of Chapter 4 and Corollary 14 is presented as follows. Probably, the most interesting part of Theorem 16 is the dependence of α_0 on ν_0 and the critical value

$$\alpha_{\mathrm{cr}} := 1 + \frac{\nu_0}{\nu}$$

of the tail parameter $\alpha \in (1,2)$, separating the α_0-stable and Gaussian limits (130) and (131). As explained later, the limit behaviour of $\widehat{p}_{\lambda,A}(H_{\nu_0})$ is closely related to the LRD property of the sectional RF

$$X_0(\boldsymbol{t}') := X(t_1, \ldots, t_{\nu_0}, 0, \ldots, 0), \quad \boldsymbol{t}' \in \mathbb{R}^{\nu_0}, \tag{132}$$

where we identify H_{ν_0} in (127) with $\mathbb{R}^{\nu_0}$ and use the notation $\boldsymbol{t} = (\boldsymbol{t}', \boldsymbol{t}'')$, $\boldsymbol{t}' := (t_1, \ldots, t_{\nu_0}) \in \mathbb{R}^{\nu_0}, \boldsymbol{t}'' = (t_{\nu_0+1}, \ldots, t_\nu) \in \mathbb{R}^{\nu-\nu_0}$. From the covariance asymptotics in (24), we see that $r_{X_0}(\boldsymbol{t}') = \mathrm{Cov}(X_0(\mathbf{0}), X_0(\boldsymbol{t}'))$ decays as

$$r_{X_0}(\boldsymbol{t}') = |\boldsymbol{t}'|^{-\nu(\alpha-1)}\left(L\left(\frac{\boldsymbol{t}'}{|\boldsymbol{t}'|}\right) + o(1)\right), \quad |\boldsymbol{t}'| \to \infty, \tag{133}$$

implying $\int_{\mathbb{R}^{\nu_0}} |r_{X_0}(\boldsymbol{t}')| \mathrm{d}\boldsymbol{t}' < \infty$ for $\nu(\alpha - 1) > \nu_0$ and $\int_{\mathbb{R}^{\nu_0}} |r_{X_0}(\boldsymbol{t}')| \mathrm{d}\boldsymbol{t}' = \infty$ for $\nu(\alpha - 1) < \nu_0$. Thus, the value $\nu(\alpha_{\mathrm{cr}} - 1) = \nu_0$ corresponds to the change from covariance SRD to covariance LRD of the sectional RF X_0 in (132). The asymptotics in (133) also suggests a α_0-stable scaling behaviour of X_0, with α_0 given by

$$\nu_0(\alpha_0 - 1) = \nu(\alpha - 1),$$

the last equation satisfied by α_0 in (129).

Sketch of the Proof of Theorem 16. Since $\lambda^{\nu_0}(\widehat{p}_{\lambda,A}(H_{\nu_0}) - p) = (\widehat{X}_{0,\lambda}(A) - \mathrm{E}\widehat{X}_{0,\lambda}(A))/\mathrm{Leb}_{\nu_0}(A)$, the theorem reduces to the limit

distribution of the integral $\widehat{X}_{0,\lambda}(\phi) = \int_{\mathbb{R}^{\nu_0}} \phi(\boldsymbol{t}'/\lambda)(X_0(\boldsymbol{t}') \wedge 1)\mathrm{d}\boldsymbol{t}'$ for $\phi = \mathbb{I}(\boldsymbol{t}' \in A) \in \Phi_0 := L^1(\mathbb{R}^{\nu_0}) \cap L^\infty(\mathbb{R}^{\nu_0})$. Then, following the proofs of Theorem 6 and Corollary 14, (130) can be deduced in two steps:
Step 1. Proof of the α_0-stable limit

$$\lambda^{-\nu_0/\alpha_0}(X_{0,\lambda}(\phi) - \mathrm{E}X_{0,\lambda}(\phi)) \xrightarrow{\mathrm{d}} L_{\alpha_0}(\phi), \quad \lambda \to \infty \tag{134}$$

for $X_{0,\lambda}(\phi) := \int_{\mathbb{R}^{\nu_0}} \phi(\boldsymbol{t}'/\lambda) X_0(\boldsymbol{t}')\mathrm{d}\boldsymbol{t}'$, $\phi \in \Phi_0$, using the characteristic function of the stochastic integral in (15) of Chapter 4.
Step 2. Extend (134) to

$$\lambda^{-\nu_0/\alpha_0}(\widehat{X}_{0,\lambda}(\phi) - \mathrm{E}\widehat{X}_{0,\lambda}(\phi)) \xrightarrow{\mathrm{d}} \mathrm{e}^{-\mu} L_{\alpha_0}(\phi), \quad \lambda \to \infty$$

using the Charlier expansion of the indicator function $x \wedge 1$.

Step 2 is relatively easy using (133) and a covariance argument as in the proof of Corollary 14. In the following, we focus on Step 1 derived using characteristic functions. Let $j_{0,\lambda}(\theta) := \log \mathrm{E}\exp\{\mathrm{i}\theta\lambda^{-\nu_0/\alpha_0}(X_{0,\lambda}(\phi) - \mathrm{E}X_{0,\lambda}(\phi))\}$, $j_0(\theta) := \log \mathrm{E}\exp\{\mathrm{i}\theta L_{\alpha_0}(\phi)\}$. Following (35) of Chapter 4, rewrite

$$\begin{aligned} j_{0,\lambda}(\theta) &= \int_{\mathbb{R}^{\nu_0}\times\mathbb{R}^{\nu-\nu_0}} \mathrm{E}\Psi\bigg(\frac{\theta}{\lambda^{\nu_0/\alpha_0}}\int_{\mathbb{R}^{\nu_0}} \phi(\boldsymbol{t}'/\lambda)\mathbb{I}((\boldsymbol{t}',\boldsymbol{0}) \\ &\quad -(\boldsymbol{s}',\boldsymbol{s}'') \in \Xi)\mathrm{d}\boldsymbol{t}'\bigg)\mathrm{d}\boldsymbol{s}'\mathrm{d}\boldsymbol{s}'' \\ &= \lambda^{\nu_0}\int_{\mathbb{R}^{\nu_0}\times\mathbb{R}^{\nu-\nu_0}} \mathrm{E}\Psi\bigg(\frac{\theta}{\lambda^{\nu_0/\alpha_0}}\int_{\mathbb{R}^{\nu_0}} \phi\Big(\frac{\boldsymbol{t}'}{\lambda}+\boldsymbol{s}'\Big) \\ &\quad \times \mathbb{I}((\boldsymbol{t}',\boldsymbol{0}) \in (\boldsymbol{0},\boldsymbol{s}'')+\Xi)\mathrm{d}\boldsymbol{t}'\bigg)\mathrm{d}\boldsymbol{s}'\mathrm{d}\boldsymbol{s}''. \end{aligned}$$

The intuitive argument leading to $j_{0,\lambda}(\theta) \to j_0(\theta)$ uses the observation that

$$\begin{aligned} &\int_{\mathbb{R}^{\nu_0}} \phi\Big(\frac{\boldsymbol{t}'}{\lambda}+\boldsymbol{s}'\Big)\mathbb{I}((\boldsymbol{t}',\boldsymbol{0}) \in (\boldsymbol{0},\boldsymbol{s}'')+\Xi)\mathrm{d}\boldsymbol{t}' \\ &\quad \to \phi(\boldsymbol{s}')\int_{\mathbb{R}^{\nu_0}} \mathbb{I}((\boldsymbol{t}',\boldsymbol{0}) \in (\boldsymbol{0},\boldsymbol{s}'')+\Xi)\mathrm{d}\boldsymbol{t}' \\ &\quad = \phi(\boldsymbol{s}')\mathrm{Leb}_{\nu_0}(\Xi_{\boldsymbol{s}''}), \end{aligned}$$

where $\Xi_{\boldsymbol{s}''} := \Xi \cap \{\boldsymbol{t} : \boldsymbol{t}'' = \boldsymbol{s}''\}$ is a section of Ξ by the hyperplane $\{\boldsymbol{t} = (\boldsymbol{t}',\boldsymbol{t}'') \in \mathbb{R}^\nu : \boldsymbol{t}'' = \boldsymbol{s}''\}$. Therefore, we first need to consider the limit of

$$\tilde{j}_{0,\lambda}(\theta) := \lambda^{\nu_0} \int_{\mathbb{R}^{\nu_0}\times\mathbb{R}^{\nu-\nu_0}} \mathrm{E}\Psi\left(\frac{\theta}{\lambda^{\nu_0/\alpha_0}}\phi(\boldsymbol{s}')\mathrm{Leb}_{\nu_0}(\Xi_{\boldsymbol{s}''})\right)\mathrm{d}\boldsymbol{s}'\mathrm{d}\boldsymbol{s}''$$

and then show that the difference $j_{0,\lambda}(\theta) - \tilde{j}_{0,\lambda}(\theta) \to 0$ is negligible. Using $\Xi = R^{1/\nu}\,\Xi^0$, we get $\mathrm{Leb}_{\nu_0}(\Xi_{\boldsymbol{s}''}) = R^{\nu_0/\nu} g^0(\boldsymbol{s}''/R^{1/\nu})$, where

$$g^0(\boldsymbol{s}'') := \mathrm{Leb}_{\nu_0}(\Xi^0 \cap \{\boldsymbol{t}'' = \boldsymbol{s}''\}) \tag{135}$$

is the ν_0-dimensional Lebesgue measure of the intersection of Ξ^0 with the hyperplane $\{(\boldsymbol{t}',\boldsymbol{t}'') \in \mathbb{R}^\nu : \boldsymbol{t}'' = \boldsymbol{s}''\}$. Since $\Xi^0 \subset \{|\boldsymbol{t}| < 1\}$ is a bounded set, $g^0(\boldsymbol{s}'') \ge 0$ in (135) is bounded and has a bounded support (which vanishes for $|\boldsymbol{s}''| > 1$). Moreover, $g^0(\boldsymbol{s}'')$ is *independent* of R. Denote

$$h(x) := x^{\alpha_0}\mathrm{E}[R^{(\nu-\nu_0)/\nu}\mathbb{I}(R > x^{\nu/\nu_0})] \quad (x > 0).$$

By change of variables,

$$\begin{aligned}
\tilde{j}_{0,\lambda}(\theta) &= \lambda^{\nu_0}\int_{\mathbb{R}^{\nu_0}} \mathrm{i}\theta\phi(\boldsymbol{s}')\mathrm{d}\boldsymbol{s}'\mathrm{E}\int_0^\infty \mathbb{I}(\mathrm{Leb}_{\nu_0}(\Xi_{\boldsymbol{s}''}) > x\lambda^{\nu_0/\alpha_0}) \\
&\quad\times\left(\mathrm{e}^{\mathrm{i}\theta\phi(\boldsymbol{s}')x} - 1\right)\mathrm{d}x \\
&= \lambda^{\nu_0}\int_{\mathbb{R}^{\nu_0}} \mathrm{i}\theta\phi(\boldsymbol{s}')\mathrm{d}\boldsymbol{s}'\int_0^\infty \left(\mathrm{e}^{\mathrm{i}\theta\phi(\boldsymbol{s}')x} - 1\right)\frac{\mathrm{d}x}{(x\lambda^{\nu_0/\alpha_0})^{\alpha_0}} \\
&\quad\times\int_{\mathbb{R}^{\nu-\nu_0}} g_0(\boldsymbol{s}'')^{\alpha_0} h\left(\frac{x\lambda^{\nu_0/\alpha_0}}{g^0(\boldsymbol{s}'')}\right)\mathrm{d}\boldsymbol{s}'' \\
&= \int_{\mathbb{R}^{\nu_0}} \mathrm{i}\theta\phi(\boldsymbol{s}')\mathrm{d}\boldsymbol{s}'\int_0^\infty \left(\mathrm{e}^{\mathrm{i}\theta\phi(\boldsymbol{s}')x} - 1\right)\frac{\mathrm{d}x}{x^{\alpha_0}} \\
&\quad\times\int_{\mathbb{R}^{\nu-\nu_0}} g_0(\boldsymbol{s}'')^{\alpha_0} h\left(\frac{x\lambda^{(\nu-1)/\alpha_0}}{g^0(\boldsymbol{s}'')}\right)\mathrm{d}\boldsymbol{s}''
\end{aligned}$$

where

$$\lim_{x\to\infty} h(x) = c_R\alpha \lim_{x\to\infty} x^{\alpha_0}\int_{x^{\nu/\nu_0}}^\infty \frac{r^{1-(\nu_0/\nu)}\mathrm{d}r}{r^{1+\alpha}} = \frac{c_R\alpha\nu}{\alpha_0\nu_0} =: h_\infty$$

is a consequence of the tail behaviour of R in (22) of Chapter 4 and $(\alpha - 1)\nu = (\alpha_0 - 1)\nu_0$. Then,

$$\tilde{\jmath}_{0,\lambda}(\theta) \to \jmath_0(\theta) = h_\infty \int_{\mathbb{R}^{\nu-\nu_0}} g_0(\boldsymbol{s}'')^{\alpha_0} \mathrm{d}\boldsymbol{s}'' \int_{\mathbb{R}^{\nu_0}} \mathrm{i}\theta\phi(\boldsymbol{s}')\mathrm{d}\boldsymbol{s}' \times \int_0^\infty \left(\mathrm{e}^{\mathrm{i}\theta\phi(\boldsymbol{s}')x} - 1\right) \frac{\mathrm{d}x}{x^{\alpha_0}}$$

can be deduced from the DCT, with $\jmath_0(\theta)$ being the log-characteristic function of the α_0-stable r.v. $L_{\alpha_0}(\phi)$ in (134). The remainder of the proof of Theorem 16, including approximation of $\jmath_{0,\lambda}(\theta)$ by $\tilde{\jmath}_{0,\lambda}(\theta)$, and the proof of the Gaussian limit in (131) resemble the proof of Theorem 6 of Chapter 4.

6 Application to Statistical Solutions of Burgers' Equation

Burgers' equation with (random) potential initial data is written as

$$\begin{aligned} \partial\vec{v}(t,\boldsymbol{x})/\partial t + \langle \vec{v}(t,\boldsymbol{x}), \nabla\rangle \vec{v}(t,\boldsymbol{x}) &= \frac{1}{2}\kappa\Delta\vec{v}(t,\boldsymbol{x}), \\ \vec{v}(0,\boldsymbol{x}) &= -\nabla\xi(\boldsymbol{x}), \end{aligned} \tag{136}$$

where $\vec{v}(t,\boldsymbol{x}) = (v_1(t,\boldsymbol{x}), \ldots, v_\nu(t,\boldsymbol{x})), (t,\boldsymbol{x}) \in \mathbb{R}_+ \times \mathbb{R}^\nu$ is a $\mathbb{R}^\nu$-valued function (velocity field) and $\xi = \{\xi(\boldsymbol{x}); \boldsymbol{x} \in \mathbb{R}^\nu\}$ is a scalar (potential) RF; $\nabla := (\partial/\partial x_1, \ldots, \partial/\partial x_\nu)$ is the gradient operator, $\langle \vec{v}(t,\boldsymbol{x}), \nabla\rangle := \sum_{i=1}^\nu v_i(t,\boldsymbol{x})\partial/\partial x_i$. The parameter $\kappa > 0$ is usually called the viscosity parameter. Burgers' equation is one of the important equations of mathematical physics. The solution $\vec{v}(t,\boldsymbol{x})$ with random initial data (known as the Burgers turbulence problem) is a (vector-valued) RF whose behavior as $t \to \infty$ and/or $\kappa \to 0$ presents considerable physical and mathematical interest and has been extensively studied in the literature. See Ref. [268] for an overview of mathematical and physical issues related to Burgers' turbulence and numerous references. For a probabilistic approach, we refer to Ref. [3] and the review paper in Ref. [256]. This chapter discusses a relatively simple case of Burgers' turbulence when $\kappa > 0$ is fixed and the initial fluctuations are linear RFs and weak in the sense that exponential moments of $\xi(\boldsymbol{x})$ exist. In this case, the results for Burgers' turbulence appear as simple corollaries of the limit theorems in Sections 3 and 5 for exponential nonlinear functions of linear RFs.

The study of Burgers' equation is facilitated by the Hopf–Cole substitution

$$\vec{v}(t,\boldsymbol{x}) = -\kappa\nabla\log u(t,\boldsymbol{x}) = -\frac{\kappa\nabla u(t,\boldsymbol{x})}{u(t,\boldsymbol{x})}, \tag{137}$$

with a scalar-valued $u(t,\boldsymbol{x})$, satisfying the heat equation $\partial u(t,\boldsymbol{x})/\partial t = \frac{1}{2}\kappa\Delta u(t,\boldsymbol{x})$ with the (exponential) initial condition $u(0,\boldsymbol{x}) = \mathrm{e}^{\xi(\boldsymbol{x})/\kappa}$, $\boldsymbol{x}\in\mathbb{R}^\nu$. Thus, (137) has an explicit representation through the heat kernel $g(t,\boldsymbol{x},\boldsymbol{y}) := (2\pi\kappa t)^{-\nu/2}\exp\{-|\boldsymbol{x}-\boldsymbol{y}|^2/2\kappa t\}$ as the ratio

$$\vec{v}(t,\boldsymbol{x}) = -\frac{\kappa\int_{\mathbb{R}^\nu}\nabla g(t,\boldsymbol{x},\boldsymbol{y})\mathrm{e}^{\xi(\boldsymbol{y})/\kappa}\mathrm{d}\boldsymbol{y}}{\int_{\mathbb{R}^\nu} g(t,\boldsymbol{x},\boldsymbol{y})\mathrm{e}^{\xi(\boldsymbol{y})/\kappa}\mathrm{d}\boldsymbol{y}}, \tag{138}$$

with a vector-valued integrand in the numerator:

$$\nabla g(t,\boldsymbol{x},\boldsymbol{y}) = -\left(\frac{\boldsymbol{x}-\boldsymbol{y}}{\kappa t}\right) g(t,\boldsymbol{x},\boldsymbol{y}). \tag{139}$$

Using the fact that $\int_{\mathbb{R}^\nu}\nabla g(t,\boldsymbol{x},\boldsymbol{y})\mathrm{d}\boldsymbol{y} = 0$, one can replace $\mathrm{e}^{\xi(\boldsymbol{y})/\kappa}$ in the numerator of (138) by $\mathrm{e}^{\xi(\boldsymbol{y})/\kappa} - \mathrm{E}\mathrm{e}^{\xi(\boldsymbol{y})/\kappa}$, provided the last expectation is finite and does not depend on $\boldsymbol{y}$. The natural parabolic scaling leads to the RF

$$\vec{v}_\lambda(t,\boldsymbol{x}) := \vec{v}(\lambda^2 t, \lambda\boldsymbol{x}),$$

and the problem concerns the limit distribution of the RF $\vec{v}_\lambda(t,\boldsymbol{x}); (t,\boldsymbol{x})\in\mathbb{R}_+\times\mathbb{R}^\nu$ as $\lambda\to\infty$. Similarly as in Ref. [3] and some other works, we ignore the meaning of the initial condition in (136) for models of ξ, discussed later, and consider only the scaling properties of the Hopf–Cole solution in (138).

Using (138), the rescaled velocity RF can be written as

$$\vec{v}_\lambda(t,\boldsymbol{x}) = -\frac{\kappa\int_{\mathbb{R}^\nu}\phi_{t,\boldsymbol{x}}(\boldsymbol{y}/\lambda)(G(\xi(\boldsymbol{y})) - \mathrm{E}G(\xi(\boldsymbol{y})))\mathrm{d}\boldsymbol{y}}{\lambda\int_{\mathbb{R}^\nu}\psi_{t,\boldsymbol{x}}(\boldsymbol{y}/\lambda)G(\xi(\boldsymbol{y}))\mathrm{d}\boldsymbol{y}}, \tag{140}$$

where $G(x) = \mathrm{e}^{x/\kappa}$ and the integrals in the numerator and denominator resemble (3) with $\phi(\boldsymbol{y}) = \phi_{t,\boldsymbol{x}}(\boldsymbol{y}) := \nabla g(t,\boldsymbol{x},\boldsymbol{y})$ and

$\phi(\boldsymbol{y}) = \psi_{t,\boldsymbol{x}}(\boldsymbol{y}) := g(t,\boldsymbol{x},\boldsymbol{y})$, respectively. Clearly, for any fixed $(t,\boldsymbol{x}) \in \mathbb{R}_+ \times \mathbb{R}^\nu$, the above ϕ's belong to Φ in (38) of Chapter 1. If ξ is ergodic stationary with $\mathrm{E}\mathrm{e}^{\xi(\mathbf{0})/\kappa} =: E_\kappa < \infty$, we have that $\mathrm{E}\int_{\mathbb{R}^\nu} \psi_{t,\boldsymbol{x}}(\boldsymbol{y}/\lambda)G(\xi(\boldsymbol{y}))\mathrm{d}\boldsymbol{y} = E_\kappa \int_{\mathbb{R}^\nu} \psi_{t,\boldsymbol{x}}(\boldsymbol{y}/\lambda)\mathrm{d}\boldsymbol{y} = \lambda^\nu E_\kappa$ and $\lambda^{-\nu}\int_{\mathbb{R}^\nu} \psi_{t,\boldsymbol{x}}(\boldsymbol{y}/\lambda)G(\xi(\boldsymbol{y}))\mathrm{d}\boldsymbol{y} \xrightarrow{\mathrm{d}} E_\kappa$ for any $(t,\boldsymbol{x})$ by the law of large numbers, reducing the limit of $\vec{v}_\lambda(t,\boldsymbol{x})$ to that of the numerator on the r.h.s. of (140), or the scaling limit of the centred exponential RF $\mathrm{e}^{\xi(\boldsymbol{y})/\kappa} - \mathrm{E}\mathrm{e}^{\xi(\boldsymbol{y})/\kappa}$.

Moving average initial data. The following corollary from Proposition 9 describes the limit of (140) when the initial data ξ is a Lévy MA RF. Recall from (83) that for such ξ, the first Appell coefficient of $\mathrm{e}^{x/\kappa}$ coincides with $E_\kappa = \exp\left\{\int_{\mathbb{R}^\nu\times\mathbb{R}}(\mathrm{e}^{va(\boldsymbol{y})} - 1 - va(\boldsymbol{y}))\pi(\mathrm{d}v)\mathrm{d}\boldsymbol{y}\right\}$.

Corollary 17. *Let*

$$\xi(\boldsymbol{x}) = \int_{\mathbb{R}^\nu} a(\boldsymbol{x}-\boldsymbol{y})M(\mathrm{d}\boldsymbol{y}), \quad \boldsymbol{x} \in \mathbb{R}^\nu$$

be a Lévy MA RF with kernel $a(\cdot)$ and the Lévy characteristics (σ,π) satisfying the conditions of Proposition 9 (particularly, π having all exponential moments). In addition, consider the following:

(i) *Let $a(\cdot)$ satisfy Assumption (A)$(d;\mathbb{R}^\nu)$, $0 < d < \nu/4$. Then,*

$$\lambda^{1+\nu-\frac{\nu+4d}{2}}\vec{v}_\lambda(t,\boldsymbol{x}) \overset{\mathrm{fdd}}{\to} W_d(\nabla g(t,\boldsymbol{x},\cdot)),$$

where the Gaussian RF W_d is defined in (14) of Chapter 3.

(ii) *Let $a(\cdot) \in L^1(\mathbb{R}^\nu)$. Then,*

$$\lambda^{1+\frac{\nu}{2}}\vec{v}_\lambda(t,\boldsymbol{x}) \overset{\mathrm{fdd}}{\to} \sigma_Y W(\nabla g(t,\boldsymbol{x},\cdot)),$$

where $W = W_0$ is defined in (14) of Chapter 3 and $\sigma_Y^2 = \int_{\mathbb{R}^\nu}\mathrm{Cov}(\mathrm{e}^{\xi(\mathbf{0})/\kappa}, \mathrm{e}^{\xi(\boldsymbol{y})/\kappa})\mathrm{d}\boldsymbol{y} > 0$ is given by $\sigma_Y^2 = E_\kappa^2\int_{\mathbb{R}^\nu}\big(\exp\{\int_{\mathbb{R}^\nu\times\mathbb{R}}(\mathrm{e}^{va(\boldsymbol{u})/\kappa} - 1)(\mathrm{e}^{va(\boldsymbol{y}+\boldsymbol{u})/\kappa} - 1)\pi(\mathrm{d}v)\mathrm{d}\boldsymbol{u}\} - 1\big)\mathrm{d}\boldsymbol{y}$; see (29) of Chapter 2.

We remark that the covariance function of the limit Gaussian RF part (ii) can be written explicitly. Indeed, by exchanging the gradient with integration and using $\int_{\mathbb{R}^\nu} g(t_1,\boldsymbol{x}_1,\boldsymbol{y})g(t_2,\boldsymbol{x}_2,\boldsymbol{y})\mathrm{d}\boldsymbol{y} = g(t_1+t_2,\boldsymbol{x}_1,\boldsymbol{x}_2)$ (the Kolmogorov–Chapman formula for Brownian transition density), we get for the components-wise covariances of

$W(\nabla g(t,\boldsymbol{x},\cdot)) = (W(\partial_{x_1} g(t,\boldsymbol{x},\cdot)),\dots,W(\partial_{x_\nu} g(t,\boldsymbol{x},\cdot)))$ that for any $\boldsymbol{x}_k = (x_{1,k},\dots,x_{\nu,k}) \in \mathbb{R}^\nu, k=1,2, t_1,t_2>0, 1\le i,j\le \nu$,

$$\begin{aligned}&\mathrm{E}W(\partial_{x_{1,i}} g(t_1,\boldsymbol{x}_1,\cdot))W(\partial_{x_{2,j}} g(t_2,\boldsymbol{x}_2,\cdot))\\ &\quad = (2\pi\kappa(t_1+t_2))^{-\nu/2}\partial_{x_{1,i}}\partial_{x_{2,j}}\mathrm{e}^{-|\boldsymbol{x}_1-\boldsymbol{x}_2|^2/2\kappa(t_1+t_2)}\end{aligned}$$

can be written as partial derivatives of the heat kernel $g(t_1 + t_2, \boldsymbol{x}_1, \boldsymbol{x}_2)$.

Random grain initial data. Non-Gaussian limits for the rescaled Hopf–Cole solution in (140) may appear when the initial data ξ is a RG RF. Assume that the initial potential RF has the form

$$\xi(\boldsymbol{y}) := M^{-1/2}(X_M(\boldsymbol{y}) - \mathrm{E}X_M(\boldsymbol{y})), \quad \boldsymbol{y}\in\mathbb{R}^\nu, \tag{141}$$

where X_M is the (aggregated) RG model as in Theorem 12, with intensity $M = \lambda^\gamma$ increasing with λ when $\gamma > 0$. Note that $\mathrm{E}\mathrm{e}^{\xi(\boldsymbol{y})/\kappa} \to \mathrm{E}\mathrm{e}^{Z_\mu/\kappa} = \mathrm{e}^{\mu/2\kappa^2}$ $(M\to\infty)$, $\mathrm{E}[\mathrm{e}^{Z_\mu/\kappa}Z_\mu]/\mathrm{E}\mathrm{e}^{Z_\mu/\kappa} = 1/\kappa$ and $\lambda^{-\nu}\int_{\mathbb{R}^\nu} g(t,\boldsymbol{x},\boldsymbol{y}/\lambda)\mathrm{e}^{\xi(\boldsymbol{y})/\kappa}\mathrm{d}\boldsymbol{y} \to \mathrm{E}\mathrm{e}^{Z_\mu/\kappa}$, $Z_\mu \sim N(0,\mu)$ in probability by the law of large numbers. The exponential function $G(x) = \mathrm{e}^{x/\kappa}, x\in\mathbb{R}$ satisfies (119), and hence condition (110) is satisfied. The application of Theorem 12 (ii) yields the following result.

Corollary 18. *Let $\vec{v}_\lambda(t,\boldsymbol{x})$ be as in* (140) *and* (141), *with X_M, M satisfying the conditions of Theorem* 12. *Then, as $\lambda\to\infty$, for any $\gamma>0$,*

$$\lambda^{1+d+\frac{\gamma}{2}-H(\gamma)}\vec{v}_\lambda(t,\boldsymbol{x}) \overset{\mathrm{fdd}}{\to} \begin{cases} B_\alpha(\nabla g(t,\boldsymbol{x},\cdot)), & \gamma > \nu(\alpha-1),\\ L_\alpha(\nabla g(t,\boldsymbol{x},\cdot)), & \gamma < \nu(\alpha-1),\\ J_\alpha(\nabla g(t,\boldsymbol{x},\cdot)), & \gamma = \nu(\alpha-1),\end{cases} \tag{142}$$

where $H(\gamma)$ and the limit RFs are the same as in (111).

For $\gamma = 0$ or $\xi(\boldsymbol{y})$ in (141) replaced by $\xi(\boldsymbol{y}) = X(\boldsymbol{y})$ in (15) of Chapter 4, we can apply Corollary 14 to conclude the following result.

Corollary 19. *Let $\vec{v}_\lambda(t,\boldsymbol{x})$ be as in* (140), *with $\xi(\boldsymbol{y}) = X(\boldsymbol{y})$ as in Theorem* 12. *Then, as $\lambda\to\infty$,*

$$\lambda^{1+d-\frac{d}{\alpha}}\vec{v}_\lambda(t,\boldsymbol{x}) \overset{\mathrm{fdd}}{\to} \kappa(\mathrm{e}^{1/\kappa}-1)L_\alpha(\nabla g(t,\boldsymbol{x},\cdot)). \tag{143}$$

7 Notes

1. The interest in limit theorems for LRD processes under Gaussian subordination was initiated by Rosenblatt [224]. Non-central limits depending on the Hermite rank of G were independently described in Dobrushin and Major [80] and Taqqu [259]. The central limit theorem for the above processes was established in Breuer and Major [47] using combinatorial formula for Hermite polynomials. Arcones [5] extended the above results to vector-valued Gaussian processes. Nualart and Peccati [190] proved that for the CLT, it is enough to prove the convergence of moments up to order 4. For MA Gaussian RFs, the simple approximation by m-dependent r.v.s, as in Theorem 1, was first applied by Giraitis *et al.* [103]. Limit theorems involving Gaussian-subordinated r.v.s appear in many scientific areas and applications. The interested reader is advised to refer to the monographs by Giraitis *et al.* [103], Peccati and Taqqu [199], Pipiras and Taqqu [215], Leonenko [161], and the references therein.

2. Non-central limit results for Appell polynomials of MA RF X were established in Surgailis [245] and Avram and Taqqu [10]. Surgailis [247] proved that the Appell rank is 'robust' w.r.t. an additive 'short memory perturbation' of a linear LRD process. The case of general G and causal MA time series X was studied in Ho and Hsing [120], Surgailis [248] and other works. Non-central and central limit theorems for more general polynomial forms in i.i.d. r.v.s can be found in hot Giraitis *et al.* [103], Bai and Taqqu [18], and elsewhere.

3. First-order reduction principle for non-causal X indexed by $\mathbb{Z}$ is discussed in Giraitis and Surgailis [102]. Results related to Theorems 5 and 6 for MA RFs in dimension $\nu \geq 2$ were obtained by Doukhan *et al.* [83]. Boissy *et al.* [46] studied the estimation of parameters for unilateral fractional autoregressive spatial processes.

4. Empirical process of causal LRD time series was studied by Dehling and Taqqu [75], Ho and Hsing [119], Wu [269,270], and others. For statistical applications, see Beran [30], Giraitis *et al.* [103], and Beran *et al.* [31]. A spatial empirical process ($\nu \geq 2$) of LRD MA RF X and some inference problems were discussed in Doukhan *et al.* [83] and Koul *et al.* [142].

5. The discussion in Section 5 is based on Surgailis [253] and Biermé *et al.* [41]. Central limit theorems for parameters of the Boolean model are discussed in Baddeley [16], Hall [111, Thm.3.5]. Mase [180], Molchanov and Stoyan [188], Molchanov [189], and elsewhere.

6. Corollary 17 extends or complements Bulinski and Molchanov [50], Albeverio *et al.* [3], Giraitis *et al.* [97], and other works. Corollary 18 is due to Surgailis [253]. Asymptotic results for Burgers' and heat equations with singular Gaussian and non-Gaussian initial data were obtained by Rosenblatt [225], Leonenko and Woyczynski [165], and Surgailis and Woyczynski [257]. An extensive account of various mathematical and physical aspects of Burgers' turbulence is given in Woyczynski [268].

7. Anh *et al.* [4] discussed limit theorems for spatio-temporal RFs

$$u(t, \boldsymbol{x}) = \int_{\mathbb{R}^\nu} h(t, \boldsymbol{x} - \boldsymbol{y}) Y(\boldsymbol{y}) \mathrm{d}\boldsymbol{y}, \quad t > 0, \tag{144}$$

where $h(t, \boldsymbol{x})$ is a deterministic kernel (e.g., the fundamental solution of a partial or fractional partial differential equation) and $Y = G(X)$ is a nonlinear function of a stationary isotropic Gaussian or a χ-square RF X admitting an orthogonal expansion in Laguerre polynomials similar to the Hermite expansion in (49) of Chapter 2.

Chapter 6

Anisotropic Scaling of RFs on $\mathbb{Z}^2$

This section discusses anisotropic scaling limits on increasing rectangles for an MA RF,

$$X(\boldsymbol{t}) = \sum_{\boldsymbol{u}\in\mathbb{Z}^2} a(\boldsymbol{t}-\boldsymbol{u})\varepsilon(\boldsymbol{u}), \quad \boldsymbol{t}\in\mathbb{Z}^2, \tag{1}$$

with non-random coefficients $a(\cdot)\in L^2(\mathbb{Z}^\nu)$ and standardised i.i.d. innovations $\varepsilon(\boldsymbol{t}), \boldsymbol{t}\in\mathbb{Z}^2$ under spectral and spatial domain assumptions on the MA coefficients. Recall from Section 2 of Chapter 1 the definition of (large-scale) γ-scaling limit:

$$d_{\lambda,\gamma}^{-1} S_{\lambda,\gamma}(\boldsymbol{x}) \xrightarrow{\text{fdd}} V_\gamma(\boldsymbol{x}), \quad \boldsymbol{x}\in\mathbb{R}_+^2, \quad \lambda\to\infty, \tag{2}$$

where

$$S_{\lambda,\gamma}(\boldsymbol{x}) = \sum_{\boldsymbol{t}\in K_{[\lambda x_1,\lambda^\gamma x_2]}} X(\boldsymbol{t}) \tag{3}$$

and $K_{[\lambda x_1,\lambda^\gamma x_2]} = \{\boldsymbol{t}\in\mathbb{Z}^2 : 1\le t_1\le\lambda x_1, 1\le t_2\le\lambda^\gamma x_2\}\subset\mathbb{Z}^2$ is a rectangle whose sides increase at rate $O(\lambda)$ and $O(\lambda^\gamma)$ as $\lambda\to\infty$. For (2), the behaviours of $a(\boldsymbol{t})$ and the spectral density $f(\boldsymbol{x})$ as $|\boldsymbol{t}|\to\infty$ and $|\boldsymbol{x}|\to 0$ are of primary importance. Several anisotropic forms of $a(\boldsymbol{t})$ and $f(\boldsymbol{x})$ are provided, allowing for different asymptotic behaviours along the coordinate axes and depending on two or more parameters. In each case, the existence or nonexistence of the scaling transition

$$V_\gamma = \begin{cases} V_+, & \gamma > \gamma_{\rm cr}, \\ V_-, & \gamma < \gamma_{\rm cr}, \\ V_{\rm cr}, & \gamma = \gamma_{\rm cr}, \end{cases} \tag{4}$$

is established, and the unbalanced scaling limits $V_\gamma^\pm$ are identified as FBS with Hurst parameters $H_i^\pm \in [0,1], i = 1,2$. Scaling transition is also demonstrated for MA RFs with an 'oblique' dependence axis and some nonlinear subordinated RFs of X in (1). The scaling diagram of a class of LRD MA RF on $\mathbb{Z}^3$ is described in detail.

1 Preliminaries: Generalised Homogeneous Functions

Recall $\mathbb{R}_0^2 := \mathbb{R}^2 \backslash \{(0,0)\}$. For $q_i > 0$, $i = 1,2$, let

$$\rho(\boldsymbol{t}) := |t_1|^{q_1} + |t_2|^{q_2}, \quad \boldsymbol{t} = (t_1,t_2) \in \mathbb{R}^2, \quad Q := \frac{1}{q_1} + \frac{1}{q_2}. \tag{5}$$

Definition 1. A measurable function $h : \mathbb{R}^2 \to \mathbb{R}$ is said to be:

(i) (q_1,q_2)-*homogeneous* if for all $\lambda > 0$,

$$\lambda h(\lambda^{1/q_1} t_1, \lambda^{1/q_2} t_2) = h(\boldsymbol{t}), \quad \boldsymbol{t} \in \mathbb{R}_0^2. \tag{6}$$

(ii) (q_1,q_2)-*invariant* if $\lambda \mapsto h(\lambda^{1/q_1} t_1, \lambda^{1/q_2} t_2)$ is a constant function on $\mathbb{R}_+$ for any $\boldsymbol{t} \in \mathbb{R}_0^2$.

Proposition 2. *(i) Any (q_1,q_2)-homogeneous function h can be represented as*

$$h(\boldsymbol{t}) = \ell(\boldsymbol{t})/\rho(\boldsymbol{t}), \quad \boldsymbol{t} \in \mathbb{R}_0^2, \tag{7}$$

where $\rho(\boldsymbol{t})$ is defined in (5) and ℓ is a (q_1,q_2)-invariant function. Moreover, ℓ can be written as

$$\ell(\boldsymbol{t}) = \ell_{\mathrm{sign}(t_2)}(t_1/\rho(\boldsymbol{t})^{1/q_1}), \quad t_2 \neq 0, \tag{8}$$

where $\ell_\pm(z) := h(z, \pm(1-|z|^{q_1})^{1/q_2})$, $z \in [-1,1]$.
(ii) Let $\kappa > 0$ and h be a (q_1,q_2)-homogeneous function. Then, h^κ is $(\kappa q_1, \kappa q_2)$-homogeneous.

Proof. (i) Relation (7) follows from (6) by taking $\lambda = 1/\rho(\boldsymbol{t})$. Then, $\ell(\boldsymbol{t}) := h(t_1/\rho(\boldsymbol{t})^{1/q_1}, t_2/\rho(\boldsymbol{t})^{1/q_2})$ is a (q_1, q_2)-invariant function. Therefore, (8) follows since $t_2/\rho(\boldsymbol{t})^{1/q_2} = \mathrm{sign}(t_2)(1 - |t_1/\rho(\boldsymbol{t})^{1/q_1}|^{q_1})^{1/q_2}$. (ii) is obvious from $\lambda^\kappa h^\kappa(\lambda^{1/q_1}t_1, \lambda^{1/q_2}t_2) = h^\kappa(\boldsymbol{t}) = \mu h^\kappa(\mu^{1/\kappa q_1}t_1, \mu^{1/\kappa q_2}t_2)$, with $\mu := \lambda^\kappa$. □

The notion of (q_1, q_2)-homogeneous function is a particular case of the generalised homogeneous function introduced in Ref. [113]. Note that $\boldsymbol{t} \mapsto (t_1, \rho(\boldsymbol{t})^{1/q_1})$ is a one-to-one transformation of the upper half-plane $\{\boldsymbol{t} \in \mathbb{R}^2 : t_2 \geq 0\}$ onto itself. Following Ref. [78], the form in (7) is called the *polar representation of h*. The two factors in (7), viz., ρ^{-1} and ℓ, are called the *radial* and *angular* functions, respectively. Note that h being strictly positive and continuous on $\mathbb{R}^2_0$ is equivalent to $\ell_\pm$ both being strictly positive and continuous on $[-1, 1]$ with $\ell_+(\pm 1) = \ell_-(\pm 1)$. Let

$$\mathcal{S}_\rho := \{\boldsymbol{t} \in \mathbb{R}^2 : \rho(\boldsymbol{t}) = 1\}, \quad \bar{\boldsymbol{t}} := \left(\frac{t_1}{\rho(\boldsymbol{t})^{1/q_1}}, \frac{t_2}{\rho(\boldsymbol{t})^{1/q_2}}\right) \in \mathcal{S}_\rho, \quad \boldsymbol{t} \in \mathbb{R}^2_0. \tag{9}$$

In what follows, *we identify a (q_1, q_2)-invariant function with its restriction on $\mathcal{S}_\rho$*:

$$\ell(\boldsymbol{t}) = \ell(\bar{\boldsymbol{t}}), \quad \boldsymbol{t} \in \mathbb{R}^2_0. \tag{10}$$

Instead of ρ in (7), a different form of the radial function can be used: $\rho_p(\boldsymbol{t}) := (|t_1|^{pq_1} + |t_2|^{pq_2})^{1/p}, \boldsymbol{t} \in \mathbb{R}^2, p > 0$. Since ρ_p is (q_1, q_2)-homogeneous, changing ρ by $\rho_p(\boldsymbol{t})$ in (7) reduces to a change in the angular function. Indeed,

$$\frac{\rho_p(\boldsymbol{t})}{\rho(\boldsymbol{t})} = \ell_p(\bar{\boldsymbol{t}}), \quad \text{with} \quad \ell_p(\boldsymbol{t}) := (|t_1|^{pq_1} + |t_2|^{pq_2})^{1/p}, \quad \boldsymbol{t} \in \mathcal{S}_\rho \tag{11}$$

and $\ell_p(\boldsymbol{t})$ is bounded from above and below on $\mathcal{S}_\rho$ in view of the inequality

$$C_1\rho(\boldsymbol{t}) \leq \rho_p(\boldsymbol{t}) \leq C_2\rho(\boldsymbol{u}), \quad \boldsymbol{u} \in \mathbb{R}^2, \tag{12}$$

with $C_i > 0$, $i = 1, 2$, independent of $\boldsymbol{t}$; see Ref. [250, (2.16)]. We will also use the elementary inequality

$$\rho(\boldsymbol{t} + \boldsymbol{u}) \leq C(\rho(\boldsymbol{t}) + \rho(\boldsymbol{u})), \quad \boldsymbol{t}, \boldsymbol{u} \in \mathbb{R}^2, \tag{13}$$

which follows from $(a+b)^p \le 2^p(a^p+b^p), a, b, p > 0$. Let $\rho_+(\boldsymbol{t}) := 1 \vee \rho(\boldsymbol{t}), \boldsymbol{t} \in \mathbb{R}^2$.

Proposition 3. *For any* $\delta > 0, w > 0$,

$$\int_{\{|\boldsymbol{t}|<\delta\}} \rho(\boldsymbol{t})^{-w} \mathrm{d}\boldsymbol{t} < \infty \quad \Longleftrightarrow \quad w < Q \tag{14}$$

and

$$\left.\begin{array}{l} \int_{\{|\boldsymbol{t}|>\delta\}} \rho(\boldsymbol{t})^{-w} \mathrm{d}\boldsymbol{t} < \infty \\ \sum_{\boldsymbol{t}\in\mathbb{Z}^2} \rho_+(\boldsymbol{t})^{-w} < \infty \end{array}\right\} \quad \Longleftrightarrow \quad w > Q. \tag{15}$$

Proof. The proof of (14)–(15) and many other similar bounds involving multiple integrals follows by repeated integration of simple power functions. We shall often use the following elementary inequalities:

$$\int_0^1 \frac{\mathrm{d}u}{(v+u^q)^w} \le C \begin{cases} 1, & 0 < q < 1/w, 0 < v < 1, \\ |\log v|, & q = 1/w, 0 < v < 1, \\ v^{(1/q)-w}, & q > 1/w, 0 < v < 1, \\ v^{-w}, & q > 0, v \ge 1, \end{cases} \tag{16}$$

$$\int_0^\infty \frac{\mathrm{d}u}{(v+u^q)^w} \le C v^{(1/q)-w}, \quad q > 1/w.$$

Consider (14), where $\delta = 1$ w.l.g., using (16) we get $\int_{\{|\boldsymbol{t}|<\delta\}} \rho(\boldsymbol{t})^{-w} \mathrm{d}\boldsymbol{t} \le C \int_0^1 \mathrm{d}t_1 \int_0^1 (t_1^{q_1} + t_2^{q_2})^{-w} \mathrm{d}t_2 \le C \int_0^1 t_1^{-q_1(w-\frac{1}{q_2})} \mathrm{d}t_1 < \infty$ for $w < Q$. The above argument can also be reversed, leading to (14). The implications in (15) follow analogously from (16). □

Convolutions. Recall the notation $[\phi_1 \star \phi_2](\boldsymbol{t}) = \sum_{\boldsymbol{u}\in\mathbb{Z}^2} \phi_1(\boldsymbol{u})\phi_2(\boldsymbol{t}+\boldsymbol{u})$ for 'the discrete' convolution of sequences $\phi_i = \{\phi_i(\boldsymbol{u}), \boldsymbol{u} \in \mathbb{Z}^2\} \in L^2(\mathbb{Z}^2), i = 1, 2$ and $(\psi_1 \star \psi_2)(\boldsymbol{t}) = \int_{\mathbb{R}^2} \psi_1(\boldsymbol{u})\psi_2(\boldsymbol{t}+\boldsymbol{u})\mathrm{d}\boldsymbol{u}$ for the 'usual' convolution of functions $\psi_i = \{\psi_i(\boldsymbol{u}), \boldsymbol{u} \in \mathbb{R}^2\}, i = 1, 2$. Note that convolution with itself is symmetric:

$$(\psi \star \psi)(\boldsymbol{t}) = (\psi \star \psi)(-\boldsymbol{t}), \quad \boldsymbol{t} \in \mathbb{R}^2 \tag{17}$$

(and similarly, $[\phi \star \phi](\boldsymbol{t}) = [\phi \star \phi](-\boldsymbol{t}), \boldsymbol{t} \in \mathbb{Z}^2$), which follows from the change of variables $\boldsymbol{u} + \boldsymbol{t} \to \boldsymbol{u}$ in the convolution integral.

Convolution naturally appears when studying the covariance function. The convolution of two (q_1, q_2)-homogeneous functions with $1 < Q < 2$ is a $(\tilde{q}_1, \tilde{q}_2)$-homogeneous function with the radial function

$$\tilde{\rho}(\boldsymbol{t}) := |t_1|^{\tilde{q}_1} + |t_2|^{\tilde{q}_2}, \quad \boldsymbol{t} \in \mathbb{R}^2, \quad \tilde{q}_i := q_i(2 - Q), \quad i = 1, 2. \tag{18}$$

When $Q \le 1$, the convolution $\rho^{-1} \star \rho^{-1}$ is not defined since $\rho(\boldsymbol{u})^{-1}$ is not integrable at the origin, referring to (14), and needs to be 'regularised' as in Proposition 4 (iii). The parametrisation by $\tilde{q}_i$ in (18) will be used in both cases $1 < Q < 2$ and $Q < 1$ in parallel with the parametrisation by q_i. We note that the mapping $(q_1, q_2) \mapsto (\tilde{q}_1, \tilde{q}_2)$ from $\{(q_1, q_2) \in \mathbb{R}^2_+ : Q < 2\}$ to $\mathbb{R}^2_+$ is one-to-one, with the inverse mapping given by

$$q_i = \frac{1}{2}\tilde{q}_i(1 + \tilde{Q}), \quad i = 1, 2, \quad \tilde{Q} := \frac{1}{\tilde{q}_1} + \frac{1}{\tilde{q}_2}, \tag{19}$$
$$\tilde{Q} = \frac{Q}{2 - Q}, \quad Q = \frac{2\tilde{Q}}{1 + \tilde{Q}}.$$

Proposition 4. (i) *Let* $w_i > 0, i = 1, 2$, $w_1 + w_2 > Q$. *Then, there exists* $C > 0$ *such that for any* $\boldsymbol{t} \in \mathbb{R}^2_0$,

$$(\rho^{-w_1} \star \rho^{-w_2})(\boldsymbol{t}) \le C\rho(\boldsymbol{t})^{Q - w_1 - w_2}, \quad w_i < Q,\ i = 1, 2, \tag{20}$$
$$(\rho_+^{-w_1} \star \rho^{-w_2})(\boldsymbol{t}) \le C\rho_+(\boldsymbol{t})^{-w_2}, \quad w_2 < Q < w_1, \tag{21}$$
$$(\rho_+^{-w_1} \star \rho_+^{-w_2})(\boldsymbol{t}) \le C\rho_+(\boldsymbol{t})^{-w_1 \wedge w_2}, \quad w_i > Q,\ i = 1, 2. \tag{22}$$

Moreover, inequalities (20)–(22) *are also valid for the 'discrete' convolution* $[\rho_+^{-w_1} \star \rho_+^{-w_2}](\boldsymbol{t}), \boldsymbol{t} \in \mathbb{Z}^2$ *with* $\rho(\boldsymbol{t})$ *on the r.h.s. of* (20) *replaced by* $\rho_+(\boldsymbol{t})$.

(ii) *Let* $1 < Q < 2$ *and* $a_i = a_i(\boldsymbol{t}), \boldsymbol{t} \in \mathbb{Z}^2, i = 1, 2$ *satisfy*

$$a_i(\boldsymbol{t}) = \rho(\boldsymbol{t})^{-1}\big(\ell_i(\boldsymbol{t}) + o(1)\big), \quad |\boldsymbol{t}| \to \infty, \tag{23}$$

where $\ell_i(\boldsymbol{u}) \not\equiv 0, \boldsymbol{u} \in \mathbb{R}^2$ *are bounded continuous* (q_1, q_2)*-invariant functions,* $i = 1, 2$. *Let* $a_{i\infty}(\boldsymbol{u}) := \rho(\boldsymbol{u})^{-1}\ell_i(\boldsymbol{u}), \boldsymbol{u} \in \mathbb{R}^2, i = 1, 2.$

Then,

$$(a_{1\infty} \star a_{2\infty})(\boldsymbol{t}) = \tilde{\rho}(\boldsymbol{t})^{-1} L_{12}(\boldsymbol{t}), \quad \boldsymbol{t} \in \mathbb{R}^2 \tag{24}$$

and

$$[a_1 \star a_2](\boldsymbol{t}) = \tilde{\rho}(\boldsymbol{t})^{-1}\big(L_{12}(\boldsymbol{t}) + o(1)\big), \quad |\boldsymbol{t}| \to \infty, \tag{25}$$

where

$$L_{12}(\boldsymbol{t}) := (a_{1\infty} \star a_{2\infty})(\boldsymbol{t}), \quad \boldsymbol{t} \in \mathcal{S}_{\tilde{\rho}} \tag{26}$$

is a bounded continuous $(\tilde{q}_1, \tilde{q}_2)$*-invariant function. Moreover, if* $\ell_1 = \ell_2 \geq 0$*, then* L_{12} *in* (26) *is strictly positive on* $\mathcal{S}_{\tilde{\rho}}$.

Proof. (i) Consider (20). By the change of variables $u_i \to \rho(\boldsymbol{t})^{1/q_i} u_i, i = 1, 2$ in the convolution integral, we get $(\rho^{-w_1} \star \rho^{-w_2})(\boldsymbol{t}) = \rho(\boldsymbol{t})^{Q-w_1-w_2} I(\boldsymbol{t})$, where $I(\boldsymbol{t}) := (\rho^{-w_1} \star \rho^{-w_2})(\bar{\boldsymbol{t}})$ and $\bar{\boldsymbol{t}} := (t_1/\rho(\boldsymbol{t})^{1/q_1}, t_2/\rho(\boldsymbol{t})^{1/q_2})$. Note $\rho(\bar{\boldsymbol{t}}) = 1$; therefore, $|\bar{\boldsymbol{t}}| > \delta$ for some $\delta > 0$ independent of $\bar{\boldsymbol{t}} \in \mathbb{R}^2$. To show that $C(\boldsymbol{t})$ is bounded, split $I(\boldsymbol{t}) = I_1(\boldsymbol{t}) + I_2(\boldsymbol{t}) + I_{12}(\boldsymbol{t})$, where

$$\begin{aligned}
I_1(\boldsymbol{t}) &:= \int_{\{|\boldsymbol{u}|<\delta\}} \rho(\boldsymbol{u})^{-w_1} \rho(\boldsymbol{u} + \bar{\boldsymbol{t}})^{-w_2} \mathrm{d}\boldsymbol{u}, \\
I_2(\boldsymbol{t}) &:= \int_{\{|\boldsymbol{u}+\bar{\boldsymbol{t}}|<\delta\}} \rho(\boldsymbol{u})^{-w_1} \rho(\boldsymbol{u} + \bar{\boldsymbol{t}})^{-w_2} \mathrm{d}\boldsymbol{u}, \\
I_{12}(\boldsymbol{t}) &:= \int_{\{|\boldsymbol{u}|>\delta\} \cup \{|\boldsymbol{u}+\bar{\boldsymbol{t}}|>\delta\}} \rho(\boldsymbol{u})^{-w_1} \rho(\boldsymbol{u} + \bar{\boldsymbol{t}})^{-w_2} \mathrm{d}\boldsymbol{u}.
\end{aligned}$$

Since $\rho(\boldsymbol{t})^{-1}$ is bounded on $\{|\boldsymbol{t}| > \delta\}$, so $I_1(\boldsymbol{t}) \leq C \int_{\mathbb{R}^2} \rho(\boldsymbol{u} + \bar{\boldsymbol{t}})^{-w_2} \mathrm{d}\boldsymbol{u} < C$ and $I_2(\boldsymbol{t}) \leq C \int_{\mathbb{R}^2} \rho(\boldsymbol{u})^{-w_1} \mathrm{d}\boldsymbol{u} < C$ according to (14) and $w_i < Q, i = 1, 2$, provided $\delta > 0$ is small enough. Next, by Hölder's inequality and (15), with $w := w_1 + w_2 > Q$,

$$I_{12}(\boldsymbol{t}) \leq \int_{\{|\boldsymbol{u}|>\delta\}} \rho(\boldsymbol{u})^{-w} \mathrm{d}\boldsymbol{u} < C, \tag{27}$$

proving (20).

Next, consider (21). Then, $(\rho_+^{-w_1} \star \rho^{-w_2})(\boldsymbol{t}) = \rho(\boldsymbol{t})^{Q-w_1-w_2}(I_1'(\boldsymbol{t}) + I_2(\boldsymbol{t}) + I_{12}(\boldsymbol{t}))$, similarly as above, where $I_2(\boldsymbol{t}) < C, I_{12}(\boldsymbol{t}) < C$ and

$I_1'(\boldsymbol{t}) := \int_{\{|\boldsymbol{u}|<\delta\}} (\frac{1}{\rho(\boldsymbol{t})} \vee \rho(\boldsymbol{u}))^{-w_1} \rho(\boldsymbol{u} + \bar{\boldsymbol{t}})^{-w_2} \mathrm{d}\boldsymbol{u} \le C \int_{\{|\boldsymbol{u}|<\delta\}} (\frac{1}{\rho(\boldsymbol{t})} \vee \rho(\boldsymbol{u}))^{-w_1} \mathrm{d}\boldsymbol{u} \le C\rho(\boldsymbol{t})^{w_1 - Q} \int_{\mathbb{R}^2} (1 \vee \rho(\boldsymbol{u}))^{-w_1} \mathrm{d}\boldsymbol{u}$, where, in the last inequality, we changed the variables $u_i \to \rho(\boldsymbol{t})^{-1/q_i} u_i, i = 1, 2$. Since $\int_{\mathbb{R}^2} (1 \vee \rho(\boldsymbol{u}))^{-w_1} \mathrm{d}\boldsymbol{u} < \infty$ when $w_1 > Q$ (see (15)), relation (21) follows.

Let us prove (22). Similar to the proof of (21), $(\rho_+^{-w_1} \star \rho_+^{-w_2})(\boldsymbol{t}) = \rho(\boldsymbol{t})^{Q - w_1 - w_2} (I_1'(\boldsymbol{t}) + I_2'(\boldsymbol{t}) + I_{12}(\boldsymbol{t}))$, where $I_1'(\boldsymbol{t}) \le C\rho(\boldsymbol{t})^{w_1 - Q}, I_{12}(\boldsymbol{t}) < C$ and $I_2'(\boldsymbol{t}) \le C\rho(\boldsymbol{t})^{w_2 - Q}$, proving (22).

Extension of (20)–(22) to 'discrete' convolution requires some routine changes related to the replacement of sums by integrals. We use the following inequality: for all $K > 0$ large enough,

$$\rho(\boldsymbol{u}_1)/2 < \rho(\boldsymbol{u}_2) < 2\rho(\boldsymbol{u}_1), \quad \forall\, |\boldsymbol{u}_i| > K, \quad i = 1, 2, \quad |\boldsymbol{u}_1 - \boldsymbol{u}_2| \le 1, \tag{28}$$

which follows by the Taylor expansion of $\rho(\boldsymbol{t})$. Consider the 'discrete' version of (20). For a large $K > 0$ and $|\boldsymbol{t}| > 2K$, we have $[\rho_+^{-w_1} \star \rho_+^{-w_2}](\boldsymbol{t}) = \sum_{|\boldsymbol{u}|<K} \rho_+(\boldsymbol{u})^{-w_1} \rho(\boldsymbol{t}+\boldsymbol{u})^{-w_2} + \sum_{|\boldsymbol{u}+\boldsymbol{t}|<K} \rho(\boldsymbol{u})^{-w_1} \rho_+(\boldsymbol{t} + \boldsymbol{u})^{-w_2} + \sum_{|\boldsymbol{u}|\ge K, |\boldsymbol{u}+\boldsymbol{t}|\ge K} \rho(\boldsymbol{u})^{-w_1} \rho(\boldsymbol{t}+\boldsymbol{u})^{-w_2} =: J_1(\boldsymbol{t}) + J_2(\boldsymbol{t}) + J_{12}(\boldsymbol{t})$, where $J_1(\boldsymbol{t}) \le C\rho(\boldsymbol{t})^{-w_2} \sum_{\boldsymbol{u}\in\mathbb{Z}^2} \rho_+(\boldsymbol{u})^{-w_1}$ and $\sum_{\boldsymbol{u}\in\mathbb{Z}^2} \rho_+(\boldsymbol{u})^{-w_1} \le C \int_{|\boldsymbol{u}|<K} \rho(\boldsymbol{u})^{-w_1} \mathrm{d}\boldsymbol{u} < \infty$ follow from (28) and (14). In a similar way, $J_2(\boldsymbol{t}) \le C\rho(\boldsymbol{t})^{-w_1} \sum_{\boldsymbol{u}\in\mathbb{Z}^2} \rho_+(\boldsymbol{u})^{-w_2} \le C\rho(\boldsymbol{t})^{-w_1}$ and $J_{12}(\boldsymbol{t}) \le C \int_{|\boldsymbol{u}|\ge K, |\boldsymbol{u}+\boldsymbol{t}|\ge K} \rho(\boldsymbol{u})^{-w_1} \rho(\boldsymbol{t} + \boldsymbol{u})^{-w_2} \mathrm{d}\boldsymbol{u} \le C\rho(\boldsymbol{t})^{Q - w_1 - w_2}$ according to (20). For $|\boldsymbol{t}| \le 2K$, the bound $[\rho_+^{-w_1} \star \rho_+^{-w_2}](\boldsymbol{t}) \le \sum_{\boldsymbol{u}\in\mathbb{Z}^2} \rho_+(\boldsymbol{u})^{-w_1 - w_2} < C$ follows as in (27). The remaining facts in (i) are easy and omitted.

(ii) Relation (24) is immediate from the change of variables $u_i \to \tilde{\rho}(\boldsymbol{t})^{1/\tilde{q}_i} u_i, i = 1, 2$ in the convolution integral. Consider (25). We have $[a_1 \star a_2](\boldsymbol{t}) = \sum_{i,j=0}^{1} [a_1^i \star a_2^j](\boldsymbol{t})$, where $a_i^0(\boldsymbol{t}) := \rho_+(\boldsymbol{t})^{-1} \ell_i(\boldsymbol{t}), a_i^1(\boldsymbol{t}) := a_i(\boldsymbol{t}) - a_i^0(\boldsymbol{t}) = o(\rho(\boldsymbol{t})^{-1}), i = 1, 2$. Equation (25) follows from

$$\lim_{|\boldsymbol{t}|\to\infty} \left|\tilde{\rho}(\boldsymbol{t})[a_1^0 \star a_2^0](\boldsymbol{t}) - L_{12}(\boldsymbol{t})\right| = 0 \tag{29}$$

and

$$[a_1^i \star a_2^j](\boldsymbol{t}) = o(\tilde{\rho}(\boldsymbol{t})^{-1}), \quad (i, j) \ne (0, 0), \quad i, j = 0, 1, \quad |\boldsymbol{t}| \to \infty. \tag{30}$$

The proof of (30) mimics that of (29) and is omitted. To prove (29), we write the 'discrete' convolution as the integral $[a_1^0 \star a_2^0](\boldsymbol{t}) = \int_{\mathbb{R}^2} a_1^0(\lceil \boldsymbol{u} \rceil) a_2^0(\lceil \boldsymbol{u} \rceil + \boldsymbol{t}) \mathrm{d}\boldsymbol{u}$, where we change the variables $u_i \to \tilde{\varrho}^{1/\tilde{q}_i} u_i, i = 1, 2, \tilde{\varrho} := \tilde{\rho}(\boldsymbol{t})$, as in the proof of (24) above. Write the above change as $\boldsymbol{u} \to R_{\tilde{\varrho}} \boldsymbol{u}$, where $R_{\tilde{\varrho}} := \mathrm{diag}(\tilde{\varrho}^{1/\tilde{q}_1} u_1, \tilde{\varrho}^{1/\tilde{q}_2} u_2)$. Then, $\tilde{\rho}(\boldsymbol{t})[a_1^0 \star a_2^0](\boldsymbol{t}) = \tilde{L}_{\varrho}(\boldsymbol{t})$, where

$$\tilde{L}_{\varrho}(\boldsymbol{t}) := \int_{\mathbb{R}^2} g_{\tilde{\varrho},\boldsymbol{z}}(\boldsymbol{u}) \mathrm{d}\boldsymbol{u}, \quad \boldsymbol{z} = R_{\tilde{\varrho}}^{-1} \boldsymbol{t},$$

where for all $\tilde{\rho} > 0$, $\boldsymbol{z} \in \mathbb{R}^2$ such that $\tilde{\rho}(\boldsymbol{z}) = 1$, $\boldsymbol{u} \in \mathbb{R}^2$,

$$g_{\tilde{\varrho},\boldsymbol{z}}(\boldsymbol{u}) := \frac{\ell_1(R_{\tilde{\varrho}}^{-1} \lceil R_{\tilde{\varrho}} \boldsymbol{u} \rceil)}{\rho(R_{\tilde{\varrho}}^{-1} \lceil R_{\tilde{\varrho}} \boldsymbol{u} \rceil) \vee \tilde{\varrho}^{-q_1/\tilde{q}_1}} \cdot \frac{\ell_2(R_{\tilde{\varrho}}^{-1} \lceil R_{\tilde{\varrho}} \boldsymbol{u} \rceil + \boldsymbol{z})}{\rho(R_{\tilde{\varrho}}^{-1} \lceil R_{\tilde{\varrho}} \boldsymbol{u} \rceil + \boldsymbol{z}) \vee \tilde{\varrho}^{-q_1/\tilde{q}_1}}, \tag{31}$$

and we used the homogeneous and invariance properties of ρ and ℓ_i, $i = 1, 2$. Clearly, $R_{\tilde{\varrho}}^{-1} \lceil R_{\tilde{\varrho}} \boldsymbol{u} \rceil \to \boldsymbol{u}$ $(\tilde{\varrho} \to \infty)$ for any $\boldsymbol{u} \neq \boldsymbol{0}$. Therefore, using the continuity of ρ and ℓ_i, $i = 1, 2$, it follows that

$$g_{\tilde{\varrho},\boldsymbol{z}}(\boldsymbol{u}) \to a_{\infty 1}(\boldsymbol{u}) a_{\infty 2}(\boldsymbol{u} + \boldsymbol{z}) \tag{32}$$

as $\tilde{\varrho} \to \infty$ or $|\boldsymbol{t}| \to \infty$ for all $\boldsymbol{u} \in \mathbb{R}_0^2$, $\boldsymbol{u} + \boldsymbol{z} \in \mathbb{R}_0^2$. Then, (29) follows from

$$\sup_{\boldsymbol{z} \in \mathbb{R}^2 : \tilde{\rho}(\boldsymbol{z}) = 1} I_{\tilde{\varrho},\boldsymbol{z}} \to 0 \quad \text{as} \quad \tilde{\varrho} \to \infty, \quad \text{where}$$

$$I_{\tilde{\varrho},\boldsymbol{z}} := \int_{\mathbb{R}^2} |g_{\tilde{\varrho},\boldsymbol{z}}(\boldsymbol{u}) - a_{\infty 1}(\boldsymbol{u}) a_{\infty 2}(\boldsymbol{u} + \boldsymbol{z})| \mathrm{d}\boldsymbol{u}. \tag{33}$$

We use the fact that

$$|g_{\tilde{\varrho},\boldsymbol{z}}(\boldsymbol{u})| + |a_{\infty 1}(\boldsymbol{u}) a_{\infty 2}(\boldsymbol{u} + \boldsymbol{z})| \leq C \rho(\boldsymbol{u})^{-1} \rho(\boldsymbol{u} + \boldsymbol{z})^{-1} \tag{34}$$

with $C > 0$ independent of $\tilde{\varrho} > 1, \boldsymbol{u} \in \mathbb{R}^2, \boldsymbol{z} \in \mathbb{R}^2, \tilde{\rho}(\boldsymbol{z}) = 1$. Indeed, with $\tilde{\boldsymbol{u}} = (\tilde{u}_1, \tilde{u}_2) := R_{\tilde{\varrho}}^{-1} \lceil R_{\tilde{\varrho}} \boldsymbol{u} \rceil$ from (13), we get $\rho(\boldsymbol{u}) \leq C(\rho(\tilde{\boldsymbol{u}}) + \rho(\boldsymbol{u} - \tilde{\boldsymbol{u}})), \rho(\boldsymbol{u} + \boldsymbol{z}) \leq C(\rho(\tilde{\boldsymbol{u}} + \boldsymbol{z}) + \rho(\boldsymbol{u} - \tilde{\boldsymbol{u}}))$, where $|\tilde{u}_i - u_i| \leq$

$\tilde{\varrho}^{-1/\tilde{q}_i}, i = 1, 2$, and hence $\rho(\boldsymbol{u} - \tilde{\boldsymbol{u}}) \le 2\tilde{\varrho}^{-q_1/\tilde{q}_1}$ since $q_1/\tilde{q}_1 = q_2/\tilde{q}_2$. Therefore,

$$\rho(\tilde{\boldsymbol{u}}) \vee \tilde{\varrho}^{-q_1/\tilde{q}_1} \ge (1/C_1)\rho(\boldsymbol{u}),$$
$$\rho(\tilde{\boldsymbol{u}} + \boldsymbol{z}) \vee \tilde{\varrho}^{-q_1/\tilde{q}_1} \ge (1/C_1)\rho(\boldsymbol{u} + \boldsymbol{z}) \tag{35}$$

for some $C_1 > 0$ independent of $\tilde{\varrho} > 1, \boldsymbol{u}, \boldsymbol{z} \in \mathbb{R}^2$, proving (34), taking into account the definition of $g_{\tilde{\varrho},\boldsymbol{z}}(\boldsymbol{u})$ in (31) and the boundedness of $L_i, i = 1, 2$.

Let us prove (33). Towards this aim, split $I_{\tilde{\varrho},\boldsymbol{z}} = I_1(\tilde{\varrho}, \boldsymbol{z}) + I_2(\tilde{\varrho}, \boldsymbol{z}) + I_{12}(\tilde{\varrho}, \boldsymbol{z})$, where

$$I_1 := \int_{\{\varrho(\boldsymbol{u})<\delta\}} \cdots, \quad I_2 := \int_{\{\varrho(\boldsymbol{u}+\boldsymbol{z})<\delta\}} \cdots,$$
$$I_{12} := \int_{\{\varrho(\boldsymbol{u})>\delta\}\cap\{\varrho(\boldsymbol{u}+\boldsymbol{z})<\delta\}} \cdots$$

and $\delta > 0$ is small enough. By (34), (35), (14), and $1 < Q < 2$, we infer that $I_1(\tilde{\varrho}, \boldsymbol{z}) \le C \int_{\{\varrho(\boldsymbol{u})<\delta\}} \rho(\boldsymbol{u})^{-1}\mathrm{d}\boldsymbol{u} \to 0$ and $I_2(\tilde{\varrho}, \boldsymbol{z}) \le C \int_{\{\varrho(\boldsymbol{u}+\boldsymbol{z})<\delta\}} \rho(\boldsymbol{u}+\boldsymbol{z})^{-1}\mathrm{d}\boldsymbol{u} \to 0$ as $\delta \to 0$ uniformly in $\tilde{\varrho} > 1, \boldsymbol{z} \in \mathbb{R}^2$. On the other hand, by (32), the integrand of $I_{12}(\tilde{\varrho}, \boldsymbol{z})$ tends to zero and is dominated by $C\rho(\boldsymbol{u})^{-1}\rho(\boldsymbol{u} + \boldsymbol{z})^{-1}$, referring to (34), which is integrable on $\{\varrho(\boldsymbol{u}) > \delta\} \cap \{\varrho(\boldsymbol{u} + \boldsymbol{z}) < \delta\}$. Hence, $I_{12}(\tilde{\varrho}, \boldsymbol{z}) \to 0$ and $I_{\tilde{\varrho},\boldsymbol{z}} \to 0$ as $\tilde{\varrho} \to \infty$, for any $\boldsymbol{z} \in \mathbb{R}^2, \tilde{\rho}(\boldsymbol{z}) = 1$. The uniform convergence in (33) follows by the continuity of $g_{\tilde{\varrho},\boldsymbol{z}}$ in (31) and the limit function in (32) in $\boldsymbol{z} \in \mathbb{R}^2, \tilde{\rho}(\boldsymbol{z}) = 1$.

It remains to show that $0 \le \ell_1 = \ell_2 =: \ell$ imply strict positivity of L_{12} in (26). For $\boldsymbol{u} = (u_1, u_2) \in \mathbb{R}^2_0$, write $\overline{\boldsymbol{u}} = (\bar{u}_1, \bar{u}_2)$, where $\bar{u}_i := u_i/\rho(\boldsymbol{u})^{1/q_i}, i = 1, 2$, as in (9). We have $L_{12}(\boldsymbol{t}) = \int_{\mathbb{R}^2} \ell(\overline{\boldsymbol{u}})\ell(\overline{\boldsymbol{u} + \boldsymbol{t}})(\rho(\boldsymbol{u})\rho(\boldsymbol{u} + \boldsymbol{t}))^{-1}\mathrm{d}\boldsymbol{u}, \boldsymbol{t} \in \mathcal{S}_{\tilde{\rho}}$. Since $\ell \not\equiv 0$, there exist $\boldsymbol{u}_0 \in \mathbb{R}^2, \rho(\boldsymbol{u}_0) = 1$ and $\delta > 0$ such that $\ell(\overline{\boldsymbol{u}}) > \delta$ on the set $U_\delta(\boldsymbol{u}_0) := \{\boldsymbol{u} \in \mathbb{R}^2_0 : |\overline{\boldsymbol{u}} - \overline{\boldsymbol{u}_0}| < \delta\}$. Moreover, since $|\overline{\boldsymbol{u} + \boldsymbol{t}} - \overline{\boldsymbol{u}}| \to 0$ uniformly in $\boldsymbol{t} \in \mathcal{S}_{\tilde{\rho}}$ as $|\boldsymbol{u}| \to \infty$, by the continuity of ℓ, we get that $\ell(\overline{\boldsymbol{u} + \boldsymbol{t}}) > \delta/2$ for any $\boldsymbol{t} \in \mathbb{R}^2$ and all $\boldsymbol{u} \in U_\delta(\boldsymbol{u}_0), |\boldsymbol{u}| > K$ provided $K = K(\delta) > 0$ is large enough. Consequently, $L_{12}(\boldsymbol{t}) \ge (\delta^2/2) \int_{U_\delta(\boldsymbol{u}_0)\cap\{|\boldsymbol{u}|>K\}} (\rho(\boldsymbol{u})\rho(\boldsymbol{u} + \boldsymbol{t}))^{-1}\mathrm{d}\boldsymbol{u} > 0$, proving part (ii) and Proposition 4. $\square$

Fourier transform. The Fourier transform

$$\widehat{a}_\infty(\boldsymbol{x}) = \int_{\mathbb{R}^2} \mathrm{e}^{\mathrm{i}\langle \boldsymbol{x}, \boldsymbol{t}\rangle} a_\infty(\boldsymbol{t}) \mathrm{d}\boldsymbol{t}, \quad \boldsymbol{x} \in \mathbb{R}^2 \tag{36}$$

of a generalised homogeneous function

$$a_\infty(\boldsymbol{t}) = \rho(\boldsymbol{t})^{-1} \ell(\boldsymbol{t}), \quad \boldsymbol{t} \in \mathbb{R}^2_0, \tag{37}$$

with $1 < Q < 2$, and a bounded $\ell(\cdot)$ can be defined by splitting $a_\infty(\boldsymbol{t}) = a_\infty(\boldsymbol{t})\mathbb{I}(\rho(\boldsymbol{t}) \le 1) + a_\infty(\boldsymbol{t})\mathbb{I}(\rho(\boldsymbol{t}) > 1) =: a^-_\infty(\boldsymbol{t}) + a^+_\infty(\boldsymbol{t})$, where $a^-_\infty \in L^1(\mathbb{R}^2), a^+_\infty \in L^2(\mathbb{R}^2)$. Hence, $\widehat{a}_\infty(\boldsymbol{x})$ is defined as a sum of a continuous function $\widehat{a}^-_\infty$ and a $L^2(\mathbb{R}^\nu)$ function $\widehat{a}^+_\infty$. The formal change of variable $\boldsymbol{t} \to (t_1/\rho(\boldsymbol{x})^{1/q_1}, t_2/\rho(\boldsymbol{x})^{1/q_2})$ in the integral (36) leads to the representation

$$\widehat{a}_\infty(\boldsymbol{x}) = \rho(\boldsymbol{x})^{1-Q} \ell_0(\boldsymbol{x}), \tag{38}$$

with the (q_1, q_2)-invariant function

$$\ell_0(\boldsymbol{x}) := \widehat{a}_\infty(\overline{\boldsymbol{x}}), \quad \overline{\boldsymbol{x}} = \left(\frac{x_1}{\rho(\boldsymbol{x})^{1/q_1}}, \frac{x_2}{\rho(\boldsymbol{x})^{1/q_2}}\right) \tag{39}$$

suggesting that $\widehat{a}_\infty$ is a $(q_1(1-Q), q_2(1-Q))$-homogeneous function; see Proposition 2 (ii). However, the above change of variables is not legitimate in general since the integral does not converge absolutely, and the validity of (38) remains open. The following proposition assumes (38) and is not used in the following text but may help in relating the spatial and spectral characteristics of anisotropic RFs.

Proposition 5. *Let $a(\cdot) \in L^2(\mathbb{R}^2) \cap L^\infty(\mathbb{R}^2)$ satisfy*

$$a(\boldsymbol{t}) = \rho(\boldsymbol{t})^{-1}\big(\ell(\boldsymbol{t}) + O(\rho(\boldsymbol{t})^{1-w})\big), \quad |\boldsymbol{t}| \to \infty, \tag{40}$$

where $\rho(\cdot)$ is as in (5), $1 < Q < 2, w > Q$, and $\ell(\cdot)$ is a continuous generalised invariant function. Assume that $\widehat{a}_\infty(\boldsymbol{x})$ in (36) is continuous on $\mathbb{R}^2_0$ and (38) holds. Then,

$$\widehat{a}(\boldsymbol{x}) = \rho(\boldsymbol{x})^{1-Q}\big(\ell_0(\boldsymbol{x}) + o(1)\big), \quad |\boldsymbol{x}| \to 0, \tag{41}$$

where

$$\ell_0(\boldsymbol{x}) := \widehat{a}_\infty(\overline{\boldsymbol{x}}), \quad \overline{\boldsymbol{x}} = \left(\frac{x_1}{\rho(\boldsymbol{x})^{1/q_1}}, \frac{x_2}{\rho(\boldsymbol{x})^{1/q_2}}\right) \tag{42}$$

is a (q_1, q_2)-invariant function. Moreover,

$$|\widehat{a}(\boldsymbol{x})|^2 = \rho_{\rm sp}(\boldsymbol{x})^{-1}\big(L_{\rm sp}(\boldsymbol{x}) + o(1)\big), \tag{43}$$

where

$$\rho_{\rm sp}(\boldsymbol{x}) := |x_1|^{\upsilon_1} + |x_2|^{\upsilon_2}, \quad \upsilon_i := 2q_i(Q-1), \quad i = 1, 2, \tag{44}$$

and $L_{\rm sp}(\boldsymbol{x}) := |\ell_0(\boldsymbol{x})|^2 \rho_{\rm sp}(\boldsymbol{x})/\rho(\boldsymbol{x})^{2(Q-1)}$ is a (υ_1, υ_2)-invariant function.

Proof. Since $a(\cdot) - a_\infty(\cdot) \in L^1(\mathbb{R}^2)$ by (40) and Proposition (3) (15), it suffices to show (41) for $a(\boldsymbol{t}) = a_\infty(\boldsymbol{t})$, which is a direct consequence of (38) and other assumptions of the proposition. The invariance property of $L_{\rm sp}(\boldsymbol{x})$ follows from (11). □

Dependence axis.

Definition 6. Let $g\colon \mathbb{R}^2 \to \mathbb{R}$ be a measurable function. We say that a line passing through the origin and given by $\{\boldsymbol{t} \in \mathbb{R}^2 : \langle \boldsymbol{a}, \boldsymbol{t}\rangle = 0\}$ with $\boldsymbol{a} \in \mathbb{R}^2_0$ is the *dependence axis of g* if for all $\boldsymbol{c} \in \mathbb{R}^2_0$ such that $a_1 c_2 \neq c_1 a_2$,

$$\liminf_{|\boldsymbol{t}|\to\infty,\, \langle \boldsymbol{c},\boldsymbol{t}\rangle = 0} \frac{\log(1/|g(\boldsymbol{t})|)}{\log|\boldsymbol{t}|} > \limsup_{|\boldsymbol{t}|\to\infty,\, \langle \boldsymbol{a},\boldsymbol{t}\rangle = 0} \frac{\log(1/|g(\boldsymbol{t})|)}{\log|\boldsymbol{t}|}. \tag{45}$$

We say that a line $\{\boldsymbol{t} \in \mathbb{R}^2 : \langle \boldsymbol{a}, \boldsymbol{t}\rangle = 0\}$ with $\boldsymbol{a} \in \mathbb{R}^2_0$ is the *dependence axis of $g : \mathbb{Z}^2 \to \mathbb{R}$* if this line is the dependence axis of $g([\boldsymbol{t}])$, $\boldsymbol{t} \in \mathbb{R}^2$.

Proposition 7. *Let $g\colon \mathbb{Z}^2 \to \mathbb{R}$ satisfy*

$$g(\boldsymbol{t}) = \rho(B\boldsymbol{t})^{-1}(\ell(B\boldsymbol{t}) + o(1)), \quad |\boldsymbol{t}| \to \infty, \tag{46}$$

where $B = (b_{ij})_{i,j=1,2}$ is a 2×2 nondegenerate matrix, $\rho(\boldsymbol{t}) = |t_1|^{q_1} + |t_2|^{q_2}$, $\boldsymbol{t} \in \mathbb{R}^2$, $q_i > 0, i = 1, 2$, and $\ell(\boldsymbol{t}) = \ell(t_1, t_2)$ is a (q_1, q_2)-invariant continuous function on $\mathcal{S}_\rho$. In addition, consider the following:

(i) *Let $q_1 < q_2$ and $|\ell(1, 0)| > 0, |\ell(-1, 0)| > 0$. Then, the dependence axis of g is $\{\boldsymbol{t} \in \mathbb{R}^2 : \langle \boldsymbol{b}_2, \boldsymbol{t}\rangle = 0\}$, with $\boldsymbol{b}_2 = (b_{21}, b_{22})$;*

(ii) *Let* $q_1 > q_2$ *and* $|\ell(0,1)| > 0$, $|\ell(0,-1)| > 0$. *Then, the dependence axis of* g *is* $\{\boldsymbol{t} \in \mathbb{R}^2 : \langle \boldsymbol{b}_1, \boldsymbol{t}\rangle = 0\}$ *with* $\boldsymbol{b}_1 = (b_{11}, b_{12})$.

Proof. It suffices to show only part (i) since (ii) is analogous. In what follows, we prove that

$$\lim_{|\boldsymbol{t}|\to\infty,\, \langle \boldsymbol{b}_2, \boldsymbol{t}\rangle = 0} |\boldsymbol{t}|^{q_1}\, |g([\boldsymbol{t}])| = \frac{|\boldsymbol{b}_2|^{q_1}}{|\det(B)|^{q_1}} \begin{cases} \ell(1,0), & \langle \boldsymbol{b}_1, \boldsymbol{t}\rangle \to +\infty, \\ \ell(-1,0), & \langle \boldsymbol{b}_1, \boldsymbol{t}\rangle \to -\infty, \end{cases} \tag{47}$$

$$\limsup_{|\boldsymbol{t}|\to\infty,\, \langle \boldsymbol{c}, \boldsymbol{t}\rangle = 0} |\boldsymbol{t}|^{q_2}\, |g([\boldsymbol{t}])| < \infty, \quad \forall \boldsymbol{c} \in \mathbb{R}^2_0, b_{21}c_2 \neq c_1 b_{22}. \tag{48}$$

Note that relation (47) implies $\lim_{|\boldsymbol{t}|\to\infty,\, \langle \boldsymbol{b}_2, \boldsymbol{t}\rangle = 0} \frac{\log(1/|g([\boldsymbol{t}])|)}{\log|\boldsymbol{t}|} = q_1$, whereas relation (48) implies $\liminf_{|\boldsymbol{t}|\to\infty,\, \langle \boldsymbol{c}, \boldsymbol{t}\rangle = 0} \frac{\log(1/|g([\boldsymbol{t}])|)}{\log|\boldsymbol{t}|} \geq q_2$, hence the statement of the proposition.

Let us prove (47). We have

$$\frac{\langle \boldsymbol{b}_1, \boldsymbol{t}\rangle}{\rho(B[\boldsymbol{t}])^{1/q_1}} = \frac{\operatorname{sign}(\langle \boldsymbol{b}_1, \boldsymbol{t}\rangle)}{(|\langle \boldsymbol{b}_1, [\boldsymbol{t}]\rangle / |\langle \boldsymbol{b}_1, \boldsymbol{t}\rangle||^{q_1} + |\langle \boldsymbol{b}_2, [\boldsymbol{t}]\rangle|^{q_2} / |\langle \boldsymbol{b}_1, \boldsymbol{t}\rangle|^{q_1})^{1/q_1}} \to \pm 1$$
$$\text{as} \quad \langle \boldsymbol{b}_1, \boldsymbol{t}\rangle \to \pm\infty$$

since $|\langle \boldsymbol{b}_2, [\boldsymbol{t}]\rangle| = O(1)$ on $\langle \boldsymbol{b}_2, \boldsymbol{t}\rangle = 0$. In a similar way, $\lim_{|\boldsymbol{t}|\to\infty,\, \langle \boldsymbol{b}_2, \boldsymbol{t}\rangle = 0} |\boldsymbol{t}|^{q_1} \rho(B[\boldsymbol{t}])^{-1} = (\frac{|b_{21}|+|b_{22}|}{|\det(B)|})^{q_1}$. Therefore, (47) follows by the asymptotic form of g and the assumption of the continuity of ℓ.

Consider (48). In view of (46) and the boundedness of ℓ, it suffices to show (48) for $\rho(B\boldsymbol{t})^{-1}$ in place of $g(\boldsymbol{t})$, $\boldsymbol{t} \in \mathbb{Z}^2$. Then, $|\boldsymbol{t}|^{q_2} \rho(B[\boldsymbol{t}])^{-1} = (\frac{|\langle \boldsymbol{b}_1, [\boldsymbol{t}]\rangle|^{q_1}}{|\boldsymbol{t}|^{q_2}} + \frac{|\langle \boldsymbol{b}_2, [\boldsymbol{t}]\rangle|^{q_2}}{|\boldsymbol{t}|^{q_2}})^{-1}$, where $\frac{|\langle \boldsymbol{b}_1, [\boldsymbol{t}]\rangle|^{q_1}}{|\boldsymbol{t}|^{q_2}} \to 0$ and $\frac{|\langle \boldsymbol{b}_2, [\boldsymbol{t}]\rangle|}{|\boldsymbol{t}|} \to \frac{|b_{21}c_2 - b_{22}c_1|}{|c_1|+|c_2|} > 0$, proving (48). □

Let us show that the dependence axis is preserved under the 'discrete' convolution $[g_1 \star g_2](\boldsymbol{t}) = \sum_{\boldsymbol{u}\in\mathbb{Z}^2} g_1(\boldsymbol{u}) g_2(\boldsymbol{u}+\boldsymbol{t})$, $\boldsymbol{t} \in \mathbb{Z}^2$.

Proposition 8. *For* $i = 1, 2$, *let* $g_i : \mathbb{Z}^2 \to \mathbb{R}$ *satisfy*

$$g_i(\boldsymbol{t}) = \rho(B\boldsymbol{t})^{-1}(\ell_i(B\boldsymbol{t}) + o(1)), \quad |\boldsymbol{t}| \to \infty, \tag{49}$$

where B *is a* 2×2 *nondegenerate matrix,* ρ *with* $Q = q_1^{-1} + q_2^{-1} \in (1, 2)$, *and* ℓ_i *are functions as in Proposition 7. For* $i = 1, 2$, *let*

$g_{i\infty}(\boldsymbol{t}) := \rho(\boldsymbol{t})^{-1}\ell_i(\boldsymbol{t})$, $\boldsymbol{t} \in \mathbb{R}^2_0$. *Then,*

$$[g_1 \star g_2](\boldsymbol{t}) = \tilde{\rho}(B\boldsymbol{t})^{-1}(\tilde{L}_{12}(B\boldsymbol{t}) + o(1)), \quad |\boldsymbol{t}| \to \infty, \tag{50}$$

where $\tilde{\rho}(\boldsymbol{t}) := |t_1|^{\tilde{q}_1} + |t_2|^{\tilde{q}_2}$, $\boldsymbol{t} \in \mathbb{R}^2$, *with* $\tilde{q}_i = q_i(2 - Q)$ *as in* (18) *and*

$$\tilde{L}_{12}(\boldsymbol{t}) := |\det(B)|^{-1}(g_{1\infty} \star g_{2\infty})(t_1/\tilde{\rho}(\boldsymbol{t})^{1/\tilde{q}_1}, t_2/\tilde{\rho}(\boldsymbol{t})^{1/\tilde{q}_2}), \quad \boldsymbol{t} \in \mathbb{R}^2_0, \tag{51}$$

is a $(\tilde{q}_1, \tilde{q}_2)$*-invariant function. Moreover, if* $\ell_1 = \ell_2 \geq 0$, *then* $\tilde{L}_{12}$ *is strictly positive.*

Proof. We follow the proof of Proposition 4 (ii). For $\boldsymbol{t} \in \mathbb{Z}^2$, split every $g_i(\boldsymbol{t})$ as a sum of $g_i^1(\boldsymbol{t}) := g_i(\boldsymbol{t}) - g_i^0(\boldsymbol{t})$ and $g_i^0(\boldsymbol{t}) := (1 \vee \rho(B\boldsymbol{t}))^{-1}\ell_i(B\boldsymbol{t})$ using the convention $g_i^0(\boldsymbol{0}) = \ell_i(\boldsymbol{0}) := 0$. Then, $[g_1 \star g_2](\boldsymbol{t}) = \sum_{k,j=0}^{1}[g_1^k \star g_2^j](\boldsymbol{t})$ and (50) follows from

$$\lim_{|\boldsymbol{t}|\to\infty} \left|\tilde{\rho}(B\boldsymbol{t})[g_1^0 \star g_2^0](\boldsymbol{t}) - \tilde{L}_{12}(B\boldsymbol{t})\right| = 0 \tag{52}$$

and

$$\tilde{\rho}(B\boldsymbol{t})[g_1^k \star g_2^j](\boldsymbol{t}) = o(1), \quad |\boldsymbol{t}| \to \infty, \quad (k, j) \neq (0, 0). \tag{53}$$

To prove (52) we write the 'discrete' convolution as the integral $[g_1^0 \star g_2^0](\boldsymbol{t}) = \int_{\mathbb{R}^2} g_1^0(\lceil \boldsymbol{u} \rceil)g_2^0(\lceil \boldsymbol{u} \rceil + \boldsymbol{t})\mathrm{d}\boldsymbol{u}$, where we change a variable: $\boldsymbol{u} \to B^{-1}R_{\tilde{\varrho}}\boldsymbol{u}$ with

$$\boldsymbol{t}' := B\boldsymbol{t}, \quad \tilde{\varrho} := \tilde{\rho}(\boldsymbol{t}'), \quad R_{\tilde{\varrho}} := \mathrm{diag}(\tilde{\varrho}^{1/\tilde{q}_1}, \tilde{\varrho}^{1/\tilde{q}_2}).$$

Then, with $\tilde{Q} := \tilde{q}_1^{-1} + \tilde{q}_2^{-1}$, we have that

$$\begin{aligned}\tilde{\rho}(B\boldsymbol{t})[g_1^0 \star g_2^0](\boldsymbol{t}) &= |\det(B)|^{-1}\tilde{\varrho}^{1+\tilde{Q}} \int_{\mathbb{R}^2} \frac{\ell_1(B\lceil B^{-1}R_{\tilde{\varrho}}\boldsymbol{u}\rceil)}{\rho(B\lceil B^{-1}R_{\tilde{\varrho}}\boldsymbol{u}\rceil) \vee 1} \\ &\quad \cdot \frac{\ell_2(B\lceil B^{-1}R_{\tilde{\varrho}}\boldsymbol{u}\rceil + \boldsymbol{t}')}{\rho(B\lceil B^{-1}R_{\tilde{\varrho}}\boldsymbol{u}\rceil + \boldsymbol{t}') \vee 1}\mathrm{d}\boldsymbol{u} \\ &= |\det(B)|^{-1} \int_{\mathbb{R}^2} g_{\tilde{\varrho},\boldsymbol{z}}(\boldsymbol{u})\mathrm{d}\boldsymbol{u}, \quad \boldsymbol{z} = R_{\tilde{\varrho}}^{-1}\boldsymbol{t}',\end{aligned}$$

where for all $\tilde{\rho} > 0$, $\boldsymbol{z} \in \mathbb{R}^2$ such that $\tilde{\rho}(\boldsymbol{z}) = 1$, $\boldsymbol{u} \in \mathbb{R}^2$,

$$g_{\tilde{\varrho},\boldsymbol{z}}(\boldsymbol{u}) := \frac{\ell_1(R_{\tilde{\varrho}}^{-1}B\lceil B^{-1}R_{\tilde{\varrho}}\boldsymbol{u}\rceil)}{\rho(R_{\tilde{\varrho}}^{-1}B\lceil B^{-1}R_{\tilde{\varrho}}\boldsymbol{u}\rceil) \vee \tilde{\varrho}^{-q_1/\tilde{q}_1}}$$
$$\cdot \frac{\ell_2(R_{\tilde{\varrho}}^{-1}B\lceil B^{-1}R_{\tilde{\varrho}}\boldsymbol{u}\rceil + \boldsymbol{z})}{\rho(R_{\tilde{\varrho}}^{-1}B\lceil B^{-1}R_{\tilde{\varrho}}\boldsymbol{u}\rceil + \boldsymbol{z}) \vee \tilde{\varrho}^{-q_1/\tilde{q}_1}},$$

and we used the generalised homogeneous and generalised invariance properties of ρ and ℓ_i, $i = 1, 2$, and the facts that $q_1/\tilde{q}_1 = q_2/\tilde{q}_2$, $1 + \tilde{Q} = 2q_1/\tilde{q}_1 = 2/(2 - Q)$. Therefore, using the continuity of ρ and ℓ_i, $i = 1, 2$, it follows that $g_{\tilde{\varrho},\boldsymbol{z}}(\boldsymbol{u}) - a_{\infty,1}(\boldsymbol{u})a_{\infty,2}(\boldsymbol{u} + \boldsymbol{z}) \to 0$ as $\tilde{\varrho} \to \infty$ or $|\boldsymbol{t}| \to \infty$ for all $\boldsymbol{u} \in \mathbb{R}^2_0$, $\boldsymbol{u} + \boldsymbol{z} \in \mathbb{R}^2_0$. Then, similarly as in Ref. [209, (7.8)], we conclude that $\sup_{\boldsymbol{z}\in\mathbb{R}^2:\rho(\boldsymbol{z})=1} | \int_{\mathbb{R}^2} g_{\tilde{\varrho},\boldsymbol{z}}(\boldsymbol{u})\mathrm{d}\boldsymbol{u} - (a_{\infty,1}\star a_{\infty,2})(\boldsymbol{z})| \to 0$, $\tilde{\varrho} \to \infty$, and (52) holds. The remaining details, including the proof of (53) are similar to those in Proposition 4. Proposition 8 is proved. □

2 Anisotropic Spatial and Spectral Dependence Characteristics. Examples

Assumption (A)$_{\mathrm{LRD},\rho}$. The coefficients in (1) have the form

$$a(\boldsymbol{t}) = \rho(\boldsymbol{t})^{-1}\big(\ell(\boldsymbol{t}) + o(1)\big), \quad |\boldsymbol{t}| \to \infty, \tag{54}$$

where $\rho(\boldsymbol{t})$ as in (5), $1 < Q < 2$, and $\ell(\boldsymbol{t}), \boldsymbol{t} \in \mathbb{R}^2_0$, is a bounded continuous generalised invariant function.

Assumption (A)$_{\mathrm{SRD}}$. The coefficients in (1) satisfy $\sum_{\boldsymbol{t}\in\mathbb{Z}^2} |a(\boldsymbol{t})| < \infty$ and $\sum_{\boldsymbol{t}\in\mathbb{Z}^2} a(\boldsymbol{t}) \neq 0$.

Assumption (A)$_{\mathrm{ND},\rho}$. The coefficients in (1) have the form as in (54) with $0 < Q < 1$ and the same $\rho(\boldsymbol{t})$ and $\ell(\boldsymbol{t})$. Moreover,

$$\sum_{\boldsymbol{t}\in\mathbb{Z}^2} a(\boldsymbol{t}) = 0. \tag{55}$$

For $(c_1, c_2) \in \mathbb{R}^2_+$, define

$$Q_{c_1,c_2} := \frac{c_1}{q_1} + \frac{c_2}{q_2}. \tag{56}$$

Then, $Q_{1,1} = Q$. Linear combinations in (56) of the reciprocal parameters q_i in Assumptions $(\mathrm{A})_{\mathrm{LRD},\rho}/(\mathrm{A})_{\mathrm{ND},\rho}$ are used in what follows to characterise the LRD/ND properties in Section 3 of Chapter 2 and scaling limits of the linear RF X in (1).

From Proposition 4 (iii), we have the following corollary.

Corollary 9. *Let X be an MA RF in* (1).

(i) *If Assumption $(A)_{\mathrm{LRD},\rho}$ holds, then*

$$r(\boldsymbol{t}) = \tilde{\rho}(\boldsymbol{t})^{-1}\big(L(\boldsymbol{t}) + o(1)\big), \quad |\boldsymbol{t}| \to \infty, \tag{57}$$

where $\tilde{\rho}(\boldsymbol{t}), L(\boldsymbol{t})$ are defined as in (18), (26), *with $a_{i\infty}(\boldsymbol{t}) \equiv a_\infty(\boldsymbol{t}) := \ell(\boldsymbol{t})/\rho(\boldsymbol{t})$.*

In addition, assume $\ell(\boldsymbol{t}) \geq 0$. Then, X has the coefficient LRD and covariance properties; moreover,

if $Q_{1,1/2} = \frac{1}{q_1} + \frac{1}{2q_2} > 1$ (or $\tilde{q}_1 < 1$), then X has the horizontal covariance LRD property;

if $Q_{1/2,1} = \frac{1}{2q_1} + \frac{1}{q_2} > 1$ (or $\tilde{q}_2 < 1$), then X has the vertical covariance LRD property.

(ii) *If Assumption $(A)_{\mathrm{SRD}}$ holds, then X has the coefficient and covariance SRD properties.*

(iii) *If Assumption $(A)_{\mathrm{ND},\rho}$ holds, then X has the coefficient and covariance ND properties.*

We remark that the 'isotropic' form of MA coefficients used in Assumption $(\mathrm{A})(d;\mathbb{Z}^2)$, Section 2 of Chapter 1, viz.,

$$a(\boldsymbol{t}) \sim \ell\left(\frac{\boldsymbol{t}}{|\boldsymbol{t}|}\right)|\boldsymbol{t}|^{2-2d}, \quad |\boldsymbol{t}| \to \infty, \tag{58}$$

is a particular case of (54) with $q_1 = q_2 = q := 2 - 2d$.

Let us turn to the spectral dependence characteristics. Let

$$\rho_{\mathrm{sp}}(\boldsymbol{x}) = |x_1|^{\upsilon_1} + |x_2|^{\upsilon_2}, \quad \boldsymbol{x} = (x_1, x_2) \in \mathbb{R}^2 \tag{59}$$

be as in (44), where $\upsilon_i > 0, i = 1, 2$, are parameters. Set

$$\Upsilon := \frac{1}{\upsilon_1} + \frac{1}{\upsilon_2}.$$

Assumption (F)$_{\mathrm{LRD},\rho}$. The spectral density f has the form

$$f(\boldsymbol{x}) = \rho_{\mathrm{sp}}(\boldsymbol{x})^{-1} L_{\mathrm{sp}}(\boldsymbol{x})\big(1 + o(1)\big), \quad |\boldsymbol{x}| \to 0, \tag{60}$$

where $\rho_{\mathrm{sp}}(\boldsymbol{x})$ is as in (59), $\Upsilon > 1$ and $L_{\mathrm{sp}}(\boldsymbol{x}) = L_{\mathrm{sp}}(-\boldsymbol{x}), \boldsymbol{x} \in \mathbb{R}^2_+$ is a strictly positive symmetric bounded, continuous generalised invariant function. Moreover, f is bounded on $\{\boldsymbol{x} \in \Pi^2 : |\boldsymbol{x}| > \delta\}$, for any $\delta > 0$.

Assumption (F)$_{\mathrm{SRD}}$. The spectral density $f(\boldsymbol{x}), \boldsymbol{x} \in \Pi^2$ is a bounded function which is continuous at $\boldsymbol{x} = \mathbf{0}$, with $f(\mathbf{0}) := \lim_{\boldsymbol{x}\to\mathbf{0}} f(\boldsymbol{x}) > 0$.

Assumption (F)$_{\mathrm{ND},\rho}$. The spectral density $f(\boldsymbol{x}), \boldsymbol{x} \in \Pi^2$ is a bounded function such that

$$f(\boldsymbol{x}) = \rho_{\mathrm{sp}}(\boldsymbol{x}) L_{\mathrm{sp}}(\boldsymbol{x})\big(1 + o(1)\big), \quad |\boldsymbol{x}| \to 0, \tag{61}$$

where $\rho_{\mathrm{sp}}(\boldsymbol{x})$ is as in (59) and $L_{\mathrm{sp}}(\boldsymbol{x}) = L_{\mathrm{sp}}(-\boldsymbol{x}), \boldsymbol{x} \in \mathbb{R}^2_+$ is a strictly positive, symmetric continuous generalised invariant function.

The *heuristic* relation

$$\upsilon_i = 2q_i|Q-1|, \quad q_i = \upsilon_i \begin{cases} \Upsilon - \frac{1}{2}, & \text{LRD}, \\ \Upsilon + \frac{1}{2}, & \text{ND} \end{cases} \quad i = 1, 2 \tag{62}$$

between the parameters q_1 and q_2 in (A)$_{\mathrm{LRD},\rho}$/(A)$_{\mathrm{ND},\rho}$ and υ_1, υ_2 in (F)$_{\mathrm{LRD},\rho}$/(F)$_{\mathrm{ND},\rho}$ is helpful to relate the scaling results, discussed in Sections 3 and 4 of Chapter 6.

Remark 10. The assumptions of strict positivity and boundedness of $L_{\mathrm{sp}}(\boldsymbol{x})$ in (60) and (61) imply that the spectral density does not vanish nor explode on the coordinate axes outside of the origin. These assumptions are important in the subsequent discussion and exclude spectral densities factorising as $f(\boldsymbol{x}) = f_1(x_1)f_2(x_2)$, as in Example 14 with $f_i, i = 1, 2$, featuring either the spectrum LRD or spectrum ND properties.

Corollary 11. *Let $X = \{X(\boldsymbol{t}); \boldsymbol{t} \in \mathbb{Z}^2\}$ be a stationary RF with spectral density $f(\boldsymbol{x}), \boldsymbol{x} \in \Pi^2$.*

(i) *If Assumption (F)$_{\mathrm{LRD},\rho}$ holds, then X has the spectrum LRD property.*

(ii) *If Assumption* $(\mathrm{F})_{\mathrm{SRD}}$ *holds, then* X *has the spectrum SRD property.*
(iii) *If Assumption* $(\mathrm{F})_{\mathrm{ND},\rho}$ *holds, then* X *has the spectrum ND property.*

Example 12. The spectral density of the fractionally integrated lattice isotropic RF X on $\mathbb{Z}^2$ in Example 18 of Chapter 3 is

$$f(\boldsymbol{x}) = (2\pi)^{-2}2^{2d}|2-\cos(x_1)) - \cos(x_2)|^{-2d}, \quad \boldsymbol{x} \in \Pi^2,$$

referring to (62), which behaves as $(2\pi)^{-2}2^{4d}(x_1^2+x_2^2)^{-2d}$ as $|\boldsymbol{x}| \to 0$. It is easy to see that the above f satisfies Assumption $(\mathrm{F})_{\mathrm{LRD},\rho}$ (respectively, $(\mathrm{F})_{\mathrm{SRD}}$ and $(\mathrm{F})_{\mathrm{ND},\rho}$) in the corresponding cases $0<d<1/2$, $d=0$ and $-1/2<d<0$, with $\upsilon_1=\upsilon_2=4|d|$ and the angular function

$$L_{\mathrm{sp}}(\boldsymbol{x}) := (2\pi)^{-2}2^{4d}\begin{cases}(x_1^{4d}+x_2^{4d})/(x_1^2+x_2^2)^{2d}, & 0<d<1/2,\\ (x_1^2+x_2^2)^{2|d|}/(x_1^{4|d|}+x_2^{4|d|}), & -1/2<d<0.\end{cases}$$

Example 13. The spectral density of the anisotropic fractionally integrated RF X on $\mathbb{Z}^2$ in Example 19 of Chapter 3 is

$$f(\boldsymbol{x}) = (2\pi)^{-2}|1-\mathrm{e}^{-\mathrm{i}x_1}(1-\theta+\theta\cos(x_2))|^{-2d}, \quad \boldsymbol{x} \in \Pi^2, \tag{63}$$

which behaves as const $\big(x_1^2+(\frac{\theta}{2})^2x_2^4\big)^{-d}$ as $|\boldsymbol{x}| \to 0$. One can show that f in (63) Assumption $(\mathrm{F})_{\mathrm{LRD},\rho}$ (respectively, $(\mathrm{F})_{\mathrm{SRD}}$ and $(\mathrm{F})_{\mathrm{ND},\rho}$) in the corresponding cases $0<d<3/4$, $d=0$ and $-3/4<d<0$, with $\upsilon_1=2|d|, \upsilon_2=4|d|$.

Example 14. The spectral density of separately fractionally integrated RF X on $\mathbb{Z}^2$ (Example 20 of Chapter 3) is given by

$$f(\boldsymbol{x}) = (2\pi)^{-2}\prod_{j=1}^{2}|1-\mathrm{e}^{-\mathrm{i}x_j}|^{-2d_j}, \quad \boldsymbol{x} \in \Pi^2, \quad -1/2<d_1,d_2<1/2, \tag{64}$$

which behaves as const $|x_1|^{-2d_1}|x_2|^{-2d_2}$ as $|\boldsymbol{x}| \to 0$. The spectral density in (64) is unbounded if $d_1 \vee d_2 > 0$ and satisfies the spectrum LRD property if $d_1 \wedge d_2 > 0$. In the case of $d_1 \wedge d_2 < 0$, the above spectral density vanishes on both coordinate axes and satisfies the spectrum ND property. We also note that for all $(d_1, d_2) \in (-1/2, 1/2)^2 \backslash \{(0,0)\}$, the product function in (64) does not satisfy Assumptions $(\mathrm{F})_{\mathrm{LRD},\rho}$ or $(\mathrm{F})_{\mathrm{ND},\rho}$.

Example 15. The spectral density

$$f(\boldsymbol{x}) = |x_2|^{\mu} \rho_{\mathrm{sp}}(\boldsymbol{x})^{-1-\frac{\mu}{\upsilon_2}} = L_{\mathrm{sp}}(\boldsymbol{x}) \rho_{\mathrm{sp}}(\boldsymbol{x})^{-1},$$
$$L_{\mathrm{sp}}(\boldsymbol{x}) := \left(\frac{|x_2|}{\rho_{\mathrm{sp}}(\boldsymbol{x})^{1/\upsilon_2}} \right)^{\mu}, \tag{65}$$

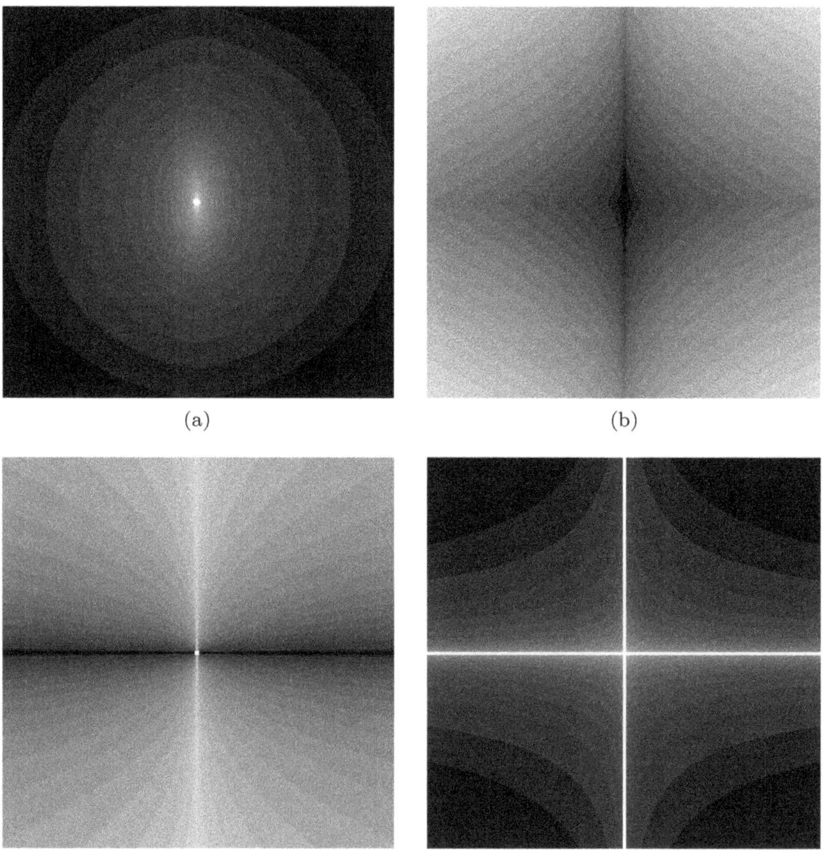

(a) (b)

(d) (c)

Figure 1. Contour graphs of spectral densities: (a) Example 12, $d > 0$, (b) Example 12, $d < 0$, (c) Example 14, $d_1, d_2 > 0$, and (d) Example 15.

depending on three parameters $\upsilon_i > 0, i = 1, 2, \mu \in (0, 1)$, vanishes on the horizontal axis $x_2 = 0$ and explodes on the vertical axis $x_1 = 0$ as $x_1 \to 0$. It may be termed 'long-range negative dependent' (LRND), different from LRD/SRD/ND or product-type densities discussed earlier. Formally, (65) can be written as an LRD density in (60), except that the generalised invariant function $L_{\mathrm{sp}}(\boldsymbol{x})$ in (65) vanishes as a power function with $x_2 \to 0$.

Typical 'low-frequency' behaviours discussed above are illustrated in Fig. 1, presenting the contour graphs of four spectral densities in Examples 12, 14, and 15. The densities in (a) and (b) satisfy, respectively, Assumptions $(\mathrm{F})_{\mathrm{LRD},\rho}$ and $(\mathrm{F})_{\mathrm{ND},\rho}$. The density in (c) factorises into a product of two functions (64). Finally, density in (d) follows Example 15 featuring the 'LRND behaviour'. As will be seen in the following section, MA RFs with spectral densities in Figs. 1(a)–1(d) display very different scaling limits.

3 Anisotropic Scaling of Linear RFs on $\mathbb{Z}^2$: Spectral Approach

Main attention is given to scaling limits under Assumptions $(\mathrm{F})_{\mathrm{LRD},\rho}$ and $(\mathrm{F})_{\mathrm{ND},\rho}$, which depend on 'memory' parameters $\upsilon_i > 0, i = 1, 2$, and the exponent $\gamma > 0$ characterising scaling anisotropy. Under these assumptions, the linear RF in (1) exhibits a scaling transition at a critical point, $\gamma_{\mathrm{cr}} > 0$, coinciding with the intrinsic anisotropy ratio υ_1/υ_2 of the spectral density, except for the very strong negative dependence ('edge effects') occurring when $\upsilon_1 \vee \upsilon_2 > 1$. The unbalanced scaling limits agree with the FBS

$$V_{\pm} = B_{H_1^{\pm}, H_2^{\pm}}, \tag{66}$$

with the Hurst parameters taking a different form in regions $F_{11}, \ldots, F_{33}$ of the parametric quadrant $\{(\upsilon_1, \upsilon_2) \in \mathbb{R}_+^2\}$, as shown in Fig. 2.

It is notable that *in all these regions, at least one of the Hurst parameters* $H_i^{\pm}, i = 1, 2$ *takes values* $1, \frac{1}{2}$, *or* 0. We may conclude that unbalanced scaling leads to either complete dependence ($H = 1$), independence ($H = 1/2$), or complete negative dependence ($H = 0$) in one of the two directions (horizontal or vertical) on the plane.

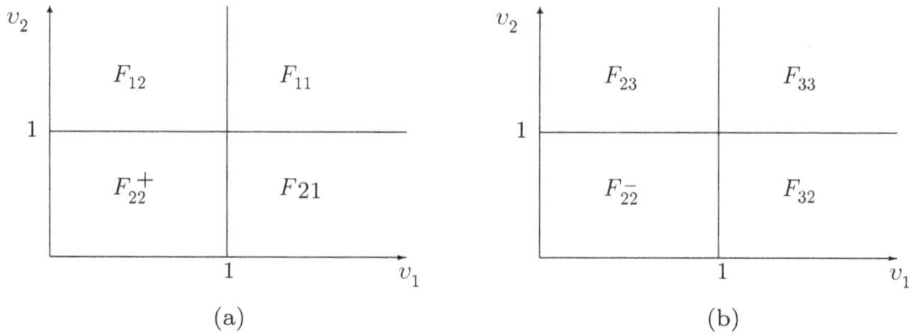

Figure 2. Parameter regions in the spectral domain: (a) $(\mathrm{F})_{\mathrm{LRD},\rho}$ and (b) $(\mathrm{F})_{\mathrm{ND},\rho}$.

Table 1. Unbalanced scaling limits $V_\pm$ in regions $F_{11}, \dots, F_{33}$ in Fig. 2.

Parameter Region	γ_{cr}	V_+	V_-	Range of Hurst Parameters
F_{11}	v_1/v_2	B_{1,H_2^+}	$B_{H_1^-,1}$	$1/2 < H_1^-, H_2^+ < 1$
F_{12}	v_1/v_2	$B_{H_1^+,1/2}$	$B_{H_1^-,1}$	$1/2 < H_1^+, H_1^- < 1$
F_{21}	v_1/v	B_{1,H_2^+}	$B_{1/2,H_2^-}$	$1/2 < H_2^+, H_2^- < 1$
F_{22}^+	v_1/v_2	$B_{H_1^+,1/2}$	$B_{1/2,H_2^-}$	$1/2 < H_1^+, H_2^- < 1$
F_{22}^-	v_1/v_2	$B_{H_1^+,1/2}$	$B_{1/2,H_2^-}$	$0 < H_1^+, H_2^- < 1/2$
F_{23}	v_1	$B_{H_1^+,1/2}$	$B_{1/2,0}$	$0 < H_1^+ < 1/2$
F_{32}	v_2	$B_{0,1/2}$	$B_{1/2,H_2^-}$	$0 < H_2^- < 1/2$
F_{33}	1	$B_{0,1/2}$	$B_{1/2,0}$	

The results for unbalanced scaling are summarised in Table 1 and exclude the boundaries $v_i = 1, i = 1, 2$ of the regions in Fig. 2. The well-balanced limit V_{cr} is also Gaussian and is determined by the limit spectral density

$$f_0(\boldsymbol{x}) := \lim_{\lambda \to 0} \lambda^{-1} f(\lambda^{-1/v_1} x_1, \lambda^{-1/v_2} x_2) \tag{67}$$

$$= \begin{cases} \rho_{sp}(\boldsymbol{x})^{-1} L_{sp}(\boldsymbol{x}), & (v_1, v_2) \in F_{11} \cup F_{12} \cup F_{21} \cup F_{22}^+, \\ \rho_{sp}(\boldsymbol{x}) L_{sp}(\boldsymbol{x}), & (v_1, v_2) \in F_{22}^-, \end{cases} \tag{68}$$

whereas for $(v_1, v_2) \in F_{23} \cup F_{32} \cup F_{33}$, the RF V_{cr} can be written as a linear combination of FBS with indices $H_1 \wedge H_2 = 0$.

Let us turn to rigorous formulations of the limit results whose proofs can be found in Section 8 of Chapter 7. The Hurst indices in Table 1 can be written in a compact way as

$$H_1^+ := \frac{1}{2}(1 + (\upsilon_1 \wedge 1)), \quad H_2^+ := \frac{1}{2}\left(1 + \upsilon_2 - \frac{\upsilon_2}{\upsilon_1 \vee 1}\right), \tag{69}$$
$$H_1^- := \frac{1}{2}\left(1 + \upsilon_1 - \frac{\upsilon_1}{\upsilon_2 \vee 1}\right), \quad H_2^- := \frac{1}{2}(1 + (\upsilon_2 \wedge 1)),$$

The normalisation $d_{\lambda,\gamma}$ in Theorems 16 and 17 is written as

$$d_{\lambda,\gamma} = \lambda^{H(\gamma)}, \tag{70}$$

with $H(\gamma) > 0$ being a linear combination of $H_i^\pm, i = 1, 2$ in (69).

With (67) in mind, we set the well-balanced limit

$$V_0(\boldsymbol{x}) := \int_{\mathbb{R}^2} \prod_{j=1}^{2} \frac{1 - \mathrm{e}^{\mathrm{i}u_j x_j}}{\mathrm{i}u_j} \sqrt{f_0(\boldsymbol{u})} Z(\mathrm{d}\boldsymbol{u}), \quad \boldsymbol{x} \in \mathbb{R}_+^2, \tag{71}$$

as a stochastic integral w.r.t. a Gaussian complex-valued white noise, $Z(\mathrm{d}\boldsymbol{u}), \boldsymbol{u} \in \mathbb{R}^2$, with zero mean and $\mathrm{E}|Z(\mathrm{d}\boldsymbol{u})|^2 = \mathrm{d}\boldsymbol{u}$.

Theorem 16. *Let X in* (3) *be a stationary linear RF on $\mathbb{Z}^2$ with zero mean and a spectral density f satisfying Assumption $(F)_{\mathrm{LRD},\rho}$, $\upsilon_i \neq 1, i = 1, 2$. Then, we have the following:*

- *The scaling limits in* (2) *exist for any $\gamma > 0$ featuring a scaling transition at $\gamma_{\mathrm{cr}} = \frac{\upsilon_1}{\upsilon_2}$ with unbalanced limits given by*

$$V_\pm \overset{\mathrm{fdd}}{=} \kappa_\pm B_{H_1^\pm, H_2^\pm}, \tag{72}$$

where $H_i^\pm, i = 1, 2$ are as in (69) *and $\kappa_\pm = \kappa_\pm(\upsilon_1, \upsilon_2) > 0$ are some constants.*

- *The well-balanced limit $V_{\mathrm{cr}} = V_0$ is given in* (71).
- *The normalisation in* (2) *is given by* (70) *with*

$$H(\gamma) = \begin{cases} H_1^+ + \gamma H_2^+, & \gamma > \gamma_{\mathrm{cr}}, \\ \frac{1}{2}(1 + \upsilon_1 + \frac{\upsilon_1}{\upsilon_2}), & \gamma = \gamma_{\mathrm{cr}}, \\ H_1^- + \gamma H_2^-, & \gamma < \gamma_{\mathrm{cr}}. \end{cases} \tag{73}$$

The asymptotic constants $\kappa_\pm$ in (72) are written in the proof of Theorem 16 (see (146) and (152)) or can be determined from these expressions by symmetry. The fact that $\kappa_\pm > 0$ is due to $L_{\rm sp}(1,0) > 0, L_{\rm sp}(0,1) > 0$. We note that the scaling exponent $H(\gamma)$ in (73) is a continuous (albeit non-analytic) function of $\gamma, \upsilon_i, i = 1,2$; particularly, $H(\gamma_{\rm cr}) = \lim_{\gamma \to \gamma_{\rm cr}} H(\gamma)$.

Next, we turn to (2) for the ND RF X satisfying Assumption $({\rm F})_{{\rm ND},\rho}$.

Theorem 17. *Let X in (3) be a stationary linear RF on $\mathbb{Z}^2$ with zero mean and a spectral density f satisfying Assumption $(F)_{{\rm ND},\rho}, \upsilon_i \neq 1, i = 1,2$. In addition, assume that f is continuous on Π^2 when $\upsilon_1 \vee \upsilon_2 > 1$. Then, we have the following:*

- *The scaling limits in (2) exist for any $\gamma > 0$ and satisfy (4), with $\gamma_{\rm cr}, H_i^\pm, i = 1,2$ in Table 1 and (69), and*

$$V_+ = \kappa_+ B_{H_1^+,1/2}, \quad V_- = \kappa_- B_{1/2,H_2^-}. \tag{74}$$

- *The well-balanced limit is given by*

$$V_{\rm cr} = \begin{cases} V_0, & H_1^+ \wedge H_2^- > 0, \\ \kappa_+ B_{H_1^+,1/2} + \kappa_- B_{1/2,H_2^-}, & H_1^+ \wedge H_2^- = 0, \end{cases} \tag{75}$$

  where $B_{H_1^+,1/2}$ and $B_{1/2,H_2^-}$ are independent FBS and V_0 is defined in (71).
- *The normalisation in (2) is given by (70) with*

$$H(\gamma) = \begin{cases} H_1^+ + (\gamma/2), & \gamma < \gamma_{\rm cr}, \\ \frac{1}{2}(1 - (\upsilon_1 \wedge 1) + \gamma_{\rm cr}) & \gamma = \gamma_{\rm cr}, \\ (1/2) + \gamma H_2^-, & \gamma > \gamma_{\rm cr}. \end{cases} \tag{76}$$

- *The RF X exhibits the scaling transition at $\gamma_{\rm cr}$ unless $\kappa_\pm = 0$.*

The constants $\kappa_\pm \geq 0$ in (74) and (75) are given in (153), (158), and (162), or they can be determined from these expressions by symmetry. In most cases, they depend on $\upsilon_i, i = 1,2$ and the angular function $L_{\rm sp}$ alone, except for the case when $H_1^+ \wedge H_2^- = 0$ when they involve the spectral density outside of the origin $\mathbf{0}$; see (158) and (162). Particularly, κ_+/κ_- may vanish when f vanishes on the horizontal/vertical axis, leading to the trivial scaling limits in respective situations. Vanishing of the spectral density on coordinate axes is a stronger form of ND, which may change the results in Theorem 17 and is left open in this book.

Remark 18. Theorems 16 and 17 extend to parameter values $v_i = 1\,(\exists i = 1,2)$ (boundaries of the regions in Fig. 1) [252]. The normalisation $d_{\lambda,\gamma}$ in (2) in such a case may contain an additional logarithmic factor and the corresponding asymptotic constants $\kappa_\pm$ take a different form. The proofs of these results are more involved and can be found in Ref. [252].

'Edge effects.' In spatial statistics, 'edge effects' refer to the situation when the contribution to a statistic from observations near the boundary of the sample region is non-negligible and may affect or even dominate the limit distribution of the statistic itself. For linear ND RFs, edge effects were studied in Refs. [145, 251]. Edge effects also appear in Theorems 17 and 21 under unbalanced scaling when the corresponding Hurst indices vanish: $H_i^\pm = 0\,(\exists i = 1,2)$. Indeed, since the normalisation $d^2_{\lambda,\gamma}$ in our theorems grows proportionally to the variance $\mathrm{Var}(S_{\lambda,\gamma}(\mathbf{1}))$, we see from (76) that when $H_1^+ = 0, \gamma < \gamma_{\mathrm{cr}}$, this variance increases as the horizontal length $[\lambda]$ of the rectangle (sampling domain), and similarly, when $H_2^- = 0, \gamma > \gamma_{\mathrm{cr}}$, it grows proportionally to the vertical length $[\lambda^\gamma]$ of the rectangle. Thus, we can speak about a 'horizontal edge effect' in the first case and a 'vertical edge effect' in the second case. Similar horizontal and vertical edge effects appear under spatial-domain ND conditions in the following section.

Other types of spectral dependence. Next, we discuss scaling limits in (2) for some other classes of linear RFs X on $\mathbb{Z}^2$, including spectrum SRD and spectral densities in Examples 14 and 15. First, consider the case of a bounded spectral density of the form

$$f(\boldsymbol{x}) = L_{\mathrm{sp}}(\boldsymbol{x})(1 + o(1)), \quad |\boldsymbol{x}| \to 0, \tag{77}$$

where L_{sp} is a strictly positive and continuous generalised invariant function, as in (60), depending on $v_i > 0, i = 1,2$. If $L_{\mathrm{sp}}(\boldsymbol{x}) \equiv \ell > 0$ is a constant function, then f belongs to the class $(\mathrm{F})_{\mathrm{SRD}}$ and

$f(\mathbf{0}) = \ell$, else f in (77) is not continuous at $\boldsymbol{x} = \mathbf{0}$ and form a more general class of spectral densities. Let $\gamma_0 := \frac{\upsilon_1}{\upsilon_2}$. We have

$$f_0(\boldsymbol{x}) := \lim_{\lambda \to 0} f(\lambda^{-1}x_1, \lambda^{-\gamma}x_2) = \begin{cases} L_{\mathrm{sp}}(1,0), & \gamma > \gamma_0, \\ L_{\mathrm{sp}}(0,1), & \gamma < \gamma_0, \\ L_{\mathrm{sp}}(\boldsymbol{x}), & \gamma = \gamma_0. \end{cases} \tag{78}$$

The following proposition shows that for $\gamma \neq \gamma_0$, the limit in (2) is the Brownian Sheet, except that the 'long-run variance' $\kappa^2\pm$ may depend on γ.

Proposition 19. *Let X in (3) be a stationary linear RF on $\mathbb{Z}^2$ with zero mean and a bounded spectral density f satisfying (77). Then, we have the following:*

- *The scaling limits in (2) exist for any $\gamma > 0$ with normalisation $d_{\lambda,\gamma} = \lambda^{(1+\gamma)/2}$ and are given by*

$$V_\gamma = \begin{cases} \kappa_+ B_{1/2,1/2}, & \gamma > \gamma_0, \\ \kappa_- B_{1/2,1/2}, & \gamma < \gamma_0, \\ V_0, & \gamma = \gamma_0, \end{cases} \tag{79}$$

where $\kappa_+^2 := L_{\mathrm{sp}}(1,0) \int_{\mathbb{R}^2} \prod_{j=1}^2 |(1-\mathrm{e}^{\mathrm{i}u_j})/u_j|^2 \mathrm{d}\boldsymbol{u}$, $\kappa_-^2 := L_{\mathrm{sp}}(0,1) \int_{\mathbb{R}^2} \prod_{j=1}^2 |(1-\mathrm{e}^{\mathrm{i}u_j})/u_j|^2 \mathrm{d}\boldsymbol{u}$ and V_0 is defined as in (71), with $f_0(\boldsymbol{x})$ given in (78).
- *The RF X does not exhibit scaling transition if $\kappa_\pm > 0$.*

Next, we consider a linear RF X on $\mathbb{Z}^2$ with spectral density

$$f(\boldsymbol{x}) = \ell \prod_{i=1}^{2} |x_i|^{-\upsilon_i} \big(1 + o(1)\big), \quad |\boldsymbol{x}| \to 0, \tag{80}$$

where $\ell > 0, \upsilon_i \in (-1,0) \cup (0,1), i = 1,2$, are parameters. The above $f(\boldsymbol{x})$ include separately fractionally integrated spectral densities of Example 14 and may explode and vanish on the coordinate axes when $|\boldsymbol{x}| \to 0$. Let

$$H_i := \frac{1}{2}(1 + \upsilon_i), \quad i = 1,2. \tag{81}$$

Proposition 20. *Let X be a stationary linear RF in* (1) *with zero mean and a spectral density f as in* (80)*. Then, the scaling limits in* (2) *exist for any $\gamma > 0$ $d_{\lambda,\gamma} = \lambda^{H(\gamma)}, H(\gamma) = H_1 + \gamma H_2$ and do not depend on γ:*

$$V_\gamma = \kappa B_{H_1,H_2},$$

where $\kappa^2 := \ell \int_{\mathbb{R}^2} \prod_{j=1}^{2} |(1 - \mathrm{e}^{\mathrm{i}u_j})/u_j|^2 |u_j|^{1-2H_j} \mathrm{d}\boldsymbol{u}$. Particularly, the above RF X does not exhibit a scaling transition.

Proposition 20 can be generalised to include a nonconstant angular function as in (78), in which case the limit V_γ may depend on γ, similarly as in Proposition 19. As noted earlier, these results heavily rely on the strict positivity condition of the angular function L_{sp}. Violation of the last condition may lead to different scaling limits, as shown in the following proposition. Consider the spectral density as in Example 15, (65). Define

$$\begin{aligned}
H_1^+ &:= \frac{1}{2}\left(1 + \left(\left(\upsilon_1 + \frac{\mu\upsilon_1}{\upsilon_2}\right) \wedge 1\right)\right), \\
H_2^+ &:= \frac{1}{2}\left(1 - \left(\mu \wedge \left(\frac{\upsilon_2}{\upsilon_1} - \upsilon_2\right)\right)\right), \\
H_1^- &:= \frac{1}{2}\left(1 + \upsilon_1 - \frac{\upsilon_1}{\upsilon_2 \vee 1}\right), \quad H_2^- := \frac{1}{2}\left(1 + (\upsilon_2 \wedge 1)\right) \\
H(\gamma) &:= \begin{cases} H_1^+ + \gamma H_2^+, & \gamma > \gamma_{\mathrm{cr}}, \\ H_1^- + \gamma H_2^-, & \gamma < \gamma_{\mathrm{cr}}, \\ 1 + \upsilon_1 + \frac{\upsilon_1}{\upsilon_2}, & \gamma = \gamma_{\mathrm{cr}}, \end{cases} \qquad \gamma_{\mathrm{cr}} := \frac{\upsilon_1}{\upsilon_2}.
\end{aligned} \tag{82}$$

Note that $H_i^- \in [\frac{1}{2}, 1], i = 1, 2$ in (82) are the same as in Theorem 16, (69), whereas the expressions for $H_i^+, i = 1, 2$, in (82) agree with those in (69) if and only if $\mu = 0$. Also, note that for $\upsilon_1 < 1$, we have $H_1^+ \in [\frac{1}{2}, 1]$ and $H_2^+ \in (0, \frac{1}{2})$. A closer inspection of (82) shows that (H_1^+, H_2^+) *can assume any values from* $(0, \frac{1}{2}) \times (\frac{1}{2}, 1)$ when υ_1, μ vary over $(0, 1)$ and $\upsilon_2 > 0$ is fixed. Finally, we remark that $H(\gamma)$ in (82) is a continuous function of $\gamma, \upsilon_i, i = 1, 2, \mu$, and its value at $\gamma = \gamma_{\mathrm{cr}}$ is the same as in Theorem 16 independently of μ.

Proposition 21. *Let X in (3) be a linear RF on $\mathbb{Z}^2$ with zero mean and a spectral density f as in Example 15, (65). In addition, suppose that $v_1 \neq 1, v_1 + \frac{v_2}{v_1} \neq 1$. Then, we have the following:*

- *The scaling limits in (2) exist for any $\gamma > 0$ under normalisation $d_{\lambda,\gamma} = \lambda^{H(\gamma)}$, featuring a scaling transition with $\gamma_{\rm cr}, H(\gamma)$ in (82) and the unbalanced limits*

$$V_+ = \kappa_+ B_{H_1^+, H_2^+}, \quad V_- = \kappa_- B_{H_1^-, H_2^-}, \tag{83}$$

where $H_i^{\pm}, i = 1, 2,$ as in (82) and $\kappa_{\pm} > 0$ are some constants.
- *The well-balanced limit $V_{\rm cr} = V_0$ is given in (71), with $f_0(\boldsymbol{u})$ replaced by $f(\boldsymbol{u}), \boldsymbol{u} \in \mathbb{R}_0^2$ in (65).*

4 Anisotropic Scaling of Linear RFs on $\mathbb{Z}^2$: Spatial Domain Approach

This section discusses scaling limits in (2) for linear RF X on $\mathbb{Z}^2$ under spatial domain assumptions, particularly Assumptions $(\mathrm{A})_{\mathrm{LRD},\rho}/(\mathrm{A})_{\mathrm{ND},\rho}$ defined in Section 2. Recall that the MA coefficients of X in these assumptions have the same asymptotic form

$$a(\boldsymbol{t}) = \rho(\boldsymbol{t})^{-1}\big(\ell(\boldsymbol{t}) + o(1)\big), \quad |\boldsymbol{t}| \to \infty, \tag{84}$$

with $\rho(\boldsymbol{t}) = |t_1|^{q_1} + |t_2|^{q_2}$ and $\ell(\boldsymbol{t})$ being a generalised invariant function, except that $Q \in (1, 2)$ in the LRD case and $Q < 1$, $\sum_{\boldsymbol{t} \in \mathbb{Z}^2} a(\boldsymbol{t}) = 0$ in the ND case, $Q = \frac{1}{q_1} + \frac{1}{q_2}$. Under the correspondence in (62) between spectral and spatial domain parameters, the scaling results are roughly similar as in Section 3 of Chapter 7, except that the proofs are different and more involved. Accordingly, the scaling limits in (2) take a generally different form in the regions $R_{11}, \ldots, R_{33}$ of the parametric quadrant $\{0 < Q < 2\} \subset \mathbb{R}_+^2$ shown in Fig. 3.

By setting

$$v_i = 2q_i|Q - 1|, \quad i = 1, 2, \tag{85}$$

as in (62), we see that the eight regions $R_{11}, \ldots, R_{33}$ in Fig. 3 correspond to the eight regions $F_{11}, \ldots, F_{33}$ in Fig. 2. The main results of this section are Theorems 22 and 23, referring to the LRD and

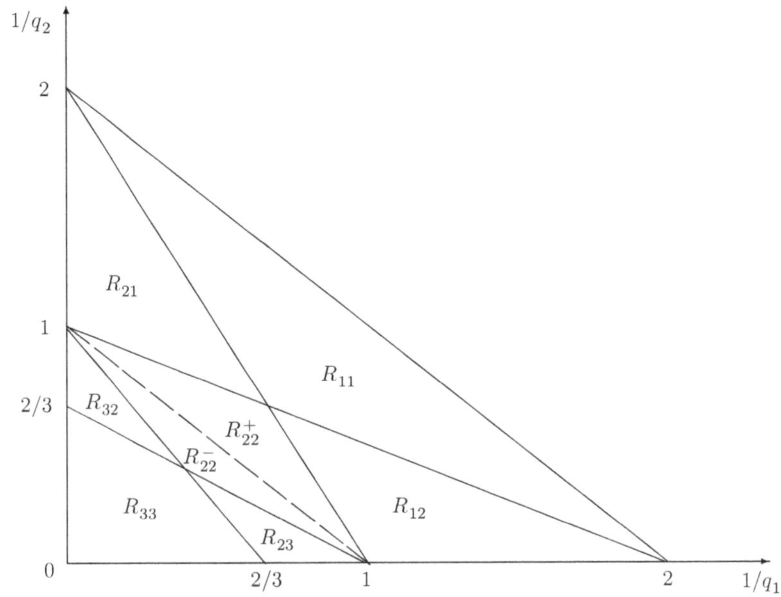

Figure 3. Partition of the parameter set $0 < Q < 2$ in Table 2. The dashed segment separates the LRD and ND regions.

ND parameter regions summarised in Table 2, where the *Hurst parameters,*

$$H_1^+ = \left(\frac{3}{2} + \frac{q_1}{q_2} - q_1\right) \wedge 1, \quad H_2^+ = \left(\frac{3}{2} + \frac{q_2}{2q_1} - q_2\right) \vee \frac{1}{2}, \tag{86}$$
$$H_1^- = \left(\frac{3}{2} + \frac{q_1}{2q_2} - q_1\right) \vee \frac{1}{2}, \quad H_2^- = \left(\frac{3}{2} + \frac{q_2}{q_1} - q_2\right) \wedge 1,$$

and other quantities completely agree with those in Table 1 under the correspondence (85). The proofs of the above results in Section 8 use the 'discrete stochastic integral' approach discussed in Section 3 of Chapter 2. The limit Gaussian RFs are written as stochastic integrals w.r.t. a real-valued Gaussian white noise and involve the generalised homogeneous function

$$a_\infty(\boldsymbol{t}) = \rho(\boldsymbol{t})^{-1}\ell(\boldsymbol{t}), \quad \boldsymbol{t} \in \mathbb{R}_0^2, \tag{87}$$

Table 2. Unbalanced scaling limits $V_\pm$ in regions $R_{11}, \ldots, R_{33}$ in Fig. 3.

Parameter Region	Critical $\gamma_{\rm cr}$	V_+	V_-	Range of Hurst Parameters
R_{11}	q_1/q_2	B_{1,H_2^+}	$B_{H_1^-,1}$	$1/2 < H_1^-, H_2^+ < 1$
R_{12}	q_1/q_2	$B_{H_1^+,1/2}$	$B_{H_1^-,1}$	$1/2 < H_1^+, H_1^- < 1$
R_{21}	q_1/q_2	B_{1,H_2^+}	$B_{1/2,H_2^-}$	$1/2 < H_2^+, H_2^- < 1$
R_{22}^+	q_1/q_2	$B_{H_1^+,1/2}$	$B_{1/2,H_2^-}$	$1/2 < H_1^+, H_2^- < 1$
R_{22}^-	q_1/q_2	$B_{H_1^+,1/2}$	$B_{1/2,H_2^-}$	$0 < H_1^+, H_2^- < 1/2$
R_{23}	$2q_1(1-Q)$	$B_{H_1^+,1/2}$	$B_{1/2,0}$	$0 < H_1^+ < 1/2$
R_{32}	$1/2q_2(1-Q)$	$B_{0,1/2}$	$B_{1/2,H_2^-}$	$0 < H_2^- < 1/2$
R_{33}	1	$B_{0,1/2}$	$B_{1/2,0}$	

Using the notation

$$Q_{c_1,c_2} := \sum_{i=1}^{2} \frac{c_i}{q_i}$$

for the weighted sum of the reciprocal parameters, the boundaries in Fig. 3 correspond to line segments $Q_{1/2,1} = 1, Q_{1,1/2} = 1, Q_{3/2,1} = 1, Q_{1,3/2} = 1$, and $Q_{1,1} = Q = 1$. Note that R_{22}^- can be equivalently defined as

$$(q_1, q_2) \in R_{22}^- \Longleftrightarrow Q_{3/2,1} \wedge Q_{1,3/2} > 1 > Q \Longleftrightarrow 0 < H_1^+, H_2^- < \frac{1}{2}.$$

Long-range dependence. We first consider the LRD case $1 < Q < 2$ or regions R_{11}, R_{12}, R_{21}, and R_{22}^+ in Table 2. Introduce a Gaussian RF written as the stochastic integral

$$V_0(\boldsymbol{x}) := \int_{\mathbb{R}^2} \left\{ \int_{]\boldsymbol{0},\boldsymbol{x}]} a_\infty(\boldsymbol{t} - \boldsymbol{u}) \mathrm{d}\boldsymbol{t} \right\} W(\mathrm{d}\boldsymbol{u}), \quad \boldsymbol{x} \in \mathbb{R}_+^2, \tag{88}$$

w.r.t. Gaussian white noise on $\mathbb{R}^2$ with zero mean and variance $\mathrm{E}W(\mathrm{d}\boldsymbol{u})^2 = \mathrm{d}\boldsymbol{u}$, whose existence is shown in the proof of Theorem 22.

Theorem 22. *Let X in (3) be a stationary linear RF on $\mathbb{Z}^2$ with zero mean, satisfying Assumption (A)$_{\mathrm{LRD},\rho}$, $Q_{1/2,1} \neq 1, Q_{1/2,1} \neq 1$. Then, we have the following:*

- *The scaling limits in* (2) *exist for any* $\gamma > 0$, *exhibiting a scaling transition at* $\gamma_{\rm cr} = \frac{q_1}{q_2}$ *with unbalanced limits given by*

$$V_\pm = \kappa_\pm B_{H_1^\pm, H_2^\pm}, \tag{89}$$

where $H_i^\pm, i = 1, 2$ *as in* (86) *and* $\kappa_\pm > 0$ *are some constants.*
- *The well-balanced limit* $V_{\rm cr} = V_0$ *is given in* (88).
- *The normalisation in* (2) *is given by* $d_{\lambda,\gamma} = \lambda^{H(\gamma)}$, *with*

$$H(\gamma) = \begin{cases} H_1^+ + \gamma H_2^+, & \gamma > \gamma_{\rm cr}, \\ H_1^- + \gamma H_2^-, & \gamma < \gamma_{\rm cr}, \\ \frac{3}{2} + \frac{3}{2}\frac{q_1}{q_2} - q_1, & \gamma = \gamma_{\rm cr}. \end{cases}$$

Negative dependence. Next, we turn to ND or the regions R_{22}^-, R_{23}, R_{32}, and R_{33} in Fig. 3. We define the well-balanced limit RF in the region R_{22}^- as the stochastic integral

$$\begin{aligned} V_0(\boldsymbol{x}) := & \int_{\mathbb{R}^2 \setminus]\mathbf{0}, \boldsymbol{x}]} \left\{ \int_{]\mathbf{0}, \boldsymbol{x}]} a_\infty(\boldsymbol{t} - \boldsymbol{u}) \mathrm{d}\boldsymbol{t} \right\} W(\mathrm{d}\boldsymbol{u}) \\ & - \int_{]\mathbf{0}, \boldsymbol{x}]} \left\{ \int_{\mathbb{R}^2 \setminus]\mathbf{0}, \boldsymbol{x}]} a_\infty(\boldsymbol{t} - \boldsymbol{u}) \mathrm{d}\boldsymbol{t} \right\} W(\mathrm{d}\boldsymbol{u}), \quad \boldsymbol{x} \in \mathbb{R}_+^2, \end{aligned} \tag{90}$$

w.r.t. the same Gaussian white noise $W(\mathrm{d}\boldsymbol{u})$ as in (88). According to Table 2,

$$\gamma_{\rm cr} := \begin{cases} q_1/q_2, & Q_{3/2,1} \wedge Q_{1,3/2} > 1, \\ 1/2q_2(1-Q), & Q_{3/2,1} < 1 < Q_{1,3/2}, \\ 2q_1(1-Q), & Q_{1,3/2} < 1 < Q_{3/2,1}, \\ 1, & Q_{3/2,1} \vee Q_{1,3/2} < 1. \end{cases}$$

Theorem 23. *Let* X *in* (3) *be a stationary MA RF on* $\mathbb{Z}^2$ *with zero mean and coefficients satisfying Assumption* $(A)_{\mathrm{ND},\rho}$, $Q_{3/2,1} \neq 1$, $Q_{1,3/2} \neq 1$. *Then, we have the following:*

- *The scaling limits in* (2) *exist for any* $\gamma > 0$, *with unbalanced limits*

$$V_+ = \kappa_+ B_{H_1^+, 1/2}, \quad V_- = \kappa_- B_{1/2, H_2^-}, \tag{91}$$

where $\gamma_{\rm cr}, H_i^{\pm}, i = 1,2$ *are as in Table 2,* (86), *and* $\kappa_{\pm} \ge 0$ *are some constants.*

- *The well-balanced limit is given by*

$$V_{\rm cr} = \begin{cases} V_0, & H_1^+ \wedge H_2^- > 0, \\ \kappa_+ B_{H_1^+,1/2} + \kappa_- B_{1/2,H_2^-}, & H_1^+ \wedge H_2^- = 0, \end{cases}$$

where $B_{H_1^+,1/2}$ *and* $B_{1/2,H_2^-}$ *are independent FBS and* V_0 *is defined in* (90) *and* $\kappa_{\pm}$ *are the same as in* (91).

- *The normalisation in* (2) *is given by* $d_{\lambda,\gamma} = \lambda^{H(\gamma)}$, *with*

$$H(\gamma) = \begin{cases} H_1^+ + \frac{\gamma}{2}, & \gamma \ge \gamma_{\rm cr}, \\ \frac{1}{2} + \gamma H_2^-, & \gamma \le \gamma_{\rm cr}. \end{cases} \tag{92}$$

- *The RF* X *exhibits scaling transition at* $\gamma_{\rm cr}$ *unless* $\kappa_{\pm} = 0$.

More general sampling regions and edge processes. An interesting and challenging question left open in this book is extending Theorems 22 and 23 to more general sampling regions or test functions as in Section 4, or in other words, replacing the sums $S_{\lambda,\gamma}(\boldsymbol{x})$ in (3) on rectangles by the integrals

$$X_{\lambda,\gamma}(\phi) := \int_{\mathbb{R}^2} X([\boldsymbol{t}])\phi(t_1/\lambda, t_2/\lambda^{\gamma})\mathrm{d}\boldsymbol{t} \tag{93}$$

for a suitable class of test functions $\phi : \mathbb{R}^2 \to \mathbb{R}$, which includes indicator functions $\phi(\boldsymbol{t}) = \mathbb{I}(\boldsymbol{t} \in D)$ of class $\mathcal{D}$ of bounded Borel sets $D \subset \mathbb{R}^2$ with rectifiable boundary ∂D. Particularly, $S_{\lambda,\gamma}(\boldsymbol{x}) = X_{\lambda,\gamma}(\mathbb{I}(\cdot \in]\mathbf{0}, \boldsymbol{x}]))$ correspond to rectangular sampling regions, which may be unrealistic in statistical applications. We conjecture that a normalised limit of $S_{\lambda,\gamma}(D) := X_{\lambda,\gamma}(\mathbb{I}(\cdot \in D))$ may exist for any $\gamma > 0, D \in \mathcal{D}$ and write as formal integrals $\int_D V_\gamma(\boldsymbol{t})\mathrm{d}\boldsymbol{t}$, or $\int_{\partial D} V_\gamma^{\partial D}(\boldsymbol{t})\mathrm{d}\boldsymbol{t}$, or a sum of such integrals, of some generalised Gaussian processes $\{V_\gamma(\boldsymbol{t}); \boldsymbol{t} \in D\}$ and $\{V_\gamma^{\partial D}(\boldsymbol{t}); \boldsymbol{t} \in \partial D\}$, with the latter interpreted as the *edge process on* ∂D of length element $\mathrm{d}\boldsymbol{t}$.

The above rather imprecise conjectures can be clarified and verified for rectangles $D =]\mathbf{0}, \boldsymbol{x}]$. The limits in Theorems 22 and 23, except for regions R_{23}, R_{32}, and R_{33}, can be written as formal integrals $\int_{]\mathbf{0},\boldsymbol{x}]} V_\gamma(\boldsymbol{t})\mathrm{d}\boldsymbol{t}$ of the generalised RF

$$V_\gamma(\boldsymbol{t}) := \begin{cases} \kappa_\pm \partial B_{H_1^\pm, H_2^\pm}(\boldsymbol{t})/\partial t_1 \partial t_2, & \gamma \neq \gamma_{\mathrm{cr}}, \\ \int_{\mathbb{R}^2} a_\infty(\boldsymbol{t}-\boldsymbol{u}) W(\mathrm{d}\boldsymbol{u}), & \gamma = \gamma_{\mathrm{cr}}, \end{cases}$$

with the convention $\int_{\mathbb{R}^2} a_\infty(\boldsymbol{t})\mathrm{d}\boldsymbol{t} = 0$ for $Q < 1$. In the regions R_{32} and R_{33}, the limit $V_\gamma = V_+, \gamma > \gamma_{\mathrm{cr}}$ can be written as $\int_{\partial D} V_+^{\partial D}(\boldsymbol{t})\mathrm{d}\boldsymbol{t}$, where

$$V_+^{\partial D}(\boldsymbol{t}) := \kappa_+ \begin{cases} \mathrm{d}B_{t_2}(t_1)/\mathrm{d}t_1, & 0 < t_1 < x_1, t_2 = x_2, \\ -\mathrm{d}B_0(t_1)/\mathrm{d}t_1, & 0 < t_1 < x_1, t_2 = 0, \\ 0, & \text{elsewhere on } \partial]\mathbf{0},\boldsymbol{x}], \end{cases}$$

where $B_{t_2} = \{B_{t_2}(t); t \geq 0\}, t_2 \geq 0$ is a family of *independent* Brownian motions. A similar interpretation through formal integrals of generalised Gaussian processes are valid for all other limits in Theorems 22 and 23. Accordingly, the edge processes $V_\gamma^{\partial D}$ can be identified with multiples of one-dimensional white noise (a formal derivative of standard Brownian motion) on the edges of the rectangle $]\mathbf{0},\boldsymbol{x}]$; they depend only on γ, as they vanish on the vertical or horizontal edges of $]\mathbf{0},\boldsymbol{x}]$ depending on whether $\gamma > \gamma_{\mathrm{cr}}$ or $\gamma < \gamma_{\mathrm{cr}}$ holds.

Edge processes can also be introduced in discrete time $\boldsymbol{t} \in \mathbb{Z}^2$ and related to the horizontal/vertical 'edge effects' discussed in Section 3 under spectral dependence assumptions. Analogous 'edge effects' appear in Theorem 23 when $H_i^\pm = 0\,(\exists i = 1,2)$ (regions R_{23}, R_{32}, and R_{33}). Of particular interest are the asymptotic constants $\kappa_+ = \sigma_{\mathrm{edge},2}$ ($H_1^+ = 0$, vertical 'edge effect') and $\kappa_- = \sigma_{\mathrm{edge},1}$ ($H_2^- = 0$, horizontal 'edge effect') written as

$$\sigma^2_{\mathrm{edge},1} := 2\sum_{u_2 \geq 0} \Bigg(\sum_{t_1 \in \mathbb{Z}, t_2 \geq 1} a(t_1, t_2 + u_2) \Bigg)^2 + 2\sum_{u_2 \leq -1} \Bigg(\sum_{t_1 \in \mathbb{Z}, t_2 \leq 0} a(t_1, t_2 + u_2) \Bigg)^2, \qquad (94)$$

$$\sigma^2_{\text{edge},2} := 2\sum_{u_1\ge 0}\left(\sum_{t_1\ge 1, t_2\in\mathbb{Z}} a(t_1+u_1,t_2)\right)^2$$
$$+2\sum_{u_1\le -1}\left(\sum_{t_1\le 0, t_2\in\mathbb{Z}} a(t_1+u_1,t_2)\right)^2$$

and involving *all* MA coefficients $a(\boldsymbol{t}), \boldsymbol{t}\in\mathbb{Z}^2$ of the RF, in contrast to other cases of Theorems 23 and 22 when $\kappa_\pm$ are expressed through the asymptotic function a_∞ in (84), (87). As explained in Section 3, 'edge effects' mean that $S_{\lambda,\gamma}(\boldsymbol{x})$ behaves as a sum of weakly dependent r.v.s indexed by points on the edges of the rectangle $K_{[\lambda x_1,\lambda^\gamma x_2]}$. Particularly, when $H_2^- = 0$ (horizontal 'edge effect'), Ref. [251] proves that

$$\mathrm{E}\left(S_{\lambda,\gamma}(\boldsymbol{x}) - \sum_{1\le t_1\le[\lambda x_1]}(U_0(t_1) - U_{[\lambda^\gamma x_2]}(t_1))\right)^2 = o(\lambda^{2H(\gamma)}),$$
$$\text{for}\quad \gamma<\gamma_{\text{cr}},$$

where, for each $t_2\in\mathbb{Z}$, the 'horizontal process' $U_{t_2} = \{U_{t_2}(t_1); t_1\in\mathbb{Z}\}$ 'located' on the line $\{(t_1,t_2)\in\mathbb{Z}^2 : t_1\in\mathbb{Z}\}\subset\mathbb{Z}^2$ is defined by

$$U_{t_2}(t_1) := \sum_{\boldsymbol{u}\in\mathbb{Z}^2} g(\boldsymbol{t}-\boldsymbol{u})\varepsilon(\boldsymbol{u}),\quad\text{where}\tag{95}$$

$$g(\boldsymbol{t}-\boldsymbol{u}) := \sum_{w\in\mathbb{Z}:\text{sgn}(w-t_2-(1/2))\ne\text{sgn}(u_2-t_2-(1/2))} \times a(t_1-u_1, w-u_2)\text{sgn}(u_2-w).$$

The process U_{t_2} is stationary and SRD, which follows from

$$\sum_{t_1\in\mathbb{Z}}|\text{Cov}(U_{t_2}(0),U_{t_2}(t_1))| = \sum_{t_1\in\mathbb{Z}}\Big|\sum_{\boldsymbol{u}\in\mathbb{Z}^2} g(\boldsymbol{u})g(\boldsymbol{u}+\boldsymbol{t})\Big|$$
$$\le \sum_{u_2\in\mathbb{Z}}\left(\sum_{(u_1,w)\in\mathbb{Z}^2}|a(u_1,w-u_2)|\mathbb{I}(\text{sgn}(w-(1/2))\right.$$
$$\left.\ne \text{sgn}(u_2-(1/2)))\right)^2 < \infty;$$

see Ref. [251]. Moreover, $\sum_{t_1\in\mathbb{Z}} \mathrm{Cov}(U_{t_2}(0), U_{t_2}(t_1)) = (1/2)\sigma^2_{\mathrm{edge},1}$ and the 'edge processes' U_0 and $U_{[\lambda^\gamma x_2]}$ in (4) 'located' on the lower and upper edges of the rectangle $K_{[\lambda x_1, \lambda^\gamma x_2]}$ become uncorrelated as $\lambda \to \infty$. Note that $\sigma_{\mathrm{edge},1}$ and/or $\sigma_{\mathrm{edge},2}$ in (94) may vanish, provided

$$\sum_{t_1\in\mathbb{Z}} a(t_1, t_2) = 0 \quad (\forall\, t_2 \in \mathbb{Z}) \quad \text{and/or} \quad \sum_{t_2\in\mathbb{Z}} a(t_1, t_2) = 0 \quad (\forall\, t_1 \in \mathbb{Z})$$

hold, implying the horizontal (respectively, vertical) ND property of Definition 16 of Chapter 1. The latter properties is a stronger form of ND which might change the scaling results in Theorem 23, including the critical γ_{cr}.

Other types of spatial dependence. The following Propositions 24 and 25 are 'spatial' versions of Propositions 19 and 20. It confirms that the scaling transition does not occur for RFs with SRD and 'hyperbolic' dependence.

Proposition 24. *Let X in* (1) *be a stationary linear RF on $\mathbb{Z}^2$ satisfying $(A)_{\mathrm{SRD}}$. Then, for any $\gamma > 0$.*

$$\lambda^{-(1+\gamma)/2} S_{\lambda,\gamma}(\boldsymbol{x}) \xrightarrow{\mathrm{fdd}} \sigma B_{1/2,1/2}(\boldsymbol{x}), \quad \boldsymbol{x} \in \mathbb{R}^2_+, \tag{96}$$

where $\sigma^2 = \big(\sum_{\boldsymbol{t}\in\mathbb{Z}^2} a(\boldsymbol{t})\big)^2 > 0$.

Next, we consider anisotropic scaling of separately fractionally integrated RFs on $\mathbb{Z}^2$ defined in Example 20 of Chapter 3. The MA coefficients of such RFs take a product form,

$$a(\boldsymbol{t}) = a_1(t_1)a_2(t_2), \quad \boldsymbol{t} = (t_1, t_2) \in \mathbb{Z}^2, \tag{97}$$

where $a_i = a_i(t), t \in \mathbb{Z}$, decay as power functions when $t \to \infty$ and vanish for $t < 0$. The corresponding linear RF has a unilateral or 'causal' representation,

$$\begin{aligned} X(\boldsymbol{t}) &= \sum_{\boldsymbol{u}\in\mathbb{Z}^2_+} a(\boldsymbol{t})\varepsilon(\boldsymbol{t} - \boldsymbol{u}) \\ &= \sum_{u_1\le t_1, u_2\le t_2} a_1(t_1 - u_1)a_2(t_2 - u_2)\varepsilon(u_1, u_2), \quad \boldsymbol{t} \in \mathbb{Z}^2. \end{aligned} \tag{98}$$

The covariance function of (98) factorises into the product of two covariance functions:

$$r(\boldsymbol{t}) = r_1(t_1)r_2(t_2), \quad \boldsymbol{t} = (t_1, t_2) \in \mathbb{Z}^2, \quad \text{where} \tag{99}$$
$$r_i(t) := \sum_{u \le 0 \wedge t} a_i(u)a_i(t-u), \quad t \in \mathbb{Z}, \quad i = 1, 2$$

are the covariance functions of two MA processes X_i indexed by integers $t \in \mathbb{Z}$ with respective MA coefficients $a_i, i = 1, 2$. The factorisation property in (99) facilitates the study of the variance and the limit distribution of sums $S_{\lambda_1,\lambda_2}(\boldsymbol{x}) := \sum_{\boldsymbol{t} \in K_{[\lambda_1 x_1, \lambda_2 x_2]}} X(\boldsymbol{t})$. Indeed,

$$\begin{aligned} \mathrm{Var}(S_{\lambda_1,\lambda_2}(\boldsymbol{x})) &= \sum_{\boldsymbol{t},\boldsymbol{s} \in K_{[\lambda_1 x_1, \lambda_2 x_2]}} r(\boldsymbol{t} - \boldsymbol{s}) \\ &= \prod_{i=1}^{2} \sum_{1 \le t_i, s_i \le [\lambda_i x_i]} r_i(t_i - s_i) \end{aligned} \tag{100}$$

factorises into the product of partial sums' variances of two stationary processes $X_i, i = 1, 2$, whose asymptotics under various LRD/SRD/ND conditions is well investigated; see e.g., Ref. [103, Chapter 3.3]. A variant of such conditions is the following assumption.

Assumption (A)$_{d_1,d_2}$. The MA coefficients in (1) have the form as in (97), where $a_i(t) = 0$ for $t \le -1, i = 1, 2$ and

$$a_i(t) = t^{d_i - 1}(A_i + o(1)), \quad t \to \infty, \quad A_i \ne 0, \quad 0 < |d_i| < 1/2, \tag{101}$$

$$\sum_{t=0}^{\infty} |a_i(t)| < \infty, \quad \sum_{t=0}^{\infty} a_i(t) =: A_i \ne 0, \quad d_i = 0,$$

$i = 1, 2$. In addition, assume that $\sum_{t=0}^{\infty} a_i(t) = 0$ for $d_i < 0, i = 1, 2$.

Conditions (101) are well known in time-series analyses and correspond to the LRD ($0 < d_i < 1/2$), SRD ($d_i = 0$), and ND ($-1/2 < d_i < 0$) concepts of the MA process X_i. Assumption (A)$_{d_1,d_2}$ is satisfied by the separately fractionally integrated RF in Example 20 of Chapter 3 and extends it to a nonparametric class of RFs.

The product form in (97)–(101) means that the arguments $t_i, i = 1,2$, can be scaled independently, and the scaling limit

$$\lim_{\lambda_i \to \infty, i=1,2} \lambda_1^{2-d_1-d_2} a(\lambda_1 t_1, \lambda_2 t_2) = \prod_{i=1}^{2} A_i t_i^{d_i - 1} =: a_\infty(\boldsymbol{t}) \qquad (102)$$

exists for $0 < |d_i| < 1/2, i = 1,2$ when $\lambda_i \to \infty, i = 1,2$ *in an arbitrary way*, in contrast to (84)

Proposition 25. *Let X be a stationary linear RF on $\mathbb{Z}^2$ in* (1) *satisfying Assumption $(A)_{d_1,d_2}$, for some $-1/2 < d_i < 1/2, i = 1,2$. Then, for any $\gamma > 0$,*

$$\lambda_1^{-H_1} \lambda_2^{-H_2} S_{\lambda_1,\lambda_2}(\boldsymbol{x}) \xrightarrow{\text{fdd}} \kappa_1 \kappa_2 B_{H_1,H_2}(\boldsymbol{x}),$$
$$\boldsymbol{x} \in \mathbb{R}_+^2, \quad \lambda_i \to \infty, \quad i = 1,2, \qquad (103)$$

where $H_i = d_i + .5, i = 1,2$ and $\kappa_i > 0$ given in (191). *As a consequence, X does not exhibit scaling transition for any $-1/2 < d_1, d_2 < 1/2$.*

5 Scaling Transition for Linear LRD RFs with 'Oblique' Dependence Axis

The dependence axis of an MA RF X on $\mathbb{Z}^2$ was defined in Section 1 of Chapter 7 as a direction on the plane along which the MA coefficients $a(\boldsymbol{t})$ decay at the smallest rate (Definition 6). Under Assumption $(\mathrm{A})_{\mathrm{LRD},\rho}$, the dependence axis of X agrees with the horizontal axis if $q_1 < q_2$ and with the vertical axis if $q_1 > q_2$. Since the scaling in (2) is parallel to the coordinate axes, we may say that in Theorem 22, *the scaling is congruous with the dependence axis of X*. This raises the question of what happens if the dependence axis of X does *not* agree with any of the two coordinate axes (the case of *incongruous scaling*), as in the case of linear RFs satisfying Assumption $\mathrm{A}_{\mathrm{LRD},\rho,\mathcal{O}}$ later. It turns out that incongruous scaling dramatically changes the scaling transition; in particular, the critical point becomes $\gamma_{\mathrm{cr}} = 1$ independently of the parameters q_1 and q_2. A similar scaling behaviour may be expected for ND RFs with 'rotated' MA coefficients in (104) but left open, partly because the concept of the dependence axis under ND is not very clear.

Assumption (A)$_{\mathrm{LRD},\rho,\mathcal{O}}$. The coefficients $a(\boldsymbol{t})$ and $\boldsymbol{t} \in \mathbb{Z}^2$ in (1) satisfy

$$a(\boldsymbol{t}) = \rho(\mathcal{O}\boldsymbol{t})^{-1}(\ell(\mathcal{O}\boldsymbol{t}) + o(1)), \quad |\boldsymbol{t}| \to \infty, \tag{104}$$

where $\mathcal{O} = (o_{ij})_{i,j=1,2}$ is a real nondegenerate matrix and $\rho(\boldsymbol{t}), \ell(\boldsymbol{t})$ are the same as in Assumption (A)$_{\mathrm{LRD},\rho}$, $1 < Q < 2$.

Assumptions (A)$_{\mathrm{LRD},\rho}$ and (A)$_{\mathrm{LRD},\rho,\mathcal{O}}$ coincide when $\mathcal{O} = I := \begin{bmatrix} 1 & 0 \\ 0 & 1 \end{bmatrix}$ is the identity matrix. Let

$$\boldsymbol{o}_1 := (o_{11}, o_{12}), \quad \boldsymbol{o}_2 := (o_{21}, o_{22}) \tag{105}$$

denote the row vectors of the matrix $\mathcal{O}$ in (104). According to Proposition 7, under the additional positivity conditions on the angular factor $\ell(\boldsymbol{t})$, the dependence axis of $a(\boldsymbol{t})$ in (104) and the corresponding RF X in (1) is given by

$$\boldsymbol{o}_2 \cdot \boldsymbol{t} = 0 \quad (q_1 < q_2) \quad \text{or} \quad \boldsymbol{o}_1 \cdot \boldsymbol{t} = 0 \quad (q_1 > q_2), \tag{106}$$

Moreover, since the dependence axis is preserved under the discrete convolution in Proposition 8, the covariance function of the linear RF X in (1) satisfies a similar asymptotics:

$$r(\boldsymbol{t}) = \tilde{\rho}(\mathcal{O}\boldsymbol{t})^{-1}(L(\mathcal{O}\boldsymbol{t}) + o(1)), \quad |\boldsymbol{t}| \to \infty, \tag{107}$$

where $\tilde{\rho}$, L are the same as in Corollary 9, (57); that is, they do not depend on the matrix $\mathcal{O}$, which enters only the argument of these functions. It follows that the dependence axes of the covariance function and the MA coefficients coincide, and we apply the same terminology to the RF X.

Table 3 summarises the main results on the scaling transition for linear RF X in (1) with MA coefficients satisfying Assumption (A)$_{\mathrm{LRD},\rho,\mathcal{O}}$. The notation $H_i^{\pm}, Q_{1/2,1}, Q_{1,1/2}$ is the same as in (86) or Theorem 22. In accordance with (106), congruous scaling in Table 3 corresponds to $o_{12}o_{21} = 0$, which do not necessarily mean that $\mathcal{O} = I$ is the identity matrix. Thus, the results in Table 3 for congruous scaling extend Table 2 for $1 < Q < 2$ and agree with the first four lines of Table 2. On the other hand, the results in the right columns of Table 3 (incongruous scaling) are different and in some sense more

Table 3. Unbalanced scaling limits $V_\pm$ of linear RF X satisfying Assumption $\mathrm{A}_{\mathrm{LRD},\rho,\mathcal{O}}$ under congruous and incongruous scaling.

	Congruous Scaling			Incongruous Scaling		
Parameter Region	$\gamma_{\rm cr}$	V_+	V_-	$\gamma_{\rm cr}$	V_+	V_-
$R_{11}: Q_{1/2,1} \wedge Q_{1,1/2} > 1$	q_1/q_2	B_{1,H_2^+}	$B_{H_1^-,1}$	1	$B_{1,H_1^- \wedge H_2^+}$	$B_{H_1^- \wedge H_2^+,1}$
$R_{12}: Q_{1/2,1} < 1 < Q_{1,1/2}$	q_1/q_2	$B_{H_1^+,1/2}$	$B_{H_1^-,1}$	1	$B_{H_1^+ \wedge H_2^-,1/2}$	$B_{1/2,H_1^+ \wedge H_2^-}$
$R_{21}: Q_{1,1/2} < 1 < Q_{1/2,1}$	q_1/q_2	B_{1,H_2^+}	$B_{1/2,H_2^-}$	1	$B_{H_1^+ \wedge H_2^-,1/2}$	$B_{1/2,H_1^+ \wedge H_2^-}$
$R_{22}^+: Q_{1/2,1} \vee Q_{1,1/2} < 1$	q_1/q_2	$B_{H_1^+,1/2}$	$B_{1/2,H_2^-}$	1	$B_{H_1^+ \wedge H_2^-,1/2}$	$B_{1/2,H_1^+ \wedge H_2^-}$

simple: the critical point $\gamma_{\rm cr} = 1$ independently of $q_i, i = 1, 2$, and the unbalanced FBS limits are symmetric w.r.t. the coordinate axes. In particular, the above conclusions apply to rotation,

$$\mathcal{O} = \begin{bmatrix} \cos\phi & \sin\phi \\ -\sin\phi & \cos\phi, \end{bmatrix}$$

of the plane by an angle $\phi \in [0, \pi/2)$. This transformation changes the original dependence axis of X in Assumption $(\mathrm{A})_{\mathrm{LRD},\rho}$, and the change in the angle ϕ can be arbitrary small; nevertheless, the results in Table 3 for $\phi \neq 0$ and $\phi = 0$ are very different.

Let us turn to rigorous formulations pertaining to Table 3. We shall restrict to the case of $q_1 \neq q_2$ or $q_1 < q_2$ by symmetry. The limit RFs $V_\pm, V_{\rm cr}$ in the subsequent theorems are Gaussian and defined as stochastic integrals w.r.t. white noise $W(\mathrm{d}\boldsymbol{u})$, as in (88) and elsewhere in this section. We use the same notation $a_\infty(\boldsymbol{u}) = \rho(\boldsymbol{u})^{-1}\ell(\boldsymbol{u})$, $\boldsymbol{u} \in \mathbb{R}^2_0$, as in (87). Define the RF

$$\widetilde{V}_D(\boldsymbol{x}) := |\det(\mathcal{O})|^{-\frac{1}{2}} \int_{\mathbb{R}^2} \left\{ \int_{]\mathbf{0},\boldsymbol{x}]} a_\infty(\widetilde{D}\boldsymbol{t} - \boldsymbol{u})\mathrm{d}\boldsymbol{t} \right\} W(\mathrm{d}\boldsymbol{u}), \quad \boldsymbol{x} \in \mathbb{R}^2_+, \tag{108}$$

where $\widetilde{D}$ is any 2×2 matrix in (109):

$$\tilde{\mathcal{O}}_{00} := \begin{bmatrix} o_{11} & 0 \\ 0 & o_{22} \end{bmatrix}, \quad \tilde{\mathcal{O}}_{20} := \begin{bmatrix} 0 & 0 \\ o_{21} & o_{22} \end{bmatrix}, \tag{109}$$

$$\tilde{\mathcal{O}}_{11} := \begin{bmatrix} o_{11} & 0 \\ 0 & 0 \end{bmatrix}, \quad \tilde{\mathcal{O}}_{21} := \begin{bmatrix} 0 & 0 \\ o_{21} & 0 \end{bmatrix}, \quad \tilde{\mathcal{O}}_{22} := \begin{bmatrix} 0 & 0 \\ 0 & o_{22} \end{bmatrix}.$$

Also, let

$$V_D(\boldsymbol{x}) := |\det(\mathcal{O})|^{-1} \int_{\mathbb{R}^2} \left\{ \int_{\mathbb{R}^2} a_\infty(\boldsymbol{t}) \mathbb{I}(D\boldsymbol{t} + \boldsymbol{u} \in]\boldsymbol{0}, \boldsymbol{x}]) \mathrm{d}\boldsymbol{t} \right\} W(\mathrm{d}\boldsymbol{u}), \quad \boldsymbol{x} \in \mathbb{R}^2_+, \tag{110}$$

where D is any 2×2 matrix in (111):

$$\mathcal{O}_{01} := \tfrac{1}{\det(\mathcal{O})} \begin{bmatrix} o_{22} & 0 \\ -o_{21} & 0 \end{bmatrix}, \quad \mathcal{O}_{11} := \tfrac{1}{\det(\mathcal{O})} \begin{bmatrix} o_{22} & 0 \\ 0 & 0 \end{bmatrix}, \tag{111}$$

$$\mathcal{O}_{21} := \tfrac{1}{\det(\mathcal{O})} \begin{bmatrix} 0 & 0 \\ -o_{21} & 0 \end{bmatrix}, \quad \mathcal{O}_{22} := \tfrac{1}{\det(\mathcal{O})} \begin{bmatrix} 0 & 0 \\ 0 & o_{11} \end{bmatrix}.$$

In (109), (111) o_{ij} are the entries of the matrix O in (104). To shorten notation, write

$$\widetilde{V}_{ij} := \widetilde{V}_{\tilde{\mathcal{O}}_{ij}}, \quad V_{ij} := V_{\mathcal{O}_{ij}}, \quad \widetilde{V}_0 := \widetilde{V}_{\mathcal{O}}.$$

Set $\tilde{\kappa}^2_{ij} := \mathrm{E}\widetilde{V}_{ij}(\mathbf{1})^2$, $\kappa^2_{ij} := \mathrm{E}V_{ij}(\mathbf{1})^2$. As explained above, we distinguish between two cases:

Case 1: $o_{21}o_{22} \neq 0$ (incongruous scaling),

Case 2: $o_{21} = 0$ (congruous scaling).

Theorem 26. *Let X in (3) be a stationary linear RF on $\mathbb{Z}^2$ with zero mean satisfying Assumption $(A)_{\mathrm{LRD},\rho,\mathcal{O}}$, $Q_{1/2,1} \neq 1, Q_{1/2,1} \neq 1$, $q_1 < q_2$. Then, the convergence in (2) holds for all $\gamma > 0$ and satisfy (4) with the limits well defined in the following parameter regions:*

(i) *Region $R_{11}(Q_{1/2,1} > 1)$:*

incongruous scaling (Case 1): $\gamma_{\mathrm{cr}} = 1$, $V_+ = \widetilde{V}_{22} \overset{\mathrm{fdd}}{=} \tilde{\kappa}_{22} B_{1,H_2^+}$, $V_- = \widetilde{V}_{21} \overset{\mathrm{fdd}}{=} \tilde{\kappa}_{21} B_{H_2^+,1}$, $V_{\mathrm{cr}} = \widetilde{V}_{20}$;

congruous scaling (Case 2): $\gamma_{\mathrm{cr}} = q_1/q_2$, $V_+ = \widetilde{V}_{22} \overset{\mathrm{fdd}}{=} \tilde{\kappa}_{22} B_{1,H_2^+}$, $V_- = \widetilde{V}_{11} \overset{\mathrm{fdd}}{=} \tilde{\kappa}_{11} B_{H_1^-,1}$, $V_{\mathrm{cr}} = \widetilde{V}_{00}$.

(ii) *Region* $R_{12}(Q_{1/2,1} < 1 < Q_{1,1/2})$:

incongruous scaling (*Case* 1): $\gamma_{\rm cr} = 1$, $V_+ = V_{11} \overset{\rm fdd}{=} \kappa_{11} B_{H_1^+,1/2}$, $V_- = V_{21} \overset{\rm fdd}{=} \kappa_{21} B_{1/2,H_1^+}$, $V_{\rm cr} = V_{01}$;

congruous scaling (*Case* 2): $\gamma_{\rm cr} = q_1/q_2$, $V_+ = V_{11} \overset{\rm fdd}{=} \kappa_{11} B_{H_1^+,1/2}$, $V_- = \widetilde{V}_{11} \overset{\rm fdd}{=} \tilde{\kappa}_{11} B_{H_1^-,1}$, $V_{\rm cr} = \widetilde{V}_{00}$.

(iii) *Region* R_{22}^+ ($Q_{1,1/2} < 1$):

incongruous scaling (*Case* 1): $\gamma_{\rm cr}$, V_+, V_-, $V_{\rm cr}$ *the same as in part (ii), Case* 1.

congruous scaling (*Case* 2): $\gamma_{\rm cr} = q_1/q_2$, $V_+ = V_{11} \overset{\rm fdd}{=} \kappa_{11} B_{H_1^+,1/2}$, $V_- = V_{22} \overset{\rm fdd}{=} \kappa_{22} B_{1/2,H_2^-}$, $V_{\rm cr} = \widetilde{V}_{00}$.

The normalisation in Theorem 26 is $d_{\lambda,\gamma} = \lambda^{H(\gamma)}$, where $H(\gamma) > 0$ in Case 2 is the same as in Theorem 22, whereas in Case 2, $H(\gamma)$ is written as a linear combination of the Hurst indices of the limit RFs. The proof of Theorem 26 uses a similar approach as Theorem 22 but is rather lengthy due to a large number of cases and subcases. A complete proof can be found in Ref. [210], which also includes the case of $q_1 = q_2$ (excluded in the formulation of Theorem 26) when the dependence axis is undefined but the presence of the matrix $\mathcal{O}$ in (104) has an affect on the variances of the limit RFs. Identification of the unbalanced limits with FBS follows by change of variables in the integral expressions of their covariances. For instance, from the definitions in (108) and (109) and a change $t_2 \to x_2 t_2$, $u_2 \to x_2 u_2$, $u_1 \to x_2^{q_2/q_1} u_1$ using the invariance property, $\lambda a_\infty(\lambda^{1/q_1} t_1, \lambda^{1/q_2} t_2) = a_\infty(\boldsymbol{t})$, we get

$$\begin{aligned} &\mathrm{E}\widetilde{V}_{22}(\boldsymbol{x})^2 \\ &= |\det(\mathcal{O})|^{-1} \int_{\mathbb{R}^2} \left| \int_{]\mathbf{0},\boldsymbol{x}]} a_\infty(-u_1, o_{22}t_2 - u_2) \mathrm{d}\boldsymbol{t} \right|^2 \mathrm{d}\boldsymbol{u} = x_1^2 x_2^{2H_2^+} \tilde{\kappa}_{22}^2, \end{aligned}$$

with $\tilde{\kappa}_{22}^2 = |\det(\mathcal{O})|^{-1} \int_{\mathbb{R}^2} \left| \int_{]\mathbf{0},\mathbf{1}]} a_\infty(-u_1, o_{22}t_2 - u_2) \mathrm{d}\boldsymbol{t} \right|^2 \mathrm{d}\boldsymbol{u} = \mathrm{E}\widetilde{V}_{22}(\mathbf{1})^2$.

What happens if MA coefficients $a(\boldsymbol{t})$ factorise as in (97) or Assumption $(\mathrm{A})_{d_1,d_2}$ and are 'rotated' by a matrix $\mathcal{O} = (o_{ij})_{i,j=1,2}$?

In other words, what can be said about the scaling limits in (2) of the corresponding RF

$$X(\boldsymbol{t}) = \sum_{\boldsymbol{u}\in\mathbb{Z}^2} b(\boldsymbol{t}-\boldsymbol{u})\varepsilon(\boldsymbol{u}), \quad \boldsymbol{t}\in\mathbb{Z}^2, \tag{112}$$

where

$$b(\boldsymbol{t}) := a([\mathcal{O}\boldsymbol{t}]) = \prod_{i=1}^{2} a_i([\boldsymbol{o}_i\cdot\boldsymbol{t}]), \quad \boldsymbol{t}\in\mathbb{Z}^2, \tag{113}$$

with $\boldsymbol{o}_i, i = 1, 2$ defined in (105). In the LRD case ($d_i \in (0, 1/2)$, $i = 1, 2$) $a(\boldsymbol{t})$ in (97) decay at rates of $O(t_1^{d_1-1})$ and $O(t_2^{d_2-1})$ along the horizontal and vertical axes, respectively, and at a much faster rate of $O(|\boldsymbol{t}|^{d_1+d_2-2})$ along any other direction. We may say that the coefficients $a(\boldsymbol{t})$ in (97) feature *two* directions or dependence axes, namely, the coordinate axes, along which they decay much slower than in any other direction on the plane. This form of MA coefficients is very different from the dependence structure in Theorem 22 and does not lead to a scaling transition as shown in Proposition 20. Obviously, the 'rotated' coefficients $b(\boldsymbol{t})$ in (113) cannot be 'separately scaled' as in (102), and we can expect that the scaling limits of (112) generally depend on γ, as it happens in Proposition 27.

Proposition 27. *Let X be a stationary linear RF in* (112)–(113), *with $a(\boldsymbol{t})$ satisfying Assumption $(A)_{d_1,d_2}$, $0 < d_i < 1/2, i = 1, 2$, and a non-degenerate 2×2 matrix $\mathcal{O}$, $o_{ii} > 0, i = 1, 2, o_{12} \neq 0$. Then, the limits in* (2) *exist for any $\gamma > 0$ and satisfy*

$$V_\gamma(\boldsymbol{x}) = \begin{cases} \kappa_{+,1}B_{H,1/2}(\boldsymbol{x}), & \gamma > 1, H < 1, \\ \kappa_{+,2}B_{1,H-1/2}(\boldsymbol{x}), & \gamma > 1, H > 1. \\ \kappa B_{H_1,H_2}(\mathcal{O}^{-1}\boldsymbol{x}), & \gamma = 1. \end{cases} \tag{114}$$

In (114), *$H := d_1 + d_2 + 1/2$, $H_i, i = 1, 2$ are the same as in* (103) *and $\kappa_{+,i}, i = 1, 2, \kappa$ are some constants.*

An analogous statement holds for $\gamma < 1$ and o_{12} replaced by o_{21} with the Hurst indices in (114) interchanged. We omit the proof of the above proposition since it resembles that of Theorem 22 and other theorems in this section. It uses Proposition 9 of Chapter 2 with the limits $V_\gamma(\boldsymbol{x})$ written as stochastic integrals w.r.t. white noise as in (108). Particularly, for $\gamma > 1$,

$$V_\gamma(\boldsymbol{x}) = \begin{cases} |\det(\mathcal{O})|^{-1} \int_{\mathbb{R}^2} \{ \int_{\mathbb{R}^2} a_\infty(\boldsymbol{t}) \mathbb{I}(\mathcal{O}_{01}\boldsymbol{t} + \boldsymbol{u} \in]\boldsymbol{0}, \boldsymbol{x}]) \mathrm{d}\boldsymbol{t} \} W(\mathrm{d}\boldsymbol{u}), & H < 1, \\ |\det(\mathcal{O})|^{-1/2} \int_{\mathbb{R}^2} \{ \int_{]\boldsymbol{0}, \boldsymbol{x}]} a_\infty(\tilde{\mathcal{O}}_{02}\boldsymbol{t} - \boldsymbol{u}) \mathrm{d}\boldsymbol{t} \} W(\mathrm{d}\boldsymbol{u}), & H > 1, \end{cases} \tag{115}$$

where $a_\infty(\boldsymbol{t}) := A_1 A_2 (t_1)_+^{d_1 - 1} (t_2)_+^{d_2 - 1}$ and $\mathcal{O}_{01}, \tilde{\mathcal{O}}_{02}$ are the same 2×2 matrices as in (111), (109) so that these RFs have a similar form as RFs $V_{01}, \tilde{V}_{02}$ in (110) and (108), except that the kernel a_∞ in (115) is different from a_∞ in (87). The integral in (115) can be more explicitly written using the identity

$$\int_{\mathbb{R}} \prod_{i=1}^{2} (o_{i1} t_1 + o_{i2} t_2)_+^{d_i - 1} \mathrm{d}t_2 = |t_1|^{d_1 + d_2 - 1} (c_+ \mathbb{I}(t_1 > 0) + c_- \mathbb{I}(t_1 < 0)), \tag{116}$$

with

$$\begin{aligned} c_+ &:= |o_{22}|^{-d_1} |o_{12}|^{-d_2} |\det(\mathcal{O})|^{d_1 + d_2 - 1} \big(B(d_2, 1 - d_1 - d_2) \\ &\qquad \times \mathbb{I}(\det(\mathcal{O}) > 0, o_{12} > 0) + B(d_1, d_2) \mathbb{I}(\det(\mathcal{O}) > 0, o_{12} < 0) \big), \\ c_- &:= |o_{22}|^{-d_1} |o_{12}|^{-d_2} |\det(\mathcal{O})|^{d_1 + d_2 - 1} \big(B(d_1, 1 - d_1 - d_2) \\ &\qquad \times \mathbb{I}(\det(\mathcal{O}) < 0, o_{12} > 0) + B(d_1, d_2) \mathbb{I}(\det(\mathcal{O}) < 0, o_{12} < 0) \big). \end{aligned}$$

Formula (116) follows by an elementary change of variables and the definition of the beta function. Consequently, (115) can be written as

$$\begin{aligned} V_\gamma(\boldsymbol{x}) = c_+ A_1 A_2 \int_{\mathbb{R} \times (0, x_2]} \\ \times \left\{ \int_0^{x_1} (c_+ (t - u_1)_+^{d_1 + d_2 - 1} \mathrm{d}t + c_- (t - u_1)_-^{d_1 + d_2 - 1} \mathrm{d}t \right\} W(\mathrm{d}\boldsymbol{u}). \end{aligned}$$

We recognise in the above expression a double-sided stochastic integral representation of FBM with one-dimensional 'time' x_1 and Hurst parameter $H = (1/2) + d_1 + d_2 \in (1/2, 1)$ (see Ref. [231]), which agrees with (114). Therefore, it also follows the expressions for $\kappa_{+,i}$ in terms of d_i and $c_\pm$:

$$\begin{aligned}
\kappa^2_{+,1} &= (A_1A_2)^2 \int_{\mathbb{R}} \left\{ \int_0^1 (c_+(t-u)_+^{d_1+d_2-2} + c_-(t-u)_-^{d_1+d_2-2}) \mathrm{d}t \right\}^2 \mathrm{d}u \\
&= (A_1A_2)^2 \left(\frac{2B(d_1+d_2, 1-2d_1-2d_2)}{(d_1+d_2)(2d_1+2d_2+1)} \right. \\
&\quad \left. + \frac{B(d_1+d_2, d_1+d_2)}{(2d_1+2d_2+1)(d_1+d_2)} \right),
\end{aligned}$$

$$\kappa^2_{+,2} = \frac{1}{(d_1+d_2)(2(d_1+d_2)-1)|\det(O)|} \prod_{i=1}^2 \frac{A_i B(d_i, 1-2d_i)}{|b_{i2}|^{1-2d_i}}.$$

6 Nonlinear LRD Random Fields

Let us discuss anisotropic scaling limits (2) and the scaling transition for a class of nonlinear subordinated RFs:

$$Y(\boldsymbol{t}) = G(X(\boldsymbol{t})), \quad \boldsymbol{t} \in \mathbb{Z}^2, \tag{117}$$

where $X = \{X(\boldsymbol{t}); \boldsymbol{t} \in \mathbb{Z}^2\}$ is a stationary MA RF in (1) satisfying Assumption $(\mathbf{A})_{\mathrm{LRD},\rho}$ and $G(x), x \in \mathbb{R}$ is a nonlinear function with $\mathrm{E}G(X(\mathbf{0}))^2 < \infty$, $\mathrm{E}G(X(\mathbf{0})) = 0$. We use the notation $r_X(\boldsymbol{t}) = \mathrm{Cov}(X(\mathbf{0}), X(\boldsymbol{t})), r_Y(\boldsymbol{t}) = \mathrm{Cov}(Y(\mathbf{0}), Y(\boldsymbol{t}))$ for the respective covariance functions and study the convergence

$$d^{-1}_{\lambda,\gamma} S^Y_{\lambda,\gamma}(\boldsymbol{x}) \xrightarrow{\mathrm{fdd}} V^Y_\gamma(\boldsymbol{x}), \quad \boldsymbol{x} \in \mathbb{R}^2_+, \quad \lambda \to \infty, \tag{118}$$

of the partial sums RF $S^Y_{\lambda,\gamma}(\boldsymbol{x}) = \sum_{\boldsymbol{t} \in K_{[\lambda x_1, \lambda^\gamma x_2]}} Y(\boldsymbol{t})$ towards a limit RF $V^Y_\gamma = \{V^Y_\gamma(\boldsymbol{x}); \boldsymbol{x} \in \mathbb{R}^2_+\}$, for each $\gamma > 0$.

Following Section 1 of Chapter 4, we consider the case when $G = A_k$ is the Appell polynomial of degree $k \geq 1$ relative to the marginal distribution of X. It is convenient to use the parametrisation

$$\tilde{q}_i = q_i(2-Q), \quad i = 1, 2, \quad \tilde{Q} = \sum_{i=1}^2 \frac{1}{\tilde{q}_i}, \quad \tilde{\rho}(\boldsymbol{t}) = \sum_{i=1}^2 |t_i|^{\tilde{q}_i}, \tag{119}$$

as in (18)–(19). Recall from Corollary 9 that the above X is LRD and its covariance satisfies

$$r_X(\boldsymbol{t}) = \tilde{\rho}(\boldsymbol{t})^{-1}\big(\tilde{L}(\boldsymbol{t}) + o(1)\big) = O(\tilde{\rho}(\boldsymbol{t})^{-1}), \quad |\boldsymbol{t}| \to \infty. \tag{120}$$

The results concerning (118) are summarised as follows:

(Y1) Subordinated RFs $Y = A_k(X), 1 \le k < \tilde{Q}$ exhibit scaling transition at the same point $\gamma^Y_{\mathrm{cr}} = \tilde{q}_1/\tilde{q}_2 = q_1/q_2$ independent of k.
(Y2) The well-balanced scaling limit V^Y_{cr} is non-Gaussian unless $k = 1$ and is given by a k-tuple Wiener–Itô integral.
(Y3) Unbalanced scaling limit V^Y_+ agrees with FBS $B_{H^+_{1k},1/2}$ with the Hurst parameter $H^+_{1k} \in (1/2, 1)$ if $k\tilde{q}_2 > 1$ and with a 'generalised Hermite slide' $V^Y_+(\boldsymbol{x}) = x_1 \mathcal{H}_{+,k}(x_2)$ if $k\tilde{q}_2 < 1$, where $\mathcal{H}_{+,k}(x_2)$ is a H^+_{2k}-SS process written as a k-tuple Wiener–Itô integral. A similar fact holds for unbalanced limits V^Y_-.
(Y4) For $k > \tilde{Q}$, the RF $Y = A_k(X)$ does not exhibit a scaling transition, and all scaling limits $V^Y_\gamma, \gamma > 0$ agree with Brownian sheet $B_{1/2,1/2}$.
(Y5) In the case of a Gaussian underlying RF X, the above conclusions hold for a general nonlinear function G in (117) and k equal to the Hermite rank of G.

The results in (Y1)–(Y5) are obtained under the following assumptions.

Assumption $(\mathbf{G})_{X,k}$. For $k \in \mathbb{N}_+$,

$$Y = \{Y(\boldsymbol{t}) := A_k(X(\boldsymbol{t})); \boldsymbol{t} \in \mathbb{Z}^2\}, \tag{121}$$

where $X = \{X(\boldsymbol{t}); \boldsymbol{t} \in \mathbb{Z}^2\}$ is a linear RF in (1) satisfying Assumption $(\mathrm{A})_{\mathrm{LRD},\rho}$ and A_k is the kth Appell polynomial relative to the (marginal) distribution of $X(\boldsymbol{t})$. Moreover, we assume that $\mathrm{E}|\varepsilon|^{2k} < \infty$ and $\ell(\boldsymbol{t}) \ge 0$ in (54).

Assumption $(\mathbf{G})_{X,\infty}$. $Y = \{Y(\boldsymbol{t}) := G(X(\boldsymbol{t})); \boldsymbol{t} \in \mathbb{Z}^2\}$, where $X = \{X(\boldsymbol{t}); \boldsymbol{t} \in \mathbb{Z}^2\}$ is a Gaussian MA RF satisfying Assumption $(\mathrm{A})_{\mathrm{LRD},\rho}$ and $G = G(x), x \in \mathbb{R}$ is an arbitrary measurable function with $\mathrm{E}G(X(\mathbf{0}))^2 < \infty, \mathrm{E}G(X(\mathbf{0})) = 0$; moreover, $\varepsilon(\mathbf{0}) \sim N(0,1), X(\mathbf{0}) \sim N(0,1)$.

As noted above, under Assumption $(\mathrm{G})_{X,\infty}$, the Appell polynomials $A_k(z)$ coincide with the Hermite polynomials $H_k(z)$. The following statement is similar in spirit to Theorem 3 of Chapter 5 and uses the decomposition $Y(\boldsymbol{t}) = X^{\bullet k}(\boldsymbol{t}) + \mathcal{Z}(\boldsymbol{t})$ in (96) of Chapter 2 into 'off-diagonal' form and a remainder term.

Proposition 28. *Let RFs X and $Y = A_k(X)$ satisfy Assumptions $(A)_{\mathrm{LRD},\rho}$ and $(G)_{X,k}$.*

(i) *Let $1 \le k < \tilde{Q}$. Then, $\sum_{\boldsymbol{t}\in\mathbb{Z}^2} |r_Y(\boldsymbol{t})| = \infty$ and*

$$r_Y(\boldsymbol{t}) = k!\tilde{\rho}(\boldsymbol{t})^{-k}\big(L(\boldsymbol{t})^k + o(1)\big), \quad r_{\mathcal{Z}}(\boldsymbol{t}) = o(\tilde{\rho}(\boldsymbol{t})^{-k}), \quad |\boldsymbol{t}| \to \infty, \tag{122}$$

with $L(\boldsymbol{t}), \tilde{\rho}(\boldsymbol{t})$ the same as in (120). Moreover, $\sum_{t_2\in\mathbb{Z}} |r_Y(0,t_2)| = \infty \iff k\tilde{q}_2 \le 1$ and $\sum_{t_1\in\mathbb{Z}} |r_Y(t_1,0)| = \infty \iff k\tilde{q}_1 \le 1$.

(ii) *Let $k > \tilde{Q}$. Then, $\sum_{\boldsymbol{t}\in\mathbb{Z}^2} |r_Y(\boldsymbol{t})| < \infty$.*

Recall the notions of vertical/horizontal LRD/SRD properties from Definition 16 of Chapter 1. From Proposition 28 (i), we see that $Y = A_k(X)$ has covariance vertical LRD property if $k\tilde{q}_2 \le 1$ and covariance vertical SRD property if $k\tilde{q}_2 > 1$. Similarly, $Y = A_k(X)$ has covariance horizontal LRD property if $k\tilde{q}_1 \le 1$ and covariance vertical SRD property if $k\tilde{q}_1 > 1$. According to (Y3), the change in these properties at $k\tilde{q}_i = 1$ corresponds to the change in the limit distribution (118). The Hurst indices of the limit MSS RFs in (Y3) are equal to

$$H_{1k}^+ := \begin{cases} 1 + \frac{\tilde{q}_1}{2\tilde{q}_2} - \frac{k\tilde{q}_1}{2}, & k\tilde{q}_2 > 1, \\ 1, & k\tilde{q}_2 < 1, \end{cases} \qquad H_{2k}^+ := \begin{cases} \frac{1}{2}, & k\tilde{q}_2 > 1, \\ 1 - \frac{k\tilde{q}_2}{2}, & k\tilde{q}_2 < 1, \end{cases}$$

$$H_{1k}^- := \begin{cases} \frac{1}{2}, & k\tilde{q}_1 > 1, \\ 1 - \frac{k\tilde{q}_1}{2}, & k\tilde{q}_1 < 1, \end{cases} \qquad H_{2k}^- := \begin{cases} 1 + \frac{\tilde{q}_2}{2\tilde{q}_1} - \frac{k\tilde{q}_2}{2}, & k\tilde{q}_1 > 1, \\ 1, & k\tilde{q}_1 < 1. \end{cases} \tag{123}$$

and agree with the corresponding indices $H_i^{\pm}$ in (86) for $k = 1, 1 < Q < 2$. The non-Gaussian processes $\mathcal{H}_{\pm,k}$ with one-dimensional time are written as k-tuple integrals:

$$\mathcal{H}_{+,k}(x_2) := \int_{\mathbb{R}^{2k}} h_{+,k}(x_2; (\boldsymbol{u})_k) \mathrm{d}^k W,$$
$$\mathcal{H}_{-,k}(x_1) := \int_{\mathbb{R}^{2k}} h_{-,k}(x_1; (\boldsymbol{u})_k) \mathrm{d}^k W, \tag{124}$$

$x_1, x_2 > 0$, where

$$h_{+.k}(x_2; (\boldsymbol{u})_k) := \int_0^{x_2} \prod_{i=1}^k a_\infty((0, t_2) - \boldsymbol{u}_i) \mathrm{d}t_2, \tag{125}$$

$$h_{-.k}(x_1; (\boldsymbol{u})_k) := \int_0^{x_1} \prod_{i=1}^k a_\infty((t_1, 0) - \boldsymbol{u}_i) \mathrm{d}t_1.$$

The 'critical' RF is written as

$$V_{\mathrm{cr}}^Y(\boldsymbol{x}) := \int_{\mathbb{R}^{2k}} h_{0,k}(\boldsymbol{x}; (\boldsymbol{u})_k) \mathrm{d}^k W, \quad \text{where}$$

$$h_{0,k}(\boldsymbol{x}; (\boldsymbol{u})_k) := \int_{]0,\boldsymbol{x}]} \prod_{i=1}^k a_\infty(\boldsymbol{t} - \boldsymbol{u}_i) \mathrm{d}\boldsymbol{t}. \tag{126}$$

The RF V_0^Y in (126) is well defined for $1 \le k < \tilde{Q}$ as a k-tuple Wiener–Itô integral and has zero mean $\mathrm{E}V_0^Y(\boldsymbol{x}) = 0$ and finite variance $\mathrm{E}(V_0^Y(\boldsymbol{x}))^2 = k!\|h_0(\boldsymbol{x};\cdot)\|_k^2$. Moreover, V_0^Y has stationary rectangular increments and is $(\gamma_{\mathrm{cr}}, H(\gamma_{\mathrm{cr}})$-SS with $H(\gamma_{\mathrm{cr}}) = 1 + \frac{\tilde{q}_1}{\tilde{q}_2} - k\tilde{q}_1/2$. The processes $\mathcal{H}_{+,k}$ and $\mathcal{H}_{-,k}$ in (124) are well defined for $1 \le k < 1/\tilde{q}_2$ and $1 \le k < 1/\tilde{q}_1$, respectively, as k-tuple Wiener–Itô–Wiener integrals. They have zero mean, finite variance, stationary increments, and are self-similar with respective indices $H_{2k}^+ \in (1/2, 1)$ and $H_{1k}^- \in (1/2, 1)$. Particularly, from the stochastic integral representation in (124), we obtain the covariance

$$\begin{aligned}
&\mathrm{E}\mathcal{H}_{+,k}(x)\mathcal{H}_{+,k}(y) \\
&= k! \int_{\mathbb{R}^{2k}} h_+(x; (\boldsymbol{u})_k) h_+(y; (\boldsymbol{u})_k) \mathrm{d}(\boldsymbol{u})_k \quad (\mathrm{d}(\boldsymbol{u})_k := \mathrm{d}\boldsymbol{u}_1 \cdots \mathrm{d}\boldsymbol{u}_k) \\
&= k! \int_0^x \int_0^y \mathrm{d}t\mathrm{d}t' \prod_{i=1}^k \int_{\mathbb{R}^2} a_\infty((t, 0) - \boldsymbol{u}_i) a_\infty((t', 0) - \boldsymbol{u}_i) \mathrm{d}\boldsymbol{u}_i
\end{aligned}$$

$$= k! \int_0^x \int_0^y \mathrm{d}t\mathrm{d}t' (a_\infty \star a_\infty)(t - t', 0)^k$$

$$= k! \int_0^x \int_0^y \mathrm{d}t\mathrm{d}t' |t - t'|^{-k\tilde{q}_1} L(1,0)^k \quad \text{(see (24), (26), (17))}$$

$$= \frac{k! L(1,0)^k}{(1 - k\tilde{q}_1)(2 - k\tilde{q}_1)} \left(x^{2-k\tilde{q}_1} + y^{2-k\tilde{q}_1} - |y - x|^{2-k\tilde{q}_1}\right), \quad x, y > 0.$$

Following the terminology in Ref. [208], the RFs $x_1 \mathcal{H}_{+,k}(x_2)$ and $x_2 \mathcal{H}_{-,k}(x_1)$ may be called a *generalised Hermite slide* since they represent a random surface 'sliding linearly to the origin' along one of the coordinate on the plane from a generalised Hermite process indexed by the other coordinate. In the case of $k = 1$, the above generalised Hermite slide is Gaussian and agrees with a FBS, B_{H_1,H_2}, where one of the two parameters H_1 and H_2 equals 1.

Rigorous statements pertaining to (Y1)–(Y5) and their proofs are given in Ref. [209]. The proofs of non-Gaussian limits in (Y2)–(Y3) use the representation of $S_k(g_{\lambda,\gamma}(\boldsymbol{x};\cdot)) = \sum_{\boldsymbol{t} \in K_{[\lambda x_1, \lambda^\gamma x_2]}} X^{\bullet k}(\boldsymbol{t})$ as an off-diagonal k-linear form in (108) of Chapter 2 with coefficients

$$g_{\lambda,\gamma}(\boldsymbol{x}; (\boldsymbol{u})_k) := \sum_{\boldsymbol{t} \in K_{[\lambda x_1, \lambda^\gamma x_2]}} a(\boldsymbol{t} - \boldsymbol{u}_1) \cdots a(\boldsymbol{t} - \boldsymbol{u}_k), \quad (\boldsymbol{u})_k \in \mathbb{Z}^{\bullet 2k},$$

together with Proposition 21 of Chapter 2. The proof of the Gaussian limits in (V3) and (V4) use approximation by m-dependent sums, as in the case of isotropic scaling (Theorem 3 (ii)) of Chapter 5. The above approximation is more delicate in the case of (V3) when the approximating terms also depend on λ and form a triangular array of m-dependent r.v.s, in which case it uses a CLT due to Ref. [193].

Remark 29 (Incongruous scaling). An interesting open problem is extending the results in Ref. [209] to subordinated RFs with an 'oblique' dependence axis. Note that if the underlying linear RF X in (121) satisfies Assumption $\mathbf{(A)}_{\mathrm{LRD},\rho,\mathcal{O}}$, then $\mathrm{Cov}(A_k(X(\mathbf{0}), A_k(X(\boldsymbol{t}))) = \tilde{\rho}(\mathcal{O}\boldsymbol{t})^{-k} L(\mathcal{O}\boldsymbol{t})^k + o(1)), |\boldsymbol{t}| \to \infty$ under suitable conditions on $k, \tilde{q}_i, i = 1, 2, L$, referring to (122) and Proposition 8 (50), and we can expect the RF $Y = A_k(X)$ to exhibit a change in the scaling transition under incongruous scaling as in the linear case of $k = 1$.

7 Anisotropic Scaling of RFs on $\mathbb{Z}^\nu$

Let $X = \{X(\boldsymbol{t}); \boldsymbol{t} \in \mathbb{Z}^\nu\}$ be a stationary RF on $\mathbb{Z}^\nu$, $\nu \geq 1$, and let

$$K_{\lambda,\boldsymbol{\gamma}}(\boldsymbol{x}) := \prod_{i=1}^{\nu}[1, [\lambda^{\gamma_i} x_i]] \subset \mathbb{Z}^\nu, \quad \boldsymbol{x} = (x_1, \ldots, x_\nu) \in \mathbb{R}_+^\nu,$$

$$\boldsymbol{\gamma} = (\gamma_1, \ldots, \gamma_\nu) \in \mathbb{R}_+^\nu,$$

be a sequence of ν-dimensional 'rectangles' whose sides grow at possibly different rates $O(\lambda^{\gamma_i}), i = 1, \ldots, \nu$ as $\lambda \to \infty$. Consider the partial sums RF

$$S_{\lambda,\boldsymbol{\gamma}}(\boldsymbol{x}) := \sum_{\boldsymbol{t} \in K_{\lambda,\boldsymbol{\gamma}}(\boldsymbol{x})} X(\boldsymbol{t}), \quad \boldsymbol{x} \in \mathbb{R}_+^\nu. \tag{127}$$

For given $\boldsymbol{\gamma} \in \mathbb{R}_+^\nu$, consider the scaling limit of the RF X along 'rectangles':

$$d_{\lambda,\boldsymbol{\gamma}}^{-1} S_{\lambda,\boldsymbol{\gamma}}(\boldsymbol{x}) \xrightarrow{\text{fdd}} V_{\boldsymbol{\gamma}}(\boldsymbol{x}), \quad \boldsymbol{x} \in \mathbb{R}_+^\nu, \tag{128}$$

where $d_{\lambda,\boldsymbol{\gamma}} \to \infty$ is a normalisation. We may ask about the existence of the scaling limits in (128) and the structure of the scaling diagram $\mathcal{V} = \{V_{\boldsymbol{\gamma}}; \boldsymbol{\gamma} \in \mathbb{R}_+^\nu\}$. A natural extension of Assumption $(\mathrm{A})_{\mathrm{LRD},\rho}$ is an MA RF,

$$X(\boldsymbol{t}) = \sum_{\boldsymbol{s} \in \mathbb{Z}^\nu} a(\boldsymbol{t} - \boldsymbol{s})\varepsilon(\boldsymbol{s}), \quad \boldsymbol{t} \in \mathbb{Z}^\nu, \tag{129}$$

with coefficients

$$a(\boldsymbol{t}) = \rho(\boldsymbol{t})^{-1}\big(\ell(\boldsymbol{t}) + o(1)\big), \quad |\boldsymbol{t}| \to \infty, \tag{130}$$

where $\rho(\boldsymbol{t}) := \sum_{i=1}^{\nu} |t_i|^{q_i}, \boldsymbol{t} = (t_1, \ldots, t_\nu) \in \mathbb{R}^\nu$, $q_i > 0, i = 1, \ldots, \nu$ are parameters and $\ell(\boldsymbol{t})$ is a generalised homogeneous function:

$$\ell(\lambda^{1/q_1} t_1, \ldots, \lambda^{1/q_\nu} t_\nu) = \ell(\boldsymbol{t}), \quad \boldsymbol{t} \in \mathbb{R}_0^\nu.$$

In the case when $\ell \geq 0$ is bounded, the coefficients in (130) satisfy $\sum_{\boldsymbol{t} \in \mathbb{Z}^\nu} |a(\boldsymbol{t})|^2 < \infty$ and $\sum_{\boldsymbol{t} \in \mathbb{Z}^\nu} |a(\boldsymbol{t})| = \infty$, provided

$$1 < Q := \sum_{i=1}^{\nu} \frac{1}{q_i} < 2.$$

The last fact is immediate by a repeated use of the identity $\int_0^\infty (u^q + w)^{-r} \mathrm{d}u = C w^{(1/q)-r}, q, r, w > 0$, $C := \int_0^\infty (u^q + 1)^{-r} \mathrm{d}u < \infty$ $(r > 1/q), := \infty$ $(r \le 1/q)$. Since $a(0, \dots, 0, t_i, 0, \dots, 0) = O(|t_i|^{-q_i})$ as $|t_i| \to \infty$, it means that when q_i are different, the MA coefficients decay at different rates in different directions of $\mathbb{Z}^\nu$. Arguing as in the case of $\nu = 2$, we may expect that the scaling limits in (128) for X in (18)–(130) crucially depend on the fulfilment or violation of the *balance conditions*:

$$\gamma_i/\gamma_j = q_j/q_i, \quad 1 \le i < j \le \nu. \tag{131}$$

This is true in the case of $\nu = 2$, where we can assume $\gamma_1 = 1, \gamma_2 = \gamma$ w.l.g., and (131) reduces to $\gamma = q_1/q_2$; see Theorem 22. The $\nu(\nu-1)/2$ linear equations in (131) describe the same number of hyperplanes in $\mathbb{R}^\nu_+$ and the partition $\mathbb{R}^\nu_+ = \bigcup_{i=1}^{\tau(\nu)} \Gamma_i$ into $\tau(\nu) \le 3^\nu$ disjoint sets of dimension $1 \le \dim(\Gamma_i) \le \nu$. On each of these sets, we can expect the same scaling limit $V_{\boldsymbol{\gamma}} \overset{\mathrm{fdd}}{=} V_i$ independent of $\boldsymbol{\gamma} \in \Gamma_i$, with the limits $V_i \ne V_j$ being generally different for $i \ne j$. Accordingly, in higher dimensions $\nu > 2$, the structure of $\mathcal{V}$ is more complicated than in the case of $\nu = 2$. Depending on $\boldsymbol{\gamma}$, these limits can be divided into three groups: the well-balanced limit arising when $\boldsymbol{\gamma}$ satisfies *all* balance conditions in (131) (group 1); 'partially unbalanced' limits arising when $\boldsymbol{\gamma}$ satisfies *some but not all* of the balance conditions in (131) (group 2), and 'completely unbalanced' limits arising when $\boldsymbol{\gamma}$ satisfies *none* of the balance conditions in (131) (group 3).

The above conjectures were verified and the structure of the scaling diagram $\mathcal{V}$ was identified in Ref. [250] in the case of $\nu = 3$. Indeed, in the latter (nontrivial) case, the Gaussian scaling limits $V_{\boldsymbol{\gamma}}$ in (128) exist for any $\boldsymbol{\gamma} \in \mathbb{R}^3_+$; moreover, 'completely unbalanced' limits (group 3) agree with the FBS B_{H_1,H_2,H_3} on $\mathbb{R}^3_+$, with *at least two* among three indices $H_i \in [\frac{1}{2}, 1], i = 1, 2, 3$ equal to $\frac{1}{2}$ or 1, while 'partially unbalanced' limits (group 2) are not FBS but have some 'partial FBS property' in one direction.

Let us describe the partition of $\mathbb{R}^3_+$ induced by balance conditions in (131) for $\nu = 3$. Set

$$\gamma_{\mathrm{cr},ij} := q_j/q_i, \quad i, j = 1, 2, 3, \ i \ne j.$$

The system in (131) comprises three equations:

$$\gamma_2/\gamma_1 = \gamma_{\mathrm{cr},21}, \quad \gamma_3/\gamma_1 = \gamma_{\mathrm{cr},31}, \quad \gamma_3/\gamma_2 = \gamma_{\mathrm{cr},32}, \tag{132}$$

among which *only two* are independent since any two of (132) imply the third one.
With $(\gamma_{\mathrm{cr},21}, \gamma_{\mathrm{cr},31}, \gamma_{\mathrm{cr},32}) \in \mathbb{R}^3_+$, $\gamma_{\mathrm{cr},32} = \gamma_{\mathrm{cr},31}/\gamma_{\mathrm{cr},21}$, we associate the partition

$$\mathbb{R}^3_+ = \bigcup_{\imath \in \wp} \Gamma_\imath$$

into 13 sets $\Gamma_\imath, \imath \in \wp$, defined as

$$\Gamma_{000} := \{\gamma \in \mathbb{R}^3_+ : \gamma_2/\gamma_1 = \gamma_{\mathrm{cr},21}, \gamma_3/\gamma_1 = \gamma_{\mathrm{cr},31}, \gamma_3/\gamma_2 = \gamma_{\mathrm{cr},32}\},$$
$$\Gamma_{011} := \{\gamma \in \mathbb{R}^3_+ : \gamma_2/\gamma_1 = \gamma_{\mathrm{cr},21}, \gamma_3/\gamma_1 > \gamma_{\mathrm{cr},31}, \gamma_3/\gamma_2 > \gamma_{\mathrm{cr},32}\},$$
$$\dots$$
$$\Gamma_{\text{-}111} := \{\gamma \in \mathbb{R}^3_+ : \gamma_2/\gamma_1 < \gamma_{\mathrm{cr},21}, \gamma_3/\gamma_1 > \gamma_{\mathrm{cr},31}, \gamma_3/\gamma_2 > \gamma_{\mathrm{cr},32}\}.$$

Thus, the set $\wp = \{\imath\}$ consists of 13 triples:

$$\wp = \{000,\ 011,\ 110,\ 10\text{-}1,\ 0\text{-}1\text{-}1,\ \text{-}1\text{-}10,\ \text{-}101,\ 111,$$
$$11\text{-}1,\ 1\text{-}1\text{-}1,\ \text{-}1\text{-}1\text{-}1,\ \text{-}1\text{-}11,\ \text{-}111\}.$$

The corresponding partition of $\mathbb{R}^2_+ = \{(\gamma_2/\gamma_1, \gamma_3/\gamma_1)\}$ is shown in Fig. 4. There, the line $\Gamma_{000} \subset \mathbb{R}^3_+$ satisfying all three balance conditions in (132) (the 'well-balanced' set) reduces to the single point $(\gamma_{\mathrm{cr},21}, \gamma_{\mathrm{cr},31}) = (q_1/q_2, q_1/q_3) \in \mathbb{R}^2_+$, the two-dimensional sets

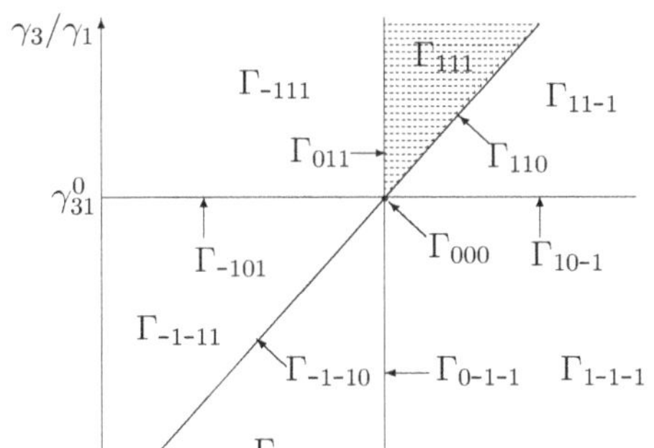

Figure 4. Partition of $\mathbb{R}^3_+$ induced by balance conditions in (132). The shaded region corresponds to the subset in (133).

$\Gamma_{011}, \Gamma_{110}, \Gamma_{10\text{-}1}, \Gamma_{0\text{-}1\text{-}1}, \Gamma_{\text{-}1\text{-}10}, \Gamma_{\text{-}101}$ satisfying only one of the balance conditions in (132) (the 'partly balanced' sets) become line segments, and the three-dimensional sets $\Gamma_{111}, \Gamma_{11\text{-}1}, \Gamma_{1\text{-}1\text{-}1}, \Gamma_{\text{-}1\text{-}1\text{-}1}, \Gamma_{\text{-}1\text{-}11}, \Gamma_{\text{-}111}$ which violate two (or all) balance conditions in (132) (the 'completely unbalanced' sets) are projected as sets of dimension 2.

The dependence of the scaling limits V_γ on the parameters q_i is different in different regions $\Gamma_\imath, \imath \in \wp$. By symmetry, it suffices to consider the region

$$\{\gamma = (\gamma_1, \gamma_2, \gamma_3) \in \mathbb{R}^3_+ : \gamma_1 q_1 \le \gamma_2 q_2 \le \gamma_3 q_3\} = \bigcup_{\imath=000,011,111,110} \Gamma_\imath \tag{133}$$

alone since $V_\imath$ for $\imath \notin \{000, 011, 111, 110\}$ can be obtained via permutation. Let

$$Q_{c_1,c_2,c_3} := \sum_{j=1}^{3} \frac{c_j}{q_j}, \quad c_1, c_2, c_3 > 0,$$

$Q = Q_{1,1,1}$. For γ satisfying (133), there are three parameter regions of $\boldsymbol{q} = (q_1, q_2, q_3)$, defined as follows:

$$\begin{aligned} \text{Region I:} &\quad Q_{1/2,1,1} < 1 < Q, \\ \text{Region II:} &\quad Q_{1/2,1/2,1} < 1 < Q_{1/2,1,1}, \\ \text{Region III:} &\quad Q_{1/2,1/2,1/2} < 1 < Q_{1/2,1/2,1}. \end{aligned} \tag{134}$$

The above regions are complemented by introducing Region 0: $Q < 1$. Since the dependence in RF X generally decreases with q_j, we may conclude that the dependence in X increases from Region 0 to Region III. The probabilistic meaning of the inequalities in (134) in terms of the summability of the covariance function $r(\boldsymbol{t}) = \mathrm{E}X(\boldsymbol{0})X(\boldsymbol{t})$ on the coordinate axes and coordinate planes in $\mathbb{Z}^3$ is provided in the following proposition.

Proposition 30. *Let $X = \{X(\boldsymbol{t}), \boldsymbol{t} \in \mathbb{Z}^3\}$ be a linear RF in* (18) *of Chapter 3 with coefficients as in* (130), *$\nu = 3$, where $\ell(\cdot) \ge 0$ is a bounded continuous function, $1 < Q < 2$, $a_\infty(\boldsymbol{t}) := \rho(\boldsymbol{t})^{-1}\ell(\boldsymbol{t})$. Then,*

$$r(\boldsymbol{t}) = \rho(\boldsymbol{t})^{Q-2}(L(\boldsymbol{t}) + o(1)), \quad |\boldsymbol{t}| \to \infty,$$

where

$$L(\boldsymbol{t}) := (a_\infty \star a_\infty)(\bar{\boldsymbol{t}}), \quad \bar{\boldsymbol{t}} := \left(\frac{t_1}{\rho(\boldsymbol{t})^{1/q_1}}, \frac{t_2}{\rho(\boldsymbol{t})^{1/q_2}}, \frac{t_3}{\rho(\boldsymbol{t})^{1/q_3}}\right)$$

is a generalised invariant function, $0 < C_1 \le L(\boldsymbol{t}) \le C_2 < \infty, \boldsymbol{t} \in \mathbb{R}^3_0$. *Moreover,* $r(\boldsymbol{t})$ *satisfies the following properties in respective parameter Regions 0–III:*

$$\begin{aligned} \text{Region 0}: &\ \textstyle\sum_{(t_1,t_2,t_3)\in\mathbb{Z}^3} |r(t_1,t_2,t_3)| < \infty, \\ \text{Region I}: &\ \textstyle\sum_{(t_1,t_2,t_3)\in\mathbb{Z}^3} |r(t_1,t_2,t_3)| = \infty, \\ &\ \textstyle\sum_{(t_2,t_3)\in\mathbb{Z}^2} |r(0,t_2,t_3)| < \infty; \\ \text{Region II}: &\ \textstyle\sum_{(t_2,t_3)\in\mathbb{Z}^2} |r(0,t_2,t_3)| = \infty, \quad \sum_{t_3\in\mathbb{Z}} |r(0,0,t_3)| < \infty; \\ \text{Region III}: &\ \textstyle\sum_{t_3\in\mathbb{Z}} |r(0,0,t_3)| = \infty. \end{aligned} \tag{135}$$

The divergence of the series in (135) in Region III can be interpreted as the (covariance) LRD property of the sectional process $\{X(0,0,t_3), t_3 \in \mathbb{Z}\}$ on the coordinate axis t_3. In Region II, this sectional process is SRD, but the sectional RF $\{X(0,t_2,t_3), (t_2,t_3) \in \mathbb{Z}^2\}$ is LRD. Similarly, in Region I, the last sectional RF is SRD, but the RF X on $\mathbb{Z}^3$ is LRD. Region 0 corresponds to the (covariance) SRD X. Conditions (134) are not symmetric w.r.t. the permutation of $q_j, j = 1,2,3$; therefore, the axes $t_j, j = 1,2,3$, there cannot generally be exchanged except for the 'isotropic' case of $q_1 = q_2 = q_3 =: q$. In the latter case, (134) reduce to $q > 3$ (Region 0), $2.5 < q < 3$ (Region I), $2 < q < 2.5$ (Region II), and $1.5 < q < 2$ (Region III), respectively.

Let us describe the scaling limits V_γ for γ satisfying (133) or belonging to the shaded region $\Gamma_{000} \cup \Gamma_{011} \cup \Gamma_{111} \cup \Gamma_{110}$ in Fig. 4.

Let $W(\mathrm{d}\boldsymbol{u})$ be a real-valued Gaussian white noise on $\mathbb{R}^3$ with zero mean and variance $\mathrm{d}\boldsymbol{u}$. Consider the following Gaussian RFs on $\mathbb{R}^3_+$, defined as stochastic integrals w.r.t. W:

$$\mathcal{Y}_1(\boldsymbol{x}) := \int_{\mathbb{R}^3} W(\mathrm{d}\boldsymbol{u}) \int_{\mathbb{R}^3} \frac{\mathbb{I}(0 < u_j < x_j, j = 2,3, 0 < t_1 < x_1)\mathrm{d}\boldsymbol{t}}{\rho(t_1 - u_1, t_2, t_3)}, \quad \text{(Region I)} \tag{136}$$

$$\mathcal{Y}_2(\boldsymbol{x}) := x_1 \int_{\mathbb{R}^3} W(\mathrm{d}\boldsymbol{u}) \int_{\mathbb{R}^2} \frac{\mathbb{I}(0 < u_3 < x_3, 0 < t_2 < x_2)\mathrm{d}t_2 \mathrm{d}t_3}{\rho(u_1, t_2 - u_2, t_3)},$$

(Region II)

$$\mathcal{Y}_3(\boldsymbol{x}) := x_1 x_2 \int_{\mathbb{R}^3} W(\mathrm{d}\boldsymbol{u}) \int_{\mathbb{R}} \frac{\mathbb{I}(0 < t_3 < x_3)\mathrm{d}t_3}{\rho(u_1, u_2, t_3 - u_3)}$$

(Region III)

$$\mathcal{Y}_{12}(\boldsymbol{x}) := \int_{\mathbb{R}^3} W(\mathrm{d}\boldsymbol{u}) \int_{\mathbb{R}^3} \frac{\mathbb{I}(0 < t_j < x_j, j = 1, 2, 0 < u_3 < x_3)\mathrm{d}\boldsymbol{t}}{\rho(t_1 - u_1, t_2 - u_2, t_3)}$$

(Regions I&II)

$$\mathcal{Y}_{23}(\boldsymbol{x}) := x_1 \int_{\mathbb{R}^3} W(\mathrm{d}\boldsymbol{u}) \int_{\mathbb{R}^2} \frac{\mathbb{I}(0 < t_j < x_j, j = 2, 3)\mathrm{d}t_2 \mathrm{d}t_3}{\rho(u_1, t_2 - u_2, t_3 - u_3)},$$

(Regions II&III)

$$\mathcal{Y}_0(\boldsymbol{x}) := \int_{\mathbb{R}^3} W(\mathrm{d}\boldsymbol{u}) \int_{\mathbb{R}^3} \frac{\mathbb{I}(0 < t_j < x_j, j = 1, 2, 3)\mathrm{d}\boldsymbol{t}}{\rho(\boldsymbol{t} - \boldsymbol{u})}.$$

(Regions I&II&III)

Proposition 31. *The Gaussian RFs in* (136) *are well defined in the indicated parameter regions. More precisely,*

(i1) $\mathcal{Y}_1$ *is well defined in Region I;*
(i2) $\mathcal{Y}_2$ *is well defined in Region II;*
(i3) $\mathcal{Y}_3$ *is well defined in Region III;*
(i4) $\mathcal{Y}_{12}$ *is well defined for* $Q_{1/2,1/2,1} < 1 < Q$, *which contains Regions I and II;*
(i5) $\mathcal{Y}_{23}$ *is well defined for* $Q/2 < 1 < Q_{1/2,1,1}$, *which contains Regions II and III;*
(i6) $\mathcal{Y}_0$ *is well defined for* $1 < Q < 2$, *which contains Regions I, II, and III.*

Moreover, the RFs $\mathcal{Y}_i, i = 1, 2, 3$ *agree, up to multiplicative constants* $\kappa_i = \mathrm{E}^{1/2}[\mathcal{Y}_i^2(\mathbf{1})]$, *with FBS having two its parameters equal to either* 1/2 *or 1. Namely,*

$$\mathcal{Y}_1 \overset{\mathrm{fdd}}{=} \kappa_1 B_{H_1,1/2,1/2}, \quad \mathcal{Y}_2 \overset{\mathrm{fdd}}{=} \kappa_2 B_{1,H_2,1/2}, \quad \mathcal{Y}_3 \overset{\mathrm{fdd}}{=} \kappa_3 B_{1,1,H_3},$$

where $H_i \in (1/2, 1), i = 1, 2, 3$ are defined as

$$H_1 := q_1 \left(Q_{\frac{3}{2},1,1} - 1 \right), \quad H_2 := q_2 \left(Q_{\frac{1}{2},\frac{3}{2},1} - 1 \right),$$
$$H_3 := q_3 \left(Q_{\frac{1}{2},\frac{1}{2},\frac{3}{2}} - 1 \right).$$

It follows that $\mathcal{Y}_i, i = 1, 2, 3$, are FBS whose smoothness increases with i and $\mathcal{Y}_0$ has a similar form as the well-balanced limit in dimension $\nu = 2$. The remaining two RFs $\mathcal{Y}_{12}$ and $\mathcal{Y}_{23}$ are not FBS but share some dependence properties of FBS along coordinate axes. Particularly, $\mathcal{Y}_{12}$ has independent increments in direction x_3, and $\mathcal{Y}_{23}$ has invariant increments in direction x_1. All RFs in (136) satisfy the operator scaling property with a diagonal matrix E: for $\mathcal{Y}_0$, we have $E = \mathrm{diag}(1/q_1, 1/q_2, 1/q_3)$ and for $\mathcal{Y}_i, i = 1, 2, 3$, $E = \mathrm{diag}(\gamma_1, \gamma_2, \gamma_3)$, where $\boldsymbol{\gamma} = (\gamma_1, \gamma_2, \gamma_3) \in \mathbb{R}^3_+$ is arbitrary. The RFs $\mathcal{Y}_{12}$ and $\mathcal{Y}_{23}$ satisfy operator scaling property with respective matrices $E = \mathrm{diag}(1/q_1, 1/q_2, \gamma_3)$ and $E = \mathrm{diag}(\gamma_1, 1/q_2, 1/q_3)$, for any $\gamma_1, \gamma_3 > 0$.

The proofs of Propositions 30 and 31 and other results in this section, including a detailed description of the scaling limits V_γ in the case of $\nu = 3$, are given in Ref. [250]. They use a similar approach as in the case of $\nu = 2$ but are longer and more cumbersome due to a larger number of different cases and subcases.

8 Some Proofs

This section contains the proofs of Theorems 16, 17, and 22 and Propositions 21, 24, and 25. We also provide a sketch of the proof of Theorem 23. A complete proof of the last theorem is given in Ref. [251].

Let us explain the main idea behind the proof of Theorems 16 and 17. Using Proposition 7 of Chapter 2, we reduce the convergence in (2) to that of the covariance functions

$$R_{\lambda,\gamma}(\boldsymbol{x}, \boldsymbol{y}) := d_{\lambda,\gamma}^{-2} \mathrm{E} S_{\lambda,\gamma}(\boldsymbol{x}) S_{\lambda,\gamma}(\boldsymbol{y}) \to \mathrm{E} V_\gamma(\boldsymbol{x}) V_\gamma(\boldsymbol{y}) =: R_\gamma(\boldsymbol{x}, \boldsymbol{y}),$$
$$\lambda \to \infty, \tag{137}$$

for any $\boldsymbol{x}, \boldsymbol{y} \in \mathbb{R}^2_+$, and a Lindeberg-type condition

$$\sup_{\boldsymbol{u} \in \mathbb{Z}^2} \left| \sum_{\boldsymbol{t} \in K_{[\lambda, \lambda^\gamma]}} a(\boldsymbol{t} - \boldsymbol{u}) \right| = o(d_{\lambda,\gamma}), \quad \lambda \to \infty. \tag{138}$$

Indeed, the finite-dimensional convergence in (2) is equivalent to the one-dimensional convergence $\mathcal{S}_{\lambda,\gamma} := d_{\lambda,\gamma}^{-1} \sum_{j=1}^m \theta_j S_{\lambda,\gamma}(\boldsymbol{x}_j) \xrightarrow{\mathrm{d}} \sum_{j=1}^m \theta_j V_\gamma(\boldsymbol{x}_j) =: \mathcal{V}_\gamma$ for any $\theta_j \in \mathbb{R}, \boldsymbol{x}_j = (x_{1j}, x_{2j}) \in \mathbb{R}^2_+$, $j = 1, \ldots, m \geq 1$. From (8), we have that $\mathrm{E}(\mathcal{S}_{\lambda,\gamma})^2 \to \sum_{j,j'=1}^m \theta_j \theta_{j'} R_\gamma(\boldsymbol{x}_j, \boldsymbol{x}_{j'}) =: \mathcal{R}_\gamma \geq 0$. Then, $\mathcal{S}_{\lambda,\gamma} = d_{\lambda,\gamma}^{-1} \sum_{\boldsymbol{u} \in \mathbb{Z}^2} g_\lambda(\boldsymbol{u}) \varepsilon(\boldsymbol{u})$ is a weighted linear form as in Proposition 7 of Chapter 2, with weights $d_\lambda(\boldsymbol{u}) := \sum_{j=1}^m \theta_j \sum_{\boldsymbol{t} \in K_{[\lambda x_{1j}, \lambda^\gamma x_{2j}]}} a(\boldsymbol{t} - \boldsymbol{u})$. Thanks to (138) and the latter proposition, $\mathcal{S}_\gamma \xrightarrow{\mathrm{d}} N(0, \mathcal{R}_\gamma)$, or, in other words, the distribution of the vector $(d_{\lambda,\gamma}^{-1} S_{\lambda,\gamma}(\boldsymbol{x}_j); j = 1, \ldots, m)$ tends to a Gaussian distribution on $\mathbb{R}^m$ with zero mean and covariance matrix $(R_\gamma(\boldsymbol{x}_j, \boldsymbol{x}_{j'}))_{1 \leq j, j' \leq m}$.

Verification of (138) being rather easy, the main attention in what follows is given to the proof of (8). We use the representation of $R_{\lambda,\gamma}(\boldsymbol{x}, \boldsymbol{y}) = d_{\lambda,\gamma}^{-2} \sum_{\boldsymbol{t} \in K_{[\lambda x_1, \lambda^\gamma x_2]}, \boldsymbol{s} \in K_{[\lambda y_1, \lambda^\gamma y_2]}} r(\boldsymbol{t} - \boldsymbol{s})$ through the spectral density:

$$R_{\lambda,\gamma}(\boldsymbol{x}, \boldsymbol{y}) = d_{\lambda,\gamma}^{-2} \int_{\Pi^2} \prod_{j=1}^2 D_{[\lambda^{\gamma_j} x_j]}(u_j) \overline{D_{[\lambda^{\gamma_j} y_j]}(u_j)} f(\boldsymbol{u}) \mathrm{d}\boldsymbol{u}, \tag{139}$$

where $\gamma_1 := 1, \gamma_2 := \gamma$ and $D_n(u), u \in \Pi$ is the *Dirichlet kernel*:

$$D_n(u) := \sum_{t=1}^n \mathrm{e}^{\mathrm{i}tu} = \frac{\mathrm{e}^{\mathrm{i}u} - \mathrm{e}^{\mathrm{i}(n+1)u}}{1 - \mathrm{e}^{\mathrm{i}u}}. \tag{140}$$

The covariance function of the (unbalanced) limiting FBS has a spectral representation: for $(H_1, H_2) \in (0,1)^2$:

$$\mathrm{E}B_{H_1,H_2}(\boldsymbol{x}) B_{H_1,H_2}(\boldsymbol{y}) = \kappa^{-2} \int_{\mathbb{R}^2} \prod_{j=1}^2 \frac{(1 - \mathrm{e}^{\mathrm{i}x_j u_j})(1 - \mathrm{e}^{-\mathrm{i}y_j u_j})}{|u_j|^{2H_j + 1}} \mathrm{d}\boldsymbol{u}, \tag{141}$$

where

$$\kappa^2 := \int_{\mathbb{R}^2} \prod_{j=1}^{2} \frac{|1-\mathrm{e}^{\mathrm{i}u_j}|^2}{|u_j|^{2H_j+1}} \mathrm{d}\boldsymbol{u} = \prod_{j=1}^{2} \frac{\pi}{H_j \Gamma(2H_j)\sin(H_j\pi)}, \tag{142}$$

which follows from the definition in (42) of Chapter 1 and the well-known spectral representation of the covariance of the FBM; see Ref. [11]. The representation in (141) is not valid when $(H_1, H_2) \notin (0,1)^2$, in which case the covariance function of B_{H_1,H_2} is directly obtained from (42) of Chapter 1. We also have that the covariance of the well-balanced limit V_0 has the spectral representation

$$\mathrm{E}V_0(\boldsymbol{x})V_0(\boldsymbol{y}) = \int_{\mathbb{R}^2} \prod_{j=1}^{2} \frac{(1-\mathrm{e}^{\mathrm{i}u_j x_j})(1-\mathrm{e}^{-\mathrm{i}u_j y_j})}{u_j^2} f_0(\boldsymbol{u})\mathrm{d}\boldsymbol{u}, \quad \boldsymbol{x}, \boldsymbol{y} \in \mathbb{R}^2_+. \tag{143}$$

Thus, except for 'degenerate' cases $(H_1, H_2) \notin (0,1)^2$, relation (8) is deduced to the convergence of the integral in (139) towards corresponding integrals in (141), (143), for any given $\boldsymbol{x}, \boldsymbol{y} \in \mathbb{R}^2_+$. The last convergence follows by suitably rescaling the integrand in (8) and using (67) together with a dominated convergence argument. The 'degenerate' cases $(H_1, H_2) \in [0,1]^2 \setminus (0,1)^2$ require somewhat different scaling argument, as seen in the following.

Proof of Theorem 16. We first prove (8) in each region of Table 1. Relation (138) is verified afterwards.

Case $\upsilon_1 \vee \upsilon_2 < 1$ (region R^+_{22} in Table 1). According to (69), $H^+_1 = (1+\upsilon_1)/2 \in (1/2, 1), H^+_2 = H^-_1 = 1/2, H^-_2 = (1+\upsilon_2)/2 \in (1/2,1)$. By change of variables, $u_1 \to u_1/\lambda, u_2 \to u_2/\lambda^\gamma$, we rewrite (139) as $R_{\lambda,\gamma}(\boldsymbol{x}, \boldsymbol{y}) = \int_{\lambda\Pi \times \lambda^\gamma \Pi} G_\lambda(\boldsymbol{u})\mathrm{d}\boldsymbol{u}$, where

$$G_\lambda(\boldsymbol{u}) := \frac{D_{[\lambda x_1]}(u_1/\lambda)\overline{D_{[\lambda y_1]}(u_1/\lambda)}}{\lambda^2} \cdot \frac{D_{[\lambda^\gamma x_2]}(u_2/\lambda^\gamma)\overline{D_{[\lambda^\gamma y_2]}(u_2/\lambda^\gamma)}}{\lambda^{2\gamma}} \cdot f_\lambda(\boldsymbol{u}), \tag{144}$$

$$f_\lambda(\boldsymbol{u}) := f(u_1/\lambda, u_2/\lambda^\gamma)/\lambda^{2H(\gamma)-1-\gamma}.$$

By (60) and the definition of $H(\gamma)$ in (73).

$$\lambda^{-1}D_{[\lambda x_1]}(u_1/\lambda) \to \frac{1-\mathrm{e}^{\mathrm{i}x_1u_1}}{-\mathrm{i}u_1}, \quad \lambda^{-\gamma}D_{[\lambda^\gamma x_2]}(u_2/\lambda^\gamma) \to \frac{1-\mathrm{e}^{\mathrm{i}x_2u_2}}{-\mathrm{i}u_2}, \tag{145}$$

$$f_\lambda(\boldsymbol{u}) \to g_\gamma(\boldsymbol{u}) := \begin{cases} f_0(\boldsymbol{u}), & \gamma = \gamma_{\mathrm{cr}}, \\ f_0(u_1,0), & \gamma > \gamma_{\mathrm{cr}}, \\ f_0(0,u_2), & \gamma < \gamma_{\mathrm{cr}}, \end{cases}$$

point-wise for any $\boldsymbol{u} \in \mathbb{R}^2_0$, where $f_0(\boldsymbol{u}) = L_{\mathrm{sp}}(\boldsymbol{u})/\rho_{\mathrm{sp}}(\boldsymbol{u})$; see (67). Particularly, $f_0(u_1,0) = |u_1|^{-\upsilon_1}L_{\mathrm{sp}}(1,0)$, $f_0(0,u_2) = |u_2|^{-\upsilon_2}L_{\mathrm{sp}}(0,1), \boldsymbol{u} = (u_1,u_2) \in \mathbb{R}^2_0$ since $L_{\mathrm{sp}}(1,0) = L_{\mathrm{sp}}(-1,0)$, $L_{\mathrm{sp}}(0,1) = L_{\mathrm{sp}}(0,-1)$. Set

$$\kappa_+^2 := L(1,0)\kappa^2 = L(1,0)\int_{\mathbb{R}^2} |u_1|^{-2H_1^+-1}|u_2|^{-2}\prod_{j=1}^{2}|1-\mathrm{e}^{\mathrm{i}u_j}|^2\mathrm{d}\boldsymbol{u} \tag{146}$$

in agreement with (142). We also see from (141), (146) that the covariance function of the limit RF is written as $\mathrm{E}V_\gamma(\boldsymbol{x})V_\gamma(\boldsymbol{y}) = \kappa_+^2\mathrm{E}B_{H_1^+,1/2}(\boldsymbol{x})B_{H_1^+,1/2}(\boldsymbol{y}) = \int_{\mathbb{R}^2}\prod_{j=1}^2\big(\frac{1-\mathrm{e}^{\mathrm{i}x_ju_j}}{\mathrm{i}u_j}\big)\big(\frac{1-\mathrm{e}^{-\mathrm{i}y_ju_j}}{-\mathrm{i}u_j}\big)|u_1|^{-2H_1^+}$ $|u_2|^{-1}\mathrm{d}\boldsymbol{u}$ and agrees with the integral $\int_{\mathbb{R}^2}G(\boldsymbol{u})\mathrm{d}\boldsymbol{u}$, where

$$G(\boldsymbol{u}) := \prod_{j=1}^{2}\left(\frac{1-\mathrm{e}^{\mathrm{i}x_ju_j}}{\mathrm{i}u_j}\right)\left(\frac{1-\mathrm{e}^{-\mathrm{i}y_ju_j}}{-\mathrm{i}u_j}\right)g_\gamma(\boldsymbol{u}) \tag{147}$$

is the product of the corresponding limit functions in (145) and $G_\lambda(\boldsymbol{u}) \to G(\boldsymbol{u}), \boldsymbol{u} \in \mathbb{R}^2_0$ according to (145). The proof of $\int_{\mathbb{R}^2}G_\lambda(\boldsymbol{u})\mathrm{d}\boldsymbol{u} \to \int_{\mathbb{R}^2}G(\boldsymbol{u})\mathrm{d}\boldsymbol{u}$ in all three cases $\gamma > \gamma_{\mathrm{cr}}, \gamma < \gamma_{\mathrm{cr}}$, and $\gamma = \gamma_{\mathrm{cr}}$ now follows from the DCT and the elementary bound $|D_n(u)| \le Cn/(1+n|u|), |u| < \pi$, implying

$$|\lambda^{-1}D_{[\lambda x]}(\tfrac{u}{\lambda})| \le C/(1+|u|), \quad |u| < \lambda\pi, \tag{148}$$

for any fixed $x \in \mathbb{R}, x \ne 0$ and any $\lambda > 1/|x|$. Hence,

$$|G_\lambda(\boldsymbol{u})| \le C\prod_{j=1}^2(1+u_j^2)^{-1}\begin{cases} \rho_{\mathrm{sp}}(\boldsymbol{u})^{-1}, & \gamma = \gamma_{\mathrm{cr}}, \\ \rho_{\mathrm{sp}}(u_1,0)^{-1}, & \gamma > \gamma_{\mathrm{cr}}, \\ \rho_{\mathrm{sp}}(0,u_2)^{-1}, & \gamma < \gamma_{\mathrm{cr}}, \end{cases} \tag{149}$$

and the r.h.s. of (149), denoted by $\bar{G}(\boldsymbol{u})$, is integrable: $\bar{I} := \int_{\mathbb{R}^2} \bar{G}(\boldsymbol{u})\mathrm{d}\boldsymbol{u} < \infty$. Indeed, $\bar{I} \le C \int_{\mathbb{R}}(1+|u_1|^2)^{-1}|u_1|^{-\upsilon_1}\mathrm{d}u_1 \times \int_{\mathbb{R}}(1+|u_2|^2)^{-1}\mathrm{d}u_2 < \infty$ when $\gamma > \gamma_{\mathrm{cr}}$, and similarly, $\bar{I} < \infty$ in the other two cases of γ.

Case $\upsilon_1 < 1 < \upsilon_2$ (region R_{12} in Table 1). For $\gamma \ge \gamma_{\mathrm{cr}}$, the limits of $R_{\lambda,\gamma}(\boldsymbol{x},\boldsymbol{y})$ and their proofs are the same as the previous case. Let $\gamma < \gamma_{\mathrm{cr}}$, in which case $H_1^- = \frac{1}{2}(1+\upsilon_1-\frac{\upsilon_1}{\upsilon_2})$, $H_2^- = 1$. In (139), change the variables as $u_1 \to u_1/\lambda, u_2 \to u_2/\lambda^{\gamma_0}$, $\gamma_0 := \gamma_{\mathrm{cr}} = \frac{\upsilon_1}{\upsilon_2}$. Then, $R_{\lambda,\gamma}(\boldsymbol{x},\boldsymbol{y}) = \int_{\lambda\Pi\times\lambda^{\gamma_0}\Pi} G_\lambda(\boldsymbol{u})\mathrm{d}\boldsymbol{u}$, where

$$
\begin{aligned}
G_\lambda(\boldsymbol{u}) &:= \frac{D_{[\lambda x_1]}(u_1/\lambda)\overline{D_{[\lambda y_1]}(u_1/\lambda)}}{\lambda^2} \\
&\quad \cdot \frac{D_{[\lambda^\gamma x_2]}(u_2/\lambda^{\gamma_0})\overline{D_{[\lambda^\gamma y_2]}(u_2/\lambda^{\gamma_0})}}{\lambda^{2\gamma}} \cdot f_\lambda(\boldsymbol{u}), \qquad (150)\\
f_\lambda(\boldsymbol{u}) &:= f(u_1/\lambda, u_2/\lambda^{\gamma_0})/\lambda^{\upsilon_1}.
\end{aligned}
$$

Contrary to (145), the second factor in the expression of $G_\lambda(\boldsymbol{u})$ in (150) now tends to x_2y_2 independent of $u_2 \ne 0$, which follows from $D_n(u/m)/n \to 1\,(n,m,m/n\to\infty)$; see the definition in (140) of the Dirichlet kernel. Therefore,

$$
G_\lambda(\boldsymbol{u}) \to G(\boldsymbol{u}) := x_2y_2\left(\frac{1-\mathrm{e}^{\mathrm{i}x_1u_1}}{\mathrm{i}u_1}\right)\left(\frac{1-\mathrm{e}^{-\mathrm{i}y_1u_1}}{-\mathrm{i}u_1}\right) f_0(\boldsymbol{u}) \qquad (151)
$$

point-wise for any $\boldsymbol{u} \in \mathbb{R}_0^2$ with $f_0(\boldsymbol{u})$ as in (67). This, together with a DCT argument as in the previous case, leads to $\int_{\mathbb{R}^2} G_\lambda(\boldsymbol{u})\mathrm{d}\boldsymbol{u} \to \int_{\mathbb{R}^2} G(\boldsymbol{u})\mathrm{d}\boldsymbol{u}$, and the proof follows by noting that the last integral equals $\kappa_-^2\, x_2y_2\mathrm{E}B_{H_1^-}(x_1)B_{H_1^-}(y_1) = \kappa_-^2\mathrm{E}B_{H_1^-,1}(\boldsymbol{x})B_{H_1^-,1}(\boldsymbol{y})$, with

$$
\begin{aligned}
\kappa_-^2 &:= \int_{\mathbb{R}^2} |(1-\mathrm{e}^{\mathrm{i}u_1})/u_1|^2 f_0(\boldsymbol{u})\mathrm{d}\boldsymbol{u} \qquad (152)\\
&= \int_{\mathbb{R}} |1-\mathrm{e}^{\mathrm{i}u_1}|^2|u_1|^{-2H_1^- -1}\mathrm{d}u_1 \int_{\mathbb{R}} L_{\mathrm{sp}}(1,u_2)\rho_{\mathrm{sp}}^{-1}(1,u_2)\mathrm{d}u_2
\end{aligned}
$$

and where we used the fact that $\int_{\mathbb{R}} f_0(\boldsymbol{u})\mathrm{d}u_2 = |u_1|^{(\upsilon_1/\upsilon_2)-\upsilon_1} \int_{\mathbb{R}} L_{\mathrm{sp}}(1,u_2)\rho_{\mathrm{sp}}^{-1}(1,u_2)\mathrm{d}u_2$ by the generalised invariant and symmetry properties of L_{sp}.

Case $\upsilon_2 < 1 < \upsilon_1$ (region R_{21} in Table 1). The proof of (8) is completely symmetric to the previous case by exchanging $\upsilon_1 \leftrightarrow \upsilon_2$, $u_1 \leftrightarrow u_2$.

Case $\upsilon_1 \wedge \upsilon_2 > 1$ (region R_{11} in Table 1). The proof of (8) for $\gamma \le \gamma_{\mathrm{cr}}$ is the same as in region R_{12}, whereas for $\gamma > \gamma_{\mathrm{cr}}$, it follows by symmetry as in region R_{21}.

It remains to check condition (138). The Cauchy–Schwarz inequality $\left|\sum_{t \in K_{[\lambda,\lambda^\gamma]}} a(\boldsymbol{t} - \boldsymbol{u})\right| \le \lambda^{(1+\gamma)/2} \times \|a\|_{L^2(\mathbb{Z}^2)}$ reduces (138) to $H(\gamma) > (1+\gamma)/2$, which holds by $H_1^\pm \wedge H_2^\pm \ge 1/2, H_1^\pm + H_2^\pm > 1$. The proof of Theorem 16 is now complete. □

Proof of Theorem 17. Similarly as in the proof of the previous theorem, we first prove (8) and then check (138).

Case $\upsilon_1 \vee \upsilon_2 < 1, \gamma_{\mathrm{cr}} = \upsilon_1/\upsilon_2$ (region R_{22}^- in Table 1). In this case, $H_i^\pm > 0, i = 1, 2$, and the proof of (8) closely resembles that of Theorem 16, case $\upsilon_1 \vee \upsilon_2 < 1$. Particularly, (145) hold with $f_0(\boldsymbol{u}) = L_{\mathrm{sp}}(\boldsymbol{u})\rho_{\mathrm{sp}}(\boldsymbol{u})$; see (67). The asymptotic constants $\kappa_\pm$ in (74) take a similar form as in (146):

$$\begin{aligned}\kappa_+^2 &:= \int_{\mathbb{R}^2} f_0(u_1, 0) \prod_{j=1}^{2} |(1 - \mathrm{e}^{\mathrm{i}u_j})/u_j|^2 \mathrm{d}\boldsymbol{u} \\ &= L_{\mathrm{sp}}(1, 0) \int_{\mathbb{R}^2} |u_1|^{\upsilon_1} \prod_{j=1}^{2} |(1 - \mathrm{e}^{\mathrm{i}u_j})/u_j|^2 \mathrm{d}\boldsymbol{u}. \end{aligned} \tag{153}$$

Case $\upsilon_1 < 1 < \upsilon_2, \gamma_{\mathrm{cr}} = \upsilon_1$ (region R_{23} in Table 1). We have three subcases, $\gamma > \upsilon_1, \gamma < \upsilon_1$, and $\gamma = \upsilon_1$, which are treated separately as follows.

Subcase $\gamma > \upsilon_1$. We shall prove that $V_+ = \kappa_+ B_{H_1^+, 1/2}$, with $d_{\lambda,\gamma} = \lambda^{H(\gamma)}, H(\gamma) = H_1^+ + \gamma/2 = (1 - \upsilon_1 + \gamma)/2$. Split $f(\boldsymbol{u}) = f_1(\boldsymbol{u}) + \tilde{f}_1(\boldsymbol{u})$, where $f_1(\boldsymbol{u}) := f(u_1, 0), \tilde{f}_1(\boldsymbol{u}) := f(\boldsymbol{u}) - f(u_1, 0)$. Accordingly, $R_{\lambda,\gamma}(\boldsymbol{x}, \boldsymbol{y}) = R_1(\boldsymbol{x}, \boldsymbol{y}) + \tilde{R}_1(\boldsymbol{x}, \boldsymbol{y})$. The convergence of the first term $R_1(\boldsymbol{x}, \boldsymbol{y}) \to \kappa_+^2 \mathrm{E} B_{H_1^+, 1/2}(\boldsymbol{x}) B_{H_1^+, 1/2}(\boldsymbol{y})$ with κ_+^2 in (153) for any $\gamma > 0$ follows as in the case $\upsilon_1 \vee \upsilon_2 < 1$ above. Therefore, it suffices to prove $\tilde{R}_1(\boldsymbol{x}, \boldsymbol{x}) \to 0$, or

$$J_\lambda := \int_{\Pi^2} |\tilde{f}_1(\boldsymbol{u})|\, |D_{[\lambda x_1]}(u_1) D_{[\lambda^\gamma x_2]}(u_2)|^2 \mathrm{d}\boldsymbol{u} = o(\lambda^{1-\upsilon_1+\gamma}). \tag{154}$$

We have $|\tilde{f}_1(\boldsymbol{u})| \le C\sum_{k=1}^{3} h_{1,k}(\boldsymbol{u})$, where $h_{1,1}(\boldsymbol{u}) := \rho_{\rm sp}(\boldsymbol{u}) - \rho_{\rm sp}(u_1,0) = |u_2|^{\upsilon_2}$, $h_{1,2}(\boldsymbol{u}) := \rho_{\rm sp}(u_1,0) \times (L_{\rm sp}(\boldsymbol{u}) - L_{\rm sp}(u_1,0))$ and $h_{1,3}(\boldsymbol{u}) := \rho_{\rm sp}(\boldsymbol{u})\delta(\boldsymbol{u})$, where $\delta(\boldsymbol{u})$ is a bounded function tending to 0 as $|\boldsymbol{u}| \to 0$. Accordingly, $J_\lambda \le C\sum_{k=1}^{3} J_{\lambda,k}$ where $J_{\lambda,k}$ is defined as in (154), with $|\tilde{f}_1(\boldsymbol{u})|$ replaced by $h_{1,k}(\boldsymbol{u})$. Here,

$$
\begin{aligned}
J_{\lambda,1} &\le \int_{\Pi^2} |u_2|^{\upsilon_2} \frac{|1-\mathrm{e}^{\mathrm{i}[\lambda x_1]u_1}|^2}{|u_1|^2|u_2|^2}\mathrm{d}\boldsymbol{u} \;\le\; \lambda \int_{\mathbb{R}\times\Pi} \frac{|1-\mathrm{e}^{\mathrm{i}u_1x_1}|^2}{|u_1|^2}|u_2|^{\upsilon_2-2}\mathrm{d}\boldsymbol{u} \\
&\le C\lambda = o(\lambda^{1-\upsilon_1+\gamma}) \qquad (155)
\end{aligned}
$$

since $\gamma > \upsilon_1$. Next,

$$
\begin{aligned}
J_{\lambda,2} \le& \int_{\Pi} |u_1|^{\upsilon_1} \frac{|1-\mathrm{e}^{\mathrm{i}[\lambda x_1]u_1}|^2}{|u_1|^2}\mathrm{d}u_1 \\
&\times \int_{\Pi} |L_{\rm sp}(u_1,u_2) - L_{\rm sp}(u_1,0)| \frac{|1-\mathrm{e}^{\mathrm{i}[\lambda^\gamma x_2]u_2}|^2}{|u_2|^2}\mathrm{d}u_2 \\
&\le C\lambda^{1-\upsilon_1+\gamma} \int_{\mathbb{R}} |u_1|^{\upsilon_1} \frac{|1-\mathrm{e}^{\mathrm{i}u_1x_1}|^2}{|u_1|^2}\mathrm{d}u_1 \\
&\times \int_{\mathbb{R}} \left|L_{\rm sp}\left(\frac{u_1}{\lambda}, \frac{u_2}{\lambda^\gamma}\right) - L_{\rm sp}\left(\frac{u_1}{\lambda}, 0\right)\right| \frac{|1-\mathrm{e}^{\mathrm{i}u_2x_2}|^2}{|u_2|^2}\mathrm{d}u_2 \\
&= o(\lambda^{1-\upsilon_1+\gamma}) \qquad (156)
\end{aligned}
$$

by the DCT and the fact that $\left|L_{\rm sp}\left(\frac{u_1}{\lambda}, \frac{u_2}{\lambda^\gamma}\right) - L_{\rm sp}\left(\frac{u_1}{\lambda}, 0\right)\right|$ is bounded and tends to 0 for any $\boldsymbol{u} \in \mathbb{R}^2_0$ by continuity of $L_{\rm sp}$ on $\{\rho_{\rm sp}(\boldsymbol{u}) = 1\}$ since $\gamma > \upsilon_1 > \upsilon_1/\upsilon_2$. A similar argument also entails $J_{\lambda,3} = o(\lambda^{1-\upsilon_1+\gamma})$, proving (154) and (8) in the subcase $\gamma > \upsilon_1$.

Subcase $\gamma < \upsilon_1$. We shall prove (8) for $V_- = \kappa_- B_{1/2,0}$, with $d^2_{\lambda,\gamma} = \lambda$ and κ_- in (158). Split $f(\boldsymbol{u}) = f_2(\boldsymbol{u}) + \tilde{f}_2(\boldsymbol{u})$, $f_2(\boldsymbol{u}) := f(0,u_2)$, $\tilde{f}_2(\boldsymbol{u}) := f(\boldsymbol{u}) - f(0,u_2)$, and accordingly, $R_{\lambda,\gamma}(\boldsymbol{x},\boldsymbol{y}) = R_2(\boldsymbol{x},\boldsymbol{y}) + \tilde{R}_2(\boldsymbol{x},\boldsymbol{y})$. It suffices to prove

$$
R_2(\boldsymbol{x},\boldsymbol{y}) \to \kappa_-^2(x_1 \wedge y_1) \times \frac{1}{2}(1 + \mathbb{I}(x_2 = y_2)) \quad \text{and} \quad \tilde{R}_2(\boldsymbol{x},\boldsymbol{x}) \to 0. \qquad (157)
$$

The first relation in (157) follows from

$$R_2(\boldsymbol{x},\boldsymbol{y}) = \frac{1}{\lambda}\int_\Pi D_{[\lambda x_1]}(u)\overline{D_{[\lambda y_1]}(u)}\mathrm{d}u \\ \times \int_\Pi f(0,v)D_{[\lambda^\gamma x_2]}(v)\overline{D_{[\lambda^\gamma y_2]}(v)}\mathrm{d}v =: J'_\lambda \times J''_\lambda,$$

where

$$J'_\lambda \to \int_{\mathbb{R}} \frac{(1-\mathrm{e}^{\mathrm{i}x_1u})(1-\mathrm{e}^{-\mathrm{i}y_1u})}{|u|^2}\mathrm{d}u = (x_1\wedge y_1)\int_{\mathbb{R}}\frac{|1-\mathrm{e}^{\mathrm{i}u}|^2}{|u|^2}\mathrm{d}u$$

follows by the DCT, and

$$J''_\lambda = \int_\Pi \frac{f(0,v)}{|1-\mathrm{e}^{\mathrm{i}v}|^2}\left(1-\mathrm{e}^{\mathrm{i}[\lambda^\gamma x_2]v}-\mathrm{e}^{-\mathrm{i}[\lambda^\gamma y_2]v}+\mathrm{e}^{\mathrm{i}([\lambda^\gamma x_2]-[\lambda^\gamma y_2])v}\right)\mathrm{d}v \\ \to \int_\Pi \frac{f(0,v)}{|1-\mathrm{e}^{\mathrm{i}v}|^2}\mathrm{d}v \times \begin{cases} 2, & x_2 = y_2, \\ 1, & x_2 \neq y_2, \end{cases}$$

by the Lebesgue–Riemann theorem [241, Thm. 1.2] and the integrability of $f(0,v)|1-\mathrm{e}^{\mathrm{i}v}|^{-2}$. Therefore, the asymptotic constant in (157) equals

$$\kappa_-^2 := 2\int_{\mathbb{R}} |(1-\mathrm{e}^{\mathrm{i}u})/u|^2\mathrm{d}u \times \int_\Pi f(0,v)|1-\mathrm{e}^{\mathrm{i}v}|^{-2}\mathrm{d}v. \tag{158}$$

To show the second relation in (157), similarly as in the subcase $\gamma > \upsilon_1$ above, write $|\tilde{f}_2(\boldsymbol{u})| \leq C\sum_{k=1}^3 h_{2,k}(\boldsymbol{u})$ and $\tilde{R}_2(\boldsymbol{x},\boldsymbol{x}) \leq C\sum_{k=1}^3 \tilde{R}_{3,k}(\boldsymbol{x},\boldsymbol{x})$ accordingly, where $h_{2,1}(\boldsymbol{u}) := \rho_{\mathrm{sp}}(\boldsymbol{u}) - \rho_{\mathrm{sp}}(0,u_2) = |u_1|^{\upsilon_1}$, $h_{2,2}(\boldsymbol{u}) := \rho_{\mathrm{sp}}(0,u_2)(L_{\mathrm{sp}}(\boldsymbol{u}) - L_{\mathrm{sp}}(0,u_2))$ and $h_{2,3}(\boldsymbol{u}) := \rho_{\mathrm{sp}}(\boldsymbol{u})\delta(\boldsymbol{u})$, where $\delta(\boldsymbol{u})$ is a bounded function tending to 0 as $|\boldsymbol{u}| \to 0$. Then, $\tilde{R}_{2,3}(\boldsymbol{x},\boldsymbol{x}) = o(1)$,

$$|\tilde{R}_{2,1}(\boldsymbol{x},\boldsymbol{x})| \leq C\lambda^{-1}\int_{\Pi^2}|u_1|^{\upsilon_1}\frac{|1-\mathrm{e}^{\mathrm{i}[\lambda x_1]u_1}|^2|1-\mathrm{e}^{\mathrm{i}[\lambda^\gamma x_2]u_2}|^2}{|1-\mathrm{e}^{\mathrm{i}u_1}|^2|1-\mathrm{e}^{\mathrm{i}u_2}|^2}\mathrm{d}\boldsymbol{u} \\ \leq C\lambda^{\gamma-\upsilon_1}\int_{\mathbb{R}}|u_1|^{\upsilon_1-2}|1-\mathrm{e}^{\mathrm{i}x_1u_1}|^2\mathrm{d}u_1 \\ \times \int_{\mathbb{R}}|u_2|^{-2}|1-\mathrm{e}^{\mathrm{i}x_2u_2}|^2\mathrm{d}u_2 \\ = O(\lambda^{\gamma-\upsilon_1}) = o(1)$$

and

$$
\begin{aligned}
|\tilde{R}_{2,2}(\boldsymbol{x},\boldsymbol{x})| &\le C\int_{\Pi}|u_2|^{\upsilon_2-2}\mathrm{d}u_2\\
&\quad\times\int_{\mathbb{R}}\left|L_{\mathrm{sp}}\left(\frac{u_1}{\lambda},u_2\right)-L_{\mathrm{sp}}(0,u_2)\right||(1-\mathrm{e}^{\mathrm{i}x_1u_1})/u_1|^2\mathrm{d}u_1\\
&= o(1), \qquad (159)
\end{aligned}
$$

proving (157) and (8).

Subcase $\gamma=\upsilon_1$. We shall prove (8) for $V_0=\kappa_+B_{H_1^+,1/2}+\kappa_-B_{1/2,0}$ equals the sum of independent limits in the subcases $\gamma>\upsilon_1$ and $\gamma<\upsilon_1$ treated above. Split $f(\boldsymbol{u})=f_1(\boldsymbol{u})+f_2(\boldsymbol{u})+\tilde{f}_{12}(\boldsymbol{u})$, $\tilde{f}_{12}(\boldsymbol{u}):=f(\boldsymbol{u})-f(u_1,0)-f(0,u_2)$ and, accordingly, $R_{\lambda,\gamma}(\boldsymbol{x},\boldsymbol{y})=R_1(\boldsymbol{x},\boldsymbol{y})+R_2(\boldsymbol{x},\boldsymbol{y})+\tilde{R}_{12}(\boldsymbol{x},\boldsymbol{y})$. Note that $H_1+\upsilon_1/2=1/2$. The convergences $R_1(\boldsymbol{x},\boldsymbol{y})\to\kappa_+^2\mathrm{E}B_{H_1,1/2}(\boldsymbol{x},\boldsymbol{y})B_{H_1^+,1/2}(\boldsymbol{x},\boldsymbol{y})$ and $R_2(\boldsymbol{x},\boldsymbol{y})\to\kappa_-^2\mathrm{E}B_{1/2,0}(\boldsymbol{x},\boldsymbol{y})B_{1/2,0}(\boldsymbol{x},\boldsymbol{y})$ were proved in the respective cases $\gamma>\gamma_{\mathrm{cr}},\upsilon_i<1,i=1,2$, and $\gamma<\gamma_{\mathrm{cr}},\upsilon_1<1<\upsilon_2$ discussed above. It remains to show

$$\tilde{R}_{12}(\boldsymbol{x},\boldsymbol{x})\to 0. \qquad (160)$$

Write $f(\boldsymbol{u})=f_0(\boldsymbol{u})T(\boldsymbol{u})$, where $f_0(\boldsymbol{u})=\rho_{\mathrm{sp}}(\boldsymbol{u})L_{\mathrm{sp}}(\boldsymbol{u})$ and $T(\boldsymbol{u}):=\frac{f(\boldsymbol{u})}{f_0(\boldsymbol{u})}$, $\boldsymbol{u}\in\Pi^2$ are continuous functions; see Assumption $(\mathrm{F})_{\mathrm{ND},\rho}$. Then, $\tilde{f}_{12}(\boldsymbol{u})=\sum_{i=1}^2\tilde{f}_{12,i}(\boldsymbol{u})$, where

$$
\begin{aligned}
\tilde{f}_{12,1}(\boldsymbol{u})&:=|u_1|^{\upsilon_1}(\widetilde{T}(\boldsymbol{u})-\widetilde{T}(u_1,0)),\\
\tilde{f}_{12,2}(\boldsymbol{u})&:=|u_2|^{\upsilon_2}(\widetilde{T}(\boldsymbol{u})-\widetilde{T}(0,u_2)),
\end{aligned}
$$

and where $\widetilde{T}(\boldsymbol{u}):=L_{\mathrm{sp}}(\boldsymbol{u})T(\boldsymbol{u})$. Accordingly, $\tilde{R}_{12}(\boldsymbol{x},\boldsymbol{x})=\sum_{i=1}^2\tilde{R}_{12,i}(\boldsymbol{x},\boldsymbol{x})$. Then,

$$
\begin{aligned}
&\tilde{R}_{12,1}(\boldsymbol{x},\boldsymbol{x})\\
&\le\int_{\mathbb{R}^2}|u_1|^{\upsilon_1}\delta_{\lambda,1}(\boldsymbol{u})|(1-\mathrm{e}^{\mathrm{i}u_1x_1})/u_1|^2\prod_{j=1}^2|(1-\mathrm{e}^{\mathrm{i}u_jx_j})/u_j|^2\mathrm{d}\boldsymbol{u}.
\end{aligned}
\qquad (161)
$$

Here, $\delta_{\lambda,1}(\boldsymbol{u}):=\left|\widetilde{T}\left(\frac{u_1}{\lambda},\frac{u_2}{\lambda^{\upsilon_1}}\right)-\widetilde{T}\left(\frac{u_1}{\lambda},0\right)\right|\mathbb{I}(\boldsymbol{u}\in\lambda\Pi\times\lambda^{\upsilon_1}\Pi)$ is bounded uniformly in $\lambda>1$, and the integral on the r.h.s. of (161) with $\delta_{\lambda,1}(\boldsymbol{u})$

replaced by 1 converges due to $v_1 < 1$. Also, note that $\delta_{\lambda,1}(\boldsymbol{u}) \to 0$ as $T(\boldsymbol{u}), \boldsymbol{u} \in \Pi^2$ is continuous and $L_{\rm sp}(\frac{u_1}{\lambda}, \frac{u_2}{\lambda^{v_1}}) - L_{\rm sp}(\frac{u_1}{\lambda}, 0) \to 0$ due to $v_2 > 1$. This proves (160) for $\tilde R_{12,1}(\boldsymbol{x},\boldsymbol{x})$ instead of $\tilde R_{12}(\boldsymbol{x},\boldsymbol{x})$. The proof of $\tilde R_{12,2}(\boldsymbol{x},\boldsymbol{x}) \to 0$ is similar to that of (159), with $\big|L_{\rm sp}(\frac{u_1}{\lambda}, u_2) - L_{\rm sp}(0,u_2)\big|$ there replaced by $\delta_{\lambda,2}(\boldsymbol{u}) := \big|\widetilde T(\frac{u_1}{\lambda}, u_2) - \widetilde T(0,u_2)\big| \to 0$ in view of the above-mentioned properties of $L_{\rm sp}(\boldsymbol{u})$ and $T(\boldsymbol{u})$, completing the proof of (8) in the parameter region R_{23}. The last relation in region R_{32} follows by symmetry.

Case $v_i > 1, i = 1,2, \gamma_{\rm cr} = 1$ (region R_{33} in Table 1). By symmetry, it suffices to consider the case $\gamma \le 1$. Split $f(\boldsymbol{u}) = f_1(\boldsymbol{u}) + f_2(\boldsymbol{u}) + \tilde f_{12}(\boldsymbol{u})$, $R_{\lambda,\gamma}(\boldsymbol{x},\boldsymbol{y}) = R_1(\boldsymbol{x},\boldsymbol{y}) + R_2(\boldsymbol{x},\boldsymbol{y}) + \tilde R_{12}(\boldsymbol{x},\boldsymbol{y})$, as in the previous case. First, let $\gamma < 1$. Then, $R_2(\boldsymbol{x},\boldsymbol{y}) \to \kappa_-^2 {\rm E} B_{1/2,0}(\boldsymbol{x}) B_{1/2,0}(\boldsymbol{y})$ as in (157), while $R_1(\boldsymbol{x},\boldsymbol{x}) \le C\lambda^{\gamma-1}\int_\Pi |u_1|^{v_1-2}{\rm d}u_1 \times \int_{\mathbb R} |(1-{\rm e}^{{\rm i}u_2x_2})/u_2|^2 {\rm d}u_2 = O(\lambda^{\gamma-1}) = o(1)$ is negligible when $\gamma < 1$; for $\gamma = 1$, we have $R_1(\boldsymbol{x},\boldsymbol{y}) \to \kappa_+^2 {\rm E} B_{0,1/2}(\boldsymbol{x}) B_{0,1/2}(\boldsymbol{y})$ analogously to (157) with

$$\kappa_+^2 := 2\int_{\mathbb R} |(1-{\rm e}^{{\rm i}u})/u|^2 {\rm d}u \times \int_\Pi f(v,0)|1-{\rm e}^{{\rm i}v}|^{-2}{\rm d}v; \tag{162}$$

c.f. (158). The proof of $\tilde R_{12}(\boldsymbol{x},\boldsymbol{x}) \to 0$ for $\gamma \le 1$ follows as in (159). This concludes the proof of (8) in all cases of Theorem 17.

It remains to show the Lindeberg criterion in (138). The relation $f(\boldsymbol{u}) = (2\pi)^2|\widehat a(\boldsymbol{u})|^2$ and the assumptions on f imply that $|\widehat a(\boldsymbol{u})| < C$ is bounded. Therefore, with $(\gamma_1,\gamma_2) := (1,\gamma)$, we get that $\sup_{\boldsymbol{s}\in\mathbb Z^2}\big|\sum_{\boldsymbol{t}\in K_{[\lambda,\lambda^\gamma]}} a(\boldsymbol{t}-\boldsymbol{s})\big| \le \int_{\Pi^2}\prod_{j=1}^2 \big|D_{[\lambda^{\gamma_j}]}(u_i)\big|\,|\widehat a(\boldsymbol{u})|{\rm d}\boldsymbol{u} \le C\int_{\Pi^2}\prod_{j=1}^2\big|D_{[\lambda^{\gamma_j}]}(u_i)\big|\,{\rm d}\boldsymbol{u} \le C(\log\lambda)^2$, and (138) holds since $H(\gamma) > 0$ in all cases of Theorem 17, completing its proof. □

Proof of Proposition 21. For $\gamma \le \gamma_{\rm cr}$, the results and their proofs completely agree with those of Theorem 16. The subsequent proof is limited to $\gamma > \gamma_{\rm cr}$ and is split into three cases as follows.

Case $v_1 + \frac{v_1}{v_2} < 1$. Then, $H_1^+ = \frac12(1 + v_1 + \frac{\mu v_1}{v_2})$, $H_2^+ = \frac12(1-\mu) > 0$. According to (65), for $\gamma > v_1/v_2, \mu < 1$, we have that

$$f(u_1/\lambda, u_2/\lambda^\gamma) \sim \frac{\lambda^{v_1}}{|u_1|^{v_1}} \times \left|\frac{\frac{u_2}{\lambda^\gamma}}{|\frac{u_1}{\lambda}|^{v_1/v_2}}\right|^\mu = \lambda^{2H(\gamma)-1-\gamma}\prod_{j=1}^2 |u_j|^{1-2H_j^+}$$

point-wise in $\boldsymbol{u} = (u_1, u_2) \in \mathbb{R}_0^2$. Then, the limit $V_+ = \kappa_+ B_{H_1^+, H_2^+}$ follows similarly as in Theorem 16, with

$$\kappa_+^2 = \prod_{j=1}^{2} \int_{\mathbb{R}} |(1 - \mathrm{e}^{\mathrm{i}u})/u|^2 |u|^{1-2H_j^+} \mathrm{d}u. \tag{163}$$

Case $\upsilon_1 < 1 < \upsilon_1 + \frac{\upsilon_1}{\upsilon_2}$. Then, $(2H_1^+, 2H_2^+) = (1 + \upsilon_1 + \frac{\mu\upsilon_1}{\upsilon_2}, 1 - \mu)$ if $\mu < \frac{\upsilon_2}{\upsilon_1} - \upsilon_2$, $= (2, 1 + \upsilon_2 - \frac{\upsilon_2}{\upsilon_1})$ if $\mu > \frac{\upsilon_2}{\upsilon_1} - \upsilon_2$. The proof of the limit $V_+ = \kappa_+ B_{H_1^+, H_2^+}$ for $\mu < \frac{\upsilon_2}{\upsilon_1} - \upsilon_2$ is the same as in the case of $\upsilon_1 + \frac{\upsilon_1}{\upsilon_2} < 1$ above and is omitted. Next, let $\mu > \frac{\upsilon_2}{\upsilon_1} - \upsilon_2$. Due to $\gamma > \upsilon_1/\upsilon_2$, we see that

$$\lambda^{-\gamma\upsilon_2} f\Big(\frac{u_1}{\lambda^{\gamma\upsilon_2/\upsilon_1}}, \frac{u_2}{\lambda^{\gamma}}\Big) \to f_0(\boldsymbol{u}) := \Big(\frac{|u_2|}{\rho_{\mathrm{sp}}(\boldsymbol{u})^{1/\upsilon_2}}\Big)^{\mu} \frac{1}{\rho_{\mathrm{sp}}(\boldsymbol{u})},$$

$$\lambda^{-1} D_{[\lambda x]}\Big(\frac{u_1}{\lambda^{\gamma\upsilon_2/\upsilon_1}}\Big) \to x, \quad \lambda^{-\gamma} D_{[\lambda^{\gamma} x]}\Big(\frac{u_2}{\lambda^{\gamma}}\Big) \to \frac{1 - \mathrm{e}^{\mathrm{i}xu_2}}{-\mathrm{i}u_2}$$

point-wise for any $u_i \neq 0, i = 1, 2, x > 0$. We also have $\int_{\mathbb{R}} f_0(u_1, u_2)\mathrm{d}u_1 = |u_2|^{1-2H_2^+} \int_{\mathbb{R}} f_0(u, 1)\mathrm{d}u$, where the last integral converges due to $f_0(u, 1) = 1/(1 + |u|^{\upsilon_1})^{1+\mu/\upsilon_2}$ and $\mu > \frac{\upsilon_2}{\upsilon_1} - \upsilon_2$. Then, $V_+ = \kappa_+ B_{1, H_2^+}$ follows as in Theorem 16. The asymptotic constant κ_+ in both cases of μ is given by

$$\kappa_+^2 = \begin{cases} \prod_{j=1}^2 \int_{\mathbb{R}} |(1 - \mathrm{e}^{\mathrm{i}u})/u|^2 |u|^{1-2H_j^+} \mathrm{d}u, & \mu < \frac{\upsilon_2}{\upsilon_1} - \upsilon_2, \\ \int_{\mathbb{R}} \big|(1 - \mathrm{e}^{\mathrm{i}v})/v\big|^2 |v|^{1-2H_2^+} \mathrm{d}v \times \int_{\mathbb{R}} f_0(u, 1)\mathrm{d}u, & \mu > \frac{\upsilon_2}{\upsilon_1} - \upsilon_2, \end{cases} \tag{164}$$

Case $\upsilon_1 > 1$. In this case, the results do not depend on μ and completely agree with those in Theorem 16, including the proof. This completes the proof of Proposition 21. □

Proof of Theorem 22. We use Proposition 9 of Chapter 2 about convergence in the distribution of linear forms in i.i.d. r.v.s towards the Gaussian white noise integral. We start with a stochastic integral representation of limit RFs:

$$V_0(\boldsymbol{x}) = \int_{\mathbb{R}^2} h_0(\boldsymbol{x}; \boldsymbol{u}) W(\mathrm{d}\boldsymbol{u}), \quad V_{\pm}(\boldsymbol{x}) = \int_{\mathbb{R}^2} h_{\pm}(\boldsymbol{x}; \boldsymbol{u}) W(\mathrm{d}\boldsymbol{u}), \quad \boldsymbol{x} \in \mathbb{R}_+^2, \tag{165}$$

where

$$h_0(\boldsymbol{x};\boldsymbol{u}) := \int_{]0,\boldsymbol{x}]} a_\infty(\boldsymbol{t}-\boldsymbol{u})\mathrm{d}\boldsymbol{t}, \tag{166}$$

$$h_+(\boldsymbol{x};\boldsymbol{u}) := \begin{cases} \mathbb{I}(u_2 \in (0,x_2]) \int_{(0,x_1]\times\mathbb{R}} a_\infty(t_1-u_1,t_2)\mathrm{d}\boldsymbol{t}, & Q_{1/2,1} < 1, \\ x_1 \int_{(0,x_2]} a_\infty(u_1,t_2-u_2)\mathrm{d}t_2, & Q_{1/2,1} > 1, \end{cases}$$

$$h_-(\boldsymbol{x};\boldsymbol{u}) := \begin{cases} \mathbb{I}(u_1 \in (0,x_1]) \int_{\mathbb{R}\times(0,x_2]} a_\infty(t_1,t_2-u_2)\mathrm{d}\boldsymbol{t}, & Q_{1,1/2} < 1, \\ x_2 \int_{(0,x_1]} a_\infty(t_1-u_1,u_2)\mathrm{d}t_1, & Q_{1,1/2} > 1, \end{cases}$$

written as integrals of the generalised invariant function a_∞ in (87). The fact that stochastic integrals in (89) are well defined and $V_\pm$ agree with FBS in (89) will be verified later in the proof of the convergence in (2). We write h_γ for h_0, h_+, and h_-, in the respective cases of $\gamma = \gamma_{\mathrm{cr}}, \gamma > \gamma_{\mathrm{cr}}$, and $\gamma < \gamma_{\mathrm{cr}}$.

The normalised partial sums on the l.h.s. of (2) can be rewritten as linear forms:

$$\lambda^{-H(\gamma)} S_{\lambda,\gamma}(\boldsymbol{x}) = S(g_{\lambda,\gamma}(\boldsymbol{x};\cdot)) = \sum_{\boldsymbol{u}\in\mathbb{Z}^2} g_{\lambda,\gamma}(\boldsymbol{x};\boldsymbol{u})\varepsilon(\boldsymbol{u}), \quad \text{with} \tag{167}$$

$$g_{\lambda,\gamma}(\boldsymbol{x};\boldsymbol{u}) := \lambda^{-H(\gamma)} \sum_{\boldsymbol{t}\in K_{[\lambda x_1,\lambda^\gamma x_2]}} a(\boldsymbol{t}-\boldsymbol{u}) \quad \boldsymbol{u}\in\mathbb{Z}^2.$$

Using Proposition 9 of Chapter 2 with appropriately chosen $\mu_i \to \infty, i = 1,2$ and the Cramér-Wold device, relation (2) follows from the $L^2(\mathbb{R}^2)$-convergence of integrands:

$$\lim_{\lambda\to\infty} \Big\| \sum\nolimits_{i=1}^m \theta_i (g^\dagger_{\lambda,\gamma}(\boldsymbol{x}_i;\cdot) - h_\gamma(\boldsymbol{x}_i;\cdot)) \Big\| = 0, \tag{168}$$

for any $m \ge 1$ and any $\theta_i \in \mathbb{R}, \boldsymbol{x}_i \in \mathbb{R}^2_+, 1 \le i \le m$, where $g^\dagger_{\lambda,\gamma}(\boldsymbol{x};\boldsymbol{u})$ is the re-normalised integrand in (34) of Chapter 2 with unit matrix A. It suffices to prove (168) for $m = \theta_1 = 1, \boldsymbol{x}_1 = \boldsymbol{x}$ since the general case follows by the norm inequality.

Consider (168) for $\gamma = \gamma_{\mathrm{cr}} = q_1/q_2$. The fact that $h_0(\boldsymbol{x};\cdot) \in L^2(\mathbb{R}^2)$ or that the RF V_0 in (88) is well defined follows from Proposition 4 (ii). In particular,

$$(a_\infty \star a_\infty)(\boldsymbol{t}) = \tilde{\rho}(\boldsymbol{t})^{-1} L(\boldsymbol{t}), \tag{169}$$

where $\tilde{\rho}(\boldsymbol{t}) = |t_1|^{\tilde{q}_1} + |t_2|^{\tilde{q}_2}$ and $\tilde{q}_i = q_i(2-Q)$; see (24). Thus, $\|h_0(\boldsymbol{x};\cdot)\|^2 = \int_{]\mathbf{0},\boldsymbol{x}]^2}(a_\infty \star a_\infty)(\boldsymbol{u}_1 - \boldsymbol{u}_2)\mathrm{d}\boldsymbol{u}_1\mathrm{d}\boldsymbol{u}_2 \le C\int_{]\mathbf{0},\boldsymbol{x}]^2}\tilde{\rho}(\boldsymbol{u}_1 - \boldsymbol{u}_2)\mathrm{d}\boldsymbol{u}_1\mathrm{d}\boldsymbol{u}_2 < \infty$ since $\tilde{Q} = \frac{1}{\tilde{q}_1} + \frac{1}{\tilde{q}_2} > 1$. For $g_{\lambda,0}(\boldsymbol{x};\boldsymbol{u}) \equiv g_{\lambda,\gamma_{\mathrm{cr}}}(\boldsymbol{x};\boldsymbol{u})$ in (167), with $\lambda' := \lambda^{\gamma_{\mathrm{cr}}}$, the corresponding $g^\dagger_{\lambda,0}(\boldsymbol{x};\boldsymbol{u}) \equiv g^\dagger_{\lambda,\gamma_{\mathrm{cr}}}(\boldsymbol{x};\boldsymbol{u})$ in (34) and $A = \mathrm{diag}(1,1)$ the unit matrix can be written as

$$g^\dagger_{\lambda,0}(\boldsymbol{x};\boldsymbol{u}) = \int_{\mathbb{R}^2} a_\lambda(\boldsymbol{t},\boldsymbol{u})\mathbb{I}\big(([\lambda t_1],[\lambda' t_2]) \in (0,\lambda x_1] \times (0,\lambda' x_2]\big)\mathrm{d}\boldsymbol{t}, \tag{170}$$

where

$$a_\lambda(\boldsymbol{t},\boldsymbol{u}) := \lambda^{q_1} a([\lambda t_1] - [\lambda u_1], [\lambda' t_2] - [\lambda' u_2]) \to a_\infty(\boldsymbol{t}-\boldsymbol{u}), \quad \lambda \to \infty \tag{171}$$

point-wise for any $\boldsymbol{t},\boldsymbol{u} \in \mathbb{R}^2, \boldsymbol{t} \ne \boldsymbol{u}$ according to Assumption $(\mathrm{A})_{\mathrm{LRD},\rho}$. To prove (168), we will need the dominating bound:

$$|a_\lambda(\boldsymbol{t},\boldsymbol{u})| \le C\rho(\boldsymbol{t}-\boldsymbol{u})^{-1}, \tag{172}$$

with $C > 0$ independent of $\boldsymbol{t},\boldsymbol{u}$, and $\lambda \ge 1$. It suffices to show (172) with a in (171) replaced by $1 \wedge \rho^{-1}$ or

$$\begin{aligned}\rho_\lambda(\boldsymbol{t},\boldsymbol{u})^{-1} &:= \lambda^{q_1}\big(1 \wedge \rho([\lambda t_1] - [\lambda u_1], [\lambda' t_2] - [\lambda' u_2])^{-1}\big) \\ &= \big(\lambda^{-q_1} \vee \rho(\boldsymbol{t}_\lambda - \boldsymbol{u}_\lambda)\big)^{-1} \;\le\; C\rho(\boldsymbol{t}-\boldsymbol{u})^{-1}, \quad \boldsymbol{t},\boldsymbol{u} \in \mathbb{R}^2.\end{aligned} \tag{173}$$

where $\boldsymbol{t}_\lambda := ([\lambda t_1]/\lambda, [\lambda' t_2]/\lambda'), \boldsymbol{u}_\lambda := ([\lambda u_1]/\lambda, [\lambda' u_2]/\lambda')$. To show (173), use the inequality

$$\rho(\boldsymbol{t}+\boldsymbol{s}) \le C(\rho(\boldsymbol{t}) + \rho(\boldsymbol{s})), \quad \boldsymbol{t},\boldsymbol{s} \in \mathbb{R}^2, \tag{174}$$

where $C = 1 \vee 2^{q_1-1} \vee 2^{q_2-1}$, which follows from $|t+s|^q \le (1 \vee 2^{q-1})(|t|^q + |s|^q)$, $t,s \in \mathbb{R}, q > 0$. Using (174) and $\rho((\boldsymbol{t}-\boldsymbol{t}_\lambda) - (\boldsymbol{u}-\boldsymbol{u}_\lambda)) \le C\lambda^{-q_1}$, we obtain

$$\begin{aligned}\rho(\boldsymbol{t}-\boldsymbol{u}) &\le C\big(\rho(\boldsymbol{t}_\lambda - \boldsymbol{u}_\lambda)) + \rho((\boldsymbol{t}-\boldsymbol{t}_\lambda) - (\boldsymbol{u}-\boldsymbol{u}_\lambda))\big) \\ &\le C\big(\rho(\boldsymbol{t}_\lambda - \boldsymbol{u}_\lambda)) + \lambda^{-q_1}\big) \le C\big(\lambda^{-q_1} \vee \rho(\boldsymbol{t}_\lambda - \boldsymbol{u}_\lambda)\big)\end{aligned}$$

and hence (173) and (172). From (172), (171), and $\mathbb{I}(([\lambda t_1], [\lambda' t_2]) \in (0, \lambda x_1] \times (0, \lambda' x_2]) \le \mathbb{I}(\boldsymbol{t} \in (0, x_1 + 1] \times (0, x_2 + 1]), \lambda \ge 1$, we obtain

$$|g^{\dagger}_{\lambda,0}(\boldsymbol{x};\boldsymbol{u})| \le \bar{g}_0(\boldsymbol{x};\boldsymbol{u}) := C \int_{\mathbb{R}^2} \rho(\boldsymbol{t}-\boldsymbol{u})^{-1}\mathbb{I}(\boldsymbol{t} \in (0, x_1+1] \times (0, x_2+1])\mathrm{d}\boldsymbol{t}, \tag{175}$$

where $\bar{g}_0(\boldsymbol{x};\cdot) \in L^2(\mathbb{R}^2)$ similarly as $h_0(\boldsymbol{x};\cdot) \in L^2(\mathbb{R}^2)$ in (167). Moreover, $g^{\dagger}_{\lambda,0}(\boldsymbol{x};\boldsymbol{u}) \to h_0(\boldsymbol{x};\boldsymbol{u})$ point-wise for any $\boldsymbol{u} \in \mathbb{R}^2$, implying (168) by (175) and the DCT. This proves (2) for $\gamma = \gamma_{\mathrm{cr}}$.

Next, we prove (2) for $\gamma > \gamma_{\mathrm{cr}}, Q_{1/2,1} > 1$, the last inequality being equivalent to $\tilde{q}_2 < 1$. Let us check that the limit V_+ given in (165)–(166) is well defined and satisfies (89). Then, $(a_\infty \star a_\infty)(0,t) = \tilde{\rho}(0,t)^{-1}L(0,\mathrm{sgn}(t)) = |t|^{-\tilde{q}_2}\tilde{L}(0,\mathrm{sgn}(t)), t \in \mathbb{R}$ is integrable at the origin; see (169) and

$$\begin{aligned}\|h_+(\boldsymbol{x};\cdot)\|^2 &= x_1^2 \int_0^{x_2}\int_0^{x_2} (a_\infty \star a_\infty)(0, t_2 - t_2')\mathrm{d}t_2\mathrm{d}t_2' \\ &= \kappa_+^2 x_1^2 x_2^{2-\tilde{q}_2} = \kappa_+^2 \mathrm{E}B_{1,H_2^+}(\boldsymbol{x})^2,\end{aligned}$$

where

$$\begin{aligned}\kappa_+^2 := \|h_+(\mathbf{1};\cdot)\|^2 &= \int_{\mathbb{R}^2}\left(\int_{(0,1]} a_\infty(u_1, t_2 + u_2)\mathrm{d}t_2\right)^2 \mathrm{d}\boldsymbol{u} \\ &= \frac{\tilde{L}(0,1) + \tilde{L}(0,-1)}{(1-\tilde{q}_2)(2-\tilde{q}_2)}.\end{aligned} \tag{176}$$

Hence, $V_+(\boldsymbol{x})$ in (165) is well defined; moreover, for a fixed x_1, the process $\{V_+(x_1,x_2); x_2 \in \mathbb{R}\}$ is a stationary-increment Gaussian process which is self-similar with index H_2^+; therefore, it agrees with FBM with variance as in (176). As a consequence, the above RF V_+ coincides with FBS in (89).

Let us prove the convergence in (168) in the above case. Using Proposition 9 of Chapter 2 with $\mu_1 = \gamma/\gamma_{\mathrm{cr}}, \mu_2 = \gamma$, this follows from

$$\|g^{\dagger}_{\lambda,\gamma}(\boldsymbol{x};\cdot) - h_+(\boldsymbol{x};\cdot)\| \to 0 \quad (\lambda \to \infty). \tag{177}$$

Then, with $\lambda' := \lambda^{\gamma}, \lambda'' := \lambda^{\gamma/\gamma_{\rm cr}}, \lambda = o(\lambda'')$,

$$
\begin{aligned}
g^{\dagger}_{\lambda,\gamma}(\boldsymbol{x};\boldsymbol{u}) = \int_{\mathbb{R}^2} a_\lambda(\boldsymbol{t},\boldsymbol{u}) \mathbb{I}\big(([\lambda t_1],[\lambda' t_2]) \in (0,\lambda x_1] \\
\times (0,\lambda' x_2]\big) \mathrm{d}\boldsymbol{t} \to \tilde{h}_+(\boldsymbol{x};\boldsymbol{u})
\end{aligned} \tag{178}
$$

point-wise for any $\boldsymbol{u} \in \mathbb{R}^2$ since the integrand

$$
a_\lambda(\boldsymbol{t},\boldsymbol{u}) := (\lambda'')^{q_1} a([\lambda t_1] - [\lambda'' u_1], [\lambda' t_2] - [\lambda' u_2]) \to a_\infty((0,t_2) - \boldsymbol{u}) \tag{179}
$$

point-wise for any $\boldsymbol{t} \neq \boldsymbol{u}$. To show the L^2-convergence in (177), we use Lemma 30 of Chapter 2 with $p = 2$. Accordingly, it suffices to check that there exist $G_\lambda, G \in L^2(\mathbb{R}^2)$ such that: (j') $\lim_{\lambda\to\infty} G_\lambda(\boldsymbol{u}) = G(\boldsymbol{u})$ point-wise in $\mathbb{R}^2$, (jj') $\lim_{\lambda\to\infty} \|G_\lambda\| = \|G\|$, and (jjj') $|g^{\dagger}_{\lambda,\gamma}(\boldsymbol{x};\boldsymbol{u}| \le C G_\lambda(\boldsymbol{u})$ with $C > 0$ independent of $\lambda > 0, \boldsymbol{u} \in \mathbb{R}^2$. To show this, define

$$
\begin{aligned}
G_\lambda(\boldsymbol{u}) &:= \int_{]\boldsymbol{0},2\boldsymbol{x}]} \rho(((\lambda/\lambda'')t_1, t_2) - \boldsymbol{u})^{-1} \mathrm{d}\boldsymbol{t}, \\
G(\boldsymbol{u}) &:= \int_{]\boldsymbol{0},2\boldsymbol{x}]} \rho((0,t_2) - \boldsymbol{u})^{-1} \mathrm{d}\boldsymbol{t},
\end{aligned} \tag{180}
$$

and note that

$$
\|G_\lambda\|^2 = \int_{]\boldsymbol{0},2\boldsymbol{x}]^2} (\rho^{-1} \star \rho^{-1})((\lambda/\lambda'')(t_1 - t_1'), t_2 - t_2') \mathrm{d}\boldsymbol{t}\mathrm{d}\boldsymbol{t}' \tag{181}
$$

$$
\to \int_{]\boldsymbol{0},2\boldsymbol{x}]^2} (\rho^{-1} \star \rho^{-1})(0, t_2 - t_2') \mathrm{d}\boldsymbol{t}\mathrm{d}\boldsymbol{t}' = \|G\|^2 < \infty \tag{182}
$$

by (20) of Proposition 4 and the condition $\tilde{q}_2 < 1$. Thus, G_λ, G satisfies (j') and (jj'). Condition (jjj') follows from the definition of a_λ in (179), the bound $|a(\boldsymbol{t})| \le C(1 \vee \rho(\boldsymbol{t})^{-1})$, and the inequality

$$
\tfrac{1}{(\lambda'')^{q_1}} \vee \rho\Big(\Big(\tfrac{[\lambda t_1]}{\lambda''}, \tfrac{[\lambda' t_2]}{\lambda'}\Big) - \Big(\tfrac{[\lambda'' u_1]}{\lambda''}, \tfrac{[\lambda' u_2]}{\lambda'}\Big)\Big) \ge c\rho((\frac{\lambda t_1}{\lambda''}, t_2) - \boldsymbol{u}), \tag{183}
$$

with $c > 0$ independent of $\boldsymbol{t}, \boldsymbol{u} \in \mathbb{R}^2$ and $\lambda \ge 1$, which follows from (173). This proves (2) or the statement of the theorem for $\gamma > \gamma_{\rm cr}, Q_{1/2,1} > 1$.

Next, we prove (2) for $\gamma > \gamma_{\rm cr}$, $Q_{1/2,1} < 1$. Note $h_+(\boldsymbol{x};\cdot) \in L^2(\mathbb{R}^2)$ and

$$\mathrm{E}(V_+(\boldsymbol{x}))^2 = \|h_+(\boldsymbol{x};\cdot)\|^2 = \kappa_+^2 x_1^{2H_1^+} x_2 = \kappa_+^2 \mathrm{E} B_{H_1^+,1/2}(\boldsymbol{x})^2,$$
$$\boldsymbol{x} \in \mathbb{R}_+^2,$$

with

$$\begin{aligned}\kappa_+^2 &:= \int_{\mathbb{R}} \left\{ \int_{(0,1]\times\mathbb{R}} a_\infty(t_1 - u_1, t_2)\mathrm{d}\boldsymbol{t} \right\}^2 \mathrm{d}u_1 \qquad (184)\\ &= \int_{(0,1]^2\times\mathbb{R}} (a_\infty \star a_\infty)(t_1 - t_1', t_2)\mathrm{d}t_1\mathrm{d}t_1'\mathrm{d}t_2\\ &= \int_{(0,1]^2} |t - t'|^{(\tilde{q}_1/\tilde{q}_2)-\tilde{q}_1}\mathrm{d}t\mathrm{d}t'\\ &\quad\times \int_{\mathbb{R}} \tilde{\rho}(\mathrm{sgn}(t-t'), s)^{-1} L(\mathrm{sgn}(t-t'), s)\mathrm{d}s < \infty\end{aligned}$$

since $\tilde{q}_1 - (\tilde{q}_1/\tilde{q}_2) = 2 - 2H_1^+ < 1$ and $\int_{\mathbb{R}} \tilde{\rho}(\pm 1, s)^{-1}\mathrm{d}s = \int_{\mathbb{R}}(1 + |s|^{\tilde{q}_2})^{-1}\mathrm{d}s < \infty, \tilde{q}_2 > 1$. The fact that $V_+ \overset{\mathrm{fdd}}{=} \kappa_+ B_{H_1^+,1/2}$ follows as in the previous case, using the scaling properties of the integrand in (166). To show the convergence in (2), or

$$S(g_{\lambda,\gamma}(\boldsymbol{x};\cdot)) \xrightarrow{\mathrm{d}} \kappa_+ B_{H_1^+,1/2}(\boldsymbol{x}), \qquad (185)$$

use Proposition 9 of Chapter 2 with $\mu_1 = 1, \mu_2 = \gamma$. Accordingly, it suffices to prove

$$\lim_{\lambda\to\infty} \|g_{\lambda,\gamma}^\dagger(\boldsymbol{x};\cdot) - h_+(\boldsymbol{x};\cdot)\| = 0, \qquad (186)$$

where

$$\begin{aligned}g_{\lambda,\gamma}^\dagger(\boldsymbol{x};\boldsymbol{u}) &= \lambda^{q_1} \int_{\mathbb{R}^2} a([\lambda t_1] - [\lambda u_1], [\lambda^{\gamma_{\rm cr}} t_2]\\ &\quad - [\lambda^\gamma u_2])\mathbb{I}(0 < [\lambda t_1] \le \lambda x_1, 0 < [\lambda^{\gamma_{\rm cr}} t_2] \le \lambda^\gamma x_2)\mathrm{d}\boldsymbol{t}\\ &= \int_{\mathbb{R}^2} a_\lambda(\boldsymbol{t};\boldsymbol{u})\mathrm{d}\boldsymbol{t}, \qquad (187)\end{aligned}$$

where

$$a_\lambda(\boldsymbol{t};\boldsymbol{u}) := \lambda^{q_1} a([\lambda t_1]-[\lambda u_1],[\lambda^{\gamma_{\rm cr}} t_2])\mathbb{I}(0<[\lambda t_1]\le \lambda x_1, 0<[\lambda^{\gamma_{\rm cr}} t_2] + [\lambda^\gamma u_2] \le \lambda^\gamma x_2) \to a_\infty(t_1-u_1,t_2)\mathbb{I}(0<t_1\le x_1, 0<u_2\le x_2) \quad (188)$$

follows since $\gamma > \gamma_{\rm cr}$ for any $\boldsymbol{t},\boldsymbol{u}\in\mathbb{R}^2, \boldsymbol{t}\ne\boldsymbol{u}, \boldsymbol{t},\boldsymbol{u}\notin\{\boldsymbol{0},\boldsymbol{x}\}$.

The dominating bound

$$\begin{aligned}|g^\dagger_{\lambda,\gamma}(\boldsymbol{u})| &\le C\int_{(0,x_1]\times\mathbb{R}} \rho(t_1-u_1,t_2)^{-1}\mathbb{I}\big(0<[\lambda' t_2]+[\lambda'' u_2]\le \lambda'' x_2\big)\mathrm{d}\boldsymbol{t}\\ &\le C\int_{(0,x_1]\times\mathbb{R}} \rho(t_1-u_1,t_2)^{-1}\mathbb{I}\big(-1<u_2+t_2\tilde\lambda<x_2+1\big)\mathrm{d}\boldsymbol{t}\\ &=: CG_\lambda(\boldsymbol{u}),\end{aligned}$$

where $\lambda' := \lambda^{\gamma_{\rm cr}}, \lambda'' := \lambda^\gamma, \tilde\lambda := \lambda'/\lambda'' \to 0$ follows using similar inequalities as in (183) and (173). Clearly, $G_\lambda(\boldsymbol{u}) \to G(\boldsymbol{u}) := \mathbb{I}(-1< u_2 < x_2+1)\int_{(0,x_1]\times\mathbb{R}}\rho(t_1-u_1,t_2)^{-1}\mathrm{d}\boldsymbol{t}$ point-wise. We also have that

$$\|G_\lambda\|^2 = \int_{((0,x_1]\times\mathbb{R})^2\times\mathbb{R}} \rho(t_1-u_1,t_2)^{-1}\rho(t_1'-u_1,t_2')^{-1}\phi_\lambda(t_2,t_2')\mathrm{d}\boldsymbol{t}\mathrm{d}\boldsymbol{t}'\mathrm{d}u_1,$$

where

$$\phi_\lambda(t_2,t_2') := \int_{\mathbb{R}} \mathbb{I}\big(-1<u_2+t_2\tilde\lambda<x_2+1, -1<u_2+t_2'\tilde\lambda < x_2+1\big)\mathrm{d}u_2 \to x_2+2$$

as $\lambda\to\infty$ and $\phi_\lambda(t_2,t_2')\le x_2+2$ for any $t_2,t_2'\in\mathbb{R}$. Moreover, since $\|G\|^2<\infty$, as seen above, it follows by the DCT that $\|G_\lambda\|^2\to\|G\|^2$. Hence, (186) follows by Lemma 30 of Chapter 2, proving (185) or (2) for $\gamma>\gamma_{\rm cr}, Q_{1/2,1}<1$. The proof of (2) for $\gamma<\gamma_{\rm cr}$ follows by symmetry: by exchanging the coordinates x_1 and x_2, together with simultaneous changes $\gamma\to 1/\gamma$ and $q_i\to q_j (i\ne j)$. Finally, the last statement of Theorem 22 is a consequence of the fact that the unbalanced scaling limits in (2) have different covariance functions. This completes the proof of Theorem 22. □

Outline of the Proof of Theorem 23. From Table 1, the FBS limits $V_\pm = B_{H_1^\pm, H_2^\pm}$ belong to two groups: either $H_1^\pm \wedge H_2^\pm > 0$ (Case 1) or $H_1^\pm \wedge H_2^\pm = 0$ (Case 2). In Case 1, the proof of (2) uses Proposition 9 of Chapter 2 and is roughly similar to that of Theorem 22, with the difference that V_0 and $V_\pm$ have a stochastic integral representation in (165) with somewhat different kernels, h_0 abd $h_\pm$; particularly, $h_\pm$ in (166) must be replaced by

$$h_+(\boldsymbol{x};\boldsymbol{u}) := \mathbb{I}(0 < u_2 \le x_2) \times \begin{cases} \int_{(0,x_1]\times\mathbb{R}} a_\infty(t_1 - u_1, t_2)\mathrm{d}\boldsymbol{t}, & u_1 \notin (0,x_1], \\ -\int_{(\mathbb{R}\setminus(0,x_1])\times\mathbb{R}} a_\infty(t_1 - u_1, t_2)\mathrm{d}\boldsymbol{t}, & u_1 \in (0,x_1], \end{cases}$$

$$h_-(\boldsymbol{x};\boldsymbol{u}) := \mathbb{I}(0 < u_1 \le x_1) \times \begin{cases} \int_{\mathbb{R}\times(0,x_2]} a_\infty(t_1, t_2 - u_2)\mathrm{d}\boldsymbol{t}, & u_2 \notin (0,x_2], \\ -\int_{\mathbb{R}\times(\mathbb{R}\setminus(0,x_2])} a_\infty(t_1, t_2 - u_2)\mathrm{d}\boldsymbol{t}, & u_2 \in (0,x_2], \end{cases}$$

which account for the ND property and the fact that the Hurst indices H_1^+, H_2^- in Theorem 23 (91) take values in the interval $(0, \frac{1}{2})$.

The above approach is not applicable in Case 2 when the limits $V_\pm$ are 'degenerated' FBS with Hurst parameters taking value 0 and do not have a stochastic integral representation. Let $\gamma < \gamma_{\mathrm{cr}} = 2q_1(1-Q)$ (region R_{23}). In this case, $V_-(\boldsymbol{x}) \overset{\mathrm{fdd}}{=} \sigma_{\mathrm{edge},1} B_{1/2,0}(\boldsymbol{x})$ and the finite-dimensional convergence in (2) reduces to the weak convergence of vector-valued random process indexed by $x_1 > 0$:

$$\begin{aligned} &\lambda^{-1/2}(S_{\lambda,\gamma}(x_1, x_{21}), \ldots, S_{\lambda,\gamma}(x_1, x_{2r})) \\ &\quad \xrightarrow{\mathrm{fdd}} \frac{\sigma_{\mathrm{edge},1}}{\sqrt{2}}(B^{(1)}(x_1) - B^{(0)}(x_1), \ldots, B^{(r)}(x_1) - B^{(0)}(x_1)) \end{aligned} \tag{189}$$

for any *fixed* $0 < x_{21} < \cdots < x_{2r}, r \ge 1$, where $B^{(i)}, i = 0, 1, \ldots, r$ are *independent* BM. By 'localising' the l.h.s. of (189) in the vicinity of horizontal lines $(x_1, 0), (x_1, \lambda^\gamma x_{21}), \ldots, (x_1, \lambda^\gamma x_{2r})$, one can reduce it to $r = 1, x_{21} = x_2$, or

$$\lambda^{-1/2} S_{\lambda,\gamma}(x_1, x_2) \xrightarrow{\mathrm{fdd}} \frac{\sigma_{\mathrm{edge},1}}{\sqrt{2}}(B^{(1)}(x_1) - B^{(0)}(x_1)) \overset{\mathrm{fdd}}{=} \sigma_{\mathrm{edge},1} B(x_1), \tag{190}$$

where B is a standard BM. Next, using the zero-sum condition $\sum_{\boldsymbol{t}\in\mathbb{Z}^2} a(\boldsymbol{t}) = 0$, we can replace $S_{\lambda,\gamma}(x_1,x_2) = S_{\lambda,\gamma}(\boldsymbol{x})$ in (190) by

$$S^0_{\lambda,\gamma}(\boldsymbol{x}) := \sum_{1\le u_1\le[\lambda x_2],u_2\in\mathbb{Z}} \varepsilon(\boldsymbol{u}) \sum_{\boldsymbol{t}\in K_{[\lambda x_1,\lambda^\gamma x_2]}} a(\boldsymbol{t}-\boldsymbol{u}),$$

which is further approximated by four sums:

$$V^{(1)}_{L,\lambda,\gamma}(\boldsymbol{x}) := \sum_{1\le u_1,t_1\le[\lambda x_1]} \sum_{-L<t_2\le 0<u_2\le L} \varepsilon(\boldsymbol{u})a(\boldsymbol{t}-\boldsymbol{u}),$$

$$V^{(2)}_{L,\lambda,\gamma}(\boldsymbol{x}) := -\sum_{1\le u_1,t_1\le[\lambda x_1]} \sum_{-L<u_2\le 0<t_2\le L} \varepsilon(\boldsymbol{u})a(\boldsymbol{t}-\boldsymbol{u}),$$

$$V^{(3)}_{L,\lambda,\gamma}(\boldsymbol{x}) := -\sum_{1\le u_1,t_1\le[\lambda x]} \sum_{-L<t_2-[\lambda^\gamma x_2]\le 0<u_2-[\lambda^\gamma x_2]\le L} \varepsilon(\boldsymbol{u})a(\boldsymbol{t}-\boldsymbol{u}),$$

$$V^{(4)}_{L,\lambda,\gamma}(\boldsymbol{x}) := \sum_{1\le u_1,t_1\le[\lambda x_1]} \sum_{-L<u_2-[\lambda^\gamma x_2]\le 0<t_2-[\lambda^\gamma x_2]\le L} \varepsilon(\boldsymbol{u})a(\boldsymbol{t}-\boldsymbol{u}),$$

which are mutually independent and have similar structure, each consisting of a finite number L of (independent) weighted sums of noise variables $\varepsilon(\boldsymbol{u}), 1 \le t_1 \le [\lambda x_1]$ in a neighbourhood of the lower ($u_2 = 1$) or upper ($u_2 = [\lambda^\gamma x_2]$) boundaries of the rectangle $K_{[\lambda x_1,\lambda^\gamma x_2]}$, behaving like a Brownian motion in x_1 independent of $x_2 > 0$.

Proof of Proposition 24. It suffices to prove (96) for $\boldsymbol{x} \in (0,1]^2$. Let $D'_\lambda := \mathbb{Z}^2 \cap [-\lambda, 2\lambda] \times [-\lambda^\gamma, 2\lambda^\gamma], D''_\lambda := \mathbb{Z}^2\backslash D'_\lambda$. Then, $\lambda^{-(1+\gamma)/2}S_{\lambda,\gamma}(\boldsymbol{x}) = S(g'_{\lambda,\gamma}(\boldsymbol{x};\cdot)) + S(g''_{\lambda,\gamma}(\boldsymbol{x};\cdot))$, where $g'_{\lambda,\gamma}(\boldsymbol{x};\boldsymbol{u}) := \lambda^{-(1+\gamma)/2}\sum_{\boldsymbol{t}\in K_{[\lambda x_1,\lambda^\gamma x_2]}} a(\boldsymbol{t}-\boldsymbol{u})\mathbb{I}(\boldsymbol{u} \in D'_\lambda)$ and $g''_{\lambda,\gamma}(\boldsymbol{x};\boldsymbol{u})$ is defined analogously with D'_λ replaced by D''_λ. Let $h''_{\lambda,\gamma}(\boldsymbol{x};\boldsymbol{u}) := \lambda^{(1+\gamma)/2}g''_{\lambda,\gamma}(\boldsymbol{x};\boldsymbol{u}) = \sum_{\boldsymbol{t}\in K_{[\lambda x_1,\lambda^\gamma x_2]}} a(\boldsymbol{t}-\boldsymbol{u})\mathbb{I}(\boldsymbol{u} \in D''_\lambda)$. By the summability property of $a(\boldsymbol{t})$, $\epsilon_\lambda := \sup_{\boldsymbol{u}\in\mathbb{Z}^2} |h''_{\lambda,\gamma}(\boldsymbol{x};\boldsymbol{u})| \to 0$, and hence

$$\begin{aligned}\mathrm{E}S(g''_{\lambda,\gamma}(\boldsymbol{x};\cdot))^2 &= \sum_{\boldsymbol{u}\in\mathbb{Z}^2} |g''_{\lambda,\gamma}(\boldsymbol{x};\boldsymbol{u})|^2 = \lambda^{-1-\gamma}\sum_{\boldsymbol{u}\in\mathbb{Z}^2} |h''_{\lambda,\gamma}(\boldsymbol{x};\boldsymbol{u})|^2 \\ &\le \epsilon_\lambda\lambda^{-1-\gamma}\sum_{\boldsymbol{u}\in\mathbb{Z}^2} |h''_{\lambda,\gamma}(\boldsymbol{x};\boldsymbol{u})|\end{aligned}$$

$$
\begin{aligned}
&\leq \epsilon_\lambda \lambda^{-1-\gamma} \sum_{1\leq t_1\leq\lambda, 1\leq t_2\leq\lambda^\gamma} \sum_{\boldsymbol{u}\in\mathbb{Z}^2} |a(\boldsymbol{t}-\boldsymbol{u})| \\
&\leq C\epsilon_\lambda \to 0.
\end{aligned}
$$

Therefore, $S(g''_{\lambda,\gamma}(\boldsymbol{x};\cdot)) = o_p(1)$ is negligible as $\lambda \to \infty$. It remains to show $S(g'_{\lambda,\gamma}(\boldsymbol{x};\cdot)) \xrightarrow{\text{fdd}} \int_{\mathbb{R}^2} h(\boldsymbol{x};\boldsymbol{u})W(\boldsymbol{u})$, where $h(\boldsymbol{x};\boldsymbol{u}) := \sigma\mathbb{I}(\boldsymbol{u} \in]\mathbf{0},\boldsymbol{x}])$. Using Proposition 9 of Chapter 2, this follows from $\|g'^{\dagger}_{\lambda,\gamma}(\boldsymbol{x};\cdot) - h(\boldsymbol{x};\cdot)\| \to 0$, where $g'^{\dagger}_{\lambda,\gamma}(\boldsymbol{x};\cdot)$ is defined as in (134) of Chapter 2, with $\mu_1 = 1, \mu_2 = \gamma$. Then, with $\gamma_1 := 1, \gamma_2 := \gamma$, we see that

$$
\begin{aligned}
g'^{\dagger}_{\lambda,\gamma}(\boldsymbol{x};\boldsymbol{u}) = & \sum_{1\leq t_i\leq[\lambda^{\gamma_i}x_i], i=1,2} a(\boldsymbol{t} - ([\lambda^{\gamma_1}x_1],[\lambda^{\gamma_2}x_2]))\mathbb{I}(-\lambda^{\gamma_i} \\
&\leq [\lambda^{\gamma_i}u_i] \leq 2\lambda^{\gamma_i}, i=1,2) \\
&\to h(\boldsymbol{x};\boldsymbol{u})
\end{aligned}
$$

point-wise for a.e. $\boldsymbol{u} \in \mathbb{R}^2$. Moreover, since $g'^{\dagger}_{\lambda,\gamma}(\boldsymbol{x};\cdot)$ is bounded and has compact support (which vanishes outside of $[-1,2]^2 \subset \mathbb{R}^2$), the last convergence extends to the convergence in $L^2(\mathbb{R}^2)$, proving $\|g'^{\dagger}_{\lambda,\gamma}(\boldsymbol{x};\cdot) - h(\boldsymbol{x};\cdot)\| \to 0$ and completing the proof of Proposition 24. □

Proof of Proposition 25. First, consider the one-dimensional convergence in (103) at a given $\boldsymbol{x} \in \mathbb{R}^2_+$. We use Proposition 7 of Chapter 2. Assumption $(\mathrm{A})_{d_1,d_2}$ implies that

$$
\lim_{n\to\infty} n^{-2d_i-1} \sum_{t,t'=1}^{n} r_i(t-t') = \kappa_i^2 := \begin{cases} A_i^2 \frac{B(d_i,1-2d_i)}{d_i(1+2d_i)}, & 0 < |d_i| < 1/2, \\ A_i^2, & d_i = 0. \end{cases} \tag{191}
$$

See Ref. [103, Prop. 3.3.1]. (Note that $B(d,1-2d)/d = \Gamma(d)\Gamma(1-2d)/d\Gamma(1-d) > 0, |d| < 1/2$.) For $-1/2 < d_i < 0$, (191) can be proved as follows. Observe the identity

$$
\begin{aligned}
&\sum_{t,t'=1}^{n} r_i(t-t') \\
&\quad = \sum_{u=0}^{\infty}\left(\sum_{t=1}^{n} a_i(t+u)\right)^2 + \sum_{u=1}^{n}\left(\sum_{t=0}^{n-u} a_i(t)\right)^2 =: I_{n1} + I_{n2},
\end{aligned}
$$

which follows by writing the l.h.s. as $\sum_{u\in\mathbb{Z}}\big(\sum_{t=1}^n \tilde a_i(t+u)\big)^2$, with $\tilde a_i(t) := a_i(t)\mathbb{I}(t\ge 0), t\in\mathbb{Z}$. Using the zero-sum condition, I_{n2} can be written as $I_{n2} = \sum_{u=1}^n(\sum_{t=n-u+1}^\infty a_i(t))^2 = \sum_{u=0}^{n-1}(\sum_{u+1}^\infty a_i(t))^2$. The relations

$$n^{-2d_i-1}I_{n1} \to A_i^2\int_0^\infty \mathrm{d}u(\int_0^1 (t+u)^{d_i-1}\mathrm{d}t)^2 =: A_i^2 I_1,$$
$$n^{-2d_i-1}I_{n2} \to A_i^2\int_0^1 \mathrm{d}u(\int_u^\infty t^{d_i-1}\mathrm{d}t)^2 =: A_i^2 I_2$$

for $-1/2 < d_i < 0$ follow by $(\mathrm{A})_{d_1,d_2}$ and the DCT, where $I_1 + I_2 = B(d_i, 1-2d_i)/d_1(1+2d_i)$; see e.g., Ref. [103, p. 545]. From (100) and (191), we immediately obtain

$$\lambda_1^{-2H_1}\lambda_2^{-2H_2}\mathrm{E}(S_{\lambda_1,\lambda_2}(\boldsymbol{x}))^2$$
$$= \prod_{i=1}^2 \lambda_i^{-1-2d_i}\sum_{t,t'=1}^{[\lambda_i x_i]} r_i(t-t') \to \prod_{i=1}^2 \kappa_i^2 x_i^{1+2d_i} \tag{192}$$

Then, $\lambda_1^{-H_1}\lambda_2^{-H_2}S_{\lambda_1,\lambda_2}(\boldsymbol{x}) \xrightarrow{\mathrm{d}} \kappa_1\kappa_2 B_{H_1,H_2}(\boldsymbol{x})$ for a fixed $\boldsymbol{x}\in\mathbb{R}_+^2$ follows by Proposition 7 of Chapter 2 since condition (14) of Chapter 2 by the factorisation property (97) reduces to

$$\sup_{u\in\mathbb{Z}}\left|\sum_{t=1}^n a_i(t-u)\right| = o(n^{d_i+(1/2)}), \quad n\to\infty, \tag{193}$$

which is obvious since the l.h.s. of (193) is bounded for $-1/2 < d_i \le 0$, and $O(n^{d_i})$ for $0 < d_i < 1/2$. The finite-dimensional convergence in (103) follows similarly by considering the linear combinations $\sum_{j=1}^m \theta_j S_{\lambda_1,\lambda_2}(\boldsymbol{x}_j)$ (the 'Cramer–Wold device'). Proposition 25 is proved. □

9 Notes

1. Most facts in Sections 1 and 2 come from Pilipauskaitė and Surgailis [209, 210]. The notion and some properties of generalised homogeneous function on $\mathbb{R}^\nu$ appeared in Hankey and Stanley [113]. The above function is a particular case of a more general class of *E-homogeneous functions* satisfying $f(\lambda^E\boldsymbol{x}) = \lambda f(\boldsymbol{x}), \boldsymbol{x}\in\mathbb{R}_0^\nu, \lambda > 0$ for a given $\nu\times\nu$ matrix E studied by Meerschaert and Scheffler [183] and Biermé *et al.* [37].

2. A general 'wisdom' in Fourier analysis says that a power decay of a function on $\mathbb{R}^\nu$ as $|\boldsymbol{t}| \to \infty$ is reflected in a power behaviour of its Fourier transform as $|\boldsymbol{x}| \to 0$, and vice versa. Rigorous statements of this kind are known as Abelian and Tauberian theorems. Such theorems play important role in the theory of distributions, time-series analysis, renewal theory, and other areas. See Bingham [43] and Feller [89]. The classical reference on Abelian/Tauberian theorems for functions decaying as a power function of the Euclidean norm $|\boldsymbol{t}|$ is Wainger [264]. An extension of the Abelian- and Tauberian-type results to the Fourier transform of $a(\boldsymbol{t})$ with anisotropic decay as in (23) is an interesting and difficult open problem. Quoting Leonenko and Olenko [162, p. 716] '...perhaps to the astonishment of some readers, multi-dimensional Abelian and Tauberian theory is more complicated than its one-dimensional counterpart for characteristic functions and most results were obtained only in the last two decades'.

3. The discussion in Sections 3–8 of Chapter 7 is based on Puplinskaitė and Surgailis [220], Pilipauskaitė and Surgailis [209, 210], and Surgailis [250–252].

4. Damarackas and Paulauskas [68, 69] discussed anisotropic scaling of MA RFs on $\mathbb{Z}^\nu$ with finite and infinite variance and MA coefficients taking various forms, leading either to the absence of scaling transition or to multiple critical points in the scaling diagram. We expect that most results in Sections 4 and 5 of Chapter 7 can be extended to MA RFs in (1) with infinite-variance innovations. General criteria for weak convergence of linear random fields with innovations in the domain of attraction of a stable law were given by Peligrad *et al.* [201]. Particularly, Proposition 25 extends to $\varepsilon \in D(\alpha), 1 < \alpha < 2$, following Damarackas and Paulauskas [68] with the limit FBS replaced by a linear fractional stable sheet introduced by Ayache *et al.* [12, 13].

5. Biermé *et al.* [39] studied anisotropic scaling limits in (128) for a binary RF X on $\mathbb{Z}^\nu$, generalising the (one-dimensional) Hammond and Sheffield [112] coalescing the random walk model. Biermé *et al.* [39] proved that these scaling limits depend on $\gamma \in \mathbb{R}^\nu_+$ and have a roughly similar structure as in Proposition 31, with unbalanced limits having degenerate dependence along at least one direction.

Chapter 7

Local Scaling Limits of Lévy-Driven Fractional RFs

A complete description of anisotropic local scaling limits of a class of fractional-type infinitely divisible RFs on $\mathbb{R}^2$ is obtained. We show that γ-tangent and γ-rectangent limits in (1) and (2) exist for any $\gamma > 0$ and exhibit a scaling transition at a critical point $\gamma_{\rm cr}$. The unbalanced rectangent limits $V_\pm$ are α-stable MSS RFs with one of the self-similarity parameters equal to 0 or 1. The unbalanced tangent limits $T_\pm$ have a similar but simpler structure, depending only on one of the two coordinates on the plane. The discussion is based on Ref. [211].

1 The Model and Main Results

Recall from Section 2 of Chapter 1 the definition of $\boldsymbol{\gamma}$-tangent and $\boldsymbol{\gamma}$-rectangent local scaling limits of RF X. In dimension $\nu = 2$, we take $\boldsymbol{\gamma} = (1, \gamma), \lambda^\Gamma = \mathrm{diag}(\lambda, \lambda^\gamma), \gamma > 0$ w.l.g. and consider the convergence

$$d_{\lambda,\gamma}^{-1}(X(\boldsymbol{t}_0 + \lambda^\Gamma \boldsymbol{t}) - X(\boldsymbol{t}_0)) \stackrel{\rm fdd}{\longrightarrow} T_\gamma(\boldsymbol{t}), \quad \lambda \to 0, \tag{1}$$

and

$$d_{\lambda,\gamma}^{-1} X(]\boldsymbol{t}_0, \boldsymbol{t}_0 + \lambda^\Gamma \boldsymbol{t}]) \stackrel{\rm fdd}{\longrightarrow} V_\gamma(\boldsymbol{t}), \quad \lambda \to 0, \tag{2}$$

for any given $\gamma > 0$. The class of RFs X, for which the above limits are identified in what follows, can be written as a stochastic integral,

$$X(\boldsymbol{t}) := \int_{\mathbb{R}^2} \{g(\boldsymbol{t}-\boldsymbol{u}) - g_1^0((t_1,0)-\boldsymbol{u}) - g_2^0((0,t_2)-\boldsymbol{u}) + g_{12}^0(-\boldsymbol{u})\} M(\mathrm{d}\boldsymbol{u}), \quad \boldsymbol{t} = (t_1,t_2) \in \mathbb{R}_+^2, \tag{3}$$

w.r.t. a Lévy random measure M on $\mathbb{R}^2$, with the deterministic integrands $g, g_1^0, g_2^0, g_{12}^0$ satisfying certain conditions guaranteeing the existence of (3). It follows from (3) that the RF X has stationary rectangular increments which do not depend on the 'initial' functions g_1^0, g_2^0, g_{12}^0:

$$X(]\boldsymbol{t}_0, \boldsymbol{t}_0+\boldsymbol{t}]) \overset{\mathrm{fdd}}{=} X(]\boldsymbol{0},\boldsymbol{t}]) = \int_{\mathbb{R}^2} g(]-\boldsymbol{u}, \boldsymbol{t}-\boldsymbol{u}]) M(\mathrm{d}\boldsymbol{u}), \quad \boldsymbol{t} \in \mathbb{R}_+^2, \tag{4}$$

where

$$g(]-\boldsymbol{u},\boldsymbol{t}-\boldsymbol{u}]) = g(t_1-u_1, t_2-u_2) - g(-u_1, t_2-u_2) - g(t_1-u_1,-u_2) + g(-u_1,-u_2)$$

in accordance with the notation of the rectangular increment. Similarly, when $g_1^0 = g_2^0 = 0$, the RF X in (3) has stationary (ordinary) increments:

$$X(\boldsymbol{t}_0+\boldsymbol{t}) - X(\boldsymbol{t}_0) \overset{\mathrm{fdd}}{=} X(\boldsymbol{t}) - X(\boldsymbol{0}) = \int_{\mathbb{R}^2} (g(\boldsymbol{t}-\boldsymbol{u}) - g(\boldsymbol{u})) M(\mathrm{d}\boldsymbol{u}), \quad \boldsymbol{t} \in \mathbb{R}^2. \tag{5}$$

In both cases, the scaling limits in (1)–(2) do not depend on $\boldsymbol{t}_0$ and depend on the scaling properties of the kernel g and the Lévy basis M, as specified in Assumptions $(\mathrm{G})_\alpha$ and $(\mathrm{M})_\alpha$. The latter assumption, defined in Section 6 of Chapter 2, roughly means that the small-scale behaviour of M is α-stable with $0 < \alpha \le 2$. Assumption $(\mathrm{G})_\alpha$ features a power-law behaviour of $g(\boldsymbol{t})$ at $\boldsymbol{t} = \boldsymbol{0}$ with generally different exponents along the coordinate axes:

$$g(\boldsymbol{t}) \sim g_0(\boldsymbol{t}) := \rho(\boldsymbol{t})^{\chi} \ell(\boldsymbol{t}), \quad \boldsymbol{t} \to \boldsymbol{0}, \quad \text{where} \quad \rho(\boldsymbol{t}) := |t_1|^{q_1} + |t_2|^{q_2}, \tag{6}$$

where $q_1 > 0, q_2 > 0, \kappa \in \mathbb{R}_0$ are parameters and $\ell(\boldsymbol{t}), \boldsymbol{t} \in \mathbb{R}^2_0$ is a generalised invariant function, resembling the behaviour of MA coefficients $a(\boldsymbol{t}), |\boldsymbol{t}| \to \infty$ in Assumption $(\mathrm{A})_{\mathrm{LRD},\rho}$. The precise definition of $(\mathrm{G})_\alpha$ is more involved and is postponed for later. The above class is quite large and contains many fractional RFs studied in Refs. [58,59,114,127,258] and elsewhere. Particularly, (5) includes a fractional Lévy RF with $g(\boldsymbol{t}) = |\boldsymbol{t}|^{H-1}, H \in (0,1)$ (Example 3) and a Matérn RF (Example 4). RFs in (3) also constitute a natural spatial generalisation of stationary-increment Lévy-driven MAs with one-dimensional time scaling to fractional linear stable motion [26]. In view of these facts, we call (3) a *fractional Lévy-driven RF*. The r.h.s. $g_0(\boldsymbol{t}) = \rho(\boldsymbol{t})^\chi \ell(\boldsymbol{t})$ in (6) vanishes or explodes at the origin $\boldsymbol{t} = \boldsymbol{0}$ depending on the sign of χ, suggesting a different scaling behaviour in (1) in the cases of $\chi > 0$ and $\chi < 0$. Actually, our limit results essentially depend on the two parameters

$$p_i := q_i(Q - \chi) > 0, \quad i = 1, 2, \tag{7}$$

alone, making the parametrisation $p_i, i = 1, 2$, in (7) more convenient than $\chi, q_i, i = 1, 2$, in (6). The parameters in (7) satisfy

$$\frac{\alpha}{1+\alpha} < P < \alpha, \quad \text{where } P := \frac{1}{p_1} + \frac{1}{p_2}. \tag{8}$$

Note that for the p_i in (7) satisfying (8), $\chi > 0$ is equivalent to $P > 1$. For $c_i > 0, i = 1, 2$, denote

$$P_{c_1,c_2} := \frac{c_1}{p_1} + \frac{c_2}{p_2}, \tag{9}$$

so that $P = P_{1,1}$. We prove that the above Lévy-driven fractional RF in (3) exhibits a small-scale scaling transition, with different unbalanced γ-rectangent RFs $V_\pm$ depending on α, p_1, p_2.

The main results are presented in Table 1 and Fig. 1, showing four sets – $\mathcal{R}_{11}, \mathcal{R}_{12}, \mathcal{R}_{21}$, and $\mathcal{R}_{22}$ – in the parameter region (8) determined by the segments $P_{1/\alpha,(1+\alpha)/\alpha} = 1, P_{(1+\alpha)/\alpha,1/\alpha} = 1$ with different γ-rectangent limits $V_\pm$. The critical or the scaling transition point in all four regions, $\mathcal{R}_{ij}$, is the same:

$$\gamma_{\mathrm{cr}} := \frac{p_1}{p_2} = \frac{q_1}{q_2}. \tag{10}$$

Table 1. Unbalanced rectangent scaling limits $V_\pm$ and their Hurst parameters in regions $\mathcal{R}_{11}, \mathcal{R}_{12}, \mathcal{R}_{21}$, and $\mathcal{R}_{22}$ in Fig. 1.

Parameter Region	V_+	V_-	Range of Hurst Parameters
$\mathcal{R}_{11}$	$\mathfrak{B}_{\mathcal{H}_1^+,1,\alpha}$	$\mathfrak{B}_{1,\mathcal{H}_2^-,\alpha}$	$0 < \mathcal{H}_1^+, \mathcal{H}_2^- < 1$
$\mathcal{R}_{12}$	$\mathfrak{B}_{\mathcal{H}_1^+,1,\alpha}$	$\mathfrak{B}_{\mathcal{H}_1^-,0,\alpha}$	$0 < \mathcal{H}_1^+, \mathcal{H}_1^- < 1$
$\mathcal{R}_{21}$	$\mathfrak{B}_{0,\mathcal{H}_2^+,\alpha}$	$\mathfrak{B}_{1,\mathcal{H}_2^-,\alpha}$	$0 < \mathcal{H}_2^+, \mathcal{H}_2^- < 1$
$\mathcal{R}_{22}$	$\mathfrak{B}_{0,\mathcal{H}_2^+,\alpha}$	$\mathfrak{B}_{\mathcal{H}_1^-,0,\alpha}$	$0 < \mathcal{H}_2^+, \mathcal{H}_1^- < 1$

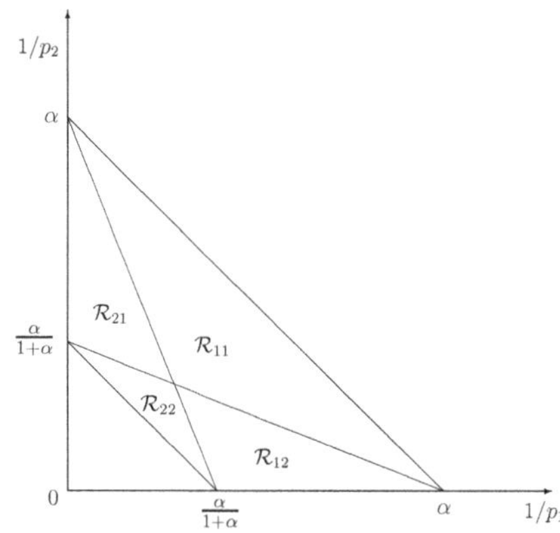

Figure 1. Regions in the parameter set $\frac{\alpha}{1+\alpha} < P < \alpha$ with different unbalanced tangent limits.

The four RFs $\mathfrak{B}_{\mathcal{H}_1^+,1,\alpha}, \mathfrak{B}_{1,\mathcal{H}_2^-,\alpha}, \mathfrak{B}_{\mathcal{H}_1^-,0,\alpha}, \mathfrak{B}_{0,\mathcal{H}_2^+,\alpha}$ in Table 1 are MSS, with one of the Hurst parameters taking the value 1 or 0 and the other parameter given by

$$\begin{aligned} \mathcal{H}_1^+ &:= \frac{1+\alpha}{\alpha} + \frac{p_1}{\alpha p_2} - p_1, \quad \mathcal{H}_1^- := \frac{1+\alpha}{\alpha}\left(1 + \frac{p_1}{p_2}\right) - p_1, \\ \mathcal{H}_2^+ &:= \frac{1+\alpha}{\alpha}\left(1 + \frac{p_2}{p_1}\right) - p_2, \quad \mathcal{H}_2^- := \frac{1+\alpha}{\alpha} + \frac{p_2}{\alpha p_1} - p_2. \end{aligned} \tag{11}$$

In the Gaussian case $\alpha = 2$, the RFs in Table 1 agree with the FBS, with the corresponding Hurst parameters being $\mathfrak{B}_{\mathcal{H}_1^\pm,\mathcal{H}_2^\pm,2} \stackrel{\text{fdd}}{=} c_\pm B_{\mathcal{H}_1^\pm,\mathcal{H}_2^\pm}$.

Let us turn to rigorous formulations of the above results. Given a function $f = f(\boldsymbol{t}), \boldsymbol{t} = (t_1, t_2) \in \mathbb{R}^2$, we use the following notation for its partial derivatives:

$$\partial_i f(\boldsymbol{t}) := \partial f(\boldsymbol{t})/\partial t_i, \quad i = 1, 2, \quad \partial_{12} f(\boldsymbol{t}) := \partial^2 f(\boldsymbol{t})/\partial t_1 \partial t_2. \tag{12}$$

Assumption $(\mathbf{G})_\alpha$. The functions g and g_0 have partial derivatives as in (12) on $\mathbb{R}^2_0$, satisfying

$$\begin{gathered} |g_0(\boldsymbol{t})| \le C\rho(\boldsymbol{t})^\chi, \quad |\partial_i g_0(\boldsymbol{t})| \le C\rho(\boldsymbol{t})^{\chi - \frac{1}{q_i}}, \\ i = 1, 2, \quad |\partial_{12} g_0(\boldsymbol{t})| \le C\rho(\boldsymbol{t})^{\chi - Q} \end{gathered} \tag{13}$$

for any $\boldsymbol{t} \in \mathbb{R}^2_0$, and

$$\begin{aligned} &g(\boldsymbol{t}) = g_0(\boldsymbol{t}) + o(\rho(\boldsymbol{t})^\chi), \quad \partial_i g(\boldsymbol{t}) = \partial_i g_0(\boldsymbol{t}) + o\big(\rho(\boldsymbol{t})^{\chi - \frac{1}{q_i}}\big), \\ &i = 1, 2, \quad \partial_{12} g(\boldsymbol{t}) = \partial_{12} g_0(\boldsymbol{t}) + o\big(\rho(\boldsymbol{t})^{\chi - Q}\big), \quad \text{as} \quad |\boldsymbol{t}| \to 0. \end{aligned} \tag{14}$$

Moreover, $g_0(\boldsymbol{t}) = \rho(\boldsymbol{t})^\chi \ell(\boldsymbol{t}), \boldsymbol{t} \in \mathbb{R}^2_0$ as in (6), where $\ell(\boldsymbol{t})$ is a generalised invariant function and the parameters $\chi \in \mathbb{R}\backslash\{0\}$, $q_i > 0$, $i = 1, 2$ satisfy

$$-\frac{1}{\alpha} Q < \chi < \left(1 - \frac{1}{\alpha}\right) Q. \tag{15}$$

Assumption $(\mathrm{G})_\alpha$ pertains to the behaviour of g alone. It is complemented by the following general condition guaranteeing the existence of the Lévy fractional RF X in (3).

Assumption $(\mathbf{G})^0_\alpha$. For any $\boldsymbol{t} = (t_1, t_2) \in \mathbb{R}^2, \delta > 0$,

$$\begin{aligned} \int_{\mathbb{R}^2} &\big| g(\boldsymbol{t} - \boldsymbol{u}) - g^0_1((t_1, 0) - \boldsymbol{u}) - g^0_2((0, t_2) - \boldsymbol{u}) \\ &+ g^0_{12}(-\boldsymbol{u}) \big|^\alpha \mathrm{d}\boldsymbol{u} < \infty \quad (0 \le \alpha \le 2) \end{aligned} \tag{16}$$

and

$$\int_{|\boldsymbol{u}| > \delta} \left(|\partial_{12} g(\boldsymbol{u})|^\alpha + \sum_{i=1}^{2} |\partial_i g(\boldsymbol{u})|^\alpha \right) \mathrm{d}\boldsymbol{u} < \infty \quad (1 \le \alpha \le 2). \tag{17}$$

Moreover, if $0 < \alpha < 1$, then there exist $\delta_0 > 0$ with the functions $\bar{g}_i(\boldsymbol{u})$, $\bar{g}_{12}(\boldsymbol{u})$, $\boldsymbol{u} = (u_1, u_2) \in \mathbb{R}^2_+$ monotone decreasing in each $u_j > 0$, $j = 1, 2$, and satisfying $|\partial_i g(\boldsymbol{u})| \le \bar{g}_i(|u_1|, |u_2|)$, $|\partial_{12} g(\boldsymbol{u})| \le \bar{g}_{12}(|u_1|, |u_2|)$, $|\boldsymbol{u}| > \delta_0$, such that (17) holds, with $\partial_i g$, $\partial_{12} g$ replaced by $\bar{g}_i, \bar{g}_{12}$, $i = 1, 2$.

Remark 1. The function $g_0(\boldsymbol{t}) = \rho(\boldsymbol{t})^\chi = (|t_1|^{q_1} + |t_2|^{q_2})^\chi$ corresponding to $\ell(\boldsymbol{t}) \equiv 1$ satisfies the bounds in (13), provided $q_i \geq 1, i = 1, 2$. The last fact follows from the expressions of the partial derivatives

$$\begin{aligned}
\partial_i \rho(\boldsymbol{t})^\chi &= \chi q_i \rho(\boldsymbol{t})^{\chi-1} |t_i|^{q_i - 1} \mathrm{sign}(t_i) = \rho(\boldsymbol{t})^{\chi - \frac{1}{q_i}} \ell_i(\boldsymbol{t}), \\
\partial_{12} \rho(\boldsymbol{t})^\chi &= \chi(\chi - 1) q_1 q_2 \rho(\boldsymbol{t})^{\chi-2} |t_1|^{q_1 - 1} |t_2|^{q_2 - 1} \mathrm{sign}(t_1 t_2) \\
&= \rho(\boldsymbol{t})^{\chi - Q} \ell_{12}(\boldsymbol{t})
\end{aligned}$$

with bounded

$$\ell_i(\boldsymbol{t}) := \chi q_i \left(\frac{|t_i|}{\rho(\boldsymbol{t})^{1/q_i}} \right)^{q_i - 1} \mathrm{sign}(t_i), \quad i = 1, 2,$$

$$\ell_{12}(\boldsymbol{t}) := \chi(\chi - 1) q_1 q_2 \prod_{i=1}^{2} \left(\frac{|t_i|}{\rho(\boldsymbol{t})^{1/q_i}} \right)^{q_i - 1} \mathrm{sign}(t_i).$$

Remark 2. Let

$$\rho_0(\boldsymbol{t}) := |t_1|^{p_1} + |t_2|^{p_2}, \quad \boldsymbol{t} = (t_1, t_2) \in \mathbb{R}^2 \tag{18}$$

for $p_i, i = 1, 2$ in (7). Note that $P = Q/(Q - \chi)$ so that $P < \alpha$ translates to $\chi < Q(1 - \frac{1}{\alpha})$ for $\chi > 0$ and $P < \alpha/(1+\alpha)$ to $\chi > -Q/\alpha$ for $\chi < 0$. Using the inequality in (13), it follows that the bounds in (13) can be respectively replaced by

$$\begin{gathered}
|g_0(\boldsymbol{t})| \leq C \rho_0(\boldsymbol{t})^{P-1}, \quad |\partial_i g_0(\boldsymbol{t})| \leq C \rho_0(\boldsymbol{t})^{P - \frac{1}{p_i} - 1}, \\
|\partial_{12} g_0(\boldsymbol{t})| \leq C \rho_0(\boldsymbol{t})^{-1}.
\end{gathered} \tag{19}$$

In the subsequent proofs, we use either (13) or (19), depending on which is more convenient. Recall that $\|f\|_\alpha = \left(\int_{\mathbb{R}^2} |f(\boldsymbol{u})|^\alpha \mathrm{d}\boldsymbol{u} \right)^{1/\alpha}$.

2 Examples of Fractional Lévy-Driven RFs

In this section, we discuss three examples of Lévy-driven fractional RFs related to fractional powers of the classical partial differential operators which are important for spatial applications. In each of these examples, Assumptions $(\mathrm{G})_\alpha$ and $(\mathrm{G})_\alpha^0$ are verified and Theorems 11 and 14 apply in the respective parameter regions.

Example 3 (Fractional Lévy RF; Example 22 of Chapter 3). Let $g(\boldsymbol{t}) = |\boldsymbol{t}|^{H-\frac{2}{\alpha}}, g_i^0(\boldsymbol{t}) = 0, i = 1, 2, g_{12}^0(\boldsymbol{t}) = -|\boldsymbol{t}|^{H-\frac{2}{\alpha}}$ for some $H \in (0,1), \alpha \in (0,2]$. The corresponding fractional Lévy RF X takes the form

$$X(\boldsymbol{t}) := \int_{\mathbb{R}^2} \left\{ |\boldsymbol{u} - \boldsymbol{t}|^{H-\frac{2}{\alpha}} - |\boldsymbol{u}|^{H-\frac{2}{\alpha}} \right\} M(\mathrm{d}\boldsymbol{u}), \quad \boldsymbol{t} \in \mathbb{R}_+^2. \tag{20}$$

If $M = W_\alpha$ is an α-stable random measure, the RF X in (20) is also called an MA fractional stable RF (a fractional Brownian RF in the case of $\alpha = 2$). It follows that the integrand in (20) satisfies Assumption $(\mathrm{G})_\alpha^0$ if and only if $H \in (0,1)$, in which case X in (20) is well defined for M, satisfying Assumption $(\mathrm{M})_\alpha$.

Let us check that $g(\boldsymbol{t}) = |\boldsymbol{t}|^{H-\frac{2}{\alpha}}$ satisfies Assumption $(\mathrm{G})_\alpha$, with $q_1 = q_2 = 2, Q = 1, \chi = (H - \frac{2}{\alpha})/2 < 0$ or $p_1 = p_2 = 2 - H + \frac{2}{\alpha} \in (1+\frac{2}{\alpha}, 2+\frac{2}{\alpha}), P = \frac{2}{2-H+\frac{2}{\alpha}} \in (\frac{\alpha}{1+\alpha}, \frac{2\alpha}{2+\alpha})$, for any $H \in (0,1)$, implying $P_{(1+\alpha)/\alpha, 1/\alpha} = P_{1/\alpha, (1+\alpha)/\alpha} < 1$. Since $g(\boldsymbol{t}) = g_0(\boldsymbol{t})$ is a homogeneous function, this follows from Remark 1 and can also be directly verified. Indeed, using $|\partial_i |\boldsymbol{t}|| = |t_i|/|\boldsymbol{t}| \le 1, |\partial_{12}|\boldsymbol{t}|| = |t_1 t_2|/|\boldsymbol{t}|^3 \le |\boldsymbol{t}|^{-1}$, it immediately follows that

$$\begin{aligned} \left|\partial_{12} |\boldsymbol{t}|^{H-\frac{2}{\alpha}}\right| &= \left| \left(H - \frac{2}{\alpha}\right)\left(H - 1 - \frac{2}{\alpha}\right) |\boldsymbol{t}|^{H-2-\frac{2}{\alpha}} \partial_1 |\boldsymbol{t}| \partial_2 |\boldsymbol{t}| \right. \\ &\quad \left. + \left(H - \frac{2}{\alpha}\right) |\boldsymbol{t}|^{H-1-\frac{2}{\alpha}} \partial_{12} |\boldsymbol{t}| \right| \\ &\le C |\boldsymbol{t}|^{H-2-\frac{2}{\alpha}} \end{aligned}$$

or the last inequality in (13), with the remaining relations being completely analogous.

Example 4 (Isotropic Matérn RF; Example 23 of Chapter 3). Let $c > 0$,

$$g(\boldsymbol{t}) := \frac{2^{1+\chi} c^{-\chi}}{\Gamma(-\chi)} \times \begin{cases} |\boldsymbol{t}|^\chi K_\chi(c|\boldsymbol{t}|), & \chi \in \left(-\frac{1}{\alpha}, 0\right), \\ |\boldsymbol{t}|^\chi K_\chi(c|\boldsymbol{t}|) - \Gamma(\chi) 2^{\chi-1} c^{-\chi}, & \chi \in \left(0, 1 - \frac{1}{\alpha}\right), \end{cases} \quad \boldsymbol{t} \in \mathbb{R}_0^2, \tag{21}$$

where K_χ denotes the modified Bessel function of the second kind. Let Assumption $(\mathrm{M})_\alpha$ hold for $0 < \alpha \le 2$ and

$$X(\boldsymbol{t}) := \begin{cases} \int_{\mathbb{R}^2} g(\boldsymbol{t}-\boldsymbol{u}) M(\mathrm{d}\boldsymbol{u}), & \chi \in \left(-\frac{1}{\alpha}, 0\right), \\ \int_{\mathbb{R}^2} (g(\boldsymbol{t}-\boldsymbol{u}) - g(-\boldsymbol{u})) M(\mathrm{d}\boldsymbol{u}), & \chi \in \left(0, 1-\frac{1}{\alpha}\right), \end{cases} \quad \boldsymbol{t} \in \mathbb{R}^2. \tag{22}$$

For $\kappa = \frac{H-1}{2} < 0, \alpha = \nu = 2$, $g(\boldsymbol{t})$ in (21) agrees with the kernel $m_{c,H}(\boldsymbol{t})$ in Example 23 (84), and X in (22) coincides with the stationary Matérn RF $\mathcal{M}_{c,H}(\boldsymbol{t})$ defined in this example, whereas for $\kappa > 0$, $X(\boldsymbol{t}) = \mathcal{M}_{c,H}(\boldsymbol{t}) - \mathcal{M}_{c,H}(\mathbf{0})$ is an increment of the Matérn RF. Clearly, the above RFs X and $\mathcal{M}_{c,H}$ have identical ordinary and rectangular increments and tangent limits.

We use the following facts about this function:

$$\begin{aligned} &K_\nu(t) = K_{-\nu}(t) \sim \Gamma(\nu) 2^{\nu-1} t^{-\nu}, \quad \nu > 0, \text{ as } t \downarrow 0, \\ &(t^\chi K_\nu(t))' = -t^\nu K_{\nu-1}(t), \quad \nu \in \mathbb{R}, \quad t > 0; \end{aligned} \tag{23}$$

see Ref. [1, Eq. (9.6.9), p. 375]. The asymptotics in (23) imply that $g(\boldsymbol{t}) \sim |\boldsymbol{t}|^{2\chi}$ $(\chi < 0), g(\boldsymbol{t}) \to 0$ $(\chi > 0)$ as $|\boldsymbol{t}| \to 0$; moreover, for $\chi = (H - \frac{2}{\alpha})/2 < 0$, we have $\lim_{c\downarrow 0} g(\boldsymbol{t}) = |\boldsymbol{t}|^{H-\frac{2}{\alpha}}$, and hence $g(\boldsymbol{t})$ in (20) can be regarded as the limiting case of (21) when $c \downarrow 0$. Equation (22) is a particular case of (3) corresponding to $g_1^0(\boldsymbol{t}) = g_2^0(\boldsymbol{t}) = g_{12}^0(\boldsymbol{t}) = 0$ $(\chi < 0)$ and $g_1^0(\boldsymbol{t}) = g_2^0(\boldsymbol{t}) = 0, g_{12}^0(\boldsymbol{t}) = -g(\boldsymbol{t})$ $(\chi > 0)$.

Proposition 5. *For any $\alpha \in (1, 2]$, the kernel $g(\boldsymbol{t})$ in (21) satisfies Assumption $(G)_\alpha$ with*

$$g_0(\boldsymbol{t}) := |\boldsymbol{t}|^{2\chi}, \quad \boldsymbol{t} \in \mathbb{R}_0^2, \tag{24}$$

$q_1 = q_2 = 2, Q = 1$, and χ as in (21), or $p_1 = p_2 = 2(1-\chi)$, $P \in (\frac{\alpha}{1+\alpha}, \alpha)$. Moreover, the integrand in (22) satisfies Assumption $(G)_\alpha^0$. As a consequence, the RF X in (22) is well defined for any M satisfying Assumption $(M)_\alpha$.

Proof. Relations (13) follow from (24) and Remark 1. With (13) in mind, let us prove a stronger version of (14):

$$\begin{aligned} g(\boldsymbol{t}) = g_0(\boldsymbol{t})(1 + o(1)), \quad \partial_i g(\boldsymbol{t}) = \partial_i g_0(\boldsymbol{t})(1 + o(1)), \\ \partial_{12} g(\boldsymbol{t}) = \partial_{12} g_0(\boldsymbol{t})(1 + o(1)) \end{aligned} \tag{25}$$

as $|\boldsymbol{t}| \to 0$. The first asymptotics in (25) follows from (23). Note the derivatives:

$$\begin{aligned} \partial_i g_0(\boldsymbol{t}) &= 2\chi |\boldsymbol{t}|^{2\chi-1} \partial_i |\boldsymbol{t}|, \quad i = 1, 2, \\ \partial_{12} g_0(\boldsymbol{t}) &= 4\chi(\chi - 1)|\boldsymbol{t}|^{2\chi-2} \prod_{i=1}^{2} \partial_i |\boldsymbol{t}|. \end{aligned} \tag{26}$$

It suffices to verify (25) for the derivatives $\partial_i \tilde{g}(\boldsymbol{t}), \partial_{12}\tilde{g}(\boldsymbol{t})$, where $\tilde{g}(\boldsymbol{t}) := \frac{2^{1+\chi}}{\Gamma(-\chi)} |\boldsymbol{t}|^\chi K_\chi(|\boldsymbol{t}|)$. Using the properties in (23), we find that

$$\partial_i \tilde{g}(\boldsymbol{t}) = -\frac{2^{1+\chi}}{\Gamma(-\chi)} |\boldsymbol{t}|^\chi K_{\chi-1}(|\boldsymbol{t}|) \partial_i |\boldsymbol{t}| = 2\chi |\boldsymbol{t}|^{2\chi-1}\big(1 + o(1)\big)\partial_i |\boldsymbol{t}|,$$

proving (25) for $\partial_i \tilde{g}(\boldsymbol{t})$. Similarly, using (23),

$$\begin{aligned} \partial_{12}\tilde{g}(\boldsymbol{t}) &= \frac{2^{1+\chi}}{\Gamma(-\chi)} \{-\partial_1(|\boldsymbol{t}|^\chi K_{\chi-1}(|\boldsymbol{t}|))\partial_2|\boldsymbol{t}| - |\boldsymbol{t}|^\chi K_{\chi-1}(|\boldsymbol{t}|)\partial_{12}|\boldsymbol{t}|\} \\ &= \frac{2^{1+\chi}}{\Gamma(-\chi)} |\boldsymbol{t}|^\chi K_{\chi-2}(|\boldsymbol{t}|) \prod_{i=1}^{2} \partial_i |\boldsymbol{t}| \\ &= 4\chi(\chi-1)|\boldsymbol{t}|^{2\chi-2}\big(1 + o(1)\big) \prod_{i=1}^{2} \partial_i |\boldsymbol{t}| \end{aligned}$$

(see (26)), proving (25). The remaining facts of Assumption $(\mathrm{G})_\alpha$ follow from the definitions in (21). Finally, Assumption $(\mathrm{G})^0_\alpha$ is guaranteed since $K_\chi(t)$ decays exponentially for any $\chi > 0$. □

Example 6 (Anisotropic fractional heat operator RF; Example 24 of Chapter 3). Let

$$g(\boldsymbol{t}) = \frac{t_1^\chi}{2^{\frac{1}{2}}(2\pi)^{\frac{3}{2}} c_2 \Gamma(\chi + \frac{3}{2})} \exp\left\{-c_1 t_1 - \frac{t_2^2}{4c_2^2 t_1}\right\} \mathbb{I}(t_1 > 0), \quad \boldsymbol{t} \in \mathbb{R}^2, \tag{27}$$

where $\chi > -\frac{3}{2}, c_1 > 0$, and $c_2 > 0$ are parameters. The parametrisation in (27) corresponds to $d = \kappa + \frac{3}{2}, \nu = 2$ in (91). Let

$$X(\boldsymbol{t}) = \int_{\mathbb{R}^2} g(\boldsymbol{t} - \boldsymbol{u}) M(\mathrm{d}\boldsymbol{u}), \tag{28}$$

and

$$\begin{aligned}
\rho(\boldsymbol{t}) &:= |t_1| + t_2^2, \quad \boldsymbol{t} \in \mathbb{R}^2, \quad q_1 := 1, \quad q_2 := 2, \quad Q = \frac{3}{2},\\
g_0(\boldsymbol{t}) &:= \rho(\boldsymbol{t})^{\chi}\ell(\boldsymbol{t}),\\
\ell(\boldsymbol{t}) &:= \frac{z^{\chi}}{2^{\frac{1}{2}}(2\pi)^{\frac{3}{2}}c_2\Gamma(\chi+\frac{3}{2})}\exp\left\{-\frac{1}{4c_2^2}\left(\frac{1}{z}-1\right)\right\}\mathbb{I}(t_1>0),\\
&\qquad \text{where } z := \frac{t_1}{\rho(\boldsymbol{t})} \in (0,1].
\end{aligned} \tag{29}$$

Note that $\ell(\boldsymbol{t})$ in (29) is a bounded generalised invariant function for any $\chi > -\frac{3}{2}$; in particular, $\ell(\boldsymbol{t}) \to 0$ $(z \downarrow 0)$. We have

$$g(\boldsymbol{t}) = \rho(\boldsymbol{t})^{\chi}\ell(\boldsymbol{t})(1+o(1)) = g_0(\boldsymbol{t})(1+o(1)), \quad \boldsymbol{t} \to \mathbf{0}. \tag{30}$$

The form of $g_0(\boldsymbol{t})$ and the asymptotics in (30) are similar to those in the lattice case; see Refs. [209, (4.8)] and [251, (4.11)].

Proposition 7. *The kernel $g(\boldsymbol{t})$ in (27) satisfies Assumptions $(G)_\alpha$ and $(G)^0_\alpha$ for any*

$$-\frac{3}{2\alpha} < \chi < \frac{3}{2}\left(1-\frac{1}{\alpha}\right), \quad \chi \neq 0, \quad \chi > -\frac{3}{2}, \quad \alpha \in \left(\frac{1}{2}, 2\right] \tag{31}$$

(equivalently, $P = \frac{3}{3-2\chi} \in \left(\frac{\alpha}{1+\alpha}, \alpha\right), P \neq 1, P > \frac{1}{2}$), with $g_0, q_i, i = 1,2$ as in (29). As a consequence, the RF X in (28) is well defined for any M satisfying Assumption $(M)_\alpha$.

Proof. It suffices to prove the proposition for $\tilde{g}(\boldsymbol{t}) := t_1^{\chi}\exp\{-t_1 - \frac{t_2^2}{t_1}\}\mathbb{I}(t_1>0), \tilde{g}_0(\boldsymbol{t}) := t_1^{\chi}\exp\{-\frac{t_2^2}{t_1}\}\mathbb{I}(t_1>0), \boldsymbol{t} \in \mathbb{R}$. Let us verify (13). It is convenient to change the variables as $u := t_1 > 0, z := t_2^2/t_1 > 0$. Then, $\rho(\boldsymbol{t}) = u(1+z), \tilde{g}_0(\boldsymbol{t}) = u^{\chi}\mathrm{e}^{-z}$ and the first bound in (13) can be written as $u^{\chi}\mathrm{e}^{-z} \le Cu^{\chi}(1+z)^{\chi}$ or $\mathrm{e}^{-z} \le C(1+z)^{\chi}, z > 0$ for any $\chi \in \mathbb{R}$. In a similar way,

$$\begin{aligned}
|\partial_1\tilde{g}_0(\boldsymbol{t})| &= u^{\chi-1}|\chi+z|\mathrm{e}^{-z} \le C(u(1+z))^{\chi-1} = C\rho(\boldsymbol{t})^{\chi-\frac{1}{q_1}},\\
|\partial_2\tilde{g}_0(\boldsymbol{t})| &= 2u^{\chi-\frac{1}{2}}\sqrt{z}\mathrm{e}^{-z} \le C(u(1+z))^{\chi-\frac{1}{2}} = C\rho(\boldsymbol{t})^{\chi-\frac{1}{q_2}},
\end{aligned}$$

$$
\begin{aligned}
\left|\partial_{12}\tilde{g}_0(\boldsymbol{t})\right| &= \left|2(1-\chi)u^{\chi-\frac{3}{2}}\sqrt{z}\mathrm{e}^{-z} + u^{\chi-\frac{1}{2}}z\mathrm{e}^{-z}\right| \le C(u(1+z))^{\chi-\frac{3}{2}} \\
&= C\rho(\boldsymbol{t})^{\chi-Q},
\end{aligned}
$$

proving (13). Relations (14) follow from (13) and

$$
\begin{aligned}
|\tilde{g}(\boldsymbol{t}) - \tilde{g}_0(\boldsymbol{t})| &= |(1-\mathrm{e}^{-t_1})\tilde{g}_0(\boldsymbol{t})| = O(|t_1|\rho(\boldsymbol{t})^{\chi}) = o(\rho(\boldsymbol{t})^{\chi}), \\
|\partial_1(\tilde{g}(\boldsymbol{t}) - \tilde{g}_0(\boldsymbol{t}))| &= |(1-\mathrm{e}^{-t_1})\partial_1\tilde{g}_0(\boldsymbol{t}) + \mathrm{e}^{-t_1}\tilde{g}_0(\boldsymbol{t})| \\
&= O(|t_1|\rho(\boldsymbol{t})^{\chi-\frac{1}{q_1}} + \rho(\boldsymbol{t})^{\chi}) = o(\rho(\boldsymbol{t})^{\chi-\frac{1}{q_1}}), \\
|\partial_2(\tilde{g}(\boldsymbol{t}) - \tilde{g}_0(\boldsymbol{t}))| &= |(1-\mathrm{e}^{-t_1})\partial_2\tilde{g}_0(\boldsymbol{t})| = O(|t_1|\rho(\boldsymbol{t})^{\chi-\frac{1}{q_2}}) \\
&= o(\rho(\boldsymbol{t})^{\chi-\frac{1}{q_2}}), \\
|\partial_{12}(\tilde{g}(\boldsymbol{t}) - \tilde{g}_0(\boldsymbol{t}))| &= |(1-\mathrm{e}^{-t_1})\partial_{12}\tilde{g}_0(\boldsymbol{t})| + O(|\partial_1\tilde{g}_0(\boldsymbol{t})| + \tilde{g}_0(\boldsymbol{t})) \\
&= o(\rho(\boldsymbol{t})^{\chi-Q}).
\end{aligned}
$$

Condition (16) of Assumption $(\mathrm{G})^0_\alpha$ in this case becomes $\|g\|_\alpha < \infty$, which follows from $\|\tilde{g}\|^\alpha_\alpha = C\int_0^\infty t_1^{\alpha\chi+\frac{1}{2}}\mathrm{e}^{-\alpha t_1}\mathrm{d}t_1 < \infty$ since $\chi > -\frac{3}{2\alpha}$. □

3 Limit RFs

In this section, we define the α-stable RFs $\mathfrak{B}_{\mathcal{H}_1^+,1,\alpha}, \mathfrak{B}_{1,\mathcal{H}_2^-,\alpha}, \mathfrak{B}_{\mathcal{H}_1^-,0,\alpha}, \mathfrak{B}_{0,\mathcal{H}_2^+,\alpha}$ in Table 1. As noted above, these RFs are MSS with one of the self-similarity parameters equal to 0 or 1. When $\alpha = 2$, these RFs are Gaussian and agree with the FBS defined in Example 11 of Chapter 1 through the covariance function. For $\alpha < 2$, the covariance does not exist, and the definition is more involved.

We start by defining two classes, $Y_{i,\alpha}, \widetilde{Y}_{i,\alpha}, i = 1, 2$, of α-stable SS SI processes with a one-dimensional time parameter. These definitions are completely analogous for $i = 1$ and $i = 2$, and the differences between the two cases essentially reduce to exchanging the coordinate axes. The processes $Y_{i,\alpha} = \{Y_{i,\alpha}(t); t \in \mathbb{R}_+\}, \widetilde{Y}_{i,\alpha} = \{\widetilde{Y}_{i,\alpha}(t); t \in \mathbb{R}_+\}, \alpha \in (0, 2]$, are defined by

$$
\begin{aligned}
Y_{i,\alpha}(t) &:= \int_{\mathbb{R}^2} h_i(t;\boldsymbol{u})W_\alpha(\mathrm{d}\boldsymbol{u}), \\
\widetilde{Y}_{i,\alpha}(t) &:= \int_{\mathbb{R}^2} \tilde{h}_i(t;\boldsymbol{u})W_\alpha(\mathrm{d}\boldsymbol{u}), \quad i = 1, 2
\end{aligned}
\tag{32}
$$

as stochastic integrals in (130) of Chapter 2 w.r.t. the α-stable random measure $W_\alpha(\mathrm{d}\boldsymbol{u})$ on $\mathbb{R}^2$ with the control measure $\mathrm{d}\boldsymbol{u}$ and the characteristic function in (124) of Chapter 2 of deterministic kernel functions

$$\begin{aligned} h_i(t;\boldsymbol{u}) &:= g_0(t\boldsymbol{e}_i - \boldsymbol{u}) - g_0(-\boldsymbol{u}), \quad i = 1,2, \\ \tilde{h}_1(t;\boldsymbol{u}) &:= \partial_2 g_0(t\boldsymbol{e}_1 - \boldsymbol{u}) - \partial_2 g_0(-\boldsymbol{u}), \\ \tilde{h}_2(t;\boldsymbol{u}) &:= \partial_1 g_0(t\boldsymbol{e}_2 - \boldsymbol{u}) - \partial_1 g_0(-\boldsymbol{u}), \\ h_0(\boldsymbol{t};\boldsymbol{u}) &:= g_0(] - \boldsymbol{u}, \boldsymbol{t} - \boldsymbol{u}]), \end{aligned} \tag{33}$$

where $t \in \mathbb{R}_+, \boldsymbol{t} \in \mathbb{R}^2_+, \boldsymbol{u} \in \mathbb{R}^2, \boldsymbol{e}_1 = (1,0), \boldsymbol{e}_2 = (0,1)$ (h_0 is used later to define the limit rectangent RF arising under a well-balanced scaling).

Recall the definition of P_{c_1,c_2} and $\rho_0(\boldsymbol{t})$ in (9) and (18). Note the equivalencies:

$$\begin{aligned} P < \alpha \ \& \ P_{\frac{1}{\alpha},\frac{1+\alpha}{\alpha}} > 1 &\iff 0 < \mathcal{H}_2^- < 1, \\ P > \tfrac{\alpha}{1+\alpha} \ \& \ P_{\frac{1}{\alpha},\frac{1+\alpha}{\alpha}} < 1 &\iff 0 < \mathcal{H}_1^- < 1, \\ P_{\frac{1}{\alpha},\frac{1+\alpha}{\alpha}} = 1 &\iff \mathcal{H}_1^- = 1 \ \& \ \mathcal{H}_2^- = 0. \end{aligned} \tag{34}$$

Similar equivalencies hold for $\mathcal{H}_1^+$ and $\mathcal{H}_2^+$ by symmetry. Also, note that $\mathcal{H}_1^+ = \mathcal{H}_2^- = 1$ when $P = \alpha$ and $\mathcal{H}_1^- = \mathcal{H}_2^+ = 0$ when $P = \frac{\alpha}{1+\alpha}$. We use (19) and properties (14)–(15) of Chapter 6 of generalised homogeneous functions in Proposition 4 of Chapter 6, particularly the inequality

$$\rho_0(\boldsymbol{t}+\boldsymbol{s}) \le C_p(\rho_0(\boldsymbol{t}) + \rho_0(\boldsymbol{s})), \quad \boldsymbol{t},\boldsymbol{s} \in \mathbb{R}^2, \quad C_p := 2^{(p_1-1)\vee(p_2-1)\vee 0}. \tag{35}$$

Proposition 8. *Let $0 < \alpha \le 2$, g_0 be as in (6) and satisfy the bounds (13); moreover, let $\frac{\alpha}{1+\alpha} < P < \alpha, P \ne 1$. Then, for any $\boldsymbol{t} \in \mathbb{R}^2_+, t \in \mathbb{R}_+$:*

- (i) $\|h_0(\boldsymbol{t};\cdot)\|_\alpha < \infty$;
- (ii) $\|h_i(t;\cdot)\|_\alpha < \infty$, *provided* $P_{1/\alpha,(1+\alpha)/\alpha} < 1$ $(i = 1)$ *or* $P_{(1+\alpha)/\alpha,1/\alpha} < 1$ $(i = 2)$ *holds;*
- (iii) $\|\tilde{h}_i(t;\cdot)\|_\alpha < \infty$, *provided* $P_{(1+\alpha)/\alpha,1/\alpha} > 1$ $(i = 1)$ *or* $P_{1/\alpha,(1+\alpha)/\alpha} > 1$ $(i = 2)$ *holds.*

Proof. It suffices to only consider $\boldsymbol{t} = \mathbf{1}$, $t = 1$, as well as the case $i = 1$ in (ii)–(iii). Let $U_r := \{\boldsymbol{u} \in \mathbb{R}^2 : \rho_0(\boldsymbol{u})^{\frac{1}{P}} < r\}$ for $r > 2^{\frac{1}{P}}$.
(i) We have $\int_{U_r} |h_0(\mathbf{1}; \boldsymbol{u})|^\alpha \mathrm{d}\boldsymbol{u} \leq C \int_{U_{2r}} |g_0(\boldsymbol{u})|^\alpha \mathrm{d}\boldsymbol{u}$, with $|g_0(\boldsymbol{u})| \leq C\rho_0(\boldsymbol{u})^{P-1}$ by (19), which satisfies $\int_{U_{2r}} \rho_0(\boldsymbol{u})^{\alpha(P-1)} \mathrm{d}\boldsymbol{u} < \infty$ by (14) of Chapter 6 since $\alpha(P-1) > -P$. On the other hand, rewriting $h_0(\mathbf{1}; \boldsymbol{u}) = \int_{[\mathbf{0},\mathbf{1}]} \partial_{12} g_0(\boldsymbol{t} - \boldsymbol{u}) \mathrm{d}\boldsymbol{t}$, with $|\partial_{12} g_0(\boldsymbol{t} - \boldsymbol{u})| \leq C\rho_0(\boldsymbol{u})^{-1}$, for all $\boldsymbol{u} \in U_r^c := \mathbb{R}^2 \backslash U_r$ by (19) and (35), we have $\int_{U_r^c} |h_0(\mathbf{1}; \boldsymbol{u})|^\alpha \mathrm{d}\boldsymbol{u} \leq C \int_{U_r^c} \rho_0(\boldsymbol{u})^{-\alpha} \mathrm{d}\boldsymbol{u} < \infty$ by (15) of Chapter 6 since $P < \alpha$.
(ii) We have $\int_{U_r} |h_1(1; \boldsymbol{u})|^\alpha \mathrm{d}\boldsymbol{u} \leq C \int_{U_{2r}} |g_0(\boldsymbol{u})|^\alpha \mathrm{d}\boldsymbol{u} < \infty$, as in the proof of (i). Next, rewriting $h_1(1; \boldsymbol{u}) = \int_0^1 \partial_1 g_0(t\boldsymbol{e}_1 - \boldsymbol{u}) \mathrm{d}t$, with $|\partial_1 g_0(t\boldsymbol{e}_1 - \boldsymbol{u})| \leq C\rho_0(\boldsymbol{u})^{\frac{1}{p_2}-1}$ on U_r^c by (19) and (35), we have $\int_{U_r^c} |h_1(1; \boldsymbol{u})|^\alpha \mathrm{d}\boldsymbol{u} \leq C \int_{U_r^c} |\rho_0(\boldsymbol{u})|^{\alpha(\frac{1}{p_2}-1)} \mathrm{d}\boldsymbol{u} < \infty$ by (15) of Chapter 6 since $P_{\frac{1}{\alpha}, \frac{1+\alpha}{\alpha}} < 1$.
(iii) We have $\int_{U_r} |\tilde{h}_1(1; \boldsymbol{u})|^\alpha \mathrm{d}\boldsymbol{u} \leq C \int_{U_{2r}} |\partial_2 g_0(\boldsymbol{u})|^\alpha \mathrm{d}\boldsymbol{u}$, with $|\partial_2 g_0(\boldsymbol{u})| \leq C\rho_0(\boldsymbol{u})^{\frac{1}{p_1}-1}$, by (19), which satisfies $\int_{U_{2r}} \rho_0(\boldsymbol{u})^{\alpha(\frac{1}{p_1}-1)} \mathrm{d}\boldsymbol{u} < \infty$ by (14) of Chapter 6 since $P_{\frac{1+\alpha}{\alpha}, \frac{1}{\alpha}} > 1$. Next, rewriting $\tilde{h}_1(1; \boldsymbol{u}) = \int_0^1 \partial_{12} g_0(t\boldsymbol{e}_1 - \boldsymbol{u}) \mathrm{d}t$, with $|\partial_{12} g_0(t\boldsymbol{e}_1 - \boldsymbol{u})| \leq C\rho(\boldsymbol{u})^{-1}$, on U_r^c by (19) and (35), we have $\int_{U_r^c} |\tilde{h}_1(1; \boldsymbol{u})|^\alpha \mathrm{d}\boldsymbol{u} \leq C \int_{U_r^c} \rho_0(\boldsymbol{u})^{-\alpha} \mathrm{d}\boldsymbol{u} < \infty$ by (15) of Chapter 6 since $P < \alpha$. Proposition 8 is proved. □

Corollary 9. *Let $g_0, \alpha, p_i, i = 1, 2$, satisfy the conditions of Proposition 8. Then, the random processes in* (32) *are well-defined, H-SSSI and have α-stable finite-dimensional distributions in the parameter regions indicated in Table 2.*

Table 2. Parameter regions and self-similarity indices of α-stable random processes $Y_{i,\alpha}, \widetilde{Y}_{i,\alpha}, i = 1, 2$, in (32).

	$\widetilde{Y}_{1,\alpha}$	$\widetilde{Y}_{2,\alpha}$	$Y_{1,\alpha}$	$Y_{2,\alpha}$
Parameter Region	$P_{\frac{1+\alpha}{\alpha}, \frac{1}{\alpha}} > 1$	$P_{\frac{1}{\alpha}, \frac{1+\alpha}{\alpha}} > 1$	$P_{\frac{1}{\alpha}, \frac{1+\alpha}{\alpha}} < 1$	$P_{\frac{1+\alpha}{\alpha}, \frac{1}{\alpha}} < 1$
H	$\mathcal{H}_1^+$	$\mathcal{H}_2^-$	$\mathcal{H}_1^-$	$\mathcal{H}_2^+$

Particularly, in the Gaussian case of $\alpha = 2$, *the processes* $Y_{i,2}, \widetilde{Y}_{i,2}, i = 1, 2$, *agree with the corresponding FBMs:*

$$Y_{1,2} \stackrel{\text{fdd}}{=} \sigma_1 B_{\mathcal{H}_1^-}, \quad Y_{2,2} \stackrel{\text{fdd}}{=} \sigma_2 B_{\mathcal{H}_2^+}, \quad \tilde{Y}_{1,2} \stackrel{\text{fdd}}{=} \tilde{\sigma}_1 B_{\mathcal{H}_1^+}, \quad \tilde{Y}_{2,2} \stackrel{\text{fdd}}{=} \tilde{\sigma}_2 B_{\mathcal{H}_2^-}, \tag{36}$$

where $\sigma_i^2 := \int_{\mathbb{R}^2} |h_i(1, \boldsymbol{u})|^2 \mathrm{d}\boldsymbol{u}, \tilde{\sigma}_i^2 := \int_{\mathbb{R}^2} |\tilde{h}_i(1, \boldsymbol{u})|^2 \mathrm{d}\boldsymbol{u}, i = 1, 2$.

Proof. The fact that the processes in (32) are well defined and have α-stable distributions follows from Proposition 8 and the general properties of stochastic integrals w.r.t. the α-stable random measure [231]. The stationarity of increments property is a consequence of the form of the integrands in (33) and the invariance of W_α w.r.t. shifts in $\mathbb{R}^2$. Similarly, the H-SS property follows from change of variables in the stochastic integral, particularly

$$\begin{aligned} Y_{1,\alpha}(\lambda t) &= \int_{\mathbb{R}^2} (g_0(\lambda t - u_1, -u_2) - g_0(-u_1, -u_2)) W_\alpha(\mathrm{d}\boldsymbol{u}) \\ &= \int_{\mathbb{R}^2} (g_0(\lambda t - \lambda u_1, -\lambda^{p_1/p_2} u_2) - g_0(-\lambda u_1, -\lambda^{p_1/p_2} u_2)) \\ &\quad \times W_\alpha(\mathrm{d}\lambda u_1, \mathrm{d}\lambda^{p_1/p_2} u_2) \\ &\stackrel{\text{fdd}}{=} \lambda^{\mathcal{H}_1^-} Y_{1,\alpha}(t), \end{aligned}$$

where we used the facts

$$g_0(\lambda t_1, \lambda^{q_1/q_2} t_2) = \lambda^{\chi q_1} g_0(\boldsymbol{t}), \quad \forall \lambda > 0,\ \boldsymbol{t} \in \mathbb{R}_0^2, \tag{37}$$

$\chi q_1 = p_1(P-1), p_1/p_2 = q_1/q_2$ (see (7)) and $W_\alpha((\mathrm{d}\lambda u_1, \mathrm{d}\lambda^{p_1/p_2} u_2) \stackrel{\text{fdd}}{=} \lambda^{(1+\frac{p_1}{p_2})/\alpha} W_\alpha(\mathrm{d}\boldsymbol{u})$. The self-similarity of $\widetilde{Y}_{1,\alpha}$ follows analogously:

$$\begin{aligned} \widetilde{Y}_{1,\alpha}(\lambda t) &= \int_{\mathbb{R}^2} (\partial_2 g_0(\lambda t - u_1, -u_2) - \partial_2 g_0(-u_1, -u_2)) W_\alpha(\mathrm{d}\boldsymbol{u}) \\ &= \int_{\mathbb{R}^2} (\partial_2 g_0(\lambda t - \lambda u_1, -\lambda^{p_1/p_2} u_2) - \partial_2 g_0(-\lambda u_1, -\lambda^{p_1/p_2} u_2)) \\ &\quad \times W_\alpha(\mathrm{d}\lambda u_1, \mathrm{d}\lambda^{p_1/p_2} u_2) \\ &\stackrel{\text{fdd}}{=} \lambda^{\mathcal{H}_1^+} \widetilde{Y}_{1,\alpha}(t). \end{aligned}$$

Finally, (36) is a consequence of the well-known characterisation of Gaussian H-SSSI processes [231]. □

Next, we define two classes of RFs on $\mathbb{R}^2_+$ following the definitions in (32). The first class of RFs is defined as a simple line extension of the corresponding random processes $\widetilde{Y}_{1,\alpha}, i = 1, 2$. For $\boldsymbol{t} = (t_1, t_2) \in \mathbb{R}^2_+$, set

$$\mathfrak{B}_{\mathcal{H}_1^+,1,\alpha}(\boldsymbol{t}) := t_2\widetilde{Y}_{1,\alpha}(t_1), \quad \mathfrak{B}_{1,\mathcal{H}_2^-,\alpha}(\boldsymbol{t}) \; := \; t_1\widetilde{Y}_{2,\alpha}(t_2). \tag{38}$$

The definition of the second class of RFs through a finite-dimensional distribution is similar to that of FBS with $H_1 \wedge H_2 = 0$ in Example 11 of Chapter 1. Let a finite collection, $\boldsymbol{t}_j = (t_{j,1}, t_{j,2}) \in \mathbb{R}^2_+, j = 1, \ldots, m, 0 < t_{1,2} < \cdots < t_{m,2}$, of points in $\mathbb{R}^2_+$ with different ordinates be given. Let $Y^{(j)}_{i,\alpha}, j = 0, 1, \ldots, m$ be independent copies of $Y_{i,\alpha}$ in (32). Define a family of finite-dimensional distributions:

$$\{\mathfrak{B}_{\mathcal{H}_1^-,0,\alpha}(\boldsymbol{t}_j);\, j = 1, \ldots, m\} \overset{\text{fdd}}{=} \{Y^{(j)}_{1,\alpha}(t_{j,1}) - Y^{(0)}_{1,\alpha}(t_{j,1});\, j = 1, \ldots, m\}. \tag{39}$$

Similarly, we define an RF $\mathfrak{B}_{0,\mathcal{H}_2^+,\alpha} = \{\mathfrak{B}_{0,\mathcal{H}_2^+,\alpha}(\boldsymbol{t}); \boldsymbol{t} \in \mathbb{R}^2_+\}$ such that for each collection $\boldsymbol{t}_j = (t_{j,1}, t_{j,2}) \in \mathbb{R}^2_+, j = 1, \ldots, m, 0 < t_{1,1} < \cdots < t_{m,1}$ of points with different abscissas,

$$\{\mathfrak{B}_{0,\mathcal{H}_2^+,\alpha}(\boldsymbol{t}_j);\, j = 1, \ldots, m\} \overset{\text{fdd}}{=} \{Y^{(j)}_{2,\alpha}(t_{j,2}) - Y^{(0)}_{2,\alpha}(t_{j,2});\, j = 1, \ldots, m\}. \tag{40}$$

These definitions can be extended to arbitrary finite collections of points in $\mathbb{R}^2_+$. Namely, let $\boldsymbol{t}_j = (t_{j,1}, t_{j,2}) \in \mathbb{R}^2_+, j = 1, \ldots, n$ be given with $t_{1,2} < t_{2,2} < \cdots < t_{m,2}, t_{j,2} \in \{t_{1,2}, \ldots, t_{m,2}\}, j = m+1, \ldots, n, n > m$. Then,

$$\begin{aligned} &\{\mathfrak{B}_{\mathcal{H}_1^-,0,\alpha}(\boldsymbol{t}_j);\, j = 1, \ldots, n\} \\ &\overset{\text{fdd}}{=} \{(Y^{(1)}_{1,\alpha}(t_{j,1}) - Y^{(0)}_{1,\alpha}(t_{j,1});\, t_{j,2} = t_{1,2}), \ldots, \\ &\qquad (Y^{(m)}_{1,\alpha}(t_{j,1}) - Y^{(0)}_{1,\alpha}(t_{j,1});\, t_{j,2} = t_{m,2})\}. \end{aligned} \tag{41}$$

Finally, let

$$V_0(\boldsymbol{t}) := \int_{\mathbb{R}^2} h_0(\boldsymbol{t};\boldsymbol{u}) W_\alpha(\mathrm{d}\boldsymbol{u}), \quad \boldsymbol{t} \in \mathbb{R}^2_+. \tag{42}$$

The following corollary summarises the properties of the introduced RFs and details their MSS indices mentioned at the beginning of this section.

Corollary 10. *Let $g_0, \alpha, p_i, i = 1, 2$ satisfy the conditions in Proposition 9. Then:*

(i) *the RF V_0 in (42) is well defined;*
(ii) *the RFs in Table 2 are well defined in the parameter regions shown in Table 2 (with $\widetilde{Y}_{1,\alpha}, \widetilde{Y}_{2,\alpha}$, $Y_{1,\alpha}, Y_{2,\alpha}$ replaced, respectively, by $\mathfrak{B}_{\mathcal{H}_1^+,1,\alpha}$, $\mathfrak{B}_{1,\mathcal{H}_2^-,\alpha}$, $\mathfrak{B}_{\mathcal{H}_1^-,0,\alpha}$, $\mathfrak{B}_{0,\mathcal{H}_2^+,\alpha}$);*
(iii) *all the RFs in (i) and (ii) have α-stable finite-dimensional distributions and stationary rectangular increments;*
(iv) *the RFs in Table 1 are (H_1, H_2)-MSS with the respective indices $(H_1, H_2) = (\mathcal{H}_1^+, 1), (1, \mathcal{H}_2^-)$, $(\mathcal{H}_1^-, 0)$, $(0, \mathcal{H}_2^+)$;*
(v) *in the Gaussian case of $\alpha = 2$, the RFs in Table 1 agree with FBS:*

$$\mathfrak{B}_{\mathcal{H}_1^+,1,2} \overset{\text{fdd}}{=} \tilde{\sigma}_1 B_{\mathcal{H}_1^+,1}, \quad \mathfrak{B}_{1,\mathcal{H}_2^-,2} \overset{\text{fdd}}{=} \tilde{\sigma}_2 B_{1,\mathcal{H}_2^-},$$
$$\mathfrak{B}_{\mathcal{H}_1^-,0,2} \overset{\text{fdd}}{=} \sigma_1 B_{\mathcal{H}_1^-,0}, \quad \mathfrak{B}_{0,\mathcal{H}_2^+,2} \overset{\text{fdd}}{=} \sigma_2 B_{0,\mathcal{H}_2^+},$$

where $\sigma_i^2, \tilde{\sigma}_i^2, i = 1, 2$ are given in (36).

Proof. Most of the facts in Corollary 10 follow from Proposition 9 and the definitions of the introduced RFs. Let us check the MSS property of $\mathfrak{B}_{\mathcal{H}_1^-,0,\alpha}$:

$$\mathfrak{B}_{\mathcal{H}_1^-,0,\alpha}(\lambda_1 t_1, \lambda_2 t_2) \overset{\text{fdd}}{=} \lambda_1^{\mathcal{H}_1^-} \mathfrak{B}_{\mathcal{H}_1^-,0,\alpha}(\boldsymbol{t}), \quad \forall\, \lambda_1 > 0,\ \lambda_2 > 0. \tag{43}$$

Let $\{\boldsymbol{t}_j = (t_{j,1}, t_{j,2}) \in \mathbb{R}^2_+, j = 1, \ldots, n\}$ be an arbitrary collection of points as in (41) with $t_{1,2} < t_{2,2} < \cdots < t_{m,2}, t_{j,2} \in \{t_{1,2}, \ldots, t_{m,2}\}, j = m+1, \ldots, n, n > m$. Then, the rescaled collection $\{\boldsymbol{t}'_j := (\lambda_1 t_{j,1}, \lambda_2 t_{j,2}), j = 1, \ldots, n\}$ satisfies the same property: $t'_{1,2} < t'_{2,2} < \cdots < t'_{m,2}, t'_{j,2} \in \{t'_{1,2}, \ldots, t'_{m,2}\}, j = m + 1, \ldots, n, n > m$.

Therefore, with $Y^{(j)}_{1,\alpha}, j = 0, 1, \ldots, m$, independent copies of $\mathcal{H}_1^-$-SS process $Y_{1,\alpha}$ in (32),

$$
\begin{aligned}
&\{\mathfrak{B}_{\mathcal{H}_1^-,0,\alpha}(\boldsymbol{t}'_j);\ j = 1, \ldots, n\} \\
&\overset{\text{fdd}}{=} \{(Y^{(1)}_{1,\alpha}(t'_{j,1}) - Y^{(0)}_{1,\alpha}(t'_{j,1});\ t'_{j,2} = t'_{1,2}), \ldots, \\
&\qquad (Y^{(m)}_{1,\alpha}(t'_{j,1}) - Y^{(0)}_{1,\alpha}(t'_{j,1});\ t'_{j,2} = t'_{m,2})\} \\
&\overset{\text{fdd}}{=} \lambda_1^{\mathcal{H}_1^-}\{(Y^{(1)}_{1,\alpha}t_{j,1}) - Y^{(0)}_{1,\alpha}(t_{j,1});\ t_{j,2} = t_{1,2}), \ldots, \\
&\qquad (Y^{(m)}_{1,\alpha}(t_{j,1}) - Y^{(0)}_{1,\alpha}(t_{j,1});\ t_{j,2} = t_{m,2})\} \\
&\overset{\text{fdd}}{=} \{\lambda_1^{\mathcal{H}_1^-}\mathfrak{B}_{\mathcal{H}_1^-,0,\alpha}(\boldsymbol{t}_j);\ j = 1, \ldots, n\},
\end{aligned}
$$

proving (43). □

4 Rectangent Limits of Lévy-Driven Fractional RFs

Theorem 11. *Let a Lévy-driven fractional RF X in* (3) *satisfy Assumptions* $(G)^0_\alpha$, $(G)_\alpha$, *and* $(M)_\alpha$; $0 < \alpha \le 2$, $\frac{\alpha}{1+\alpha} < P < \alpha$, $P \ne 1$, $P_{\frac{1}{\alpha},\frac{1+\alpha}{\alpha}} \ne 1$, $P_{\frac{1+\alpha}{\alpha},\frac{1}{\alpha}} \ne 1$. *Then, the γ-rectangent RF in* (2) *exists for any $\gamma > 0$, $\boldsymbol{t}_0 \in \mathbb{R}^2_+$ and satisfies the trichotomy*

$$
V_\gamma = \begin{cases} V_+, & \gamma > \gamma_{\rm cr}, \\ V_-, & \gamma < \gamma_{\rm cr}, \\ V_0, & \gamma = \gamma_{\rm cr}, \end{cases}
$$

with $\gamma_{\rm cr} = \frac{q_1}{q_2} = \frac{p_1}{p_2}$, V_0 *defined in* (42), *and*

$$
V_- := \begin{cases} \mathfrak{B}_{1,\mathcal{H}_2^-,\alpha}, & P_{\frac{1}{\alpha},\frac{1+\alpha}{\alpha}} > 1, \\ \mathfrak{B}_{\mathcal{H}_1^-,0,\alpha}, & P_{\frac{1}{\alpha},\frac{1+\alpha}{\alpha}} < 1, \end{cases} \qquad V_+ := \begin{cases} \mathfrak{B}_{\mathcal{H}_1^+,1,\alpha}, & P_{\frac{1+\alpha}{\alpha},\frac{1}{\alpha}} > 1, \\ \mathfrak{B}_{0,\mathcal{H}_2^+,\alpha}, & P_{\frac{1+\alpha}{\alpha},\frac{1}{\alpha}} < 1. \end{cases}
$$

The normalisation $d_{\lambda,\gamma} = \lambda^{H(\gamma)}$ in (2) *is defined in the proof of Theorem 11.*

Proof. Case $\gamma = \gamma_{\rm cr}$. Let $H(\gamma_{\rm cr}) := (1+\gamma_{\rm cr})\frac{1+\alpha}{\alpha} - p_1 = (1+\gamma_{\rm cr})\frac{1+\alpha}{\alpha} - q_1(Q-\chi)$, $\Gamma_{\rm cr} := \mathrm{diag}(1, \gamma_{\rm cr})$. Then, $\lambda^{-H(\gamma_{\rm cr})} X(]\mathbf{0}, \lambda^{\Gamma_{\rm cr}} \boldsymbol{t}]) = \int_{\mathbb{R}^2} f_\lambda(\boldsymbol{t}; \boldsymbol{u}) M(\mathrm{d}\boldsymbol{u})$, where

$$f_\lambda(\boldsymbol{t}; \boldsymbol{u}) := \lambda^{-H(\gamma_{\rm cr})} g(] - \boldsymbol{u}, \lambda^{\Gamma_{\rm cr}} \boldsymbol{t} - \boldsymbol{u}]), \quad \boldsymbol{u} \in \mathbb{R}^2.$$

It suffices to show that for any $\theta_i \in \mathbb{R}$, $\boldsymbol{t}_i \in \mathbb{R}^2_+$, $i = 1, \ldots, m$, $m \geq 1$,

$$\sum_{i=1}^m \theta_i \int_{\mathbb{R}^2} f_\lambda(\boldsymbol{t}_i; \boldsymbol{u}) M(\mathrm{d}\boldsymbol{u}) \xrightarrow{\mathrm{d}} \sum_{i=1}^m \theta_i \int_{\mathbb{R}^2} h_0(\boldsymbol{t}_i; \boldsymbol{u}) W_\alpha(\mathrm{d}\boldsymbol{u}), \quad (44)$$

where $h_0(\boldsymbol{t}; \boldsymbol{u}) = g_0(] - \boldsymbol{u}, \boldsymbol{t} - \boldsymbol{u}])$ as in (33). Using Proposition 24 of Chapter 2 with $\lambda_1 = \lambda, \lambda_2 = \lambda^\gamma, f_{\boldsymbol{\lambda}} = f_\lambda$ this follows from

$$\left\| \sum_{i=1}^m \theta_i (f^\dagger_\lambda(\boldsymbol{t}_i; \cdot) - h_0(\boldsymbol{t}_i; \cdot)) \right\|_\alpha \to 0. \quad (45)$$

It suffices to prove (45) for $m = \theta_1 = 1$, $\boldsymbol{t}_1 = \boldsymbol{t}$. We have $f^\dagger_\lambda(\boldsymbol{t}; \boldsymbol{u}) = \lambda^{-\chi q_1} g_0(] - \lambda^{\Gamma_{\rm cr}} \boldsymbol{u}, \lambda^{\Gamma_{\rm cr}}(\boldsymbol{t} - \boldsymbol{u})]) + o(1) = h_0(\boldsymbol{t}; \boldsymbol{u}) + o(1)$ for any $\boldsymbol{u} \neq \mathbf{0}$, $(0, t_2), (t_1, 0), \boldsymbol{t}$, by (14). Consequently, we have (45) by the DCT using (14), (13), similarly as in the proof of Proposition 8 (i). Indeed, the above relations imply the existence of $\epsilon > 0$ such that for all $0 < |\boldsymbol{u}| < \epsilon$,

$$\begin{aligned} |g(\boldsymbol{u})| \leq C\rho(\boldsymbol{u})^\chi, \quad |\partial_i g(\boldsymbol{u})| \leq C\rho(\boldsymbol{u})^{\chi - \frac{1}{q_i}}, \quad i = 1, 2, \\ |\partial_{12} g(\boldsymbol{u})| \leq C\rho(\boldsymbol{u})^{\chi - Q}. \end{aligned} \quad (46)$$

Letting $\boldsymbol{t} = \mathbf{1}$ and using (46), we have that for all $\lambda > 0$ small enough, $|\boldsymbol{u}| < \epsilon$,

$$\begin{aligned} |f^\dagger_\lambda(\mathbf{1}; \boldsymbol{u})| &\leq C\lambda^{-\chi q_1}(\rho(\lambda^{\Gamma_{\rm cr}}(\mathbf{1} - \boldsymbol{u}))^\chi + \rho(\lambda^{\Gamma_{\rm cr}}(\boldsymbol{e}_2 - \boldsymbol{u}))^\chi \\ &\quad + \rho(\lambda^{\Gamma_{\rm cr}}(\boldsymbol{e}_1 - \boldsymbol{u}))^\chi + \rho(-\lambda^{\Gamma_{\rm cr}} \boldsymbol{u})^\chi) \\ &= C\big(\rho(\mathbf{1} - \boldsymbol{u})^\chi + \rho(\boldsymbol{e}_2 - \boldsymbol{u})^\chi + \rho(\boldsymbol{e}_1 - \boldsymbol{u})^\chi + \rho(-\boldsymbol{u})^\chi\big), \end{aligned}$$

where the dominating function is $L^\alpha(\mathbb{R}^2)$-integrable on $|\boldsymbol{u}| < \epsilon$; see (14). Similarly, for all $|\lambda^{\Gamma_{\rm cr}} \boldsymbol{u}| < \epsilon \leq |\boldsymbol{u}|$,

$$\begin{aligned} |f^\dagger_\lambda(\mathbf{1}; \boldsymbol{u})| &= \left| \lambda^{-\chi q_1} \int_{]\mathbf{0}, \mathbf{1}]} \partial_{12} g(\lambda^{\Gamma_{\rm cr}}(\boldsymbol{s} - \boldsymbol{u})) \mathrm{d}\boldsymbol{s} \right| \\ &\leq C \int_{]\mathbf{0}, \mathbf{1}]} \rho(\boldsymbol{s} - \boldsymbol{u})^{\chi - Q} \mathrm{d}\boldsymbol{s}, \end{aligned}$$

where the dominating function is $L^\alpha(\mathbb{R}^2)$-integrable on $|\boldsymbol{u}| \geq \epsilon$. Finally, we have $\int_{|\lambda^{\Gamma_{\rm cr}}\boldsymbol{u}|\geq\epsilon} |f^\dagger_\lambda(\boldsymbol{1};\boldsymbol{u})|^\alpha \mathrm{d}\boldsymbol{u} = \lambda^{\alpha(1+\gamma_{\rm cr}-H(\gamma_{\rm cr}))} I_\lambda = o(1)$ because $1+\gamma_{\rm cr} - H(\gamma_{\rm cr}) = q_1(Q(1-\frac{1}{\alpha}) - \chi) = \frac{p_1}{\alpha}(\alpha - P) > 0$ and

$$
\begin{aligned}
I_\lambda &= \int_{|\boldsymbol{u}|\geq\epsilon} \Big| \int_{]\boldsymbol{0},\boldsymbol{1}]} \partial_{12} g(\lambda^{\Gamma_{\rm cr}}\boldsymbol{s} - \boldsymbol{u}) \mathrm{d}\boldsymbol{s} \Big|^\alpha \mathrm{d}\boldsymbol{u} \\
&\leq \int_{|\boldsymbol{u}|\geq\epsilon} \int_{]\boldsymbol{0},\boldsymbol{1}]} |\partial_{12} g(\lambda^{\Gamma_{\rm cr}}\boldsymbol{s} - \boldsymbol{u})|^\alpha \mathrm{d}\boldsymbol{s}\mathrm{d}\boldsymbol{u} \leq \int_{|\boldsymbol{u}|\geq\frac{\epsilon}{2}} |\partial_{12} g(\boldsymbol{u})|^\alpha \mathrm{d}\boldsymbol{u} < \infty
\end{aligned}
$$

for $1 \leq \alpha \leq 2$ follows by Jensen's inequality from Assumption $(\mathrm{G})^0_\alpha$; for $0 < \alpha < 1$, the relation $I_\lambda = O(1)$ is a consequence of the monotonicity of the dominating function $\bar{g}_{12}$ in (17), Assumption $(\mathrm{G})^0_\alpha$. This completes the proof of (45).

Case $\gamma < \gamma_{\rm cr}, P_{\frac{1}{\alpha},\frac{1+\alpha}{\alpha}} < 1$. Let $H(\gamma) = \mathcal{H}^-_1$. It suffices to prove the convergence

$$
\sum_{i=1}^n \theta_i \lambda^{-\mathcal{H}^-_1} X(]\boldsymbol{0}, \lambda^\Gamma \boldsymbol{t}_i]) \stackrel{\mathrm{d}}{\longrightarrow} \sum_{i=1}^n \theta_i \mathfrak{B}_{\mathcal{H}^-_1,0,\alpha}(\boldsymbol{t}_i) \tag{47}
$$

for any $n \geq 1$, $\theta_i \in \mathbb{R}, i = 1,\ldots,n$, and any collection $\{\boldsymbol{t}_i = (t_{i,1}, t_{i,2}), i = 1,\ldots,n,\}$ of points of $\mathbb{R}^2_+$. As in (41), write the last collection as a union of m sub-collections with different ordinates: $t_{1,2} < t_{2,2} < \cdots < t_{m,2}, t_{j,2} \in \{t_{1,2},\ldots,t_{m,2}\}, j = m+1,\ldots,n, n > m$. First, assume $m = 1$, or $t_{1,2} = \cdots = t_{n,2} =: t_2$. Split $\lambda^{-\mathcal{H}^-_1} X(]\boldsymbol{0}, \lambda^\Gamma \boldsymbol{t}]) = \bar{X}^{(1)}_\lambda(t_1) - \bar{X}^{(0)}_\lambda(t_1)$, where

$$
\begin{aligned}
\bar{X}^{(0)}_\lambda(t_1) &:= \int_{\mathbb{R}^2} f_\lambda(t_1;\boldsymbol{u}) M(\mathrm{d}\boldsymbol{u}), \\
\bar{X}^{(1)}_\lambda(t_1) &:= \int_{\mathbb{R}^2} f_\lambda(t_1;\boldsymbol{u} - (0,\lambda^\gamma t_2)) M(\mathrm{d}\boldsymbol{u}),
\end{aligned}
\tag{48}
$$

with $f_\lambda(t_1;\boldsymbol{u}) = \lambda^{-\mathcal{H}^-_1}(g((\lambda t_1, 0) - \boldsymbol{u}) - g(-\boldsymbol{u}))$, $\boldsymbol{u} \in \mathbb{R}^2$. The two processes in (48) (indexed by $t_1 \in \mathbb{R}_+$) are the first-order differences mentioned above; they are identically distributed and asymptotically independent, as follows from the subsequent argument. Let us prove that as $\lambda \downarrow 0$,

$$(\bar{X}_\lambda^{(0)}(t_1), \bar{X}_\lambda^{(1)}(t_1)) \xrightarrow{\text{fdd}} (Y_{1,\alpha}^{(0)}(t_1), Y_{1,\alpha}^{(1)}(t_1)), \tag{49}$$

where $Y_{1,\alpha}^{(j)}$, $j = 0, 1$, are independent copies of $Y_{1,\alpha}$ in (32). For this purpose, write $f_\lambda(t_1; \boldsymbol{u})$ as the sum of

$$\begin{aligned} f_{\lambda,0}(t_1; \boldsymbol{u}) &= f_\lambda(t_1; \boldsymbol{u})\mathbb{I}(|u_2| < \epsilon\lambda^\gamma, |\boldsymbol{u}| < \epsilon), \\ f_{\lambda,1}(t_1; \boldsymbol{u}) &= f_\lambda(t_1; \boldsymbol{u})\mathbb{I}(|u_2| \ge \epsilon\lambda^\gamma, |\boldsymbol{u}| < \epsilon), \\ f_{\lambda,2}(t_1; \boldsymbol{u}) &= f_\lambda(t_1; \boldsymbol{u})\mathbb{I}(|\boldsymbol{u}| \ge \epsilon) \end{aligned} \tag{50}$$

for all $\boldsymbol{u} \in \mathbb{R}^2$ and some small enough $\epsilon > 0$. Then, using the fact that the processes $\{\int_{\mathbb{R}^2} f_{\lambda,0}(t_1; \boldsymbol{u})M(\mathrm{d}\boldsymbol{u}); t_1 \in \mathbb{R}_+\}$ and $\{\int_{\mathbb{R}^2} f_{\lambda,0}(t_1; \boldsymbol{u} - (0, \lambda^\gamma t_2))M(\mathrm{d}\boldsymbol{u}); t_1 \in \mathbb{R}_+\}$ are independent and have the same distribution, relation (49) follows if we show that, with h_1 in (33),

$$\begin{aligned} \int_{\mathbb{R}^2} f_{\lambda,0}(t_1; \boldsymbol{u})M(\mathrm{d}\boldsymbol{u}) &\xrightarrow{\text{fdd}} \int_{\mathbb{R}^2} h_1(t_1; \boldsymbol{u})W_\alpha(\mathrm{d}\boldsymbol{u}) = Y_{1,\alpha}(t_1) \quad \text{and} \\ \|f_{\lambda,i}(t_1; \cdot)\|_\alpha &= o(1), \quad i = 1, 2. \end{aligned} \tag{51}$$

The first relation in (51) follows by Proposition 24 with $\lambda_1 = \lambda$, $\lambda_2 = \lambda^{\gamma_{\mathrm{cr}}}$ from $\|f_{\lambda,0}^\dagger(t_1; \cdot) - h_1(t_1; \cdot)\|_\alpha = o(1)$ for each t_1 fixed (e.g., $t_1 = 1$), as follows. Indeed, since $\gamma < \gamma_{\mathrm{cr}}$, we have that

$$\begin{aligned} f_{\lambda,0}^\dagger(1; \boldsymbol{u}) &= \lambda^{-q_1\chi}(g(\lambda^{\Gamma_{\mathrm{cr}}}(\boldsymbol{e}_1 - \boldsymbol{u})) - g(-\lambda^{\Gamma_{\mathrm{cr}}}\boldsymbol{u})) \\ &\quad \times \mathbb{I}(|u_2| < \epsilon\lambda^{\gamma-\gamma_{\mathrm{cr}}}, |\lambda^{\Gamma_{\mathrm{cr}}}\boldsymbol{u}| < \epsilon) \\ &\sim \lambda^{-q_1\chi}(g_0(\lambda^{\Gamma_{\mathrm{cr}}}(\boldsymbol{e}_1 - \boldsymbol{u})) - g_0(-\lambda^{\Gamma_{\mathrm{cr}}}\boldsymbol{u})) \\ &= g_0(\boldsymbol{e}_1 - \boldsymbol{u}) - g_0(-\boldsymbol{u}) = h_1(1; \boldsymbol{u}) \end{aligned} \tag{52}$$

using (14) and $\lambda^{-q_1\chi}o(\rho(\lambda^{\Gamma_{\mathrm{cr}}}\boldsymbol{u})^\chi) = o(1)$ for any $\boldsymbol{u} \ne \mathbf{0}, \boldsymbol{e}_1$. To get the dominating function for $f_{\lambda,0}^\dagger(1; \cdot)$ in (52), note that $|\lambda^{\Gamma_{\mathrm{cr}}}(\boldsymbol{e}_1 - \boldsymbol{u})| < 2\epsilon$ for $|\lambda^{\Gamma_{\mathrm{cr}}}\boldsymbol{u}| < \epsilon$ and all $\lambda > 0$ small enough, so that we can use (14), (13) or (19) to bound the terms on the r.h.s. of (52). Indeed, using

$\lambda^{-q_1\chi}(|g(\lambda^{\Gamma_{\rm cr}}(\boldsymbol{e}_1-\boldsymbol{u}))|+|g(-\lambda^{\Gamma_{\rm cr}}\boldsymbol{u}))|) \le C(\rho(\boldsymbol{e}_1-\boldsymbol{u})^{\chi}+\rho(\boldsymbol{u})^{\chi}) \le C(\rho_0(\boldsymbol{e}_1-\boldsymbol{u})^{P-1}+\rho_0(\boldsymbol{u})^{P-1})$ for $|\boldsymbol{u}|<2$ and $\lambda^{-q_1\chi}|g(\lambda^{\Gamma_{\rm cr}}(\boldsymbol{e}_1-\boldsymbol{u}))-g(-\lambda^{\Gamma_{\rm cr}}\boldsymbol{u}))| = \lambda^{1-\chi q_1}|\int_0^1 \partial g_1(\lambda^{\Gamma_0}(s\boldsymbol{e}_1-\boldsymbol{u})\mathrm{d}s| \le C\int_0^1 \rho_0(s\boldsymbol{e}_1-\boldsymbol{u})^{\frac{1}{p_2}-1}\mathrm{d}s$ for $|\boldsymbol{u}|\ge 2$, we obtain

$$|f^{\dagger}_{\lambda,0}(1;\boldsymbol{u})| \le C\Bigg((\rho_0(\boldsymbol{e}_1-\boldsymbol{u})^{P-1}+\rho_0(-\boldsymbol{u})^{P-1})\mathbb{I}(|\boldsymbol{u}|<2) + \int_0^1 \rho_0(s\boldsymbol{e}_1-\boldsymbol{u})^{\frac{1}{p_2}-1}\mathrm{d}s\,\mathbb{I}(|\boldsymbol{u}|\ge 2)\Bigg), \qquad (53)$$

where the dominating function on the r.h.s. of (53) is in $L^{\alpha}(\mathbb{R}^2)$; see the proof of Proposition 8 (ii). The above dominating argument applies to $f^{\dagger}_{\lambda,1}(1;\cdot)$ as well, except that we now have

$$f^{\dagger}_{\lambda,1}(1;\boldsymbol{u}) = \lambda^{-q_1\chi}(g(\lambda^{\Gamma_{\rm cr}}(\boldsymbol{e}_1-\boldsymbol{u}))-g(-\lambda^{\Gamma_{\rm cr}}\boldsymbol{u})) \times \mathbb{I}(|u_2|\ge \epsilon\lambda^{\gamma-\gamma_{\rm cr}}, |\lambda^{\Gamma_{\rm cr}}\boldsymbol{u}|<\epsilon) = o(1)$$

point-wise (since $\lambda^{\gamma-\gamma_{\rm cr}} \to \infty$), resulting in $\|f_{\lambda,1}(1;\cdot)\|_{\alpha} = \|f^{\dagger}_{\lambda,1}(1;\cdot)\|_{\alpha} = o(1)$. Finally, rewriting $\int_{\mathbb{R}^2}|f_{\lambda,2}(t_1;\boldsymbol{u})|^{\alpha}\mathrm{d}\boldsymbol{u} = \int_{|\boldsymbol{u}|\ge\epsilon}|f_{\lambda}(t_1;\boldsymbol{u})|^{\alpha}\mathrm{d}\boldsymbol{u}$, with $f_{\lambda}(t_1;\boldsymbol{u}) = \lambda^{1-\mathcal{H}_1^-}\int_0^{t_1}\partial_1 g(\lambda s\boldsymbol{e}_1-\boldsymbol{u})\mathrm{d}s$, we get $\|f_{\lambda,2}(t_1;\cdot)\|_{\alpha} = o(1)$ since $1-\mathcal{H}_1^- = p_1(1-P_{\frac{1}{\alpha},\frac{1+\alpha}{\alpha}})>0$, and $\int_{|\boldsymbol{u}|\ge\epsilon}|\int_0^{t_1}\partial_1 g(\lambda s\boldsymbol{e}_1-\boldsymbol{u})\mathrm{d}s|^{\alpha}\mathrm{d}\boldsymbol{u}<C$ follows from Assumption $(\mathrm{G})^0_{\alpha}$, similarly as in the case $\gamma=\gamma_{\rm cr}$ above. This completes the proof of (49) or (47) for $m=1$.

The above argument extends to the general case of $1\le m\le n$ in (47). Let $1\le j\le m$. For $\boldsymbol{t}=(t_1,t_2)\in\mathbb{R}^2_+$ such that $t_2 = t_{j,2}$, consider a similar decomposition as in (48): $\lambda^{-\mathcal{H}_1^-}X(]\boldsymbol{0},\lambda^{\Gamma}\boldsymbol{t}]) = \bar{X}^{(j)}_{\lambda}(t_1)-\bar{X}^{(0)}_{\lambda}(t_1)$, where $\bar{X}^{(0)}_{\lambda}(t_1)$ is the same as in (48) and

$$\bar{X}^{(j)}_{\lambda}(t_1) := \int_{\mathbb{R}^2} f_{\lambda}(t_1;\boldsymbol{u}-(0,\lambda^{\gamma}t_{j,2}))M(\mathrm{d}\boldsymbol{u}).$$

Then, split $f_{\lambda}(t_1;\boldsymbol{u}) = \sum_{i=0}^2 f_{\lambda,i}(t_1;\boldsymbol{u}), \boldsymbol{u}\in\mathbb{R}^2$ as in (50). If $\epsilon>0$ is small enough, then $m+1$ processes $\{\int_{\mathbb{R}^2} f_{\lambda,0}(t_1;\boldsymbol{u}-(0,\lambda^{\gamma}t_{j,2}))M(\mathrm{d}\boldsymbol{u}); t_1\in\mathbb{R}_+\}$, $j=0,\dots,m$, are independent;

moreover,

$$\int_{\mathbb{R}^2} f_{\lambda,0}(t_1; \boldsymbol{u} - (0, \lambda^\gamma t_{j,2}))M(\mathrm{d}\boldsymbol{u}) \xrightarrow{\text{fdd}} Y^{(j)}_{1,\alpha}(t_1), \quad j = 0, \dots, m,$$

whereas other terms are negligible as in (51), which proves (47).
Case $\gamma < \gamma_{\mathrm{cr}}, P_{\frac{1}{\alpha}, \frac{1+\alpha}{\alpha}} > 1$. Let $H(\gamma) := 1 + \gamma\mathcal{H}_2^-$. It suffices to prove the one-dimensional convergence

$$\lambda^{-1-\gamma\mathcal{H}_2^-} X(]\mathbf{0}, \lambda^\Gamma \boldsymbol{t}]) \xrightarrow{\mathrm{d}} t_1 \tilde{Y}_{\alpha,2}(t_2) = t_1 \int_{\mathbb{R}^2} \tilde{h}_2(t_2; \boldsymbol{u}) W_\alpha(\mathrm{d}\boldsymbol{u}) \quad (54)$$

for arbitrary $\boldsymbol{t} = (t_1, t_2) \in \mathbb{R}^2_+$, with $\tilde{h}_2(t; \boldsymbol{u}) = \partial_1 g_0(t\boldsymbol{e}_2 - \boldsymbol{u}) - \partial_1 g_0(-\boldsymbol{u})$, $t \in \mathbb{R}_+$, $\boldsymbol{u} \in \mathbb{R}^2$, in (33). Since the l.h.s. of (54) can be written as $\int_{\mathbb{R}^2} f_\lambda(\boldsymbol{t}; \boldsymbol{u}) M(\mathrm{d}\boldsymbol{u})$, then using Proposition 24 of Chapter 2 with $\lambda_1 := \lambda^{\frac{\gamma}{\gamma_{\mathrm{cr}}}}$, $\lambda_2 := \lambda^\gamma$, it suffices to prove

$$\|f^\dagger_\lambda(\boldsymbol{t}; \cdot) - t_1 \tilde{h}_2(t_2; \cdot)\|_\alpha = o(1), \quad (55)$$

where, with $\Gamma'_{\mathrm{cr}} = \mathrm{diag}(\frac{\gamma}{\gamma_{\mathrm{cr}}}, \gamma)$, $\lambda' = \lambda^{1-\frac{\gamma}{\gamma_{\mathrm{cr}}}} \downarrow 0$,

$$\begin{aligned} f^\dagger_\lambda(\boldsymbol{t}; \boldsymbol{u}) &= \lambda^{\frac{1}{\alpha}(\frac{\gamma}{\gamma_{\mathrm{cr}}} + \gamma) - 1 - \gamma\mathcal{H}_2^-} g(] - \lambda^{\Gamma'_{\mathrm{cr}}} \boldsymbol{u}, \lambda^\Gamma \boldsymbol{t} - \lambda^{\Gamma'_{\mathrm{cr}}} \boldsymbol{u}]) \\ &= \lambda^{\frac{\gamma}{\gamma_{\mathrm{cr}}} - \gamma\chi q_2} \int_0^{t_1} \{\partial_1 g(\lambda^{\Gamma'_{\mathrm{cr}}}(\lambda' s\boldsymbol{e}_1 + t_2\boldsymbol{e}_2 - \boldsymbol{u})) \\ &\quad - \partial_1 g(\lambda^{\Gamma'_{\mathrm{cr}}}(\lambda' s\boldsymbol{e}_1 - \boldsymbol{u}))\}\mathrm{d}s \end{aligned}$$

satisfies

$$\begin{aligned} f^\dagger_\lambda(\boldsymbol{t}; \boldsymbol{u}) &\sim \lambda^{\frac{\gamma}{\gamma_{\mathrm{cr}}} - \gamma\chi q_2} \int_0^{t_1} \{\partial_1 g_0(\lambda^{\Gamma'_{\mathrm{cr}}}(\lambda' s\boldsymbol{e}_1 + t_2\boldsymbol{e}_2 - \boldsymbol{u})) \\ &\quad - \partial_1 g_0(\lambda^{\Gamma'_{\mathrm{cr}}}(\lambda' s\boldsymbol{e}_1 - \boldsymbol{u}))\}\mathrm{d}s \\ &= \int_0^{t_1} \{\partial_1 g_0(\lambda' s\boldsymbol{e}_1 + t_2\boldsymbol{e}_2 - \boldsymbol{u}) - \partial_1 g_0(\lambda' s\boldsymbol{e}_1 - \boldsymbol{u})\}\mathrm{d}s \\ &\sim t_1 \tilde{h}_2(t_2; \boldsymbol{u}) \end{aligned}$$

for almost all $\boldsymbol{u} \in \mathbb{R}^2_+$. The DCT argument to show (55) for $\boldsymbol{t} = \mathbf{1}$ proceeds as follows. Split $f^\dagger_\lambda(\mathbf{1}; \boldsymbol{u}) = f^\dagger_\lambda(\mathbf{1}; \boldsymbol{u})\mathbb{I}(\rho_0(\boldsymbol{u}) > 2C_p(\lambda')^{p_1})$, $\rho_0(\boldsymbol{u} - \boldsymbol{e}_2) > 2C_p(\lambda')^{p_1}) + f^\dagger_\lambda(\mathbf{1}; \boldsymbol{u})\mathbb{I}(\rho_0(\boldsymbol{u}) \le 2C_p(\lambda')^{p_1}) + f^\dagger_\lambda(\mathbf{1}; \boldsymbol{u})$

$\mathbb{I}(\rho_0(\boldsymbol{u}-\boldsymbol{e}_2) \le 2C_p(\lambda')^{p_1}) =: \sum_{i=1}^{3} f^{\dagger}_{\lambda,i}(\boldsymbol{u})$, where C_p is as in (13) of Chapter 6. It suffices to prove

$$|f^{\dagger}_{\lambda,1}(\boldsymbol{u})| \le \bar{f}(\boldsymbol{u}), \quad \bar{f} \in L^{\alpha}(\mathbb{R}^2) \quad \text{and} \tag{56}$$

$$\|f^{\dagger}_{\lambda,i}\|_{\alpha} \to 0, \quad i = 2, 3. \tag{57}$$

To get the dominated function in (56), it is convenient to use $\chi q_2 = p_2(P-1)$ and $|\partial_1 g_0(\boldsymbol{u})| \le C\rho_0(\boldsymbol{u})^{\frac{1}{p_2}-1}$, referring to (19). Therefore, for $\boldsymbol{t} = \boldsymbol{1}$, $|\boldsymbol{u}| < 2$, we obtain

$$|f^{\dagger}_{\lambda,1}(\boldsymbol{u})| \le C\int_0^1 (\rho_0(\lambda' s\boldsymbol{e}_1 + \boldsymbol{e}_2 - \boldsymbol{u})^{\frac{1}{p_2}-1} + \rho_0(\lambda' s\boldsymbol{e}_1 - \boldsymbol{u})^{\frac{1}{p_2}-1})\mathrm{d}s, \tag{58}$$

where $\rho_0(\lambda' s\boldsymbol{e}_1 - \boldsymbol{u}) \ge C_p^{-1}\rho_0(\boldsymbol{u}) - \rho_0(\lambda' s\boldsymbol{e}_1) = \frac{1}{2C_p}\rho_0(\boldsymbol{u}) + (\frac{1}{2C_p}\rho_0(\boldsymbol{u}) - (\lambda')^{p_1}) \ge \frac{1}{2C_p}\rho_0(\boldsymbol{u})$ and, similarly, $\rho_0(\lambda' s\boldsymbol{e}_1 + \boldsymbol{e}_2 - \boldsymbol{u}) \ge \frac{1}{2C_p}\rho_0(\boldsymbol{e}_2 - \boldsymbol{u})$ for $\rho_0(\boldsymbol{u}) \ge 2C_p(\lambda')^{p_1}, \rho_0(\boldsymbol{u}-\boldsymbol{e}_2) > 2C_p(\lambda')^{p_1}$. Therefore, $|f^{\dagger}_{\lambda,1}(\boldsymbol{u})|\mathbb{I}(|\boldsymbol{u}|) < 2) \le C(\rho(\boldsymbol{e}_2 - \boldsymbol{u})^{\frac{1}{p_2}-1} + \rho(\boldsymbol{u})^{\frac{1}{p_2}-1}) = \bar{f}(\boldsymbol{u})$, where $\int_{|\boldsymbol{u}|\le 2} \bar{f}(\boldsymbol{u})^{\alpha}\mathrm{d}\boldsymbol{u} \le C\int_{|\boldsymbol{u}|\le 2}\rho_0(\boldsymbol{u})^{\alpha(\frac{1}{p_2}-1)}\mathrm{d}\boldsymbol{u} < \infty$, since $\alpha(1-\frac{1}{p_2}) < P$ is equivalent to $P_{\frac{1}{\alpha},\frac{1+\alpha}{\alpha}} > 1$, referring to (14) of Chapter 6. For $|\boldsymbol{u}| \ge 2$, we write

$$\begin{aligned} |f^{\dagger}_{\lambda,1}(\boldsymbol{u})| &= \lambda^{\gamma p_2}\left|\int_{]\boldsymbol{0},\boldsymbol{1}]} \partial_{12} g(\lambda^{\Gamma'_{\mathrm{cr}}}(\lambda' s_1\boldsymbol{e}_1 + s_2\boldsymbol{e}_2 - \boldsymbol{u}))\mathrm{d}\boldsymbol{s}\right| \\ &\le C\int_{]\boldsymbol{0},\boldsymbol{1}]} \rho_0(\lambda' s_1\boldsymbol{e}_1 + s_2\boldsymbol{e}_2 - \boldsymbol{u})^{-1}\mathrm{d}\boldsymbol{s} \le C\rho_0(\boldsymbol{u})^{-1}, \end{aligned}$$

with $\int_{|\boldsymbol{u}|\ge 2}\rho_0(\boldsymbol{u})^{-\alpha}\mathrm{d}\boldsymbol{u} < \infty$, referring to as in (15) of Chapter 6, also the proof of Proposition 8 (iii).

It remains to show (57) (where it suffices to consider $i = 2$ only). Using (58), (14), and the bound on $|g_0|$ in (19), we get

$$\begin{aligned} |f^{\dagger}_{\lambda,2}(\boldsymbol{u})| &\le C\int_0^1 \rho_0(\lambda' s\boldsymbol{e}_1 + \boldsymbol{e}_2 - \boldsymbol{u})^{\frac{1}{p_2}-1}\mathrm{d}s + C(\rho_0(\lambda'\boldsymbol{e}_1 - \boldsymbol{u})^{P-1} \\ &\quad + \rho_0(-\boldsymbol{u})^{P-1})/\lambda' \\ &\le C(1 + (\lambda')^{-1}(\rho_0(\lambda'\boldsymbol{e}_1 - \boldsymbol{u})^{P-1} + \rho_0(-\boldsymbol{u})^{P-1}), \end{aligned}$$

so that (57) follows from $\int_{\rho_0(\boldsymbol{u})\le C(\lambda')^{p_1}} \rho_0(\boldsymbol{u})^{\alpha(P-1)} \mathrm{d}\boldsymbol{u} = C(\lambda')^{1+\gamma_{\mathrm{cr}}+p_1\alpha(P-1)}$ since $1+\gamma_{\mathrm{cr}}+p_1\alpha(P-1) > \alpha$ is equivalent to $P_{1/\alpha,(1+\alpha)/\alpha} > 1$. This proves (55) and completes the proof of Theorem 11. □

Extending Theorem 11, we may ask the following two questions: (q1) what happens in the region $P > \alpha$? (q2) when the (anisotropic) γ-rectangent limits in (1) agree with the Lévy sheet $W_\alpha = \{W_\alpha(]\mathbf{0}, \boldsymbol{x}]); \boldsymbol{x} \in \mathbb{R}^2_+\}$ in (132) of Chapter 2? (Obviously, the limits V_γ in Theorem 11 *never* agree with W_α.)

Concerning (q1), recall Table 1, where in the parameter region $\mathcal{R}_{11}$, the unbalanced limits $V_\pm$ in Theorem 11 are MSS RFs with indices $(H_1, H_2) \to (1,1)$ as $P \to \alpha$. The last fact is an easy observation from the formulas in (11). One may expect that for $P > \alpha$, all scaling limits are (1,1)-MSS RFs of the form $\{t_1 t_2 V, (t_1,t_2) \in \mathbb{R}^2_+\}$, where V is an r.v. See Ref. [230, Prop. 8.2.10] for a related fact in the case of one-parameter processes. As shown in Proposition 12, this is true indeed, using the fact that

$$\|f_\lambda - f\|_\alpha \to 0 \quad \text{implies} \quad \int_{\mathbb{R}^2} f_\lambda(\boldsymbol{u}) M(\boldsymbol{u}) \xrightarrow{\mathrm{d}} \int_{\mathbb{R}^2} f(\boldsymbol{u}) M(\boldsymbol{u}) \tag{59}$$

for M satisfying $(\mathrm{M})_\alpha$ (where the last condition can be relaxed [211]). Let $\Lambda = \mathrm{diag}(\lambda_1, \lambda_2)$.

Proposition 12. *Let X be a Lévy-driven RF given by (3), where M is a Lévy random measure on $\mathbb{R}^2$ satisfying Assumption $(M)_\alpha$, $0 < \alpha \le 2$. Let (16) hold, where the kernel g admits the $L^\alpha(\mathbb{R}^2)$-derivative $\partial_{12} g$ in the sense that*

$$\lim_{t_i \downarrow 0,\, i=1,2} \int_{\mathbb{R}^2} \left|(t_1 t_2)^{-1} g(]\boldsymbol{u}, \boldsymbol{u}+\boldsymbol{t}]) - \partial_{12} g(\boldsymbol{u})\right|^\alpha \mathrm{d}\boldsymbol{u} = 0. \tag{60}$$

Then, as $\lambda_i \downarrow 0, i = 1,2$,

$$(\lambda_1\lambda_2)^{-1} X(]\mathbf{0}, \Lambda \boldsymbol{t}]) \xrightarrow{\mathrm{fdd}} t_1 t_2 V, \quad \textit{where } V := \int_{\mathbb{R}^2} \partial_{12} g(\boldsymbol{u}) M(\mathrm{d}\boldsymbol{u}).$$

Proof. The proof follows from (4), (60), and the criterion in (59). □

Condition (60) is not directly related to the asymptotic form of $g(\boldsymbol{t})$ at $\mathbf{0}$, as specified in Assumption $(\mathrm{G})_\alpha$. On the other hand, if

g satisfies the latter assumption, then from (14), (13), (19), and (14) of Chapter 6, we see that $\|\partial_{12} g\|_\alpha < \infty$ and (60) hold when $P > \alpha$, relating Proposition 12 to question (q1). Concerning question (q2), in order for the rectangent limits to be given by a Lévy sheet, we replace the power-law behaviour of g in Assumption $(\mathrm{G})_\alpha$ by the condition that g is bounded but *discontinuous at* $\mathbf{0}$. Let $\mathbb{R}^2_{ij} := \{\boldsymbol{t} \in \mathbb{R}^2 : \mathrm{sgn}(t_1) = i, \mathrm{sgn}(t_2) = j\}$, $i, j = \pm 1$, denote the four open quadrants in $\mathbb{R}^2$.

Assumption $(\mathbf{G})'_\alpha$. The function g has finite limits, $g_{ij} = \lim_{|\boldsymbol{t}|\to 0,\, \boldsymbol{t}\in\mathbb{R}^2_{ij}} g(\boldsymbol{t})$, $i, j \in \{1, -1\}$, on each quadrant such that

$$g[\mathbf{0}] := \sum_{i,j\in\{1,-1\}} ij\, g_{ij} \neq 0. \tag{61}$$

Moreover,

$$\int_{\mathbb{R}^2} \left|g(]-\boldsymbol{u}, \boldsymbol{t} - \boldsymbol{u}]) - g[\mathbf{0}]\mathbb{I}(\boldsymbol{u} \in]\mathbf{0}, \boldsymbol{t}])\right|^\alpha \mathrm{d}\boldsymbol{u} = o(t_1 t_2),$$
$$t_i \downarrow 0,\ i = 1, 2. \tag{62}$$

Note that condition (61) and the existence of the *limites quadrantales* g_{ij} is typical of distribution functions in $\mathbb{R}^2$ with an atom at $\mathbf{0}$ or, more generally, functions from the Skorohod space $D(\mathbb{R}^2)$ (see Ref. [36]), which are discontinuous at the origin. The simplest example of g satisfying (61) and (62) is the function taking constant values on each quadrant, that is,

$$g(\boldsymbol{t}) = \sum_{i,j\in\{1,-1\}} g_{ij}\mathbb{I}(\boldsymbol{t} \in \mathbb{R}^2_{ij}), \quad \boldsymbol{t} \in \mathbb{R}^2. \tag{63}$$

For g in (63), $g(]-\boldsymbol{u}, \boldsymbol{t} - \boldsymbol{u}]) = g[\mathbf{0}]\mathbb{I}(\boldsymbol{u} \in]\mathbf{0}, \boldsymbol{t}])$, implying

$$(\lambda_1\lambda_2)^{-\frac{1}{\alpha}} X(]\mathbf{0}, \Lambda\boldsymbol{t}]) = g[\mathbf{0}](\lambda_1\lambda_2)^{-\frac{1}{\alpha}} M(]\mathbf{0}, \lambda^\Gamma\boldsymbol{t}]) \xrightarrow{\mathrm{fdd}} g[\mathbf{0}]W_\alpha(]\mathbf{0}, \boldsymbol{t}]), \tag{64}$$

referring to (132) of Chapter 2. A similar result holds for more general g, satisfying Assumption $(\mathrm{G})'_\alpha$.

Proposition 13. *Let X be a Lévy-driven RF in* (3) *satisfying Assumptions $(G)^0_\alpha$, $(G)'_\alpha$, and $(M)_\alpha$, for some $0 < \alpha \le 2$. Then, as $\lambda_i \downarrow 0, i = 1, 2$,*

$$(\lambda_1\lambda_2)^{-\frac{1}{\alpha}} X(]\mathbf{0}, \Lambda\boldsymbol{t}]) \xrightarrow{\mathrm{fdd}} g[\mathbf{0}]W_\alpha(]\mathbf{0}, \boldsymbol{t}]). \tag{65}$$

Proof. Split $X(]\mathbf{0},\boldsymbol{t}]) = X'(]\mathbf{0},\boldsymbol{t}]) + X''(]\mathbf{0},\boldsymbol{t}])$, where $X'(]\mathbf{0},\boldsymbol{t}]) := g[\mathbf{0}]M(]\mathbf{0},\boldsymbol{t}])$, $X''(]\mathbf{0},\boldsymbol{t}]) := \int_{\mathbb{R}^2}(g(]-\boldsymbol{u},\boldsymbol{t}-\boldsymbol{u}]) - g[\mathbf{0}]\mathbb{I}(\boldsymbol{u} \in]\mathbf{0},\boldsymbol{t}]))$ $M(\mathrm{d}\boldsymbol{u})$. Then, $(\lambda_1\lambda_2)^{-\frac{1}{\alpha}}X'(]\mathbf{0},\Lambda\boldsymbol{t}]) \xrightarrow{\mathrm{fdd}} g[\mathbf{0}]W_\alpha(]\mathbf{0},\boldsymbol{t}])$ according to (64), while $(\lambda_1\lambda_2)^{-\frac{1}{\alpha}}X''(]\mathbf{0},\Lambda\boldsymbol{t}]) \xrightarrow{\mathrm{fdd}} 0$ in view of Proposition 24 of Chapter 2 and condition (62). □

5 Tangent RFs of Lévy-Driven Fractional RFs

The (ordinary) increment $X(\boldsymbol{t}+\boldsymbol{t}_0) - X(\boldsymbol{t}_0)$ of a Lévy-driven fractional RF X in (3) depends on three kernels, g, g_1^0, and g_2^0, in contrast to the rectangular increment $X(]\boldsymbol{t}_0,\boldsymbol{t}_0+\boldsymbol{t}])$ depending on g alone. However, if $g_1^0 = g_2^0 = 0, -g_{12}^0 =: g^0$, then ordinary increments of RF,

$$X(\boldsymbol{t}) = \int_{\mathbb{R}^2} \{g(\boldsymbol{t}-\boldsymbol{u}) - g^0(-\boldsymbol{u})\} M(\mathrm{d}\boldsymbol{u}), \quad \boldsymbol{t} \in \mathbb{R}^2, \tag{66}$$

depend on g alone. The class of Lévy-driven RFs in (66) contains Lévy fractional RFs in Examples 3–6. In this section, we identify γ-tangent limits of (66) for any $\gamma > 0$ under (non-isotropic) assumptions on g, which are similar but simpler than in Assumption $(\mathrm{G})_\alpha$. The following Theorem 14 shows that the family $\{T_\gamma; \gamma > 0\}$ of γ-tangent RFs exhibits a scaling transition at the same point $\gamma_{\mathrm{cr}} = q_1/q_2$ as in Theorem 11 with unbalanced $T_\gamma^\pm$ having a rather simple structure.

Theorem 14. *Let X be a Lévy-driven fractional RF in* (66), *where M satisfies Assumption $(M)_\alpha$, $0 < \alpha \le 2$, and $\int_{\mathbb{R}^2} |g(\boldsymbol{t}-\boldsymbol{u}) - g_{12}^0(-\boldsymbol{u})|^\alpha \mathrm{d}\boldsymbol{u} < \infty$ for all $\boldsymbol{t} \in \mathbb{R}^2$. Let $g(\boldsymbol{t}) = g_0(\boldsymbol{t})(1+o(1))$ as $|\boldsymbol{t}| \to 0$, where g, g_0 have partial derivatives ∂_i, $i = 1,2$, in* (12) *on $\mathbb{R}_0^2$ such that*

$$\begin{aligned} &|g(\boldsymbol{t})| + |g_0(\boldsymbol{t})| \le C\rho(\boldsymbol{t})^\chi, \\ &|\partial_i g(\boldsymbol{t})| + |\partial_i g_0(\boldsymbol{t})| \le C\rho(\boldsymbol{t})^{\chi - \frac{1}{q_i}}, \quad i = 1,2, \end{aligned} \tag{67}$$

for all $\boldsymbol{t} \in \mathbb{R}_0^2$. Moreover, $g_0(\boldsymbol{t}) = \rho(\boldsymbol{t})^\chi \ell(\boldsymbol{t})$, $\boldsymbol{t} \in \mathbb{R}_0^2$, as in (6), *where ℓ is a generalised invariant function and the parameters $\chi < 0$, $q_i > 0$,*

$i = 1, 2$, with $Q = \frac{1}{q_1} + \frac{1}{q_2}$ satisfy

$$-\frac{1}{\alpha}Q < \chi < \frac{1}{q_1 \vee q_2} - \frac{1}{\alpha}Q. \tag{68}$$

Then, the γ-tangent limits T_γ in (1) of X exist for any $\gamma > 0$ under the normalisation $d_{\lambda,\gamma} = \lambda^{H(\gamma)}$, $H(\gamma) = (1 \wedge \frac{\gamma}{\gamma_{\mathrm{cr}}})H(\gamma_{\mathrm{cr}})$, $H(\gamma_{\mathrm{cr}}) = \frac{1+\gamma_{\mathrm{cr}}}{\alpha} + \chi q_1$ and are given by

$$T_\gamma = \begin{cases} T_+, & \gamma > \gamma_{\mathrm{cr}}, \\ T_-, & \gamma < \gamma_{\mathrm{cr}}, \\ T_0, & \gamma = \gamma_{\mathrm{cr}}, \end{cases} \tag{69}$$

where $\gamma_{\mathrm{cr}} = \frac{q_1}{q_2}$ and

$$T_0(\boldsymbol{t}) = \int_{\mathbb{R}^2} \{g_0(\boldsymbol{t} - \boldsymbol{u}) - g_0(-\boldsymbol{u})\} W_\alpha(\mathrm{d}\boldsymbol{u}), \quad \boldsymbol{t} \in \mathbb{R}^2_+. \tag{70}$$

W_α is an α-stable random measure, as in Theorem 11, and the 'unbalanced' γ-tangent RFs

$$T_+(\boldsymbol{t}) = T_0(t_1, 0), \quad T_-(\boldsymbol{t}) = T_0(0, t_2), \quad \boldsymbol{t} = (t_1, t_2) \in \mathbb{R}^2_+, \tag{71}$$

depend on only one coordinate on the plane.

Remark 15. We can rewrite the condition (68) in terms of p_i, $i = 1, 2$, in (7) as $P_{\frac{1}{\alpha}, \frac{1+\alpha}{\alpha}} < 1$, $P_{\frac{1+\alpha}{\alpha}, \frac{1}{\alpha}} < 1$, $P > \frac{\alpha}{1+\alpha}$, which corresponds to the parameter region $\mathcal{R}_{22}$ in Fig. 1. The limits $T_\pm$ in (71) are nothing but α-stable processes in (32), viz., $T_+(\boldsymbol{t}) \overset{\mathrm{fdd}}{=} Y_{1,\alpha}(t_1), T_-(\boldsymbol{t}) \overset{\mathrm{fdd}}{=} Y_{2,\alpha}(t_2)$ used to define unbalanced rectangent limits $V_\pm$; see (39)–(40). Although $T_\pm$ and $V_\pm$ are very different, both are (H_1, H_2)-MSS RFs with the same indices (H_1, H_2), $H_1 \wedge H_2 = 0$. These observations relate Theorems 14 and 11 and raise the question about the existence of γ-tangent limits in other parameter regions in Fig. 1.

Remark 16. For $\alpha = 2$ in Example 1 of Chapter 2, we have $g(\boldsymbol{t}) = g_0(\boldsymbol{t}) = |\boldsymbol{t}|^{H-1}$, with $q_1 = q_2 = 2$, $Q = 1$, $\chi = \frac{H-1}{2}$, and conditions (68) are equivalent to $\frac{1-H}{2} < \frac{1}{2} < 1 - \frac{H}{2}$ or $0 < H < 1$. Condition (67) in this example is also satisfied for all $\boldsymbol{t} \in \mathbb{R}^2_0$ since $\partial_i |\boldsymbol{t}|^{H-1} = (H-1) |\boldsymbol{t}|^{H-2} \partial_i |\boldsymbol{t}|$ with $\partial_i |\boldsymbol{t}| = \frac{t_i}{|\boldsymbol{t}|}$, $i = 1, 2$.

Proof of Theorem 14. Let us check that the stochastic integral in (70) is well defined for any $\boldsymbol{t} \in \mathbb{R}^2$ or $\|h_0(\boldsymbol{t};\cdot)\|_\alpha < \infty$, where $h_0(\boldsymbol{t};\boldsymbol{u}) := g_0(\boldsymbol{t}-\boldsymbol{u}) - g_0(-\boldsymbol{u})$. Let $q := \max\{q_1, q_2, 1\}$. Then, using (67), we get $\int_{\rho(\boldsymbol{u})<2^q\rho(\boldsymbol{t})} |h_0(\boldsymbol{t};\boldsymbol{u})|^\alpha \mathrm{d}\boldsymbol{u} \le C\int_{\rho(\boldsymbol{u})<3^q\rho(\boldsymbol{t})} \rho(\boldsymbol{u})^{\alpha\chi}\mathrm{d}\boldsymbol{u} < \infty$ as $Q > -\alpha\chi$, referring to (14) of Chapter 6. Next, for $\rho(\boldsymbol{u}) \ge 2^q\rho(\boldsymbol{t})$, we rewrite $h_0(\boldsymbol{t};\boldsymbol{u}) = \int_0^{t_1} \partial_1 g_0(s-u_1, t_2-u_2)\mathrm{d}s + \int_0^{t_2} \partial_2 g_0(-u_1, s-u_2)\mathrm{d}s$, where $\partial_1 g_0(s-u_1,t_2-u_2) \le C\rho(\boldsymbol{u})^{\chi-\frac{1}{q_1}}$, $\partial_2 g_0(-u_1, s-u_2) \le C\rho(\boldsymbol{u})^{\chi-\frac{1}{q_1}}$ by (67), (35). We get $\int_{\rho(\boldsymbol{u})\ge 2^q\rho(\boldsymbol{t})} |h_0(\boldsymbol{t};\boldsymbol{u})|^\alpha \mathrm{d}\boldsymbol{u} \le C\int_{\rho(\boldsymbol{u})\ge 2^q\rho(\boldsymbol{t})} \rho(\boldsymbol{u})^{\alpha(\chi-\frac{1}{q_1\vee q_2})}\mathrm{d}\boldsymbol{u} < \infty$. Hence, $\|h_0(\boldsymbol{t};\cdot)\|_\alpha < \infty$.

To prove the convergence in (1), we use Proposition 24 of Chapter 2. It suffices to consider the case $\boldsymbol{t}_0 = \boldsymbol{0}$. First, let $\gamma = \gamma_{\mathrm{cr}}$, with $H(\gamma_{\mathrm{cr}}) = \frac{1}{\alpha}(1+\gamma_{\mathrm{cr}}) + \chi q_1$. Then, $\lambda^{-H(\gamma_{\mathrm{cr}})}X(\lambda^{\Gamma_{\mathrm{cr}}}\boldsymbol{t}) = \int_{\mathbb{R}^2} f_\lambda(\boldsymbol{t};\boldsymbol{u})M(\mathrm{d}\boldsymbol{u})$, where $f_\lambda(\boldsymbol{t};\boldsymbol{u}) := \lambda^{-H(\gamma_{\mathrm{cr}})}(g(\lambda^{\Gamma_{\mathrm{cr}}}\boldsymbol{t}-\boldsymbol{u}) - g(-\boldsymbol{u}))$ satisfies

$$f_\lambda^\dagger(\boldsymbol{t};\boldsymbol{u}) = \lambda^{-\chi q_1}(g(\lambda^{\Gamma_{\mathrm{cr}}}(\boldsymbol{t}-\boldsymbol{u})) - g(-\lambda^{\Gamma_{\mathrm{cr}}}\boldsymbol{u})) \to h_0(\boldsymbol{t};\boldsymbol{u})$$

for all $\boldsymbol{u} \ne \boldsymbol{0}, \boldsymbol{t}$ according to $g(\boldsymbol{t}) = g_0(\boldsymbol{t})(1+o(1))$, $|\boldsymbol{t}| \to 0$. The domination argument implying $f_\lambda^\dagger(\boldsymbol{t};\cdot) \to h_0(\boldsymbol{t};\cdot)$ in $L^\alpha(\mathbb{R}^2)$ uses (67) and follows as in the proof $\|h_0(\boldsymbol{t};\cdot)\|_\alpha < \infty$ above.

Next, let $\gamma < \gamma_{\mathrm{cr}}$, with $H(\gamma) = \frac{\gamma}{\gamma_{\mathrm{cr}}}H(\gamma_{\mathrm{cr}}) = \frac{1}{\alpha}(\gamma + \frac{\gamma}{\gamma_{\mathrm{cr}}}) + \gamma\chi q_2$. Then, we have $\lambda^{-H(\gamma)}X(\lambda^{\Gamma}\boldsymbol{t}) = \int_{\mathbb{R}^2} f_{\lambda,\gamma}(\boldsymbol{t};\boldsymbol{u})M(\mathrm{d}\boldsymbol{u})$, where using Proposition 24 of Chapter 2 with $\lambda_1 = \lambda^{\frac{\gamma}{\gamma_{\mathrm{cr}}}}$, $\lambda_2 = \lambda^\gamma$, we see that

$$\begin{aligned} f_{\lambda,\gamma}^\dagger(\boldsymbol{t};\boldsymbol{u}) &= \lambda^{-\gamma\chi q_2}(g(\lambda t_1 - \lambda^{\frac{\gamma}{\gamma_{\mathrm{cr}}}}u_1, \lambda^\gamma t_2 - \lambda^\gamma u_2) \\ &\quad - g(-\lambda^{\frac{\gamma}{\gamma_{\mathrm{cr}}}}u_1, -\lambda^\gamma u_2)) \to h_0((0,t_2);\boldsymbol{u}) \end{aligned}$$

for all $\boldsymbol{u} \ne \boldsymbol{0}, (0,t_2)$. The $L^\alpha(\mathbb{R}^2)$-convergence $f_{\lambda,\gamma}^\dagger(\boldsymbol{t};\cdot) \to h_0((0,t_2);\cdot)$ in the case of $\gamma < \gamma_{\mathrm{cr}}$, and the proof of (1) for $\gamma > \gamma_{\mathrm{cr}}$ follows analogously, so we omit the details. Theorem 14 is proved. □

6 Notes

1. The notion of *tangent RF* appeared in Falconer [88]. It corresponds to that of **1**-tangent RFs (Definition 7 of Chapter 1, with (30) of

Chapter 1) replaced by a stronger functional convergence. Tangent RFs of Lévy fractional RFs in Example 3 in an arbitrary dimension $\nu \geq 1$ were studied by Benassi *et al.* [29] and Cohen [58].

2. *Anisotropic scaling of ambit RFs.* It of interest to extend the class of RFs X in (3) by including a *random volatility RF* (independent of M) as in the case of *ambit RF*; see Barndorff-Nielsen *et al.* [22]. Particularly, one may consider local scaling limits for the RF X with the representation

$$X(\boldsymbol{t}) := \int_{\mathbb{R}^2} \{g(\boldsymbol{t}-\boldsymbol{u}) - g_1^0((t_1,0)-\boldsymbol{u}) - g_2^0((0,t_2)-\boldsymbol{u}) \\ + g_{12}^0(-\boldsymbol{u})\}A(\boldsymbol{u})M(\mathrm{d}\boldsymbol{u}), \quad \boldsymbol{t} \in \mathbb{R}_+^2, \tag{72}$$

where $M, g, g_1^0, g_2^0, g_{12}^0$ are the same as in (3) and $A = \{A(\boldsymbol{u}); \boldsymbol{u} \in \mathbb{R}^2\}$ is a 'volatility' RF independent of M. If RF A is sufficiently smooth, results as in Theorems 11 and 14 for X in (72) can be expected, with the difference that the limit RFs include $A(\mathbf{0})$ as a multiplicative factor.

3. *Incongruous scaling.* The asymptotic results in Theorems 11–14 depend on the parameters $q_i, i = 1, 2$, which characterise the local behaviour of $g(\boldsymbol{t})$ at $\boldsymbol{t} = \mathbf{0}$ along the coordinate axis. The local behaviour of the 'rotated' kernel $g(\mathcal{O}\boldsymbol{t})$ and the corresponding Lévy-driven RF $X(\mathcal{O}\boldsymbol{t})$ with a 2×2 matrix $\mathcal{O}$ can be very different, as it happens in the large-scale limit in Theorem 26 of Chapter 6. We may expect that an 'incongruous' local anisotropic scaling leads to a simpler scaling diagram with critical $\gamma_{\mathrm{cr}} = 1$ independent of $q_i, i = 1, 2$, as in Table 1.

4. *Power variations.* We expect that our results can be applied for the statistical estimation and identification of fractional parameters, $p_i, i = 1, 2$, based on observations of X in (3) on a dense discrete mesh on the plane. For isotropic Lévy fractional RF in Example 3 and a dyadic mesh, this question is treated in Benassi *et al.* [29] and Cohen [58] using log-variations of (ordinary) increments of X. There is considerable limit theory for increment-based statistics of fractional RFs driven by Gaussian white noise; see Cohen and Istas [59], Lee [150], Loh [170], Pakkanen [195, 196], and the references therein. On the other hand, for α-stable and related stationary increments, Lévy-driven moving average processes with one-dimensional time parameter (directly related to Lévy-driven fractional RF in (3)),

Barndorff-Nielsen *et al.* [23] and Basse-O'Connor *et al.* [26] provide a detailed study of the asymptotic behaviour of *power variations* computed from observations on $i/n, i = 1, \ldots, n$. For a class of Lévy-driven fractional isotropic RFs, Basse-O'Connor *et al.* [27] discuss the first-order asymptotics of power variations computed from rectangular increments of RF at points of a dense square grid.

In the case of anisotropic RFs, the number of fractional parameters is more than one, and a fixed (square or non-square) observation grid seems insufficient. Indeed, power variations computed from a grid of size $(1/n, 1/n^\gamma)$ for a fractional RF X, as in this chapter, are supposed to estimate the normalising exponent $H(\gamma)$ in Theorem 11, which does not determine the fractional parameters p_1 and p_2 and takes a different form in different parameter regions in Fig. 1. It seems that consistent estimation of $p_i, i = 1, 2$, would require power variations from *two* square grids corresponding to different values $\gamma' < \gamma''$, which should estimate the Hurst parameters $\mathcal{H}_i^{\pm}, i = 1, 2$ in (11), leading to the estimates of $p_i, i = 1, 2$. Finally, defining and studying spatial analogues of the *increment ratio* statistics (Bardet and Surgailis [21]), measuring the 'mean incremental roughness' of RF along a given direction, is an interesting and open question.

5. *Scaling with a non-diagonal matrix.* The scaling in (1) and (2) with a diagonal matrix Γ is quite particular, raising the question about these limits for a general 2×2 matrix Γ with positive real parts of the eigenvalues; c.f. Biermé *et al.* [37]. This question is interesting and open. We expect that a possible answer depends on Γ or how the point $\lambda^\Gamma \boldsymbol{t}$ tends to $\mathbf{0}$ as $\lambda \to 0$. In the most general case, the trajectory $\{\lambda^\Gamma \boldsymbol{t}; 0 < \lambda < 1\}$ may wind up around $\mathbf{0}$ infinitely many times, and the limits in (1) and (2) probably do not exist. In some other cases (e.g., a diagonalisable matrix Γ), the trajectory of $\lambda^\Gamma \boldsymbol{t}$ may approach $\mathbf{0}$ along some 'oblique' direction, resembling the scaling of RFs with 'oblique' dependence axis discussed in Section 3 of Chapter 6, and the limits in (1) and (2) may exist.

Chapter 8

Aggregation of Shot-Noise Inputs

This chapter discusses joint spatio-temporal scaling limits of sums $A_{\lambda,\gamma}$ (indexed by $(x, y) \in \mathbb{R}^2_+$) of a large number $O(\lambda^\gamma)$ of independent integrated LRD shot-noise input processes at time scale λ. Conditions on the input process are provided in order that the normalised RFs $A_{\lambda,\gamma}$ tend to an α-stable Lévy sheet ($1 < \alpha < 2$) if $\gamma < \gamma_{\rm cr}$, to an FBS if $\gamma > \gamma_{\rm cr}$, and to an *intermediate* RF if $\gamma = \gamma_{\rm cr}$. A general class of intermediate RFs on $\mathbb{R}^2$ satisfying the (H, γ)-SS property is introduced and studied in detail.

1 Joint Spatio-Temporal Scaling of Broadband Network Traffic

Broadband network traffic exhibits three distinctive properties – heavy tails, self-similarity, and long-range dependence [186, 197, 267]. To explain these empirical facts, several mathematical models of network traffic were proposed and studied [128, 130, 185, 186, 197, 260]. The usual framework can be described as joint temporal-spatial scaling (aggregation) of independent 'sources' or 'inputs', each described by a stationary process $X = \{X(t); t \in \mathbb{R}\}$ with LRD. The aggregation procedure involves the summation of M independent integrated copies $X_i, i = 1, \dots, M$ of X at time scale T when T and M increase jointly to infinity, possibly at different rates. Accordingly, we may consider the limit distribution of

$$A(T,M) := \sum_{i=1}^{M} \int_0^T X_i(s)\mathrm{d}s, \quad \text{where } T = x\lambda,$$

$$M = [y\lambda^\gamma], \quad (x,y) \in \mathbb{R}^2_+, \tag{1}$$

increase jointly with $\lambda \to \infty$ and $\gamma > 0$ is an arbitrary fixed number. Then, $A_{\lambda,\gamma}(x,y) := A(x\lambda, [y\lambda^\gamma])$ is an RF indexed by $(x,y) \in \mathbb{R}^2_+$, and the limit problem becomes (large-scale) anisotropic scaling of RFs:

$$d^{-1}_{\lambda,\gamma}\{A_{\lambda,\gamma}(x,y) - \mathrm{E}A_{\lambda,\gamma}(x,y)\} \xrightarrow{\text{fdd}} V_\gamma(x,y). \tag{2}$$

(2) fits into the framework of anisotropic scaling of RFs on $\mathbb{R}^2$ or scaling with aggregation of random processes with a one-dimensional time parameter. We also use the notation $(x,y) \in \mathbb{R}^2$ since the arguments in (2) are of different nature in a temporal-spatial setting. In the literature, the limit in (2) is often restricted to the case $y = 1$ and to a limit random process with one-dimensional time $x \in \mathbb{R}_+$; however, these results can be easily extended to the RF setup in (2) using the fact that the summands in (1) are independent.

A 'typical' result in the heavy-tailed aggregated traffic research (see, e.g., Refs. [93, 130, 185]) states that there exists a $\gamma_{\mathrm{cr}} > 0$ such that the limit in (2) is α-stable for $\gamma < \gamma_{\mathrm{cr}}$ and Gaussian for $\gamma > \gamma_{\mathrm{cr}}$, or, more precisely,

$$V_\gamma(x,y) = \begin{cases} L_\alpha(x,y), & \gamma < \gamma_{\mathrm{cr}}, \\ B_{H_1,1/2}(x,y), & \gamma > \gamma_{\mathrm{cr}}, \\ J(x,y), & \gamma = \gamma_{\mathrm{cr}}, \end{cases} \tag{3}$$

where L_α is an α-stable Lévy sheet and $B_{H_1,1/2}, H_1 \in (\frac{1}{2}, 1)$ is an FBS. Obviously, L_α and $B_{H_1,1/2}$ are different RFs, with the latter having finite variance and the former having infinite variance. For $\gamma = \gamma_{\mathrm{cr}}$, the limit J in (3) was proved in Refs. [92, 93] to be an infinitely divisible 'intermediate' RF which is neither Gaussian nor stable. In the context of Section 3 of Chapter 1, the trichotomy in (3) represents a scaling transition arising under particular dependence conditions – horizontal LRD and vertical SRD (independence). Following Ref. [185], we may interpret the exponent $\gamma > 0$ in (1)–(2) as the *connection rate*, indicating how fast the number M of independent sources grows with T, and refer to the cases $\gamma > \gamma_{\rm cr}$ and $\gamma < \gamma_{\rm cr}$ as *fast* and *slow* growth for the connection rate, respectively.

In Section 4, the limits in (2)–(3) are derived for a general class of LRD shot-noise 'inputs',

$$X(t) := \sum_{j\in\mathbb{Z}} W_j(t - T_j), \quad t \in \mathbb{R}, \tag{4}$$

where $\{T_j; j \in \mathbb{Z}\}$ is a homogeneous Poisson point process with unit rate and $\{W_j; j \in \mathbb{Z}\}$ is a sequence of independent copies of a 'pulse process' $W = \{W(t); t \in \mathbb{R}\}$ with trajectories in $L^1(\mathbb{R})$. Chapter 9 deals with a related but distinct class of regenerative X, with a random pulse starting anew at points of a heavy-tailed renewal process. The above classes contain the most widely studied inputs, namely, the M/G/∞ queue and the ON/OFF process. The mathematical treatment of these two classes is rather involved and different from that in the previous chapters.

2 Aggregation of M/G/∞ Queue

We start with the simplest shot-noise 'input' X in (4) – the M/G/∞ queue. In the *M/G/∞* queue, the number of customers, or active sources at time $t \in \mathbb{R}$, is given by

$$X(t) := \sum_j \mathbb{I}(T_j < t \le T_j + R_j), \tag{5}$$

where $\cdots < T_{-1} < 0 \le T_0 < T_1 < \cdots$ are points (arrival times) of a stationary Poisson process with unit rate and $\{R_j; j \in \mathbb{Z}\}$ (service, or transmission durations) are strictly positive i.i.d. r.v.s with a finite mean, independent of $\{T_j; j \in \mathbb{Z}\}$. Accordingly, $\{(T_j, R_j)\}$ form a Poisson point process on $\mathbb{R} \times \mathbb{R}_+$ with intensity $\mathrm{d}uF(\mathrm{d}r)$, where F is the distribution of a generic duration R. Using a Poisson random measure $\mathcal{N}(A) = \sum_j \mathbb{I}((T_j, R_j) \in A)$ defined for any (bounded) Borel subset $A \subset \mathbb{R} \times \mathbb{R}_+$, with mean $\mathrm{E}\mathcal{N}(A) = \int_A \mathrm{d}uF(\mathrm{d}r)$, the process in (5) can be written as a Poisson stochastic integral,

$$X(t) = \int_{\mathbb{R}\times\mathbb{R}_+} \mathbb{I}(u < t \le u + r)\mathcal{N}(\mathrm{d}u, \mathrm{d}r). \tag{6}$$

The aggregation of independent inputs of (5) is also known as the *infinite-source Poisson model* [185], with the integral $\int_0^T X_i(s)\mathrm{d}s$ in

(1) representing the workload from the ith source coming to the network during the time interval $[0, T]$. Clearly, (5) is a particular case of (2) of Chapter 4 or (4), corresponding to a rectangular pulse process, $W(t) = \mathbb{I}(0 < t \le R)$. From (6) and the properties of the Poisson integral, we see that X is a strictly stationary process with mean $\mathrm{E}X(t) = \int_{\mathbb{R}\times\mathbb{R}_+} \mathbb{I}(u < t \le u + r)\mathrm{d}uF(\mathrm{d}r) = \int_{\mathbb{R}_+} rF(\mathrm{d}r) = \mathrm{E}R < \infty$ and covariance function

$$\begin{aligned}\mathrm{Cov}(X(0), X(t)) &= \int_{\mathbb{R}\times\mathbb{R}_+} \mathbb{I}(u < 0 < t \le u + r)\mathrm{d}uF(\mathrm{d}r)\\ &= \mathrm{E}(R - t)_+ = \int_t^\infty \mathrm{P}(R > r)\mathrm{d}r, \quad t \ge 0. \qquad (7)\end{aligned}$$

If the tail d.f. $\mathrm{P}(R > r)$ regularly varies with exponent $\alpha \in (1, 2)$ as $r \to \infty$, the above covariance function regularly varies with exponent $\alpha - 1 \in (0, 1)$ as $t \to \infty$, and hence it is not integrable on the real line and $\{X(t)\}$ has the covariance LRD property.

The infinite-source Poisson model of (5) is highly explicit and can serve as a particularly simple illustration of the following sections where the trichotomy in (3) is established for a much larger class of input processes. Actually, (5) represents an RG model in dimension $\nu = 1$ with a generic grain $\Xi =]0, R]$, and the limits in (3) can be rather easily derived from Theorem 10 of Chapter 4. Indeed, following (22) of Chapter 4, assume that the distribution $F(\mathrm{d}r)$ has a probability density $f(r), r > 0$, such that

$$f(r) \sim c_f\, r^{-1-\alpha}, \quad r \to \infty \quad (\exists\, 1 < \alpha < 2,\ c_f > 0). \qquad (8)$$

The random set $\Xi =]0, R] \subset \mathbb{R}$ satisfies the conditions of Proposition 5 of Chapter 4, $\mathrm{Cov}(X(0), X(t)) \sim (c_f/\alpha)t^{-(\alpha-1)}$, and Theorem 10 of Chapter 4 applies. For $\phi(u) = \mathbb{I}(u \in]0, x])$, the aggregate in (1) coincides in distribution with the stochastic integral

$$\begin{aligned}A(\lambda x, M) &= X_{\lambda,M}(\mathbb{I}(]0, x])\\ &\overset{\mathrm{fdd}}{=} \int_{\mathbb{R}\times\mathbb{R}_+} \left\{ \int_0^{\lambda x} \mathbb{I}(u < t \le u + r)\mathrm{d}t \right\} \mathcal{N}_M(\mathrm{d}u, \mathrm{d}r) \qquad (9)\end{aligned}$$

with respect to a Poisson random measure $\mathcal{N}_M(\mathrm{d}\boldsymbol{u}, \mathrm{d}r)$, $\mathrm{E}\mathcal{N}_M(\mathrm{d}\boldsymbol{u}, \mathrm{d}r) = M\mathrm{E}\mathcal{N}(\mathrm{d}\boldsymbol{u}, \mathrm{d}r)$. The limits (51) of Chapter 4 occurring in Theorem 10 of Chapter 4 for the above RG model are given by the stochastic integrals

$$\mathcal{B}_\alpha(x) := \int_{\mathbb{R}\times\mathbb{R}_+} \left\{ \int_0^x \mathbb{I}(u < t \le u + r)\mathrm{d}t \right\} W_\alpha(\mathrm{d}u, \mathrm{d}r), \quad (10)$$

$$J_\alpha(x) := \int_{\mathbb{R}\times\mathbb{R}_+} \left\{ \int_0^x \mathbb{I}(u < t \le u + r)\mathrm{d}t \right\} \widetilde{\mathcal{N}}_\alpha(\mathrm{d}u, \mathrm{d}r),$$

$x > 0$, which formally resemble (9), except that they do not depend on λ, and the random measure W_α (respectively, $\widetilde{\mathcal{N}}_\alpha$) is Gaussian (respectively, centred Poisson) with the same variance:

$$\mathrm{E}W_\alpha^2(\mathrm{d}u, \mathrm{d}r) = \mathrm{E}J_\alpha^2(\mathrm{d}u, \mathrm{d}r) = \mathrm{d}u\, \nu_\alpha(\mathrm{d}u),$$
$$\text{where } \nu_\alpha(\mathrm{d}r) := c_f\, r^{-\alpha-1}\mathrm{d}r, \ r \in \mathbb{R}_+. \quad (11)$$

The fact that (10) are well defined for $\alpha \in (1, 2)$ and have finite variance,

$$\mathrm{E}\mathcal{B}_\alpha^2(x) = \mathrm{E}J_R^2(x)$$
$$= c_f \int_0^\infty \frac{\mathrm{d}r}{r^{1+\alpha}} \int_{\mathbb{R}} \left\{ \int_0^x \mathbb{I}(u < t \le u + r)\mathrm{d}t \right\}^2 \mathrm{d}u < \infty, \quad (12)$$

can be directly verified by noting that $\int_0^x \mathbb{I}(u < t \le u + r)\mathrm{d}t \le r \wedge x$ and $\int_0^x \mathbb{I}(u < t \le u+r)\mathrm{d}t = \mathrm{Leb}_1((0, x] \cap (u, u+r]) \ne 0$ implies $|u| \le r \vee x$. Hence, $\int_1^\infty r^{-1-\alpha}\mathrm{d}r \int_{\mathbb{R}}\{\cdots\}^2\mathrm{d}u \le x^2 \int_1^\infty r^{-1-\alpha}\mathrm{d}r \int_{|u|\le r} \mathrm{d}u$ $= 2x^2 \int_1^\infty r^{-\alpha}\mathrm{d}r < \infty$ and $\int_0^1 r^{-1-\alpha}\mathrm{d}r \int_{\mathbb{R}}\{\cdots\}^2\mathrm{d}u \le \int_0^1 r^{1-\alpha}\mathrm{d}r \times$ $\int_{|u|\le x} \mathrm{d}u = 2x \int_0^1 r^{1-\alpha}\mathrm{d}r < \infty$. By the change of variables $u \to xu, r \to xr$ in the integral in (12), we get that

$$\mathrm{E}\mathcal{B}_\alpha^2(x) = \mathrm{E}J_\alpha^2(x) = C_\alpha^2 x^{2H_1}, \quad \text{where } H_1 := \frac{3-\alpha}{2},$$
$$C_\alpha^2 := \mathrm{E}B_\alpha^2(1) = \mathrm{E}J_R^2(1). \quad (13)$$

Furthermore, the two processes in (10) have stationary increments. Since for any X with stationary increments and finite variance, we have $\mathrm{Cov}(X(t), X(s)) = (1/2)(\mathrm{Var}(X(t)) + \mathrm{Var}(X(s)) - \mathrm{Var}(X(t) - X(s)))$, relation (13) implies that for any $x, y > 0$,

$$\mathrm{E}B_\alpha(x)B_\alpha(y) = \mathrm{E}J_\alpha(x)J_\alpha(y) = \frac{C_\alpha^2}{2}\big(x^{2H} + y^{2H} - |x-y|^{2H}\big).$$

In other words, the Gaussian process $\mathcal{B}_\alpha \overset{\text{fdd}}{=} C_\alpha B_{H_1}$ coincides with the FBM B_{H_1} up to a multiplicative constant C_α. The process J_α having the same second-order properties with B_α is known as the *Telecom process*. Further properties of the Telecom process and its extension – the Telecom RF – are discussed in Section 3.

The third process $L_\alpha(x) \overset{\text{fdd}}{=} L_\alpha(]0, x])$ in Theorem 10 of Chapter 4, case $\nu = 1$, is a classical homogeneous asymmetric Lévy process with independent increments and the characteristic function

$$\begin{aligned} \mathrm{E}\mathrm{e}^{\mathrm{i}\theta L_\alpha(x)} &= \exp\left\{x \int_0^\infty \Psi(\theta r)\nu_\alpha(\mathrm{d}r)\right\} \\ &= \exp\big\{ -x|\theta|^\alpha \sigma^\alpha (1 - \operatorname{sgn}(\theta)\tan(\pi\alpha/2))\big\}, \quad \theta \in \mathbb{R}, \end{aligned} \tag{14}$$

where $\sigma^\alpha = c_f \cos(\pi\alpha/2)\Gamma(2-\alpha)/(1-\alpha)$; see (125) of Chapter 4 and (11).

We formulate the corollary from Theorem 10 of Chapter 4, pertaining to the aggregated workload from the M/G/∞ queue.

Corollary 1. *Let $A_{\lambda,\gamma}(x) := \sum_{i=1}^{[\lambda^\gamma]} \int_0^{\lambda x} X_i(s)\mathrm{d}s, x > 0$ be the aggregated workload from independent M/G/∞ queues as in* (5)*, with the service time distribution of R satisfying* (8)*. Then, as $\lambda \to \infty$,*

$$\begin{aligned} &\lambda^{-H(\gamma)}(A_{\lambda,\gamma}(x) - \mathrm{E}A_{\lambda,\gamma}(x)) \\ &\quad \overset{\text{fdd}}{\longrightarrow} \begin{cases} \mathcal{B}_\alpha(x), & \gamma > \alpha - 1, \ H(\gamma) = \frac{3-\alpha+\gamma}{2}, \\ L_\alpha(x), & \gamma < \alpha - 1, \ H(\gamma) = \frac{1+\gamma}{\alpha}, \\ J_\alpha(x), & \gamma = \alpha - 1, \ H(\gamma) = 1, \end{cases} \end{aligned} \tag{15}$$

where $\mathcal{B}_\alpha \overset{\text{fdd}}{=} C_\alpha B_{H_1}, L_\alpha, J_\alpha$ are as in (10) *and* (14).

Contrary to the first two limit processes $\mathcal{B}_\alpha$ and L_α in (15), which are self-similar with the respective indices $H_1 = \frac{3-\alpha}{2}$ and $\frac{1}{\alpha}$, the Telecom process is *not* self-similar but satisfies a 'bridge property' between B_α and L_α, termed *asymptotic local* and *global* self-similarity, as discussed in what follows. Roughly, it says that $J_\alpha(x)$ resembles $\mathcal{B}_\alpha(x)$ for 'small' x and $L_\alpha(x)$ for 'large' x.

Proposition 2. *The Telecom process* $J_\alpha, 1 < \alpha < 2$ *satisfies the following two properties:*

- $\lambda^{-H} J_\alpha(\lambda x) \xrightarrow{\text{fdd}} \mathcal{B}_\alpha(x)$ *as* $\lambda \to 0$ *(asymptotic local self-similarity);*
- $\lambda^{-1/\alpha} J_\alpha(\lambda x) \xrightarrow{\text{fdd}} L_\alpha(x)$ *as* $\lambda \to \infty$ *(asymptotic global self-similarity).*

Proposition 2 is part of Theorem 7. Corollary 1 easily extends to RF convergence in (2)–(3), with limit RFs given by similar stochastic integrals to (10) and having independent increments in the vertical direction. We give a formal definition of the RF extension of the Telecom process.

Definition 3. The Telecom RF $J_\alpha = \{J_\alpha(x,y); (x,y) \in \mathbb{R}^2_+\}$ is defined for $\alpha \in (1,2)$ as the stochastic integral

$$J_\alpha(x,y) := \int_{\mathbb{R}\times\mathbb{R}_+} \left\{ \int_0^x \mathbb{I}(u < t \le u + r)\mathrm{d}t \right\} \widetilde{\mathcal{N}}_\alpha(\mathrm{d}u, \mathrm{d}v, \mathrm{d}r) \tag{16}$$

w.r.t. the centred Poisson random measure $\widetilde{\mathcal{N}}_\alpha$ on $\mathbb{R}\times\mathbb{R}_+\times\mathbb{R}_+$ with variance $\mathrm{d}u\mathrm{d}v\nu_\alpha(\mathrm{d}r) = c_f \mathrm{d}u\mathrm{d}v r^{-\alpha-1}\mathrm{d}r$.

3 A Class of Intermediate RFs

We define a large class of RFs, $J_\nu = \{J_\nu(x,y); (x,y) \in \mathbb{R}^2_+\}$, generalising the Telecom RF and depending on an infinite measure ν on $L^1(\mathbb{R}_+)$ with a scaling property. Such RFs will appear in Section 4 in the 'intermediate' limit for general shot-noise input processes. The above-mentioned RFs have a rather special structure and share the 'bridge property' of the Telecom process in Proposition 2.

Let ν be a σ-finite measure on $\mathbb{W} := \{w \in L^1(\mathbb{R}) : w(t) = 0$ for $t < 0\}$, equipped with the σ-algebra $\mathcal{B}(\mathbb{W})$ of Borel subsets of $\mathbb{W}$ and such that for any $x > 0$,

$$\int_{\mathbb{R}\times\mathbb{W}} \left(\left| \int_0^x w(t-u)\mathrm{d}t \right| \wedge \left| \int_0^x w(t-u)\mathrm{d}t \right|^2 \right) \mathrm{d}u\, \nu(\mathrm{d}w) < \infty. \tag{17}$$

With the above ν, we can associate a centred Poisson random measure $\widetilde{\mathcal{N}}_\nu(\mathrm{d}u, \mathrm{d}v, \mathrm{d}w)$ on $\mathbb{R} \times \mathbb{R}_+ \times \mathbb{W}$ with control measure $\mathrm{d}u\,\mathrm{d}v\,\nu(\mathrm{d}w)$. Recall from Section 2 of Chapter 2 that a stochastic integral $\int f \mathrm{d}\widetilde{\mathcal{N}}_\nu \equiv \int_{\mathbb{R}\times\mathbb{R}_+\times\mathbb{W}} f(u,v,w)\widetilde{\mathcal{N}}_\nu(\mathrm{d}u, \mathrm{d}v, \mathrm{d}w)$ is well defined for any (measurable) $f = f(u,v,w), (u,v,w) \in \mathbb{R} \times \mathbb{R}_+ \times \boldsymbol{W}$ with $\int_{\mathbb{R}\times\mathbb{R}_+\times\mathbb{W}} \left(|f(u,v,w)| \wedge |f(u,v,w)|^2\right) \mathrm{d}u\,\mathrm{d}v\,\nu(\mathrm{d}w) < \infty$, and its characteristic function is given by

$$\mathrm{E}\mathrm{e}^{\mathrm{i}\theta \int f \mathrm{d}\widetilde{\mathcal{N}}_\nu} = \exp\left\{ \int_{\mathbb{R}\times\mathbb{R}_+\times\mathbb{W}} (\mathrm{e}^{\mathrm{i}\theta f(u,v,w)} - 1 - \mathrm{i}\theta f(u,v,w))\mathrm{d}u\,\mathrm{d}v\,\nu(\mathrm{d}w)\right\}, \quad \theta \in \mathbb{R}. \tag{18}$$

Moreover, $\mathrm{E}|\int f \mathrm{d}\widetilde{\mathcal{N}}_\nu| < \infty$ and $\mathrm{E}\int f \mathrm{d}\widetilde{\mathcal{N}}_\nu = 0$.

Given $\lambda > 0$ and $H \in \mathbb{R}$, we associate a one-to-one mapping $\phi_{\lambda,H}$ on $\mathbb{W}$, with its inverse $\phi^{-1}_{\lambda,H} = \phi_{\lambda^{-1},H}$ defined as

$$\phi_{\lambda,H} w(t) := \lambda^{H-1} w(\lambda^{-1}t), \quad t \in \mathbb{R},\ w \in \mathbb{W}. \tag{19}$$

Note that $\{\phi_{\lambda,H}; \lambda > 0\}$ form a group of scaling transformations on $\mathbb{W}$.

Definition 4. Let $(H,\gamma) \in \mathbb{R} \times \mathbb{R}_+$.

(i) A σ-finite measure ν on $\mathbb{W}$ will be called (H,γ)-scaling if

$$\lambda^{1+\gamma} \nu \circ \phi_{\lambda,H} = \nu, \quad \forall \lambda > 0, \tag{20}$$

where $\nu \circ \phi_{\lambda,H}(B) := \nu(\phi_{\lambda,H}(B))$ for $B \in \mathcal{B}(\mathbb{W})$.

(ii) An RF $J = \{J(x,y); (x,y) \in \mathbb{R}^2_+\}$ will be called (H,γ)-*SS* if

$$\{J(\lambda x, \lambda^\gamma y); (x,y) \in \mathbb{R}^2_+\} \stackrel{\mathrm{fdd}}{=} \{\lambda^H J(x,y); (x,y) \in \mathbb{R}^2_+\}, \quad \forall \lambda > 0. \tag{21}$$

Some comments regarding Definition 4 are in order. The (H,γ)-SS property was defined earlier in (22) of Chapter 1. The notion of (H,γ)-scaling measure applies to infinite measures; indeed, (20) implies $\lambda^{1+\gamma}\nu(\mathbb{W}) = \nu(\mathbb{W})\,(\forall \lambda > 0)$, meaning that ν cannot be a finite measure. Such measures are used to construct 'intermediate' (H,γ)-SS RFs J_ν through the Poisson stochastic integrals in (18);

see (22) later. To avoid any confusion with the scaling setup in (2), we note that in Definition 4 (i), $\gamma > 0$ is *fixed*: the above-mentioned RFs J_ν appear as 'intermediate' limits in (2) for a particular $\gamma = \gamma_{\rm cr}$ only. On the other hand, the (H, γ)-SS property is a general property shared by anisotropic scaling limits in (2) for any $\gamma > 0$ (Theorem 8 of Chapter 1). Somewhat surprisingly, a given measure ν can be (H, γ)-scaling with different (H, γ), as seen in Example 9, but then the corresponding RF J_ν lacks the important 'intermediate' property of Theorem 7 and apparently cannot appear as an 'intermediate' scaling limit in (2).

Given a (H, γ)-scaling measure ν satisfying (17), we define an RF $J_\nu = \{J_\nu(x, y); (x, y) \in \mathbb{R}^2_+\}$ as the stochastic integral

$$J_\nu(x, y) := \int_{\mathbb{R}\times(0,y]\times\mathbb{W}} \left\{ \int_0^x w(t-u)\mathrm{d}t \right\} \widetilde{\mathcal{N}}_\nu(\mathrm{d}u, \mathrm{d}v, \mathrm{d}w) \tag{22}$$

w.r.t. a (centred) Poisson random measure in (18). We extend the definition in (22) to $\bar{\mathbb{R}}^2_+$ by setting $J_\nu(x, y) := 0$ for $x \wedge y = 0$. The Telecom RF in (16) corresponds to $\nu = \nu_\alpha$ in (11) concentrated on the set of indicator functions $\{\mathbb{I}(\cdot \in]0, r]); r \in \mathbb{R}_+\} \subset \mathbb{W}$, which is identified with $\mathbb{R}_+$. Note that ν_α is a σ-finite measure, which is (H, γ)-scaling for $\gamma = \alpha - 1$, $H = 1$, and satisfies (17). The Telecom RF satisfies the (H, γ)-SS property but is not (H_1, H_2)-MSS for any $H_i, i = 1, 2$. Indeed, the latter property would imply that the restriction $\{J_\alpha(1, \lambda^{H_2} y); y \in \mathbb{R}_+\} \overset{\mathrm{fdd}}{=} \{\lambda^{H_2} J_\alpha(1, y); y \in \mathbb{R}_+\}$ to the vertical line $x = 1$ is a H_2-SS process with independent increments, or, in other words, a stable process, but this is not true; see, e.g., Ref. [92].

The following proposition gives some properties of RF J_ν in (22).

Proposition 5. *Let ν be a (H, γ)-scaling measure on $\mathbb{W}$ satisfying* (17), *and let J_ν be as in* (22). *Then:*

(i) *$J_\nu(x, y)$ is well defined for any $(x, y) \in \mathbb{R}^2_+$ as a stochastic integral w.r.t. the Poisson random measure $\widetilde{\mathcal{N}}_\nu$ and has infinitely divisible finite-dimensional distributions and zero mean $\mathrm{E}J_\nu(x, y) = 0$. Moreover, the RF J_ν has stationary rectangular increments and independent rectangular increments in the vertical direction;*
(ii) *RF J_ν is (H, γ)-SS.*

(iii) *If, in addition,*

$$C_\nu^2 := \int_{\mathbb{R}\times\mathbb{W}} \left(\int_0^1 w(t-u)\mathrm{d}t\right)^2 \mathrm{d}u\,\nu(\mathrm{d}w) < \infty, \tag{23}$$

then $\mathrm{E}J_\nu^2(x,y) < \infty$, $(x,y) \in \mathbb{R}_+^2$, *and*

$$\mathrm{E}J_\nu(x_1,y_1)J_\nu(x_2,y_2) = \frac{C_\nu^2}{2}(x_1^{2H_1} + x_2^{2H_1} - |x_1-x_2|^{2H_1}) \times (y_1 \wedge y_2), \quad (x_i,y_i) \in \mathbb{R}_+^2,\ i=1,2, \tag{24}$$

where $H_1 := H - \frac{\gamma}{2}$.

Proof. (i) is obvious from definition (22) and the properties of the stochastic integral $\int f\mathrm{d}\widetilde{\mathcal{N}}_\nu$.

(ii) According to (18), for any $\theta_j \in \mathbb{R}$, $(x_j,y_j) \in \mathbb{R}_+^2$, $j=1,\ldots,m$, $m \in \mathbb{N}$, the ch.f. of J_ν can be written as $\mathrm{E}\exp\{\mathrm{i}\sum_{j=1}^m \theta_j J_\nu(\lambda x_j, \lambda^\gamma y_j)\} = \mathrm{e}^{I_\lambda}$, where

$$I_\lambda = \int_{\mathbb{R}\times\mathbb{R}_+\times\mathbb{W}} \Psi\left(\sum_{j=1}^m \theta_j \mathbb{I}(v \le \lambda^\gamma y_j) \int_0^{\lambda x_j} w(t-u)\mathrm{d}t\right) \mathrm{d}u\,\mathrm{d}v\,\nu(\mathrm{d}w).$$

In the above integral I_λ, change the variables $w \to \phi_{\lambda,H}w'$, $u \to \lambda u', v \to \lambda^\gamma v'$ and observe that $\int_0^{\lambda x_j}(\phi_{\lambda,H}w')(t-\lambda u')\mathrm{d}t = \lambda^H \int_0^{x_j} w'(t-u')\mathrm{d}t$ and $\lambda^{\gamma+1}(\nu\circ\phi_{\lambda,H})(\mathrm{d}w') = \nu(\mathrm{d}w')$ according to (20). This yields the equality of characteristic functions: $\mathrm{E}\exp\{\mathrm{i}\sum_{j=1}^m \theta_j J_\nu(\lambda x_j, \lambda^\gamma y_j)\} = \mathrm{E}\exp\{\mathrm{i}\sum_{j=1}^m \theta_j \lambda^H J_\nu(x_j,y_j)\}$.

(iii) Using (20) with $\lambda = x$, similarly as in part (ii), we get that

$$\begin{aligned}\mathrm{E}J_\nu^2(x,y) &= y\int_{\mathbb{R}\times\mathbb{W}} \left(\int_0^x w(t-u)\mathrm{d}t\right)^2 \mathrm{d}u\,\nu(\mathrm{d}w) \\ &= y\int_{\mathbb{R}\times\mathbb{W}} \left(\int_0^x (\phi_{x,H}w)(t-u)\mathrm{d}t\right)^2 \mathrm{d}u\,(\nu\circ\phi_{x,H})(\mathrm{d}w) \\ &= yx^{2H-\gamma}\int_{\mathbb{R}\times\mathbb{W}} \left(\int_0^1 w(t-u)\mathrm{d}t\right)^2 \mathrm{d}u\,\nu(\mathrm{d}w) = C_\nu^2 x^{2H_1} y.\end{aligned} \tag{25}$$

Then, $\mathrm{E}J_\nu(x_1,y_1)J_\nu(x_2,y_1) = (C_\nu^2/2)(x_1^{2H_1} + x_2^{2H_1} - |x_2-x_1|^{2H_1})y_1$ follows from $\mathrm{E}|J_\nu(x_2,y_1) - J_\nu(x_1,y_1)|^2 = \mathrm{E}|J_\nu(x_2-x_1,y_1)|^2$, which holds by $J_\nu(x_i,0) = 0$, $i = 1,2$, and the stationarity of rectangular increments. Finally, we get $\mathrm{E}J_\nu(x_1,y_1)J_\nu(x_2,y_2) = \mathrm{E}J_\nu(x_1,y_1)J_\nu(x_2,y_1)$, $y_1 \le y_2$, by (22) and the independence property of the Poisson random measure on disjoint sets. □

A natural question is what additional conditions the exponent H should satisfy in order that ν and J_ν in Proposition 5 exist. From (24), it is clear that H cannot be arbitrary in general.

Proposition 6 ([159]). *Let ν and J_ν be as in Proposition 5. Then, $0 \le H \le 1+\gamma$. Moreover, if ν satisfies (23), then $0 \le H_1 = H - \frac{\gamma}{2} \le 1$.*

We remark that the proof of the lower bound $H \ge 0$ of Proposition 6 uses the argument from Ref. [230, Lemma 8.2.2] and the property $\mathrm{E}\sup_{0<y<K}|J_\nu(1,y)| < \infty$ $(\forall K > 0)$ of the homogeneous Lévy process [232, Remark 25.19]. The second statement of the proposition is a consequence of (24) and the classical Schoenberg theorem concerning covariance functions [59, Cor. 2.1.1].

Theorem 7 establishes the asymptotic local and global self-similarity of the random process $J_\nu(x) := \{J_\nu(x,1); x > 0\}$ under some additional conditions on (H,γ)-scaling measure ν. Denote $\mathcal{W}(w) := \int_0^\infty w(t)\mathrm{d}t, w \in \mathbb{W}$.

Theorem 7 ([159]). *Let ν be a (H,γ)-scaling measure on $\mathbb{W}$ satisfying (17).*

(i) *Let ν satisfy (23) and $\frac{\gamma}{2} < H \le 1 + \frac{\gamma}{2}$. Then,*

$$\lambda^{-H_1} J_\nu(\lambda x) \xrightarrow{\mathrm{fdd}} C_\nu B_{H_1}(x), \quad \lambda \to 0, \tag{26}$$

where B_{H_1} is an FBM with $H_1 = H - \frac{\gamma}{2}$.

(ii) *Let $\frac{1+\gamma}{2} < H < 1+\gamma$, $\alpha := \frac{1+\gamma}{H}$ and*

$$\int_{\mathbb{W}} \left(|\mathcal{W}(w)| \wedge |\mathcal{W}(w)|^2\right) \nu(\mathrm{d}w) \; < \; \infty. \tag{27}$$

In addition, assume that

$$\int_{\mathbb{W}} |w(t)|^{\alpha'} \nu(\mathrm{d}w) \le C t^{-(\alpha'+\delta)}, \quad t > 0,$$

$$\left(\exists \alpha' \in [1, \alpha),\ 1 > \delta > 1 - \frac{\alpha'}{\alpha}\right). \tag{28}$$

Then,

$$\lambda^{-1/\alpha} J_\nu(\lambda x) \xrightarrow{\mathrm{fdd}} L_\nu(x), \quad \lambda \to \infty, \tag{29}$$

where $L_\nu = \{L_\nu(x); x > 0\}$ *is an* α*-stable Lévy process with ch.f.*

$$\mathrm{E}\mathrm{e}^{\mathrm{i}\theta L_\nu(x)} = \mathrm{e}^{x((\theta)_+^\alpha D_\nu^+ + (\theta)_-^\alpha D_\nu^-)}, \quad \theta \in \mathbb{R},$$

$$D_\nu^\pm := \int_{\mathbb{W}} \{\mathrm{e}^{\pm \mathrm{i}\mathcal{W}(w)} - 1 \mp \mathrm{i}\mathcal{W}(w)\} \nu(\mathrm{d}w). \tag{30}$$

Sketch of the proof of Theorem 7. The proof of part (i) using the characteristic function in (18) is rather easy. Particularly, to show the one-dimensional convergence at a fixed $x > 0$, consider

$$\begin{aligned} I_\lambda(x;\theta) &:= \log \mathrm{E} \exp\{\mathrm{i}\theta\lambda^{-H_1} J_\nu(\lambda x)\} \\ &= \int_{\mathbb{R}\times\mathbb{W}} \Psi\left(\theta\lambda^{-H_1} \int_0^{\lambda x} w(t-u)\mathrm{d}t\right) \mathrm{d}u\, \nu(\mathrm{d}w). \end{aligned}$$

In view of the scaling property in (21), $I_\lambda(x;\theta) = \lambda^{-\gamma} \int_{\mathbb{R}\times\mathbb{W}} \Psi(\theta\lambda^{\gamma/2}$ $\int_0^x w(t-u)\mathrm{d}t)\mathrm{d}u\, \nu(\mathrm{d}w)$, where $\lambda^{-\gamma}\Psi(\lambda^{\gamma/2} z) \to -z^2/2\, (\lambda \to 0)$ and (25),
together with $|\Psi(z)| \le |z|^2/2$ and (Pratt's) lemma 30 of Chapter 2, imply $I_\lambda(x;\theta) \to -(1/2)\int_{\mathbb{R}\times\mathbb{W}} \{\theta \int_0^x w(t-u)\mathrm{d}t\}^2 \mathrm{d}u\, \nu(\mathrm{d}w) = \log \mathrm{E}$ $\exp\{\mathrm{i}\theta C_\nu B_{H_1}(x)\}$.

The proof of part (ii) also uses characteristic functions but is more involved. The central observation is the equality

$$\int_{\mathbb{W}} \Psi(\theta\mathcal{W}(w))\nu(\mathrm{d}w) = (\theta)_+^\alpha D_\nu^+ + (\theta)_-^\alpha D_\nu^- = \log \mathrm{E}\mathrm{e}^{\mathrm{i}\theta L_\nu(1)}, \quad \theta \in \mathbb{R}, \tag{31}$$

relating the above integral to the log ch.f. of the α-stable r.v. $L_\alpha(1)$ in (30), which holds by the (H,γ)-scaling property of ν. Indeed, let

$\theta > 0$. Then, $\theta\mathcal{W}(w) = \mathcal{W}(\phi_{\theta^{1/H},H}w)$ and $\int_{\mathbb{W}}\Psi(\theta\mathcal{W}(w))\nu(\mathrm{d}w) = \int_{\mathbb{W}}\Psi(\mathcal{W}(\phi_{\theta^{1/H},H}w))\nu(\mathrm{d}w) = \theta^{\alpha}\int_{\mathbb{W}}\Psi(\mathcal{W}(w))\nu(\mathrm{d}w) = \theta^{\alpha}D^{+}_{\nu}$ follows by (20); for $\theta < 0$, relation (31) follows analogously. Turning now to the one-dimensional convergence in (29), consider $I_\lambda(x;\theta) := \log\mathrm{E}\exp\{\mathrm{i}\theta\lambda^{-1/\alpha}J_\nu(\lambda x)\} = \int_{\mathbb{R}\times\mathbb{W}}\Psi\big(\theta\lambda^{-1/\alpha}\int_0^{\lambda x}w(t-u)\mathrm{d}t\big)\mathrm{d}u\,\nu(\mathrm{d}w)$. We approximate $I_\lambda(x;\theta)$ by

$$\begin{aligned}\tilde{I}_\lambda(x;\theta) &:= \int_{]0,\lambda x]\times\mathbb{W}}\Psi\left(\theta\lambda^{-1/\alpha}\int_0^\infty w(t)\mathrm{d}t\right)\mathrm{d}u\,\nu(\mathrm{d}w)\\ &= x\int_{\mathbb{W}}\Psi\big(\theta\mathcal{W}(w)\big)\nu(\mathrm{d}w),\end{aligned}$$

where the last equality follows from (31) with θ replaced by $\theta\lambda^{-1/\alpha}$. Particularly, $\tilde{I}_\lambda(x;\theta) = \log\mathrm{E}\mathrm{e}^{\mathrm{i}\theta L_\alpha(x)}$ does not depend on λ, and (29) reduces to the approximation

$$I_\lambda(x;\theta) - \tilde{I}_\lambda(x;\theta) \to 0 \quad (\lambda\to\infty),$$

whose proof in Ref. [159] uses the additional condition (28).

Example 8. The conditions of Theorem 7 can be easily verified for the Telecom process J_α in (10), corresponding to the measure ν_α in (11), for $\gamma = \alpha - 1, H = 1$. By elementary integration, $C^2_{\nu_\alpha} = C\int_{[0,1]^2}|t_1 - t_2|^{1-\alpha}\mathrm{d}t_1\mathrm{d}t_2 < \infty$. Hence, (23) and the assumptions of part (i) are satisfied. Since $\mathcal{W}(w) = \int_0^\infty w(t)\mathrm{d}t = r$ for $w(t) = \mathbb{I}(t \le r)$, (27) is satisfied as well, while (28) holds with $\alpha' = 1, \delta = \alpha - 1$ since $\alpha - 1 > 1 - \frac{1}{\alpha}$. We also have that $D^{\pm}_{\nu_\alpha} = c'\mathrm{e}^{\mp\frac{1}{2}\mathrm{i}\pi\alpha}, c' := c_f\Gamma(2-\alpha)/(\alpha-1)$, and hence the characteristic function in (30) in this example agrees with (14).

Example 9. Let $\nu(\mathrm{d}w)$ be concentrated on 'rectangular' functions, $w(t) = a\,\mathbb{I}(0 < t \le r), (a,r) \in \mathbb{R}^2_+$, and given by

$$\nu(\mathrm{d}a,\mathrm{d}r) := c_\nu\, a^{-\kappa-1}r^{-\varrho-1}\,\mathrm{d}a\,\mathrm{d}r, \quad (a,r)\in\mathbb{R}^2_+, \tag{32}$$

where $c_\nu > 0, \kappa > 0, \varrho > 0$ are parameters satisfying

$$1 < \varrho < \kappa < 2. \tag{33}$$

Condition (17) for the above ν follows by elementary integration. As a consequence, the corresponding RF $J_\nu = \{J_\nu(x,y);(x,y)\in\mathbb{R}^2_+\}$

in (22) is well defined. It turns out that J_ν has κ-stable finite-dimensional distributions and a stochastic integral representation:

$$J_\nu(x,y) \overset{\mathrm{fdd}}{=} \int_{\mathbb{R}\times(0,y]\times\mathbb{R}_+} \left\{ \int_0^x \mathbb{I}(u < t \le r+u)\mathrm{d}t \right\} \times Z_\kappa(\mathrm{d}u, \mathrm{d}v, \mathrm{d}r), \quad (x,y) \in \mathbb{R}^2_+, \tag{34}$$

where $Z_\kappa(\mathrm{d}u, \mathrm{d}v, \mathrm{d}r)$ is a κ-stable random measure on $\mathbb{R} \times \mathbb{R}^2_+$ with ch.f.

$$\mathrm{E}\mathrm{e}^{\mathrm{i}\theta Z_\kappa(B)} = \exp\left\{ ((\theta)^\kappa_+ D^+_\kappa + (\theta)^\kappa_- D^-_\kappa) \int_B \frac{\mathrm{d}u\, \mathrm{d}v\, \mathrm{d}r}{r^{1+\varrho}} \right\}, \tag{35}$$

where $D^\pm_\kappa := c_\nu \int_{\mathbb{R}_+} (\mathrm{e}^{\pm \mathrm{i}a} - 1 \mp ia) a^{-1-\kappa} \mathrm{d}a$. By changing the variables $t \to \lambda_1^{H_1} t$, $u \to \lambda_1^{H_1} u$, $r \to \lambda_1^{H_1} r$, $v \to \lambda_2^{H_2} v$ in the characteristic function of the κ-stable integral on the r.h.s. of (34), with arbitrary $\lambda_i > 0, i = 1, 2$ and

$$H_1 := \frac{1+\kappa-\varrho}{\kappa}, \qquad H_2 := \frac{1}{\kappa},$$

we see that the RF $J_\nu \overset{\mathrm{fdd}}{=} \tilde{J}_\nu$ is (H_1, H_2)-MSS and, hence, also (H,γ)-SS for *any* $\gamma > 0$ with $H = H_1 + \gamma H_2$. Particularly, the random process $\{J_\nu(x,1), x > 0\}$ is H_1-SS and κ-stable. Obviously, this J_ν does not satisfy Theorem 7 (neither (i) nor (ii)) and may serve as a justification for the necessity of additional conditions such as (23) and (27) in this theorem. Example 9 is interesting since it provides a negative answer to the question about the uniqueness of the pair (H,γ) in Definition 4 (i).

4 Limit Results

Consider the shot-noise input process X in (4), with $\{T_j\}, \{W_j\}$ satisfying the stated conditions. Moreover, we assume that $\{T_j\}$ and $\{W_j\}$ are independent, and

$$\int_0^\infty (\mathrm{E}|W(t)| + \mathrm{E}|W(t)|^2)\mathrm{d}t < \infty. \tag{36}$$

Then, $X(t)$ in (4) can be written as

$$X(t) = \int_{\mathbb{R}\times\mathbb{W}} w(t-u)\mathcal{N}(\mathrm{d}u, \mathrm{d}w), \quad t \in \mathbb{R}, \tag{37}$$

where $\mathcal{N}(\mathrm{d}u, \mathrm{d}w)$ is a Poisson random measure on $\mathbb{R}\times\mathbb{W}$ with mean $\mathrm{d}u\mathrm{P}_W(\mathrm{d}w)$. It follows that X is a stationary process with

$$\mathrm{E}X(t) = \int_0^\infty \mathrm{E}W(u)\mathrm{d}u,$$

$$\mathrm{Cov}(X(0), X(t)) = \int_0^\infty \mathrm{E}W(u)W(t+u)\mathrm{d}u, \quad t \geq 0. \tag{38}$$

The centred aggregated workload in (2) is written as the stochastic integral

$$\begin{aligned} &A_{\lambda,\gamma}(x,y) - \mathrm{E}A_{\lambda,\gamma}(x,y) \\ &\quad = \int_{\mathbb{R}\times(0,[\lambda^\gamma y]]\times\mathbb{W}} \left\{\int_0^{\lambda x} w(t-u)\mathrm{d}t\right\} \widetilde{\mathcal{N}}(\mathrm{d}u, \mathrm{d}v, \mathrm{d}w) \end{aligned} \tag{39}$$

w.r.t. the centred Poisson random measure $\widetilde{\mathcal{N}}(\mathrm{d}u, \mathrm{d}v, \mathrm{d}w)$ on $\mathbb{R}\times\mathbb{R}_+\times\mathbb{W}$ with variance $\mathrm{d}u\mathrm{d}v\mathrm{P}_W(\mathrm{d}w)$. We have that

$$\begin{aligned} \mathrm{Var}(A_{\lambda,\gamma}(x,y)) &= [\lambda^\gamma y]\int_{\mathbb{R}} \mathrm{E}\left\{\int_0^{\lambda x} W(t-u)\mathrm{d}t\right\}^2 \mathrm{d}u \\ &= [\lambda^\gamma y]\int_0^{\lambda x}\int_0^{\lambda x} \mathrm{d}t_1\mathrm{d}t_2 \int_{\mathbb{R}} \mathrm{d}u\mathrm{E}W(t_1-u)W(t_2-u) \\ &= 2[\lambda^\gamma y]\int_0^{\lambda x} (\lambda x - t)\,\mathrm{Cov}(X(0), X(t))\mathrm{d}t, \end{aligned} \tag{40}$$

where the last equality follows from (38).

The main result of this section is the following theorem.

Theorem 10 ([159]). *Consider $A_{\lambda,\gamma}(x,y)$, represented as in (39), for some $\gamma > 0$ and a shot-noise process X in (37) satisfying (36).*

(i) *Let*

$$\mathrm{Cov}(X(0), X(t)) \sim c_X t^{-2(1-H_1)}, \quad t \to \infty$$
$$\left(\exists\, c_X > 0,\ H_1 \in \left(\frac{1}{2}, 1\right)\right), \tag{41}$$

and

$$\lambda^{1+\gamma} \int_{\mathbb{R}} \mathrm{E}\left|\lambda^{1-H} \int_0^1 W(\lambda(t-u))\mathrm{d}t\right|^{2+\delta} \mathrm{d}u \to 0,$$
$$\lambda \to \infty \quad (\exists\, \delta > 0), \tag{42}$$

where $H := H_1 + \frac{\gamma}{2}$. *Then, the convergence in* (2) *holds with* $d_{\lambda,\gamma} = \lambda^H$ *and* $V_\gamma = C_W B_{H_1,\frac{1}{2}}$, *where* $B_{H_1,\frac{1}{2}}$ *is an FBS,* $C_W^2 := \frac{c_X}{(2H_1-1)H_1}$.

(ii) *Let* $\alpha \in (1,2)$, *and let the distribution of* $\mathcal{W} := \int_0^\infty W(t)\mathrm{d}t$ *satisfy*

$$\mathrm{P}(\mathcal{W} > x) = (c_+ + o(1))x^{-\alpha},$$
$$\mathrm{P}(\mathcal{W} \le -x) = (c_- + o(1))x^{-\alpha}, \quad x \to \infty, \tag{43}$$

for some $c_\pm \ge 0$, $c_+ + c_- > 0$. *Moreover, let*

$$\mathrm{E}|W(t)|^{\alpha'} \le C(1 \vee t)^{-(\alpha'+\delta)}, \quad t > 0$$
$$\left(\exists\, \alpha' \in [1,\alpha),\ 1 \wedge \delta > \left(1 - \frac{\alpha'}{\alpha}\right)(1+\gamma)\right). \tag{44}$$

Then, the convergence in (2) *holds with* $d_{\lambda,\gamma} = \lambda^H$, $H := \frac{1+\gamma}{\alpha}$ *and* $V_\gamma = L_\alpha$, *where* L_α *is an* α*-stable Lévy sheet.*

(iii) *Assume that there exists a* (H,γ)*-scaling measure* ν_W *on* $\mathbb{W}$ *satisfying* (17) *and such that for any* $\lambda > 1$, *the measure* $\lambda^{1+\gamma}\mathrm{P}_W \circ \phi_{\lambda,H}$ *is absolutely continuous w.r.t.* ν_W *with a bounded Radon–Nikodym derivative,* $g_{\lambda,\gamma,H}(w) := \frac{\mathrm{d}(\lambda^{1+\gamma}\mathrm{P}_W \circ \phi_{\lambda,H})}{\mathrm{d}\nu_W}(w)$, *tending to 1 as* $\lambda \to \infty$:

$$g_{\lambda,\gamma,H}(w) \to 1, \quad |g_{\lambda,\gamma,H}(w)| \le C \quad \text{for } \nu_W - \text{a.e. } w \in \mathbb{W}. \tag{45}$$

Then, the convergence in (2) *holds with* $d_{\lambda,\gamma} = \lambda^H$ *and* $V_\gamma = J_{\nu_W}$, *where the RF* J_{ν_W} *is defined in* (22).

Let us comment on the conditions and proof of Theorem 10. Equation (41) is typical of the LRD of X. Equation (42) is a Lyapunov-type condition, implying asymptotic normality of the sum $A_{\lambda,\gamma}(x,y)$ of $[y\lambda^\gamma]$ i.i.d. inputs. If $H > 1$ and the pulse process is bounded, $\sup_{t\ge 0}|W(t)| \le C$, where $C < \infty$ is non-random, then (42) is automatically satisfied in view of the inequality $(H-1)(2+\delta) > 1+\gamma$ and the definition of H. Hence, the proof of part (i) of Theorem 10 is short and simple.

The proof of part (ii) essentially reduces to an approximation of $A_{\lambda,\gamma}(x,y) - \mathrm{E}A_{\lambda,\gamma}(x,y)$ by the rescaled Lévy sheet $M(\lambda x, \lambda^\gamma y)$ with the Lévy measure $\pi(\mathrm{d}z) = \mathrm{P}(\mathcal{W} \in \mathrm{d}z)$ in the domain of attraction of the α-stable law; see (43). The convergence $\lambda^{-(1+\gamma)/\alpha}M(\lambda x, \lambda^\gamma y) \xrightarrow{\text{fdd}} L_\alpha(x,y)$ is classical, following Ref. [122, proof of Thm. 2.2.2]. The above-mentioned approximation uses characteristic functions and reduces to the convergence

$$I_{\lambda,\gamma}(x) - \tilde{I}_{\lambda,\gamma}(x) \to 0, \quad \lambda \to \infty. \tag{46}$$

for $I_{\lambda,\gamma}(x) := \log \mathrm{E}\exp\{\mathrm{i}\theta\lambda^{-H}(A_{\lambda,\gamma}(x,1) - \mathrm{E}A_{\lambda,\gamma}(x,1)\}$, $\tilde{I}_{\lambda,\gamma}(x) := \lambda^{1+\gamma}x\mathrm{E}\Psi(\theta\lambda^{-H}\mathcal{W})$. Relation (46) is verified by writing $I_{\lambda,\gamma}(x)$ as a sum of two integrals:

$$\begin{aligned} I_{\lambda,\gamma}(x) &= [\lambda^\gamma]\left(\int_0^{\lambda x} + \int_{-\infty}^0\right)\mathrm{E}\Psi\left(\theta\lambda^{-H}\int_0^{\lambda x} W(t-u)\mathrm{d}t\right)\mathrm{d}u \\ &=: I^+_{\lambda,\gamma}(x) + I^-_{\lambda,\gamma}(x) \end{aligned}$$

and showing that $I^-_{\lambda,\gamma}(x)$ is negligible, whereas $I^+_{\lambda,\gamma}(x)$ satisfies (46), using conditions (43) and (44). We note that the last conditions are easily verifiable in concrete situations, as described in Section 5. The examples in Section 5 show that the bounds on α', δ in (44) are sharp and cannot be replaced by smaller quantities in general.

Finally, part (iii) of Theorem 10 is rather simple since the expectation in $I_{\lambda,\gamma}(x)$ can be replaced by an integral over the (H,γ)-scaling measure ν_W, leading to a log-characteristic function of the intermediate RF J_{ν_W}. The conditions in (iii) (particularly, (45)) are much stronger and require a certain scaling property of W.

A natural question is: can we recognise the trichotomy in (3) from Theorem 10? The limits in (i) and (ii) clearly agree with the first two limits in (3), while the limit in (iii) is apparently related to the third limit in (3). Moreover, under the second-moment assumption (23), all three limit RFs in (i)–(iii) are different. (Indeed, the limit in (ii) is different from those in (i) and (iii), as the latter have finite variances, while the limits in (i) and (iii) are different since the distribution in (18) is non-Gaussian by the uniqueness of the Lévy–Khinchine representation.) Obviously, for a given γ, only one of the three possibilities (i)–(iii) may occur. This leads to the two following questions: (a) can all three limits (i)–(iii) occur for the same shot-noise process X in different regions of γ? and if the answer to (a) is affirmative, (b) does there exist a $\gamma_{\rm cr} > 0$ separating these regions as in (3)? For the examples treated in the following section, the answer to (a) and (b) is positive. The following Propositions 11 and 12 show that a positive answer to (b) holds in the general case of Theorem 10 under some additional conditions. Note that conditions (42) and (44) are 'monotone' in γ: indeed, if (42) holds for some $\gamma > 0$, then it is also satisfied for any $\gamma' > \gamma$, as $1+\gamma+(1-H_1-\frac{\gamma}{2})(2+\delta)$ decreases with γ, whereas (44) extends to any $\gamma' < \gamma$ as $(1-\frac{\alpha'}{\alpha})(1+\gamma)$ decreases with γ. Let

$$\gamma^+_{\rm cr} := \inf\{\gamma > 0 :\ (42) \text{ holds}\}, \quad \gamma^-_{\rm cr} := \sup\{\gamma > 0 : (44) \text{ holds}\},$$

with the infimum taken for a fixed $H_1 \in (\frac{1}{2}, 1)$ and some $\delta > 0$ and the supremum for a given $\alpha \in (1,2)$ and some α', δ satisfying (44). Then, $\gamma^-_{\rm cr} \le \gamma^+_{\rm cr}$, and a positive answer to (b) follows if

$$\gamma^+_{\rm cr} = \gamma^-_{\rm cr}. \tag{47}$$

Proposition 11 ([159]). *Let X be a shot-noise process in* (37) *satisfying* (36), (41), (43), *and H_1, α related by*

$$2(1-H_1) = \alpha - 1. \tag{48}$$

In addition, assume that $\sup_{t\ge 0}|W(t)| \le C$ is bounded by a non-random constant, $C < \infty$, and

$$\mathrm{E}|W(t)| \ \le\ C(1 \vee t)^{-\alpha}, \quad t > 0. \tag{49}$$

Then, (47) *is true. As a consequence, the statements (i) and (ii) of Theorem 10 hold for any $\gamma > \gamma_{\rm cr}$ and $\gamma < \gamma_{\rm cr}$, respectively, where $\gamma_{\rm cr} := \gamma^+_{\rm cr} = \gamma^-_{\rm cr} = \alpha - 1$.*

We remark that the additional conditions regarding (49) do not contradict (41) in view of the inequality $|\mathrm{Cov}(X(0), X(t))| = |\int_0^\infty \mathrm{E}[W(u)W(t+u)]\mathrm{d}u| \le C\int_0^\infty \mathrm{E}|W(t+u)|\mathrm{d}u \le C(1\vee t)^{-(\alpha-1)} = C(1\vee t)^{-2(1-H_1)}$, in agreement with (41). Condition (48) holds in the following Examples 13 and 15 and does not hold in Examples 14 and 17, where $\gamma_{\mathrm{cr}} > \alpha - 1$; see Section 5. It seems that Proposition 11 strongly relies on the boundedness of the pulse process W, in which case long memory in X and a stable limit for the integral of X are essentially due to the long duration of the pulse. Proposition 12 applies to the unbounded pulse process yielding (47) and a positive answer to (b) provided no other limits except for (i)–(iii) may occur.

Proposition 12 ([159]). *Assume that X and $(H, \gamma, \nu), \nu := \nu_W$ satisfy Theorem 10 (iii). Moreover, let H, γ, ν satisfy* (23), (27), *and*

$$\frac{1+\gamma}{2} < H < 1 + \frac{\gamma}{2}. \tag{50}$$

Then, such (H, γ) is unique; in other words, the intermediate limit $V_\gamma = J_\nu$ in Theorem 10 (iii) occurs at a single point, $\gamma_{\mathrm{cr}} = \gamma$.

5 Examples

We present four examples of a pulse process, W, satisfying the conditions of Theorem 10.

Example 13 ('Independent transmission rate and duration', or 'continuous flow rewards' model; see Ref. [130]). Let

$$W(t) = A\mathbb{I}(0 < t \le R), \tag{51}$$

where the r.v. $R > 0$ has a density $f(r)$ in (8) and $A > 0$ is an r.v. with $\mathrm{E}A^{2+\delta} < \infty (\exists \delta > 0)$, independent of R. The M/G/∞ queue corresponds to $A = 1$. As in Corollary 1, we have

$$\gamma_{\mathrm{cr}} := \alpha - 1, \quad H := \begin{cases} \frac{3+\gamma-\alpha}{2}, & \gamma > \gamma_{\mathrm{cr}}, \\ 1, & \gamma = \gamma_{\mathrm{cr}}, \\ \frac{1+\gamma}{\alpha}, & 0 < \gamma < \gamma_{\mathrm{cr}}. \end{cases}$$

The above pulse process in (51) satisfies Theorem 10, with $H_1 := \frac{3-\alpha}{2}$ as in (13) and the measure $\nu_W(\mathrm{d}a, \mathrm{d}r) = \mathrm{P}(A \in \mathrm{d}a)\nu_\alpha(\mathrm{d}r)$ concentrated on the 'rectangular' functions $w(t) = a\mathbb{I}(0 < t \le r)$ in Example 9. Note that (50) holds for $\gamma = \gamma_{\mathrm{cr}}$, so that the intermediate limit is unique. The RF J_{ν_W} in this example has a similar form to the Telecom RF in (16) and satisfies Theorem 7 following Example 8. We remark that a violation of condition $\mathrm{E}A^2 < \infty$ may lead to drastically different limits in (2) (a trichotomy different from (3)) [130].

Example 14 ('Deterministically related transmission rate and duration' model, see Ref. [208]). Let

$$W(t) = R^{1-p}\mathbb{I}(0 < t < R^p), \tag{52}$$

where $R > 0$ is an r.v. satisfying (8) and $p \in (0, 1]$ is a (shape) parameter. (For $p = 1$, (52) coincides with (51), $A = 1$.) Let

$$1 < \alpha < 2, \quad \gamma_{\mathrm{cr}} := \frac{\alpha}{p} - 1, \quad H := \begin{cases} \frac{\gamma}{2} + \frac{2+p-\alpha}{2p}, & \gamma > \gamma_{\mathrm{cr}}, \\ \frac{1}{p}, & \gamma = \gamma_{\mathrm{cr}}, \\ \frac{1+\gamma}{\varrho}, & 0 < \gamma < \gamma_{\mathrm{cr}}. \end{cases}$$

Condition (36) holds if $2 - p < \alpha < 2$. The conditions of Theorem 10 can be verified in this example in respective parameter regions. Particularly, for $\gamma > \gamma_{\mathrm{cr}}$, condition (41) is satisfied with $H_1 = \frac{p+2-\varrho}{2p} \in (\frac{1}{2}, 1)$, as

$$\mathrm{Cov}(X(0), X(t)) = t \int_0^\infty \mathrm{E}R^{2(1-p)}\mathbb{I}(R^p > t(1+u))\mathrm{d}u \sim c_X t^{\frac{2-p-\varrho}{p}},$$

where $c_X = \frac{c_\varrho \varrho}{\varrho - 2(1-p)} \int_0^\infty (1+u)^{\frac{2(1-p)-\varrho}{p}} \mathrm{d}u < \infty$. Since $\mathcal{W} = R$ for $W(t)$ in (52), condition (43) holds by (22) of Chapter 4. Condition (44) for any $\gamma < \gamma_{\mathrm{cr}}$ and $1 \le \alpha' < \alpha$ sufficiently close to α holds in view of $\mathrm{E}|W(t)|^{\alpha'} = \mathrm{E}R^{\alpha'(1-p)}\mathbb{I}(R^p > t) \le C(1 \vee t)^{(\alpha'(1-p)-\alpha)\frac{1}{p}}$ since $\delta = \frac{\alpha-\alpha'}{p} > (1+\gamma)(1 - \frac{\alpha'}{\alpha})$ is equivalent to $\gamma < \gamma_{\mathrm{cr}}$. Relation (45) follows as in the previous example, with $\nu_W = \nu_\alpha$ given by (11).

Example 15 ('Exponentially damped transmission rate' model). Let

$$W(t) = \mathrm{e}^{-At}\mathbb{I}(0 < t \le R), \tag{53}$$

where $R > 0$ and $A > 0$ are independent r.v.s with

$$\mathrm{P}(R > r) \sim c_\varrho r^{-\varrho} \quad (r \to \infty), \quad \mathrm{P}(A \le a) \sim c_\kappa a^\kappa \quad (a \to 0) \tag{54}$$

for some positive exponents $\varrho > 0$, $\kappa > 0$ and asymptotic constants $c_\varrho > 0$, $c_\kappa > 0$. Let

$$1 < \varrho + \kappa < 2, \quad \gamma_{\mathrm{cr}} := \varrho + \kappa - 1, \quad H := \begin{cases} \frac{\gamma+3-\varrho-\kappa}{2}, & \gamma > \gamma_{\mathrm{cr}}, \\ 1, & \gamma = \gamma_{\mathrm{cr}}, \\ \frac{1+\gamma}{\varrho+\kappa}, & 0 < \gamma < \gamma_{\mathrm{cr}}. \end{cases} \tag{55}$$

The corresponding distribution P_W is identified with the distribution induced by the mapping $(a, r) \mapsto J(t; a, r) := \mathrm{e}^{-at}\mathbb{I}(0 < t \le r)$, $t > 0$, from $(\mathbb{R}^2_+, P_A \times P_R)$ to $L^1(\mathbb{R}_+)$.

Proposition 16. *Let W be as in (53) and (54), where $\varrho + \kappa \in (1, 2)$. Moreover, for $\gamma = \gamma_{\mathrm{cr}}$, assume that P_R, P_A have densities $f_R(r)$ and $f_A(a)$ satisfying $f_R(r) \le C a^{\varrho-1}, r > 0$, $f_A(a) \le C a^{\kappa-1}, a > 0$, $\lim_{a\to 0} a^{1-\kappa} f_A(a) = \kappa c_\kappa$. Then:*

(i) *conditions (41) and (42) hold for any $\gamma > \gamma_{\mathrm{cr}}$ with $2(1 - H_1) = \varrho + \kappa - 1$ and*

$$\begin{aligned} c_X &= \Gamma(\kappa + 1) c_\kappa c_\varrho c_{\kappa,\varrho}, \\ c_{\kappa,\varrho} &:= \int_0^\infty (1 + z)^{-\varrho}(1 + 2z)^{-\kappa}\mathrm{d}z = \frac{{}_2F_1(\kappa, 1; \kappa + \varrho; -1)}{\kappa + \varrho - 1}, \end{aligned}$$

where ${}_2F_1(a, b; c; z)$ denotes the hypergeometric function;

(ii) *conditions (43) and (44) hold for any $0 < \gamma < \gamma_{\mathrm{cr}}$ with $\alpha = \varrho + \kappa$ and*

$$c_+ := \kappa c_\varrho c_\kappa \int_0^1 (1 - u)^{\kappa+\varrho-1} \left(\log \frac{1}{u}\right)^{-\varrho} \mathrm{d}u < \infty, \quad c_- = 0; \tag{56}$$

(iii) *conditions (17) and (45) hold for $\gamma = \gamma_{\rm cr}$ with the measure ν_W induced by the mapping $J(\cdot;a,r)$ from $(\mathbb{R}^2_+, \nu_A \times \nu_R)$ to $L^1(\mathbb{R}_+)$, with $\nu_R(\mathrm{d}r) = \varrho c_\varrho r^{\varrho-1}\mathrm{d}r$, $\nu_A(\mathrm{da}) := \kappa c_\kappa a^{\kappa-1}\mathrm{da}$;*

(iv) *the above measure ν_W ($\gamma = \gamma_{\rm cr}$) satisfies the conditions in (i)–(ii) of Theorem 7.*

Proof. (i) Let $\alpha = \varrho + \kappa$. Using (54) and (55), we get

$$
\begin{aligned}
&\mathrm{Cov}(X(0), X(t))t^{\alpha-1} \\
&\quad = t^{\alpha-1}\int_0^\infty \mathrm{P}(R > u+t)\mathrm{E}\mathrm{e}^{-A(2u+t)}\mathrm{d}u \\
&\quad = t^{\alpha-1}\int_0^\infty (2u+t)\mathrm{P}(R > u+t)\int_0^\infty \mathrm{P}(A \le z)\mathrm{e}^{-z(2u+t)}\mathrm{d}z\mathrm{d}u \\
&\quad = t^{\alpha}\int_0^\infty (2u+1)\mathrm{P}(R > t(u+1))\int_0^\infty \mathrm{P}(A \le t^{-1}z)\mathrm{e}^{-z(2u+1)}\mathrm{d}z\mathrm{d}u \\
&\quad \sim c_\varrho c_\kappa \int_0^\infty (2u+1)(u+1)^{-\varrho}\int_0^\infty z^\kappa \mathrm{e}^{-z(2u+1)}\mathrm{d}z\mathrm{d}u = c_X
\end{aligned}
$$

from the DCT. Since $|W(t)| \le 1$ in (53), condition (42) is satisfied when $H > 1$ or $\gamma > \gamma_{\rm cr}$.

(ii) Note the integral in (56) converges since $(1-u)^{\kappa+\varrho-1}(\log(1/u))^{-\varrho} \sim (1-u)^{\kappa-1}$ $(u \nearrow 1)$, where $\kappa > 0$. Consider the tail of the d.f. of $\mathcal{W} = \int_0^R \mathrm{e}^{-At}\mathrm{d}t = (1-\mathrm{e}^{-AR})/A$. Then, assuming that $1/x > 0$ is a continuity point of the d.f. of A, we can write

$$
\begin{aligned}
J(x) &:= x^{\varrho+\kappa}\mathrm{P}(\mathcal{W} > x) \\
&= x^{\varrho+\kappa}\int_0^{1/x} \mathrm{P}\left(R > \frac{1}{a}\log\frac{1}{1-ax}\right)\mathrm{dP}(A \le a)
\end{aligned}
$$

as a sum of

$$
J_1(x,\epsilon) := x^{\varrho+\kappa}c_\varrho \int_{\epsilon/x}^{(1-\epsilon)/x} a^\varrho\left(\log\frac{1}{1-ax}\right)^{-\varrho}\mathrm{dP}(A \le a),
$$

$$
J_2(x,\epsilon) := x^{\varrho+\kappa}\int_0^{\epsilon/x} \mathrm{P}\left(R > \frac{1}{a}\log\frac{1}{1-ax}\right)\mathrm{dP}(A \le a),
$$

$$
J_3(x,\epsilon) := x^{\varrho+\kappa}\int_{(1-\epsilon)/x}^{1/x} \mathrm{P}\left(R > \frac{1}{a}\log\frac{1}{1-ax}\right)\mathrm{dP}(A \le a),
$$

and $J_4(x,\epsilon) := J(x) - \sum_{i=1}^{3} J_i(x,\epsilon)$ for a small $\epsilon > 0$. Let us prove that

$$\begin{aligned} \lim\nolimits_{\epsilon\to 0} \limsup\nolimits_{x\to\infty} J_i(x,\epsilon) = 0, \quad i = 2, 3, \\ \lim\nolimits_{x\to\infty} J_4(x,\epsilon) = 0 \quad (\forall\, \epsilon > 0) \end{aligned} \tag{57}$$

and

$$\begin{aligned} J(\epsilon) &:= \lim_{x\to\infty} J_1(x,\epsilon) \\ &= \kappa c_\varrho c_\kappa \int_\epsilon^{1-\epsilon} a^{\kappa+\varrho-1} \left(\log \frac{1}{1-a}\right)^{-\varrho} \mathrm{d}a \qquad (58) \\ &\to \kappa c_\varrho c_\kappa \int_0^1 a^{\kappa+\varrho-1} \left(\log \frac{1}{1-a}\right)^{-\varrho} \mathrm{d}a =: J \quad \text{as } \epsilon \to 0, \end{aligned}$$

where $J = c_+ < \infty$, referring to (56). Relations (57) and (58) imply condition (43) with $c_- = 0$.

Consider $J_2(x,\epsilon)$. Using $\log(1/(1-ax)) \ge ax$ $(0 \le ax < 1)$ and the bounds $\mathrm{P}(R > x) \le Cx^{-\varrho}$ $(\forall x > 0)$, $\mathrm{P}(A \le x) \le Cx^{\kappa}$ $(\forall x > 0)$ (which follow from (54)), we obtain

$$J_2(x,\epsilon) \le Cx^\kappa \mathrm{P}\left(A \le \frac{\epsilon}{x}\right) \le C\epsilon^\kappa \to 0 \quad (\epsilon \to 0).$$

Similarly, using $\log(1/(1-ax)) \ge \log(1/\epsilon)$ $((1-\epsilon)/x < a < 1/x)$, we obtain

$$\begin{aligned} J_3(x,\epsilon) &\le (-\log \epsilon)^{-\varrho} x^{\varrho+\kappa} \int_0^{1/x} a^\varrho \mathrm{dP}(A \le a) \\ &\to (-\log \epsilon)^{-\varrho} \frac{c_\kappa \kappa}{\kappa + \varrho} \quad (x \to \infty), \end{aligned}$$

and subsequently, let $\epsilon \to 0$, thus proving the first relation in (57). The second relation in (57) follows similarly using the uniform convergence $\lim_{x\to\infty} \sup_{a\in[c,d]} x^\varrho |\mathrm{P}(R > ax) - c_\varrho (ax)^{-\varrho}| = 0$ on each compact interval $[c,d] \subset (0,\infty)$.

Let us prove (58). Using condition (54) and integrating by parts, we infer the existence of the following limit as $x \to \infty$:

$$\begin{aligned} J_1(x,\epsilon) &= x^\kappa c_\varrho \Bigg[\mathrm{P}\left(A \le \frac{1-\epsilon}{x}\right) \frac{(1-\epsilon)^\varrho}{\log^\varrho(1/\epsilon)} - \mathrm{P}\left(A \le \frac{\epsilon}{x}\right) \frac{\epsilon^\varrho}{\log^\varrho(1/(1-\epsilon))} \\ &\quad - \int_\epsilon^{1-\epsilon} \mathrm{P}\left(A \le \frac{a}{x}\right) \left(a^\varrho \left(\log \frac{1}{1-a}\right)^{-\varrho}\right)'_a \mathrm{d}a \Bigg] \\ &\to c_\kappa c_\varrho \Bigg[(1-\epsilon)^{\kappa+\varrho} \left(\log \frac{1}{\epsilon}\right)^{-\varrho} - \epsilon^{\kappa+\varrho} \left(\log \frac{1}{1-\epsilon}\right)^{-\varrho} \\ &\quad - \int_\epsilon^{1-\epsilon} a^\kappa \left(a^\varrho \left(\log \frac{1}{1-a}\right)^{-\varrho}\right)'_a \mathrm{d}a \Bigg] = J(\epsilon). \end{aligned}$$

Then, (58) follows by the convergence of the integral J. This proves (43). Relation (44) follows from $\mathrm{E}|W(t)| = \mathrm{E}\mathrm{e}^{-At}\mathbb{I}(R > t) \le C(1 \vee t)^{-\varrho-\kappa}$ since $\delta = \alpha - 1 > 1 + \gamma - H$ is equivalent to $\gamma < \gamma_{\mathrm{cr}}$.

(iii) Follows similarly as in Examples 8 and 13 above.

(iv) As in Example 8, we find that $C^2_{\nu_W} = C \int_{[0,1]^2} |t_1 - t_2|^{1-\kappa-\varrho} \mathrm{d}t_1 \mathrm{d}t_2 < \infty$, and hence (23) holds. Since $H(\gamma_{\mathrm{cr}}) = 1$, we see that all conditions of part (i) are satisfied. Next, since $\mathcal{W}(w) = (1 - \mathrm{e}^{-ar})/a \le (1 \wedge ar)/a$ and $\int_{\mathbb{W}} |w(t)| \nu_W(\mathrm{d}w) = C \int_{\mathbb{R}^2_+} \mathrm{e}^{-at}\mathbb{I}(t \le r) a^{\kappa-1} r^{-\varrho-1} \mathrm{d}a\mathrm{d}r = Ct^{-\kappa-\varrho}$ for $w(t) = \mathrm{e}^{-at}\mathbb{I}(0 < t \le r)$, conditions (27) and (28) hold with $\alpha = \varrho + \kappa \in (1,2), \alpha' = 1$ and $\delta = \alpha - 1 > 1 - \frac{1}{\alpha}$. □

Example 17 ('Brownian pulse' model). Let

$$W(t) = B(t)\mathbb{I}(0 < t \le R), \tag{59}$$

where $B = \{B(t); t \ge 0\}$ is a standard Brownian motion and $R > 0$ is an r.v. independent of B and satisfying (54). Consider the measure space $C(\mathbb{R}_+) \times \mathbb{R}_+$ equipped with $P_B \times P_R$, where P_B is the Wiener measure and P_R is the distribution of the r.v. R. Then, the distribution P_W on $L^1(\mathbb{R}_+)$ can be identified with the distribution induced by the mapping $J : C(\mathbb{R}_+) \times \mathbb{R}_+ \to L^1(\mathbb{R}_+)$, $J(g,r)(t) := g(t)\mathbb{I}(0 < t \le r)$, $t > 0$, $(g,r) \in C(\mathbb{R}_+) \times \mathbb{R}_+$.

Proposition 18 ([159]). *Let W be as in (59), where $2 < \varrho < 3$. Let $\alpha := \frac{2\varrho}{3}$, $H_1 := 2 - \frac{\varrho}{2}$,*

$$\gamma_{\rm cr} := \varrho - 1, \qquad H := \begin{cases} \frac{\gamma}{2} + H_1, & \gamma < \gamma_{\rm cr}, \\ \frac{3}{2}, & \gamma = \gamma_{\rm cr}, \\ \frac{1+\gamma}{\alpha}, & \gamma < \gamma_{\rm cr}. \end{cases} \tag{60}$$

Moreover, for $\gamma = \gamma_{\rm cr}$ assume that P_R has a density of $f_R(r)$, as in Proposition 16. Then:

(i) *conditions* (41) *and* (42) *hold for any $\gamma > \gamma_{\rm cr}$, with $c_X = \frac{c_\varrho}{(\varrho-1)(\varrho-2)}$;*
(ii) *conditions* (43) *and* (44) *hold for any $0 < \gamma < \gamma_{\rm cr}$;*
(iii) *conditions* (17) *and* (45) *hold for $\gamma = \gamma_{\rm cr}$, with the measure ν_W induced by the mapping J on $C(\mathbb{R}_+) \times \mathbb{R}_+$ equipped with the measure $P_B \times \nu_R$, with $\nu_R(\mathrm{d}r)$, $r > 0$, as in* (11)*;*
(iv) *the above measure ν_W ($\gamma = \gamma_{\rm cr}$) satisfies the conditions in (i)–(ii) of Theorem 7.*

6 Notes

1. The beginnings of extensive statistical and probabilistic research on self-similar network traffic can be traced to the seminal work of Leland *et al.* [160]. Further progress in this research area can be found in Park and Willinger [197].

2. A more general version of Corollary 1, case $\gamma \neq \alpha - 1$, appeared in Mikosch *et al.* [185]. The convergence to the Telecom process (case $\gamma = \alpha - 1$) was established in Kaj and Taqqu [130]. The asymptotic self-similarity properties of the Telecom process (Proposition 2) were shown in Gaigalas [92]. The normalised Telecom process $C_\alpha^{-1} J_\alpha$ in (10) is also termed *fractional Poisson motion*, not to be confused with other fractional Poisson processes discussed in the literature (Beghin and Orsinger [28]).

3. Most results in Sections 3 and 4 come from Leipus *et al.* [159]. See the review paper by Colombani *et al.* [60]. The κ-stable RF in (34) appeared in Kaj and Taqqu [130] in the fast connection rate limit for the 'continuous flow reward' model (Example 13), with the reward

distribution $A > 0$ varying regularly at infinity with exponent κ. Pipiras and Taqqu [214] provide useful criteria for the convergence of stochastic integrals with respect to the Poisson random measures, which can also be applied to show the convergence towards the intermediate limit in Theorem 10 (iii).

4. The α-stable limit in Theorem 10 (ii) extends to $\gamma = 0$ or the convergence

$$\lambda^{-1/\alpha} \int_0^{\lambda t} (X(s) - \mathrm{E}X(s))\mathrm{d}s \xrightarrow{\mathrm{fdd}} L_\alpha(\boldsymbol{t})$$

of the shot-noise input process X in (4) towards an α-stable Lévy process L_α with independent increments. The central condition (43) for the above convergence implies that $\mathcal{W} = \int_0^\infty W(s)\mathrm{d}s$ is an r.v. with a heavy-tailed distribution. Gaussian and non-Gaussian scaling limits of LRD shot-noise processes with a deterministic or other pulse processes were discussed in several works; see Giraitis and Surgailis [100, 101], Cioczek-Georges and Mandelbrot [64, 65], Klueppelberg and Kuehn [135], Klueppelberg and Matsui [136], Iksanov *et al.* [124], and the references therein.

Chapter 9

Aggregation of Regenerative Inputs

This chapter deals with the aggregation problem in (1)–(2) of Chapter 8 for regenerative input processes X having the form

$$X(t) = \sum_{j} W_j(t - T_{j-1})\mathbb{I}(T_{j-1} < t < T_j), \quad t \in \mathbb{R}, \tag{1}$$

somewhat similar to the shot-noise input in (4) of Chapter 8, except that $\{T_j\}$ in (1) is a stationary renewal process with heavy-tailed inter-renewal intervals $Z_j = T_j - T_{j-1}$ and 'pulses' W_j in (1) are generally not independent of Z_j. The crucial regeneration property means that the 'cycles' $\{X(t + T_j); 0 \le t < Z_j\} = \{W_j(t); 0 \le t < Z_j\}$ between random times T_j are independent and identically distributed, so that X 'regenerates anew' at each T_j. Regeneration occurs in many queueing systems at times T_j when a new arrival finds the system empty. Dependence and distributional properties of regenerative inputs in (1) are more delicate and difficult to study than those of shot-noise inputs since they involve both Z_j and $\{W_j(t)\}_{t\in[0,Z_j)}$. On the other hand, the former class of inputs is more interesting for applications since it contains many queueing models with limited service capacity.

Let us briefly describe the contents of this chapter. As in most related works, we assume that the generic distribution $Z \overset{\mathrm{d}}{=} Z_j$ of inter-renewal intervals is heavy-tailed:

$$\mathrm{P}(Z > x) \sim c_Z x^{-\alpha}, \qquad x \to \infty, \tag{2}$$

with some constants $\alpha \in (1, 2), c_Z > 0$. The condition $\alpha > 1$ guarantees that $\mathrm{E}Z < \infty$, which is necessary for the stationarity of $\{T_j\}$, while $\alpha < 2$ implies that $\{T_j\}$ is LRD [93, 185]. The most widely studied case of such processes X, from the point of view of the limits in (2), is the ON/OFF model with $W(t) = \mathbb{I}(t < Z^{\mathrm{on}})$ and $Z = Z^{\mathrm{on}} + Z^{\mathrm{off}}$, where Z^{on} and Z^{off} are (random) durations of ON and OFF intervals, respectively. As noted in Ref. [185], the limit results for the heavy-tailed ON/OFF and infinite-source Poisson inputs have striking similarities, with the limits in (2) of Chapter 8 being virtually the same for both models, including the critical γ_{cr} and the 'intermediate' Telecom RF arising at $\gamma = \gamma_{\mathrm{cr}}$. Related results were obtained in Refs. [81, 130]. For more general pulse processes as in the current chapter, some differences between limit results for classes (4) and (1) of inputs appear, which are especially notable in Examples 14 of Chapter 8 and 23. The main results in this chapter are Theorems 10, 14, and 15, providing the conditions on the regenerative inputs for fast/slow/intermediate limits in (3) of Chapter 8, together with examples of regenerative processes satisfying all these conditions.

1 Regenerative Process: Definition, Stationarity, and Covariance Function

Let $Z > 0$ be an r.v. and $\{W(t)\}_{t \in [0, Z)}$ be a real-valued process with $\int_0^Z |W(t)| \mathrm{d}t < \infty$ a.s., defined on the same probability space. Let

$$(Z_j, \{W_j(t)\}_{t \in [0, Z_j)}), \quad j = 1, 2, \ldots, \tag{3}$$

be consecutive cycles (i.i.d. copies of a generic cycle $(Z, \{W(t)\}_{t \in [0, Z)})$). We shall refer to Z_j as the length (duration) and $\{W_j(t)\}_{t \in [0, Z_j)}$ as the pulse (cycle) process of the jth cycle. Throughout this section, we assume

$$\mu := \mathrm{E}Z < \infty. \tag{4}$$

Let the initial cycle $(Z_0, \{W_0(t)\}_{t \in [0, Z)})$ be independent of (3). Define $T_0 := Z_0, T_n := \sum_{i=0}^n Z_i, n \geq 1$, and consider a regenerative process $X = \{X(t); t \geq 0\}$, defined as

$$X(t) := W_0(t)\mathbb{I}(t < T_0) + \sum_{j=1}^{\infty} W_j(t - T_{j-1})\mathbb{I}(T_{j-1} \leq t < T_j), \quad t \geq 0. \tag{5}$$

Thus, X is composed of independent cycles and regenerates at each time T_j of the renewal process $\{T_j; j \geq 0\}$ with inter-renewal distribution Z and delay time $T_0 \geq 0$. Let $X^0 = \{X^0(t); t \geq 0\}$ denote the zero-delayed process defined as in (5) with $T_0 = 0$. The regeneration property of (5) implies that for any $n = 1, 2, \ldots,$

$$\{X(t + T_n); t \geq 0\} \stackrel{\text{fdd}}{=} X^0$$

and

$$\{X(t + T_n); t \geq 0\} \quad \text{is independent of} \quad \{X(t); 0 \leq t < T_n\}.$$

As noted before, the class of regenerative processes is very large and includes many queueing models. We present three examples of such processes discussed in the literature.

Example 1 (ON/OFF process). Let $Z = Z^{\text{on}} + Z^{\text{off}}, W(t) = \mathbb{I}(Z^{\text{on}} > t)$, where $Z^{\text{on}} > 0$ and $Z^{\text{off}} > 0$ are the respective durations of ON (busy) and OFF (idle) intervals, which can be mutually dependent. See Refs. [115,185]. This process arises in queueing models with a single server, where $X(t)$ represents the number of customers served at time t.

Example 2 (Workload process). Let $Z = Z^{\text{on}} + Z^{\text{off}}$ as in the ON/OFF process and $W(t) := (Z^{\text{on}} - t)_+$. Then, $X(t)$ in (5) is the forward recurrence time of the busy period. In the G/G/1/0 queue, the process $X(t)$ represents the current workload in the system.

Example 3 (Renewal–reward process). Let $W(t) = W\mathbb{I}(t \in [0, Z))$, where $W \in \mathbb{R}$ is a (random) constant that can be dependent or independent of duration $Z > 0$. See Ref. [213] and the references therein.

Let us turn to the properties of the regenerative process. Since we are dealing with stationary input processes, the first question is about the stationarity of (5), or the distribution of the initial cycle $(Z_0, \{W_0(t)\}_{t\in[0,Z_0)})$, guaranteeing the stationarity of (5). Somewhat surprisingly, this question admits a simple and elegant answer. In the literature, this 'initial distribution' is usually obtained by a random shift of a 'length-biased' generic distribution $(Z, \{W(t)\}_{t\in[0,Z)})$; see the monograph by [263, Chapter 10.2.1]. In what follows, we use a differential form of the 'initial distribution', namely

$$\begin{aligned} &\mathrm{P}(W_0(\cdot) \in A, Z_0 \in \mathrm{d}s_0) \\ &\quad := \mu^{-1} \int_{-\infty}^{0} \mathrm{d}s_{-1} \mathrm{P}(W(\cdot - s_{-1}) \in A, Z + s_{-1} \in \mathrm{d}s_0), \end{aligned} \tag{6}$$

for any $s_0 > 0$ and any (Borel measurable) set $A \subset L^1([0, s_0))$.

From (6), other probabilities involving random Z_0 can be obtained by integrating over s_0. Particularly, (6) yields the well-known distribution of $T_0 = Z_0$ in a stationary renewal process:

$$\mathrm{P}(T_0 > t) = \mu^{-1} \int_t^{\infty} \mathrm{P}(Z > s)\mathrm{d}s, \quad t \geq 0. \tag{7}$$

The proof of (6) and other results in this section use the renewal process

$$N(t) := \sum_{j=0}^{\infty} \mathbb{I}(T_j \in [0, t]), \quad t \geq 0, \tag{8}$$

and the *renewal function*

$$U(t) := \mathrm{E}[N(t)|T_0 = 0] = \mathrm{E}N_0(t), \quad t \geq 0, \tag{9}$$

where

$$N_0(t) := \sum_{j=0}^{\infty} \mathbb{I}(S_j \leq t), \quad S_j := \sum_{k=1}^{j} Z_k, S_0 := 0$$

is the pure renewal process starting at 0. The renewal function can be extended on $\mathbb{R}$ by setting $U(t) := 0\,(t < 0)$, as non-decreasing and identifiable with an (infinite) measure on $\mathbb{R}$. The convolution $G_1 \star G_2$ of two right-continuous non-decreasing functions (measures) $G_i, i = 1, 2$, is defined as $(G_1 \star G_2)(t) = \int_{\mathbb{R}} G_1(t-s)G_2(\mathrm{d}s)$, or the measure $(G_1 \star G_2)(A) = \int_{\mathbb{R}^2} \mathbb{I}(t+s \in A)G_1(\mathrm{d}t)G_2(\mathrm{d}s), A \in \mathcal{B}(\mathbb{R})$. With this notation, the renewal function is written as

$$U(t) = \sum_{j=0}^{\infty} F^{\star j}(t), \quad t \geq 0, \tag{10}$$

where $F^{\star j}(t), t \geq 0$, denote the jth-fold convolution of the d.f. $F(t) = \mathrm{P}(Z \leq t)$. Since U has an atom at $t = 0$ $U(\mathrm{d}0) = U(0) - U(0-) = 1$ and may have other atoms in $(0, \infty)$, the (Lebesgue–Stieltjes) integral $\int_A h(s)U(\mathrm{d}s)$ needs some care; in particular, the reader should distinguish between integrals over $A = [0, t]$ and $A = (0, t]$. According to (10), the increment $U(\mathrm{d}s)$ can be interpreted as the probability that N_0 has a point $T_j \in \mathrm{d}s$: $U(\mathrm{d}s) = \mathrm{P}(\exists j \geq 0 : T_j \in \mathrm{d}s) = \sum_{j=0}^{\infty} \mathrm{P}(T_j \in \mathrm{d}s)$. The well-known fact

$$\mathrm{E}N(t) = \mu^{-1}t \tag{11}$$

for the stationary renewal process in (8) follows from (10) and (7); see Remark 4.

Proof of (6). Let $\theta_\tau x := \{x(\tau + s); s \in [0, \infty)\}$ denote the shift of a real-valued measurable function, $x = \{x(s); s \in [0, \infty)\}$, and we write $\theta_\tau B$ for the corresponding shift of a measurable subset B in the space of measurable functions $x : \mathbb{R}_+ \to \mathbb{R}$. In view of the regeneration property, it suffices to prove that any shift by $\tau > 0$ does not change the distribution of the initial cycle given by (6) or that

$$\mathrm{P}(\theta_\tau B) = \mathrm{P}(B) = \mu^{-1} \int_0^{\infty} \mathrm{P}(W(\boldsymbol{t} + s) \in A, Z > t_0 + s)\mathrm{d}s, \tag{12}$$

for any cylinder set $B = \{W(\boldsymbol{t}) \in A, T_0 > t_0\}$, where $W(\boldsymbol{t}) := (W(t_1), \ldots, W(t_m))$, $A \subset \mathcal{B}(\mathbb{R}^m)$, $\boldsymbol{t} := (t_1, \ldots, t_m)$, $0 \leq t_1 < \cdots < t_m < t_0 < \infty$; $W(\boldsymbol{t}+s) := (W(t_1+s), \ldots, W(t_m+s))$, $s \geq 0$. To show (12), split $\mathrm{P}(\theta_\tau B) = \mathrm{P}(W(\boldsymbol{t} + \tau) \in A, T_0 > \tau + t_0) + \mathrm{P}(W(\boldsymbol{t} + \tau) \in$

$A, T_0 \le \tau + t_0) =: p_0(\tau) + p_1(\tau)$. According to (6),

$$
\begin{aligned}
p_0(\tau) &= \mu^{-1} \int_{-\infty}^{0} \mathrm{P}(W(\boldsymbol{t} + \tau - s_{-1}) \in A, Z > t_0 + \tau - s_{-1}) \mathrm{d}s_{-1} \\
&= \mu^{-1} \int_{\tau}^{\infty} \mathrm{P}(W(\boldsymbol{t} + s) \in A, Z > t_0 + s) \mathrm{d}s. \qquad (13)
\end{aligned}
$$

Using the regeneration property, (7), and a change of variables, it follows that

$$
\begin{aligned}
p_1(\tau) &= \sum_{j=1}^{\infty} \mathrm{P}(W(\boldsymbol{t} + \tau) \in A, T_0 - \tau > t_0, N(0, \tau] = j) \\
&= \sum_{j=1}^{\infty} \int_{0 < s_0 \le s_1 \le \tau} \mathrm{P}(T_0 \in \mathrm{d}s_0) F^{\star(j-1)}(\mathrm{d}s_1 - s_0) \\
&\quad \times \mathrm{P}(W(\boldsymbol{t} + \tau - s_1) \in A, Z > t_0 + \tau - s_1) \\
&= \mu^{-1} \int_{0 < s_0 \le s_1 \le \tau} \mathrm{d}s_0 \, \mathrm{P}(Z > s_0) \, U(\mathrm{d}s_1 - s_0) \\
&\quad \times \mathrm{P}(W(\boldsymbol{t} + \tau - s_1) \in A, Z > t_0 + \tau - s_1) \\
&= \mu^{-1} \int_{0}^{\tau} \mathrm{P}(W(\boldsymbol{t} + \tau - x) \in A, Z > t_0 + \tau - x) \mathrm{d}x \\
&\quad \times \int_{[0,x]} U(\mathrm{d}y) \mathrm{P}(Z > x - y), \qquad (14)
\end{aligned}
$$

where

$$
\int_{[0,x]} U(\mathrm{d}y) \mathrm{P}(Z > x - y) = \mathrm{P}(N_0(x) \ge 1) = 1 \quad (\forall\, x > 0); \qquad (15)
$$

see Ref. [89, Chapter 11, (1.13)] and Remark 4 later. From (14) and (15), we conclude that

$$
p_1(\tau) = \mu^{-1} \int_{0}^{\tau} \mathrm{P}(W(\boldsymbol{t} + s) \in A, Z > t_0 + s) \mathrm{d}s. \qquad (16)
$$

Clearly, (12) follows from (13) and (16). □

Remark 4. (15) is a complete probability formula:

$$\begin{aligned}
&\int_{[0,x]} U(\mathrm{d}y)\mathrm{P}(Z > x - y) \\
&\quad = \mathrm{P}(Z > x) + \int_{0<y\leq x} \mathrm{P}(Z \in \mathrm{d}y)\mathrm{P}(Z > x - y) \\
&\qquad + \int_{0<y_1<y_2\leq x} \mathrm{P}(Z_1 \in \mathrm{d}y_1, Z_2 \in \mathrm{d}y_2 - y_1)\mathrm{P}(Z > x - y_2) + \cdots \\
&\quad = \mathrm{P}(N_0(x) \geq 1) = 1.
\end{aligned}$$

The change of variables used in (14) can be written more explicitly as follows (with any bounded functions $h(s), g(s), s \geq 0$ in place of the probabilities in the integral in (14)):

$$\begin{aligned}
&\int_{0<s_0\leq s_1<\tau} \mathrm{d}s_0\, h(s_0)\, U(\mathrm{d}s_1 - s_0) g(\tau - s) \\
&\quad = \int_0^\tau h(\tau - x)\mathrm{d}x \int_{[0,x]} U(\mathrm{d}y) g(x - y) \\
&\quad = \int_{[0,\tau]} U(\mathrm{d}y) \int_y^\tau h(\tau - x) g(x - y)\mathrm{d}x \qquad \left(\int_\tau^\tau \cdots \mathrm{d}x = 0\right) \\
&\quad = \int_{[0,\tau]} U(\mathrm{d}y) \int_y^\tau h(x' - y) g(\tau - x')\mathrm{d}x \qquad (x' := \tau + y - x) \\
&\quad = \int_0^\tau g(\tau - x)\mathrm{d}x \int_{[0,x]} U(\mathrm{d}y) h(x - y).
\end{aligned}$$

Applying the above equality with $h(s) = \mathrm{P}(Z > s), g(s) = 1$ and using (7), (9), (10), and (15), we get that

$$\begin{aligned}
\mathrm{E}N(t) &= \mu^{-1} \int_0^t \mathrm{d}s_0 U(t - s_0) \\
&= \mu^{-1} \int_{0<s_0\leq s_1\leq t} \mathrm{d}s_0 \mathrm{P}(Z > s_0) U(\mathrm{d}s_1 - s_0) \\
&= \mu^{-1} \int_0^t \mathrm{d}x \int_{[0,x]} U(\mathrm{d}y)\mathrm{P}(Z > x - y) = \mu^{-1} t.
\end{aligned}$$

Let us turn now to the dependence properties of a regenerative process. The following lemma gives the expressions for the moments and the covariance function of a stationary process in (5) in terms of a generic distribution $(Z, \{W(t)\}_{t\in[0,Z)})$. Define

$$\begin{aligned} G^0(t) &:= \mathrm{E}[W(Z-t)\mathbb{I}(t<Z)], \quad G^1(t) := \mathrm{E}[W(t)\mathbb{I}(t<Z)], \\ z(t) &:= \int_0^t G^0(s)G^1(t-s)\mathrm{d}s, \quad t\ge 0. \end{aligned} \tag{17}$$

Lemma 5. *Let $X = \{X(t); t\ge 0\}$ be a stationary regenerative process in (5)–(4) with initial distribution as in (6).*
(i) For any $p>0, t\ge 0$, we have that

$$\begin{aligned} \mathrm{E}|X(t)|^p &= \mu^{-1}\int_0^\infty \mathrm{E}[|W(s)|^p\mathbb{I}(s<Z)]\mathrm{d}s \\ &= \mu^{-1}\int_0^\infty \mathrm{E}[|W(Z-s)|^p\mathbb{I}(s<Z)]\mathrm{d}s, \end{aligned} \tag{18}$$

provided the r.h.s. of (18) is finite. Particularly, if (18) holds with $p=1$, then

$$\mathrm{E}X(t) = \mu^{-1}\int_0^\infty G^1(s)\mathrm{d}s = \mu^{-1}\int_0^\infty G^0(s)\mathrm{d}s. \tag{19}$$

(ii) Assume additionally that

$$\mathrm{E}|X(t)|^2 = \mu^{-1}\int_0^\infty \mathrm{E}[|W(s)|^2\mathbb{I}(s<Z)]\mathrm{d}s < \infty. \tag{20}$$

Then,

$$\mathrm{Cov}(X(0), X(t)) = R(t) + h(t), \quad t\ge 0, \tag{21}$$

where

$$\begin{aligned} R(t) &:= \mathrm{E}[X(0)X(t)\mathbb{I}(t<T_0)] \\ &= \mu^{-1}\int_0^\infty \mathrm{E}[W(s)W(s+t)\mathbb{I}(Z>t+s)]\mathrm{d}s, \\ h(t) &:= \mu^{-1}\int_{[0,t]} z(t-s)U(\mathrm{d}s) - (\mathrm{E}X(0))^2, \end{aligned} \tag{22}$$

and U *is the renewal function in* (9). *Moreover,*

$$m := \int_0^\infty z(s)\mathrm{d}s = (\mu \mathrm{E}X(0))^2. \tag{23}$$

Proof. (i) By stationarity and (6), $\mathrm{E}|X(t)|^p = \mathrm{E}|X(0)|^p = \mathrm{E}|W_0(0)|^p = \mu^{-1}\int_0^\infty \mathrm{E}|W(s)|^p \mathbb{I}(s < Z)\mathrm{d}s$, proving (18) and (19) being completely analogous.

(ii) Relations (20) and (18) imply that $\int_0^\infty |G^i(s)|^p \mathrm{d}s < \infty, p = 1, 2, i = 0, 1$; in particular,

$$\int_0^\infty |G^1(s)|^2 \mathrm{d}s = \int_0^\infty (\mathrm{E}[W(s)\mathbb{I}(s < Z)])^2 \mathrm{d}s$$
$$\leq \int_0^\infty \mathrm{E}[|W(s)|^2 \mathbb{I}(s < Z)\mathrm{d}s = \mu \mathrm{E}|X(0)|^2.$$

Therefore, the functions $z(t)$ and $R(t)$ in (22) are well defined and bounded, i.e.,

$$|z(t)| \leq \mu \mathrm{E}|X(0)|^2, \quad |R(t)| \leq \mathrm{E}|X(0)|^2, \quad t \geq 0.$$

Relation (23) follows from (19). Note that $\mathrm{E}[W_0(0)\mathbb{I}(T_0 \in \mathrm{d}s)] = G^0(s)\mathrm{d}s$ by (6). Then, using the regeneration property or independence of 'tours' in (3),

$$\begin{aligned}
\mathrm{E}[X(0)X(t)] &= \mathrm{E}[X(0)X(t)\mathbb{I}(t < T_0)] \\
&\quad + \sum_{j=0}^\infty \mathrm{E}[X(0)X(t)\mathbb{I}(T_j \leq t < T_{j+1})] \\
&= R(t) + \sum_{j=0}^\infty \int_{0<s_0\leq s_1<t} \mathrm{E}[W_0(0)\mathbb{I}(T_0 \in \mathrm{d}s_0)]F^{\star j} \\
&\quad \times (\mathrm{d}s_1 - s_0)\mathrm{E}[W(t-s_1)\mathbb{I}(s_1 < t)] \\
&= R(t) + \int_{0<s_0\leq s_1<t} G^0(s_0)\mathrm{d}s_0 U(\mathrm{d}s_1 - s_0)G^1(t - s_1) \\
&= R(t) + \int_{[0,t]} z(t-s)U(\mathrm{d}s),
\end{aligned}$$

where the last equality follows from the change of variable as in (17). □

Example 6 (Covariance function of ON/OFF process; Example 1). Denote $F^{\rm on}(t) := {\rm P}(Z^{\rm on} \le t)$, $F^{\rm off}(t) := {\rm P}(Z^{\rm on} \le t)$, $F(t) = {\rm P}(Z \le t)$, $Z = Z^{\rm on} + Z^{\rm off}$, $\bar F^{\rm on}(t) := 1 - F^{\rm on}(t)$, $\bar F^{\rm off}(t) := 1 - F^{\rm off}(t)$, $\bar F(t) := 1 - F(t)$, $t \ge 0$, $\mu = \mu_{\rm on} + \mu_{\rm off}$, $\mu_{\rm on} := {\rm E}Z^{\rm on}$, $\mu_{\rm off} := {\rm E}Z^{\rm off}$. Then, $G^1(t) = \bar F^{\rm on}(t)$ and $G^0(t) = {\rm P}(Z^{\rm off} < t < Z^{\rm on} + Z^{\rm off}) = {\rm P}(Z > t) - {\rm P}(Z > t, Z^{\rm off} \ge t) = \bar F(t) - {\rm P}(Z^{\rm off} \ge t) = \bar F(t) - \bar F^{\rm off}(t)$ at each continuity point $t \ge 0$ of the p.d.f. $F^{\rm on}(t)$ (i.e., a.e. on $\mathbb{R}_+$). Therefore,

$$z(t) = \int_0^t \bar F^{\rm on}(s) \bar F(t-s){\rm d}s - \tilde z(t), \quad \text{where}$$

$$\tilde z(t) := \int_0^t \bar F^{\rm off}(s) \bar F^{\rm on}(t-s){\rm d}s. \tag{24}$$

The generic distribution $(Z, \{W(t)\}_{t\in[0,Z)})$ in this example can be identified with the joint distribution of $(Z^{\rm on}, Z), 0 < Z^{\rm on} < Z < \infty$ and the initial distribution with the distribution $(Z_0, \{W_0(t)\}_{t\in[0,Z_0)})$, where the distribution of $(Z^{\rm on}_0, Z_0), 0 \le Z^{\rm on}_0 < Z_0 < \infty$ is given by

$$\begin{aligned}
&{\rm P}(Z^{\rm on}_0 = 0, Z_0 \in {\rm d}s_0) = \mu^{-1} \int_0^\infty {\rm P}(Z^{\rm on} \le s, Z \in {\rm d}s_0 + s){\rm d}s, \\
&{\rm P}(Z^{\rm on}_0 \in {\rm d}s^{\rm on}_0, Z_0 \in {\rm d}s_0) \\
&\quad = \mu^{-1} \int_0^\infty {\rm P}(Z^{\rm on} \in {\rm d}s^{\rm on}_0 + s, Z \in {\rm d}s_0 + s){\rm d}s, \quad 0 < s^{\rm on}_0 < s_0.
\end{aligned}$$

Accordingly, ${\rm P}(Z^{\rm on}_0 \in {\rm d}s^{\rm on}_0) = \mu^{-1}\int_0^\infty {\rm P}(Z^{\rm on} \in {\rm d}s^{\rm on}_0 + s){\rm d}s$, $(s^{\rm on}_0 > 0)$, ${\rm P}(Z^{\rm on}_0 > t) = \mu^{-1}\int_0^\infty {\rm P}(Z^{\rm on} > t + s){\rm d}s$ $(t > 0)$, ${\rm P}(Z^{\rm on}_0 = 0) = \mu^{-1}\int_0^\infty {\rm P}(Z^{\rm on} \le s, Z > s){\rm d}s$, where ${\rm P}(Z^{\rm on} \le s, Z > s) = F^{\rm on}(s) - F(s) = \bar F(s) - \bar F^{\rm on}(s)$ and

$${\rm P}(Z^{\rm on}_0 = 0) = \mu^{-1}\int_0^\infty (\bar F(s) - \bar F^{\rm on}(s)){\rm d}s = \mu^{-1}(\mu - \mu_{\rm on}) = \frac{\mu_{\rm off}}{\mu}.$$

Hence, the initial or stationary distribution of the ON/OFF process is given by

$$\begin{aligned}
{\rm P}(X(0) = 0) &= {\rm P}(Z^{\rm on} = 0) = \frac{\mu_{\rm off}}{\mu}, \\
{\rm P}(X(0) = 1) &= {\rm P}(Z^{\rm on} > 0) = \frac{\mu_{\rm on}}{\mu}.
\end{aligned}$$

Therefore, $\mathrm{E}X(0) = \mu_{\text{on}}/\mu$ and $m = (\mu \mathrm{E}X(0))^2 = \mu_{\text{on}}^2$. We also obtain that

$$R(t) = \mathrm{P}(Z_0^{\text{on}} > t) = \mu^{-1} \int_0^\infty \bar{F}^{\text{on}}(t+s)\mathrm{d}s, \quad t \geq 0. \qquad (25)$$

Using (24) and (25), equation (21) for the covariance function of the ON/OFF process can be rearranged in a more convenient form as follows:

$$\mathrm{Cov}(X(0), X(t)) = \frac{\mu_{\text{on}}\mu_{\text{off}}}{\mu^2} - \frac{1}{\mu} \int_{[0,t]} \tilde{z}(t-s)U(\mathrm{d}s), \quad t \geq 0. \qquad (26)$$

Indeed, the last observation amounts to checking the identity

$$R(t) + \frac{1}{\mu} \int_{[0,t]} \left\{ \int_0^{t-s} \bar{F}^{\text{on}}(t-s-u)\bar{F}(u)\mathrm{d}u \right\} U(\mathrm{d}s)$$

$$= \frac{\mu_{\text{on}}\mu_{\text{off}}}{\mu^2} + \frac{\mu_{\text{on}}^2}{\mu^2} = \frac{\mu_{\text{on}}}{\mu}, \quad t \geq 0$$

or

$$\int_{[0,t]} \left\{ \int_0^{t-s} \bar{F}^{\text{on}}(t-s-u)\bar{F}(u)\mathrm{d}u \right\} U(\mathrm{d}s) = \int_0^t \bar{F}^{\text{on}}(s)\mathrm{d}s. \qquad (27)$$

The l.h.s. of (27) can be written as $(\bar{F}^{\text{on}} \star \bar{F} \star U)(t)$, where $(\bar{F} \star U)$ $(t) = 1$ (see (17)), implying (27) and (26). Note that *the expression* (26) *is symmetric w.r.t.the exchange of ON and OFF distributions*, a fact which also follows from the observation that $\tilde{X}(t) := 1 - X(t), t \geq 0$ is a stationary ON/OFF process with $\tilde{Z}^{\text{on}} := Z^{\text{off}}, \tilde{Z}^{\text{off}} := Z^{\text{on}}$ and has the same covariance function as the process X.

Further examples of covariance functions of regenerative processes are discussed in the following section.

2 Gaussian Convergence

This section studies Gaussian (FBS) limits ($\gamma > \gamma_{\text{cr}}$) in (2) of Chapter 8 for the stationary regenerative inputs in (5) satisfying additional conditions. Essentially, these conditions ensure the convergence of the variance or the regular decay of the covariance function in (21), as in the case of shot-noise inputs. From (21), we see that $\mathrm{Cov}(X(0), X(t)) = R(t) + h(t)$ is the sum of two terms and its LRD property may be due to one (either $R(t)$ or $h(t)$ in (22)) or to both of them. The term $R(t)$ is simpler than that in the shot-noise case and similar to it, essentially reflecting the LRD property due to the long initial renewal interval $T_0 > t$. The term $h(t)$ reflects the behaviour after T_0 and has a more complicated structure involving the renewal function $U(t)$. Under the regular variation condition (2) of Z, the behaviour of $h(t)$ can be analysed through renewal methods. The *elementary renewal theorem* states that $U(t) \sim \mu^{-1}t$ $(t \to \infty)$, $\mu = \mathrm{E}Z < \infty$, and its refinement (the *key renewal theorem*) extends this result to integrals $\int_{[0,t]} z(t-s)U(\mathrm{d}s)$, viz.,

$$\int_{[0,t]} z(t-s)U(\mathrm{d}s) \to \mu^{-1} \int_0^\infty z(s)\mathrm{d}s \quad (t \to \infty), \tag{28}$$

under additional assumptions on $z(t)$ and $F(t)$; see Refs. [6,89]. The key renewal theorem is one of the fundamental facts of probability theory. Application of (28) to $h(t)$ in (22) yields $h(t) \to 0$, which is insufficient for our purposes. The rate of convergence in (28) in the case of heavy-tailed (2) was obtained in Ref. [115], which is used in the subsequent discussion. Following Ref. [115, Thm. 3.1], we impose the following regularity condition on $z(t)$ and $F(t)$:

$$z(t) \text{ is non-negative, right continuous, and has bounded variation on } [0,\infty),\ \lim_{t\to\infty} z(t) = 0 \tag{29}$$

and

$$F^{j\star}(t) \text{ is nonsingular for some } j \geq 1. \tag{30}$$

Theorem 7 ([115]). *Let* (2), (29), *and* (30) *hold. In addition, suppose that there exists the limit*

$$\lim_{t\to\infty} t^\alpha z(t) = c_z^* \in \mathbb{R}. \tag{31}$$

Then,

$$\int_{[0,t]} z(t-s)U(\mathrm{d}s) - \frac{m}{\mu} \sim \frac{1}{(\alpha-1)\mu}\Big(\frac{c_Z m}{\mu} - c_z^*\Big)t^{-(\alpha-1)}, \quad t \to \infty, \tag{32}$$

where $m := \int_0^\infty z(s)\mathrm{d}s$. (32) *holds when* $c_z^* = 0$ *as well, in which case* (31) *becomes*

$$z(t) = o(t^{-\alpha}), \quad t \to \infty. \tag{33}$$

Corollary 8. *Let* $\{X(t); t \geq 0\}$ *be a stationary regenerative process in* (5) *with finite variance and covariance as in* (21), *satisfying conditions* (2), (29), (30), *and* (31).
(i) ('R(t) dominates'). Assume additionally that

$$R(t) \sim c_R t^{-\beta}, \quad t \to \infty \quad (\exists\, 0 < \beta < \alpha - 1,\ c_R > 0). \tag{34}$$

Then,

$$\mathrm{Cov}(X(0), X(t)) \ \sim\ R(t)\ \sim\ c_R t^{-\beta}, \quad t \to \infty. \tag{35}$$

(ii) ('h(t) dominates') Assume additionally $R(t) = o(t^{-(\alpha-1)})$, $t \to \infty$. *Then,*

$$\mathrm{Cov}(X(0), X(t)) \ \sim\ h(t) \ \sim\ \frac{1}{(\alpha-1)\mu^2}\Big(\frac{c_Z m}{\mu} - c_z^*\Big)t^{-(\alpha-1)}, \quad t \to \infty. \tag{36}$$

Proof. According to Theorem 7, relations (2), (29), and (31) imply the asymptotics of $h(t)$ as in (32). Then, part (i) follows by (22) and (21). Part (ii) follows directly from (32) and $R(t) = o(t^{-(\alpha-1)})$. □

Remark 9. While the conditions of the right continuity of $z(t)$ and $\lim_{t\to\infty} z(t) = 0$ in (29) are mild, the two other conditions in (29) are more restrictive. Clearly, $z(t) \geq 0$ holds if either $G^i(t) \geq 0, i = 1, 2$ for all $t \geq 0$ or $G^1(t) = \mathrm{E}[W(t)\mathbb{I}(t < Z)] = 0$ $(\forall t > 0)$ holds. The bounded variation condition in (29) holds if $G^0(t)$ is integrable and $G^1(t)$ has bounded variation on $[0, \infty)$, $\lim_{t\to\infty} G^1(t) = 0$, or vice versa [159].

Theorem 10 ([159]). *Let $X_i = \{X_i(t); t \ge 0\}$ in (1) be i.i.d. copies of the stationary regenerative process in (5) with finite variance and a covariance function:*

$$\mathrm{Cov}(X(0), X(t)) \sim c_X t^{-(\alpha-1)}, \quad t \to \infty \quad (\exists\, c_X > 0,\ 1 < \alpha < 2). \tag{37}$$

Assume, in addition, that

$$\sup_{t\in[0,Z)} |W(t)| \le C \tag{38}$$

with a nonrandom constant $C < \infty$. Then, for any $\gamma > \alpha - 1$,

$$\lambda^{-H(\gamma)}(A_{\lambda,\gamma}(x,y) - \mathrm{E}A_{\lambda,\gamma}(x,y)) \overset{\mathrm{fdd}}{\longrightarrow} C_W B_{H,1/2}(x,y), \tag{39}$$

where $H = \frac{3-\alpha}{2}$ and $H(\gamma) = H - \frac{\gamma}{2}, C_W$ and the limit in (39) are the same as in Theorem 1 (i) of Chapter 8.

The proof of Theorem 10 is similar to that of Theorem 1 (i) of Chapter 8. Namely, relation (10) entails the convergence of the variance of the l.h.s. of (39) to C_W^2 and hence the normalisation $\lambda^{H(\gamma)}$. Then, the Gaussian limit in (39) is a consequence of the Lyapunov criterion (Proposition 3 of Chapter 2), using the boundedness of pulse and input processes.

Remark 11. The boundedness condition (38) on a pulse process is quite restrictive and is not satisfied in Example 2 (workload process) unless Z^{on} is bounded. (The same condition is also imposed in Theorems 14 and 15.) In the case of a renewal–reward process (Example 3), the boundedness of the reward variable W is required. The proof in Ref. [213] of the Gaussian limit for a renewal–reward process when W is unbounded and has finite variance uses approximation by bounded W, suggesting that a similar approximation might help to weaken (38) in Theorem 1 of Chapter 8.

We end this section with a discussion of the covariance LRD property for the regenerative processes in Examples 1 and 2

Corollary 12. *Let $\{X(t); t \ge 0\}$ be a stationary ON/OFF process (notation as in Example 6) with*

$$\bar{F}^{\rm on}(t) \sim c_{\rm on} t^{-\alpha_{\rm on}} \ (\exists\ c_{\rm on} > 0, \quad \alpha_{\rm on} \in (1,2)), \qquad (40)$$
$$\bar{F}^{\rm off}(t) = o(\bar{F}^{\rm on}(t)), \quad t \to \infty.$$

In addition, assume that $F(t) = \mathrm{P}(Z^{\rm on} + Z^{\rm off} \le t)$ satisfies (30). Then,

$$\mathrm{Cov}(X(0), X(t)) \sim \frac{c_{\rm on}\mu_{\rm off}^2}{(\alpha_{\rm on}-1)\mu^3} t^{-(\alpha_{\rm on}-1)}, \quad t \to \infty. \qquad (41)$$

Proof. Let us first show that (40) implies $\mathrm{P}(Z > t) \sim \bar{F}^{\rm on}(t) \sim c_{\rm on} t^{-\alpha_{\rm on}}$, $Z = Z^{\rm on} + Z^{\rm off}$. Indeed, for small $\epsilon > 0$, we have $\mathrm{P}(Z^{\rm on} > t) \le \mathrm{P}(Z^{\rm on} + Z^{\rm off} > t) \le \mathrm{P}(Z^{\rm off} > \epsilon t) + \mathrm{P}(Z^{\rm on} > (1-\epsilon)t)$, implying $1 \le \liminf_{t\to\infty} \mathrm{P}(Z > t)/\bar{F}^{\rm on}(t) \le \limsup_{t\to\infty} \bar{F}^{\rm on}((1-\epsilon)t)/\bar{F}^{\rm on}(t) \le (1-\epsilon)^{-\alpha_{\rm on}}$, and hence the above equivalence as $\epsilon > 0$ is arbitrary.

Since $R(t) = \mathrm{P}(Z^{\rm on} > t) \sim c_{\rm on} t^{-\alpha_{\rm on}}$, Corollary 8 is not applicable, but we can use (26) and derive (41) directly from Theorem 7. Towards this goal, it suffices to show that

$$\tilde{z}(t) = \int_0^t \bar{F}^{\rm off}(s)\bar{F}^{\rm on}(t-s)\mathrm{d}s \ \sim\ c_z^* t^{-\alpha_{\rm on}}, \quad t \to \infty,$$

where $c_z^* = c_Z \int_0^\infty \bar{F}^{\rm off}(s)\mathrm{d}s = c_Z\mu_{\rm off}$. This follows by splitting $\tilde{z}(t) = \int_0^{t/2} \cdots + \int_{t/2}^t \cdots =: z'(t) + z''(t)$, where $z''(t) = o(t^{-\alpha_{\rm on}})$ according to the second relation in (40), while $t^{\alpha_{\rm on}} z'(t) \to c_Z\mu_{\rm off}$ follows from (40) and the DCT. $\square$

Next, consider the workload process in Example 2:

$$X(t) = (Z_0^{\rm on} - t)_+ + \sum_{j=0}^{\infty} (T_j + Z_{j+1}^{\rm on} - t)_+ \mathbb{I}(T_j \le t < T_{j+1}), \quad t \ge 0, \qquad (42)$$

with $W(t) = (Z^{\rm on} - t)_+, Z = Z^{\rm on} + Z^{\rm off}$ and possibly dependent intervals $Z^{\rm on} > 0, Z^{\rm off} > 0$. According to Lemma 5, $\mathrm{E}X(t)^2 = \mu^{-1}$

$\int_0^\infty \mathrm{E}(Z^{\mathrm{on}} - s)_+^2 \mathrm{d}s = \mu^{-1}\mathrm{E}(Z^{\mathrm{on}})^3/3 < \infty$ if and only if $\mathrm{E}(Z^{\mathrm{on}})^3 < \infty$, in which case $\mathrm{E}X(t) = \mu^{-1}\mathrm{E}(Z^{\mathrm{on}})^2/2$. We also have

$$\begin{aligned} G^0(t) &= \mathrm{E}[(t - X^{\mathrm{off}})_+ \mathbb{I}(t < Z^{\mathrm{on}} + Z^{\mathrm{off}})], \quad G^1(t) = \mathrm{E}(Z^{\mathrm{on}} - t)_+, \\ R(t) &= \mu^{-1}\int_t^\infty u(u-t)\bar{F}^{\mathrm{on}}(u)\mathrm{d}u. \end{aligned} \tag{43}$$

Proposition 13. *Let $\{X(t); t \ge 0\}$ be a stationary workload process in (42). Assume (30),*

$$\begin{aligned} &\bar{F}^{\mathrm{off}}(t) \sim c_{\mathrm{off}} t^{-\alpha_{\mathrm{off}}}, \quad t \to \infty \quad (\exists\, c_{\mathrm{off}} > 0,\ \alpha_{\mathrm{off}} \in (1,2)), \quad \text{and} \\ &\mathrm{E}[(Z^{\mathrm{on}})^3 | Z^{\mathrm{off}}] < C, \end{aligned} \tag{44}$$

where $C > 0$ is a non-random constant.

(i) Assume additionally that

$$\bar{F}^{\mathrm{on}}(t) \sim c_{\mathrm{on}} t^{-\alpha_{\mathrm{on}}}, \quad t \to \infty \quad (\exists\, c_{\mathrm{on}} > 0,\ 3 < \alpha_{\mathrm{on}} < \alpha_{\mathrm{off}} + 2). \tag{45}$$

Then,

$$\mathrm{Cov}(X(0), X(t)) \sim R(t) \sim \frac{c_{\mathrm{on}}}{(\alpha_{\mathrm{on}} - 2)(\alpha_{\mathrm{on}} - 3)\mu} t^{-(\alpha_{\mathrm{on}} - 3)}, \quad t \to \infty. \tag{46}$$

(ii) Assume additionally that

$$\bar{F}^{\mathrm{on}}(t) = o(t^{-\alpha_{\mathrm{off}} - 2}), \quad t \to \infty. \tag{47}$$

Then,

$$\mathrm{Cov}(X(0), X(t)) \sim h(t) \sim \frac{c_{\mathrm{off}} m}{(\alpha_{\mathrm{off}} - 1)\mu^3} t^{-(\alpha_{\mathrm{off}} - 1)}, \quad t \to \infty, \tag{48}$$

where $h(t), m$ are as in Corollary 8.

Proof. We use Corollary 8 and need to verify its conditions. As in the proof of the previous proposition, (44) imply $\mathrm{P}(Z > t) = \mathrm{P}(Z^{\mathrm{on}} + Z^{\mathrm{off}} > t) \sim c_{\mathrm{off}} t^{-\alpha_{\mathrm{off}}}$, and hence (2) holds with $\alpha = \alpha_{\mathrm{off}}$.

Let us check (31) or (33). According to (43),

$$\begin{aligned} G^0(t) &= \mathrm{E}[\mathrm{P}(Z^{\mathrm{on}} > t - Z^{\mathrm{off}} | Z^{\mathrm{off}})(t - Z^{\mathrm{off}})_+ \mathbb{I}(Z^{\mathrm{off}} \le t(1-\epsilon))] \\ &\quad + \mathrm{E}[\mathrm{P}(Z^{\mathrm{on}} > t - Z^{\mathrm{off}} | Z^{\mathrm{off}})(t - Z^{\mathrm{off}})_+ \mathbb{I}(Z^{\mathrm{off}} > t(1-\epsilon))] \\ &=: G'(t) + G''(t). \end{aligned}$$

By (44), $\mathrm{P}(Z^{\mathrm{on}} > t - Z^{\mathrm{off}}|Z^{\mathrm{off}}) \le (t - Z^{\mathrm{off}})_+^{-3}\mathrm{E}[(Z^{\mathrm{on}})^3|Z^{\mathrm{off}}] \le C(t - Z^{\mathrm{off}})_+^{-3}$ and, similarly, $P(Z^{\mathrm{on}} > t - Z^{\mathrm{off}}|Z^{\mathrm{off}}) \le (t - Z^{\mathrm{off}})_+^{-1}\mathrm{E}[Z^{\mathrm{on}}|Z^{\mathrm{off}}] \le C(t - Z^{\mathrm{off}})_+^{-1}$. Hence, $G'(t) \le C(\epsilon t)^{-2} = o(t^{-\alpha_{\mathrm{off}}})$, for any given $\epsilon > 0$ and $G''(t) \le C\mathrm{P}(t(1-\epsilon) < Z^{\mathrm{off}} \le t) = \bar{F}^{\mathrm{off}}(t(1-\epsilon)) - \bar{F}^{\mathrm{off}}(t)$, implying

$$\limsup_{t\to\infty} t^{\alpha_{\mathrm{off}}} G''(t) \le c_{\mathrm{off}} \left(\frac{1}{(1-\epsilon)^{\alpha_{\mathrm{off}}}} - 1 \right),$$

where the r.h.s. tends to 0 with $\epsilon \to 0$. Therefore, $\lim_{t\to\infty} t^{\alpha_{\mathrm{off}}} G^0(t) = 0$, proving (33). Also note that $G^i(t) \ge 0, i = 1, 2$, $G^1(t)$ is monotone, and $G^0(t)$ is integrable, so that $z(t)$ satisfies (29) in view of Remark 9.

Next, we verify the additional conditions in (i) and (ii) of Corollary 8. From (43) and (45), we conclude that $R(t)$ satisfies the asymptotics in (46) or (34) with $\beta = \alpha_{\mathrm{on}} - 3 < \alpha_{\mathrm{off}} - 1$. Hence, part (i) of Corollary 8 applies, yielding the asymptotics of $\mathrm{Cov}(X(0), X(t))$, as in (46). Similarly, from (47), we conclude that $R(t) = o(t^{-(\alpha_{\mathrm{off}}-1)})$, and hence (48) follows from (36). □

3 Stable and Intermediate Convergences

In this section, we discuss the convergence in (2) of Chapter 8 towards α-stable and intermediate limits for aggregated sums of regenerative processes in (5). Given a stationary regenerative process in (5) with a generic pair $(Z, \{W(t)\}_{t\in[0,Z)})$, denote

$$\mathcal{W}_Z := \int_0^Z W(s)\mathrm{d}s, \quad \mu_{\mathcal{W}} := \mathrm{E}\mathcal{W}_Z, \quad \widetilde{\mathcal{W}}_Z := \mathcal{W}_Z - (\mu_{\mathcal{W}}/\mu)Z, \quad (49)$$

provided the expectation $\mu_{\mathcal{W}}$ exists. Note that $\mathrm{E}\widetilde{\mathcal{W}}_Z = 0$.

Theorem 14 ([159]). *Let $X_i = \{X_i(t); t \ge 0\}$ in (1) be i.i.d. copies of the stationary regenerative process in (5). Assume that $(Z, \{W(t)\}_{t\in[0,Z)})$ satisfy (2), (38), and*

$$\mathrm{P}(\widetilde{\mathcal{W}}_Z > x) = (c_+ + o(1))\, x^{-\alpha}, \quad \mathrm{P}(\widetilde{\mathcal{W}}_Z < -x) = (c_- + o(1))\, x^{-\alpha}, \qquad x \to \infty, \tag{50}$$

for some $c_\pm \ge 0$, $c_+ + c_- > 0$. Then,

$$\lambda^{-(1+\gamma)/\alpha}(A_{\lambda,\gamma}(x,y) - \mathrm{E}A_{\lambda,\gamma}(x,y)) \xrightarrow{\mathrm{fdd}} L_\alpha(x,y), \quad \forall\, 0 < \gamma < \alpha - 1,$$

where L_α is an α-stable Lévy sheet.

Despite the similarities between the results and conditions of Theorem 1 (ii) of Chapter 8 and Theorem 14, the proof of the latter theorem is much more technical and involved. An obvious reason is that dealing with regenerative and renewal processes is generally harder than dealing with stochastic integrals and deterministic integrands. The main idea behind the proof of (14) can be traced back to Ref. [185] and consists in the approximation of $A_{\lambda,\gamma}(x,y)$ by a sum of i.i.d. r.v.s with distribution $\widetilde{W}_Z$ of (50) in the domain of attraction of the α-stable law. The aforementioned approximation relies on the (the number of regeneration points $T_i \in [0,t]$) decomposition of the integrated input process as

$$\begin{aligned}
\int_0^\lambda X(t)\mathrm{d}t &= \int_0^{\lambda\wedge T_0} W_0(t)\mathrm{d}t + \left(\int_0^{T_{N(\lambda)}} - \int_\lambda^{T_{N(\lambda)}}\right) \\
&\quad \times \sum_{i=1}^{\infty} W_i(t - T_{i-1})\mathbb{I}(T_{i-1} \le t < T_i)\mathrm{d}t \\
&= \int_0^{\lambda\wedge T_0} W_0(t)\mathrm{d}t + \sum_{i=1}^{N(\lambda)} \int_0^{Z_i} W_i(t)\mathrm{d}t \\
&\quad - \int_\lambda^{T_{N(\lambda)}} W_{N(\lambda)}(t - T_{N(\lambda)-1})\mathrm{d}t,
\end{aligned}$$

where $N(t)$ is the renewal process in (8) (the number of regeneration points $T_i \in [0,t]$). Analogously, we decompose the interval $[0,\lambda)$:

$$\lambda = \lambda \wedge T_0 + \sum_{i=1}^{N(\lambda)} Z_i - (T_{N(\lambda)} - \lambda).$$

Then, since $\mathrm{E}X := \mathrm{E}X(t) = \mu_{\mathcal{W}}/\mu$, referring to (19), so $\int_0^\lambda (X(t) - \mathrm{E}X)\mathrm{d}t = I_1(\lambda) + I_2(\lambda) - I_3(\lambda)$, where

$$I_1(\lambda) := \int_0^{\lambda\wedge T_0} W_0(t)\mathrm{d}t - (\mathrm{E}X)(\lambda \wedge T_0),$$

$$I_2(\lambda) := \sum_{i=1}^{N(\lambda)} \widetilde{\mathcal{W}}_{Z,i}, \quad \text{with } \widetilde{\mathcal{W}}_{Z,i} := \int_0^{Z_i} W_i(s)\mathrm{d}s - (\mathrm{E}X)Z_i,$$

$$I_3(\lambda) := \int_\lambda^{T_{N(\lambda)}} W_{N(\lambda)}(t - T_{N(\lambda)-1})\mathrm{d}t - (\mathrm{E}X)(T_{N(\lambda)} - \lambda), \tag{51}$$

with the convention that $I_2(\lambda) = I_3(\lambda) := 0$ if $N(\lambda) = 0$ (or $T_0 > \lambda$). Thus,

$$A_{\lambda,\gamma}(x,1) - \mathrm{E}A_{\lambda,\gamma}(x,1) = A_1(\lambda x) + A_2(\lambda x) - A_3(\lambda x),$$

$$A_k(\lambda) := \sum_{j=1}^{[\lambda^\gamma]} I_k^{(j)}(\lambda), \quad k = 1,2,3,$$

where $(I_1^{(j)}(\lambda), I_2^{(j)}(\lambda), I_3^{(j)}(\lambda))$, $j = 1,2,\dots$ are *independent copies* of $(I_1(\lambda), I_2(\lambda), I_3(\lambda))$ in (51). The approximation is defined by replacing the random number $N(\lambda x)$ of summands in $I_2(\lambda x)$ by its mean $\mathrm{E}N(\lambda x) = \lambda x/\mu$:

$$\widetilde{A}_{\lambda,\gamma}(x) := \sum_{j=1}^{[\lambda^\gamma]} \widetilde{I}_k^{(j)}(\lambda x), \quad \text{where} \quad \widetilde{I}_k^{(j)}(\lambda) := \sum_{i=1}^{[\lambda/\mu]} \widetilde{\mathcal{W}}_{Z,i}^{(j)},$$

where $\widetilde{\mathcal{W}}_{Z,i}^{(j)}$, $1 \le i \le [\lambda x/\mu]$, $1 \le j \le [\lambda^\gamma]$ are i.i.d. copies of $\widetilde{\mathcal{W}}_Z$.

With Theorems 10 and 14 in mind, and in analogy with the results in the shot-noise case (Chapter 8), it is natural to expect an intermediate limit for the aggregate $A_{\lambda,\gamma}(x,y)$ at a critical $\gamma = \gamma_{\mathrm{cr}} = \alpha - 1$, under general conditions on the pulse process W. Somewhat surprisingly, the intermediate convergence for regenerative inputs is quite difficult. The result in Theorem 15 is not very satisfactory, as explained in what follows. Observe from (50) and the decomposition in (51) that the stable limit in Theorem 14 is due to the α-tail behaviour of the difference $\widetilde{\mathcal{W}}_Z = \mathcal{W}_Z - Z(\mathrm{E}\mathcal{W}_Z/\mu)$. Since Z is assumed to have the same α-tail, it follows that $\mathrm{P}(|\mathcal{W}_Z| > x) = O(|x|^{-\alpha})$, but of course this does not exclude $\mathrm{P}(|\mathcal{W}_Z| > x) = o(|x|^{-\alpha})$ (condition (52)), as in the following theorem.

Theorem 15 ([159]). *Let $X_i = \{X_i(t); t \geq 0\}$ in (1) be i.i.d. copies of the stationary regenerative process in (5). Assume that $(Z, \{W(t)\}_{t\in[0,Z)})$ satisfy (2), (38), and*

$$\mathrm{P}(|\mathcal{W}_Z| > x) \leq Cx^{-\alpha-\delta} \quad \text{and} \quad \mathrm{E}|W(t)|\mathbb{I}(t < Z) \leq Ct^{-\alpha-\delta}$$
$$\forall\, x, t \geq 1 \quad (\exists\, \delta, C > 0). \tag{52}$$

Then, with $\gamma_{\mathrm{cr}} = \alpha - 1$,

$$\lambda^{-1}(A_{\lambda,\gamma_{\mathrm{cr}}}(x, y) - \mathrm{E}A_{\lambda,\gamma_{\mathrm{cr}}}(x, y)) \xrightarrow{\mathrm{fdd}} -(\mu_{\mathcal{W}}/\mu)J(x, y), \tag{53}$$

where $J = \{J(x, y); (x, y) \in \mathbb{R}^2_+\}$ is the Telecom RF in (16), (11) of Chapter 8.

Thus, the intermediate limit in (53) is restricted to the situation when $\mathcal{W}_Z$ *has a lighter tail than Z does, and the tail behaviour of $\widetilde{\mathcal{W}}_Z$ is determined by the term $-Z(\mu_{\mathcal{W}}/\mu)$*. It relies on the following principal result from Ref. [93], whose proof uses advanced renewal techniques and is not included in this book.

Theorem 16 ([93]). *Let $N(x) := \sum_{n=0}^{\infty} \mathbb{I}(T_n \in [0, x])$, $\cdots < T_{-1} < 0 \leq T_0 < T_1 < \cdots$, be a stationary renewal process as in (5), with the generic duration distribution Z satisfying (2) and $\{N^{(j)}(x); x \geq 0\}, j = 1, 2, \ldots$, being independent copies of the stationary renewal process $N = \{N(x); x \geq 0\}$. Then,*

$$\lambda^{-1} \sum_{j=1}^{[\lambda^{\alpha-1}]} (N^{(j)}(\lambda x) - \mathrm{E}N^{(j)}(\lambda x)) \xrightarrow{\mathrm{fdd}} -\mu^{-1}J(x), \tag{54}$$

where the limit Telecom process J is defined in (10), with c_f replaced by $\alpha c_Z/\mu$.

To explain the relation between the above-mentioned theorems, consider the following decomposition for $\int_0^\lambda (X(t) - \mathrm{E}X(t))\mathrm{d}t = \int_0^\lambda X(t)\mathrm{d}t - \lambda(\mu_{\mathcal{W}}/\mu)$:

$$\int_0^\lambda X(t)\mathrm{d}t - \lambda(\mu_{\mathcal{W}}/\mu) = \sum_{k=1}^4 \bar{I}_k(\lambda), \quad \text{where}$$

$$\bar{I}_1(\lambda) := \mu_{\mathcal{W}}(N(\lambda) - \mathrm{E}N(\lambda)), \quad \bar{I}_2(\lambda) := \sum_{i=1}^{N(\lambda)} (\mathcal{W}_{Z,i} - \mu_{\mathcal{W}}),$$

$$\bar{I}_3(\lambda) := \int_0^{\lambda \wedge T_0} W_0(t)\mathrm{d}t, \quad \bar{I}_4(\lambda) := -\int_\lambda^{T_{N(\lambda)}} W_{N(\lambda)}(t - T_{N(\lambda)-1})\mathrm{d}t. \tag{55}$$

Recall that $T_{N(\lambda)-1} < \lambda \le T_{N(\lambda)}$ are the renewal points before and after time λ. Equation (55) is similar to but different from (51). Essentially, Theorem 15 follows by showing that $\bar{I}_1(\lambda)$ in (55) is the main term, while the remaining terms $\bar{I}_k(\lambda), k = 2, 3, 4$, are negligible under intermediate scaling at $\gamma = \gamma_{\mathrm{cr}} = \alpha - 1$.

What happens if Z and $\mathcal{W}_Z$ have the same α-tail as in Theorem 14? In this case, the term $\bar{I}_2(\lambda)$ may also contribute to the limit of $A_{\lambda,\gamma_{\mathrm{cr}}}$, which should depend on the distribution of the pulse process W. The question about the limit distribution of aggregated sums $\sum_{j=1}^{[\lambda^{\alpha-1}]} \bar{I}_2^{(j)}(\lambda)$ of i.i.d. copies of $\bar{I}_2(\lambda)$ is open and does not seem easy. It is interesting that the summands $\mathcal{W}_{Z,i}, i = 1, \ldots, N(\lambda)$ when $N(\lambda) = n$ is fixed are *exchangeable*, or, in other words, the sequences $\{\mathcal{W}_{Z,i}; i = 1, \ldots, n\}$ and $\{\mathcal{W}_{Z,\pi(i)}; i = 1, \ldots, n\}$ have the same distribution for any permutation $(\pi(1), \ldots, \pi(n))$ of $(1, \ldots, n)$.

Remark 17. Marginal convergence in (54) for a fixed $x > 0$ was proved in Ref. [159] using large deviations for sums of heavy-tailed i.i.d. r.v.s [187].

4 Examples

In this section, we present several examples of regenerative processes satisfying the conditions of Theorems 10, 14, and 15. The basic example is an ON/OFF process (Corollary 18) whose covariance LRD property was earlier discussed in Corollary 12. As noted above, the ON/OFF model may represent the input from a queueing system having a single server with given service and arrival interval distributions F_σ and F_τ. Verification of the conditions of Theorems 10–15 in terms of F_σ, F_τ requires some effort. For the G/G/1/0 and G/G/1/∞ queueing system, this is done in Examples 19 and 21. Further examples of regenerative processes include the renewal–reward process (Example 22) and the exponentially damped pulse process (Example 23).

Corollary 18 (ON/OFF process). *Let $A_{\lambda,\gamma}(x,y)$ be the aggregated RF from an ON/OFF input process $X = \{X(t); t \ge 0\}$ satisfying the conditions of Corollary 12, particularly $\mathrm{P}(Z^{\mathrm{off}} > t) = o(\mathrm{P}(Z^{\mathrm{on}} > t))$. Then:*

(i) For any $\gamma > \alpha_{\mathrm{on}} - 1$, $\lambda^{-H(\gamma)}(A_{\lambda,\gamma}(x,y) - \mathrm{E}A_{\lambda,\gamma}(x,y)) \xrightarrow{\mathrm{fdd}} C_W B_{H_1,1/2}(x,y)$ holds, with $H(\gamma) = H_1 + \frac{\gamma}{2}, H_1 = \frac{3-\alpha_{\mathrm{on}}}{2} \in (\frac{1}{2}, 1)$ and $C_W^2 := \frac{c_X}{(2H-1)H}, c_X = \frac{c_{\mathrm{on}}\mu_{\mathrm{off}}^2}{(\alpha_{\mathrm{on}}-1)\mu^3}$ (see (41)).

(ii) For any $0 < \gamma < \alpha_{\mathrm{on}} - 1$, $\lambda^{-(1+\gamma)/\alpha_{\mathrm{on}}}(A_{\lambda,\gamma}(x,y) - \mathrm{E}A_{\lambda,\gamma}(x,y)) \xrightarrow{\mathrm{fdd}} L_{\alpha_{\mathrm{on}}}(x,y)$, where $L_{\alpha_{\mathrm{on}}}$ is the Lévy sheet in (14) with parameters $c_+ = c_{\mathrm{on}}(\mu/\mu_{\mathrm{off}})^\alpha, c_- = 0$.

(iii) For $\gamma = \gamma_{\mathrm{cr}} = \alpha_{\mathrm{on}} - 1$, suppose $\mathrm{P}(Z^{\mathrm{off}} > t) \le Ct^{-\alpha_{\mathrm{on}}-\delta}, t > 0$ $(\exists C, \delta > 0)$. Then, $\lambda^{-1}(A_{\lambda,\gamma_{\mathrm{cr}}}(x,y) - \mathrm{E}A_{\lambda,\gamma_{\mathrm{cr}}}(x,y)) \xrightarrow{\mathrm{fdd}} -\frac{\mu_{\mathrm{on}}}{\mu_{\mathrm{on}}+\mu_{\mathrm{off}}} J(x,y)$, where J is the Telecom RF in (53).

Proof. Part (i) is a direct consequence from Theorem 10 and Corollary 12. Part (ii) follows from Theorem 14 by noting that $\widetilde{\mathcal{W}}_Z = \mu^{-1}(\mu_{\mathrm{off}} Z^{\mathrm{on}} - \mu_{\mathrm{on}} Z^{\mathrm{off}})$ and (45) imply (2) and (50), with $\alpha = \alpha_{\mathrm{on}}, c_Z = c_{\mathrm{on}}$ and c_+, c_- as in part (i). To prove part (iii), we use Theorem 15 with $X(t)$ replaced by $-X'(t)$, where $X'(t) := 1 - X(t), t \ge 0$, is a stationary ON/OFF process with the ON and OFF intervals exchanged. It suffices to verify (52) for $\mathcal{W}'_Z := Z^{\mathrm{off}}, W'(t) := \mathbb{I}(t < Z^{\mathrm{off}})$, which clearly follow from the additional conditions on the d.f. of Z^{off} assumed in part (iii). □

Next, we discuss examples of input processes from the stationary G/G/1 queue with a single server and general inter-arrival and service distributions. Let $\tau, \tau_i > 0, i = 1, 2, \ldots$ (inter-arrival times), be i.i.d. r.v.s with $F_\tau(x) := \mathrm{P}(\tau \le x)$ and $\sigma, \sigma_i > 0, i = 1, 2, \ldots$ (service times), be i.i.d. r.v.s with $F_\sigma(x) := \mathrm{P}(\sigma \le x)$. We assume that the sequences $\{\tau, \tau_i, i \ge 1\}$ and $\{\sigma, \sigma_i, i \ge 1\}$ are independent and $\mu_\tau := \mathrm{E}\tau < \infty, \mu_\sigma := \mathrm{E}\sigma < \infty$. Let $X(t) \in \{0, 1\}$ be the number of customers served at time t. Then, $\{X(t)\}$ is a regenerative ON/OFF process which regenerates at each moment the new arrival finds the system empty. The usual terminology in queueing theory is *busy period* for the ON interval and *idle period* for the OFF interval, with the regeneration interval $Z = Z^{\mathrm{on}} + Z^{\mathrm{off}}$ termed the *busy cycle*. In order to exclude the possibility that this moment coincides with the end of service of the previous customer (in which case the OFF intervals may have zero length with positive probability), we assume that the d.f. F_τ is continuous.

The queueing systems G/G/1 are further distinguished according to service disciplines or rules which arriving customers follow when the system is busy. See Refs. [6, 57]. We consider two types of G/G/1 queues, namely the G/G/1/0 queue without a waiting room (customers arriving when the system is busy are rejected) and the G/G/1/∞ queue with an infinite waiting room (all arriving customers wait until being served). Our aim is to verify the conditions of Corollaries 18 or 12 in terms of the d.f. $F_\tau(x)$ and $F_\sigma(x)$. Clearly, these conditions should be different for G/G/1/0 and G/G/1/∞.

Example 19 (G/G/1/0 queue). In this case, the joint distribution of the ON and OFF intervals can be identified as

$$(Z^{\mathrm{on}}, Z^{\mathrm{off}}) \stackrel{\mathrm{d}}{=} \left(\sigma_1, \sum_{j=1}^{\mathrm{t}_0^S} \tau_j - \sigma_1\right), \quad \text{where}$$

$$\mathrm{t}_0^S := \inf\left\{n \ge 1 : \sum_{j=1}^{n} \tau_j \ge \sigma_1\right\} \tag{56}$$

is the first time the process $S_n := \sigma_1 - \sum_{j=1}^n \tau_j, n \ge 1$ hits $(-\infty, 0]$. The fact that this time has finite mean $\mathrm{E}\mathrm{t}_0^S < \infty$ can be verified with the help of the renewal function $U_\tau(t) := 1 + \sum_{n=1}^\infty \mathrm{P}(\sum_{j=1}^n \tau_j \le t)$

of the arrival process. Indeed, from the elementary renewal theorem we have that $U_\tau(t) \le C(1 \vee t), t \ge 0$; therefore,

$$\infty > \mathrm{E}U_\tau(\sigma) = 1 + \sum_{n=1}^{\infty} \mathrm{P}\left(\sum_{j=1}^{n} \tau_j < \sigma_1\right)$$

$$= 1 + \sum_{n=1}^{\infty} \mathrm{P}(\mathrm{t}_0^S > n) = \mathrm{E}\mathrm{t}_0^S, \tag{57}$$

where we used $\mathrm{E}\sigma < \infty$, the independence of σ_1 and $\sum_{j=1}^{n} \tau_j$, and the fact that the d.f. of $\sum_{j=1}^{n} \tau_j$ is continuous.

Proposition 20. *In the above G/G/1/0 queueing system, we have*

$$\bar{F}^{\mathrm{on}}(x) = \bar{F}_\sigma(x) \qquad \text{and} \qquad \bar{F}^{\mathrm{off}}(x) \le q\bar{F}_\tau(x), \quad x \ge 0, \tag{58}$$

where

$$q := \mathrm{E}\mathrm{t}_0^S = \frac{\mu_\sigma + \mu_{\mathrm{off}}}{\mu_\tau}. \tag{59}$$

Proof. The first relation in (58) is obvious from (56). Let us prove the second relation in (58). Using (56), we can write

$$\begin{aligned}\bar{F}^{\mathrm{off}}(x) &= \mathrm{P}(\tau_1 > x + \sigma_1) \\ &\quad + \sum_{n=1}^{\infty} \mathrm{P}\left(\sum_{j=1}^{n} \tau_j < \sigma_1, \tau_{n+1} > x + \sigma_1 - \sum_{j=1}^{n} \tau_j\right) \\ &\le \bar{F}_\tau(x)\left(1 + \mathrm{P}\left(\sum_{j=1}^{n} \tau_j < \sigma_1\right)\right) = \mathrm{E}\mathrm{t}_0^S \bar{F}_\tau(x),\end{aligned} \tag{60}$$

referring to (57). This proves (58) with q as in (59). The second equality in (59) follows from (56) and Wald's identity: $\mu_{\mathrm{off}} = \mathrm{E}\sum_{j=1}^{\mathrm{t}_0^S} \tau_j - \mu_\sigma = \mathrm{E}\mathrm{t}_0^S \mu_\tau - \mu_\sigma = q\mu_\tau - \mu_\sigma$. □

Returning to the question of convergence of aggregated inputs from G/G/1/0 queues, we see from (58) that the crucial conditions in (40) on p.d.f.s of the ON/OFF intervals for the validity of Corollary 18 can be replaced by their analogues on the service/arrival d.f.s:

$$\begin{aligned} &\bar{F}_\sigma(t) \sim c_\sigma t^{-\alpha_\sigma} \quad (\exists\, c_\sigma > 0,\ \alpha_\sigma \in (1,2)), \\ &\bar{F}_\tau(t) = o(\bar{F}_\sigma(t)), \quad t \to \infty. \end{aligned} \tag{61}$$

It follows that under (61) and regularity condition (30), statements (i) and (ii) of Corollary 18 apply with $\alpha_{\text{on}} = \alpha_\sigma, c_{\text{on}} = c_\sigma$. The 'intermediate' part (iii) of Corollary 18 also holds for this model in view of (61), provided $\bar{F}_\tau(t) \le Ct^{-\alpha_\sigma-\delta}, t > 0$ for some $\delta > 0$.

Example 21 (G/G/1/∞ queue). In this system, all arriving customers wait to be served, and the system regenerates at each time a new arrival finds the server empty. The existence of a stationary regime ('steady state') requires that the *traffic intensity* $\rho := \mu_\sigma/\mu_\tau < 1$, or $\mu_\sigma < \mu_\tau$. In such a case, the joint distribution of the ON and OFF intervals can be written as

$$(Z^{\text{on}}, Z^{\text{off}}) \stackrel{\text{d}}{=} \left(\sum_{j=1}^{\mathrm{t}_0^S} \sigma_j, \sum_{j=1}^{\mathrm{t}_0^S} \tau_j - \sum_{j=1}^{\mathrm{t}_0^S} \sigma_j \right), \quad \text{where}$$

$$\mathrm{t}_0^S := \inf\{n \ge 1 : S_n \le 0\} \tag{62}$$

is the first time the random walk $S_n := \sum_{j=1}^n (\sigma_j - \tau_j), n \ge 1$ hits $(-\infty, 0]$. The question about the distribution of the busy period in the G/G/1/∞ queue is non-trivial and much harder than in (56). The above problem is classical in queueing theory; see Refs. [6, 7, 184, 275], and the references therein. It was proved in Ref. [184] that under the tail assumption on $\bar{F}_\sigma(t)$ in (61) (and $\mu_\tau > \mu_\sigma$),

$$\mathrm{P}(Z^{\text{on}} > x) \sim c_{\text{on}}\, x^{-\alpha}, \quad x \to \infty, \quad \text{where} \quad c_{\text{on}} := \frac{c_\sigma}{(\mu_\tau - \mu_\sigma)^{1+\alpha}}. \tag{63}$$

Following (60), we can derive a similar upper bound on the tail d.f. of the OFF interval:

$$\begin{aligned}\bar{F}^{\rm off}(x) &= \mathrm{P}(\tau_1 > x + \sigma_1) \\ &\quad + \sum_{n=1}^{\infty} \mathrm{P}\left(\sum_{j=1}^{n} \tau_j < \sum_{j=1}^{n} \sigma_j, \tau_{n+1} > x + \sum_{j=1}^{n} (\sigma_j - \tau_j)\right) \\ &\le \bar{F}_\tau(x) \left(1 + \sum_{n=1}^{\infty} \mathrm{P}\left(\sum_{j=1}^{n} \tau_j < \sum_{j=1}^{n} \sigma_j\right)\right) = \mathrm{E}t_0^S\, \bar{F}_\tau(x),\end{aligned} \tag{64}$$

From (63)–(64), assumptions (61), and Corollary 18, we obtain exactly the same conclusions for the aggregated RF $A_{\lambda,\gamma}(x, y)$, $(x, y) \in \mathbb{R}^2$ from the $G/G/1/\infty$ inputs as in Example 19 above.

Example 22 (Renewal–reward process; Example 3). Gaussian and stable convergences of the aggregate $A(T, M)$ in (1) of Chapter 8 for discrete-time renewal–reward inputs were investigated in Ref. [213], with particular emphasis on the infinite-variance reward case. The above-mentioned work additionally assumes that W and Z are independent and $\mathrm{E}W = 0$ when $\mathrm{E}|W| < \infty$. In contrast, the result in Ref. [159, Corollary 5.16] applies to the renewal–reward model with bounded $|W| < \infty$, but the independence of W and Z is not required (more precisely, it is replaced by some sort of asymptotic independence as $Z \to \infty$). The intermediate limit in Ref. [159] for renewal–reward inputs is obtained under condition (52) of Theorem 15, which is *not* satisfied for independent W and Z, suggesting that the latter limit, if it exists, may generally be different from the Telecom RF.

Example 23 (A 'regenerative version' of exponentially damped shot-noise in Example 14 of Chapter 8). Let $\{X(t)\}$ be a stationary regenerative process corresponding to $Z = R, W(t)\mathbb{I}(t < Z) = \mathrm{e}^{-At}\mathbb{I}(t < R)$, where the r.v.s $A > 0$ and $R > 0$ are independent and distributed as in (54) of Chapter 8. The following proposition shows that the above 'regenerative version' behaves differently from the 'shot-noise version' of Example 14 of Chapter 8, in the sense of scaling limits of the aggregated inputs. Particularly, the critical point $\gamma_{\rm cr}$ separating fast and slow growth regimes and the parameters H_1, α of the limit distributions in (3) of Chapter 8 are different in Examples 23 and 14 of Chapter 8 under the same distributional assumptions on (A, R).

Proposition 24. *Let $\{X(t)\}$ be a stationary regenerative process with $Z = R, W(t) = \mathrm{e}^{-At}\mathbb{I}(t < R)$, where $A > 0$ and $R > 0$ are independent r.v.s satisfying (54) of Chapter 8, with*

$$1 < \rho < \rho + \kappa < 2. \tag{65}$$

Let $\gamma_{\rm cr} := \rho - 1, \alpha := \rho$. Then:
(i) Conditions (37) and (38) of Theorem 10 hold for any $\gamma > \gamma_{\rm cr}$, and hence

$$\lambda^{-(3-\alpha+\gamma)/2}(A_{\lambda,\gamma}(x,y) - \mathrm{E}A_{\lambda,\gamma}(x,y)) \xrightarrow{\rm fdd} C_W B_{H_1,1/2}(x,y),$$
$$\forall\, \gamma > \gamma_{\rm cr}, \tag{66}$$

where $H_1 = (3-\alpha)/2$ and $C_W^2 = \frac{c_X}{(2H_1-1)H_1}$, with c_X as given in (70).
(ii) Conditions (2), (38), and (50) of Theorem 14 with $\alpha = \rho$ hold for any $0 < \gamma < \gamma_{\rm cr}$, and hence

$$\lambda^{-(1+\gamma)/\alpha}(A_{\lambda,\gamma}(x,y) - \mathrm{E}A_{\lambda,\gamma}(x,y)) \xrightarrow{\rm fdd} L_\alpha(x,y) \quad \forall\, \gamma < \gamma_{\rm cr}, \tag{67}$$

where L_α is a completely asymmetric α-stable Lévy sheet with the parameters $c_+ = 0, c_- > 0$ in (71).
(iii) Conditions (2), (38), and (52) of Theorem 15 hold for $\gamma = \gamma_{\rm cr}$, and hence

$$\lambda^{-1}(A_{\lambda,\gamma_{\rm cr}}(x,y) - \mathrm{E}A_{\lambda,\gamma_{\rm cr}}(x,y)) \xrightarrow{\rm fdd} -(\mu_{\mathcal{W}}/\mu)J(x,y), \tag{68}$$

where $J(x,y)$ is the Telecom RF.

Proof. (i) Consider (37) ((38) is obvious). Recall (21) and (17). Here, $R(t) = \mu^{-1}\int_0^\infty \mathrm{E}[W(s) \times W(s+t)\mathbb{I}(Z > t+s)]\mathrm{d}s = \mu^{-1}\int_0^\infty \mathrm{P}(R > s+t)\mathrm{E}\mathrm{e}^{-A(2s+t)}\mathrm{d}s = O(t^{-(\rho+\kappa-1)}) = o(t^{-(\rho-1)})$ $(t \to \infty)$ follows from Proposition 15 (i) of Chapter 8. We also see that

$G^1(t) = \mathrm{E}[\mathrm{e}^{-At}]\mathrm{P}(R > t)$ is monotone and has bounded variation on $[0, \infty)$. Let us check

$$G^1(t) \sim c_\rho c_\kappa \Gamma(\kappa+1) t^{-\rho-\kappa} \quad \text{and} \quad G^0(t) = o(t^{-\rho}) \quad (t \to \infty). \tag{69}$$

The first relation in (69) is easy from (54) of Chapter 8. To show the second relation in (69), split $G^0(t) = \mathrm{E}[\mathrm{e}^{-A(R-t)}\mathbb{I}(t < R)] = \mathrm{E}[\mathrm{e}^{-A(R-t)}\mathbb{I}(t < R, A > \epsilon)] + \mathrm{E}[\mathrm{e}^{-A(R-t)}\mathbb{I}(t < R, A \le \epsilon)] =: G'_\epsilon(t) + G''_\epsilon(t)$, where $t^\rho G''_\epsilon(t) \le \mathrm{P}(A \le \epsilon)\, t^\rho\, \mathrm{P}(R > t)$ can be made arbitrarily small uniformly in $t \ge 1$ by an appropriate choice of $\epsilon > 0$. On the other hand, integrating by parts, we see that for any $\epsilon > 0$,

$$\begin{aligned} t^\rho G'_\epsilon(t) &\le t^\rho \mathrm{E}[\mathrm{e}^{-\epsilon(R-t)}\mathbb{I}(t < R)] \\ &= t^\rho \mathrm{P}(R > t) - \epsilon \int_0^\infty \frac{t^\rho}{(t+z)^\rho}(t+z)^\rho \mathrm{P}(R > t+z)\mathrm{e}^{-\epsilon z}\mathrm{d}z \\ &\to c_\rho - c_\rho\, \epsilon \int_0^\infty \mathrm{e}^{-\epsilon z}\mathrm{d}z = 0, \end{aligned}$$

as $t \to \infty$ according to (54) of Chapter 8 and the DCT, proving (69). Therefore, $0 \le z(t) \le G^1(t/2)(\int_0^{t/2} G^0(s)\mathrm{d}s + C\int_{t/2}^t G^0(s)(t-s)^{-\rho-\nu}\mathrm{d}s = o(t^{-\rho})$. Therefore and by Remark 9, we see that $z(t)$ in this example satisfies all the conditions of Proposition 8 with $\alpha = \rho$, and part (ii) of the latter proposition applies, yielding $\mathrm{Cov}(X(0), X(t)) \sim c_X t^{-(\rho-1)}$, $t \to \infty$, as in (36) or (37), with

$$\begin{aligned} c_X &= \frac{c_Z m}{(\rho-1)\mu^2}, \\ m &:= \int_0^\infty G^0(s)\mathrm{d}s \int_0^\infty G^1(t)\mathrm{d}t = (\mathrm{E}[(1-\mathrm{e}^{-AR})/A])^2. \end{aligned} \tag{70}$$

This proves part (i).

(ii) From Proposition 15 (ii) of Chapter 8, we have that $\mathcal{W}_Z = \int_0^R \mathrm{e}^{-At}\mathrm{d}t$ satisfies $\mathrm{P}(\mathcal{W}_Z > x) \sim c_\mathcal{W} x^{-\rho-\kappa} = o(x^{-\rho})$, $x \to \infty$. Then, the tail behaviour of $\widetilde{\mathcal{W}}_Z$ in (49) is determined by $-(\mu_\mathcal{W}/\mu)R$:

$$\mathrm{P}(\widetilde{\mathcal{W}}_Z \le -x) \ \sim\ \mathrm{P}(R > (\mu/\mu_\mathcal{W})x) \ \sim\ c_\rho(\mu_\mathcal{W}/\mu)^\rho x^{-\rho}, \quad x \to \infty.$$

$\mathrm{P}(\widetilde{\mathcal{W}}_Z > x) = o(x^{-\rho})$. Hence, $\widetilde{\mathcal{W}}_Z$ satisfies (50) with

$$c_+ = 0, \quad c_- = c_\rho(\mu_\mathcal{W}/\mu)^\rho = c_\rho\left(\frac{\mathrm{E}[(1-\mathrm{e}^{-AR})/A]}{\mathrm{E}R}\right)^\rho. \tag{71}$$

(iii) Relations (2) and (38) are obvious. The first relation in (52) follows from the proof of part (ii) since $\kappa > 0$. The second one follows from (69) since $\mathrm{E}[|W(t)|\mathbb{I}(t < Z)] = G^1(t)$. □

5 Notes

1. The pioneering work on regenerative processes is by Smith [240]. The reader is advised to refer to the monographs by Asmussen [6] and Thorisson [263] for an in-depth treatment of the subject. Lemma 5 appeared in Leipus and Surgailis [154].

2. Most of the discussion in Sections 2–5 of Chapter 10 is based on Leipus *et al.* [159]. The proofs of the Gaussian and stable limits (Theorem 10) use the same approach as in Mikosch *et al.* [185] for ON/OFF processes.

3. There is considerable literature on scaling limits (with or without aggregation) of cumulative input and related processes. Pipiras *et al.* [213] obtained joint fast, slow, and intermediate limits for renewal–reward inputs when rewards have finite or infinite variance. Iterated limits in the above model were discussed by Levy and Taqqu [166]. Mikosch and Samorodnitsky [186] studied aggregated scaling limits of cumulative inputs described by a marked point process (T_j, Z_j), with marks Z_j possibly dependent on the arrival times T_j. An overview of limit results for stochastic models that involve heavy tails, LRD, and aggregation is given in Kaj [128]. Iksanov [123] and Iksanov *et al.* [124] studied scaling limits (without aggregation) for a large class of random processes, which include integrated Poisson shot-noises and regenerative processes.

Chapter 10

Aggregation of Random-Coefficient AR(1) Processes

This chapter discusses aggregation of independent and dependent random-coefficient autoregressive (1) (RCAR(1)) processes $X = \{X(t); t \in \mathbb{Z}\}$. We consider properties of the limit processes $\mathcal{X}^{\mathrm{id}}(t)$ and $\mathcal{X}^{\mathrm{co}}(t)$ with discrete time $t \in \mathbb{Z}$, arising from contemporaneous aggregation of the RCAR(1) process X with independent (idiosyncratic) and common innovations. As noted in the Introduction, the 'memory mechanism' of the RCAR(1) model is very different from that of network traffic models discussed in Chapters 8 and 9: contrary to the latter models, the RCAR(1) process is non-ergodic (where each individual $\{X_i(t)\}$ picks a random value a and sticks to it forever), the long memory being a consequence of a high concentration of a's near the unit root 1. In contrast, in Chapters 8 and 9, each individual input $X = \{X(t)\}$ is 'innovated' at the arrival times T_j of new pulses W_j, and the long memory of X is essentially due to a few 'large contributions' from these 'innovations'.

1 The Aggregated RCAR(1) Process

Many macroeconomic time series are aggregates of microeconomic data representing the time evolution of individual 'microagents'. For instance, we may want to model monthly changes of average household 'wealth' in a population comprising $N = 1$ million households. A simple model for individual household 'wealth' dynamics is the autoregressive equation

$$Y_i(t) = a_i Y_i(t-1) + \varepsilon_i(t) + \eta(t), \quad i = 1, \dots, N, \tag{1}$$

where $\varepsilon_i(t)$ and $\eta(t)$ are random processes (idiosyncratic and common 'shocks'), accounting for a change of 'wealth' at time t, and $a_i \in [0,1]$ is a nonrandom quantity representing the portion of the past 'wealth' $Y_i(t-1)$. It is sensible to assume that the dynamics in (1) is heterogeneous and that the autoregressive coefficients $a_i, i = 1, \dots, N$ vary in the population. Since tracing a huge number N of individual households is unrealistic, we may randomly sample a much smaller number M (say, $M = 1000$) of them and average the data from the sampled evolutions

$$X_j(t) = a_j X_j(t-1) + \varepsilon_j(t) + \eta(t), \quad j = 1, \dots, M, \tag{2}$$

where $a_j, j = 1, \dots, M$ are i.i.d. r.v.s with the probability distribution equal to the population distribution of $a_i, i = 1, \dots, N$. The above rather simplistic econometric discourse leads to the aggregation problem for sums $\mathcal{X}_M(t) = \sum_{j=1}^M X_j(t)$ of RCAR(1) processes in (2) with i.i.d. coefficients $\{a_j\}$ and i.i.d. innovations $\{\varepsilon_j(s)\}, \{\eta(s)\}$.

Section 1 discusses properties of the limit-aggregated dynamics $\mathcal{X}(t) = \operatorname{fdd-lim}_{M\to\infty} d_M^{-1} \mathcal{X}_M(t), t \in \mathbb{Z}$ under various conditions on the distribution of random coefficients and innovations, including the case of infinite variance. We start with the simplest stationary first-order autoregressive (AR(1)) sequence, defined as a solution of the difference equation

$$X(t) = aX(t-1) + \varepsilon(t), \quad t \in \mathbb{Z}, \tag{3}$$

where $\{\varepsilon, \varepsilon(t); t \in \mathbb{Z}\}$ is an i.i.d. sequence and $a \in [0,1)$ is an autoregressive coefficient. By iterating (3), one arrives at its solution:

$$X(t) = \varepsilon(t) + a\varepsilon(t-1) + a^2\varepsilon(t-2) + \cdots = \sum_{s \le t} a^{t-s}\varepsilon(s). \tag{4}$$

The RCAR(1) process corresponds to (4), where $a \in [0,1)$ is an r.v., independent of $\{\varepsilon(t)\}$. The above series converges under mild assumptions on the innovation distribution ε, as seen in what follows. Following Section 2 of Chapter 1, we write $\xi \in (D)_\alpha$ $(0 < \alpha \le 2)$ if the distribution of the r.v. ξ satisfies Assumption $(D)_\alpha$; in other words, ξ belongs to the domain of normal attraction of the α-stable law.

Proposition 1. (*i*) *Assume that* $\mathrm{E}|\varepsilon|^p < \infty$ *for some* $0 < p \le 2$ *and* $\mathrm{E}\varepsilon = 0\,(p \ge 1)$. *Then, there exists a unique strictly stationary solution to* (3) *given by the series in* (4). *The latter series converges conditionally a.s. and in* L^p *for a.e.* $a \in [0, 1)$. *Moreover, if*

$$\mathrm{E}\left[\frac{1}{1-a}\right] < \infty, \tag{5}$$

then the series in (4) *converges unconditionally in* L^p.

(*ii*) *Let* $\varepsilon \in (D)_\alpha$ $(\exists\, 0 < \alpha < 2)$ *and* (5) *hold; for* $\alpha = 1$, *we additionally assume* $\mathrm{E}\big[\frac{1}{(1-a)^{1+\delta}}\big] < \infty$ $(\exists\, \delta > 0)$. *Then,* $X(t) \in (D)_\alpha$ $(\forall\, t \in \mathbb{Z})$.

Proof. (i) Let us first prove that (3) admits a unique stationary solution. Let $\{X_i(t)\}, i = 1, 2$ be two such solutions. By iteration, we have that for any $n > 0$, $X_i(0) = \varepsilon(0) + a\varepsilon(-1) + \cdots + a^{n-1}\varepsilon(-n+1) + a^n X_i(-n)$, $i = 1, 2$. Hence, $X_1(0) - X_2(0) = a^n(X_1(-n) - X_2(-n))$ or $|X_1(0) - X_2(0)| \le a^n|X_1(-n) - X_2(-n)|$. For any $\epsilon > 0, 0 < \delta < 1$, and $K > 0$, we can write

$$\begin{aligned}\mathrm{P}(|X_1(0) - X_2(0)| > \epsilon) \le \mathrm{P}(a > 1 - \delta) + \sum_{i=1}^{2} \mathrm{P}(|X_i(-n)| > K)\\ + \mathrm{P}(2(1-\delta)^n K > \epsilon).\end{aligned}$$

Since $a < 1$ a.s., $\mathrm{P}(a > 1 - \delta)$ can be made arbitrarily small by a suitable choice of δ. Next, $\mathrm{P}(|X_i(-n)| > K) = \mathrm{P}(|X_i(0)| > K), i = 1, 2$, do not depend on n by stationarity and can be made arbitrarily small by choosing K large enough. Clearly, $\mathrm{P}(2(1-\delta)^n K > \epsilon) = 0$ for n large enough. This proves $\mathrm{P}(|X_1(0) - X_2(0)| > \epsilon) = 0$. To show the unconditional convergence, we use the von Bahr–Esséen inequality. Then,

$$\mathrm{E}\left[\left|\sum_{k=0}^{\infty} a^k \varepsilon(t-k)\right|^p \Bigg| a\right] \le C\mathrm{E}|\varepsilon|^p \sum_{k=0}^{\infty} a^{kp} = \frac{C\mathrm{E}|\varepsilon|^p}{1 - a^p}. \tag{6}$$

This proves the conditional convergence in L^p of the series in (4). The a.s. convergence of (4) follows from (6). Clearly, (6) and (5) imply that (4) converges unconditionally in L^p.
(ii) Follows from Ref. [121, Thm. 3.1]. □

Under the second-moment condition $\sigma^2 := \mathrm{E}\varepsilon^2 < \infty$ and

$$\mathrm{E}\left[\frac{1}{1-a}\right] < \infty, \tag{7}$$

we see that the RCAR(1) process $\{X(t)\}$ in (4) has zero mean and covariance function

$$\mathrm{Cov}(X(0), X(t)) = \sigma^2 \mathrm{E}\left[\frac{a^t}{1-a^2}\right], \quad t \geq 0, \tag{8}$$

and a spectral density

$$f(x) = \frac{\sigma^2}{2\pi}\mathrm{E}\frac{1}{|1-a\mathrm{e}^{\mathrm{i}x}|^2}, \qquad x \in \Pi.$$

Using the fact that $\int_\Pi \frac{\mathrm{d}x}{|1-a\mathrm{e}^{\mathrm{i}x}|^2} = \frac{2\pi}{1-a^2}, |a| < 1$, we see that $\int_\Pi f(x)\mathrm{d}x = \sigma^2\mathrm{E}\left[\frac{1}{1-a^2}\right] < \infty$ is equivalent to (7). The covariance in (8) may decay as $t \to \infty$ very slowly provided the *mixing distribution* $\Phi(\mathrm{d}u) := \mathrm{P}(a \in \mathrm{d}u)$ is sufficiently concentrated near the 'unit root' $u = 1$. In most of this section, we assume that $\Phi(\mathrm{d}u)$ is absolutely continuous and the mixing density

$$\phi(u) := \Phi(\mathrm{d}u)/\mathrm{d}u, \quad u \in [0,1),$$

regularly varies at $u = 1$:

$$\phi(1-u) \sim (c_\phi + o(1))u^\beta, \quad u \to 0+, \tag{9}$$

for some $c_\phi > 0, \ -1 < \beta < 1$.

Proposition 2. *Let* $\sigma^2 = \mathrm{E}\varepsilon^2 < \infty$ *and* (7) *hold. Then:*

(*i*) $\mathrm{Cov}(X(0), X(t)) \geq 0$ *and* $\sum_{t\in\mathbb{Z}} \mathrm{Cov}(X(0),\ X(t)) = \infty$ *is equivalent to*

$$\mathrm{E}\left[\frac{1}{(1-a)^2}\right] = \infty. \tag{10}$$

Moreover, (10) *implies* $\lim_{x\to 0} f(x) = \infty$.

(*ii*) *In addition, assume the existence of a regularly varying density* ϕ *as in* (9), *where* $c_\phi > 0,\ 0 < \beta < 1$. *Then,*

$$\mathrm{Cov}(X(0), X(t)) \ \sim \ \frac{\sigma^2 c_\phi \Gamma(\beta)}{2} \cdot \frac{1}{t^\beta}, \quad t \to \infty, \tag{11}$$

and

$$f(x) \ \sim \ \frac{\sigma^2 c_\phi}{4\cos(\pi\beta/2)} \frac{1}{|x|^{1-\beta}}, \quad x \to 0. \tag{12}$$

Proof. (i) From (8), we get $\sum_{t\in\mathbb{Z}} \mathrm{Cov}(X(0), X(t)) = \sigma^2 \mathrm{E}\big[\frac{1}{(1-a)^2}\big]$; therefore, the equivalence of (10) and the divergence of the covariance series is obvious. The statement about spectral density follows by Fatou's lemma: $\liminf_{x\to 0} \mathrm{E}\frac{1}{|1-a\mathrm{e}^{\mathrm{i}x}|^2} \ge \mathrm{E}\liminf_{x\to 0}\frac{1}{|1-a\mathrm{e}^{\mathrm{i}x}|^2} = \mathrm{E}\frac{1}{(1-a)^2} = \infty$.
(ii) The statements can be derived from the classical Tauberian theorem [89, Chapter 13] or through a more elementary computation as follows. By a change of variable,

$$\mathrm{Cov}(X(0), X(t)) = \sigma^2 \int_0^1 (1-u)^t \cdot \frac{u^\beta (c_\phi + o(u))}{u(2-u)} \mathrm{d}u \ = \ \sigma^2 C_t t^{-\beta},$$

where

$$C_t := \int_0^t \Big(1 - \frac{u}{t}\Big)^t \cdot \frac{u^\beta (c_\phi + o(\frac{u}{t}))}{u(2-\frac{u}{t})} \mathrm{d}u \ \to \ (c_\phi/2)\int_0^\infty \mathrm{e}^{-u} u^{\beta-1}\mathrm{d}u$$
$$= c_\phi \Gamma(\beta)/2, \quad t \to \infty,$$

follows from the DCT. Similarly, for $x > 0$,

$$f(x) = \frac{\sigma^2}{2\pi}\int_0^1 \frac{u^\beta (c_\phi + o(u))\mathrm{d}u}{|1 - \mathrm{e}^{\mathrm{i}x} + u\mathrm{e}^{\mathrm{i}x}|^2} \ = \ \frac{\sigma^2}{2\pi} C_x x^{\beta-1},$$

where

$$C_x := \int_0^{1/x} \frac{u^\beta (c_\phi + o(ux))\mathrm{d}u}{|(1-\mathrm{e}^{\mathrm{i}x})/x + u\mathrm{e}^{\mathrm{i}x}|^2} \ \to \ c_\phi \int_0^\infty \frac{u^\beta \mathrm{d}u}{|u - \mathrm{i}|^2}$$
$$= \frac{c_\phi \pi}{2\cos(\pi\beta/2)}, \quad x \to 0,$$

where the last integral is found from Ref. [106, 3.241]. □

For special Beta-type mixing densities, the covariance function can be explicitly found.

Example 3 ([107]). Let

$$\phi(u) = \frac{2}{B(p,q)} u^{2p-1}(1-u^2)^{q-1}, \quad u \in [0,1),$$

where $p, q > 0$, $B(p,q) = \Gamma(p)\Gamma(q)/\Gamma(p+q)$. Then,

$$\begin{aligned}\mathrm{Cov}(X(0), X(t)) &= \frac{2\sigma^2}{B(p.q)} \int_0^1 u^{t+2p-1}(1-u^2)^{q-2} \mathrm{d}x \\ &= \frac{\sigma^2\Gamma(q-1)}{B(p,q)} \frac{\Gamma(p+\frac{t}{2})}{\Gamma(p+\frac{t}{2}+q-1)} \\ &\sim \frac{\sigma^2 2^{q-1}\Gamma(q-1)}{B(p,q)} \frac{1}{t^{q-1}}, \quad t \to \infty,\end{aligned}$$

where we used $\Gamma(t)/\Gamma(t+b) \sim t^{-b}$, $t \to \infty$. We see that for the above ϕ, the RCAR(1) process has the covariance LRD property when $1 < q \le 2$.

Example 4 ([53]). Let

$$\phi(u) = C(d)u^{d-1}(1-u)^{1-2d}(1+u), \quad u \in [0,1),$$

where $d \in (0, \frac{1}{2})$, $C(d) := \Gamma(3-d)/2\Gamma(d)\Gamma(2-2d)$. Then,

$$\begin{aligned}\mathrm{Cov}(X(0), X(t)) &= \sigma^2 C(d) \int_0^1 u^{t+d-1}(1-u)^{-2d} \mathrm{d}x \\ &= \sigma^2 C(d) B(t+d, 1-2d)\end{aligned}$$

agrees with the ARFIMA$(0,d,0)$ covariance function in (28) of Chapter 3 up to a constant multiplicative factor.

The RCAR(1) process in (2) writes as a sum, $X_j(t) = X_j^{\mathrm{id}}(t) + X_j^{\mathrm{co}}(t)$, of two RCAR(1) processes with idiosyncratic and common innovations. Thus, $\mathcal{X}_M(t) = \mathcal{X}_M^{\mathrm{id}}(t) + \mathcal{X}_M^{\mathrm{co}}(t)$ can be written as a sum of two aggregates,

$$\mathcal{X}_M^{\mathrm{id}}(t) := \sum_{j=1}^M X_j^{\mathrm{id}}(t) \quad \text{and} \quad \mathcal{X}_M^{\mathrm{co}}(t) := \sum_{j=1}^M X_j^{\mathrm{co}}(t), \quad t \in \mathbb{Z}, \quad (13)$$

referred to in what follows as the *aggregated RCAR(1) process with independent (idiosyncratic) innovations* and the *aggregated RCAR(1) process with common innovations.* Here, $X_j^{\mathrm{id}}, j \geq 1$ are independent copies of the RCAR(1) process X in (4), whereas

$$X_j^{\mathrm{co}}(t) := \sum_{s \leq t} a_j^{t-s} \eta(s), \quad j \geq 1, \tag{14}$$

where $\{a_j\}$ are i.i.d. copies of the r.v. a and the innovation process $\{\eta(s)\}$ is the same for all j. We are interested in the limit-aggregated processes:

$$\mathcal{X}^{\mathrm{id}}(t) := \operatorname*{fdd-lim}_{M \to \infty} \quad d_{M,\mathrm{id}}^{-1} \mathcal{X}_M^{\mathrm{id}}(t) \quad \text{and}$$

$$\mathcal{X}^{\mathrm{co}}(t) := \operatorname*{fdd-lim}_{M \to \infty} \quad d_{M,\mathrm{co}}^{-1} \mathcal{X}_M^{\mathrm{co}}(t), \tag{15}$$

where $d_{M,\mathrm{id}}, d_{M,\mathrm{co}}$ are normalisations. Under the second-moment conditions of Proposition 5, the first limit with $d_{M,\mathrm{id}} = M^{1/2}$ is identified by the classical CLT as a Gaussian process with the covariance function in (8), whereas the second limit (with $d_{M,\mathrm{co}} = M$) is an MA process with MA coefficients $\bar{a}_j = \mathrm{E} a^j, j \geq 0$.

Proposition 5. *(i) Let $\sigma^2 = \mathrm{E}\varepsilon^2 < \infty$ and (7) hold. Then,*

$$M^{-1/2} \mathcal{X}_M^{\mathrm{id}}(t) \overset{\mathrm{fdd}}{\to} \mathcal{X}^{\mathrm{id}}(t), \tag{16}$$

where $\{\mathcal{X}^{\mathrm{id}}(t); t \in \mathbb{Z}\}$ is a stationary Gaussian process with zero mean and covariance function

$$\mathrm{Cov}(\mathcal{X}^{\mathrm{id}}(0), \mathcal{X}^{\mathrm{id}}(t)) = \sigma^2 \mathrm{E}\left[\frac{a^t}{1-a^2}\right].$$

(ii) Let $\mathrm{E}\eta^2 < \infty$ and

$$\mathrm{E}\left[\frac{1}{(1-a)^{1/2}}\right] < \infty \tag{17}$$

hold. Then,

$$M^{-1} \mathcal{X}_M^{\mathrm{co}}(t) \overset{\mathrm{fdd}}{\to} \mathcal{X}^{\mathrm{co}}(t), \tag{18}$$

where

$$\mathcal{X}^{\mathrm{co}}(t) := \sum_{s \leq t} \bar{a}_{t-s} \eta(s), \quad t \in \mathbb{Z}, \tag{19}$$

is an MA process with coefficients $\bar{a}_j = \mathrm{E} a^j, j \geq 0$. The last process is LRD if and only if $\mathrm{E}[\frac{1}{1-a}] = \infty$.

Proof. As noted above, part (i) is a simple consequence of the i.i.d. property of the summands and the (multidimensional) CLT.

(ii) Let us check that the MA process in (19) is well defined or that $\bar{A} := \sum_{s=0}^{\infty}(\mathrm{E}a^s)^2 < \infty$. By Minkowski's inequality, $\bar{A}^{1/2} \le \mathrm{E}\big(\sum_{s=0}^{\infty} a^{2s}\big)^{1/2} = \mathrm{E}(1-a^2)^{-1/2} \le \mathrm{E}(1-a)^{-1/2} < \infty$ due to (17).

Consider (18). Let $\mathcal{Y}_M(t) := M^{-1}\mathcal{X}_M^{\mathrm{co}}(t) - \mathcal{X}^{\mathrm{co}}(t) = \sum_{s\le t}\{M^{-1}\sum_{j=1}^{M}(a_j^{t-s} - \mathrm{E}a_j^{t-s})\}\eta(s)$. We shall prove that for each $t \in \mathbb{Z}$,

$$\mathrm{E}\big[(\mathcal{Y}_M(t))^2\big|\mathcal{A}\big] \to_p 0, \quad M \to \infty, \tag{20}$$

where $\mathcal{A} := \sigma\{a_1, a_2, \ldots\}$ is the σ-algebra generated by random coefficients. Equation (20) implies $\mathrm{P}\big[|\mathcal{Y}_M(t)| > \epsilon\big|\mathcal{A}\big] \le \epsilon^{-2}\mathrm{E}\big[(\mathcal{Y}_M(t))^2\big|\mathcal{A}\big] \to_p 0$, and hence $\mathrm{P}(|\mathcal{Y}_M(t)| > \epsilon) = \mathrm{EP}\big[|\mathcal{Y}_M(t)| > \epsilon\big|\mathcal{A}\big] \to 0$, for each $\epsilon > 0$. Thus, (20) implies $M^{-1}\mathcal{X}_M^{\mathrm{co}}(t) \to_p \mathcal{X}^{\mathrm{co}}(t)$ and (18).

To show (20), split $\Sigma_M := \mathrm{E}\big[(\mathcal{Y}_M(t))^2\big|\mathcal{A}\big] = \sigma^2\big(\sum_{s=0}^{K} + \sum_{s=K+1}^{\infty}\big)\big\{M^{-1}\sum_{j=1}^{M}(a_j^s - \mathrm{E}a_j^s)\big\}^2 =: \sigma^2(\Sigma_{M,1} + \Sigma_{M,2})$, where $K > 0$ is a large integer. Since $a_j^s - \mathrm{E}a_j^2, j \ge 1$, are zero mean i.i.d., we get that for each $K < \infty$, $\Sigma_{M,1} \to_p 0\,(M \to \infty)$ by the law of large numbers. Thus, (20) follows from

$$\lim_{K\to\infty}\sup_{M\ge 1}\mathrm{E}(\Sigma_{M,2})^{1/2} = 0. \tag{21}$$

Next,

$$\Sigma''_{M,2} \le 2M^{-2}\sum_{s=K+1}^{\infty}\left\{\sum_{j=1}^{M} a_j^s\right\}^2 + 2\sum_{s=K+1}^{\infty}(\mathrm{E}a^s)^2 =: 2(\Sigma'_{M,2} + \Sigma''_{M,2}),$$

where $\Sigma''_{M,2}$ is nonrandom and vanishes with $K \to \infty$, as shown above. Then, it suffices to show (21) with $\Sigma_{M,2}$ replaced by $\Sigma'_{M,2}$. By the Minkowski inequality,

$$\sum_{s=K+1}^{\infty}\left\{\sum_{j=1}^{M} a_j^s\right\}^2 \le \left\{\sum_{j=1}^{M}\left(\sum_{s=K+1}^{\infty} a_j^{2s}\right)^{1/2}\right\}^2$$

$$= \left\{\sum_{j=1}^{M}\left(\frac{a^{2(K+1)}}{1-a_j^2}\right)^{1/2}\right\}^2.$$

Hence, using (17), we conclude that $\mathrm{E}(\Sigma'_{M,2})^{1/2} \le \mathrm{E}\left[\frac{a^{K+1}}{(1-a^2)^{1/2}}\right] \to 0\,(K \to \infty)$, proving (20) and (18). Finally, since $\sum_{s=0}^{\infty} \bar{a}_s = \mathrm{E}[\frac{1}{1-a}]$, the aggregated process $\mathcal{X}^{\mathrm{co}}$ is LRD if and only if the last expectation is infinite. □

We note that the limit processes in Proposition 5 *cannot* have the ND property since their covariance functions are nonnegative. In the case when the mixing distribution has a density as in (9) with $-1/2 < \beta < 0$, $\mathcal{X}^{\mathrm{co}}$ is LRD, and its MA coefficients decay regularly with exponent $1 + \beta \in (\frac{1}{d}, 1)$:

$$\begin{aligned}\bar{a}_j &= \int_0^1 (1-u)^j \cdot u^\beta (c_\phi + o(u))\,\mathrm{d}u \\ &= C_j j^{-1-\beta} \sim c_\phi \Gamma(\beta+1) j^{-1-\beta},\end{aligned} \tag{22}$$

where

$$\begin{aligned}C_j &:= \int_0^j \left(1 - \frac{u}{j}\right)^j \cdot u^\beta \left(c_\phi + o\left(\frac{u}{j}\right)\right) \mathrm{d}u \to c_\phi \int_0^\infty \mathrm{e}^{-u} u^\beta \mathrm{d}u \\ &= c_\phi \Gamma(\beta+1) \quad j \to \infty,\end{aligned}$$

as in the proof of Proposition 2.

What can be said about the existence of the aggregated RCAR(1) processes $\mathcal{X}^{\mathrm{id}}$ and $\mathcal{X}^{\mathrm{co}}$ when conditions (7) and (17) are violated? The last condition seems close to necessary for the existence of the MA process in (19). Indeed, for ϕ in (9), we have that $\mathrm{E}(1-a)^{-1/2} = \infty$ and $\sum_{j=0}^{\infty} (\bar{a}_j)^2 = \infty$ when $\beta \le -1/2$; see (22). On the other hand, it turns out that for $\beta < 0$ (which excludes (7)), the aggregated process $\mathcal{X}^{\mathrm{id}}$ exists but is a random constant independent of t, which has a symmetric $2(1+\beta)$-stable distribution.

Proposition 6 ([219]). *Let* $\sigma^2 = \mathrm{E}\varepsilon^2 < \infty$ *and* (9) *hold,* $-1 < \beta < 0$. *Then,*

$$M^{-1/2(1+\beta)} \mathcal{X}^{\mathrm{id}}_M(t) \xrightarrow{\mathrm{fdd}} \mathcal{Z}^{\mathrm{id}}, \tag{23}$$

with $\mathcal{Z}^{\mathrm{id}}$ *being an r.v. with ch.f.* $\mathrm{E}\mathrm{e}^{\mathrm{i}\theta \mathcal{Z}^{\mathrm{id}}} = \mathrm{e}^{-K|\theta|^{2(1+\beta)}}$, *where* $K = (\sigma^2/4)^{1+\beta} c_\phi \int_0^\infty (1 - \mathrm{e}^{-z}) z^{-\beta-2} \mathrm{d}z$.

Intuitively, the last result can be explained as follows. It is clear that under (9) with β decreasing, the dependence increases in the RCAR(1) process X, as well as in the limiting process $\mathcal{X}^{\mathrm{id}}$. According to (11), for $\beta > 0$, the covariance function of these processes decays as $t^{-\beta}$, and the exponent tends to 0 with $\beta \downarrow 0$, suggesting that for $\beta < 0$, the dependence in the aggregated process $\mathcal{X}^{\mathrm{id}}_M$ is extremely strong, and the limit process in (23) is completely dependent and degenerate.

Let us turn to the aggregation of the RCAR(1) process with infinite-variance innovations. The existence of $\mathcal{X}^{\mathrm{co}}$ in (18) given by (19) in the infinite variance case is rather straightforward.

Proposition 7 ([218]). *Assume* $\mathrm{E}|\eta|^p < \infty$ *for some* $0 < p \le 2$ *and* $\mathrm{E}\eta = 0\, (p \ge 1)$, *as in Proposition* 1. *In addition, let*

$$\mathrm{E}\left[\frac{1}{(1-a)^{1/p}}\right] < \infty \quad \text{if} \quad p \ge 1,$$

$$\sum_{j=0}^{\infty} \bar{a}_j^p < \infty \qquad\qquad \text{if} \quad 0 < p < 1$$

hold, where $\bar{a}_j = \mathrm{E}a^j$. *Then,* $M^{-1}\mathcal{X}^{\mathrm{co}}_M(t) \overset{\mathrm{fdd}}{\to} \mathcal{X}^{\mathrm{co}}(t)$, *where* $\mathcal{X}^{\mathrm{co}}$ *is the MA process in* (19).

Consider the existence of the aggregated RCAR(1) process $\mathcal{X}^{\mathrm{id}}$ with idiosyncratic infinite variance innovations. If $\varepsilon \in (D)_\alpha\, (0 < \alpha < 2)$, then $X(t) \in (D)_\alpha$ for each t; therefore, the marginal distribution of $M^{1/\alpha}\mathcal{X}_M(t)$ tends to the α-stable limit $\mathcal{X}^{\mathrm{id}}(t)$, according to the classical CLT. We may expect that the latter convergence extends to finite-dimensional convergence and ask about the dependence structure of the limit α-stable process.

The last question has an elegant answer as a stochastic integral:

$$\mathcal{X}^{\mathrm{id}}(t) := \sum_{s \le t} \int_{[0,1)} u^{t-s} W_\alpha(s; \mathrm{d}u), \tag{24}$$

where $W_\alpha(s;\cdot), s \in \mathbb{Z}$ are i.i.d. copies of an α-stable random measure W_α on $[0,1)$, with control measure equal to the mixing distribution Φ. The above random measure has ch.f. $\mathrm{E}\mathrm{e}^{\mathrm{i}\theta W_\alpha(A)} = \mathrm{e}^{-\Phi(A)|\theta|^\alpha \omega_\alpha(\theta)}$, $A \subset [0,1)$, as in (123)–(124) of Chapter 2, with $\mathrm{Leb}_\nu(A)$ replaced by

$\Phi(A)$. (The parameters $c_\pm$ of $\omega_\alpha(\theta)$ are the same as in $(D)_\alpha$.) We note that the sum in (24) can be written as a stochastic integral w.r.t. the α-stable random measure over $\mathbb{Z} \times [0,1)$ with integrand $g(s,u) = u^{t-s}\mathbb{I}(s \le t)$ and control measure equal to the Cartesian product of the counting measure on $\mathbb{Z}$ and the mixing measure Φ on $[0,1)$. The existence of (24) under (7) follows from the fact that $\sum_{\mathbb{Z}} \int_{[0,1)} |g(s,u)|^\alpha \Phi(\mathrm{d}u) = \sum_{s\ge 0} \mathrm{E} a^{s\alpha} = \mathrm{E}[\frac{1}{1-a^\alpha}] \le C\mathrm{E}[\frac{1}{1-a}] < \infty$. The process $\mathcal{X}^{\mathrm{id}}$ in (24) is strictly stationary, ergodic, and has α-stable finite-dimensional distributions. It is generally different from the usual α-stable MA processes and belongs to the vast class of *mixed stable MA processes* studied in Ref. [255]. For $\alpha = 2$, (24) agrees with the limit Gaussian process in Proposition 5 (i) since, from

$$\begin{aligned}\mathrm{Cov}(\mathcal{X}^{\mathrm{id}}(0), \mathcal{X}^{\mathrm{id}}(t)) &= (\sigma^2/2) \sum_{s \le t} \int_{[0,1)} u^{t-2s} \Phi(\mathrm{d}u) \\ &= (\sigma^2/2)\mathrm{E}\left[\frac{a^t}{1-a^2}\right],\end{aligned}$$

c.f. (8).

Theorem 8 ([219]). *Assume* (7) *and* $\varepsilon \in (D)_\alpha$ $(0 < \alpha \le 2)$. *Then,*

$$M^{-1/\alpha} \mathcal{X}^{\mathrm{id}}_M(t) \overset{\mathrm{fdd}}{\to} \mathcal{X}^{\mathrm{id}}(t), \tag{25}$$

where the limit process is given in (24).

Distributional LRD property of the limit aggregated processes $\mathcal{X}^{\mathrm{id}}$ *and* $\mathcal{X}^{\mathrm{co}}$ *with infinite variance.* Distributional LRD/SRD/ND properties were defined in Section 2 of Chapter 1 in terms of the partial sum limits.

For $\mathcal{X}^{\mathrm{co}}$ under power-law mixing density assumption in (9), this issue is a rather simple consequence from Proposition 7, Theorem 29 of Chapter 2, and the asymptotics of $\bar{a}_j$ in (22).

Corollary 9. *Let* $\mathcal{X}^{\mathrm{co}}$ *be the aggregated MA process in* (19), (9) *with i.i.d. innovations* $\eta(s), \eta \in D(\alpha), 0 < \alpha \le 2$.

(*i*) *Let* $1 < \alpha \le 2$ *and* $0 > \beta > \frac{1}{\alpha} - 1$. *Then,*

$$n^{-H} \sum_{t=1}^{[n\tau]} \mathcal{X}^{\mathrm{co}}(t) \overset{\mathrm{fdd}}{\to} -(c_\phi \Gamma(\beta+1)/\beta) L_{\alpha,-\beta}(\tau),$$

where $H = \frac{1}{\alpha} - \beta \in (\frac{1}{\alpha}, 1)$ *and* $\{L_{\alpha,-\beta}(\tau); \tau > 0\}$ *is the fractional linear stable motion in* (150) *of Chapter 2 with* $d = -\beta$.

(*ii*) *Let* $0 < \alpha \le 2$ *and* $\beta > 0$. *Then,*

$$n^{-\frac{1}{\alpha}} \sum_{t=1}^{[n\tau]} \mathcal{X}^{\mathrm{co}}(t) \overset{\mathrm{fdd}}{\to} \bar{A} W_\alpha(\tau),$$

where $\{W_\alpha(\tau); \tau > 0\}$ *is a homogeneous* α*-stable Lévy process as in Theorem 29* (*ii*) *of Chapter 2,* $\bar{A} := \sum_{s=0}^{\infty} \mathrm{E} a^s = \mathrm{E}(1-a)^{-1}$.

A more interesting limit termed *fractional Ornstein–Uhlenbeck (OU) stable process* arises in the partial sum limit of the aggregated process $\mathcal{X}^{\mathrm{id}}$ in (24). Introduce an α-stable random measure $W_\alpha(\mathrm{d}u, \mathrm{d}s)$ on $\mathbb{R}_+ \times \mathbb{R}$ with control measure $\nu(\mathrm{d}u, \mathrm{d}s) := c_\phi u^\beta \mathrm{d}u\, \mathrm{d}s$ and the ch.f. $\mathrm{E}e^{\mathrm{i}\theta W_\alpha(A)} = \exp\{-\nu(A)|\theta|^\alpha \omega_\alpha(\theta)\}, \theta \in \mathbb{R}, A \subset \mathbb{R}_+ \times \mathbb{R}$, $\omega_\alpha(\theta)$ given in (124).

Definition 10. The fractional OU stable process $\Lambda_{\alpha,\beta} = \{\Lambda_{\alpha,\beta}(\tau); \tau > 0\}$, depending on parameters $1 < \alpha \le 2$ and $0 < \beta < \alpha - 1$, is defined as the stochastic integral

$$\Lambda_{\alpha,\beta}(\tau) := \int_{\mathbb{R}_+ \times \mathbb{R}} (\mathfrak{f}(u, \tau - s) - \mathfrak{f}(u, -s)) W_\alpha(\mathrm{d}u, \mathrm{d}s), \quad \tau \ge 0, \quad (26)$$

w.r.t. the α-stable random measure $W_\alpha(\mathrm{d}u, \mathrm{d}s)$ and deterministic kernel

$$\mathfrak{f}(u, t) := \begin{cases} (1 - \mathrm{e}^{-ut})/u, & \text{if } u > 0 \text{ and } t > 0, \\ 0, & \text{otherwise.} \end{cases} \quad (27)$$

Proposition 11. (*i*) *The fractional OU stable process* $\Lambda_{\alpha,\beta}$ *in* (26) *is well defined for any* $1 < \alpha \le 2, 0 < \beta < \alpha - 1$. *It has* α*-stable finite-dimensional distributions and stationary increments. Moreover,* $\Lambda_{\alpha,\beta}$ *is* H*-SS with* $H = 1 - \frac{\beta}{\alpha} \in (\frac{1}{\alpha}, 1)$.

(*ii*) $\Lambda_{2,\beta}(\tau) \overset{\mathrm{fdd}}{=} C_\beta^{1/2} B_H(\tau)$ *agrees with the FBM* $B_H, H = 1 - \frac{\beta}{2}$, *with* C_β *in* (29) *as follows.*

Proof. (i) The existence of the stable integral in (26) follows from

$$I := \int_{\mathbb{R}_+\times\mathbb{R}} |\mathfrak{f}(u,\tau-s)-\mathfrak{f}(u,-s)|^\alpha \nu(\mathrm{d}u,\mathrm{d}s) < \infty, \tag{28}$$

where we can take $\tau = 1$ w.l.g. Using $|\mathrm{e}^z - 1| \le 2(1\wedge|z|)$, $|\mathrm{e}^z-1-z| \le 2(|z|\wedge z^2)$, we get

$$\begin{aligned}
&\int_{\mathbb{R}} |\mathfrak{f}(u,1-s)-\mathfrak{f}(u,-s)|^\alpha \mathrm{d}s\\
&= u^{-\alpha}\int_0^\infty |\mathrm{e}^{-us}(1-\mathrm{e}^{-u})|^\alpha \mathrm{d}s + u^{-\alpha}\int_0^1 |1-\mathrm{e}^{-us}|^\alpha \mathrm{d}s\\
&= \frac{(1-\mathrm{e}^{-u})^\alpha}{\alpha u^{1+\alpha}} + u^{-\alpha}\int_0^1 |1-\mathrm{e}^{-us}|^\alpha \mathrm{d}s\\
&\le C\left(\frac{1}{u}\wedge\frac{1}{u^{\alpha+1}} + \frac{1}{u^\alpha}\wedge 1\right).
\end{aligned}$$

Thus, $I \le C(\int_0^1 u^{\beta-1}\mathrm{d}u + \int_1^\infty u^{\beta-\alpha}\mathrm{d}u) < \infty$, proving (28).

The remaining facts in (i) follow from the general properties of the stable integral. Particularly, the H-SSSI property is obtained by change of variables in the stochastic integral. Namely, $\Lambda_{\alpha,\beta}(\tau+v) - \Lambda_{\alpha,\beta}(v) = \int_{\mathbb{R}_+\times\mathbb{R}}(\mathfrak{f}(u,\tau+v-s)-\mathfrak{f}(u,v-s))W_\alpha(\mathrm{d}u,\mathrm{d}s) = \int_{\mathbb{R}_+\times\mathbb{R}}(\mathfrak{f}(u,\tau-s)-\mathfrak{f}(u,-s))W_\alpha(\mathrm{d}u,\mathrm{d}(s-v))$, where the shifted random measure $\{W_\alpha(\mathrm{d}u,\mathrm{d}(s-v)); (u,s)\in\mathbb{R}_+\times\mathbb{R}\}$ has the same distribution as $\{W_\alpha(\mathrm{d}u,\mathrm{d}s); (u,s)\in\mathbb{R}_+\times\mathbb{R}\}$ for any shift $v\in\mathbb{R}$, implying the SI property of $\Lambda_{\alpha,\beta}$. In a similar way,

$$\begin{aligned}
\Lambda_{\alpha,\beta}(\lambda\tau) &= \int_{\mathbb{R}_+\times\mathbb{R}} (\mathfrak{f}(u,\lambda\tau-s)-\mathfrak{f}(u,-s))W_\alpha(\mathrm{d}u,\mathrm{d}s)\\
&= \lambda^H \int_{\mathbb{R}_+\times\mathbb{R}} (\mathfrak{f}(u,\tau-s)-\mathfrak{f}(u,-s))\lambda^{\beta/\alpha}W_\alpha(\mathrm{d}u/\lambda,\mathrm{d}\lambda s),
\end{aligned}$$

where the rescaled α-stable random measure $\{\lambda^{\beta/\alpha}W_\alpha(\mathrm{d}u/\lambda,\mathrm{d}\lambda s); (u,s)\in\mathbb{R}_+\times\mathbb{R}\}$ has the same distribution as $\{W_\alpha(\mathrm{d}u,\mathrm{d}s); (u,s)\in\mathbb{R}_+\times\mathbb{R}\}$ for any $\lambda>0$, resulting in the H-SS property of $\Lambda_{\alpha,\beta}$.

(ii) follows by Gaussianity and the H-SSSI property of $\Lambda_{2,\beta}$ in part (i) and the computation of the variance:

$$\begin{aligned}
&\mathrm{Var}(\Lambda_{2,\beta}(\tau)) \\
&= c_\phi \sigma^2 \int_{\mathbb{R}_+ \times \mathbb{R}} (\mathfrak{f}(u, \tau - s) - \mathfrak{f}(u, -s))^2 u^\beta \mathrm{d}u \mathrm{d}s \\
&= c_\phi \sigma^2 \int_{\mathbb{R}_+} u^{\beta-2} \mathrm{d}u \left\{ \int_0^\infty \mathrm{e}^{-2us}(1 - \mathrm{e}^{-u\tau})^2 \mathrm{d}s \right. \\
&\quad \left. + \int_0^\tau (1 - \mathrm{e}^{-us})^2 \mathrm{d}s \right\} \\
&= (C_\beta/2)\tau^{2-\beta}, \quad C_\beta := c_\phi \sigma^2 \Gamma(\beta)/(2-\beta)(1-\beta). \qquad (29)
\end{aligned}$$

Proposition 11 is proved. □

Let

$$\begin{aligned}
z_\alpha(\tau; u) &:= \int_{\mathbb{R}} (\mathfrak{f}(u, \tau - s) - \mathfrak{f}(u, -s)) \mathrm{d}\zeta_\alpha(s; u) \qquad (30) \\
&= \int_0^\tau \mathrm{d}t \int_{-\infty}^t \mathrm{e}^{-u(t-s)} \mathrm{d}\zeta_\alpha(s; u), \quad \tau, u > 0
\end{aligned}$$

be a family of integrated α-stable OU processes, $\{ \int_{-\infty}^t \mathrm{e}^{-u(t-s)} \mathrm{d}\zeta_\alpha(s; u); t \in \mathbb{R}\}$, driven by an α-stable Lévy process, $\{\zeta_\alpha(s; u); s \in \mathbb{R}\}$, with stationary and independent increments and the ch.f.

$$\mathrm{E}\mathrm{e}^{\mathrm{i}\theta\zeta_\alpha(s;u)} = \mathrm{e}^{-\sigma_\alpha^\alpha(u)|\theta|^\alpha \omega_\alpha(\theta)}, \quad \theta \in \mathbb{R}, \qquad (31)$$

where $\sigma_\alpha(u) > 0$ is a scale parameter. According to the definition of the stochastic integral,

$$\begin{aligned}
\Lambda_{\alpha,\beta}(\tau) &= \operatorname*{fdd-lim}_{n\to\infty} \sum_{i=1}^n z_\alpha(\tau; i/n), \qquad \text{where} \\
z_\alpha(\tau; i/n) &:= \sum_{i/n \in [0,1)} \int_{\mathbb{R}} (\mathfrak{f}(i/n, \tau - s) - \mathfrak{f}(i/n, -s)) \\
&\quad \times W_\alpha(]i/n, (i+1)/n], \mathrm{d}s)
\end{aligned}$$

are the integrated OU in (30) with $\zeta_\alpha(s; i/n) = W_\alpha(]i/n, (i+1)/n] \times]0, s])$, which are *independent for different* $i = 1, \ldots, n$.

Therefore, the fractional OU stable process can be regarded as a superposition of independent integrated OU stable processes with scale parameters $\sigma_\alpha^\alpha(i/n) := \nu(]i/n,(i+1)/n] \times]0,1])$. As noted above, this process arises in contemporaneous aggregation of independent RCAR(1) time series.

Theorem 12 ([219]). *Let $\mathcal{X}^{\mathrm{id}}$ be the aggregated process in* (24). *Suppose $\varepsilon \in (D)_\alpha$ $(0 < \alpha \le 2)$ as in Theorem* 8, *and the mixing distribution satisfies* (9), *with $0 < \beta < 1$.*

(*i*) *Let $1 < \alpha < 2$ and $0 < \beta < \alpha - 1$. Then,*

$$n^{-H} \sum_{t=1}^{[n\tau]} \mathcal{X}^{\mathrm{id}}(t) \stackrel{\mathrm{fdd}}{\to} \Lambda_{\alpha,\beta}(\tau),$$

where $H = 1 - \frac{\beta}{\alpha}$ and the limit process is given in (26).

(*ii*) *Let $0 < \alpha < 2$ and $0 \vee (\alpha - 1) < \beta < 1$. Then,*

$$n^{-1/\alpha} \sum_{t=1}^{[n\tau]} \mathcal{X}^{\mathrm{id}}(t) \stackrel{\mathrm{fdd}}{\to} K^{1/\alpha} W_\alpha(\tau),$$

where $K := \mathrm{E}(1-a)^{-\alpha}$ and $\{W_\alpha(\tau); \tau > 0\}$ is a homogeneous α-stable Lévy process, as in Theorem 29 (*ii*) *of Chapter 2.*

Since $\Lambda_{\alpha,\beta}$ has dependent increments and L_α has independent increments, we conclude from Theorem 12 that the aggregated process $\mathcal{X}^{\mathrm{id}}$ with mixing density as in (9) has the distributional LRD $0 < \beta < \alpha - 1$ and distributional SRD for $\beta > 0 \vee (\alpha - 1)$.

2 Joint Temporal and Contemporaneous Aggregation of RCAR(1) Processes

Let us turn to the study of joint temporal and contemporaneous aggregates:

$$A_{\lambda,\gamma}^{\mathrm{id}}(x,y) := \sum_{j=1}^{[\lambda^\gamma y]} \sum_{t=1}^{[\lambda x]} X_j^{\mathrm{id}}(t) \quad \text{and} \quad A_{\lambda,\gamma}^{\mathrm{co}}(x,y) := \sum_{j=1}^{[\lambda^\gamma y]} \sum_{t=1}^{[\lambda x]} X_j^{\mathrm{co}}(t), \tag{32}$$

where $X_j^{\mathrm{id}}(t), X_j^{\mathrm{co}}(t)$ are as in (13). Thus, $A_{\lambda,\gamma}^{\mathrm{id}}(x,y)$ is a sum of independent and identically distributed partial sum processes

$\{\sum_{t=1}^{[\lambda x]} X_j^{\mathrm{id}}(t); x \ge 0\}$, whereas $A_{\lambda,\gamma}^{\mathrm{co}}(x,y)$ sums over identically distributed but mutually dependent partial sum processes $\{\sum_{t=1}^{[\lambda x]} X_j^{\mathrm{co}}(t); x \ge 0\}$. As in the previous sections, we are interested in the limit behavior of the two RFs in (32) (indexed by points $(x,y) \in \mathbb{R}_+^2$) when $\lambda \to \infty$, for any given exponent $\gamma > 0$, or, more precisely, in the trichotomies

$$\lambda^{-H(\gamma)} A_{\lambda,\gamma}^{\mathrm{id}}(x,y) \xrightarrow{\mathrm{fdd}} \begin{cases} V_+^{\mathrm{id}}(x,y), & \gamma > \gamma_{\mathrm{cr}}^{\mathrm{id}}, \\ V_-^{\mathrm{id}}(x,y), & \gamma < \gamma_{\mathrm{cr}}^{\mathrm{id}}, \\ V_{\mathrm{cr}}^{\mathrm{id}}(x,y), & \gamma = \gamma_{\mathrm{cr}}^{\mathrm{id}} \end{cases} \tag{33}$$

and

$$\lambda^{-H(\gamma)} A_{\lambda,\gamma}^{\mathrm{co}}(x,y) \xrightarrow{\mathrm{fdd}} \begin{cases} V_+^{\mathrm{co}}(x,y), & \gamma > \gamma_{\mathrm{cr}}^{\mathrm{co}}, \\ V_-^{\mathrm{co}}(x,y), & \gamma < \gamma_{\mathrm{cr}}^{\mathrm{co}}, \\ V_{\mathrm{cr}}^{\mathrm{co}}(x,y), & \gamma = \gamma_{\mathrm{cr}}^{\mathrm{co}}, \end{cases} \tag{34}$$

which can be interpreted as a scaling transition of the aggregated RFs in (32) arising at critical exponents $\gamma_{\mathrm{cr}}^{\mathrm{id}} > 0, \gamma_{\mathrm{cr}}^{\mathrm{co}} > 0$. We shall impose the same condition (9) on the mixing distribution of the RCAR(1) process and consider the finite-variance case $\alpha = 2$ of the RCAR(1) process alone (see Notes at the end of this section on extensions to the infinite-variance case $\alpha < 2$). It is natural to expect that the limits in (33) and (34) depend on β and are different for the two RFs in (32). On the other hand, the scaling transition points are the same in both cases and equal to $1 + \beta$:

$$\gamma_{\mathrm{cr}} = \gamma_{\mathrm{cr}}^{\mathrm{id}} = \gamma_{\mathrm{cr}}^{\mathrm{co}} = 1 + \beta. \tag{35}$$

Schematic scaling diagrams of the two RFs in (32) are presented in Figs. 1 and 2. Let us turn to rigorous definitions of the RFs in Figs. 1 and 2 and the main results for aggregated sums in (32) (Theorems 19 and 20).

Stable Lévy slide. For $-1 < \beta < 0$, let $V_\beta = \{V_\beta(y); y \ge 0\}$ be a $2(1+\beta)-$stable homogeneous Lévy process with ch.f. $\mathrm{E}\mathrm{e}^{\mathrm{i}\theta V_\beta(y)} = \mathrm{e}^{-yK_\beta|\theta|^{2(1+\beta)}}$ and scale parameter

$$K_\beta := c_\phi \int_0^\infty (1 - \mathrm{e}^{-1/4u}) u^\beta \mathrm{d}u = \frac{c_\phi \Gamma(-\beta)}{4^{1+\beta}(1+\beta)}.$$

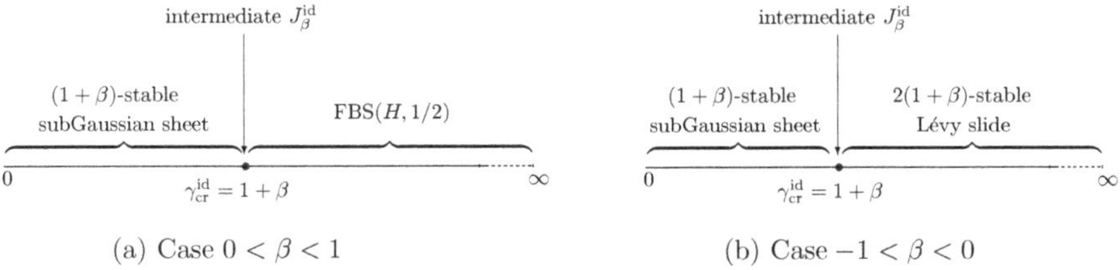

Figure 1. Scaling diagram of RCAR(1) RF $A^{\rm id}_{\lambda,\gamma}$ in (32) with mixing density (9).

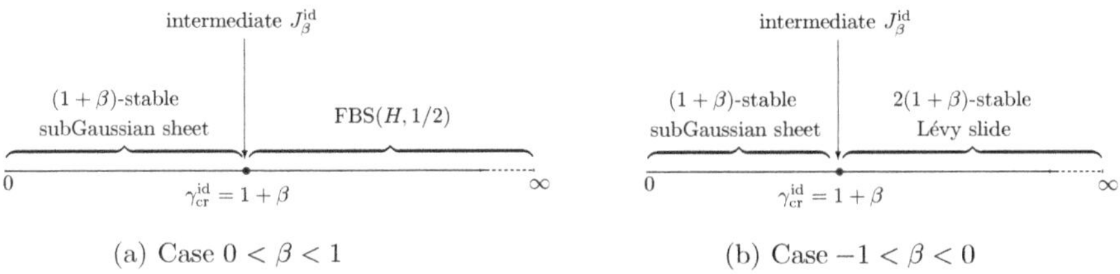

Figure 2. Scaling diagram of RCAR(1) RF $A^{\rm co}_{\lambda,\gamma}$ in (32) with mixing density (9).

The process V_β is nonnegative and has non-decreasing paths, with $V_\beta(1) \overset{\rm d}{=} \mathcal{Z}^{\rm id}$ having the same distribution as the $2(1+\beta)$-stable r.v. $\mathcal{Z}^{\rm id}$ in Proposition 6, (23) for $c_\phi = \sigma = 1$. The RF

$$xV_\beta(y), \quad (x, y) \in \mathbb{R}^2_+ \tag{36}$$

appearing in Theorem 19 is termed a $2(1+\beta)$*-stable Lévy slide* in accordance with the terminology in Section 6 of Chapter 6.

SubGaussian sheet. For $-1 < \beta < 1$, we define the $(1+\beta)$-stable subGaussian sheet $\mathcal{W}_{1+\beta} = \{\mathcal{W}_{1+\beta}(x, y); (x, y) \in \mathbb{R}^2_+\}$ as the stochastic integral

$$\mathcal{W}_{1+\beta}(x, y) = \int_{]0,y]\times C(\mathbb{R}_+)} B(x) W_{1+\beta}(\mathrm{d}v, \mathrm{d}B) \tag{37}$$

w.r.t. a symmetric $(1+\beta)$-stable random measure $W_{1+\beta}$ on the product space $\mathbb{R}_+ \times C(\mathbb{R}_+)$ equipped with the product measure $\mu(\mathrm{d}v, \mathrm{d}B) = \mathrm{d}v \mathrm{P}_B(\mathrm{d}B)$, where P_B is the Wiener measure on $C(\mathbb{R}_+)$ induced by BM B. The characteristic function of $W_{1+\beta}$ is given by

$$\mathrm{E}\mathrm{e}^{\mathrm{i}\theta W_{1+\beta}(A)} = \mathrm{e}^{-|\theta|^{1+\beta}\mu(A)}, \ \theta \in \mathbb{R},$$

for any measurable set $A \subset \mathbb{R}_+ \times C(\mathbb{R}_+)$ of finite measure μ. Recall from Section 5 of Chapter 2, Proposition 23 of Chapter 2, that the stochastic integral $\int_{\mathbb{R}_+ \times C(\mathbb{R}_+)} g(v, B) W_{1+\beta}(\mathrm{d}v, \mathrm{d}B)$ is well defined for any $g \in L^{1+\beta}(\mathbb{R}_+ \times C(\mathbb{R}_+); \mu)$ and has $(1+\beta)$-stable distribution.

Proposition 13. *(i) The sub-Gaussian sheet $\mathcal{W}_{1+\beta}$ in (37) is well defined for any $-1 < \beta < 1$. It has stationary rectangular increments, which are independent in the vertical direction. Moreover, $\mathcal{W}_{1+\beta}$ has symmetric $(1+\beta)$-stable finite-dimensional distributions and is (H_1, H_2)-MSS with $H_1 = 1/2, H_2 = 1/(1+\beta)$:*

$$\mathcal{W}_{1+\beta}(\lambda_1 x, \lambda_2 y) \stackrel{\mathrm{fdd}}{=} \lambda_1^{1/2} \lambda_2^{1/(1+\beta)} \mathcal{W}_{1+\beta}(x, y), \quad \forall \lambda_1, \lambda_2 > 0.$$

(ii) For any fixed $y > 0$,

$$\mathcal{W}_{1+\beta}(x, y) \stackrel{\mathrm{fdd}}{=} (W_{(1+\beta)/2}(y))^{1/2} B(x), \quad x > 0, \tag{38}$$

where B is a standard Brownian motion and $W_{(1+\beta)/2}(y) > 0$ is a $(1+\beta)/2$-stable r.v., independent of B and having the Laplace transform $\mathrm{E}e^{-\theta W_{(1+\beta)/2}(y)} = \exp\{-\kappa_\beta \theta^{(1+\beta)/2} y\}, \theta \geq 0$, where $\kappa_\beta := 2^{(1+\beta)/2} \mathrm{E}|Z|^{1+\beta}, Z \sim N(0,1)$.

Proof. (i) Since $\int_{\mathbb{R}_+ \times C(R_+)} |B(x)|^{1+\beta} \mathbb{I}(0 < v \leq y) \mu(\mathrm{d}v, \mathrm{d}B) = y \mathrm{E}_B |B(x)|^{1+\beta} < \infty$, the stochastic integral in (37) is well defined and

$$\begin{aligned} &\mathrm{E}\exp\left\{\mathrm{i}\sum_{j=1}^m \theta_j \mathcal{W}_{1+\beta}(x_j, y_j)\right\} \\ &\quad = \exp\left\{-\int_{\mathbb{R}_+} \mathrm{d}v\, \mathrm{E}_B \Big|\sum_{j=1}^m \theta_j \mathbb{I}(v \leq y_j) B(x_j)\Big|^{1+\beta}\right\}; \end{aligned}$$

thus, the scaling and rectangular increments properties in part (i) easily follow.

(ii) For any $y > 0, 0 < x_1 < \cdots < x_m, \theta_j \in \mathbb{R}, j = 1, \ldots, m \in \mathbb{N}_+$, we have that

$$\begin{aligned} \mathrm{E}\exp\left\{\mathrm{i}\sum_{j=1}^m \theta_i \mathcal{W}_{1+\beta}(x_i, y)\right\} &= \exp\left\{-y\, \mathrm{E}_B \Big|\sum_{j=1}^m \theta_j B(x_j)\Big|^{1+\beta}\right\} \\ &= \mathrm{e}^{-y\, \mathrm{E}|Z|^{1+\beta} \sigma^{1+\beta}}, \end{aligned} \tag{39}$$

where $Z \sim N(0,1)$ and $\sum_{j=1}^m \theta_j B(x_j) \stackrel{\mathrm{d}}{=} \sigma Z \sim N(0,\sigma^2)$, with

$$\sigma^2 := \mathrm{E}_B \Big| \sum_{j=1}^m \theta_j B(x_j)\Big|^2 = \sum_{k=1}^m \Big(\sum_{j=1}^k \theta_j\Big)^2 (x_k - x_{k-1}) \tag{40}$$

On the other hand, by the definition of the r.v. $W_{(1+\beta)/2}(y)$, the multivariate ch.f. of the random process on the r.h.s. of (38) agrees with the r.h.s. of (39):

$$\begin{aligned}&\mathrm{E}\exp\left\{ \mathrm{i}(W_{(1+\beta)/2}(y))^{1/2} \sum_{j=1}^m \theta_j B(x_j)\right\}\\ &\quad = \mathrm{E}\mathrm{e}^{-(\sigma^2/2)W_{(1+\beta)/2}(y)} = \exp\{-y\kappa_\beta 2^{-(1+\beta)/2}\sigma^{1+\beta}\}. \qquad \square\end{aligned}$$

Intermediate RFs J_β^{id} and J_β^{co}. Introduce a Poisson random measure $\mathcal{N}(\mathrm{d}u, \mathrm{d}v, \mathrm{d}B)$ on the product space $\mathbb{R}_+^2 \times C(\mathbb{R})$ with mean

$$\mu(\mathrm{d}u, \mathrm{d}v, \mathrm{d}B) \;=\; \mathrm{E}\mathcal{N}(\mathrm{d}u, \mathrm{d}v, \mathrm{d}B) \;=\; c_\phi u^\beta \mathrm{d}u\mathrm{d}v \times \mathrm{P}_B(\mathrm{d}B), \tag{41}$$

where P_B is the Wiener measure as in (37) above, and a family

$$\begin{aligned} z(x;u,B) &:= \int_0^x \mathrm{d}t \int_{-\infty}^t \mathrm{e}^{-u(t-s)}\mathrm{d}B(s)\\ &= \int_{\mathbb{R}} \big(\mathfrak{f}(u, x-s) - \mathfrak{f}(u,-s)\big)\mathrm{d}B(s) \quad x \in \mathbb{R}, \end{aligned} \tag{42}$$

of 'elementary' integrated OU processes defined on the same probability space, with kernel $\mathfrak{f}$ in (27). For each $u > 0$, the process $\{z(x;u,B); x \in \mathbb{R}\}$ is a.s. continuously differentiable on $\mathbb{R}$, and its derivative $z'(x;u,B) = \mathrm{d}z(x;u,B)/\mathrm{d}x$ satisfies the Langevin equation:

$$\mathrm{d}z'(x;u,B) = -uz'(x;u,B)\mathrm{d}x + \mathrm{d}B(x).$$

Accordingly, the joint ch.f. of the above Gaussian family of 'elementary' processes $z(x_j;u_j,\cdot)$, $x_j \in \mathbb{R}$, $u_j \in \mathbb{R}_+$, $j = 1,\dots,m \ge 1$ on the

Wiener space $(C(\mathbb{R}_+), \mathrm{P}_B)$ is given by

$$\mathrm{E}_B \exp\left\{\mathrm{i}\sum_{j=1}^{m} \theta_j z(x_j; u_j, B)\right\}$$
$$= \exp\left\{-(1/2)\int_{\mathbb{R}}\left(\sum_{j=1}^{m}\theta_j\left(\mathfrak{f}(u_j, x_j - s) - \mathfrak{f}(u_j, -s)\right)\right)^2 \mathrm{d}s\right\}. \tag{43}$$

Definition 14. The *intermediate RF* J^{id}_β, depending on the parameter $\beta \in (-1, 1)$, is defined as a stochastic integral,

$$\begin{aligned} J^{\mathrm{id}}_\beta(x, y) := &\int_{(0,1)\times(0,y]\times C(\mathbb{R})} z(x; u, B)(\mathcal{N}(\mathrm{d}u, \mathrm{d}v, \mathrm{d}B) \\ &- \mathrm{E}\mathcal{N}(\mathrm{d}u, \mathrm{d}v, \mathrm{d}B)\mathbb{I}(-1/2 < \beta < 1)) \\ &+ \int_{[1,\infty)\times(0,y]\times C(\mathbb{R})} z(x; u, B)\widetilde{\mathcal{N}}(\mathrm{d}u, \mathrm{d}v, \mathrm{d}B), \\ &\times (x, y) \in \mathbb{R}^2_+, \end{aligned} \tag{44}$$

w.r.t. the above Poisson measure with mean μ in (41) and $\widetilde{\mathcal{N}} = \mathcal{N} - \mu$ is the centred Poisson measure.

For $-1/2 < \beta < 1$, the two integrals in (44) can be combined into a single one:

$$\begin{aligned} J^{\mathrm{id}}_\beta(x, y) &= \int_{\mathbb{R}_+\times(0,y]\times C(\mathbb{R})} z(x; u, B)\widetilde{\mathcal{N}}(\mathrm{d}u, \mathrm{d}v, \mathrm{d}B) \\ &= \int_{\mathbb{R}_+\times(0,y]\times C(\mathbb{R})} \left\{\int_{\mathbb{R}} (\mathfrak{f}(u, x - s) - \mathfrak{f}(u, -s))\mathrm{d}B(s)\right\} \\ &\quad\times \widetilde{\mathcal{N}}(\mathrm{d}u, \mathrm{d}v, \mathrm{d}B). \end{aligned} \tag{45}$$

Proposition 15. (*i*) *The intermediate RF J^{id}_β in (44) is well defined for any $-1 < \beta < 1$. It has infinitely divisible finite-dimensional distributions and stationary rectangular increments, which are independent in the vertical direction. Moreover, J^{id}_β is (H, γ)-SS with $H = 3/2$, $\gamma = 1 + \beta$:*

$$J^{\mathrm{id}}_\beta(\lambda x, \lambda^{1+\beta} y) \overset{\mathrm{fdd}}{=} \lambda^{3/2} J^{\mathrm{id}}_\beta(x, y), \quad \forall\, \lambda > 0. \tag{46}$$

(*ii*) *If* $0 < \beta < 1$, *then* $\mathrm{E}(J^{\mathrm{id}}_\beta(x,y))^2 < \infty$ *and*

$$\begin{aligned}
&\mathrm{Cov}\big(J^{\mathrm{id}}_\beta(x_1,y_1), J^{\mathrm{id}}_\beta(x_2,y_2)\big)\\
&\quad = \frac{\Gamma(\beta)\psi_1}{2(2-\beta)(1-\beta)}(y_1\wedge y_2)\big(x_1^{2-\beta} + x_2^{2-\beta} - |x_1 - x_2|^{2-\beta}\big).
\end{aligned} \tag{47}$$

Proof. (i) It suffices to prove the existence of (44) for $y = 1$. Consequently (see (22) of Chapter 2), it suffices to check that $I_1 := \int_{(0,1)\times(0,1)\times C(\mathbb{R})} \mu(\mathrm{d}u,\mathrm{d}v,\mathrm{d}B) = C\int_0^1 u^\beta \mathrm{d}u < \infty$ and $I_2 := \int_{[1,\infty)\times(0,1)\times C(\mathbb{R})} |z(x;u,B)|^2\ \mu(\mathrm{d}u,\mathrm{d}v,\mathrm{d}B) = C\int_1^\infty \mathrm{E}_B|z(x;u,B)|^2\ u^\beta \mathrm{d}u < \infty$. Here, $\mathrm{E}_B|z(x;u,B)|^2 = \sigma^2(x;u)$, where

$$\begin{aligned}
\sigma^2(x;u) &:= \int_{-\infty}^x \big(\mathfrak{f}(u,x-s) - \mathfrak{f}(u,-s)\big)^2 \mathrm{d}s\\
&\le C\left(\frac{(ux)\wedge(ux)^3}{r^3} + \frac{(1-\mathrm{e}^{-ux})^2}{u^3}\right)\\
&\le C\left(\frac{(ux)\wedge(ux)^3}{r^3} + \frac{1\wedge(ux)^2}{u^3}\right) \le C\frac{x}{u^2}\big(1\wedge(xu)\big),
\end{aligned}$$

as in the proof of Proposition 11. Thus, $I_2 < \infty$ when $\beta < 1$. The remaining facts in (i) are easy to prove from the definition of J^{id}_β and its (multivariate) ch.f., so we omit the details.

(ii) From (45) and (42), we have that

$$\begin{aligned}
&\int_{\mathbb{R}_+\times(0,y]\times C(\mathbb{R}_+)} z(x;u,B)^2\mu(\mathrm{d}u,\mathrm{d}v,\mathrm{d}B)\\
&\quad = y\psi_1\int_{\mathbb{R}_+\times\mathbb{R}} (\mathfrak{f}(u,x-s) - \mathfrak{f}(u,-s))^2 u^\beta \mathrm{d}u\mathrm{d}s\\
&\quad = \frac{y\Gamma(\beta)\psi_1}{(2-\beta)(1-\beta)}x^{2-\beta};
\end{aligned}$$

see (29). The computation of the covariance in (47) follows as in Proposition 11. $\square$

The intermediate RF $J^{\rm co}_\beta$ is defined as a *double* stochastic integral,

$$\int_{\mathbb{R}_+\times\mathbb{R}_+\times\mathbb{R}} h(s,u,v)\mathrm{d}B(s)\mathcal{N}(\mathrm{d}u,\mathrm{d}v) =: I(h), \tag{48}$$

with a deterministic kernel, $h = h(s,u,v)$, depending on (x,y) and having a similar form as in $J^{\rm id}_\beta$; see (50) later. The above stochastic integral is somewhat unusual and is different from other stochastic integrals in this book. The definition of the integral in (48) and its basic properties are provided in what follows.

Let B be a standard Brownian motion on $\mathbb{R}$ and $\mathcal{N} = \mathcal{N}(\mathrm{d}u,\mathrm{d}v)$ be a Poisson random measure on $\mathbb{R}_+\times\mathbb{R}_+$ with intensity $c_\phi u^\beta \mathrm{d}u\mathrm{d}v =: \mu(\mathrm{d}u,\mathrm{d}v)$. We assume that B and $\mathcal{N}$ are mutually *independent* and write $\mathrm{E} = \mathrm{E}_B\times\mathrm{E}_{\mathcal{N}}$, where $\mathrm{E}_B, \mathrm{E}_{\mathcal{N}}$ refer to expectations w.r.t. $B, \mathcal{N}$ only. For $1\le p\le 2$, let $\mathcal{L}^p \equiv \mathcal{L}^p(\mathbb{R}\times\mathbb{R}_+\times\mathbb{R}_+)$ denote the Banach space of all measurable real-valued functions $h = h(s,u,v)$, $(s,u,v)\in \mathbb{R}\times\mathbb{R}_+\times\mathbb{R}_+$ such that

$$\|h\|_{\mathcal{L}^p} := \left(\int_{\mathbb{R}_+\times\mathbb{R}_+}\left\{\int_{\mathbb{R}} h^2(s,u,v)\mathrm{d}s\right\}^{p/2}\mu(\mathrm{d}u,\mathrm{d}v)\right)^{1/p}$$
$$+\int_{\mathbb{R}_+\times\mathbb{R}_+}\left\{\int_{\mathbb{R}} h^2(s,u,v)\mathrm{d}s\right\}^{1/2}\mu(\mathrm{d}u,\mathrm{d}v) \ < \ \infty.$$

Let $\mathcal{L}_0$ consist of all (step) functions $h = h(s,u,v)$ taking a finite number of non-zero values $h_n(i,j,k)$ on 'squares' $]i/n,(i+1)/n]\times]j/n,(j+1)/n]\times]k/n,(k+1)/n]\subset\mathbb{R}\times\mathbb{R}_+\times\mathbb{R}_+, i\in\mathbb{Z}, j,k\in\mathbb{N}$ for some $n=1,2,\ldots$. For such $h\in\mathcal{L}_0$, define the double stochastic integral $I(h)$ in (48) as a sum:

$$I(h) \ := \ \textstyle\sum_{i,j,k} h_n(i,j,k)B\big(]\frac{i}{n},\frac{i+1}{n}]\big)\mathcal{N}\big(]\frac{j}{n},\frac{j+1}{n}]\times]\frac{k}{n},\frac{k+1}{n}]\big), \tag{49}$$

where $B(]i/n,(i+1)/n]) := B((i+1)/n) - B(i/n)$.

Proposition 16 ([207]). *For any $1\le p\le 2$, the double stochastic integral $I(h)$ in (49) extends to any $h\in\mathcal{L}^p$, by continuity in L^p, and satisfies the inequality*

$$\mathrm{E}|I(h)|^p \ \le \ C\|h\|^p_{\mathcal{L}^p},$$

with $C > 0$ independent of $h \in \mathcal{L}^p$. Moreover, for any $h \in \mathcal{L}^p$ and any $\theta \in \mathbb{R}$,

$$\begin{aligned} \mathrm{E}\mathrm{e}^{\mathrm{i}\theta I(h)} &= \mathrm{E}_B \exp\left\{\int_{\mathbb{R}_+} (\mathrm{e}^{\mathrm{i}\theta \int_{\mathbb{R}} h(s,u,v)\mathrm{d}B(s)} - 1)\mu(\mathrm{d}u,\mathrm{d}v)\right\} \\ &= \mathrm{E}_{\mathcal{N}} \exp\left\{-\frac{\theta^2}{2}\int_{\mathbb{R}}\left(\int_{\mathbb{R}_+\times\mathbb{R}_+} h(s,u,v)\mathcal{N}(\mathrm{d}u,\mathrm{d}v)\right)^2 \mathrm{d}s\right\}. \end{aligned}$$

In particular, $I(h)$ has a mixed Gaussian distribution with 'random variance' $\int_{\mathbb{R}} \big(\int_{\mathbb{R}_+\times\mathbb{R}_+} h(s,u,v)\mathcal{N}(\mathrm{d}u,\mathrm{d}v)\big)^2 \mathrm{d}s$.

Proposition 17. *The intermediate RF*

$$J^{\mathrm{co}}_\beta(x,y) := \int_{\mathbb{R}_+\times]0,y]\times\mathbb{R}} \big(\mathfrak{f}(u,x-s) - \mathfrak{f}(u,-s)\big)\mathrm{d}B(s)\mathcal{N}(\mathrm{d}u,\mathrm{d}v), \\ \times (x,y) \in \mathbb{R}^2_+ \tag{50}$$

is well defined for any $\beta \in (-1/2, 0)$, as a double stochastic integral of Proposition 16, and satisfies $\mathrm{E}|J^{\mathrm{co}}_\beta(x,y)|^p < \infty$ for any $p \in [1, 2(1+\beta)) \subset [1,2)$. Moreover, J^{co}_β has stationary rectangular increments, satisfying the (H,γ)-SS property with $H = \frac{3}{2}, \gamma = 1+\beta$:

$$J^{\mathrm{co}}_\beta(\lambda x, \lambda^{1+\beta} y) \overset{\mathrm{fdd}}{=} \lambda^{3/2} J^{\mathrm{co}}_\beta(x,y), \quad \forall\, \lambda > 0.$$

Proof. It suffices to show the existence of (50) for $y = 1$. By Proposition 16, $\mathcal{J}^{\mathrm{co}}_\beta(x,1) = I(h_x)$, $h_x(s,u,v) := (\mathfrak{f}(u,x-s) - \mathfrak{f}(u,-s))\mathbb{I}(v \le 1)$ is well defined, provided $\|h_x\|_{\mathcal{L}^p} < \infty$ for some $1 \le p \le 2$. From (48),

$$\sigma^2(x;u,v) := \int_{\mathbb{R}} h^2_x(s,u,v)\mathrm{d}s \;\le\; C\frac{x}{u^2}(1 \wedge (xu))\mathbb{I}(v \le 1),$$

and hence

$$\begin{aligned} &\int_{\mathbb{R}_+} \left\{\int_{\mathbb{R}} h^2_x(s,u,v)\mathrm{d}s\right\}^{p/2} \mu(\mathrm{d}u,\mathrm{d}v) \\ &\quad \le C x^{p/2} \int_0^\infty \left\{\frac{1}{u^2}(1 \wedge (xu))\right\}^{p/2} u^\beta \mathrm{d}u \le\; C x^{(3p/2)-1-\beta} < \infty \end{aligned}$$

for $1 + \beta < 1 \le p < 2(1+\beta)$. Therefore,

$$\mathrm{E}|\mathcal{J}_\beta^{\mathrm{co}}(x,1)|^p \le C\|h_x\|_{\mathcal{L}^p}^p \le C(x^{(3p/2)-1-\beta} + x^{(1-2\beta)(p/2)}) < \infty, \quad x > 0, \quad 1 \le p < 2(1+\beta).$$

The stationarity of rectangular increments property can be easily proved due to the translation invariance properties of the integrand and random measure in (50).

The (H, γ)-SS property of $\mathcal{J}_\beta^{\mathrm{co}}$ can be derived from the characteristic function of this RF given in Proposition 16. A more direct and intuitively simpler way is to use the invariance of the integrand and the random measure in (50) under automorphism $(s, u, v) \to (\lambda s, u/\lambda, \lambda^\gamma v)$ of $\mathbb{R}_+ \times \mathbb{R}_+ \times \mathbb{R}$. We have that $J_\beta^{\mathrm{co}}(\lambda x, \lambda^\gamma y) = \lambda \int_{\mathbb{R}_+ \times]0,y] \times \mathbb{R}} \big(\mathfrak{f}(u, x-s) - \mathfrak{f}(u,-s)\big) \mathrm{d}B(\lambda s) \mathcal{N}(\mathrm{d}u/\lambda, \mathrm{d}\lambda^\gamma v)$, where

$$\mathrm{d}B(\lambda s)\mathcal{N}(\mathrm{d}u/\lambda, \mathrm{d}\lambda^\gamma v) \overset{\mathrm{fdd}}{=} \lambda^{1/2}\mathrm{d}B(s)\mathcal{N}(\mathrm{d}u, \mathrm{d}v), \quad \gamma = 1+\beta$$

by the (1/2)-SS property of B and the invariance $\mu(\mathrm{d}u/\lambda, \mathrm{d}\lambda^\gamma v) = \mu(\mathrm{d}u, \mathrm{d}v)$ of the mean of the Poisson measure $\mathcal{N}$. Hence, $J_\beta^{\mathrm{co}}(\lambda x, \lambda^\gamma y) \overset{\mathrm{fdd}}{=} \lambda^{3/2} J_\beta^{\mathrm{co}}(x,y)$. □

Remark 18. Contrary to the RF J_β^{id}, the RF J_β^{co} does *not* have independent rectangular increments in the vertical direction. Since both RFs appear as scaling limits of the aggregated sums in (32), the lack of this property can be explained by the fact that the summands X_j^{co} are not independent like the summands X_j^{id}. On the other hand, the (H, γ)-SS property holds for general scaling limits, referring to Theorem 8 of Chapter 1, and is not related to the independence of vertical increments.

Further properties of RFs J_β^{id} and J_β^{co} can be found in Refs. [206, 207]. They include the existence of moments, path properties, and the asymptotic self-similarity or 'bridge' properties of random processes $\{J_\beta^{\mathrm{id}}(x,1); x \in \mathbb{R}\}$ and $\{J_\beta^{\mathrm{co}}(x,1); x \in \mathbb{R}\}$ with one-dimensional 'time' $x \in \mathbb{R}$, linking different scaling limits of the aggregated sums in (32) and analogous to the 'bridge processes' of intermediate RFs in Chapters 8 and 9.

Limit results. We first consider anisotropic scaling limits of the aggregated sums $A_{\lambda,\gamma}^{\mathrm{id}}(x,y)$ in (32) of independent RCAR(1) processes with standardised i.i.d. innovations.

Theorem 19. *Let* $\mathrm{E}\varepsilon^2 = 1$ *and the mixing distribution satisfy assumption* (9), *with* $-1 < \beta < 1$ *and* $\gamma_{\mathrm{cr}} = 1 + \beta$.

(*i*) *Let* $0 < \beta < 1$. *Then, the convergence in* (33) *holds for* $\gamma > \gamma_{\mathrm{cr}}$ *with*

$$V_+^{\mathrm{id}}(x, y) = c_+ B_{H_1,H_2}(x, y), \quad H(\gamma) = 1 + \frac{\gamma - \beta}{2},$$

where $c_+^2 := \frac{\Gamma(\beta)c_\phi}{(2-\beta)(1-\beta)}$ *and* B_{H_1,H_2} *is FBS with* $H_1 = 1 - \frac{\beta}{2}, H_2 = \frac{1}{2}$.

(*ii*) *Let* $-1 < \beta < 0$. *Then, the convergence in* (33) *holds for* $\gamma > \gamma_{\mathrm{cr}}$, *with*

$$V_+^{\mathrm{id}}(x, y) = x V_\beta(y), \quad H(\gamma) = 1 + \frac{\gamma}{2(1 + \beta)},$$

the limit given by $2(1 + \beta)$*-stable Lévy slide as defined in* (36).

(*iii*) *Let* $-1 < \beta < 1$. *Then, the convergence in* (33) *holds for* $0 < \gamma < \gamma_{\mathrm{cr}}$ *with*

$$V_-^{\mathrm{id}}(x, y) = c_- \mathcal{W}_{1+\beta}(x, y), \quad H(\gamma) = \frac{1}{2} + \frac{\gamma}{1 + \beta},$$

where $\mathcal{W}_{1+\beta}$ *is the sub-Gaussian sheet in* (37) *and the constant* $c_- > 0$ *is defined by*

$$c_-^{1+\beta} := \frac{2c_\phi \Gamma(\frac{1-\beta}{2})}{2^{(1+\beta)/2}(\beta + 1)\mathrm{E}|Z|^{1+\beta}}, \qquad Z \sim N(0, 1).$$

(*iv*) *Let* $-1 < \beta < 1$. *Then, the convergence in* (33) *holds for* $\gamma = \gamma_{\mathrm{cr}}$, *with*

$$V_{\mathrm{cr}}^{\mathrm{id}}(x, y) \;=\; J_\beta^{\mathrm{id}}(x, y), \quad H(\gamma_{\mathrm{cr}}) = \frac{3}{2},$$

where J_β^{id} *is the intermediate RF in* (45).

Proof. By independence of the summands, it suffices to show (33) for a fixed $y > 0$, or the convergence of characteristic functions:

$$U^{\mathrm{id}}_{\lambda,\gamma}(\boldsymbol{\theta}) := \mathrm{E}\exp\left\{\mathrm{i}\lambda^{-H(\gamma)}\sum_{j=1}^{m}\theta_j A^{\mathrm{id}}_{\lambda,\gamma}(x_j,y)\right\}$$

$$\to \mathrm{E}\exp\left\{\mathrm{i}\sum_{j=1}^{m}\theta_j V^{\mathrm{id}}_{\gamma}(x_j,y)\right\} =: U^{\mathrm{id}}_{\gamma}(\boldsymbol{\theta}), \qquad (51)$$

as $\lambda \to \infty$, for any given $0 < x_1 < \cdots < x_m$, $\boldsymbol{\theta} = (\theta_1,\ldots,\theta_m) \in \mathbb{R}^m, m \geq 1$. Let

$$\vartheta_\lambda(s,a) := \sum_{j=1}^{m}\theta_j\sum_{t=1}^{[\lambda x_j]} a^{t-s}\mathbb{I}(s \leq t). \qquad (52)$$

Then,

$$U^{\mathrm{id}}_{\lambda,\gamma}(\boldsymbol{\theta}) = \left(\mathrm{E}\exp\left\{\mathrm{i}\lambda^{-H(\gamma)}\sum_{s\in\mathbb{Z}}\vartheta_\lambda(s,a)\varepsilon(u)\right\}\right)^{[y\lambda^\gamma]}$$

$$= \left(1 - \frac{\Psi^{\mathrm{id}}_{\lambda,\gamma}(\boldsymbol{\theta})}{[y\lambda^\gamma]}\right)^{[y\lambda^\gamma]},$$

where

$$\Psi^{\mathrm{id}}_{\lambda,\gamma}(\boldsymbol{\theta}) := [y\lambda^\gamma]\left(1 - \mathrm{E}\left[\prod_s \mathrm{E}\big[\mathrm{e}^{\mathrm{i}\lambda^{-H(\gamma)}\vartheta(s,a)\varepsilon(s)}\big|a\big]\right]\right). \qquad (53)$$

On the other hand, since the limit characteristic function $U_\gamma(\boldsymbol{\theta})$ on the r.h.s. of (51) is infinitely divisible in all cases (i)–(iv), it can be represented as

$$U^{\mathrm{id}}_{\gamma}(\boldsymbol{\theta}) = \mathrm{e}^{-\Psi_\gamma(\boldsymbol{\theta})} \qquad (54)$$

with some $\Psi_\gamma(\boldsymbol{\theta})$, which will be identified in each case of Theorem 19. Thus, the convergence in (51) follows from

$$\Psi^{\mathrm{id}}_{\lambda,\gamma}(\boldsymbol{\theta}) \to \Psi_\gamma(\boldsymbol{\theta}), \qquad \forall\, \boldsymbol{\theta} \in \mathbb{R}^m. \qquad (55)$$

Case (i) $0 < \beta < 1, \gamma > \gamma_{\mathrm{cr}}$. We use the representation in (55) of Chapter 1 of the ch.f. of a standardised r.v. ε with finite variance:

there exists an $\epsilon > 0$ such that

$$\chi(\theta) \ := \ \mathrm{E}\mathrm{e}^{\mathrm{i}\theta\varepsilon} = \mathrm{e}^{-(1/2)\theta^2 h(\theta)}, \qquad |\theta| < \epsilon, \tag{56}$$

where $h(\theta) \to 1$ $(\theta \to 0)$. From Proposition 15 (ii), the limit Gaussian RF has a representation,

$$V_+^{\mathrm{id}}(x,y) = \int_{\mathbb{R}_+ \times (0,y] \times C(\mathbb{R})} \left\{ \int_{\mathbb{R}} (\mathfrak{f}(u, x-s) - \mathfrak{f}(u,-s))\mathrm{d}B(s) \right\} \times W(\mathrm{d}u, \mathrm{d}v, \mathrm{d}B),$$

as a stochastic integral w.r.t. Gaussian white noise $W(\mathrm{d}u, \mathrm{d}v, \mathrm{d}B)$ with variance $\mu(\mathrm{d}u, \mathrm{d}v, \mathrm{d}B)$ in (41); c.f. (44). Therefore, the ch.f. in (54) of this limit Gaussian RF can be written as

$$U_\gamma^{\mathrm{id}}(\boldsymbol{\theta}) \ = \ \exp\Big\{ - y(c_\phi/2) \int_0^\infty K(u) u^\beta \mathrm{d}u \Big\},$$
$$\text{where} \quad K(u) := \int_{\mathbb{R}} \left(\sum_{j=1}^m \theta_j \big(\mathfrak{f}(u, x_j - s) - \mathfrak{f}(u, -s)\big) \right)^2 \mathrm{d}s. \tag{57}$$

Since $\sup_{a\in[0,1), s\in\mathbb{Z}} |\vartheta_\lambda(s,a)| = O(\lambda)$ and $H(\gamma) > 1$, $\lambda^{-H(\gamma)}|\vartheta(s,a)| \to 0$ $(\lambda \to \infty)$ uniformly in $a \in [0,1), s \in \mathbb{Z}$. Hence, using (56), $\Psi_{\lambda,\gamma}^{\mathrm{id}}(\boldsymbol{\theta})$ in (53) can be rewritten as

$$\Psi_{\lambda,\gamma}^{\mathrm{id}}(\boldsymbol{\theta}) = [y\lambda^\gamma] \left(1 - \mathrm{E}\exp\left\{ -\frac{1}{2\lambda^{2H(\gamma)}} \sum_{s\in\mathbb{Z}} (\vartheta_\lambda(s,a))^2 \times \ h\big(\lambda^{-H(\gamma)}\vartheta_\lambda(s,a)\big) \right\} \right), \tag{58}$$

where h can be ignored (i.e., we can put $h(\theta) = 1$ w.l.g. in the subsequent proof). Then, after a change of variables, we obtain

$$\Psi_{\lambda,\gamma}^{\mathrm{id}}(\boldsymbol{\theta}) \ = \ \frac{[y\lambda^\gamma]}{\lambda^{1+\beta}} \int_0^\lambda \left(1 - \exp\left\{ -\frac{\lambda^{1+\beta}}{2\lambda^\gamma} K_{\lambda,\gamma}(u) \right\} \right) u^\beta \phi\Big(1 - \frac{u}{\lambda}\Big) \mathrm{d}r,$$

where

$$K_{\lambda,\gamma}(u) := \frac{1}{\lambda^3}\sum_{s\in\mathbb{Z}}\left(\vartheta_n\left(s, 1-\frac{u}{\lambda}\right)\right)^2$$

$$= \int_{\mathbb{R}}\left(\sum_{j=1}^m \theta_j \int_0^{[\lambda x_j]/\lambda}\left(1-\frac{u}{\lambda}\right)^{\lceil\lambda t\rceil-\lceil\lambda s\rceil}\mathbb{I}(s\le t)\mathrm{d}t\right)^2 \mathrm{d}s$$

is written as a Riemann sum. By the DCT, it follows that

$$K_{\lambda,\gamma}(u) \to \int_{\mathbb{R}}\left(\sum_{j=1}^m \theta_j \int_0^{x_j} \mathrm{e}^{-u(t-s)}\mathbb{I}(s\le t)\mathrm{d}t\right)^2 \mathrm{d}s \;=\; K(u)$$

$$\text{for each} \quad u>0,$$

where $K(u)$ is the same as in (57). Moreover, the inequality $1-z \le \mathrm{e}^{-z}$, $z\ge 0$, yields $(1-u/\lambda)^{\lceil\lambda t\rceil-\lceil\lambda s\rceil} \le \mathrm{e}^{-u(\lceil\lambda t\rceil-\lceil\lambda s\rceil)/\lambda} \le C\mathrm{e}^{-u(t-s)}$, $u\ge 0$, $t\ge s$, and hence

$$|K_{\lambda,\gamma}(u)| \le C\sum_{j=1}^m \int_{\mathbb{R}}\left(\int_0^{x_j}\mathrm{e}^{-u(t-s)}\mathbb{I}(s\le t)\mathrm{d}t\right)^2\mathrm{d}s =: \bar{K}(u),$$

with $\bar{K}(u)$ independent of λ and satisfying $\int_0^\infty \bar{K}(u)u^\beta \mathrm{d}u<\infty$. Since $\lim_{\lambda\to\infty}\lambda^{\gamma-1-\beta}\left(1-\exp\{-\lambda^{1+\beta-\gamma}(1/2)K_{\lambda,\gamma}(u)\}\right) = -(1/2)K(u)$, we conclude that (55) holds with

$$\Psi^{\mathrm{id}}_\gamma(\boldsymbol{\theta}) \;=\; -y(c_\phi/2)\int_0^\infty K(u)u^\beta\mathrm{d}u = \log U^{\mathrm{id}}_\gamma(\boldsymbol{\theta}),$$

as in (57).

Case (ii) $-1<\beta<0, \gamma>\gamma_{\mathrm{cr}}$. We prove (55) with $\Psi^{\mathrm{id}}_\gamma(\boldsymbol{\theta}) = -yK_\beta|\sum_{j=1}^m \theta_j x_j|^{2(1+\beta)}$; see (36). With (58) in mind, split the sum $\sum_{s\in\mathbb{Z}}(\vartheta_n(s,a))^2 = R_0(a)+R_1(a)$ into two terms over $s\le 0$ and $1\le s\le[\lambda x_m]$, where

$$R_0(a) := (1-a^2)^{-1}\left(\sum_{j=1}^m \theta_j \sum_{t=1}^{[\lambda x_j]} a^t\right)^2,$$

$$R_1(a) := \sum_{s=1}^{[\lambda x_m]} \left(\sum_{j=1}^{m} \theta_j \sum_{t=s}^{[\lambda x_j]} a^{t-s} \mathbb{I}(s \le t) \right)^2, \quad \text{for} \quad 0 \le a < 1. \tag{59}$$

The simple bound $R_1(a) \le C \min\{\lambda^3, \frac{\lambda}{(1-a)^2}\}$ for any $0 \le a < 1$ will suffice to show that $R_1(a)$ is negligible. As in the proof of (i) above, we may take $h(\theta) = 1$ in (56) and $\phi(u) = c_\phi(1-u)^\beta$ to simplify the derivations. Then, after a change of variable $1 - a = u/\lambda', \lambda' := \lambda^{\gamma/(1+\beta)} \to \infty, \lambda = o(\lambda')$, we get

$$\Psi^{\mathrm{id}}_{\lambda,\gamma}(\boldsymbol{\theta}) = c_\phi \int_0^{\lambda'} \Big(1 - \exp\Big\{ -(1/2)\big(\tilde{R}_0(u) + \tilde{R}_1(u)\big)\Big\}\Big) u^\beta \mathrm{d}u, \tag{60}$$

where, for any $u > 0$,

$$\begin{aligned}\tilde{R}_0(u) &:= \frac{1}{u\,(2 - \frac{u}{\lambda'})} \left(\sum_{j=1}^{m} \theta_j \left(\frac{1}{\lambda} \sum_{t=1}^{[\lambda x_j]} (1 - \frac{u}{\lambda'})^t \right) \right)^2 \\ &\to \frac{1}{2u} \left(\sum_{j=1}^{m} \theta_j x_j \right)^2 =: \tilde{R}(u)\end{aligned}$$

and

$$\tilde{R}_1(u) := \frac{1}{\lambda'\lambda^2} R_1\left(1 - \frac{u}{\lambda'}\right) \le C \min\left\{ \frac{\lambda}{\lambda'}, \frac{\lambda'}{u^2 \lambda} \right\} \to 0. \tag{61}$$

Write $\Psi^{\mathrm{id}}_{\lambda,\gamma}(\boldsymbol{\theta}) = \sum_{i=1}^{3} \tilde{\Psi}_i(\boldsymbol{\theta})$, where

$$\begin{aligned}\tilde{\Psi}_1(\boldsymbol{\theta}) &:= c_\phi \int_0^{\lambda'} \big(1 - \mathrm{e}^{-(1/2)\tilde{R}(u)}\big) u^\beta \mathrm{d}u, \\ \tilde{\Psi}_2(\boldsymbol{\theta}) &= c_\phi \int_0^{\lambda'} \big(\mathrm{e}^{-(1/2)\tilde{R}(u)} - \mathrm{e}^{-(1/2)\tilde{R}_0(u)}\big) u^\beta \mathrm{d}u, \\ \tilde{\Psi}_3(\boldsymbol{\theta}) &= c_\phi \int_0^{\lambda'} \big(\mathrm{e}^{-(1/2)\tilde{R}_0(u)} - \mathrm{e}^{-(1/2)(\tilde{R}_0(u)+\tilde{R}_1(u))}\big) u^\beta \mathrm{d}u.\end{aligned}$$

Here, $\tilde{\Psi}_1(\boldsymbol{\theta}) \to c_\phi \int_0^\infty (1 - \mathrm{e}^{-(1/2)\tilde{R}(u)}) u^\beta \mathrm{d}u = K_\beta |\sum_{j=1}^m \theta_j x_j|^{2(1+\beta)}$ follows by the DCT. The relation $\tilde{\Psi}_2(\boldsymbol{\theta}) \to 0$ follows in a similar way since $H(u) := \mathrm{e}^{-(1/2)\tilde{R}(u)} - \mathrm{e}^{-(1/2)\tilde{R}_0(u)} \to 0$ and $|H(u)| \le |1 - \mathrm{e}^{-(1/2)\tilde{R}(u)}| + |1 - \mathrm{e}^{-(1/2)\tilde{R}_0(u)}| \le C(|\tilde{R}(u)| + |\tilde{R}_0(u)|) \le C\min(1, 1/u) =: \bar{H}(u)$, with $\int_0^\infty \bar{H}(u)\ u^\beta \mathrm{d}u < \infty$. Finally, $\tilde{\Psi}_3(\boldsymbol{\theta}) \to 0$ follows from the bound (61) since $|\tilde{\Psi}_3(\boldsymbol{\theta})| \le c_\phi \int_0^{\lambda'} |1 - \mathrm{e}^{-(1/2)\tilde{R}_1(u)}| u^\beta \mathrm{d}u \le C \int_0^{\lambda'} |\tilde{R}_1(u)| u^\beta \mathrm{d}u \le C(\int_0^{\lambda'/\lambda} + \int_{\lambda'/\lambda}^\infty) |\tilde{R}_1(u)| u^\beta \mathrm{d}u = O((\lambda'/\lambda)^\beta) = o(1)$. This proves (55).

Case (iii) $-1 < \beta < 1, \gamma < \gamma_{\mathrm{cr}}$. We make the simplifications $h(\theta) = 1$ and $\phi(u) = c_\phi (1-u)^\beta$, as in the proof of (ii) above, and use the same decomposition $\sum_{s\in\mathbb{Z}} (\vartheta_n(s,a))^2 = R_0(a) + R_1(a)$ as in (59). We will show that now (for $\gamma < \gamma_{\mathrm{cr}}$), the term $R_0(a)$ is negligible, whereas $R_1(a)$ is dominant and yields the required limit. Indeed, using the obvious bound $R_0(a) \le C(1-a)^{-1} (\sum_{t=1}^\infty a^t)^2 \le C(1-a)^{-3}$ and the same change of variable $1 - a = u/\lambda', \lambda' := \lambda^{\gamma/(1+\beta)}, \lambda' = o(\lambda)$ results in (60), where

$$\tilde{R}_0(u) \le \lambda^{-2H(\gamma)} R_0(u/\lambda') \ \le\ C\left(\frac{\lambda'}{\lambda}\right) u^{-3} \ =\ o(1)$$

and

$$\begin{aligned}\tilde{R}_1(u) &:= \frac{1}{u^2} \cdot \frac{1}{\lambda} \sum_{s=1}^{[\lambda x_m]} \left(\sum_{j=1}^m \theta_j \left(1 - \left(1 - \frac{u}{\lambda'}\right)^{1+[\lambda x_j]-s}\right) \mathbb{I}(s \le [\lambda x_j]) \right)^2 \\ &=: \frac{1}{u^2} \cdot R_1'(u),\end{aligned}$$

where $(1 - \frac{u}{\lambda'})^{1+[\lambda x_j]-s} \le \mathrm{e}^{-([\lambda x_j]-s)/\lambda'} \to 0$ for $0 \le [\lambda x_j] - s = O(\lambda)$. Therefore, by the DCT, we get that

$$R_1'(u) \to \int_0^{x_m} \left(\sum_{j=1}^m \theta_j \mathbb{I}(s \le x_j) \right)^2 \mathrm{d}s = \sum_{k=1}^m \left(\sum_{j=1}^k \theta_j \right)^2 (x_k - x_{k-1})$$

and

$$\Psi^{\mathrm{id}}_{\lambda,\gamma}(\boldsymbol{\theta}) \to y c_\phi \int_0^\infty \left(1 - \exp\left\{-(1/2)u^{-2}\sum_{k=1}^m \left(\sum_{j=1}^k \theta_j\right)^2 \times\ (x_k - x_{k-1})\right\}\right) u^\beta \mathrm{d}u$$

$$= y k_\beta^{1+\beta} \mathrm{E}_B \Big|\sum_{j=1}^m \theta_j B(x_j)\Big|^{1+\beta} \ =\ \Psi^{\mathrm{id}}_\gamma(\boldsymbol{\theta});$$

see (39) and (40), where the last integral is found from $\int_0^\infty (1 - \mathrm{e}^{-z}) z^{-(\beta+3)/2}\mathrm{d}z = \frac{2}{\beta+1}\Gamma\big(\frac{1-\beta}{2}\big)$ $(|\beta| < 1)$ [106, 3.434].

Case (iv) $-1 < \beta < 1, \gamma = \gamma_{\mathrm{cr}}$. From the definitions in (45) and (43) and the ch.f. of a Poisson stochastic integral, we find that

$$\Psi^{\mathrm{id}}_\gamma(\boldsymbol{\theta}) := -\log \mathrm{E} \exp\left\{\mathrm{i}\sum_{j=1}^m \theta_j J^{\mathrm{id}}_\beta(x_j, y)\right\}$$

$$= c_\phi \int_0^\infty \left(1 - \exp\left\{-\frac{1}{2}\int_{\mathbb{R}} \left(\sum_{j=1}^m \theta_j \big(\mathfrak{f}(u, x_j - s) - \mathfrak{f}(u, -s)\big)\right)^2 \mathrm{d}s\right\}\right) u^\beta \mathrm{d}u.$$

Similarly to the previous cases, we prove (55) using the decomposition $\sum_{s\in\mathbb{Z}} (\vartheta_n(s, a))^2 = R_0(a) + R_1(a)$ and the change of variable $1 - a = u/\lambda, \lambda' = \lambda$, with *both* terms relevant when $\gamma = \gamma_{\mathrm{cr}}$. Indeed, in this case, we have (60) with

$$\tilde{R}_0(u) := \frac{\left(1 - \frac{u}{\lambda}\right)^2}{u^3\left(2 - \frac{u}{\lambda}\right)} \left(\sum_{j=1}^m \theta_j \left(1 - \left(1 - \frac{u}{\lambda}\right)^{[\lambda x_j]}\right)\right)^2,$$

$$\tilde{R}_1(u) := \frac{1}{u^2\lambda}\sum_{s=1}^{[\lambda x_m]} \left(\sum_{j=1}^m \theta_j \left(1 - \left(1 - \frac{u}{\lambda}\right)^{[\lambda x_j] - s + 1}\right) \mathbb{I}(s \le [\lambda x_j])\right)^2.$$

It is easy to verify that for each $u > 0$, $\tilde{R}_0(u) \to K_0(u)$, $\tilde{R}_1(u) \to K_1(u)$, where

$$K_0(u) := \frac{1}{2u^3}\left(\sum_{j=1}^m \theta_j(1 - \mathrm{e}^{-ux_j})\right)^2$$

$$= \int_{-\infty}^0 \left(\sum_{j=1}^m \theta_j(\mathfrak{f}(u, x_j - s) - \mathfrak{f}(u, -s))\right)^2 \mathrm{d}s,$$

$$K_1(x) := u^{-2}\int_0^{x_m} \left(\sum_{j=1}^m \theta_j(1 - \mathrm{e}^{-u(x_j - s)})\mathbb{I}(s \le x_j)\right)^2 \mathrm{d}s$$

$$= \int_0^{x_m} \left(\sum_{j=1}^m \theta_j \mathfrak{f}(u, x_j - s)\right)^2 \mathrm{d}s$$

and $\Psi^{\mathrm{id}}_\gamma(\boldsymbol{\theta}) = c_\phi \int_0^\infty (1 - \exp\{-(1/2)(K_0(u) + K_1(u))\})u^\beta \mathrm{d}u$. The convergence (55) now follows by the DCT using a similar argument as in the proof of Theorem 8. □

Theorem 20 discusses scaling limits in (34) of the aggregated RCAR(1) RF $A^{\mathrm{co}}_{\lambda,\gamma}(x, y)$ with common innovations. It turns out that when $-1/2 < \beta < 0$, these limits are sub-Gaussian RFs, $V^{\mathrm{co}}_\pm(x, y)$ taking a product form along directions x and y. Particularly, the 'subcritical' limit

$$V^{\mathrm{co}}_-(x, y) = L_{1+\beta}(y)B(x), \quad (x, y) \in \mathbb{R}^2_+ \tag{62}$$

can be written as a product of two *mutually independent* processes, namely, a standard BM $\{B(x); x \ge 0\}$ and a homogeneous completely asymmetric $(1+\beta)$-stable Lévy process $\{L_{1+\beta}(y); y \ge 0\}$ with non-decreasing trajectories and the Laplace transform

$$\mathrm{E}\mathrm{e}^{-\theta L_{1+\beta}(y)} = \mathrm{e}^{-yK_\beta\theta^{1+\beta}}, \quad \theta \ge 0,$$

where $K_\beta := c_\phi \int_0^\infty (1 - \mathrm{e}^{-1/u})u^\beta \mathrm{d}u = c_\phi \Gamma(-\beta)/(1+\beta)$. Since both $L_{1+\beta}$ and B are SS processes, the product RF in (62) is (H_1, H_2)-MSS with $H_1 = 1/2, H_2 = 1/(1+\beta)$:

$$B(\lambda_1 x)L_{1+\beta}(\lambda_2 y) \stackrel{\text{fdd}}{=} \lambda_1^{1/2} B(x)L_{1+\beta}(y), \qquad \forall \lambda_1, \lambda_2 > 0.$$

The above RF has an infinite variance and stationary rectangular increments since

$$\begin{aligned} &B(x_2)L_{1+\beta}(y_2) - B(x_1)L_{1+\beta}(y_2) - B(x_2)L_{1+\beta}(y_1) + B(x_1)L_{1+\beta}(y_1) \\ &= (B(x_2) - B(x_1))(L_{1+\beta}(y_2) - L_{1+\beta}(y_1)) \stackrel{\text{fdd}}{=} B(x_2 - x_1)L_{1+\beta}(y_2 - y_1); \end{aligned}$$

however, finite-dimensional distributions of (62) are not infinitely divisible and seem rather peculiar.

Theorem 20 ([207]). *Let* $\mathrm{E}\eta^2 = 1$ *and the mixing distribution satisfy assumption* (9), *with* $-1/2 < \beta < 0$, $\gamma_{\text{cr}} = 1 + \beta$. *Then:*

(*i*) *The convergence in* (34) *holds for* $\gamma > \gamma_{\text{cr}}$, *with*

$$V_+^{\text{co}}(x, y) = c_+ y B_H(x), \quad H(\gamma) = \gamma - \beta + \frac{1}{2},$$

where B_H *is a standard FBM with* $H = \frac{1}{2} - \beta$ *and* $c_+ = -c_\phi \Gamma(\beta) C_\beta, C_\beta^2 := \int_{-\infty}^{1}((1-s)^{-\beta} - (-s)_+^{-\beta})^2 \mathrm{d}s = (-\beta)\Gamma(\beta)^2 B(-\beta, 1+2\beta)/(1-2\beta)$; *see Ref. [103, p. 545].*

(*ii*) *The convergence in* (34) *holds for* $0 < \gamma < \gamma_{\text{cr}}$ *with* $V_-^{\text{co}}(x, y)$ *in* (62) *and* $H(\gamma) = \frac{1}{2} + \frac{\gamma}{1+\beta}$.

(*iii*) *The convergence in* (34) *holds for* $\gamma = \gamma_{\text{cr}}$, *with*

$$V_{\text{cr}}^{\text{co}}(x, y) = J_\beta^{\text{co}}(x, y), \quad H(\gamma_{\text{cr}}) = \frac{3}{2},$$

where J_β^{co} *is the intermediate RF in* (50).

Outline of the Proof of Theorem 20. Assume $\eta \sim N(0,1)$ is a standard normal r.v., $\mathrm{E}\mathrm{e}^{\mathrm{i}\theta\eta} = \mathrm{e}^{-\theta^2/2}$. Then, $A_{\lambda,\gamma}^{\text{co}}$ is a sub-Gaussian RF having a conditional Gaussian distribution, given $a_1, a_2, \ldots$, whose ch.f. for a fixed $y > 0$ can be written as

$$\mathrm{E}\exp\left\{\mathrm{i}\lambda^{-H(\gamma)}\sum_{j=1}^{m}\theta_j A^{\mathrm{co}}_{\lambda,\gamma}(x_j,y)\right\}$$

$$= \mathrm{E}\exp\left\{-\frac{1}{2\lambda^{2H(\gamma)}}\sum_{s\in\mathbb{Z}}\left(\sum_{i=1}^{[y\lambda^\gamma]}\vartheta_\lambda(a_i,s)\right)^2\right\}, \tag{63}$$

where $\vartheta_\lambda(a,s)$ is defined at (52). Note that $\{\vartheta_\lambda(a_i,s); s\in\mathbb{Z}\}, i = 1,2,\dots$ are i.i.d. processes, and we need to determine the behavior of the sum in the exponent on the r.h.s. of (63), which can be written as the integral $\sum_{s\in\mathbb{Z}}\left(\lambda^{-H(\gamma)}\sum_{i=1}^{[y\lambda^\gamma]}\vartheta_\lambda(a_i,s)\right)^2 = \int_{\mathbb{R}}(G_\lambda(s;\boldsymbol{\theta}))^2\mathrm{d}s$ of the squared process

$$G_\lambda(s;\boldsymbol{\theta}) := \frac{\lambda^\beta}{\lambda^\gamma}\sum_{i=1}^{[y\lambda^\gamma]}\vartheta_\lambda(a_i,[\lambda s]) = \sum_{j=1}^{m}\theta_j g_\lambda(s;x_j), \quad \text{where}$$

$$g_\lambda(s;x) := \frac{\lambda^\beta}{\lambda^\gamma}\sum_{i=1}^{[y\lambda^\gamma]}\sum_{t=1}^{[\lambda x]}a_i^{t-s}\mathbb{I}(s\le t), \quad s\in\mathbb{R}.$$

The main step in the proof consists of identifying the limit $g(s;x)$ of $g_\lambda(s;x)$, which is nonrandom for $\gamma > \gamma_{\mathrm{cr}}$ and random for $\gamma = \gamma_{\mathrm{cr}}$:

$$g_\lambda(s;x) \overset{\mathrm{fdd}}{\to} g(s,x)$$

$$:= \begin{cases} -c_\phi\Gamma(\beta)((x-s)_+^{-\beta} - (-s)_+^{-\beta}), & \gamma > \gamma_{\mathrm{cr}}, \\ \int_{\mathbb{R}_+}\left\{\int_0^x \mathrm{e}^{-u(t-s)}\mathbb{I}(s\le t)\mathrm{d}t\right\}\mathcal{N}(\mathrm{d}u), & \gamma = \gamma_{\mathrm{cr}}, \end{cases} \tag{64}$$

where $\mathcal{N}(\mathrm{d}u) := \int_{]0,y]}\mathcal{N}(\mathrm{d}u,\mathrm{d}v)$ is the Poisson random measure in (50) integrated over $v \in]0,y]$. The proof of (64) uses characteristic functions and the fact that $\{a_i\}$ are i.i.d. with distribution as in (9), $-1/2 < \beta < 0$. From the r.h.s. of (64), one can recognise the expressions of the characteristic functions of the limit RFs in parts (i) and (iii) of Theorem 20.

For $\gamma < \gamma_{\rm cr}$, the proof of part (ii) of Theorem 20 is more straightforward thanks to the approximation that $A^{\rm co}_{\lambda,\gamma}(x,y) \approx \sum_{s=1}^{[\lambda x]} \eta(s) \times \sum_{i=1}^{[\lambda^\gamma y]} \frac{1}{1-a_i}$ is the product of two mutually independent sums of i.i.d. r.v.s. Condition (9) implies that the distribution of the r.v. $\frac{1}{1-a}$ belongs to the domain of attraction of $(1+\beta)$-stable law with infinite mean. Thus, $\lambda^{-1/2} \sum_{s=1}^{[\lambda x]} \eta(s) \overset{\rm fdd}{\to} B(x)$ and $\lambda^{-1/(1+\beta)} \sum_{i=1}^{[\lambda^\gamma y]} \frac{1}{1-a_i} \overset{\rm fdd}{\to} L_{1+\beta}(y)$, leading to the product limit RF in (62).

One may ask about the limit behaviour of $A^{\rm id}_{\lambda,\gamma}(x,y)$ and $A^{\rm co}_{\lambda,\gamma}(x,y)$ for other values of the parameter β in Theorems 19 and 20. The limit of $A^{\rm id}_{\lambda,\gamma}(x,y)$ for $\beta = 0, \gamma > \gamma_{\rm cr}$ is treated in Ref. [206], resulting in $V^{\rm id}_{+} = c_+ B_{1,1/2}$. The derivations in the above case are more delicate and involve an additional logarithmic normalising factor. On the other hand, the cases $\beta > 1$ in Theorem 19 and $\beta > 0$ in Theorem 20 are relatively easy, leading to the classical BS limit $B_{1/2,1/2}$, independently of $\gamma > 0$ [206, 207].

Remark 21 (Infinite variance RCAR(1) process). A natural question arises concerning the joint limits in (33) and (34) for RCAR(1) 'inputs' with infinite-variance innovations belonging to the domain of attraction of α-stable law, $0 < \alpha < 2$. For independent RCAR(1) inputs $X^{\rm id}_j, j \ge 1$ with generic innovation $\varepsilon \in D(\alpha)$ and a mixing density as in Theorem 19, this question was studied in Ref. [205]. Very roughly, the behaviour of $A^{\rm id}_{\lambda,\gamma}$ in the above case depends on both parameters $\alpha \in (0,2)$ and $-1 < \beta < 1$ but is quite different in the regions (i) $\beta < \alpha - 1$ and (ii) $\beta > \alpha - 1$. In region (i) ('low-variability inputs'), the behaviour of $A^{\rm id}_{\lambda,\gamma}$ resembles that in Theorem 19, with $\gamma_{\rm cr} = 1 + \beta$ independent of α, except that the limits $V^{\rm id}_{\pm}, V^{\rm id}_{\rm cr}$ depend on α and have a similar structure as in the 'Gaussian' case $\alpha = 2$. Particularly, the intermediate RF $J^{\rm id}_{\alpha,\beta}$ arising at $\gamma = 1 + \beta$ has a similar Poisson integral representation as $J^{\rm id}_{\beta}$ in (44), with the difference that the Gaussian OU process $z(x; u, B)$ is replaced by its α-stable counterpart $z_\alpha(x; u)$ in (30), and the Poisson random measure $\mathcal{N}(\mathrm{d}u, \mathrm{d}v, \mathrm{d}\zeta)$ is defined on the product space $\mathbb{R}^2_+ \times D(\mathbb{R})$ involving the probability measure P_α of an α-stable Lévy process instead of the Wiener measure P_B in (44). In the region

$0 > \beta > \alpha - 1$ (which can be described as having a 'very strong dependence' ($\beta < 0$) and 'even stronger variability' ($\alpha < 1 + \beta$) of RCAR(1) 'inputs'), the RF $A^{\rm id}_{\lambda,\gamma}$ 'chooses' between two different limit behaviours: an $\alpha(\beta+1)$-stable Lévy slide $xV_{\alpha(\beta+1)}(y)$ and a Lévy sheet $W_\alpha(x,y)$ satisfying the respective MSS properties: for any $\lambda_1, \lambda_2 > 0$,

$$
\begin{aligned}
(\lambda_1 x)V_{\alpha(\beta+1)}(\lambda_2 y) &\overset{\rm fdd}{=} \lambda_1 \lambda_2^{1/\alpha(\beta+1)} x V_{\alpha(\beta+1)}(y), \\
W_\alpha(\lambda_1 x, \lambda_2 y) &\overset{\rm fdd}{=} \lambda_1^{1/\alpha} \lambda_2^{1/\alpha} W_\alpha(x,y),
\end{aligned}
\tag{65}
$$

The critical $\gamma^{\rm id}_{\rm cr}$ in the above case can be determined by setting $\lambda_1 = \lambda, \lambda_2 = \lambda^\gamma$ and equating the corresponding SS indices in (65): $1 + \frac{\gamma}{\alpha(\beta+1)} = \frac{1+\gamma}{\alpha}$ or

$$
\gamma^{\rm id}_{\rm cr} = \frac{(1+\beta)(1-\alpha)}{(-\beta)} \in (1+\beta, \infty).
$$

The critical $V^{\rm id}_{\rm cr}$ arising at $\gamma = \gamma^{\rm id}_{\rm cr}$ can be written as a sum of mutually independent RFs in (65) and has an unusual distribution as a sum of independent $\alpha(1+\beta)$-stable and α-stable RFs with different stability indices [205].

Remark 22 (Higher-order RCAR 'input' process). The results in this section clearly benefit from the detailed structure of the RCAR(1) process in (3), raising the question of their robustness and extension to higher- order RCAR models, making suitable assumptions about the mixing distribution. In general, the LRD property of a stationary RCAR(p) process depends on the roots of the autoregressive polynomial and can be quite complicated [192]. A rather forthright higher-order version of the RCAR(1) equation in (3) seems to be the RCAR(p) model with real positive roots:

$$
(1 - a_1 B) \cdots (1 - a_p B) X(t) = \varepsilon(t), \quad t \in \mathbb{Z}, \tag{66}
$$

where $a_i \in [0,1)$, $i = 1, \dots, p$, are independent r.v.s and $BX(t) = X(t-1)$ is the backward shift. The stationary solution of (66) is written as an MA process, $X(t) = \sum_{s \le t} b(t-s)\varepsilon(s)$, $t \in \mathbb{Z}$, where

$b(t) := \sum_{0\le s_1\le\cdots\le s_{p-1}\le t} a_1^{s_1} a_2^{s_2-s_1}\cdots a_p^{t-s_{p-1}}$ satisfy

$$\sum_{t=0}^{\infty} b(t) = \prod_{i=1}^{p}(1-a_i)^{-1} =: A.$$

Assuming $\varepsilon \in D(\alpha)$ $(0<\alpha\le 2)$, from Theorem 29 (ii) of Chapter 2, we see that given $a_1,\ldots,a_p$ *conditionally*,

$$\lambda^{-1/\alpha}\sum_{t=1}^{[\lambda\tau]} X(t) \overset{\text{fdd}}{\to} AW_\alpha(\tau). \tag{67}$$

In the case when each $(1-a_i)^{-1}$ has a heavy-tailed distribution with the tail parameter $1+\beta_i>0$ as in the case of the (random) factor A in (67), has a heavy-tailed distribution with parameter $1+\beta_{\min} := 1+\min_{1\le i\le p}\beta_i$ (see Ref. [87, p. 255]), we conjecture that some or many results in Sections 2 and 3 can be extended to the RCAR(p) process in (66) with β replaced by $\beta_{\min}$.

3 Notes

1. The seminal work of Granger [107] notes that 'aggregation of dynamic equations, that is, equations involving lagged dependent variables, can lead to a class of model that has fundamentally different properties to those in current use in econometrics'. The results in Ref. [107] dealing with a special beta-type distributed mixing distribution were extended by Gonçalves and Gouriéroux [105], Zaffaroni [272], Oppenheim and Viano [192], and others to more general RCAR models. Lewbel [167], Linden [168], Chong [56], and Celov *et al.* [53] discuss spectral density and other properties of the aggregated process for parametric mixing density. Contemporaneous aggregation of RCAR(1) series with infinite variance innovations is treated in Puplinskaitė and Surgailis [218, 219]. The discussion of joint temporal and contemporaneous aggregation in Section 2 follows Pilipauskaitė and Surgailis [206, 207] and Pilipauskaitė *et al.* [205].

2. *Contemporaneous aggregation of ARCH-type and stochastic volatility models.* Linear dynamic equations are *homoscedastic* (having a

constant conditional variance), which is a drawback in modelling of financial data, centring on nonlinear ARCH-type and other heteroscedastic processes featuring *random volatility.* Contemporaneous aggregation of heteroscedastic processes was investigated by Ding and Granger [77], Kazakevičius *et al.* [133], Zaffaroni [273, 274], Giraitis *et al.* [96], and elsewhere. The existence of the limit-aggregated stochastic volatility process with LRD was established in Refs. [96, 273]. Probably, the most interesting aspect of Giraitis *et al.* [96] from scaling point of view is the nonlinear limit process

$$Y_G(\tau) = \int_{-\infty}^{\tau} \left\{ \int_{0\vee s}^{\tau} G\left(\frac{B(t)-B(s)}{\sqrt{t-s}}\right) \frac{\mathrm{d}t}{(t-s)^{\frac{3}{2}-H}} \right\} \mathrm{d}B(s), \quad \tau \geq 0, \tag{68}$$

arising as the partial sum limit of the limit-aggregated random coefficient *linear ARCH* (LARCH) process, in a similar way as FBM arises in Corollary 9 (ii), $\alpha = 2$, from the limit aggregated RCAR(1) process $\mathcal{X}^{\mathrm{co}}$ with common innovations. In (68), G is a (nonlinear) function with $\mathrm{E}G^2(Z) < \infty, Z \sim N(0,1)$, and B is a standard Brownian motion, $H \in (\frac{1}{2}, 1)$. The process in (68) is well defined as a backward Itô stochastic integral and satisfies $\mathrm{E}Y_G(\tau) = 0$ and

$$\begin{aligned}
\mathrm{E}Y_G^2(\tau) &= \int_{-\infty}^{\tau} \mathrm{E}\left[\int_{0\vee s}^{\tau} G\left(\frac{B(t)-B(s)}{\sqrt{t-s}}\right) \frac{\mathrm{d}s}{(t-s)^{\frac{3}{2}-H}}\right]^2 \mathrm{d}s \\
&\leq \int_{-\infty}^{\tau} \left\{ \int_{0\vee s}^{\tau} \mathrm{E}^{1/2}\left[G\left(\frac{B(t)-B(s)}{\sqrt{t-s}}\right)^2\right] \frac{\mathrm{d}s}{(t-s)^{\frac{3}{2}-H}} \right\}^2 \mathrm{d}s \\
&= \mathrm{E}G^2(Z) \int_{-\infty}^{\tau} \left\{ \int_{0\vee s}^{\tau} \frac{\mathrm{d}t}{(t-s)^{\frac{3}{2}-H}} \right\}^2 \mathrm{d}s \\
&= c_H^2 \tau^{2H} \mathrm{E}G^2(Z), \quad c_H^2 := B\left(H - \frac{1}{2}, 2-H\right) \Big/ H(2H-1),
\end{aligned}$$

where we used the Minkowski inequality and the fact that $(B(t) - B(s))/\sqrt{t-s} \overset{\mathrm{d}}{=} Z$ $(\forall s \leq t)$. The process $\{Y_G(\tau); \tau \geq 0\}$ in (68) is H-SS and has stationary increments; for constant G, it agrees with an FBM with variance $c_H^2 \mathrm{E}G^2(Z)$. As shown in Giraitis *et al.* [96], using the Hermite expansion of G, $Y_G(\tau)$ can be written as an infinite sum of *Hermite processes* (a particular case of Hermite RFs in (8) of Chapter 5 corresponding to $\nu = 1, \phi(t) = \mathbb{I}(t \in]0, \tau])$). We note that Gaussian subordinated SS RFs given by an infinite Wiener–Itô expansion are described in Dobrushin [79], but (68) seems to be the first instance when such a process appears in limit theorems of probability theory.

3. *Aggregation of RCINAR(1) process.* Barczy *et al.* [19, 20] discuss aggregation of a random-coefficient INAR(1) (RCINAR(1)) process, $X^{\mathrm{IN}} = \{X^{\mathrm{IN}}(t); t \in \mathbb{Z}\}$, defined as in Remark 19, (80) of Chapter 2, by assuming that the autoregressive coefficient $a \in [0, 1)$ is random and independent of the Poisson process $\mathcal{N}(s; \mathrm{d}u)$. Although the covariance functions of AR(1) and INAR(1) processes with *nonrandom a* take a similar form (c.f. (8) and (81)) of Chapter 2, the latter process has a non-zero mean and the covariance function of RCINAR(1) does not vanish with the lag:

$$\begin{aligned}\mathrm{Cov}(X^{\mathrm{IN}}(0), X^{\mathrm{IN}}(t)) &= c\mathrm{E}\left[\frac{a^t}{1-a}\right] + c^2\mathrm{Var}\left(\frac{1}{1-a}\right)\\ &\to c^2\mathrm{Var}\left(\frac{1}{1-a}\right)\end{aligned}$$

as $t \to \infty$. Barczy *et al.* [19] consider iterated limits ($M \to \infty$ followed by $\lambda \to \infty$, or vice versa) of the aggregated sums $A^{\mathrm{IN}}_{\lambda,M}(x) := \sum_{j=1}^{M} \sum_{t=1}^{[\lambda x]} \widetilde{X}^{\mathrm{IN}}_j(t)$ of i.i.d. copies of the *conditionally centred* RCINAR(1) process $\widetilde{X}^{\mathrm{IN}}(t) := X^{\mathrm{IN}}(t) - \mathrm{E}[X^{\mathrm{IN}}(t)|a]$ under the same assumption (9) on the distribution of a, with similar results as in the RCAR(1) case [206]. Note that $\widetilde{X}^{\mathrm{IN}}(t) \overset{\mathrm{fdd}}{=} \widetilde{\mathcal{N}}(A_t)$ has a similar representation to that of $X^{\mathrm{IN}}(t)$ in (80) of Chapter 2, with $\mathcal{N}$ replaced by the centred Poisson random measure $\widetilde{\mathcal{N}}$. Joint limits of the above $A^{\mathrm{IN}}_{\lambda,M}(x)$, in the spirit of Theorem 19, were investigated by Barczy *et al.* [20].

4. *Regime switching AR(1) models.* In econometric literature, a widely used explanation of the long-memory phenomenon is regime switching, where the duration of regime has a heavy-tailed distribution. We refer to Liu [171], Diebold and Inoue [76], Davidson and Sibbertsen [70], Leipus *et al.* [152], and the references therein. Probably, the simplest example of such models is a zero mean i.i.d. sequence, $\{\varepsilon(t); t \in \mathbb{Z}\}$, 'perturbed' by a renewal–reward process, $\{m(t); t \in \mathbb{Z}\}$

(see Example 22 of Chapter 9), or

$$X(t) = \varepsilon(t) + m(t), \tag{69}$$

where $\{\varepsilon(t)\}$ and $\{m(t)\}$ are mutually independent. The 'mean switching' process in (69) regenerates anew at renewal times of $\{m(t)\}$ and the scaling results of Chapter 9 (Theorems 10–15 of Chapter 9) apply to aggregated sums of (69) with little changes. Equation (69) is a particular case of the autoregressive equation

$$X(t) = m(t) + a(t)X(t-1) + \sigma(t)\varepsilon(t), \tag{70}$$

with renewal switching in 'levels' $\{m(t)\}$, 'slope' $\{a(t)\}$, and/or 'volatility' $\{\sigma(t)\}$. LRD properties and partial sums limits of (70) were studied by Leipus and Surgailis [151] and Leipus *et al.* [152, 153], but the latter works do not discuss contemporaneous and joint aggregation.

4. *Statistical inference in RCAR(1) models.* Aggregation by itself inflicts a considerable loss of information about the evolution of individual 'micro-agents', the latter being largely determined by the mixing distribution. The *disaggregation problem* involves recovering the 'lost information', or the mixing distribution from observations of RCAR(1) panel data $\{X_j(t); 1 \le t \le n, 1 \le j \le M\}$ comprising M samples of the RCAR(1) processes in (2), or from the limiting aggregated process $\{\mathcal{X}(t); 1 \le t \le n\}$.

The above problem was investigated in several works. Leipus *et al.* [155] and Celov *et al.* [54] discussed nonparametric estimation of the mixing density $\phi(u)$ using empirical covariances of the limit-aggregated process. Parametric estimation of the mixing density for panel RCAR(1) data is discussed in Robinson [223], Chong [56], Beran *et al.* [32], and other works. In the nonparametric context, Leipus *et al.* [156, 158] studied estimation of the empirical d.f. of a and the tail parameter β in (9) from panel RCAR(1) observations and derived their asymptotic properties as M, n increase to infinity, possibly at different rates. Leipus *et al.* [157] obtained the limit distribution of the sample (cross) covariances:

$$\begin{aligned}\widehat{r}_{M,n}(t,s) := \frac{1}{Mn} \sum_{1\le j, j+s \le M} \sum_{1 \le k, k+t \le n} &(X_j^{\mathrm{id}}(k) - \bar{X}_{M,n}^{\mathrm{id}}) \\ &\times (X_{j+s}^{\mathrm{id}}(k+t) - \bar{X}_{M,n}^{\mathrm{id}}),\end{aligned} \tag{71}$$

computed from an idiosyncratic RCAR(1) panel, $\{X^{\rm id}_j(k); k = 1,\dots,n,\ j = 1,\dots,M\}$, satisfying similar conditions as in Theorem 19. In (71), $\bar X^{\rm id}_{M,n} := \frac{1}{Mn}\sum_{1\le j\le M, 1\le k\le n} X^{\rm id}_j(k)$ is the panel sample mean and $\widehat r_{M,n}(t,s)$ estimate the 'true' covariances $r(t,s) = {\rm E}X^{\rm id}(0)X^{\rm id}(t)\mathbb{I}(s=0)$ for any fixed lag $(t,s)\in\mathbb{Z}^2$. Needless to say, for $s=0$, the last covariance cannot be recovered from a single realisation, $X^{\rm id}(k), k = 1,\dots,n$, of the nonergodic RCAR(1) process. Similarly as in Theorem 19, M and n in Leipus *et al.* [157] increase jointly to infinity, focusing on the power growth rate $M = O(n^\gamma)$. The results in Leipus *et al.* [157] are similar in spirit to those in Theorem 19, exhibiting a trichotomy of the limit distribution of (71) at some $\gamma = \gamma_{\rm cr}$, except that they are more technical and involved and refer to one-dimensional distributions (which correspond to $x = y = 1$ in Theorem 19). We remark that these limit results are different for 'isosectional' $(s = 0)$ and 'cross-sectional' $(s \ne 0)$ covariance estimates in (71) and are also different from Theorem 19 when $M = O(n^\gamma)$, including the critical $\gamma = \gamma_{\rm cr}$ and the 'intermediate' limits at $\gamma = \gamma_{\rm cr}$.

5. *Aggregation of RCAR RFs.* Azomahou [15], Lavancier [148], Leonenko and Taufer [164], and Puplinskaitė and Surgailis [221] discussed aggregation and dependence properties of some RCAR RFs on $\mathbb{Z}^\nu$. Particularly, Ref. [221] studied aggregation of independent copies of stationary solution of nearest-neighbor RCAR equation on $\mathbb{Z}^2$:

$$X(\boldsymbol{t}) = \sum_{|\boldsymbol{u}|=1} a(\boldsymbol{u})X(\boldsymbol{t}+\boldsymbol{u}) + \varepsilon(\boldsymbol{t}), \quad \boldsymbol{t}\in\mathbb{Z}^2, \tag{72}$$

where $\varepsilon(\boldsymbol{t}), \varepsilon, \boldsymbol{t} \in \mathbb{Z}^2$ are i.i.d. r.v.s, $\varepsilon \in D(\alpha)\,(0 < \alpha \le 2)$, and $\boldsymbol{a} := \{a(\boldsymbol{t}); |\boldsymbol{t}| = 1\}$ are random coefficients independent of $\{\varepsilon(\boldsymbol{t})\}$ and satisfying

$$a(\boldsymbol{t}) \ge 0, \quad A := \sum_{|\boldsymbol{t}|=1} a(\boldsymbol{t}) \in (0,1).$$

The stationary solution of (72) obtained by inverting the equation can be written as a (random-coefficient) MA RF:

$$X(\boldsymbol{t}) = \sum_{\boldsymbol{u}\in\mathbb{Z}^2} g(\boldsymbol{t}-\boldsymbol{u}; \boldsymbol{a})\varepsilon(\boldsymbol{u}), \tag{73}$$

where

$$g(\boldsymbol{t};\boldsymbol{a}) = \sum_{k=0}^{\infty} a^{\star k}(\boldsymbol{t}) = \sum_{k=0}^{\infty} A^k p_k(\boldsymbol{t};\boldsymbol{a})$$

is the (random) Green function expressed through the k-step probabilities of the nearest-neighbor random walk $\{S_k; k \geq 0\}$ with one-step probabilities $p_1(\boldsymbol{t};\boldsymbol{a}) := a(\boldsymbol{t})/A$, $|\boldsymbol{t}| = 1$. Reference [221] obtained the conditions on $\boldsymbol{a}, \varepsilon$ for the existence of X in (72)–(73) and the aggregated RF

$$\begin{aligned}\mathcal{X}(\boldsymbol{t}) &:= \operatorname*{fdd-lim}_{M\to\infty} M^{-1/\alpha} \sum_{j=1}^{M} X_j(\boldsymbol{t}) \\ &= \sum_{\boldsymbol{u}\in\mathbb{Z}^2} \int_{[0,1)^4} g(\boldsymbol{t}-\boldsymbol{u};\boldsymbol{a}) W_\alpha(\boldsymbol{u}, \mathrm{d}\boldsymbol{a}), \end{aligned} \tag{74}$$

which extend Proposition 1 and Theorem 8 to dimension $\nu = 2$ and where W_α is an α-stable random measure on $\mathbb{Z}^2 \times [0,1)^4$, similarly as in (24). The main attention in Ref. [221] is given to the dependence and scaling properties of two aggregated RCRFs in (74) corresponding to the random coefficients

$$a(\boldsymbol{t}) = \begin{cases} \dfrac{a}{3}, & |\boldsymbol{t}| = 1, \boldsymbol{t} \neq (-1,0), \\ 0, & \boldsymbol{t} = (-1,0), \end{cases} \quad \text{and} \quad a(\boldsymbol{t}) = \frac{a}{4}, \quad |\boldsymbol{t}| = 1, \tag{75}$$

where $a \in [0,1)$ is an r.v. satisfying the same condition (9) as in the RCAR(1) case. For $\{\mathcal{X}(\boldsymbol{t}); \boldsymbol{t} \in \mathbb{Z}^2\}$ in (74) and (75) (referred to as the 3N and 4N models, respectively), the anisotropic scaling limits in (2) are obtained in Ref. [221] for each $\gamma > 0$, including the scaling transition at $\gamma_{\mathrm{cr}} = 1/2$ (3N model) and $\gamma_{\mathrm{cr}} = 1$ (4N model). Similarly as in Chapters 6 and 7, the unbalanced limits $V_\pm$ of the above 3N and 4N models have a degenerate dependence structure (either invariant or independent rectangular increments in one direction on the plane); for $\alpha = 2$, they agree with FBS $B_{H_1^\pm, H_2^\pm}$, with one of the Hurst indices $H_i^\pm, i = 1,2$ equal to 1 or 1/2.

References

[1] Abramowitz, M. and Stegun, I.A. (1972). *Handbook of Mathematical Functions with Formulas, Graphs, and Mathematical Tables.* Washington, D.C.: U.S. Government printing office.

[2] Adler, R.J. and Taylor, J.E. (2007). *Random Fields and Geometry.* New York: Springer.

[3] Albeverio, S., Molchanov, S.A. and Surgailis, D. (1994). Stratified structure of the Universe and Burgers' equation - a probabilistic approach. *Probab. Theory Relat. Fields*, 100, pp. 457–484.

[4] Anh, V.V., Leonenko, N.N. and Ruiz-Medina, M.D. (2013). Macroscaling limit theorems for filtered spatiotemporal random fields. *Stoch. Anal. Appl.*, 31, pp. 460–508.

[5] Arcones, M. (1994). Limit theorems for nonlinear functions of a stationary Gaussian sequence of vectors. *Ann. Probab.*, 22, pp. 2242–2274.

[6] Asmussen, S. (2003). *Applied Probability and Queues.* 2nd edition. New York: Springer.

[7] Asmussen, S. and Foss, S. (2018). Regular variation in a fixed-point problem for single- and multiclass branching processes and queues. *J. Appl. Probab.*, 50, pp. 47–61.

[8] Astrauskas, A. (1983). Limit theorems for sums of linearly generated random variables. *Lithuanian Math. J.*, 23, pp. 127–134.

[9] Astrauskas, A., Levy, J.B. and Taqqu, M.S. (1991). The asymptotic dependence structure of the linear fractional Lévy motion. *Lithuanian Math. J.*, 31, pp. 1–28.

[10] Avram, F. and Taqqu, M.S. (1987). Noncentral limit theorems and Appell polynomials. *Ann. Probab.*, 15, pp. 767–775.

[11] Ayache, A., Leger, S. and Pointer, M. (2002). Drap brownien fractionnaire. *Potential Anal.*, 17, pp. 41–53.

[12] Ayache, A., Roueff, F. and Xiao, Y. (2007). Local and asymptotic properties of linear fractional stable sheets. *C. R. Acad. Sci. Paris Ser. I*, 344, pp. 389–394.
[13] Ayache, A., Roueff, F. and Xiao, Y. (2009). Linear fractional stable sheets: Wavelet expansion and sample path properties. *Stoch. Process. Appl.*, 119, pp. 1168–1197.
[14] Ayache, A. and Xiao, Y. (2005). Asymptotic properties and Hausdorff dimensions of fractional Brownian sheets. *J. Fourier Anal. Appl.*, 11, pp. 407–439.
[15] Azomahou, T.T. (2009). Memory properties and aggregation of spatial autoregressive models. *J. Stat. Plan. Inf.*, 139, pp. 2581–2597.
[16] Baddeley, A. (1980). A limit theorem for statistics of spatial data. *Adv. Appl. Probab.*, 12, pp. 447–461.
[17] von Bahr, B. and Esséen, C.-G. (1965). Inequalities for the rth absolute moment of the sum of random variables, $1 \le r \le 2$. *Ann. Math. Stat.*, 36, pp. 299–303.
[18] Bai, S. and Taqqu, M.S. (2014). Generalized Hermite processes, discrete chaos and limit theorems. *Stoch. Process. Appl.*, 124, pp. 1710–1739.
[19] Barczy, M., Nedényi, F. and Pap, G. (2017). Iterated scaling limits for aggregation of randomized INAR(1) processes with idiosyncratic Poisson innovations. *J. Math. Anal. Appl.*, 451, pp. 524–543.
[20] Barczy, M., Nedényi, F. and Pap, G. (2021). On simultaneous limits for aggregation of stationary randomized INAR(1) processes with Poisson innovations. *Math. Slovaca*, 71, pp. 1241–1268.
[21] Bardet, J.-M. and Surgailis, D. (2011). Measuring the roughness of random paths by increment ratios. *Bernoulli*, 17, pp. 749–780.
[22] Barndorff-Nielsen, O.E., Benth, F.E. and Veraart, A.E.D. (2018). *Ambit Stochastics*. New York: Springer.
[23] Barndorff-Nielsen, O.E., Concuera, J.M. and Podolskij, M. (2011). Multipower variation for Brownian semistationary processes. *Bernoulli*, 17, pp. 1159–1194.
[24] Barndorff-Nielsen, O.E., Lunde, A., Shepard, N. and Veraart, A.D. (2014). Integer-valued trawl processes: A class of stationary infinitely divisible processes. *Scand. J. Stat.*, 41, pp. 693–724.
[25] Basse-O'Connor, A., Graversen, S.-E. and Pedersen, J. (2012). Multiparameter processes with stationary increments: Spectral representation and integration. *Electron. J. Probab.*, 17, pp. 1–21.
[26] Basse-O'Connor, A., Lachièze-Rey, R. and Podolskij, M. (2017). Power variation for a class of stationary increments Lévy driven moving averages. *Ann. Probab.*, 45, pp. 4477–4528.

[27] Basse-O'Connor, A., Pilipauskaitė, V. and Podolskij, M. (2021). Power variations for fractional type infinitely divisible random fields. *Electron. J. Probab.*, 26, pp. 1–35.

[28] Beghin, L. and Orsinger, E. (2009). Fractional Poisson processes and related planar random motions. *Electronic J. Probab.*, 14, pp. 1790–1826.

[29] Benassi, A., Cohen, S. and Istas, J. (2004). On roughness indices for fractional fields. *Bernoulli*, 10, pp. 357–373.

[30] Beran, J. (1994). *Statistics for Long-Memory Processes.* Monographs on Statistics and Applied Probability, Vol. 61. New York: Chapman and Hall.

[31] Beran, J., Feng, Y., Gosh, S. and Kulik, R. (2013). *Long Memory Processes: Probabilistic Properties and Statistical Methods.* Springer: Heidelberg.

[32] Beran, J., Schützner, M. and Ghosh, S. (2010). From short to long memory: Aggregation and estimation. *Comput. Stat. Data Anal.*, 54, pp. 2432–2442.

[33] Berger, Q. (2019). Strong renewal theorems and local large deviations for multivariate random walks and renewals. *Electron. J. Probab.*, 24, pp. 1–47.

[34] Besag, J. (1974). Spatial interaction and the statistical analysis of lattice systems (with discussion). *J. R. Stat. Soc. B*, 36, pp. 192–236.

[35] Besag, J. and Kooperberg, C. (1995). On conditional and intrinsic autoregressions. *Biometrika*, 82, 733–746.

[36] Bickel, P.J. and Wichura, M.J. (1971). Convergence criteria for multiparameter stochastic processes and some applications. *Ann. Math. Stat.*, 42, pp. 1656–1670.

[37] Biermé, H., Meerschaert, M.M. and Scheffler, H.P. (2007). Operator scaling stable random fields. *Stoch. Process. Appl.*, 117, pp. 312–332.

[38] Biermé, H., Estrade, A. and Kaj, I. (2010). Self-similar random fields and rescaled random balls models. *J. Theor. Probab.*, 23, pp. 1110–1141.

[39] Biermé, H., Durieu, O. and Wang, Y. (2017). Invariance principles for operator-scaling Gaussian random fields. *Ann. Appl. Probab.*, 27, pp. 1190–1234.

[40] Biermé, H., Durieu, O. and Wang, Y. (2018). Generalized operator-scaling random ball model. *ALEA, Lat. Am. J. Probab. Math. Stat.*, 15, pp. 1401–1429.

[41] Biermé, H., Durieu, O. and Surgailis, D. (2025). Limit distribution of the sample volume fraction of Boolean set. *Preprint* http://arxiv.org/abs/2505.13340.

[42] Billingsley, P. (1968). *Convergence of Probability Measures.* New York: Wiley.

[43] Bingham, N.H., Goldie, C.M. and Teugels, J.L. (1989). *Regular Variation.* Cambridge: Cambridge University Press.

[44] Bolin, D. (2014). Spatial Matérn fields driven by non-Gaussian noise. *Scand. J. Stat.*, 41, pp. 557–579.

[45] Bolthausen, E. (1982). On the central limit theorem for stationary mixing random fields. *Ann. Probab.*, 10, pp. 1047–1050.

[46] Boissy, Y., Bhattacharyya, B.B., Li, X. and Richardson, G.D. (2005). Parameter estimates for fractional autoregressive spatial processes. *Ann. Stat.*, 33, pp. 2533–2567.

[47] Breuer, P. and Major, P. (1983). Central limit theorems for non-linear functionals of Gaussian fields. *J. Multivar. Anal.*, 13, pp. 425–441.

[48] Brockwell, P.J. and Davis, R.A. (1991). *Time Series: Theory and Methods.* New York: Springer.

[49] Bružaitė, K., Surgailis, D. and Vaičiulis, M. (2007). Time-varying fractionally integrated processes with finite or infinite variance and nonstationary long memory. *Acta Appl. Math.*, 96, pp. 99–118.

[50] Bulinski, A. and Molchanov, S.A. (1991). Asymptotic normality of a solution of Burgers' equation with random initial data. *Theory Probab. Appl.*, 36, pp. 217–235.

[51] Bulinski, A., Spodarev, E. and Timmermann, F. (2012). Central limit theorems for the excursion sets volumes of weakly dependent random fields. *Bernoulli*, 18, pp. 100–118.

[52] Burgers, J.M. (1974). *The Nonlinear Diffusion Equation.* New York: Reidel.

[53] Celov, D., Leipus, R. and Philippe, A. (2007). Time series aggregation, disaggregation and long memory. *Lithuanian Math. J.*, 47, pp. 379–393.

[54] Celov, D., Leipus, R. and Philippe, A. (2010). Asymptotic normality of the mixture density estimator in a disaggregation scheme. *J. Nonparametric Stat.*, 22, pp. 425–442.

[55] Cieselski, Z. and Kamont, A. (1995). Lévy's fractional Brownian random field and function spaces. *Acta Sci. Math.* (Szeged), 60, pp. 99–118.

[56] Chong, T.T.-L. (2006). The polynomial aggregated AR(1) model. *Econom. J.*, 9, pp. 98–122.

[57] Cohen, J.W. (1982). *The Single Server Queue.* 2nd edition. Amsterdam: North Holland.

[58] Cohen, S. (2012). Fractional Lévy fields. In *Lévy Matters II.* Berlin: Springer, pp. 1–95.

[59] Cohen, S. and Istas, J. (2013). *Fractional Fields and Applications.* Mathématiques et Applications, Vol. 73. Berlin: Springer.

[60] Colombani, L., Durieu, O., Giraudo, D., Pilipauskaite, V. and Redding, L. (2025). Limit theorems: Some recent results. *ESAIM: Proc. Surv.*, 80, pp. 26–50.

[61] Clausel, M. and Vedel, B. (2011). Explicit construction of operator scaling Gaussian random fields. *Fractals*, 19, pp. 101–111.

[62] Cox, D.R. (1984). Long-range dependence: A review. In David, H.A. and David, H.T. (Eds.), *Statistics: An Appraisal.* Iowa: Iowa State University Press, pp. 55–74.

[63] Cressie, N.A.C. (1993). *Statistics for Spatial Data.* New York: Wiley.

[64] Cioczek-Georges, R. and Mandelbrot, B.B. (1995). A class of micropulses and antipersistent fractional Brownian motion. *Stoch. Process. Appl.*, 60, pp. 1–8.

[65] Cioczek-Georges, R. and Mandelbrot, B.B. (1996). Alternative micropulses and fractional Brownian motion. *Stoch. Process. Appl.*, 64, pp. 143–152.

[66] Daley, D.J. and Vesilo, R.A. (1997). Long range dependence of point processes, with queueing examples. *Stoch. Process. Appl.*, 70, pp. 265–282.

[67] Daley, D.J. and Vesilo, R.A. (2000). Long range dependence of inputs and outputs of some classical queues. *Fields Inst. Commun.*, 28, pp. 179–186.

[68] Damarackas, J. and Paulauskas, V. (2017). Spectral covariance and limit theorems for random fields with infinite variance. *J. Multivar. Anal.*, 153, pp. 156–175.

[69] Damarackas, J. and Paulauskas, V. (2021). On Lamperti type limit theorem and scaling transition for random fields. *J. Math. Anal. Appl.*, 497(1), p. 124852.

[70] Davidson, J. and Sibbertsen, P. (2005). Generating schemes for long memory processes: Regimes, aggregation and linearity. *J. Econom.*, 128, pp. 253–282.

[71] Davydov, Y.A. (1970). The invariance principle for stationary processes. *Theor. Probab. Appl.*, 15, pp. 487–498.

[72] Davydov, Y. and Paulauskas, V. (2019). Lamperti type theorems for random fields. processes. *Theor. Probab. Appl.*, 63, pp. 520–544.

[73] Dedecker, J., Doukhan, P., Lang, G., León, J.R., Louhichi, S. and Prieur, C. (2007). *Weak Dependence. With Examples and Applications.* Lecture Notes Statistics, Vol. 190. New York: Springer.

[74] Dehling, H. and Philipp, W. (2002). Empirical process techniques for dependent data. In Dehling, H., Mikosch, T. and Sørensen, M. (Eds.), *Empirical Process Techniques for Dependent Data.* Boston: Birkhäuser, pp. 1–113.

[75] Dehling, H. and Taqqu, M.S. (1989). The empirical process of some long range dependent sequences with an application to U-statistics. *Ann. Stat.*, 17, pp. 1767–1783.
[76] Diebold, F.X. and Inoue, A. (2001). Long memory and regime switching. *J. Econom.*, 105, pp. 131–159.
[77] Ding, Z. and Granger, C.W.J. (1996). Modeling volatility persistence of speculative returns: A new approach. *J. Econom.*, 73, pp. 185–215.
[78] Dobrushin, R.L. (1979). Gaussian and their subordinated self-similar random generalized fields. *Ann. Probab.*, 7, pp. 1–28.
[79] Dobrushin, R.L. (1980). Automodel generalized random fields and their renormgroup. In Dobrushin, R.L. and Sinai, Y.G. (Eds.), *Multicomponent Random Systems*. New York: Dekker, pp. 153–198.
[80] Dobrushin, R.L. and Major, P. (1979). Non-central limit theorems for non-linear functionals of Gaussian fields. *Z. Wahrscheinlichkeitstheor. Verw. Geb.*, 50, pp. 27–52.
[81] Dombry, C. and Kaj, I. (2011). The on-off network traffic under intermediate scaling. *Queueing Syst.*, 69, pp. 29–44.
[82] Doukhan, P., Oppenheim, G. and Taqqu, M.S. (Eds.) (2003). *Theory and Applications of Long-Range Dependence*. Boston: Birkhäuser.
[83] Doukhan, P., Lang, G. and Surgailis, D. (2002). Asymptotics of weighted empirical processes of linear random fields with long range dependence. *Ann. Inst. Henri Poincaré*, 38, pp. 879–896.
[84] Doukhan, P., Lang, G. and Surgailis, D. (2007). Randomly fractionally integrated processes. *Lithuanian Math. J.*, 47, pp. 3–28.
[85] Doukhan, P., Jakubowski, A., Lopes, S.R.C. and Surgailis, D. (2019). Discrete-time trawl processes. *Stoch. Process. Appl.*, 129, pp. 1326–1348.
[86] Ehm, W. (1981). Sample function properties of multi-parameter stable processes. *Z. Wahrscheinlichkeitstheor. Verw. Geb.*, 56, pp. 195–228.
[87] Embrechts, P. and Goldie, C.M. (1980). On closure and factorization properties of subexponential and related distributions. *J. Aust. Math. Soc. (Ser. A)*, 29, pp. 243–256.
[88] Falconer, K.J. (2002). Tangent fields and the local structure of random fields. *J. Theor. Probab.*, 15, pp. 731–750.
[89] Feller, W. (1966). *An Introduction to Probability Theory and Its Applications*, Vol. 2. New York: Wiley.
[90] Ferreti, A., Ippoliti, L., Valentini, P. and Bhansali, R.J. (2023). Long memory conditional random fields on regular lattices. *Environmetrics*, 34(5), p. e2817.
[91] Fyodorov, Y.V., Khoruzhenko, B.A. and Simm, N.J. (2016). Fractional Brownian motion with Hurst index $H = 0$ and the Gaussian unitary ensemble. *Ann. Probab.*, 44, pp. 2980–3031.

[92] Gaigalas, R. (2006). A Poisson bridge between fractional Brownian motion and stable Lévy motion. *Stoch. Process. Appl.*, 116, pp. 447–462.

[93] Gaigalas, R. and Kaj, I. (2003). Convergence of scaled renewal processes and a packet arrival model. *Bernoulli*, 9, pp. 671–703.

[94] Gel'fand, I.M. and Vilenkin, N.Y. (1964). *Generalized Functions - Vol. 4: Applications of Harmonic Analysis.* New York: Academic Press.

[95] Genton, M.G., Perrin, O. and Taqqu, M.S. (2007). Self-similarity and Lamperti transformation for random fields. *Stoch. Models*, 23, pp. 397–411.

[96] Giraitis, L., Leipus, R. and Surgailis, D. (2010). Aggregation of random coefficient GLARCH(1,1) process. *Econom. Theory*, 26, pp. 406–425.

[97] Giraitis, L., Molchanov, S.A. and Surgailis, D. (1992). Long memory shot noises and limit theorems with application to Burgers' equation. In Brillinger, D., *et al.* (Eds.), *New Directions in Time Series Analysis, Part II.* IMA Volumes in Mathematics and its Applications, Vol. 46. New York: Springer, pp. 153–176.

[98] Giraitis, L. and Surgailis, D. (1985). CLT and other limit theorems for functionals of Gaussian processes. *Z. Wahrscheinlichkeitstheor. Verw. Geb.*, 70, pp. 191–212.

[99] Giraitis, L. and Surgailis, D. (1986). Multivariate Appell polynomials and the central limit theorem. In Eberlein, E. and Taqqu, M.S. (Eds.), *Dependence in Probability and Statistics.* Boston: Birkhäuser, pp. 21–71.

[100] Giraitis, L. and Surgailis, D. (1990). On shot noise processes with long range dependence. In Grigelionis, B., *et al.* (Eds.), *Probability Theory and Mathematical Statistics.* Vilnius: Mokslas-VSP, pp. 401–408.

[101] Giraitis, L. and Surgailis, D. (1991). On shot noise processes attracted to fractional Lévy motion. In Cambanis, S., Samorodnitsky, G. and Taqqu, M.S. (Eds.), *Stable Processes and Related Topics.* Boston: Birkhäuser, pp. 261–273.

[102] Giraitis, L. and D. Surgailis, D. (2002). The reduction principle for the empirical process of long memory linear sequence. In Dehling, H., Mikosch, T. and Sorensen, M. (Eds.), *Empirical Process Techniques for Dependent Data.* Boston: Birkhäuser, pp. 241–255.

[103] Giraitis, L., Koul, H.L. and Surgailis, D. (2012). *Large Sample Inference for Long Memory Processes.* London: Imperial College Press.

[104] Gnedenko, B.V. and Kolmogorov, A.N. (1954). *Limit Distributions for Sums of Independent Random Variables.* Cambridge, Mass.: Addison-Wesley.

[105] Gonçalves, E. and Gouriéroux, C. (1988). Aggrégation de processus autoregressifs d'ordre 1. *Ann. Econ. Stat.*, 12, pp. 127–149.
[106] Gradshteyn, I.S. and Ryzhik, I.M. (2000). *Tables of Integrals, Series and Products.* New York: Academic Press.
[107] Granger, C.W.J. (1980). Long memory relationship and the aggregation of dynamic models. *J. Econom.*, 14, pp. 227–238.
[108] Guyon, (1995). *Random Fields on a Network: Modeling, Statistics, and Applications.* New York: Springer-Verlag.
[109] Guttorp, P. and Gneiting, T. (2006). Studies in the history of probability and statistics. XLIX. On the Matérn correlation family. *Biometrika*, 93, pp. 989–995.
[110] Hall, P. and Heyde, C.C. (1980). *Martingale Limit Theory and Applications.* New York: Academic Press.
[111] Hall, P. (1988) *Introduction to the Theory of Coverage Processes.* New York: Wiley.
[112] Hammond, A. and Sheffield, S. (2013). Power-law Pólya's urn and fractional Brownian motion. *Probab. Theory Relat. Fields*, 157, pp. 691–719.
[113] Hankey, A. and Stanley, H.E. (1972). Systematic application of generalized homogeneous functions to static scaling, dynamic scaling, and universality. *Phys. Rev. B*, 6, pp. 3515–3542.
[114] Hansen, L.V. and Thorarinsdottir, T.L. (2013). A note on moving average models for Gaussian random fields. *Stat. Probab. Lett.*, 83, pp. 850–855.
[115] Heath, D., Resnick, S. and Samorodnitsky, G. (1998). Heavy tails and long range dependence in ON/OFF processes and associated fluid models. *Math. Oper. Res.*, 23, pp. 145–165.
[116] Heyde, C.C. and Yang, Y. (1997). On defining long-range dependence. *J. Appl. Probab.*, 34, pp. 939–944.
[117] Hida, T. (1970). *Stationary Stochastic Processes.* Princeton: Princeton University Press.
[118] Hille, E. (1939) Notes on linear transformations, II. Analyticity of semigroups. *Ann. Math.*, 40, 1–47.
[119] Ho, H.-C. and Hsing, T. (1996). On the asymptotic expansion of the empirical process of long memory moving averages. *Ann. Stat.*, 24, pp. 992–1024.
[120] Ho, H.-C. and Hsing, T. (1997). Limit theorems for functionals of moving averages. *Ann. Probab.*, 25, pp. 1636–1669.
[121] Hult, H. and Samorodnitsky, G. (2008). Tail probabilities for infinite series of regularly varying random vectors. *Bernoulli*, 14, pp. 838–864.

[122] Ibragimov, I.A. and Linnik, Y.V. (1971). *Independent and Stationary Sequences of Random Variables.* Groningen: Wolters-Noordhoff.

[123] Iksanov, A. (2016). *Renewal Theory for Perturbed Random Walks and Similar Processes.* Boston: Birkhäuser.

[124] Iksanov, A., Marynych, A. and Meiners, M. (2017). Asymptotics of random processes with immigration I: Scaling limits. *Bernoulli*, 23, pp. 1233–1278.

[125] Itô, K. (1951). Multiple Wiener integral. *J. Math. Soc. Jpn.*, 3, pp. 157–164.

[126] Itô, K. (1956). Spectral type of shift transformations of differential process with stationary increments. *Trans. Am. Math. Soc.*, 81, pp. 253–263.

[127] Jónsdóttir, K.Y., Rønn-Nielsen, A., Mouridsen, K. and Vendel Jensen, E.B. (2013). Lévy-based modelling in brain imaging. *Scand. J. Stat.*, 40, pp. 511–529.

[128] Kaj, I. (2005). Limiting fractal random processes in heavy-tailed systems. In Levy-Vehel, J. and Lutton, E. (Eds.), *Fractals in Engineering, New Trends in Theory and Applications.* London: Springer, pp. 199–218.

[129] Kaj, I., Leskelä, L., Norros, I. and Schmidt, V. (2007). Scaling limits for random fields with long-range dependence. *Ann. Probab.*, 35, pp. 528–550.

[130] Kaj, I. and Taqqu, M.S. (2008). Convergence to fractional Brownian motion and to the Telecom process: the integral representation approach. In Vares, M.E. and Sidoravicius, V. (Eds.), *In and Out of Equilibrium 2.* Progress in Probability, Vol. 60. Basel: Birkhäuser, pp. 383–427.

[131] Kamont, A. (1996). On the fractional anisotropic Wiener field. *Probab. Math. Stat.*, 16, pp. 85–96.

[132] Kasahara, M. and Maejima, M. (1988). Weighted sums of i.i.d. random variables attracted to integrals of stable processes. *Probab. Theory Relat. Fields*, 78, pp. 75–96.

[133] Kazakevičius, V., Leipus, R. and Viano, M.-C. (2004). Stability of random coefficient ARCH models and aggregation schemes. *J. Econom.*, 120, pp. 139–158.

[134] Kelbert, M.Y., Leonenko, N.N. and Ruiz-Medina, M.D. (2005). Fractional random fields associated with stochastic fractional heat equations. *Adv. Appl. Prob.*, 37, pp. 108–133.

[135] Klueppelberg, C. and Kuehn, C. (2004). Fractional Brownian motion as a weak limit of Poisson shot noise processes - with application to finance. *Stoch. Process. Appl.*, 113, pp. 333–351.

[136] Klueppelberg. C. and Matsui, M. (2016). Generalized fractional Lévy processes with fractional Brownian motion limit. *Adv. Appl. Prob.*, 47, pp. 1108–1131.

[137] Kolmogorov, A.N. (1940). Wiener spiral and some other interesting curves in Hilbert space. *Dokl. Akad. Nauk SSSR*, 26, pp. 115–118.

[138] Kolmogorov, A.N. (1941). The local structure of turbulence in an incompressible fluid at very high Reynolds numbers. *Dokl. Akad. Nauk SSSR*, 30, pp. 209–303.

[139] Kostić, M. (2024). Multidimensional fractional calculus: Theory and applications. *Axioms*, 13, p. 623.

[140] Koul, H.L. (2002). *Weighted Empirical Processes in Dynamic Nonlinear Models.* Lecture Notes in Statistics, Vol. 166. New York: Springer-Verlag.

[141] Koul, H.L. and Surgailis, D. (2001). Asymptotics of empirical processes of long memory moving averages with infinite variance. *Stoch. Process. Appl.*, 91, pp. 309–336.

[142] Koul, H.L., Mimoto, N. and Surgailis, D. (2016). Goodness-of-fit tests for marginal distribution of linear random fields with long memory. *Metrika*, 79, pp. 165–193.

[143] Kuensch, H. (1981). Almost sure entropy and the variational principle for random fields with unbounded state space. *Z. Wahrscheinlichkeitstheor. Verw. Geb.*, 58, pp. 69–85.

[144] Kulik, R. and Spodarev, E. (2021). Long-range dependence of heavy-tailed random functions. *J. Appl. Probab.*, 58, pp. 569–593.

[145] Lahiri, S.N. and Robinson, P.M. (2016). Central limit theorems for long range dependent spatial linear processes. *Bernoulli*, 22, pp. 345–375.

[146] Lamperti, J. (1962). Semi-stable stochastic processes. *Trans. Am. Math. Soc.*, 104, pp. 62–78.

[147] Lavancier, F. (2007). Invariance principles for non-isotropic long memory random fields. *Inference Stoch. Process.*, 10, pp. 255–282.

[148] Lavancier, F. (2011). Aggregation of isotropic autoregressive fields. *J. Stat. Plan. Inf.*, 141, pp. 3862–3866.

[149] Lawler, G.F. and Limic, V. (2012). *Random Walk: A Modern Introduction.* Cambridge: Cambridge University Press.

[150] Lee, J. (2021). Hurst estimation for operator scaling random fields. *Stat. Probab. Lett.*, 178, p. 109188.

[151] Leipus, R. and Surgailis, D. (2003). Random coefficient autoregression, regime switching and long memory. *Adv. Appl. Probab.*, 35, pp. 1–18.

[152] Leipus, R., Paulauskas, V. and Surgailis, D. (2005). Renewal regime switching and stable limit laws. *J. Econom.*, 129, pp. 299–327.

[153] Leipus, R., Paulauskas, V. and Surgailis, D. (2006). On a random-coefficient AR(1) process with heavy-tailed renewal switching coefficient and heavy-tailed noise. *Adv. Appl. Probab.*, 43, pp. 421–440.

[154] Leipus, R. and Surgailis, D. (2007). On long-range dependence in regenerative processes based on a general on/off scheme. *Adv. Appl. Probab.*, 44, pp. 378–391.

[155] Leipus, R., Oppenheim, G., Philippe, A. and Viano, M.-C. (2006). Orthogonal series density estimation in a disaggregation scheme. *J. Stat. Plan. Inf.*, 136, pp. 2547–2571.

[156] Leipus, R., Philippe, A., Pilipauskaitė, V. and Surgailis, D. (2017). Nonparametric estimation of the distribution of the autoregressive coefficient from panel random-coefficient AR(1) data. *J. Multivar. Anal.*, 153, pp. 121–135.

[157] Leipus, R., Philippe, A., Pilipauskaitė, V. and Surgailis, D. (2019). Sample autocovariances of random-coefficient AR(1) panel model. *Electron. J. Stat.*, 13, pp. 4527–4572.

[158] Leipus, R., Philippe, A., Pilipauskaitė, V. and Surgailis, D. (2020). Estimating long memory in panel random-coefficient AR(1) data. *J. Time Ser. Anal.*, 41, pp. 520–535.

[159] Leipus, R., Pilipauskaitė, V. and Surgailis, D. (2023). Aggregation of network traffic and anisotropic scaling of random fields. *Theory Probab. Math. Stat.*, 108, pp. 77–126.

[160] Leland, W.E., Taqqu, M.S., Willinger, W. and Wilson, D.V. (1994). On the self-similar nature of Ethernet traffic (extended version). *IEEE/ACM Trans. Netw.*, 2, pp. 1–15.

[161] Leonenko, N.N. (1999). *Random Fields with Singular Spectrum.* Dordrecht: Kluwer.

[162] Leonenko, N.N. and Olenko, A. (2012). Tauberian and Abelian theorems for long-range dependent random fields. *Methodol. Comput. Appl. Probab.*, 14, pp. 1–28.

[163] Leonenko, N.N., Ruiz-Medina, M.D. and Taqqu, M.S. (2011). Fractional elliptic, hyperbolic and parabolic random fields. *Electron. J. Probab.*, 16, pp. 1134–1172.

[164] Leonenko, N.N. and Taufer, E. (2013). Disaggregation of spatial autoregressive processes. *Spatial Stat.*, 3, pp. 1–20.

[165] Leonenko, N.N. and Woyczynski, W.A. (1998). Scaling limits of solutions of the heat equation for singular non-Gaussian data. *J. Stat. Phys.*, 91, pp. 423–438.

[166] Levy, J.B. and Taqqu, M.S. (2000). Renewal reward processes with heavy-tailed interrenewal times and heavy-tailed rewards. *Bernoulli*, 6, pp. 23–44.

[167] Lewbel, A. (1994). Aggregation of simple dynamics. *Am. Econ. Rev.*, 84, pp. 905–918.

[168] Linden, M. (1999). Time series properties of aggregated AR(1) processes with uniformly distributed coefficients. *Econ. Lett.*, 64, pp. 31–36.

[169] Lodhia, A., Scheffield, S., Sun, X. and Watson, S.S. (2016). Fractional Gaussian fields: A survey. *Probab. Surv.*, 13, pp. 1–56.

[170] Loh, W.-L. (2015). Estimating the smoothness of a Gaussian random field from irregularly spaced data via higher-order quadratic variations. *Ann. Stat.*, 43, pp. 2766–2794.

[171] Liu, M. (2000). Modeling long memory in stock market volatility. *J. Econom.*, 99, pp. 139–171.

[172] Major, P. (1981). *Multiple Wiener-Itô Integrals.* Lecture Notes in Mathematics, Vol. 849. Berlin: Springer.

[173] Malyshev, V.A. and Minlos, R.A. (1991). *Gibbs Random Fields.* Dordrecht: Kluwer.

[174] Makogin, V. and Mishura, Y. (2015). Example of a Gaussian self-similar field with stationary rectangular increments that is not a fractional Brownian sheet. *Stoch. Anal. Appl.*, 33, pp. 413–428.

[175] Makogin, V. and Mishura, Y. (2019). Gaussian self-similar random fields with distinct stationarity properties of their rectangular increments. *Stoch. Models*, 35, pp. 391–428.

[176] Makogin, V. and Spodarev, E. (2025). A short memory condition for infinitely divisible random fields. *Stat. Probab. Lett.*, 221, p. 110385.

[177] Mandelbrot, B.B. (1982). *The Fractal Geometry of Nature.* New York: Freeman and Co.

[178] Mandelbrot, B.B. and Van Ness, J.W. (1968). Fractional Brownian motions, fractional noises and applications. *SIAM Rev.*, 10, pp. 422–437.

[179] Martinez, C. and Sanz, M. (2001) *The Theory of Fractional Powers of Operators.* North-Holland Math. Studies 187, Amsterdam.

[180] Mase, S. (1982). Asymptotic properties of stereological estimators. *J. Appl. Prob.*, 19, pp. 111–126.

[181] Matheron, G. (1975). *Random Sets and Integral Geometry.* New York: Wiley.

[182] McKenzie, E. (2003). Discrete variate time series. In Shanbhag, D.N. and Rao, C.R. (Eds.), *Handbook of Statistics*, Vol. 21. Amsterdam: Elsevier, pp. 573–606.

[183] Meerschaert, M.M. and Scheffler, H.P. (2001). *Limit Distributions for Sums of Independent Random Vectors: Heavy Tails in Theory and Practice.* New York: Wiley-Interscience.

[184] de Meyer, A. and Teugels, J.L. (1980). On the asymptotic behaviour of the distributions of the busy period and service time in M/G/1. *J. Appl. Probab.*, 17, pp. 802–813.

[185] Mikosch, T., Resnick, S., Rootzén, H. and Stegeman, A. (2002). Is network traffic approximated by stable Lévy motion or fractional Brownian motion? *Ann. Appl. Probab.*, 12, pp. 23–68.

[186] Mikosch, T. and Samorodnitsky, G. (2007). Scaling limits for cumulative input processes. *Math. Oper. Res.*, 32, pp. 890–918.

[187] Mikosch, T. and Wintenberger, O. (2013). Precise large deviations for dependent regularly varying sequences. *Probab. Theory Relat. Fields*, 156, pp. 851–887.

[188] Molchanov, I. and Stoyan, D. (1994). Asymptotic properties of estimators for parameters of the boolean model. *Adv. Appl. Probab.*, 26, pp. 301–323.

[189] Molchanov, I. S. (1997). *Statistics of the Boolean Model for Practitioners and Mathematicians.* New York: Wiley.

[190] Nualart, D. and Peccati, G. (2005). Central limit theorems for sequences of multiple stochastic integrals. *Ann. Probab.*, 33, pp. 177–193.

[191] Oldham, K. and Spanier, J. (1974). *The Fractional Calculus: Theory and Application of Differentiation and Integration to Arbitrary Order.* New York: Acad. Press.

[192] Oppenheim, G. and Viano, M.-C. (2004). Aggregation of random parameters Ornstein-Uhlenbeck or AR processes: some convergence results. *J. Time Ser. Anal.*, 25, pp. 335–350.

[193] Orey, S. (1958). A central limit theorem for m-dependent random variables. *Duke Math. J.*, 25, pp. 543–546.

[194] Ortigueira, M.D., Valério, D. and Machado, J.T. (2019). Variable order fractional systems. *Commun. Nonlinear Sci. Numer. Simulat.*, 71, pp. 231–243.

[195] Pakkanen, M.S. (2014). Limit theorems for power variations of ambit fields driven by white noise. *Stoch. Process. Appl.*, 124, pp. 1942–1973.

[196] Pakkanen, M.S. and Revéillac, A. (2016). Functional limit theorems for generalized variations of the fractional Brownian sheet. *Bernoulli*, 22, pp. 1671–1708.

[197] Park, K. and Willinger, W. (Eds.) (2000). *Self-Similar Network Traffic and Performance Evaluations.* New York: Wiley.

[198] Paulauskas, V. (2016). Some remarks on the definitions of memory for stationary random processes and fields. *Lithuanian Math. J.*, 56, pp. 229–250.

[199] Peccati, G. and Taqqu, M.S. (2011). *Wiener Chaos: Moments, Cumulants and Diagrams.* New York: Springer.

[200] Pecknold, S., Lovejoy, S. and Schertzer, D. (1992). The morphology and texture of anisotropic multifractals using generalized scale invariance. In Brillinger, D., *et al.* (Eds.), *New Directions in Time Series Analysis, Part II.* IMA Volumes in Mathematics and its Applications, Vol. 46. New York: Springer, pp. 269–311.

[201] Peligrad, M., Sang, H., Xiao, Y. and Yang, G. (2022). Limit theorems for linear random fields with innovations in the domain of attraction of a stable law. *Stoch. Process. Appl.*, 150, pp. 596–621.

[202] Philippe, A., Surgailis, D. and Viano, M.-C. (2006). Invariance principle for a class of non-stationary processes with long memory. *C. R. Acad. Sci. Paris Ser.*, 1 342, pp. 269–274.

[203] Philippe, A., Surgailis, D. and Viano, M.-C. (2006). Almost periodically correlated processes with long memory. In Bertail, P., Doukhan, P. and Soulier, P. (Eds.), *Dependence in Probability and Statistics.* Lecture Notes in Statistics, Vol. 187. New York: Springer.

[204] Philippe, A., Surgailis, D. and Viano, M.-C. (2008). Time-varying fractionally integrated processes with nonstationary long memory. *Theory Probab. Appl.*, 52, pp. 651–673.

[205] Pilipauskaitė, V., Skorniakov, V. and Surgailis, D. (2020). Joint temporal and contemporaneous aggregation of random-coefficient AR(1) processes with infinite variance. *Adv. Appl. Probab.*, 52, pp. 237–265.

[206] Pilipauskaitė, V. and Surgailis, D. (2014). Joint temporal and contemporaneous aggregation of random-coefficient AR(1) processes. *Stoch. Process. Appl.*, 124, pp. 1011–1035.

[207] Pilipauskaitė, V. and Surgailis, D. (2015). Joint aggregation of random-coefficient AR(1) processes with common innovations. *Stat. Probab. Lett.*, 101, pp. 73–82.

[208] Pilipauskaitė, V. and Surgailis, D. (2016). Anisotropic scaling of random grain model with application to network traffic. *J. Appl. Probab.*, 53, pp. 857–879.

[209] Pilipauskaitė, V. and Surgailis, D. (2017). Scaling transition for nonlinear random fields with long-range dependence. *Stoch. Process. Appl.*, 127, pp. 2751–2779.

[210] Pilipauskaitė, V. and Surgailis, D. (2021). Scaling limits of linear random fields on $\mathbb{Z}^2$ with general dependence axis. In Vares, M.E., Fernandez, R., Fontes, L.R. and Newman, C.M. (Eds.), *In and Out of Equilibrium 3: Celebrating Vladas Sidoravicius.* Progress in Probability. Basel: Birkhäuser, pp. 683–710.

[211] Pilipauskaitė, V. and Surgailis, D. (2022). Local scaling limits of Lévy driven fractional random fields. *Bernoulli*, 28, pp. 2833–2861.

[212] Pilipauskaitė, V. and Surgailis, D. (2024). Fractional operators and fractionally integrated random fields on Z^ν. *Fractal Fract.*, 8, p. 353.

[213] Pipiras, V., Taqqu, M.S. and Levy, L.B. (2004). Slow, fast, and arbitrary growth conditions for renewal reward processes when the renewals and the rewards are heavy-tailed. *Bernoulli*, 10, pp. 121–163.

[214] Pipiras, V. and Taqqu, M.S. (2008). Small and large scale asymptotics of some Lévy stochastic integrals. *Methodol. Comput. Appl. Probab.*, 10, pp. 299–314.

[215] Pipiras, V. and Taqqu, M.S. (2017). *Long-Range Dependence and Self-Similarity.* Cambridge: Cambridge University Press.

[216] Pratt, J.W. (1960). On interchanging limits and integrals. *Ann. Math. Stat.*, 31, pp. 74–77.

[217] Preston, C. (1976). *Random Fields.* Lecture Notes in Mathematics, Vol. 534. Berlin: Springer-Verlag.

[218] Puplinskaitė, D. and Surgailis, D. (2009). Aggregation of random coefficient AR(1) process with infinite variance and common innovations. *Lithuanian Math. J.*, 49, pp. 446–463.

[219] Puplinskaitė, D. and Surgailis, D. (2010). Aggregation of random coefficient AR1(1) process with infinite variance and idiosyncratic innovations. *Adv. Appl. Probab.*, 42, pp. 509–527.

[220] Puplinskaitė, D. and Surgailis, D. (2015). Scaling transition for long-range dependent Gaussian random fields. *Stoch. Process. Appl.*, 125, pp. 2256–2271.

[221] Puplinskaitė, D. and Surgailis, D. (2016). Aggregation of autoregressive random fields and anisotropic long-range dependence. *Bernoulli*, 22, pp. 2401–2441.

[222] Rajput, B.S. and Rosinski, J. (1989). Spectral representations of infinitely divisible processes. *Probab. Theory Relat. Fields*, 82, pp. 451–487.

[223] Robinson, P.M. (1978). Statistical inference for a random coefficient autoregressive model. *Scand. J. Stat.*, 5, pp. 163–168.

[224] Rosenblatt, M. (1961). Independence and dependence. In *Proceedings of 4th Berkeley Symposium on Mathematical Statistics and Probability.* Berkeley: University of California Press, pp. 411–443.

[225] Rosenblatt, M. (1987). Scale normalization and random solutions of the Burgers equation. *J. Appl. Probab.*, 24, pp. 328–338.

[226] Roueff, F., Samorodnitsky, G. and Soulier, P. (2012). Function-indexed empirical processes based on an infinite source Poisson transmission stream. *Bernoulli*, 18, pp. 783–802.

[227] Samko, S.G., Kilbas, A.A. and Marichev, O.I. (1993). *Fractional Integrals and Derivatives: Theory and Applications.* Yverdon: Gordon and Breach.

[228] Samko, S.G., (2013). Fractional integration and differentiation of variable order: An overview. *Nonlinear Dyn.*, 71, pp. 653–662.

[229] Samko, S.G., and Ross, B. (1993). Integration and differentiation to a variable fractional order. *Integral Transforms Spec. Funct.*, 1, pp. 277–300.

[230] Samorodnitsky, G. (2016). *Stochastic Processes and Long Range Dependence.* New York: Springer.

[231] Samorodnitsky, G. and Taqqu, M.S. (1994). *Stable Non-Gaussian Random Processes.* New York: Chapman and Hall.

[232] Sato, K.-I. (1999). *Lévy Processes and Infinitely Divisible Distributions.* Cambridge: Cambridge University Press.

[233] Schertzer, D. and Lovejoy, S. (1985). Generalised scale invariance in turbulent phenomena. *Phys.-Chem. Hydrodyn. J.*, 6, pp. 623–635.

[234] Serra, J.-P. (1982). *Image Analysis and Mathematical Morphology.* London: Academic Press.

[235] Shilov, G.E. and Gurevich, B.L. (1977). *Integral, Measure and Derivative.* New York: Dover.

[236] Shiryaev, A.N. (1996). *Probability.* New York: Springer.

[237] Shohat, J. (1936). The relation of the classical orthogonal polynomials to the polynomials of Appell. *Am. J. Math.*, 58, pp. 453–464.

[238] Sinai, Y.G. (1976). Self-similar probability distributions. *Theory Probab. Appl.*, 21, pp. 64–80.

[239] Sinai, Y.G. (1982). *Theory of Phase Transitions: Rigorous Results.* Amsterdam: Elsevier.

[240] Smith, W.L. (1955). Regenerative stochastic processes. *Proc. R. Soc. Ser. A*, 232, pp. 6–31.

[241] Stein, E.M and Weiss, G. (1971). *Introduction to Fourier Analysis on Euclidean Spaces.* Princeton: Princeton University Press.

[242] Stoyan, D., Kendall, W.S. and Mecke, J. (1989). *Stochastic Geometry and Its Applications.* Berlin: Akademie-Verlag.

[243] Surgailis, D. (1981). On infinitely divisible self-similar random fields. *Z. Wahrscheinlichkeitstheor. Verw. Geb.*, 58, pp. 453–477.

[244] Surgailis, D. (1981). On L^2 and non-L^2 multiple stochastic integration. In Arató, M., Vermes, D. and Balakrishnan, A.V. (Eds.), *Stochastic Differential Systems.* Lecture Notes in Control and Information Sciences, Vol. 36. Berlin: Springer-Verlag, pp. 212–226.

[245] Surgailis, D. (1982). Zones of attraction of self-similar multiple integrals. *Lithuanian Math. J.*, 22, pp. 185–201.

[246] Surgailis, D. (1984). On multiple Poisson stochastic integrals and associated Markov semigroups. *Probab. Math. Stat.*, 3, pp. 217–239.

[247] Surgailis, D. (2000). Long range dependence and Appell rank. *Ann. Probab.*, 28, pp. 478–497.

[248] Surgailis, D. (2004). Stable limits of sums of bounded functions of long memory moving averages with finite variance. *Bernoulli*, 10, pp. 327–355.

[249] Surgailis, D. (2008). Nonhomogeneous fractional integration and multifractional processes. *Stoch. Process. Appl.*, 118, pp. 171–198.

[250] Surgailis, D. (2019). Anisotropic scaling limits of long-range dependent linear random fields on $\mathbb{Z}^3$. *J. Math. Anal. Appl.*, 472, pp. 328–351.

[251] Surgailis, D. (2020). Scaling transition and edge effects for negatively dependent linear random fields on $\mathbb{Z}^2$. *Stoch. Process. Appl.*, 130, pp. 7518–7546.

[252] Surgailis, D. (2023). Scaling transition for singular linear random fields: Spectral approach. *Preprint* http://arxiv.org/abs/2301.02015.

[253] Surgailis, D. (2024). Scaling limits of nonlinear functions of random grain model, with application to Burgers' equation. *Stoch. Process. Appl.*, 174, p. 104390.

[254] Surgailis, D. (2026). On some open problems in spatial fractional integration. *Fractal Fract.*, 10, 100.

[255] Surgailis, D., Rosinski, J., Mandrekar, V. and Cambanis, S. (1993). Stable mixed moving averages. *Probab. Theory Relat. Fields*, 97, pp. 543–558.

[256] Surgailis, D. and Woyczynski, W.A. (2003). Limit theorems for the Burgers equation initialized by data with long-range dependence. In Doukhan, P., Oppenheim, G. and Taqqu, M.S. (Eds.), *Long Range Dependence: Theory and Applications.* Boston: Birkhäuser, pp. 507–523.

[257] Surgailis, D. and Woyczynski, W.A. (1994). Scaling limits of solutions of Burgers' equation with singular Gaussian initial data. In Houdré, C. and Pérez-Abreu, V. (Eds.), *Chaos Expansions, Multiple Wiener-Itô Integrals and Their Applications.* Boca Raton: CRC Press, pp. 145–161.

[258] Takenaka, S. (1991). Integral-geometric construction of self-similar stable processes. *Nagoya Math. J.*, 123, pp. 1–12.

[259] Taqqu, M.S. (1979). Convergence of integrated processes of arbitrary Hermite rank. *Z. Wahrscheinlichkeitstheor. Verw. Geb.*, 50, pp. 53–83.

[260] Taqqu, M.S., Willinger, W. and Sherman, R. (1997). Proof of a fundamental result in self-similar traffic modeling. *Comput. Commun. Rev.*, 27, pp. 5–23.

[261] Taqqu, M.S. (2003). Fractional Brownian motion and long-range dependence. In Doukhan, P., Oppenheim, G. and Taqqu, M.S. (Eds.), *Long Range Dependence: Theory and Applications.* Boston: Birkhäuser, pp. 5–38.

[262] Teyssiérre, G. and Kirman, A.P. (Eds.) (2007). *Long Memory in Economics.* Berlin: Springer.

[263] Thorisson, H. (2000). *Coupling, Stationarity, and Regeneration.* New York: Springer-Verlag.

[264] Wainger, S. (1965). *Special Trigonometric Series in k Dimensions.* Memoirs of the American Mathematical Society, Vol. 59, Providence, Rhode Island: American Mathematical Society.

[265] Wallin, J. and Bolin, D. (2015). Geostatistical modelling using non-Gaussian Matérn fields. *Scand. J. Stat.*, 42, pp. 872–890.

[266] Wiener, N. (1938). The homogeneous chaos. *Am. J. Math.*, 60, pp. 897–936.

[267] Willinger, W., Taqqu, M.S., Leland, M. and Wilson, D. (1997). Self-similarity through high-variability: Statistical analysis of Ethernet LAN traffic at the source level. *IEEE/ACM Trans. Netw.*, 5, pp. 71–86.

[268] Woyczynski, W.A. (1998). *Burgers-KPZ Turbulence.* Lecture Notes in Mathematics, Vol. 1700. New York: Springer.

[269] Wu, W.B. (2003). Empirical processes of long memory sequences. *Bernoulli*, 9, pp. 809–831.

[270] Wu, W.B. (2007). M-estimation of linear models with dependent errors. *Ann. Stat.*, 35, pp. 495–521.

[271] Yaglom, A.M. (1957). Some classes of random fields in n-dimensional space related to stationary random processes. *Theory Probab. Appl.*, 2, pp. 292–338.

[272] Zaffaroni, P. (2004). Contemporaneous aggregation of linear dynamic models in large economies. *J. Econom.*, 120, pp. 75–102.

[273] Zaffaroni, P. (2007). Aggregation and memory of models of changing volatility. *J. Econom.*, 136, pp. 237–249.

[274] Zaffaroni, P. (2007). Contemporaneous aggregation of GARCH processes. *J. Time Ser. Anal.*, 28, pp. 521–544.

[275] Zwart, B. (2001). Tail asymptotics for the busy period in the GI/G/1 queue. *Math. Oper. Res.*, 26, pp. 485–493.

Index

G

H

I

www.ingramcontent.com/pod-product-compliance
Lightning Source LLC
LaVergne TN
LVHW050807010926
846999LV00009B/34

* 9 7 8 9 8 1 1 2 4 9 4 1 9 *